INFINITE WORDS

Automata, Semigroups, Logic and Games

This is volume 141 in the PURE AND APPLIED MATHEMATICS series
Founding Editors: Paul A. Smith and Samuel Eilenberg

INFINITE WORDS

Automata, Semigroups, Logic and Games

Dominique Perrin and Jean-Éric Pin

Université de Marne-la-Vallée, France

and

CNRS, Université Paris VII, France

ELSEVIER
ACADEMIC
PRESS

Amsterdam Boston Heidelberg London New York Oxford
Paris San Diego San Francisco Singapore Sydney Tokyo

ELSEVIER B.V.
Sara Burgerhartstraat 25
P.O. Box 211, 1000 AE Amsterdam,
The Netherlands

ELSEVIER Ltd
The Boulevard, Langford Lane
Kidlington, Oxford OX5 1GB, UK

ELSEVIER Inc.
525 B Street, Suite 1900
San Diego, CA 92101-4495, USA

ELSEVIER Ltd
84 Theobalds Road
London WC1X 8RR UK

First edition 2004

ISBN: 0-12-532111-2
ISBN 13: 978-0-12-532111-2

ISSN: 0079-8169

Transferred to Digital Printing 2007

CONTENTS

PREFACE

The theory of automata arose as an interdisciplinary field, with roots in several scientific domains including pure mathematics, electronics and computer science. This diversity is reflected in the material presented in this book which covers topics related to computer science, algebra, logic, topology and game theory.

The elementary theory of automata allows both the *specification* and the *verification* of simple properties of finite sequences of symbols. The possible practical applications include lexical analysis, text processing and software verification.

There are at least two possible extensions of this theory.

The theory of *formal series* is one of them. Words are replaced by functions associating to each word some numerical value. This value can be an integer counting the number of paths labeled by this word in an automaton, such as the integer represented by this word in some basis. It can also be a real number corresponding to some probability.

The other possible extension is the subject of this book: Finite sequences of symbols are replaced by infinite sequences.

The motivation for this generalization originates in the early work of Richard Büchi in the sixties. Working on weak logical theories of the integers, he was lead to consider the monadic second-order theory of the successor function on the integers. He was able to prove the decidability of this theory. He actually showed that all properties of the integers expressible in this logic can also be defined in terms of finite automata.

Later on, Robert McNaughton proved the equivalence of deterministic and non-deterministic automata, a natural extension of the corresponding result for finite words. This difficult result had been conjectured by David Muller while working on questions related to oscillating circuits.

Many other results have appeared since then and interest in the theory has increased considerably, motivated by applications to problems in computer science. The notion of an infinite sequence is of interest in modelling the behavior of systems which are supposed to work indefinitely, for example operating systems.

This book presents a comprehensive treatment of all aspects of this theory. It gathers for the first time basic results with advanced ones. Although several surveys have appeared on infinite words, this book is the first manual devoted to the topic. All proofs are given in detail, with a few, duly mentioned, exceptions.

The book is intended for researchers or advanced students in mathematics or computer science. No particular background is required to read it, except for a standard mathematical culture. The dependence between chapters is not too strong, making it possible to read some chapters independently from other ones.

The book can be used to lecture and the authors have used the manuscript for several years for graduate courses in computer science. It is unlikely that all the material would be covered in a single course, but a selection with emphasis either on topology, or on logic, or on automata and semigroups, is possible.

The book is organized as follows. The first chapter contains the definitions of rational expressions, Büchi and Muller automata and recognizable sets. It covers the necessary elements of the theory of automata on finite words such as Kleene's theorem. A proof of McNaughton's theorem is given, using Safra's determinization algorithm. Although this construction is rather involved, we have chosen to place it at the very beginning of the book because it is straightforward. Other proofs of McNaughton's theorem are given later on.

In the second chapter, we shift to a more algebraic point of view. The key idea of this chapter is to give a purely algebraic definition of recognizable sets. This point of view will be adopted quite often in the sequel. Our main tools are finite semigroups and their counterpart for infinite words, called ω-semigroups.

The third chapter introduces the topological aspects of the theory. It is really an excursion into the field known as descriptive set theory, situated at the border between analysis and logic. We show that the main notions introduced so far have a natural translation in terms of topology.

The fourth chapter is devoted to games. These games are two player mathematical games which are used as a tool to prove some results on infinite words and automata. For instance, in this chapter, games are used to prove the Büchi-Landweber theorem. Some particular games, such as Wadge games or Fraïssé-Ehrenfeucht games, are used in further chapters.

In the fifth chapter, we present a classification of recognizable sets of infinite words known as the Wagner hierarchy. It emphasizes once again the importance of finite semigroups in this theory.

Chapters VI and VII present the theory of varieties for infinite words. This is an extension of the so-called Eilenberg variety theory, which associates sets of finite words with families of finite semigroups. The families of semigroups are actually varieties of finite semigroups. The extension to infinite words leads to the notion of varieties of ω-

semigroups. The classical result of Schützenberger on star-free sets and aperiodic semigroups is generalized by means of an appropriate notion of aperiodic ω-semigroups.

Logic enters in Chapter VIII. The main point is that there is a close connexion between the concepts of automata theory and those of logic, as was the case with topology. Thus, recognizability is equivalent to monadic second-order definability while aperiodicity is equivalent to first-order definability.

The last two chapters deal with two natural extensions of infinite words. The first is concerned with two-sided infinite words, for which all notions generalize in a natural way. The second deals with infinite trees. This case is important because of its role in the applications. The situation is very different with trees instead of words and, in particular, Büchi and Muller automata are no longer equivalent. The main result is Rabin's theorem, which states the equivalence between recognizability by tree automata and monadic second-order definability.

We wish to express our gratitude to all those who have helped us during the long maturation of this book. We thank Danièle Beauquier, Olivier Carton and Wolfgang Thomas for their numerous suggestions on preliminary versions. We are also indebted to the numerous students and colleagues who have read our manuscripts and indicated errors of all sorts: André Arnold, Marie-Pierre Béal, Martin Beaudry, Nicolas Bedon, Jean Berstel, Ahmed Bouabdallah, Mario Branco, Véronique Bruyère, Hugues Calbrix, Christian Choffrut, Joëlle Cohen, Pierre-Louis Curien, Volker Diekert, Paul Gastin, Hugo Gimbert, Dietrich Kuske, Bertrand Le Saec, Igor Litovski, James Martin, Anca Muscholl, Bruno Petazzoni, Antoine Petit, Christophe Prieur, Roman R. Redziejowski, Jean-François Rey, Laurent Rosaz, Pierre-Yves Schobbens, Ludwig Staiger, Istvan Toth, Sébastien Veigneau, Klaus Wagner, Pascal Weil and Thomas Wilke.

The original manuscript was realized in AMS-LaTeX. The figures were realized using the packages `pstricks` and GasTeX.

I

AUTOMATA AND INFINITE WORDS

1 Introduction

This first chapter constitutes an introduction to the theory of automata on infinite words. It includes some basic definitions, such as ω-rational sets of infinite words. These sets are defined starting from the elements of the alphabet by making use of the four operations of *union*, *concatenation*, *finite iteration* and *infinite iteration*. These operations are denoted $\cup$, ., $*$, ω as indicated in Table 1.1. The expressions obtained are often called ω-rational expressions. The name of rational expressions is reserved for those expressions that do not use the symbol ω and thus define sets of finite words. The classical theorem of Kleene establishes the equivalence between rational expressions and finite automata. Its extension to infinite words is one of the results contained in this chapter. We shall see that a set is ω-rational if and only if it can be recognized by a finite automaton (Theorem 5.4). Recognizing infinite words with finite automata requires a convention. The simplest one, introduced by Büchi, consists in considering an infinite

Operation	**Symbol**
union	$\cup$
concatenation	$\cdot$
finite iteration	$*$
infinite iteration	ω

Table 1.1. The operations used in ω-rational expressions.

Acceptance modes	Definition
Büchi	$\mathrm{Inf}(p) \cap F \neq \emptyset$
Muller	$\mathrm{Inf}(p) \in \mathcal{T}$

Table 1.2. Büchi and Muller's acceptance modes.

path in the automaton as successful if it starts in an initial state and passes infinitely often through a terminal state.

Another basic result of the theory of finite automata is given by the *determinization* algorithm which allows one to replace any finite automaton by an equivalent deterministic one. A consequence of this is the closure of the class of rational sets under complementation. The extension of these results is fraught with difficulties. The solution requires first the introduction of a more powerful acceptance mode than Büchi's one, since one has to specify the set of states met infinitely often on an infinite path. The acceptance mode, called Muller's mode, declares an infinite path p to be successful if the set $\mathrm{Inf}(p)$ of states met infinitely often on p belongs to a prescribed set $\mathcal{T}$ of sets of states. This constitutes a more constrained acceptance mode than Büchi's one, for which it suffices to check whether the set $\mathrm{Inf}(p)$ meets the set F of final states (see Table 1.2). The basic result of this theory, due to R. McNaughton, states that any finite automaton is equivalent on infinite words to a deterministic Muller automaton (Theorem 7.1). This implies in particular that the class of ω-rational sets is closed under complementation, a result proved for the first time by Büchi in a direct way (see the notes at the end of the chapter).

This chapter is organized as follows. Section 2 introduces our notation. Section 3 introduces ω-rational sets of infinite words. It contains the result that characterizes them as the finite unions of sets of the form XY^ω where X and Y are rational sets of finite words. Sections 4, 5, 6, 7 and 8 introduce the definitions of the various acceptance modes: Büchi's mode, Muller's mode, Rabin's mode and a transition mode. Section 9 contains a proof of McNaughton's theorem. This proof, discovered in 1989 by S. Safra has the advantage, compared to other possible ones, of being direct and of providing a better algorithm: starting from a nondeterministic n-state Büchi automaton, one obtains a deterministic Muller automaton with $O(n^n)$ states (the other constructions lead to a double exponential). We shall however see other proofs in the following chapters, especially one using ω-semigroups which makes various generalizations possible. The last section (Section 10) deals with computational issues concerning the transformations between various possible representations of ω-rational sets and the operations on them.

2 Words and trees

In this book, we are going to consider possibly infinite sequences of elements of a set called an *alphabet*. The elements of this set are called *letters* or also *symbols*. Most often, in the examples, the alphabet will be finite or even reduced to two elements. We shall however also consider countable alphabets. Recall that a set A is *countable* if there exists an injective map from A to $\mathbb{N}$.

A finite sequence of elements of A is called a *finite word* on A, or just a *word*. We denote by mere juxtaposition

$$a_0a_1\cdots a_n$$

the sequence $(a_0, a_1, \ldots, a_n)$. The set of words is endowed with the operation of *concatenation product* also called *product*, which associates with two words $x = a_0a_1\cdots a_p$ and $y = b_0b_1\cdots b_q$ the word $xy = a_0a_1\cdots a_pb_0b_1\cdots b_q$. This operation is associative. It has a neutral element, the *empty word*, denoted by 1 or ε and which is the empty sequence.

We denote by A^* the set of words on A and by A^+ the set of nonempty words. The set A^* (resp. A^+), equipped with the concatenation product is thus a monoid with neutral element 1 (resp. a semigroup). The set A^* is called the *free monoid* on A and A^+ the *free semigroup* on A. This terminology will be justified later.

If u is a word and a a letter, we denote by $|u|_a$ the number of occurrences of a in u. Thus, if $A = \{a, b\}$ and $u = abaab$, we have $|u|_a = 3$ and $|u|_b = 2$. The sum

$$|u| = \sum_{a\in A} |u|_a$$

is the *length* of the word u. Thus $|abaab| = 5$.

An *infinite word* on the alphabet A is an infinite sequence of elements of A, which we also denote by juxtaposition

$$u = a_0a_1\cdots a_n\cdots$$

This notation represents the mapping from $\mathbb{N}$ into A defined, for all $n \in \mathbb{N}$ by $u(n) = a_n$.

We also denote by $u[r, s]$ the word $u(r)u(r+1)\cdots u(s) = a_r\cdots a_s$. This notation is also used for finite words.

We denote indifferently by $A^{\mathbb{N}}$ or by A^{ω} the set of infinite words over the alphabet A and we let

$$A^{\infty} = A^* \cup A^{\omega}$$

which is thus the set of finite or infinite words on the alphabet A. The product of a finite word $u = a_0a_1\cdots a_n$ from A^* with an infinite word $v = b_0b_1\cdots$ of A^{ω} is the infinite word

$$uv = a_0a_1\cdots a_nb_0b_1\cdots$$

Let $u = a_0a_1\cdots a_n$ be a word in A^*. A word $x \in A^*$ is a *factor* of u if there exist integers r and s such that $0 \leq r \leq s \leq n$ and $x = u[r, s]$. This is equivalent to saying that there exist words v and w in A^* such that $u = vxw$. In the same way, we say that

x is a *left factor* or a *prefix* of u if there exists a word w in A^* such that $u = xw$,

x is a *right factor* or a *suffix* of u if there exists a word v in A^* such that $u = vx$,

x is a *proper factor* of u if there exist words v and w not both empty such that $u = vxw$,

x is *strict factor* of u if there exist words v and w in A^+ such that $u = vxw$.

For example, if $u = abaabab$, aba is a prefix, ab is a suffix, $abaab$ is a proper factor and $baaba$ is a strict factor u.

The relation "to be a left factor of" is a partial order on words called "prefix order" and sometimes denoted $\leq$. Thus,

$$1 \leq a \leq ab \leq abb \leq abba$$

If one starts with an ordered alphabet, a total order relation can be defined on words, called *lexicographic order* and denoted $\leq_{lex}$. It is the usual order in a dictionary. In formal terms, one has $u \leq_{lex} v$ if $u \leq v$ or if $u = xau'$, $v = xbv'$ with $x, u', v' \in A^*$, $a, b \in A$ and $a <_{lex} b$.

The notions of a factor and of a left factor extend without difficulty to infinite words: a word $x \in A^*$ is a *factor* of an infinite word u if there exist integers r and s such that $0 \leq r \leq s$ and $x = u[r, s]$, or in other terms, if there exist a finite word v and an infinite word w such that $u = vxw$. We say that x is a *left factor* or a *prefix* of u if there exists an infinite word w such that $u = xw$.

If x is a prefix of u, we denote by $x^{-1}u$ the unique word v such that $xv = u$. The prefix order can be extended to a partial order on A^∞ by setting for u and v in A^∞, $u \leq v$ if $u = v$ or if u is a finite prefix of v.

An integer $p > 0$ is a *period* of a finite word $u = a_0a_1 \cdots a_n$ if, for all k such that $k + p \leq n$, we have $a_k = a_{k+p}$. The smallest period of u is called *the period* of u. For instance, the period of the word $abcaabcaab$ is 4, since $abcaabcaab = (abca)^2ab$.

The notion of a period extends without difficulty to infinite words. We say that an infinite word is *periodic* if the set of its periods is not empty.

An integer $p > 0$ is an *ultimate period* of an infinite word $u = a_0a_1 \cdots$ if there is a $k_0 \geq 0$ such that for all $k \geq k_0$ we have $a_k = a_{k+p}$. If p and q are two ultimate periods of u, their gcd is still a ultimate period of u. The smallest ultimate period of an infinite word u is called *the ultimate period* of u. An infinite word which admits a ultimate period is called *ultimately periodic*. We know, for example, that rational integers are those real numbers that admit, in a given base, an ultimately periodic expansion.

The set A^ω is equipped with an internal operation called the *shift*, which is defined as the mapping $\sigma : A^\omega \to A^\omega$, associating to each infinite word $u \in A^\omega$ the infinite word $\sigma(u)$ defined, for all $n \in \mathbb{N}$, by

$$\sigma(u)(n) = u(n + 1).$$

Thus, for $u = a_0a_1 \cdots$, one has $\sigma(u) = a_1a_2 \cdots$, which means that the action of σ consists in shifting one place to the left all symbols of u. It is therefore a surjective application, which is not injective as soon as $\text{Card}(A) \geq 2$.

If X is a set of infinite words, we let

$$\sigma(X) = \{\sigma(u) \mid u \in X\}$$

We say that X is *stable* (resp. *shift invariant*) if $\sigma(X) \subset X$ (resp. $\sigma(X) = X$).

It is easy to verify that an integer p is a period of an infinite word u if and only if $\sigma^p(u) = u$ and that u is ultimately of period p if and only if there is an integer q such that $\sigma^{p+q}(u) = \sigma^q(u)$.

A subset of A^* is said to be *prefix-free* or simply *prefix* if its elements are incomparable for the prefix order. For instance, if $A = \{a, b\}$, the set $\{a^n b \mid n \geq 0\}$ is prefix-free. As a dual definition, a subset of A^* is called *prefix-closed* if it contains the prefixes of all of its elements. In particular, if X is a subset of A^+ or of A^ω, the set Pref(X) of all prefixes of the elements of X is prefix-closed. The complement of a prefix-closed set is a *right ideal*: a subset R of A^* is indeed a right ideal if for all $u \in R$ and for all $v \in A^*$, one has $uv \in R$.

There is consequently a bijection between finite prefix-free sets and finite prefix-closed sets defined as follows: we associate with each prefix-free set P, the prefix-closed set formed of the prefixes of P; conversely, we associate with every prefix-closed set T, the prefix-free set formed by the elements of T which are maximal for the prefix ordering.

These notions are closely related with the notion of a *tree* that we introduce now. A tree is a tuple (N, r, p), where N is a nonempty set whose elements are called the *nodes* of the tree, r is a distinguished element of N, called the *root* of the tree and

$$p : N \setminus \{r\} \to N$$

is a mapping associating with each node distinct from the root a unique node called its *parent* and such that for each node n in N there is an integer $k \geq 0$ such that

$$p^k(n) = r$$

The terms employed for indicating kinship are commonly used for trees. Thus, we say that n is a *child* of $p(n)$. The notions of an *ancestor* and of a *descendant* are clear also. Formally, n is an ancestor of n' if there is an integer $k \geq 0$ such that $n = p^k(n')$. The node n' is then a descendant of n.

If n is a node in a tree (N, r, p), the mapping p induces a mapping

$$p_n : N_n \setminus \{n\} \to N_n,$$

where N_n is the set of descendants of n. The triple (N_n, n, p_n) is by definition the subtree of N rooted at n.

One associates in a natural way a tree with each nonempty prefix-closed set T using the elements of T as nodes, the empty word as a root and by defining for each word u

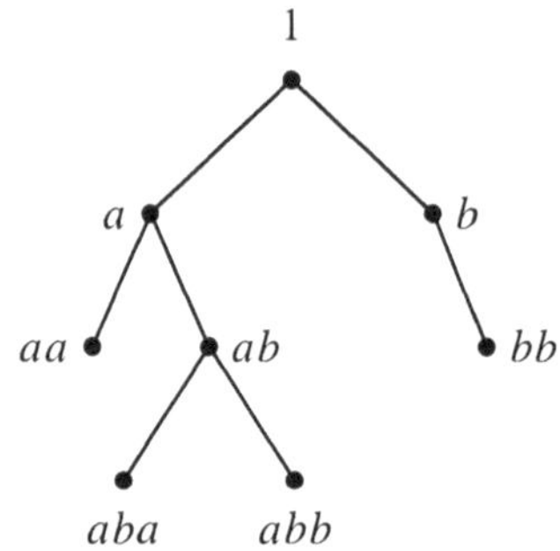

Figure 2.1. A prefix-closed set.

and each letter a such that $ua \in T$, $p(ua) = u$. Conversely, one may associate with each tree (N, r, p) a prefix-closed set T on the alphabet N by

$$T = \{n_1 \cdots n_k \mid k \geq 0,\ p(n_1) = r \text{ and for every } i \in \{2, \ldots, k\},\ p(n_i) = n_{i-1}\}$$

A node u is thus an ancestor of v if u is a prefix of v. And the subtree rooted at a node u is the set $u^{-1}T = \{v \in A^* \mid uv \in T\}$. This allows a convenient graphical representation of prefix-closed sets. For instance, if $A = \{a, b\}$ and if $T = \{1, a, aa, ab, aba, abb, b, bb\}$, the set T is represented in Figure 2.1. The number of children of a node is the *arity* of this node. The *arity of a tree* is the maximal arity of its nodes. In particular, the tree associated with a prefix-closed set on an alphabet with k elements is a tree of arity at most k. The free monoid A^* itself, which is a prefix-closed set, is associated with a tree represented in Figure 2.2. We shall have occasions to use three variants of trees. First of all, an *oriented tree* is a tree in which an order relation denoted $\leq$ is defined on the set of children of each node. An oriented tree can therefore be defined by a function

$$f : N \to N^*$$

associating to each node the ordered list of its children.

We say that two oriented trees $T = (N, r, f)$ and $T' = (N', r', f')$ are *equivalent* if there is a bijection σ from N onto N' (extending to a bijection from N^* onto N'^*) such that

$$\sigma(r) = r', \text{ and for every } n \in N, \sigma(f(n)) = f'(\sigma(n))$$

which means that σ preserves the order on the nodes (i.e. if $n_1 < n_2$ in T, then $\sigma(n_1) < \sigma(n_2)$). For example, the trees represented in Figure 2.3 are equivalent.

A *planar tree* is an equivalence class of this relation. Planar trees can be represented by a figure without mention of the set N. For example, the planar tree which is the equivalence class of the trees of Figure 2.3 is represented in Figure 2.4. It is relatively easy to compute the number of planar trees with n nodes.

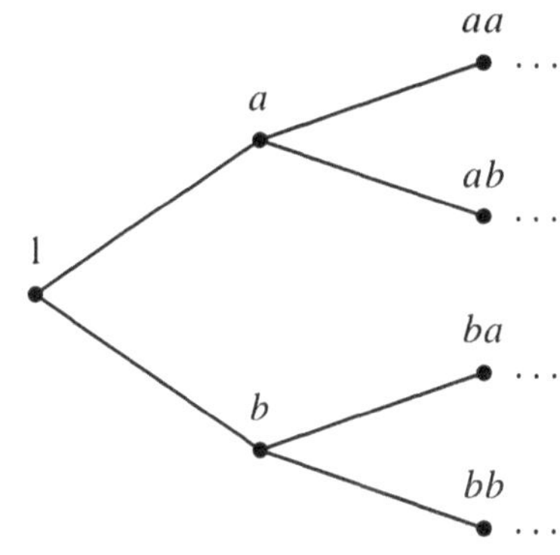

Figure 2.2. The tree of the free monoid $\{a, b\}^*$.

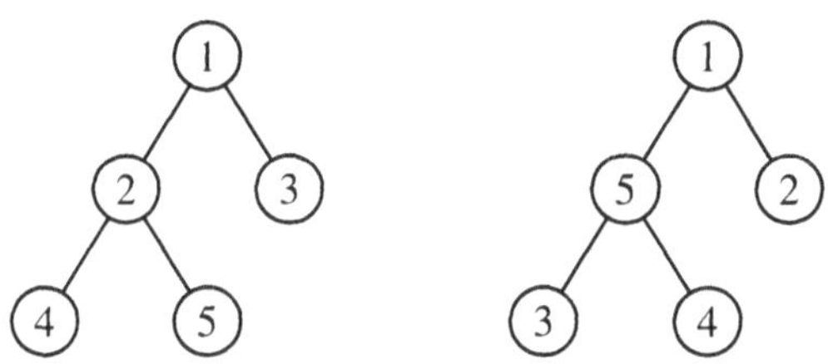

Figure 2.3. Two equivalent trees.

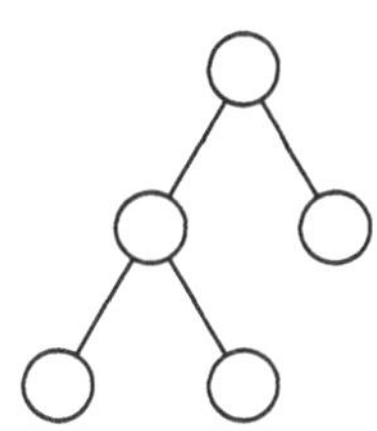

Figure 2.4. A planar tree.

Proposition 2.1. *For each $n > 0$, the number of planar trees with n nodes is the Catalan number $C_n = \dfrac{(2n-2)!}{n!(n-1)!}$.*

Proof. Let P_n be the number of planar trees with n nodes and let, by convention $P_0 = 0$. Let (N, r, f) be an oriented tree with at least two nodes and let z be the leftmost child of r. We obtain a partition of N by considering on one hand the set N_z of descendants of z and, on the other hand, the set $N' = N \setminus N_z$. These two sets define oriented trees

Figure 2.5. Decomposition of a tree.

(see Figure 2.5). Conversely given an oriented tree (N_z, z, f_z) with k nodes and an oriented tree (N', r, f') with $n - k$ nodes, we can, supposing N_z and N' disjoint, build an oriented tree with n nodes (N, r, f), where $N = N_z \cup N'$ and where

$$f(x) = \begin{cases} f_z(x) & \text{if } x \in N_f, \\ f'(x) & \text{if } x \in N_r \setminus \{r\}, \\ zf'(r) & \text{if } x = r. \end{cases}$$

Since this operation preserves the equivalence of trees we have for all $n \geq 2$:

$$P_n = \sum_{1 \leq k \leq n-1} P_k P_{n-k}$$

Thus, the generating series

$$P(X) = \sum_{n \geq 0} P_n X^n$$

satisfies the equation

$$P^2 - P + X = 0$$

The desired formula is obtained by expanding $\frac{1}{2}(1 - (1 - 4X)^{\frac{1}{2}})$. □

Finally a *labeled tree* is a pair (T, e) of a tree T and a function

$$e : N \to V$$

from N into a set V of labels associating with each node $n \in N$ its label $e(n)$.

In particular, we may label the tree associated with a prefix-closed set T on an alphabet A by setting, for all $u \in T$, $e(u) = u$. In this case, a node u is at the left of a node v if $u \leq_{lex} v$ and if u is not an ancestor of v.

In a tree, a *path* from a node n_0 to a node n_k is a sequence $(n_0, \ldots, n_k)$ of nodes such that, for $0 \leq i \leq k-1$, n_i is the parent of n_{i+1}. In a similar way, an *infinite path* starting at n_0 is a sequence $(n_0, n_1, \ldots)$ of nodes such that for all $i \geq 0$, n_i is the parent of n_{i+1}.

The following result, known as *König's lemma*, is actually a compactness property (see Exercise III.6).

Proposition 2.2. *An infinite tree in which every node has finite arity contains an infinite path.*

Proof. Let (N, r, f) be an infinite tree in which every node is of finite arity. We build an infinite path starting at the root

$$(n_0, n_1, \ldots)$$

and such that for each k, the subtree rooted at n_k is infinite. Let first $n_0 = r$. Suppose then that n_k has already been chosen for $k \geq 0$. Since n_k is of finite arity, the set of its children is finite. There has to be at least one child such that the corresponding subtree is infinite. We choose this child for the node n_{k+1}. □

3 Rational sets of infinite words

In this section, we define the notion of an ω-rational set. It extends the corresponding notion of rational sets of finite words. Let us first recall that one defines on the subsets of A^* the rational operations as

(1) the set union $X \cup Y$,

(2) the set product

$$XY = \{xy \mid x \in X \text{ and } y \in Y\}$$

(3) the star operation defined by

$$X^* = \{x_1 \cdots x_n \mid n \geq 0 \text{ and } x_1, \ldots, x_n \in X\}$$

Thus X^* is the submonoid of A^* generated by X.

We also define the "plus" operation by

$$L^+ = \{u_1 \cdots u_n \mid n > 0 \text{ and } u_1, \ldots, u_n \in L\}$$

Thus L^+ is the subsemigroup generated by L. Obviously, $L^* = L^+ \cup \{1\}$ and $L^+ = LL^*$.

The class of *rational subsets* of A^* is the smallest class $\mathcal{R}$ of subsets of A^* such that

(a) $\mathcal{R}$ contains the empty set and for each $a \in A$, the singleton set $\{a\}$,

(b) $\mathcal{R}$ is closed under finite union, finite product and the operation $L \to L^*$.

It is an immediate consequence of the definitions that all finite subsets of A^* are rational. Indeed, since rational sets are closed under finite union, it suffices to verify that every singleton $\{u\}$ is rational. If $u = 1$, we observe that $\{1\} = \emptyset^*$, and if $u = a_1 \cdots a_n$, $\{u\} = \{a_1\} \cdots \{a_n\}$.

Rational sets are described by *rational expressions*, which are expressions using symbols from the alphabet A and the symbols $\cup$, $\cdot$, $^+$, *. To simplify the notation, we often denote a singleton set u instead of $\{u\}$ and the union $X \cup Y$ is also denoted $X + Y$. For instance, the expression $(ab + 1)^+$ is a shorthand for $(\{a\} \cdot \{b\} \cup \{1\})^+$.

Example 3.1. The set of words on the alphabet $A = \{a, b\}$ in which every a is always followed by a b can be described by the rational expression $b^*(ab^+)^*$. It is thus a rational subset of A^*. It is also described by the rational expression $(b + ab)^*$. This example shows that a rational set can be described by several rational expressions.

The definition of the product of two sets $X, Y \subset A^*$ can be extended to a more general case by setting, for $X \subset A^*$ and $Y \subset A^\infty$,

$$XY = \{xy \mid x \in X \text{ and } y \in Y\}$$

Then a restricted form of associativity holds: for each $X, Y \subset A^*$ and $Z \subset A^\infty$,

$$(XY)Z = X(YZ)$$

We now introduce an additional operation expressing infinite iteration. For each $X \subset A^*$, let

$$X^\omega = \{x_0x_1 \cdots \mid \text{for all } i \geq 0, x_i \in X \setminus \{1\}\}$$

and

$$X^\infty = X^* \cup X^\omega.$$

Thus, X^ω is the set of infinite words obtained by concatenating an infinite sequence of nonempty words of X. In particular, if $u = a_0a_1 \cdots a_n$ and if $X = \{u\}$, we have $X^\omega = \{u^\omega\}$, where u^ω is the infinite word

$$a_0a_1 \cdots a_na_0a_1 \cdots a_na_0a_1 \cdots a_na_0a_1 \cdots$$

obtained by repeating u an infinity of times.

The following proposition puts together some useful identities, whose formal proof is left to the reader.

Proposition 3.1. *For all $X, Y \subset A^*$, we have*

(1) $(X + Y)^\omega = (X^*Y)^\omega + (X + Y)^*X^\omega$,
(2) $(XY)^\omega = X(YX)^\omega$,
(3) *for all* $n > 0$, $(X^n)^\omega = (X^+)^\omega = X^\omega$,
(4) $XX^\omega = X^+X^\omega = X^\omega$.

Identity (1) is to be compared with the identity between subsets of A^*:

$$(X + Y)^* = (X^*Y)^*X^*$$

Identity (2) relates the operator $^\omega$ to the product. It is the counterpart of the identity

$$(XY)^* = 1 + X(YX)^*Y$$

which may also be written

$$(XY)^+ = X(YX)^*Y$$

Identity (3) expresses the fact that infinite iteration rules out multiplicities, in contrast with the analogous identity

$$X^* = (1 + X + X^2 + \cdots + X^{n-1})(X^n)^*$$

We can now give the definition of ω-rational subsets of A^∞. The class of *ω-rational* subsets of A^∞ is the smallest set $\mathcal{R}$ of subsets of A^∞ such that

(a) $\emptyset \in \mathcal{R}$ and for all $a \in A$, $\{a\} \in \mathcal{R}$,

(b) $\mathcal{R}$ is closed under finite union,

(c) for each subset X of A^* and for each subset Y of A^∞, $X \in \mathcal{R}$ and $Y \in \mathcal{R}$ imply $XY \in \mathcal{R}$,

(d) for every subset X of A^*, $X \in \mathcal{R}$ implies $X^* \in \mathcal{R}$ and $X^\omega \in \mathcal{R}$.

As a summary, the class of ω-rational subsets of A^∞ is the smallest class of subsets of A^∞ containing the finite subsets of A^∞ and closed under finite union, finite product and the operations $X \to X^*$ and $X \to X^\omega$.

In the sequel, we shall be especially interested in the ω-rational subsets of A^∞ which are contained in A^ω and which will be called ω-rational subsets of A^ω. There is a simple characterization of these subsets, which can also be used as a definition.

Theorem 3.2. *A subset of A^ω is ω-rational if and only if it is a finite union of sets of the form XY^ω where X and Y are rational subsets of A^*.*

Proof. We denote by $\mathcal{R}at(A^\omega)$ the class of subsets defined in the statement. It is clear that every element of $\mathcal{R}at(A^\omega)$ is an ω-rational subset of A^ω. To prove the converse, we establish a slightly more precise statement: if X is a rational subset of A^∞, then

(1) $X \cap A^*$ is a rational subset of A^*.

(2) $X \cap A^\omega \in \mathcal{R}at(A^\omega)$.

This property reduces obviously to $X \in \mathcal{R}at(A^\omega)$ when $X \subset A^\omega$. Let $\mathcal{E}$ be the class of subsets of A^∞ satisfying (1) and (2). We have successively:

(a) $\emptyset \in \mathcal{E}$ and $\{a\} \in \mathcal{E}$ for every $a \in A$.

(b) $\mathcal{E}$ is closed under finite union.

(c) $\mathcal{E}$ is closed under product. In fact, if $X \subset A^*$ and $Y \subset A^\infty$, then $(XY \cap A^*) = X(Y \cap A^*)$, which is rational since Y satisfies condition (1). Also, $(XY \cap A^\omega) = X(Y \cap A^\omega)$, which is in $\mathcal{R}at(A^\omega)$ since, by condition (2), $Y \cap A^\omega \in \mathcal{R}at(A^\omega)$.

(d) $\mathcal{E}$ is closed under the operation $X \to X^*$.

(e) $\mathcal{E}$ is closed under the operation $X \to X^\omega$.

As a result, $\mathcal{E}$ contains the class of rational subsets of A^∞, which proves the theorem. □

Example 3.2. The set X of infinite words on the alphabet $\{a, b\}$ with only a finite number of occurrences of the symbol b is given by the ω-rational expression $X = (a + b)^*a^\omega$. The complement of X in A^ω, which is the set of words with an infinite number of occurrences of b is given by the expression $(a^*b)^\omega$, and is therefore also ω-rational.

Example 3.2 actually presents a particular case of a general result: the set $\mathcal{R}at(A^\omega)$ is closed under all boolean operations. This result will be proved later. The delicate point is the complement since, as it can be seen on the previous example, given an ω-rational expression for a set, it is not easy to find an ω-rational expression for its complement. Computing the intersection of two ω-rational sets is easier and can be done directly (see Exercise 7).

Let A and B be two alphabets. Any application $\varphi : A \to B^+$ defines a unique semigroup morphism $\varphi : A^+ \to B^+$, obtained by setting for a word $u = a_0a_1 \cdots a_k$, $\varphi(u) = \varphi(a_0)\varphi(a_1) \cdots \varphi(a_k)$. It can also be turned into a monoid morphism from A^* into B^* by setting $\varphi(1) = 1$. It also extends to a mapping $\varphi : A^\infty \to B^\infty$, also called morphism and defined for an infinite word $u = a_0a_1 \cdots$, by $\varphi(u) = \varphi(a_0)\varphi(a_1) \cdots$.

A semigroup morphism $\varphi : A^+ \to B^+$ will be called *alphabetic* if, for every $a \in A$, $\varphi(a) \in B$. If X is a subset of A^∞, we set

$$\varphi(X) = \{\varphi(u) \mid u \in X\}$$

Then the following formulas hold, where X is a subset of A^+ and where X_1 and X_2 are subsets of A^∞:

$$\begin{aligned} \varphi(X_1 \cup X_2) &= \varphi(X_1) \cup \varphi(X_2), \\ \text{if } X_1 \subset A^+, \quad \varphi(X_1X_2) &= \varphi(X_1)\varphi(X_2), \\ \varphi(X^+) &= \varphi(X)^+, \\ \varphi(X^\omega) &= \varphi(X)^\omega. \end{aligned}$$

It follows immediately that ω-rational sets are stable under morphism.

Proposition 3.3. *Let A and B be two alphabets and let $\varphi : A^+ \to B^+$ be a morphism. If X is an ω-rational subset of A^∞ (resp. of A^*, A^ω), then $\varphi(X)$ is an ω-rational subset of B^∞ (resp. of B^*, B^ω).*

4 Automata

An *automaton* on the alphabet A is given by a set Q, called the set of *states*, and a subset E of $Q \times A \times Q$, called the set of *edges* or *transitions*. Some additional components may be added, and in particular a subset $I \subset Q$ of *initial* states and a subset $F \subset Q$ of *final* or *terminal* states. The automaton is often denoted as a tuple $\mathcal{A} = (Q, A, E)$ or $\mathcal{A} = (Q, A, E, I, F)$ if I and F are specified. Part of the components may always be omitted. In particular, we sometimes denote the automaton merely (E, I, F) when Q and A are unambiguously defined. Moreover, if $I = \{i\}$, or if $F = \{f\}$, the automaton is denoted (E, i, F) or (E, I, f).

An automaton is said to be *finite* (resp. *countable*) if both its alphabet and the sets of its states are finite (resp. countable).

Two transitions (p, a, q) and (p', a', q') are called *consecutive* if $q = p'$. A *path* in the automaton $\mathcal{A}$ is a finite sequence of consecutive transitions

$$e_0 = (q_0, a_0, q_1), \quad e_1 = (q_1, a_1, q_2), \quad \ldots \quad , \quad e_{n-1} = (q_{n-1}, a_{n-1}, q_n)$$

also denoted

$$q_0 \xrightarrow{a_0} q_1 \xrightarrow{a_1} q_2 \cdots q_{n-1} \xrightarrow{a_{n-1}} q_n \qquad \text{or} \qquad q_0 \xrightarrow{a_0 \cdots a_{n-1}} q_n.$$

The state q_0 is the *origin* of the path and the state q_{n+1} its *end*. One says that the path *passes through* (or *visits*) the states $q_0, q_1, \ldots, q_{n+1}$. The word $x = a_0 a_1 \cdots a_n$ is the *label* of the path and the integer $n+1$ its *length*. The set $\{q_0, q_1, \ldots, q_{n+1}\}$ is the *content* of the path.

It is convenient to introduce, for each state $q \in Q$, an empty path with origin and end equal to q. Its label is the empty word and its length is 0.

An *infinite path* in the automaton $\mathcal{A}$ is an infinite sequence p of consecutive transitions

$$e_0 = (q_0, a_0, q_1), \quad e_1 = (q_1, a_1, q_2), \quad \ldots$$

also denoted

$$q_0 \xrightarrow{a_0} q_1 \xrightarrow{a_1} q_2 \quad \cdots$$

The state q_0 is the *origin* of the infinite path and the infinite word $a_0 a_1 \cdots$ is its *label*. We say that the path p *passes infinitely often* through a state q (or that p *visits* q infinitely often, or yet that q is *infinitely repeated* in p) if there are infinitely many integers n such that $q_n = q$. The set of *infinitely repeated* states in p is denoted by $\mathrm{Inf}(p)$.

We usually specify for each automaton a set of *successful* finite or infinite paths. This will be done for infinite paths in various ways in the next sections. For finite paths, there is just one notion.

A finite path in $\mathcal{A}$ is *initial* if its origin is in I and *final* if its end is in F. A path is *successful* if it is both initial and final. The set of words *recognized* by the automaton $\mathcal{A}$ is the set, denoted by $L^*(\mathcal{A})$, of all labels of successful paths in $\mathcal{A}$. We also set $L^+(\mathcal{A}) = L^*(\mathcal{A}) \setminus \{1\}$.

A set of finite words X is *recognizable* if there exists a finite automaton $\mathcal{A}$ such that $X = L^*(\mathcal{A})$.

Example 4.1. Let $\mathcal{A} = (Q, A, E, I, F)$ where $Q = \{1, 2\}$, $A = \{a, b\}$ and

$$E = \{(1, a, 1), (2, b, 1), (1, a, 2), (2, b, 2)\}, \quad I = \{1\}, \quad F = \{2\}$$

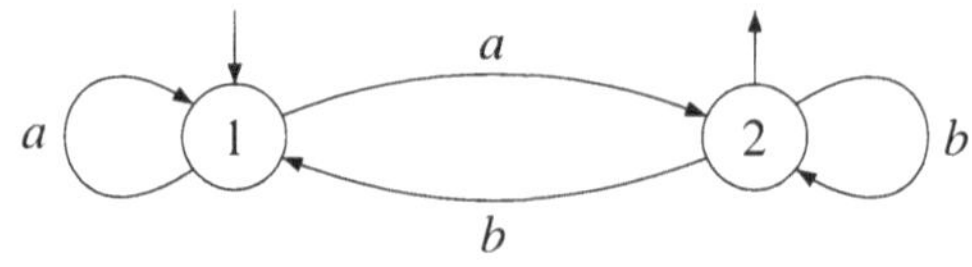

Figure 4.1. The automaton $\mathcal{A}$.

This automaton is represented in Figure 4.1. According to a convention used in all the sequel, the initial states are indicated by an incoming arrow and the final states by an outgoing one. We have $L^+(\mathcal{A}) = a\{a, b\}^*$, which is the set of finite words beginning with an a. In this automaton, every finite word is the label of exactly two paths. Indeed, every letter determines the state it comes from and conversely, the transitions going out of a given state all have the same label. This automaton can thus be interpreted as a mechanism predicting the next symbol. Every infinite word is the label of exactly one path.

An automaton $\mathcal{A} = (Q, A, E, I, F)$ is said to have *deterministic transitions*, if, for every state $q \in Q$ and every letter $a \in A$, there is at most one state q' such that (q, a, q') is a transition. It is *deterministic* if it has deterministic transitions and if I is a singleton. It is *complete* if, for every state $q \in Q$ and every letter $a \in A$, there is at least one state q' such that (q, a, q') is a transition. If q_0 is the unique initial state, we adopt the notation (Q, A, E, q_0, F) instead of $(Q, A, E, \{q_0\}, F)$.

For instance, the automaton of Example 4.1 is neither complete nor deterministic. In contrast, the automaton represented in Figure 4.2 is complete and deterministic.

Example 4.2. The automaton given by Figure 4.2 is deterministic. Each word is the label of exactly two paths, one going out of state 1 and the other from state 2. The automaton can be considered as "remembering the last symbol" since each state is accessible only by one symbol.

If $\mathcal{A} = (Q, A, E)$ is deterministic, we define a partial function from $Q \times A$ into Q by associating with each pair (q, a) in $Q \times A$ the unique state $q \cdot a$ (when it exists), such that $(q, a, q \cdot a) \in E$. If there is no q' such that $(q, a, q') \in E$, the image of (q, a) is not defined. The partial function $(q, a) \to q \cdot a$ thus defined is the *transition function*

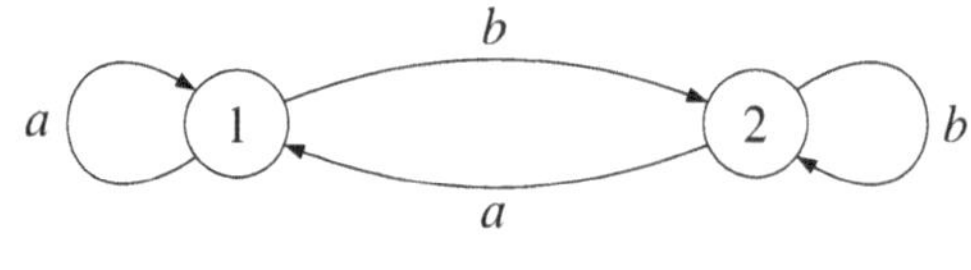

Figure 4.2. A complete deterministic automaton.

of the automaton. It is clear that a deterministic automaton is defined by its transition function.

The transition function can be extended to a partial function from $Q \times A^*$ into Q by setting $q \cdot 1 = q$ and, for every word $u \in A^+$ and for every symbol $a \in A$, $q \cdot (ua) = (q \cdot u) \cdot a$ if $(q \cdot u)$ and $(q \cdot u) \cdot a$ are defined. For example, if $\mathcal{A}$ is the automaton of Example 4.2, we have $1 \cdot bbaba = 2 \cdot bbaba = 1$.

One may also define the *transition function* of a nondeterministic automaton $\mathcal{A} = (Q, A, E)$. It is the function from $\mathcal{P}(Q) \times A^*$ into $\mathcal{P}(Q)$, traditionally denoted δ, defined by the following formula, where $S \subset Q$ and $u \in A^*$,

$$\delta(S, u) \quad = \quad \{q \in Q \mid q \text{ is the end of a path in } \mathcal{A} \text{ with label } u \text{ going out from some state of } S\}$$

It is clear that knowing E or δ is the same. We shall use one of either notation in the sequel. This transition function allows one to define the *deterministic version* of an automaton (Q, A, E, I, F) as the complete deterministic automaton

$$(\mathcal{P}(Q), A, \delta, I, \mathcal{F}) \text{ with } \mathcal{F} = \{P \subset Q \mid P \cap F \neq \emptyset\}$$

This construction is motivated by the following result, which shows that, for finite words, deterministic automata have the same expressive power as non deterministic ones. We shall see in the next sections that this elementary result does not extend easily to infinite words.

Proposition 4.1. *An automaton and its deterministic version recognize the same set of finite words.*

Proof. Let $\mathcal{A} = (Q, A, E, I, F)$ and let $\mathcal{B}$ be its deterministic version. If $u = a_0 \cdots a_{n-1}$ is recognized by $\mathcal{A}$, there is a successful path

$$c : q_0 \xrightarrow{a_0} q_1 \xrightarrow{a_1} q_2 \quad \cdots \quad q_{n-1} \xrightarrow{a_{n-1}} q_n$$

Define a sequence of subsets of Q by setting $P_0 = I$, $P_1 = P_0 \cdot a_0, \ldots, P_n = P_{n-1} \cdot a_{n-1}$. Since c is a successful path, $q_0 \in I = P_0$ and $q_n \in F$. Suppose, by induction, that $q_{i-1} \in P_{i-1}$. Then since $q_{i-1} \xrightarrow{a_i} q_i$ is a transition, $q_i \in P_i$. In particular, $q_n \in P_n \cap F$. Therefore $P_n \cap F \neq \emptyset$ and $P_n \in \mathcal{F}$. Consequently, u is accepted by $\mathcal{B}$.

Conversely, let $u = a_0 \cdots a_{n-1}$ be a word accepted by $\mathcal{B}$. Set, as above,

$$P_0 = I, \; P_1 = P_0 \cdot a_0, \; \ldots, \; P_n = P_{n-1} \cdot a_{n-1}.$$

Since $P_n \in \mathcal{F}$, one may choose an element q_n in $P_n \cap F$, and, for $0 \leq i \leq n$, an element $q_i \in P_i$ such that $q_i \xrightarrow{a_i} q_{i+1}$ is a transition in $\mathcal{A}$. Since $q_0 \in I$ and $q_n \in F$, the path $q_0 \xrightarrow{a_0} q_1 \xrightarrow{a_1} q_2 \cdots q_{n-1} \xrightarrow{a_{n-1}} q_n$ is successful, and u is accepted by $\mathcal{A}$. □

An important consequence of Proposition 4.1 is that recognizable sets are closed under complementation. Actually, they are closed under any boolean operation, but we need an auxiliary definition before proving this result. The *product* of two automata $\mathcal{A} = (Q, A, E)$ and $\mathcal{A}' = (Q', A, E')$ is the automaton

$$\mathcal{A} \times \mathcal{A}' = (Q \times Q', A, P)$$

with

$$P = \{((p, p'), a, (q, q')) \mid (p, a, q) \in E \text{ and } (p', a, q') \in E'\}.$$

Proposition 4.2. *Recognizable subsets of A^* are closed under finite union, finite intersection and set difference.*

Proof. Let X and X' be two recognizable subsets of A^*. By Proposition 4.1, we may assume that $X = L^*(\mathcal{A})$ and $X' = L^*(\mathcal{A}')$ for some deterministic complete automata $\mathcal{A} = (Q, A, \cdot, i, F)$ and $\mathcal{A}' = (Q', A, \cdot, i', F')$. Let $\mathcal{B}$ be the product of $\mathcal{A}$ and $\mathcal{A}'$. Equipped with the initial state (i, i'), $\mathcal{B}$ is a deterministic automaton. Let R be a set of final states for $\mathcal{B}$. If $R = (F \times Q') \cup (Q \times F')$, we have $L^*(\mathcal{B}) = X \cup X'$. If $R = F \times F'$, we have $L^*(\mathcal{B}) = X \cap X'$. Finally, if $R = F \times (Q \setminus F')$, we have $L^*(\mathcal{B}) = X \setminus X'$. □

Let $\mathcal{A} = (Q, A, E)$ be an automaton. The *reversed* automaton of $\mathcal{A}$ is the automaton $\mathcal{A}^r = (Q, A, E^r)$, where

$$E^r = \{(q, a, p) \mid (p, a, q) \in E\},$$

obtained by reverting the arrows of $\mathcal{A}$. An automaton $\mathcal{A}$ has *co-deterministic* transitions if the automaton $\mathcal{A}^r$ has deterministic transitions. Thus, the automaton of Figure 4.1 has co-deterministic transitions.

An automaton $\mathcal{A} = (Q, A, E)$ is *unambiguous* if, for each pair of states $(p, q) \in Q \times Q$ and for each word x, there is at most one path from p to q with label x. It is easy to verify that any automaton with deterministic or co-deterministic transitions is unambiguous, but the converse is not true as shown by the following example (see Figure 4.3). Let $\mathcal{A} = (Q, A, E)$ be an automaton. We define its *transition matrix* as follows. It is the $Q \times Q$-matrix T with coefficients in $\mathcal{P}(A^*)$ defined by

$$T_{p,q} = \{a \in A \mid (p, a, q) \in E\}.$$

The set $\mathcal{P}(A^*)$ of subsets of A^* comes with two operations: the union, which we shall denote additively and the product

$$XY = \{xy \mid x \in X \text{ and } y \in Y\}$$

The set $\mathcal{P}(A^*)^{Q \times Q}$ of $Q \times Q$-matrices with coefficients in $\mathcal{P}(A^*)$ comes in turn with an addition and a product by setting, for any $M, N \in \mathcal{P}(A)^{Q \times Q}$ and $p, q \in Q$

$$(M + N)_{p,q} = M_{p,q} + N_{p,q}$$

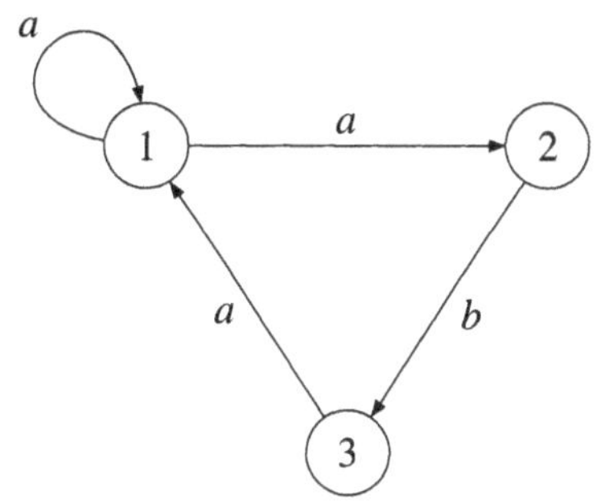

Figure 4.3. An unambiguous automaton.

and

$$(MN)_{p,q} = \sum_{r \in Q} M_{p,r} N_{r,q}$$

The fact that the above union may have an infinite number of terms is not an obstacle, since infinite unions are well-defined. We note that the identity matrix defined as

$$1_{p,q} = \begin{cases} \{1\} & \text{if } p = q, \\ \emptyset & \text{otherwise} \end{cases}$$

is the neutral element of the product.

Given a transition matrix T, we define a new matrix T^* by

$$T^* = \sum_{n \geq 0} T^n = 1 + T + T^2 + \cdots$$

The reader can verify that

$$T^*_{p,q} = \{u \in A^* \mid \text{there exists a path from } p \text{ to } q \text{ with label } u\}.$$

It can be shown that if a matrix T has all its coefficients in $\mathcal{P}(A^+)$, as it is the case here, then $X = T^*$ is the unique solution of the equation

$$X = 1 + XT. \tag{4.1}$$

We recall the following statement, which is classical in automata theory.

Proposition 4.3. *If $\mathcal{A}$ is a finite automaton, the coefficients of T^* are rational subsets of A^*.*

Proof. We use an induction on the number n of states of $\mathcal{A}$. The case $n = 0$ is trivial. If $n = 1$, we have $T^*_{1,1} = B^*$, with

$$B = \{a \mid (1, a, 1) \in E\}.$$

Since E is finite, B is finite and $T_{1,1}$ is a rational subset of A^*. If $n > 1$, let us consider a partition of the matrix T into blocks

$$\begin{pmatrix} U & V \\ W & Z \end{pmatrix}$$

where U and Z are square matrices. Let, for the same partition,

$$T^* = \begin{pmatrix} U' & V' \\ W' & Z' \end{pmatrix}.$$

Then

$$\begin{aligned} U' &= (U + VZ^*W)^* & V' &= U'VZ^* \\ W' &= Z'WU^* & Z' &= (Z + WU^*V)^* \end{aligned}$$

since these formulas allow to verify that T^* is a solution of Equation 4.1. As a result, the coefficients of the matrices U', V', W', Z' are rational, and thus so are those of T^*. □

Example 4.3. The matrices T and T^* corresponding to Example 4.2 are

$$T = \begin{pmatrix} a & b \\ a & b \end{pmatrix} \quad \text{and}$$

$$T^* = \begin{pmatrix} (a + bb^*a)^* & (a + bb^*a)^*bb^* \\ (b + aa^*b)^*aa^* & (b + aa^*b)^* \end{pmatrix} = \begin{pmatrix} (b^*a)^* & (b^*a)^*bb^* \\ (a^*b)^*aa^* & (a^*b)^* \end{pmatrix}$$

We now recall the statement of Kleene and give a sketch of the proof.

Theorem 4.4 (Kleene). *A set of finite words is recognizable if and only if it is rational.*

Proof. By Proposition 4.3 we already know that every recognizable set is rational. To prove the converse, we prove a lemma which will also be used for the case of infinite words.

An automaton is said to be *normalized* if it has only one initial state i and only one final state f and if no transition ends in i or starts from f. The following lemma shows that, for subsets of A^+, one may always replace an automaton by a normalized one.

Lemma 4.5. *For any finite automaton $\mathcal{A}$, there is a finite normalized automaton $\mathcal{A}'$ such that $L^+(\mathcal{A}) = L^+(\mathcal{A}')$.*

Proof. Let $\mathcal{A} = (Q, A, E, I, F)$ and $\mathcal{A}' = (Q \cup \{i', f'\}, A, E', \{i'\}, \{f'\})$, where i' and f' are two new states and where $E' = E \cup E_0 \cup E_1 \cup E_2$ with

$$E_0 = \{(i', a, f') \mid \text{ there exists } i \in I \text{ and } f \in F \text{ such that } (i, a, f) \in E\}$$
$$E_1 = \{(i', a, q) \mid \text{ there exists } i \in I \text{ such that } (i, a, q) \in E\}$$
$$E_2 = \{(q, a, f') \mid \text{ there exists } f \in F \text{ such that } (q, a, f) \in E\}$$

Then $\mathcal{A}'$ is normalized by construction. Moreover, if a nonempty word $u = a_1 \cdots a_n$ is recognized by $\mathcal{A}$, there exists a nonempty path

$$q_0 \xrightarrow{a_1} q_1 \xrightarrow{a_2} q_2 \ \cdots \ q_{n-1} \xrightarrow{a_n} q_n$$

starting in I and ending in F. Thus,

$$i' \xrightarrow{a_1} q_1 \xrightarrow{a_2} q_2 \ \cdots \ q_{n-1} \xrightarrow{a_n} f'$$

is a path in $\mathcal{A}'$, which shows that u is recognized by $\mathcal{A}'$. Conversely, if u is recognized by $\mathcal{A}'$, there exists a path in $\mathcal{A}'$ of the form

$$i' \xrightarrow{a_1} q_1 \xrightarrow{a_2} q_2 \ \cdots \ q_{n-1} \xrightarrow{a_n} f'$$

Thus there exist $i \in I$ and $f \in F$ such that

$$i \xrightarrow{a_1} q_1 \xrightarrow{a_2} q_2 \ \cdots \ q_{n-1} \xrightarrow{a_n} f$$

is a path in $\mathcal{A}$ and u is recognized by $\mathcal{A}$. □

Consider now two subsets X and X' of A^*, recognized respectively by the normalized finite automata $\mathcal{A} = (E, i, f)$ and $\mathcal{A}' = (E', i', f')$. We may suppose that E and E' are disjoint. Let then $\mathcal{B}$ be the automaton $\mathcal{B} = (E \cup E', \{i, i'\}, \{f, f'\})$ represented in Figure 4.4. This automaton recognizes $X \cup X'$. To build an automaton $\mathcal{C}$ recognizing the product XX' we merge f and i', as shown in Figure 4.5. We finally build an automaton recognizing X^* by merging i and f in the automaton $\mathcal{A}$, as represented in Figure 4.6. Lemma 4.5 shows that every recognizable set of A^+ is recognized by a normalized

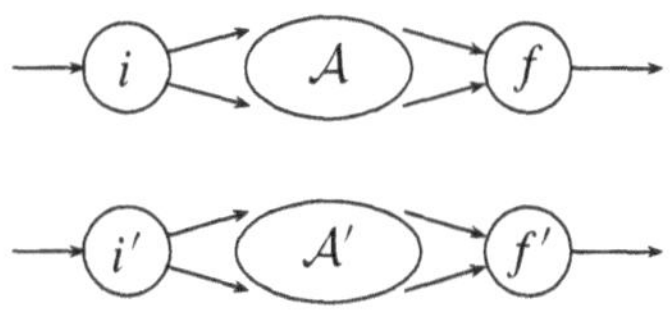

Figure 4.4. An automaton recognizing the union.

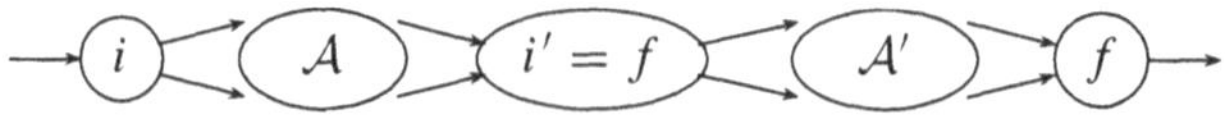

Figure 4.5. An automaton recognizing the product.

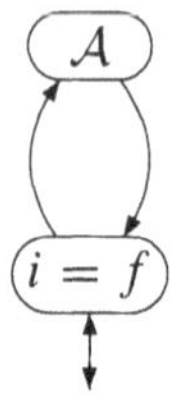

Figure 4.6. An automaton recognizing X^*.

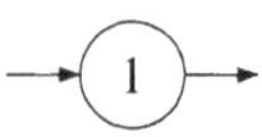

Figure 4.7. An automaton recognizing $\{1\}$.

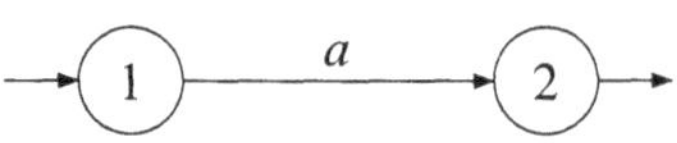

Figure 4.8. An automaton recognizing $\{a\}$.

automaton. Furthermore, we have seen that the union, the product and the star of such sets is still recognizable.

Working with normalized automata simplified the previous constructions. However, the price to pay is a special treatment for the empty word. We first observe that the set $\{1\}$ is recognized by the automaton with a single state, both initial and final, and no transition, represented in Figure 4.7.

It follows, by Proposition 4.2, that a set X of finite words is recognizable if and only if $X \setminus \{1\}$ is recognizable. We can now conclude that the class of recognizable sets of A^* is closed under finite union, product and star. For instance, for the product it follows from the following elementary formula, in which $\varepsilon(X) = X \cap \{1\}$:

$$XX' = (X \cap A^+)(X' \cap A^+) \cup \varepsilon(X)(X' \cap A^+) \cup (X \cap A^+)\varepsilon(X') \cup \varepsilon(X)\varepsilon(X')$$

Finally, the empty set is recognized by the empty automaton and if a is a letter, the set $\{a\}$ is recognized by the automaton (E, I, F) with $E = \{(1, a, 2)\}$, $I = \{1\}$ and $F = \{2\}$, represented in Figure 4.8. Thus any rational subset of A^* is recognizable and this concludes the proof of Kleene's theorem. □

5 Büchi automata

We now introduce Büchi automata, which correspond to the simplest recognizing mode for infinite words. A *Büchi automaton* is a 5-tuple $\mathcal{A} = (Q, A, E, I, F)$ where

(1) (Q, A, E) is an automaton,

(2) I and F are subsets of Q, called resp. set of *initial states* and set of *final states*.

Let $\mathcal{A} = (Q, A, E, I, F)$ be a Büchi automaton. We say that an infinite path in $\mathcal{A}$ is *initial* if its origin is in I and *final* if it visits F infinitely often. It is *successful* if it is initial and final. The set of infinite words *recognized* by $\mathcal{A}$ is the set, denoted by $L^\omega(\mathcal{A})$, of labels of infinite successful paths in $\mathcal{A}$. In the case where F is *finite* and in particular if $\mathcal{A}$ is a finite automaton, $L^\omega(\mathcal{A})$ is also the set of labels of infinite initial paths p in $\mathcal{A}$ and such that $\mathrm{Inf}(p) \cap F \neq \emptyset$.

A set X of infinite words is *recognizable* if there exists a finite Büchi automaton $\mathcal{A}$ such that $X = L^\omega(\mathcal{A})$.

A Büchi automaton $\mathcal{A} = (Q, A, E, I, F)$ is said to be *deterministic* if it has deterministic transitions and if I is a singleton, i.e. if $\mathcal{A}$ contains exactly one initial state i. In this case, every word in A^+ (resp. A^ω) is the label of at most one initial path. In particular, every word in $L^\omega(\mathcal{A})$ is the label of exactly one initial path.

A Büchi automaton $\mathcal{A} = (Q, A, E, I, F)$ is called *co-deterministic* if it has co-deterministic transitions and if any word in A^ω is the label of at most one final path. It is *co-complete* if any word in A^ω is the label of at least one final path.

More generally, a Büchi automaton $\mathcal{A} = (Q, A, E, I, F)$ is said to be *ω-unambiguous* if every word in A^ω is the label of at most one infinite successful path. In particular, every word in $L^\omega(\mathcal{A})$ defines a unique successful infinite path of which it is the label. It is clear that any deterministic or co-deterministic Büchi automaton is ω-unambiguous, but the converse is not true. The various terms are summarized in Table 5.1.

Example 5.1. Let $\mathcal{A}$ be the Büchi automaton obtained from the automaton of Example 4.1. We have $L^\omega(\mathcal{A}) = a(a^*b)^\omega$, which is the set of infinite words starting with a and containing an infinite number of occurrences of b.

Example 5.2. Let $\mathcal{A}$ be the Büchi automaton represented in Figure 5.2. We have now $L^+(\mathcal{A}) = \{a, b\}^+$ and $L^\omega(\mathcal{A}) = \{a, b\}^* a^\omega$. This automaton is also co-deterministic.

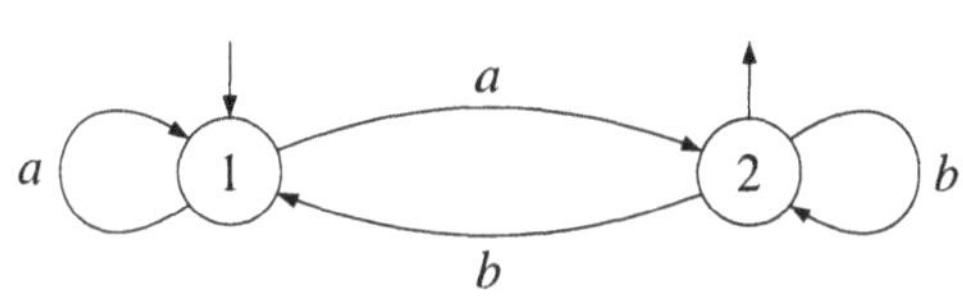

Figure 5.1. A co-deterministic Büchi automaton.

<table>
<tr><th>Det. transitions</th><th>Co-det. transitions</th><th>Unambiguous</th></tr>
<tr><td>Forbidden configuration:
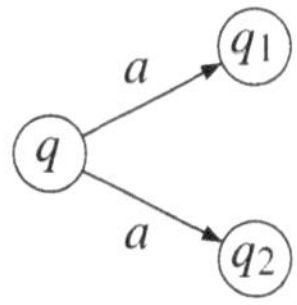

where a is a letter.</td><td>Forbidden configuration:
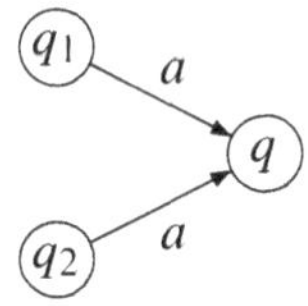

where a is a letter.</td><td>Forbidden configuration:
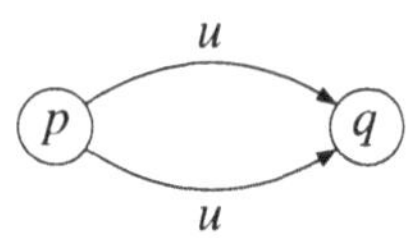

where u is a word.</td></tr>
<tr><th>Deterministic</th><th>Co-deterministic</th><th>ω-unambiguous</th></tr>
<tr><td>Deterministic transitions and exactly one initial state</td><td>Co-det. transitions and two final paths with the same label are equal</td><td rowspan="3">Every infinite word is the label of at most one successful path</td></tr>
<tr><th>Complete</th><th>Co-complete</th></tr>
<tr><td>Every word is the label of some initial path</td><td>Every word is the label of some final path</td></tr>
</table>

Table 5.1. Summary of the definitions.

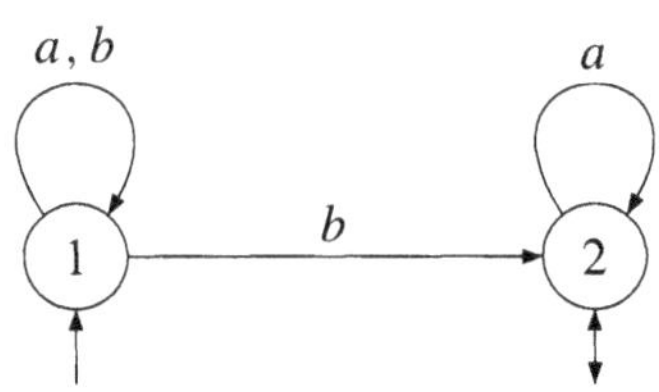

Figure 5.2. Another co-deterministic Büchi automaton.

Example 5.3. The Büchi automaton represented in Figure 5.3 is ω-unambiguous but it is neither deterministic nor co-deterministic. It is not deterministic since it has two initial states and it is not co-deterministic since the infinite word $(ab)^\omega$ is the label of two final paths, one starting at state 2, the other one at state 3.

We shall see later on that a recognizable set of infinite words is completely determined by its ultimately periodic words. The following lemma proves the easy part of this result.

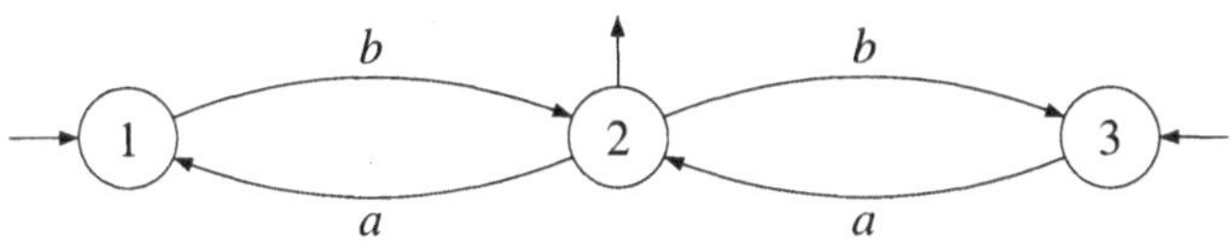

Figure 5.3. An ω-unambiguous automaton.

Lemma 5.1. *Any nonempty recognizable subset of A^ω contains an ultimately periodic word.*

Proof. Let X be a nonempty recognizable subset of A^ω recognized by a Büchi automaton $\mathcal{A} = (E, I, F)$. Since X is nonempty, there exists a path of the form $p = p_0 p_1 p_2 \cdots$, where p_0 starts in I, ends in a state $q \in F$, and where $p_1, p_2, \cdots$ are paths from q to q. The path $p_0 p_1 p_1 p_1 \cdots$ is thus also a successful path and its label is an ultimately periodic word. □

Let $\mathcal{A} = (E, I, F)$ be a Büchi automaton. A state q is called *accessible* if there is a (possibly empty) finite initial path in $\mathcal{A}$ ending in q. A state q is called *coaccessible* if there exists an infinite final path starting at q. Finally, $\mathcal{A}$ is *trim* if all its states are both accessible and coaccessible.

Suppressing all "useless" states of a Büchi automaton always gives a trim automaton. More formally, one has the following result.

Proposition 5.2. *With any Büchi automaton $\mathcal{A}$, one may associate a trim Büchi automaton $\mathcal{A}'$ such that*

(a) *The automata $\mathcal{A}$ and $\mathcal{A}'$ recognize the same subset of A^ω,*
(b) *if $\mathcal{A}$ is deterministic, so is $\mathcal{A}'$,*
(c) *if $\mathcal{A}$ is finite, so is $\mathcal{A}'$.*

Proof. (a) Let $\mathcal{A} = (Q, A, E, I, F)$ be a Büchi automaton and let P be the set of states of $\mathcal{A}$ that are both accessible and coaccessible. Let $\mathcal{A}' = (P, A, E', I \cap P, F \cap P)$ where $E' = E \cap (P \times A \times P)$. It is clear that $L^\omega(\mathcal{A}') \subset L^\omega(\mathcal{A})$. Conversely, let $u = a_0 a_1 \cdots \in L^\omega(\mathcal{A})$. There is a final path

$$p = q_0 \xrightarrow{a_0} q_1 \xrightarrow{a_1} q_2 \cdots$$

with label u such that $q_0 \in I$. The states $q_0, q_1, \ldots$ are then both accessible and coaccessible. Thus, p is a path in $\mathcal{A}'$ and $u \in L^\omega(\mathcal{A}')$. We conclude that $L^\omega(\mathcal{A}) = L^\omega(\mathcal{A}')$.

(b) If $\mathcal{A}$ is deterministic, the automaton (P, A, E') is deterministic. Moreover, if the set recognized by $\mathcal{A}$ is nonempty, there exists a final path starting at the unique initial state. Thus, the initial state is coaccessible and $I \cap P$ is a singleton. Therefore $\mathcal{A}'$ is deterministic.

(c) Finally it is clear that these constructions preserve the finiteness of the automaton. □

A Büchi automaton can, in the same way, be made complete.

Proposition 5.3. *A subset of A^ω recognized by a Büchi automaton $\mathcal{A}$ can be recognized by a complete Büchi automaton $\mathcal{A}'$ such that if $\mathcal{A}$ is finite (resp. deterministic), then $\mathcal{A}'$ is also finite (resp. deterministic).*

Proof. Let $\mathcal{A} = (Q, A, E, I, F)$ be a Büchi automaton recognizing a subset X of A^ω. If $\mathcal{A}$ is not complete, we add a new state p and we create a transition (q, a, p) if there is no transition of the form (q, a, q') in $\mathcal{A}$. More formally, let $\mathcal{A}' = (Q \cup \{p\}, A, E', I, F)$, where p is a new state and $E' = E \cup E_1 \cup E_2$ with

$$E_1 = \{(p, a, p) \mid a \in A\}$$
$$E_2 = \{(q, a, p) \mid q \in Q, a \in A \text{ and } (\{q\} \times \{a\} \times Q) \cap E = \emptyset\}$$

The automaton $\mathcal{A}'$ still recognizes X, is complete and it is deterministic (resp. finite) if $\mathcal{A}$ is deterministic (resp. finite). □

Example 5.4. The Büchi automaton of Example 5.2 can be completed as is shown in Figure 5.4.

We now prove the analogue of Kleene's theorem for infinite words.

Theorem 5.4. *A subset of A^ω is recognizable if and only if it is ω-rational.*

Proof. First consider a recognizable subset X of A^ω and let $\mathcal{A} = (E, I, F)$ be a finite Büchi automaton recognizing X. Then

$$X = L^\omega(E, I, F) = \bigcup_{i \in I} \bigcup_{f \in F} L^*(E, i, f)\big(L^+(E, f, f)\big)^\omega$$

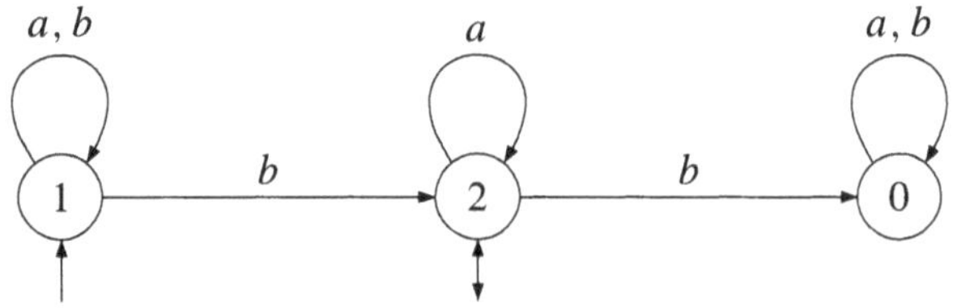

Figure 5.4. Completion of the automaton of Figure 5.2.

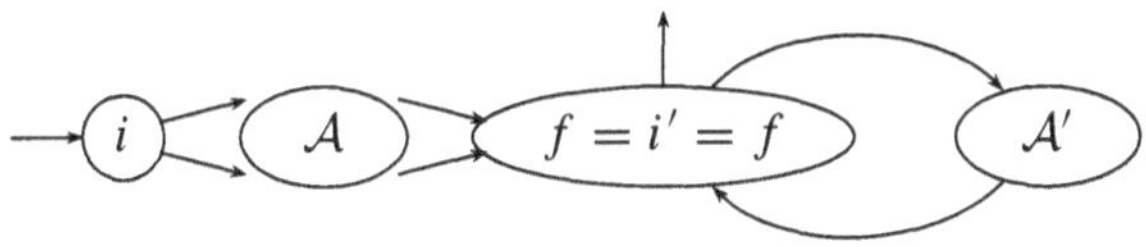

Figure 5.5. An automaton recognizing $X(X')^\omega$.

which shows that X is ω-rational, since, by Kleene's theorem, the sets $L^*(E, i, f)$ and $L^+(E, f, f)$ are rational subsets of A^*.

Conversely, let us first consider an ω-rational set Y of the form $X(X')^\omega$ with X, X' rational subsets of A^*. Let $\mathcal{A} = (Q, A, E, i, f)$ and $\mathcal{A}' = (Q', A, E', i', f')$ be two normalized automata such that $X = L^+(\mathcal{A})$ and $X' = L^+(\mathcal{A}')$. We build an automaton $\mathcal{B}$ by merging the states f, i' and f' as indicated in Figure 5.5. Formally, we have $\mathcal{B} = ((Q \cup Q') \setminus \{i', f'\}, T, i, f)$, where $T = E \cup E_0 \cup E_1 \cup E_2$ with

$$E_0 = \{(f, a, f) \mid (i', a, f') \in E'\}$$
$$E_1 = \{(f, a, q) \mid q \in Q' \setminus \{i', f'\} \text{ and } (i', a, q) \in E'\}$$
$$E_2 = \{(q, a, f) \mid q \in Q' \setminus \{i', f'\} \text{ and } (q, a, f') \in E'\}$$

This shows that $X(X')^\omega$ is recognizable.

To complete the proof of the theorem, we only have to prove that the class of recognizable subsets of A^ω is stable under finite union.

Let then Y and Y' be two recognizable subsets of A^ω, recognized resp. by the finite Büchi automata $\mathcal{A} = (Q, A, E, I, F)$ and $\mathcal{A}' = (Q', A, E', I', F')$. We may suppose that Q and Q' are disjoint and thus we may identify E and E' with subsets of $(Q \cup Q') \times A \times (Q \cup Q')$. With this convention, we have the formula

$$Y \cup Y' = L^\omega(E, I, F) \cup L^\omega(E', I', F') = L^\omega(E \cup E', I \cup I', F \cup F')$$

and thus $Y \cup Y'$ is recognized by the finite automaton $(Q \cup Q', A, E \cup E', I \cup I', F \cup F')$. □

We conclude this section with an additional closure property of the class of recognizable sets of A^ω. We have already seen (Proposition 3.3) that the class of ω-rational sets is closed under morphism. We consider now inverse morphisms. Let $\varphi : A \to B^+$ be a function and let $\varphi : A^\omega \to B^\omega$ be the morphism induced by φ.

Proposition 5.5. *If X is a recognizable subset of B^ω, then $\varphi^{-1}(X)$ is a recognizable subset of A^ω.*

Proof. Let $\mathcal{B} = (Q, B, E, I, F)$ be a Büchi automaton recognizing X. Let

$$\mathcal{A} = (Q \times \{0, 1\}, A, E', I \times \{0\}, Q \times \{1\}),$$

where $E' = E_1 \cup E_2$, with

$$E_1 = \big\{((q,i), a, (q',0)) \mid i \in \{0,1\} \text{ and there exists a path of } \mathcal{B} \text{ from } q \text{ to } q', \text{ labeled } \varphi(a), \text{ visiting no state of } F\big\}$$

and

$$E_2 = \big\{((q,i), a, (q',1)) \mid i \in \{0,1\} \text{ and there exists a path of } \mathcal{B} \text{ from } q \text{ to } q', \text{ labeled } \varphi(a), \text{ visiting a state of } F\big\}.$$

Let $u = a_0a_1 \cdots$ be a word in A^ω and let for all $n \geq 0$, $v_n = \varphi(a_n)$. Then u is accepted by $\mathcal{A}$ if and only if there exists a successful path in $Q \times \{0,1\}$ and thus a path

$$(q_0, 0) \xrightarrow{a_0} (q_1, \varepsilon_1) \xrightarrow{a_1} (q_2, \varepsilon_2) \cdots$$

labeled u passing infinitely often in $Q \times \{1\}$. By the definition of E', this means that there exists a sequence of consecutive paths in $\mathcal{B}$:

$$q_0 \xrightarrow{v_0} q_1 \xrightarrow{v_1} q_2 \cdots$$

such that an infinite number of the paths $q_n \xrightarrow{v_n} q_{n+1}$ visits a state of F. This is equivalent to the fact that the infinite path $q_0 \xrightarrow{v_0} q_1 \xrightarrow{v_1} q_2 \cdots$ is successful in $\mathcal{B}$, i.e. that $\varphi(u)$ is in X. □

Example 5.5. Let $\varphi : \{a,b,c\}^+ \to \{a,b\}^+$ be the semigroup morphism defined by $\varphi(a) = a$, $\varphi(b) = ab$, $\varphi(c) = babaab$ and let $X = (ababa)^\omega$. The set X is recognized by the Büchi automaton represented in Figure 5.6. Applying the construction described above, we obtain after deletion of some useless states, the following automaton recognizing $\varphi^{-1}(X) = \{acba, bba\}^\omega$.

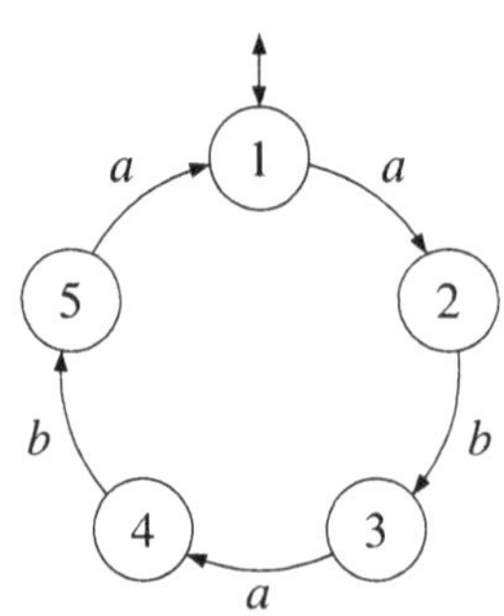

Figure 5.6. An automaton recognizing $(ababa)^\omega$.

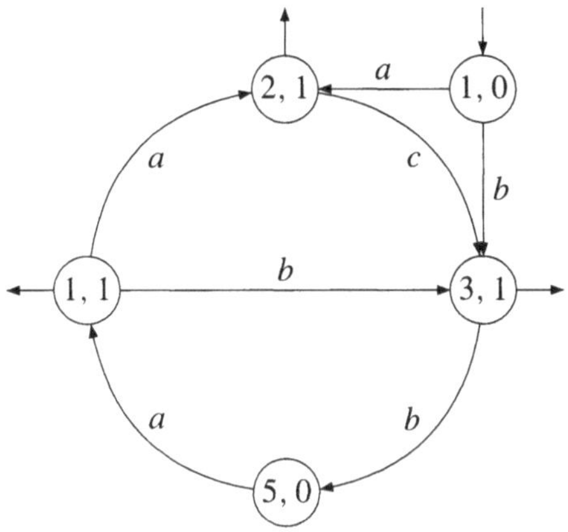

Figure 5.7. An automaton recognizing $\varphi^{-1}((ababa)^\omega)$.

6 Deterministic Büchi automata

It is well known that a set of finite words recognized by a finite automaton can also be recognized by a deterministic one. The situation is quite different for infinite words and we will see that (as soon as A has at least two symbols) there are recognizable subsets of A^ω which cannot be recognized by a finite deterministic Büchi automaton. Actually, this difference between deterministic and nondeterministic automata even holds for countable automata.

The description of the subsets of A^ω recognized by deterministic Büchi automata comes with the introduction of a new operator. For a subset L of A^*, let

$$\overrightarrow{L} = \{u \in A^\omega \mid u \text{ has infinitely many prefixes in } L\}.$$

The operator $L \to \overrightarrow{L}$ plays a role similar to that of the operator $L \to L^\omega$, since it allows one to define infinite words from finite ones. Comparing the properties of these two operators takes an important part in what follows.

The following example gives the value of $\overrightarrow{L}$ for simple sets L and it can help the reader to get more familiar with this operator.

Example 6.1.

(a) If $L = a^*b$, then $\overrightarrow{L} = \emptyset$.

(b) If $L = (ab)^+$, then $\overrightarrow{L} = (ab)^\omega$.

(c) If $L = (a^*b)^+ = (a+b)^*b$, that is if L is the set of words ending with b, then $\overrightarrow{L} = (a^*b)^\omega$, which is the set of infinite words containing an infinity of occurrences of b.

We now give a simple example showing that not every set of words can be written in the form $\overrightarrow{L}$.

Example 6.2. The set $X = (a+b)^*a^\omega$ of words with a *finite* number of occurrences of b is not of the form $\overrightarrow{L}$. Otherwise, the word ba^ω would have a prefix $u_1 = ba^{n_1}$ in L, the word $ba^{n_1}ba^\omega$ would have a prefix $u_2 = ba^{n_1}ba^{n_2}$ in L, etc. and the infinite word $u = ba^{n_1}ba^{n_2}ba^{n_3}\cdots$ would have an infinity of prefixes in L. This word would therefore be in $\overrightarrow{L}$, which is impossible, since u contains infinitely many b's.

The following statement shows that the operator $L \to \overrightarrow{L}$, just as the operator $L \to L^\omega$, preserves the class of recognizable sets.

Proposition 6.1. *Let $\mathcal{A}$ be a deterministic Büchi automaton. Then $L^\omega(\mathcal{A}) = \overrightarrow{L^+(\mathcal{A})}$.*

Proof. Let $\mathcal{A} = (Q, A, E, i, F)$. If $u \in L^\omega(\mathcal{A})$, then u is the label of a path

$$p = (q_0, a_0, q_1)(q_1, a_1, q_2)\cdots$$

such that $q_0 = i$ and such that there exists a subsequence $n_0 < n_1 < \cdots$ satisfying $q_{n_0}, q_{n_1}, \ldots \in F$. By construction, the words $u_k = a_0a_1\cdots a_{n_{k-1}}$ are in $L^+(\mathcal{A})$ and are prefixes of u. Thus $L^\omega(\mathcal{A}) \subset \overrightarrow{L^+(\mathcal{A})}$.

Conversely, if $u \in \overrightarrow{L^+(\mathcal{A})}$, u has infinitely many prefixes in $L^+(\mathcal{A})$. And since $\mathcal{A}$ is deterministic, we deduce that u is the label of an initial path passing infinitely often in F. Thus $u \in L^\omega(\mathcal{A})$. □

We now study the subsets of A^ω which can be recognized by a deterministic Büchi automaton, which will be called *deterministic*. Informally, a subset X of A^ω is deterministic if testing whether a word belongs to X can be made in one left-to-right pass. A first characterization of deterministic sets is given below. We shall give others later, in particular in connection with topology.

Theorem 6.2. *Let X be a subset of A^ω. The following conditions are equivalent:*

(1) *X can be recognized by a deterministic Büchi automaton,*

(2) *there exists a set L of A^+ such that $X = \overrightarrow{L}$.*

If, moreover, the alphabet A is countable, these conditions are equivalent to

(3) *X can be recognized by a countable deterministic Büchi automaton.*

Proof. If $\mathcal{A}$ is a deterministic Büchi automaton recognizing X, then $X = L^\omega(\mathcal{A}) = \overrightarrow{L^+(\mathcal{A})}$ by Proposition 6.1. Thus (1) implies (2).

Let now L be a subset of A^+ such that $X = \overrightarrow{L}$. Then L is recognized by the deterministic automaton $\mathcal{A} = (A^*, A, \cdot, 1, L)$, where the transition function is defined, for all $u \in A^*$ and for every $a \in A$, by $u.a = ua$. By Proposition 6.1, we have $L^\omega(\mathcal{A}) = \overrightarrow{L^+(\mathcal{A})} = \overrightarrow{L} = X$. Thus (2) implies (1). If, moreover, the alphabet A is countable, $\mathcal{A}$ is a countable automaton. Thus, in this case, (2) implies (3), which establishes the equivalence of the three conditions since (3) obviously implies (1). □

The analogous statement for recognizable sets is given below.

Corollary 6.3. *Let X be a subset of A^ω. The following conditions are equivalent:*

(1) *X is recognized by a finite deterministic Büchi automaton.*

(2) *There exists a recognizable subset L of A^+ such that $X = \overrightarrow{L}$.*

Proof. The proof of Theorem 6.2 can be adapted for finite automata. In the proof that (1) implies (2), it suffices to observe that if $\mathcal{A}$ is a finite automaton, then $L^+(\mathcal{A})$ is recognizable. For (2) implies (1), we choose for $\mathcal{A}$ a finite deterministic automaton recognizing L. □

The class of deterministic sets is closed under finite union and under finite intersection. We shall see that it is not closed under complementation.

Proposition 6.4. *Any finite union (resp. intersection) of deterministic sets is deterministic.*

Proof. For finite union, the result follows from the formula

$$\overrightarrow{\bigcup_{1 \le i \le n} L_i} = \bigcup_{1 \le i \le n} \overrightarrow{L_i}$$

and one may apply Theorem 6.2.

For intersection, we need a special construction. Consider two deterministic Büchi automata $\mathcal{A}_1 = (Q_1, A, E_1, i_1, F_1)$ and $\mathcal{A}_2 = (Q_2, A, E_2, i_2, F_2)$. Let

$$Q = Q_1 \times Q_2 \times \{0, 1\}$$

and let $\mathcal{A} = (Q, A, E, (i_1, i_2, 1), F)$ be the deterministic automaton defined by

$$E = \big\{((p_1, p_2, s), a, (q_1, q_2, t)) \mid (p_1, a, q_1) \in E_1, (p_2, a, q_2) \in E_2 \text{ and}$$
$$t = 0 \text{ if and only if } (s = 1, p_2 \in F_2 \text{ and } q_1 \notin F_1) \text{ or } (s = 0 \text{ and } q_1 \notin F_1)\big\}$$

and

$$F = Q_1 \times F_2 \times \{1\}.$$

Intuitively, the component s of a state (q_1, q_2, s) memorizes the fact that the current path has visited a state of F_1 since its last visit to F_2. In fact, if $(p_1, p_2, s) \xrightarrow{a} (q_1, q_2, t)$ is a transition, and if $p_2 \in F_2$, we have

$$t = \begin{cases} 1 & \text{if } q_1 \in F_1 \\ 0 & \text{if } q_1 \notin F_1 \end{cases}$$

But, if $p_2 \notin F_2$, we have

$$t = \begin{cases} 1 & \text{if } s = 1 \text{ (“a state in } F_1 \text{ has already been visited”) or} \\ & \text{if } q_1 \in F_1 \text{ (“first visit of a state in } F_1\text{”)} \\ 0 & \text{if } s = 0 \text{ and if } q_1 \notin F_1 \text{ (“no state in } F_1 \text{ has been visited”).} \end{cases}$$

We shall show that $L^\omega(\mathcal{A}) = L^\omega(\mathcal{A}_1) \cap L^\omega(\mathcal{A}_2)$. Let

$$c = (q_{0,1}, q_{0,2}, s_0) \xrightarrow{a_0} (q_{1,1}, q_{1,2}, s_1) \xrightarrow{a_1} (q_{2,1}, q_{2,2}, s_2) \cdots$$

be a path in $\mathcal{A}$ labeled $u = a_0a_1a_2\cdots$. By projection, c defines a path c_1 of $\mathcal{A}_1$ and a path c_2 of $\mathcal{A}_2$:

$$c_1 = q_{0,1} \xrightarrow{a_0} q_{1,1} \xrightarrow{a_1} q_{2,1} \cdots$$
$$c_2 = q_{0,2} \xrightarrow{a_0} q_{1,2} \xrightarrow{a_1} q_{2,2} \cdots$$

Suppose that c is a successful path of $\mathcal{A}$. Then $q_{0,1} = i_1$, $q_{0,2} = i_2$ and $s_0 = 1$. Moreover c passes infinitely often by a state of F. Thus, there exists an infinite sequence $(n_k)_{k \geq 0}$ such that $q_{n_k,2} \in F_2$ and $s_{n_k} = 1$. As an immediate consequence, c_2 is a successful path of $\mathcal{A}_2$. The path c_1 is also successful since condition $s_{n_k} = 1$ indicates that, since the last visit to a state of F_2, at least one state of F_1 has been visited. Thus $u \in L^\omega(\mathcal{A}_1) \cap L^\omega(\mathcal{A}_2)$.

Conversely, if $u \in L^\omega(\mathcal{A}_1) \cap L^\omega(\mathcal{A}_2)$, u is the label of a successful path c_1 of $\mathcal{A}_1$ and c_2 of $\mathcal{A}_2$, say

$$c_1 = q_{0,1} \xrightarrow{a_0} q_{1,1} \xrightarrow{a_1} q_{2,1} \cdots$$
$$c_2 = q_{0,2} \xrightarrow{a_0} q_{1,2} \xrightarrow{a_1} q_{2,2} \cdots$$

The word u defines thus an initial path c in $\mathcal{A}$. Now, if $q_{n,2} \in F_2$, we have $q_{n',1} \in F_1$ for some $n' \geq n$. If n'' is the smallest integer strictly larger than n' such that $q_{n'',2} \in F_2$, we have $(i_1, i_2, 1) \cdot a_0a_1 \cdots a_{n''} = (q_{n'',1}, q_{n'',2}, 1) \in F$, showing that c is a successful path and that $u \in L^\omega(\mathcal{A})$. □

Example 6.2 shows that the set $X = (a + b)^*a^\omega$ is not of the form $\overrightarrow{L}$. But X is recognized by the Büchi automaton of Example 5.2. Thus, there exist recognizable subsets of A^ω which cannot be recognized by a deterministic Büchi automaton. This contrasts with the case of finite words.

Corollary 6.3 leaves open the problem of whether there exists a recognizable set which could be recognized by a deterministic Büchi automaton but not by a finite one (or, equivalently, if there exist recognizable sets of the form $\overrightarrow{L}$ for which L cannot be chosen recognizable). We shall see later that this cannot happen (Theorem 9.9).

7 Muller and Rabin automata

7.1 Muller automata. Contrary to what happens for finite words, deterministic Büchi automata are not able to recognize all recognizable subsets of A^ω. This is the motivation for introducing Muller automata which are, by definition, deterministic, but with a more powerful acceptance mode. The result that all recognizable sets can be recognized by Muller automata (McNaughton's theorem) will be established in the next section.

A *Muller automaton* is a 5-tuple $\mathcal{A} = (Q, A, E, i, \mathcal{T})$ where (Q, A, E) is a *finite deterministic* automaton, i is the initial state and $\mathcal{T}$ is a set of subsets of Q, called the *table* of the automaton. We often denote a Muller automaton by $\mathcal{A} = (E, i, \mathcal{T})$.

A path p in $\mathcal{A}$ is *initial* if it starts in the initial state and *final* if $\mathrm{Inf}(p) \in \mathcal{T}$, that is, if the set of infinitely visited states is an element of the table. In this way, one may both specify the infinitely visited states and those that are not. A path is *successful* if it is both initial and final. The set of infinite words recognized by $\mathcal{A}$ is the set, denoted by $L^\omega(\mathcal{A})$, of labels of infinite successful paths in $\mathcal{A}$.

Example 7.1. The automaton represented in Figure 7.1, with the initial state $i = 1$ and the table $\mathcal{T} = \{\{2\}\}$, recognizes the set $(a+b)^*b^\omega$ of infinite words with a finite number of a's.

We first show that Muller automata are not more powerful than finite Büchi automata. The proof does not make use of the fact that Muller automata are, by definition, supposed to be deterministic (see Exercise 15).

Theorem 7.1. *The set of infinite words recognized by a Muller automaton is ω-rational.*

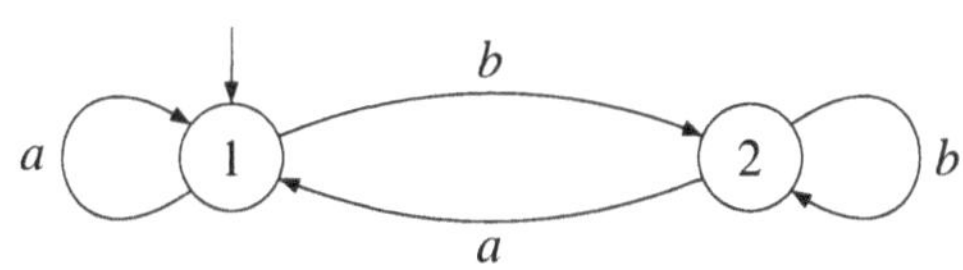

Figure 7.1. A Muller automaton.

Proof. Let $\mathcal{A} = (Q, A, E, i, \mathcal{T})$ be a Muller automaton. Since

$$L^\omega(\mathcal{A}) = \bigcup_{T \in \mathcal{T}} L^\omega(E, i, \{T\})$$

and since the class of recognizable sets is closed under finite union, we may assume that $\mathcal{T}$ is a singleton. Let thus $\mathcal{T} = \{T\}$ where $T = \{t_0, t_1, \ldots, t_k\}$. Let $D = E \cap (T \times A \times T)$, $X = L^+(E, i, t_0)$, $Y_0 = L^+(D, \{t_0\}, \{t_1\})$, $Y_1 = L^+(D, \{t_1\}, \{t_2\})$, $\ldots$, $Y_k = L^+(D, \{t_k\}, \{t_0\})$. Thus, X is the set of labels of the paths going from the initial state to t_0. Next, Y_i, for $0 \leq i < k$ (resp. $i = k$) is the set of labels of paths going from t_i to t_{i+1} (resp. from t_k to t_0) and visiting only states in T. We claim that $L^\omega(\mathcal{A}) = X(Y_0 Y_1 \cdots Y_k)^\omega$, which proves that $L^\omega(\mathcal{A})$ is ω-rational.

First, if $u \in X(Y_0 Y_1 \cdots Y_k)^\omega$, then u is the label of an initial path p such that $\mathrm{Inf}(p) = T$. Consequently $u \in L^\omega(\mathcal{A})$.

Conversely, let $u \in L^\omega(\mathcal{A})$. Then u is the label of an initial path

$$p = (q_0, a_0, q_1)(q_1, a_1, q_2) \cdots$$

such that $\mathrm{Inf}(p) = T$. In particular, there is an integer n_0 such that $q_{n_0} = t_0$ and such that, for all $n \geq n_0$, one has $q_n \in T$. One may therefore find an infinite sequence of integers $n_0 < n_1 < n_2 \cdots$ such that the sequence $q_{n_0}, q_{n_1}, \ldots$ is equal to the periodic sequence $t_0, t_1, \ldots, t_k, t_0, t_1, \ldots, t_k, t_0, \ldots$. We have then $a_0 a_1 \cdots a_{n_0 - 1} \in X$ and for all $r \geq 0$,

$$a_{n_r} a_{n_r + 1} \cdots a_{n_{r+1} - 1} \in Y_{\bar{r}},$$

where $\bar{r}$ is the rest of the division of r by $k + 1$. This shows that $u \in X(Y_0 \ldots Y_k)^\omega$, proving the claim. □

Example 7.2. Consider again the automaton of Figure 7.1, but this time with $\mathcal{T} = \{\{1, 2\}\}$. The set X recognized by this automaton is formed of all infinite words having an infinite number of occurrences of a and of b. The method used in the proof of Theorem 7.1 leads to the rational expression

$$X = (a + b)^* a\big((a + b)^* b(a + b)^* a\big)^\omega.$$

We now consider the various reductions and modifications that one may operate on a Muller automaton. Let $\mathcal{A} = (Q, A, E, i, \mathcal{T})$ be a Muller automaton. A state q is called *accessible* if there exists a finite initial path (possibly empty) ending in q. A state q is called *coaccessible* if there exists an infinite final path starting in q. A subset T of Q is called *admissible* if there exists an infinite initial path p such that $\mathrm{Inf}(p) = T$. Finally, $\mathcal{A}$ is *trim* if all its states are accessible and coaccessible and if all the elements of $\mathcal{T}$ are admissible. In practice, it is easy to verify if a subset of Q is admissible. One computes first the graph G with vertex set Q and with edges

$$R = \{(q, q') \mid \text{there exists } a \in A \text{ such that } (q, a, q') \in E\}.$$

Then a subset T of Q is admissible if the restriction of G to T is strongly connected and accessible from i, or equivalently if there exists a path between two arbitrary elements of T, and a path from i to some vertex in T.

We now establish two equivalence results for Muller automata. The first statement shows that one may always suppose a Muller automaton to be trim.

Proposition 7.2. *Any nonempty subset of A^ω recognized by a Muller automaton can be recognized by a trim Muller automaton.*

Proof. Let $\mathcal{A} = (Q, A, E, i, \mathcal{T})$ be a Muller automaton recognizing a nonempty subset X of A^ω. Let P be the set of states of $\mathcal{A}$ that are both accessible and coaccessible. Since X is nonempty, the initial state i is in P. Let $\mathcal{B} = (P, A, E', i, \mathcal{T}')$ where $\mathcal{T}'$ is the set of admissible T in $\mathcal{T}$ and let $E' = E \cap (P \times A \times P)$. It is clear that $L^\omega(\mathcal{B}) \subset L^\omega(\mathcal{A})$. Conversely, let $u = a_0 a_1 \cdots \in L^\omega(\mathcal{A})$. There exists then an initial path

$$p = q_0 \xrightarrow{a_0} q_1 \xrightarrow{a_1} q_2 \quad \cdots$$

labeled u such that $\mathrm{Inf}(p) \in \mathcal{T}$. The states $q_0, q_1, \ldots$ are both accessible and coaccessible and $\mathrm{Inf}(p)$ is admissible. Thus p is a path in $\mathcal{B}$ and $u \in L^\omega(\mathcal{B})$. Finally $L^\omega(\mathcal{A}) = L^\omega(\mathcal{B})$. □

The second statement shows that one may always suppose a Muller automaton to be complete.

Proposition 7.3. *Any subset of A^ω recognized by a Muller automaton can be recognized by a complete Muller automaton.*

Proof. Let $\mathcal{A} = (Q, A, E, i, \mathcal{T})$ be a Muller automaton recognizing a subset X of A^ω. If $\mathcal{A}$ is not complete, we add a new state p and we "complete" the automaton by creating a transition (q, a, p) if there is no transition of the form(q, a, q') in $\mathcal{A}$. More formally, let $\mathcal{A}' = (Q \cup \{p\}, A, E', i, \mathcal{T})$, where p is a new state and $E' = E \cup E_1 \cup E_2$ with

$$E_1 = \{(p, a, p) \mid a \in A\} \tag{7.1}$$
$$E_2 = \{(q, a, p) \mid q \in Q, a \in A \text{ and } (\{q\} \times \{a\} \times Q) \cap E = \emptyset\}. \tag{7.2}$$

The Muller automaton $\mathcal{A}'$ is complete and recognizes X. □

The following two results, combined with McNaughton's theorem to be proved in Section 9, constitute basic properties of recognizable subsets of A^ω.

Proposition 7.4. *Let $\mathcal{A} = (E, i, \mathcal{T})$ be a complete Muller automaton. Then the automaton $\mathcal{B} = (E, i, \mathcal{P}(Q) \setminus \mathcal{T})$, obtained by changing $\mathcal{T}$ into its complement, recognizes $A^\omega \setminus L^\omega(\mathcal{A})$.*

Proof. Since $\mathcal{A}$ is complete, every infinite word u is the label of one and only one initial path in $\mathcal{A}$. If $\mathrm{Inf}(p) \in \mathcal{T}$, then $u \in L^\omega(\mathcal{A})$, and otherwise $u \in A^\omega \setminus L^\omega(\mathcal{A})$. □

Proposition 7.5. *Let $\mathcal{A}_1 = (Q_1, A, E_1, i_1, \mathcal{T}_1)$ and $\mathcal{A}_2 = (Q_2, A, E_2, i_2, \mathcal{T}_2)$ be two Muller automata. Let π_1 and π_2 be the projections from $Q_1 \times Q_2$ onto Q_1 (resp. Q_2). The automaton $\mathcal{A} = (Q_1 \times Q_2, A, E, (i_1, i_2), \mathcal{T})$, with*

$$\mathcal{T} = \{R \subset Q_1 \times Q_2 \mid \pi_1(R) \in \mathcal{T}_1 \text{ and } \pi_2(R) \in \mathcal{T}_2\}$$
$$E = \Big\{\big((q_1, q_2), a, (q_1', q_2')\big) \mid (q_1, a, q_1') \in E_1 \text{ and } (q_2, a, q_2') \in E_2\Big\}$$

recognizes $L^\omega(\mathcal{A}_1) \cap L^\omega(\mathcal{A}_2)$.

Proof. The projections $\pi_1 : Q_1 \times Q_2 \to Q_1$ and $\pi_2 : Q_1 \times Q_2 \to Q_2$ induce a function, also denoted π_1 (resp. π_2) from the set of paths in $\mathcal{A}$ to the set of paths in $\mathcal{A}_1$ (resp. $\mathcal{A}_2$). Let u be the label of an infinite path p in $\mathcal{A}$. Let us show that $\pi_1(\mathrm{Inf}(p)) = \mathrm{Inf}(\pi_1(p))$ (resp. $\pi_2(\mathrm{Inf}(p)) = \mathrm{Inf}(\pi_2(p))$). In fact, if $q_1 \in \pi_1(\mathrm{Inf}(p))$, there exists a state $q_2 \in Q_2$ such that $(q_1, q_2) \in \mathrm{Inf}(p)$ and thus $q_1 \in \mathrm{Inf}(\pi_1(p))$. Conversely, if $q_1 \in \mathrm{Inf}(\pi_1(p))$, there exists an infinite sequence of states $(q_{2,n})_{n \geq 0}$ of Q_2 such that for every $n \geq 0$,the path p passes by $(q_1, q_{2,n})$. Since Q_2 is finite, there exists $q_2' \in Q_2$ such that $(q_1, q_2') \in \mathrm{Inf}(p)$ and thus $q_1 \in \pi_1(\mathrm{Inf}(p))$. This proves the equality $\pi_1(\mathrm{Inf}(p)) = \mathrm{Inf}(\pi_1(p))$ and the equality corresponding to π_2 can be proved in the same way.

Now, if p is a successful path, it starts in (i_1, i_2) and there exists $R \in \mathcal{T}$ such that $\mathrm{Inf}(p) = R$. We have then, by definition of $\mathcal{T}$, $\pi_1(R) \in \mathcal{T}_1$ and $\pi_2(R) \in \mathcal{T}_2$. Thus $\mathrm{Inf}(\pi_1(p)) = \pi_1(\mathrm{Inf}(p)) = \pi_1(R) \in \mathcal{T}_1$ and $\mathrm{Inf}(\pi_2(p)) \in \pi_2(R)$ whence $\pi_1(p)$ (resp. $\pi_2(p)$) is a successful path in $\mathcal{A}_1$ (resp. $\mathcal{A}_2$). Finally $u \in L^\omega(\mathcal{A}_1) \cap L^\omega(\mathcal{A}_2)$.

Conversely, let $u \in L^\omega(\mathcal{A}_1) \cap L^\omega(\mathcal{A}_2)$. The word u is then the label of a successful path p_1 (resp. p_2) of $\mathcal{A}_1$ (resp. $\mathcal{A}_2$). Let p be the unique path in $\mathcal{A}$ such that $\pi_1(p) = p_1$ and $\pi_2(p) = p_2$. This path starts in (i_1, i_2) and is labeled u, whence, by the above discussion $\pi_1(\mathrm{Inf}(p)) = \mathrm{Inf}(\pi_1(p)) \in \mathcal{T}_1$ and $\pi_2(\mathrm{Inf}(p)) = \mathrm{Inf}(\pi_2(p)) \in \mathcal{T}_2$. This implies $\mathrm{Inf}(p) \in \mathcal{T}$, showing that p is a successful path and that $u \in L^\omega(\mathcal{A})$. □

Proposition 7.6. *For any subset X of A^ω, the following conditions are equivalent.*

(1) *X is recognizable by a Muller automaton.*

(2) *X is of the form*

$$X = \bigcup_{1 \leq i \leq n} (U_i \setminus V_i) \tag{7.3}$$

where the U_i and the V_i are subsets of A^ω recognizable by finite deterministic Büchi automata.

(3) *X is a finite boolean combination of subsets of A^ω recognizable by finite Büchi deterministic automata.*

Proof. (1) $\Rightarrow$ (2). Let $\mathcal{A} = (Q, A, E, i, \mathcal{T})$ be a Muller automaton. Since

$$L^{\omega}(\mathcal{A}) = \bigcup_{T \in \mathcal{T}} L^{\omega}(E, i, \{T\})$$

we may suppose that $\mathcal{T}$ contains just one element T. Now, the connection between Muller's and Büchi's acceptance conditions is summarized in the formula

$$L^{\omega}(E, i, \{T\}) = \bigcap_{t \in T} L^{\omega}(E, i, t) \setminus \bigcup_{t \notin T} L^{\omega}(E, i, t)$$

Since, by Proposition 6.4, the class of deterministic sets is closed under finite union and under finite intersection, $L^{\omega}(\mathcal{A})$ is of the form 7.3.

(2) $\Rightarrow$ (3) is obvious.

(3) $\Rightarrow$ (1). If X is recognized by a finite deterministic Büchi automaton $\mathcal{A} = (Q, A, E, i, F)$, it is also recognized by the Muller automaton $(Q, A, E, i, \mathcal{T})$ with

$$\mathcal{T} = \{T \subset Q \mid T \cap F \neq \emptyset\}.$$

Furthermore, Propositions 7.4 and 7.5 show that the class of sets recognized by a Muller automaton is a boolean algebra. Thus any finite boolean combination of sets recognized by a Muller automaton is recognized by a Muller automaton. □

7.2 Rabin automata. The expression of a recognizable set as a union of differences as in Condition (2) of Proposition 7.6 motivates the introduction of an additional class of automata. A *Rabin automaton* is a tuple $\mathcal{A} = (Q, A, E, i, \mathcal{R})$ where (Q, A, E) is deterministic automaton, i is the initial state and $\mathcal{R} = \{(L_j, U_j) \mid j \in J\}$ is a family of pairs of sets of states. An infinite path p is successful if there exists an index $j \in J$ such that p visits U_j infinitely often and visits L_j only finitely often. When the automaton is finite, this is equivalent to

$$\mathrm{Inf}(p) \cap L_j = \emptyset \quad \text{and} \quad \mathrm{Inf}(p) \cap U_j \neq \emptyset.$$

An infinite word is recognized by $\mathcal{A}$ if it is the label of a successful path in $\mathcal{A}$.

Any deterministic Büchi automaton (Q, i, F) can be considered as a particular Rabin automaton with $\mathcal{R} = \{(\emptyset, F)\}$. Muller automata are equivalent to finite Rabin automata, as shown in the following proposition, in which the term "recognizable deterministic set" means a subset of A^{ω} recognized by a finite deterministic Büchi automaton.

Proposition 7.7. *Any finite Rabin automaton is equivalent to a Muller automaton. Conversely, any Muller automaton is equivalent to a finite Rabin automaton. Moreover, a subset of A^{ω} is recognized by a Rabin automaton with n pairs if and only if it is a union of n differences of recognizable deterministic subsets of A^{ω}.*

Proof. Let $\mathcal{A} = (Q, A, E, i, \mathcal{R})$ be a finite Rabin automaton and let

$$\mathcal{T} = \{T \subset Q \mid T \cap L = \emptyset \text{ and } T \cap U \neq \emptyset \text{ for some } (L, U) \in \mathcal{R}\}. \tag{7.4}$$

Then the Muller automaton $(Q, A, E, i, \mathcal{T})$ recognizes the same set of infinite words as $\mathcal{A}$. Thus every finite Rabin automaton is equivalent to a Muller automaton.

Let $\mathcal{E}_n$ be the class of subsets of A^ω of the form $\bigcup_{1 \leq i \leq n}(U_i \setminus V_i)$ where the U_i and the V_i are recognizable deterministic subsets of A^ω. We have to show that $\mathcal{E}_n$ is also the class of sets recognized by a Rabin automaton with n pairs. We shall obtain as a consequence that any Muller automaton is equivalent to a finite Rabin automaton, since Proposition 7.6 shows that any subset recognized by a Muller automaton belongs to some $\mathcal{E}_n$.

Let $\mathcal{A} = (Q, A, E, i, \mathcal{R})$ be a Rabin automaton with n pairs, say $\mathcal{R} = \{(L_j, U_j) \mid 1 \leq j \leq n\}$. On the one hand

$$L^\omega(\mathcal{A}) = \bigcup_{1 \leq j \leq n} L^\omega(\mathcal{A}_j) \tag{7.5}$$

where $\mathcal{A}_j = (Q, A, E, i, \mathcal{R}_j)$ is the Rabin automaton defined by $\mathcal{R}_j = \{(L_j, U_j)\}$ and on the other hand

$$L^\omega(\mathcal{A}_j) = L^\omega(Q, A, E, i, U_j) \setminus L^\omega(Q, A, E, i, L_j)$$

showing that $L^\omega(\mathcal{A}) \in \mathcal{E}_n$.

Let now X_1 and X_2 be two recognizable deterministic subsets recognized respectively by the finite deterministic Büchi automata $\mathcal{A}_1 = (Q_1, A, E_1, i_1, F_1)$ and $\mathcal{A}_2 = (Q_2, A, E_2, i_2, F_2)$. By Proposition 5.3, we may suppose that $\mathcal{A}_1$ and $\mathcal{A}_2$ are complete. Then $X_1 \setminus X_2$ is recognized by the one pair Rabin automaton $\mathcal{A} = (Q_1 \times Q_2, A, E, (i_1, i_2), \mathcal{R})$, with

$$E = \{((q_1, q_2), a, (q_1', q_2')) \mid (q_1, a, q_1') \in E_1 \text{ and } (q_2, a, q_2') \in E_2\}.$$
$$\mathcal{R} = \{((Q_1 \times F_2), (F_1 \times Q_2))\}.$$

This shows that $\mathcal{E}_1$ is equal to the class of sets recognized by a one pair Rabin automaton.

Let us now consider a set X in $\mathcal{E}_n$. By definition, X is the union of n sets X_j from $\mathcal{E}_1$ and by the above, every X_j is recognized by a one pair Rabin automaton $\mathcal{A}_j = (Q_j, A, E_j, i_j, \mathcal{R}_j)$, with $\mathcal{R}_j = \{(L_j, U_j)\}$. Then the n pair Rabin automaton

$$\mathcal{A} = (Q_1 \times Q_2 \times \cdots \times Q_n, A, E, (i_1, i_2, \ldots, i_n), \mathcal{R})$$

defined by

$$E = \{((q_1, q_2, \ldots, q_n), a, (q_1', q_2', \ldots, q_n')) \mid (q_1, a, q_1') \in E_1,\ (q_2, a, q_2') \in E_2, \ldots, (q_n, a, q_n') \in E_n\}$$

$$\mathcal{R} = \{(Q_1 \times \cdots \times Q_{j-1} \times L_j \times Q_{j+1} \times \cdots \times Q_n,\ Q_1 \times \cdots \times Q_{j-1} \times U_j \times Q_{j+1} \times \cdots \times Q_n) \mid 1 \leq j \leq n, (L_j, U_j) \in \mathcal{R}_j\}$$

recognizes X. Let in fact

$$p = (q_{0,1}, \ldots, q_{0,n}) \xrightarrow{a_0} (q_{1,1}, \ldots, q_{1,n}) \xrightarrow{a_1} (q_{2,1}, \ldots, q_{2,n}) \cdots$$

be a path in $\mathcal{A}$ labeled $u = a_0a_1a_2 \cdots$. By projection, p defines, for $1 \leq j \leq n$, a path p_j in $\mathcal{A}_j$

$$p_j = q_{0,j} \xrightarrow{a_0} q_{1,j} \xrightarrow{a_1} q_{2,j} \cdots$$

Let $(L, U) = (Q_1 \times \cdots \times Q_{j-1} \times L_j \times Q_{j+1} \times \cdots \times Q_n, Q_1 \times \cdots \times Q_{j-1} \times U_j \times Q_{j+1} \times \cdots \times Q_n)$ be a pair of $\mathcal{R}$. We have then $(q_{i,1}, \ldots, q_{i,n}) \notin L$ if and only if $q_{i,j} \notin L_j$ and $(q_{i,1}, \ldots, q_{i,n}) \in U$ if and only if $q_{i,j} \in U_j$. Thus p is a successful path in $\mathcal{A}$ if and only if one among the p_j is a successful path in $\mathcal{A}_j$. $\square$

A direct construction of a Rabin automaton equivalent to a given Muller automaton is also possible (see Exercise 17).

The notion of Streett automaton is the dual to that of Rabin automaton. A *Streett automaton* is a tuple

$$\mathcal{A} = (Q, A, E, i, \mathcal{S})$$

where (Q, A, E) is a deterministic automaton, $i \in Q$ is the initial state and $\mathcal{S} = \{(L_j, U_j) \mid j \in J\}$ is a family of pairs of sets of states. An infinite path p is successful if it starts in the initial state and if, for each $j \in J$, either it visits L_j a finite number of times or it visits U_j an infinite number of times.

Negating the acceptance condition shows that the complement of a set recognized by a Streett automaton can be recognized by a Rabin automaton with the same number of pairs.

7.3 Muller automata with a full table. We end this section by an important result characterizing those Muller automata which are equivalent to a finite deterministic Büchi automaton. Let $\mathcal{A} = (Q, A, E, i, \mathcal{T})$ be a Muller automaton. We say that the table $\mathcal{T}$ is *full* if for every admissible set $T \in \mathcal{T}$ and for every admissible set T' containing T, we have $T' \in \mathcal{T}$.

Theorem 7.8. *A Muller automaton is equivalent to a finite deterministic Büchi automaton if and only if its table is full.*

The 'if' part is a consequence of the following statement.

Proposition 7.9. *Any Muller automaton with a full table is equivalent to a finite deterministic Büchi automaton.*

Proof. Let $\mathcal{A}' = (Q, A, E, i, \mathcal{T})$ be a Muller automaton with a full table. We define a finite deterministic Büchi automaton

$$\mathcal{A} = (\mathcal{P}(Q) \times Q, (\emptyset, i), F)$$

where $F = \{(\emptyset, q) \mid q \in Q\}$ and where, for every $(S, q) \in \mathcal{P}(Q) \times Q$ and for every $a \in A$, the transition function is given by the following formula.

$$(S, q) \cdot a = \begin{cases} (\emptyset, q \cdot a) & \text{if } S \cup \{q \cdot a\} \text{ contains an element of } \mathcal{T} \\ (S \cup \{q \cdot a\}, q \cdot a) & \text{otherwise.} \end{cases}$$

Additionally, $(S, q) \cdot a$ is undefined if $q \cdot a$ is undefined.

Let u be a word accepted by $\mathcal{A}$. Then u defines in $\mathcal{A}$ an initial path p. Let $T \in \mathcal{T}$ be such that all its states are repeated infinitely often in p. Thus $T \subset \mathrm{Inf}(p)$ and since the table of $\mathcal{A}'$ is full, we have also $\mathrm{Inf}(p) \in \mathcal{T}$. Thus u is accepted by $\mathcal{A}'$.

Conversely, if u is accepted by $\mathcal{A}'$, u defines in $\mathcal{A}'$ a (unique) initial path p' such that $\mathrm{Inf}(p') \in \mathcal{T}$. Let p be the initial path defined by u in $\mathcal{A}$. This path is final. Indeed, on the contrary, there would exist a prefix v of u, such that, if vw is a prefix of u, the state $(\emptyset, i) \cdot vw$ is not a final state of $\mathcal{A}$. Let $(\emptyset, i) \cdot v = (S, q)$ and let us choose w in such a way that the path of $\mathcal{A}'$ starting at q and defined by w visits all the states of $\mathrm{Inf}(p)$. We have then $(\emptyset, i) \cdot vw = (T, q \cdot w)$ where T is a set containing $\mathrm{Inf}(p)$. But $\mathrm{Inf}(p) \in \mathcal{T}$, whence $T \in \mathcal{T}$, which is impossible. □

The construction described in the proof above can be paraphrased intuitively as follows. To simulate a Muller automaton with a full table by a finite deterministic Büchi automaton, one builds a new automaton which sends signals. It accumulates in a box the states met since the last signal. As soon as the box contains an element of the table, it sends a signal and empties the box. We shall use a similar idea for the proof of McNaughton's theorem in Section 9.

Example 7.3. The automaton of Example 7.2 has a full table. Applying the above construction, one obtains the automaton of Figure 7.2.

The necessity of the condition in Theorem 7.8 follows from the next result. Note that the finiteness of the automaton is not used in the proof. We shall return to this point in a while (Theorem 9.9).

Proposition 7.10. *A Muller automaton equivalent to a deterministic Büchi automaton has a full table.*

Proof. Let $\mathcal{A} = (Q, A, E, i, \mathcal{T})$ be a Muller automaton. Suppose that the set $L^\omega(\mathcal{A})$ can be recognized by a deterministic Büchi automaton $\mathcal{B}$. Then there exist a subset L of A^+ such that $X = \overrightarrow{L}$. Let T and T' be two admissible subsets of Q such that $T \in \mathcal{T}$ and $T \subset T'$. We have to show that $T' \in \mathcal{T}$. Since T is admissible, there exists an infinite path p starting in i such that $\mathrm{Inf}(p) = T$. We fix a state $t \in T$. There exists a finite word u such that $i \cdot u = t$ and an infinite word v such that the infinite path starting at t defined by v does not leave T but visits each state of T infinitely often. We

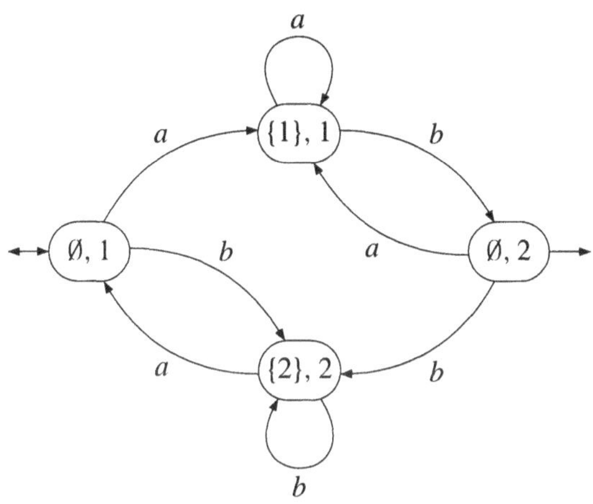

Figure 7.2. A Büchi automaton.

build then by induction two sequences of words $u_0, u_1, \cdots$ and $v_0, v_1, \cdots$ satisfying for every $k \geq 0$, the following properties:

(a) $u_0 v_0 u_1 v_1 \cdots u_k v_k \in L$,

(b) if $q_k = i.u_0 v_0 u_1 v_1 \cdots u_{k-1} v_{k-1}$, then $q_k u_k = t$ and the path starting at q_k defined by u_k passes at least once by each state of T' and stays in T'.

For this, let $u_0 = u$ and knowing the sequences up to rank $k - 1$, we choose u_k in such a way that (b) is satisfied – which is possible, since T' is admissible. The word $u_0 v_0 u_1 v_1 \cdots u_{k-1} v_{k-1} u_k v$ is accepted by $\mathcal{A}$ and there exists a prefix v_k of v such that condition (a) is satisfied.

Finally, let $w = u_0 v_0 u_1 v_1 \cdots$. We have $w \in \overrightarrow{L}$ by construction, but simultaneously, the initial path s defined by w satisfies $\text{Inf}(s) = T'$. We thus obtain that $T' \in \mathcal{T}$. □

The notion of a Muller automaton with a full table appears as the first level of a hierarchy of Muller automata using the notion of a chain. This leads to the notion of a *Rabin chain automaton* (see Chapter V).

8 Transition automata

It is sometimes convenient to use a variant of automata in which a set of final transitions is specified, instead of the usual set of final states.

Formally, a *transition Büchi automaton* is a 5-tuple $\mathcal{A} = (Q, A, E, I, F)$ where (Q, A, E) is a finite automaton, $I \subseteq Q$ is the set of initial states and $F \subseteq E$ is the set of final transitions. A path is *initial* if it starts in some initial state, *final* if it goes through

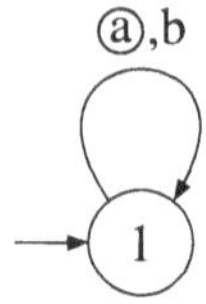

Figure 8.1. A transition automaton recognizing $(\{a, b\}^*a)^\omega$.

F infinitely often and *successful* if it is initial and final. The notions of complete (resp. co-complete) automata are defined in the usual way.

Example 8.1. The transition automaton represented in Figure 8.1 recognizes the set of infinite words containing infinitely many a's. The final transitions are circled.

Similarly, a *transition Muller automaton* is a 5-tuple $\mathcal{A} = (Q, A, E, I, \mathcal{T})$ where (Q, A, E) is a finite deterministic automaton, i is the initial state and $\mathcal{T}$ is a set of subsets of E, called the *table* of the automaton. If p is an infinite path, the set of transitions which occur infinitely often in p is denoted by $\mathrm{Inf}_T(p)$. A path p is *successful* if it is an initial path and if $\mathrm{Inf}_T(p) \in \mathcal{T}$, that is, if the set of transitions occurring infinitely often in p is an element of the table.

We now establish the equivalence of these notions with the standard ones. Recall that a Büchi automaton is co-deterministic (resp. co-complete) if any infinite word labels at most (resp. at least) one final path.

Proposition 8.1. *Every Büchi automaton is equivalent to a transition Büchi automaton. Conversely, every transition Büchi automaton is equivalent to a Büchi automaton. Furthermore, if one of these automata is co-deterministic (resp. co-complete) one can choose the other one to have the same property.*

Proof. Let $\mathcal{A} = (Q, A, E, I, F)$ be a Büchi automaton. Then $\mathcal{A}$ is equivalent to the transition Büchi automaton $\mathcal{A}' = (Q, A, E, I, F')$ where $F' = \{(p, a, q) \in E \mid p \in F\}$. Clearly, if $\mathcal{A}$ is deterministic (resp. co-deterministic, complete, co-complete), then so is $\mathcal{A}'$.

Conversely, let $\mathcal{A} = (Q, A, E, I, F)$ be a transition Büchi automaton. Then $\mathcal{A}$ is equivalent to the Büchi automaton $\mathcal{A}' = (Q', A, E', I', F')$ defined by $Q' = Q \times \{0, 1\}$, $I' = I \times \{0, 1\}$ and

$$F' = \{(p, 1) \in Q' \mid (p, a, q) \in F \text{ for some } q \in Q \text{ and } a \in A\}$$
$$E' = \big\{((p, 1), a, (q, \varepsilon)) \mid (p, a, q) \in F,\ \varepsilon \in \{0, 1\}\big\}$$
$$\cup \big\{((p, 0), a, (q, \varepsilon)) \mid (p, a, q) \in E \setminus F,\ \varepsilon \in \{0, 1\}\big\}$$

If $\mathcal{A}$ is co-deterministic (resp. co-complete), then so is $\mathcal{A}'$. □

Proposition 8.2. *Every Muller automaton is equivalent to a transition Muller automaton. Conversely, every transition Muller automaton is equivalent to a Muller automaton.*

Proof. Let $\mathcal{A} = (Q, A, E, i, \mathcal{T})$ be a Muller automaton. We claim that $\mathcal{A}$ is equivalent to the transition Muller automaton $\mathcal{A}' = (Q, A, E, j, \mathcal{T}')$, where j is a new state $Q' = E \times \{j\}$, $\mathcal{T}' = \mathcal{T}$ and

$$E' = \{(j, a, (i, a, q)) \mid (i, a, q) \in E\}$$
$$\cup \{((q, a, q'), b, (q', b, q'')) \mid (q, a, q') \in E \text{ and } (q', a, q'') \in E\}$$

Indeed, to any infinite path of $\mathcal{A}$ starting at i

$$p = (i, a_0, q_1)(q_1, a_1, q_2)\cdots$$

corresponds an infinite path of $\mathcal{A}'$ with the same label starting at j

$$p' = (j, a_0, (i, a_0, q_1))((i, a_0, q_1), a_1, (q_1, a_1, q_2))\cdots$$

and conversely, every infinite path of $\mathcal{A}'$ starting at j arises this way. Furthermore, the transition (q, a, q') occurs in p if and only if p' visits the state (q, a, q'). Therefore $\mathrm{Inf}_T(p) = \mathrm{Inf}(p')$ and $\mathcal{A}$ and $\mathcal{A}'$ are equivalent. □

9 McNaughton's theorem

The aim of this section is to prove the following result, due to R. McNaughton.

Theorem 9.1. *Any recognizable subset of A^ω can be recognized by a Rabin automaton.*

The proof that we are going to present relies on a determinization algorithm due to S. Safra. It computes a Rabin automaton equivalent to a given Büchi automaton. Using Proposition 7.7 allows one to obtain the conclusion. The states of the Rabin automaton are labeled trees.

We first give an informal description of the construction. Consider a Büchi automaton

$$\mathcal{A} = (Q, A, E, I, F)$$

and a successful path p labeled $u \in A^\omega$. There exist states $i \in I$ and $f \in F$ and a factorization $p = p_0 p_1 p_2 \cdots$ where p_0 is a path from i to f labeled u_0 and for every $n > 0$, p_n is a path from f to f labeled u_n.

$$i \xrightarrow{u_0} f \xrightarrow{u_1} f \xrightarrow{u_2} f \ \cdots$$

The usual determinization algorithm does not work with infinite words. One obtains indeed a deterministic automaton $\mathcal{B} = (\mathcal{P}(Q), A, \cdot, \{I\}, \mathcal{F})$, where $\mathcal{F} = \{P \mid P \cap F \neq \emptyset\}$ with transitions given by

$$S.a = \delta(S, a)$$

for $S \subset Q$ and $a \in A$. Processing u in $\mathcal{B}$ gives a path p'

$$i \xrightarrow{u_0} S_0 \xrightarrow{u_1} S_1 \xrightarrow{u_2} S_2 \cdots$$

Each S_i contains f, but this is not enough to make sure that such a path is successful, since nothing says that the state f appearing in S_i comes from the state f appearing in S_{i-1}. Thus, one cannot define as a table $\mathcal{T} = \{P \subset Q \mid P \cap F \neq \emptyset\}$. For example, if $\mathcal{A}$ is the automaton of Example 5.2, recognizing the set of infinite words containing a finite number of b's,

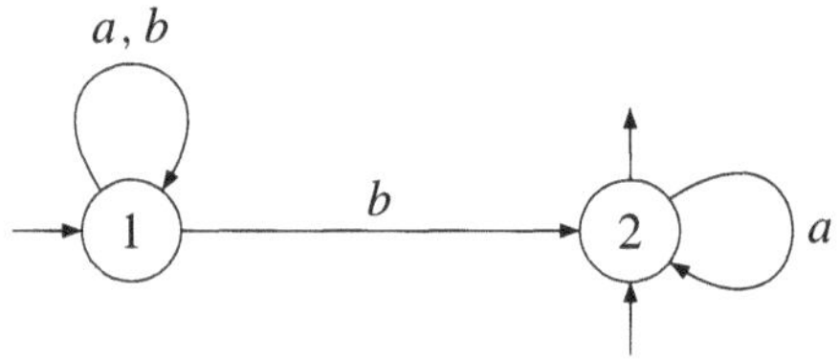

the path $\{1\} \xrightarrow{b} \{1,2\} \xrightarrow{b} \{1,2\} \xrightarrow{b} \{1,2\} \cdots$ would be successful in $\mathcal{B}$, although b^ω is not recognized by $\mathcal{A}$. Actually, the automaton obtained by this algorithm, once made trim, contains only one state and thus recognizes A^ω whatever be the acceptance mode (see Figure 9.1).

The idea is to look for a path $I \xrightarrow{u_0} S_0 \xrightarrow{u_1} S_1 \xrightarrow{u_2} S_2 \cdots$ such that the following two conditions are satisfied:

(1) $S_0 \subset \delta(I, u_0)$, and, for every $n \geq 0$, $S_{n+1} \subset \delta(S_n, u_{n+1})$.

(2) For every $n \geq 0$ and every $q \in S_{n+1}$, there is a state $p \in S_n$ and a path $p \xrightarrow{u_{n+1}} q$ in $\mathcal{A}$ *passing through a final state.*

To find such a path, we are going to build an automaton memorizing the occurrences of final states. The states of this automaton are oriented trees whose nodes are labeled by the sets S_i mentioned above. We then apply the usual determinization algorithm, taking care of adding the new final states that appear as the label of a new child of the vertex (see Figure 9.2). When all the states in the label S of a vertex have already visited a final state, that is when they all appear in the children of the node, this node is marked and all its descendants disappear.

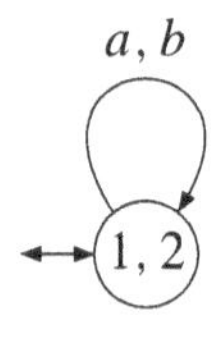

Figure 9.1. The automaton obtained by determinization.

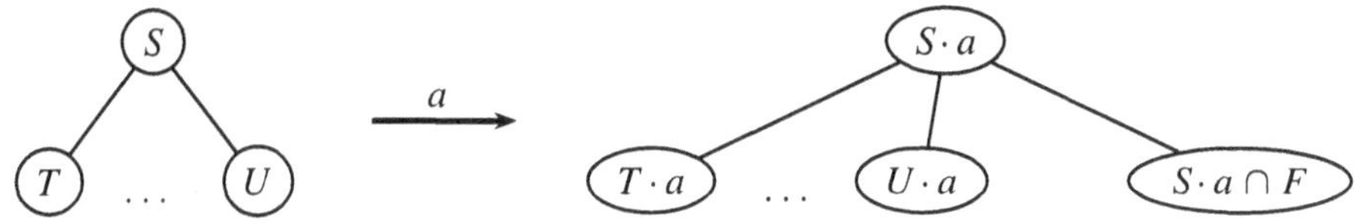

Figure 9.2. The action of letter a.

We now proceed to the formal description of the construction. Let $\mathcal{A} = (Q, A, E, I, F)$ be a finite Büchi automaton with

$$Q = \{1, 2, \dots, n\} \quad \text{and} \quad V = \{1, 2, \dots, 2n\}.$$

We build a deterministic Rabin automaton $\mathcal{D}$ as follows. Its *states* are labeled oriented trees with marks on some nodes. Formally the states are tuples (T, f, e, M) where

(1) the set of nodes T is a subset of V,

(2) $f : T \to T^*$ is a function mapping each node on the ordered sequence of its children,

(3) e is a function from T into the set of nonempty subsets of Q, mapping each node to its *label*,

(4) $M \subset T$ is the set of *marked nodes*.

These trees should also satisfy the following conditions:

(5) The root of the tree is 1.

(6) The marked nodes have to be leaves in the tree.

(7) For every node v, the union of the labels of its children is a strict subset of $e(v)$.

(8) If v is not an ancestor of w and if w is not an ancestor of v, then $e(v) \cap e(w) = \emptyset$.

The set $\mathcal{T}_n$ of all trees defined in this way is finite. More precisely, the following result holds:

Proposition 9.2. *A tree in $\mathcal{T}_n$ has at most n nodes.*

Proof. We associate with each node $v \in T$, the set

$$r(v) = e(v) \setminus \bigcup_{w \text{ child of } v} e(w)$$

By condition (7), $r(v)$ is not empty and, if v_1 and v_2 are distinct, we have $r(v_1) \cap r(v_2) = \emptyset$. This follows from condition (7) if v_1 is an ancestor of v_2 and from condition (8) in the other cases, since $r(v) \subset e(v)$. The sets $r(v)$ are thus pairwise distinct and we obtain

$$\text{Card}(T) = \sum_{v \in T} 1 \leq \sum_{v \in T} \text{Card}(r(v)) \leq \text{Card}(Q) = n$$

establishing the proposition. □

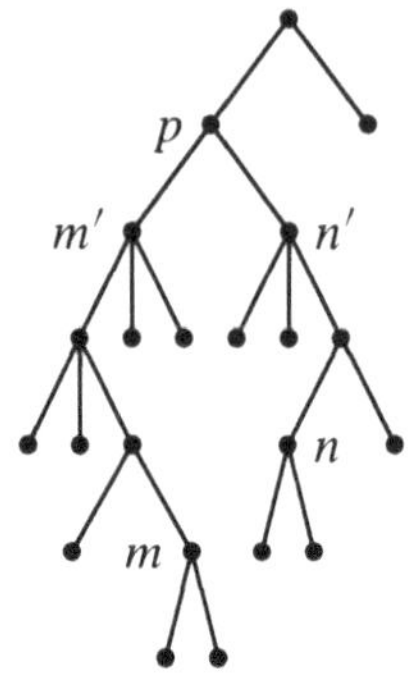

Figure 9.3. The node m is on the left of n.

In an oriented tree, the children of a given node are ordered. These local orders can be extended to a partial order on the set of nodes as follows. Given two nodes m and n which are not ancestor of one another, let p be their least common ancestor and let m' (resp. n') be the child of p which is an ancestor of m (resp. n). We say that m is *on the left* of n if $m' < n'$, as illustrated in Figure 9.3.

We return to the construction of the automaton $\mathcal{D}$. The set of its states is thus $\mathcal{T}_n$ and its transition function Δ is defined as follows. Let $R = (T, f, e, M)$ be a tree in $\mathcal{T}_n$ and let a be a letter from A. The state $\Delta(R, a)$ is obtained by the following steps.

(1) We perform the transition by a on the labels of each node and we erase the marks. For this, we build the tree (T, f, e_1, M_1), with $M_1 = \emptyset$, and, for each $v \in T$,

$$e_1(v) = \delta(e(v), a)$$

(2) We add to each node v a new child placed at the right of all children of v and labeled $e(v) \cap F$. This new node is marked and taken arbitrarily among the available nodes (in practice, we take the smallest available node). Formally, we choose an injection from T into $V \setminus T$ associating with each node $v \in T$ a node denoted $\overline{v}$. This is possible since T has at most n elements. Let $\overline{T} = \{\overline{v} \mid v \in T\}$ and consider the tree (T_2, f_2, e_2, M_2) where

$$T_2 = T \cup \overline{T}, \quad M_2 = \overline{T}$$

and, for every $v \in T$,

$$f_2(v) = f(v)\overline{v}, \qquad f_2(\overline{v}) = \varepsilon$$
$$e_2(v) = e_1(v), \qquad e_2(\overline{v}) = e_1(v) \cap F.$$

(3) In the label of each node v, we suppress the states appearing in the label of a node at the left of v. For this, we build e_3 defined for each node $v \in T_2$ by,

$$e_3(v) = e_2(v) \setminus \bigcup_{w \text{ to the left of } v} e_2(w).$$

Figure 9.4. Suppressing a node with an empty label.

(4) We suppress the nodes with an empty label and we update the function f and the marks accordingly. This operation is represented in Figure 9.4. Formally, we change to the tree (T_4, f_4, e_4, M_4) where

$$T_4 = \{v \in T_2 \mid e_3(v) \neq \emptyset\},$$

$M_4 = M_2 \cap T_4$, $e_4(v)$ is the restriction of $e_3(v)$ to T_4 and, for each node $v \in T_4$, the word $f_4(v)$ is obtained by erasing the symbols of $T_2 \setminus T_4$.

(5) We mark all nodes with a label equal to the union of the labels of their children, i.e. such that

$$e(v) = \bigcup_{w \text{ child of } v} e(w).$$

and we suppress all their descendants.

We finally obtain a state $(T_5, f_5, e_5, M_5) = \Delta(R, a)$ which is an element of $\mathcal{T}_n$.

The initial state of $\mathcal{D}$ is the tree reduced to an unmarked node labeled I if $I \cap F = \emptyset$, to a marked node labeled I if $I \subset F$ and to a node labeled I with a marked child labeled $I \cap F$ in all other cases.

There remains to specify the set $\mathcal{R}$ defining the acceptance condition. Let

$$\mathcal{R} = \{(L_v, U_v) \mid v \in V\}$$

where

$$L_v = \{R \in \mathcal{T}_n \mid v \text{ is not a node of } R\}$$
$$U_v = \{R \in \mathcal{T}_n \mid v \text{ is a marked node of } R\}.$$

Thus, a path in $\mathcal{D}$ is successful if there exists an element $v \in V$ such that, ultimately, the path uses states in which v is a node and infinitely often states in which v is marked.

Before proving that this Rabin automaton recognizes the same set of infinite words as the automaton we started from, we are going to illustrate the construction by some examples. In these examples, the states are represented by labeled oriented trees and marked nodes are indicated by a double circle. An arrow of the form $\xrightarrow{(i)}$ indicates that step i of the algorithm has been performed.

Example 9.1. Consider the automaton represented in Figure 9.5, which recognizes the set of words having a finite nonzero number of b's. We detail the steps of Safra's algorithm. The initial state is the tree with a single node of Figure 9.6. The action of the letters a and b on the initial state are represented in Figure 9.7 and Figure 9.8, respectively. A new state has now been created. The actions of the letters a and b on this new state are represented in Figure 9.9 and Figure 9.10, respectively. Thus another new state has been created. The action of the letters a and b on this new state are easily derived from the ones represented in Figures 9.9 and 9.10 by exchanging the names 2 and 3 in every place. After renaming the states, we obtain the automaton of Figure 9.11.

We have $L_1 = \emptyset$, $L_2 = \{1, 3\}$, $L_3 = \{1, 2\}$, $U_1 = \emptyset$, $U_2 = \{2\}$, $U_3 = \{3\}$. Thus the accepting pairs are $(\{1, 3\}, \{2\})$ and $(\{1, 2\}, \{3\})$. Note that, by Formula (7.4), these Rabin pairs are equivalent to the table $\mathcal{T} = \{\{2\}, \{3\}\}$.

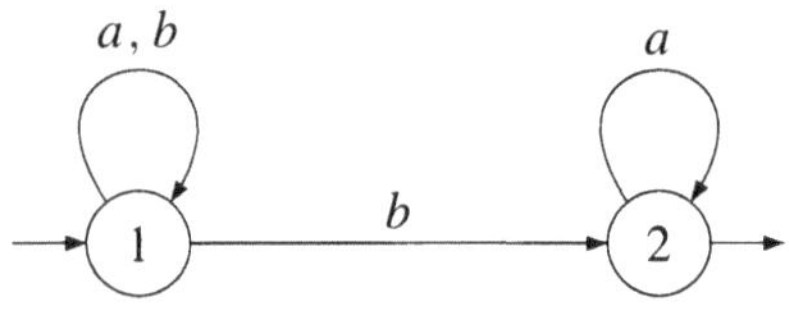

Figure 9.5. A Büchi automaton.

Figure 9.6. The initial state.

Figure 9.7. The action of a on the initial state.

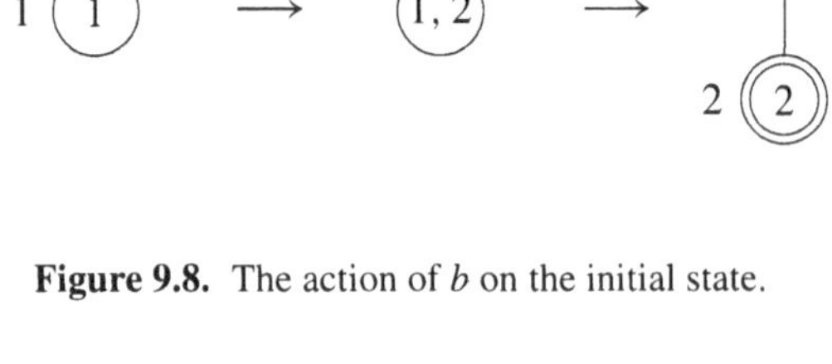

Figure 9.8. The action of b on the initial state.

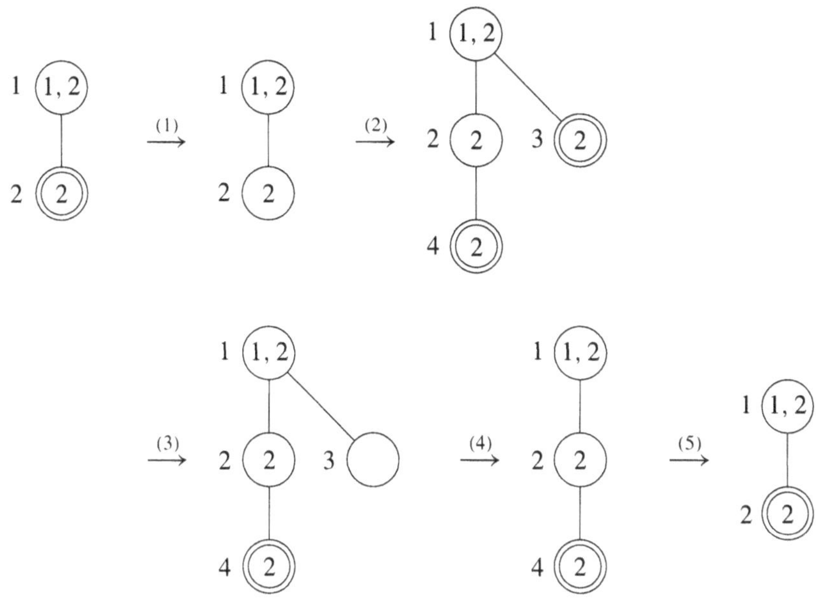

Figure 9.9. The action of a on the new state.

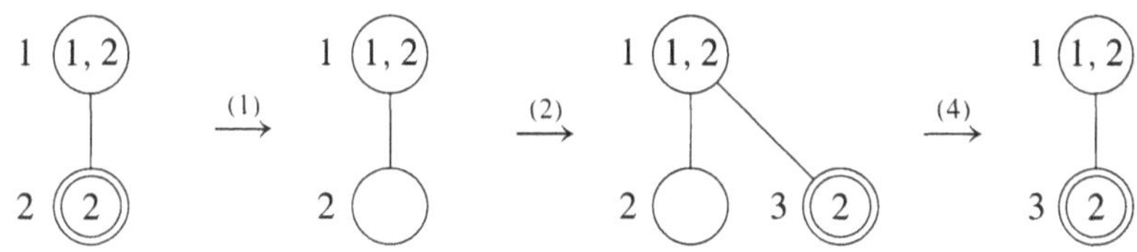

Figure 9.10. The action of b on the new state.

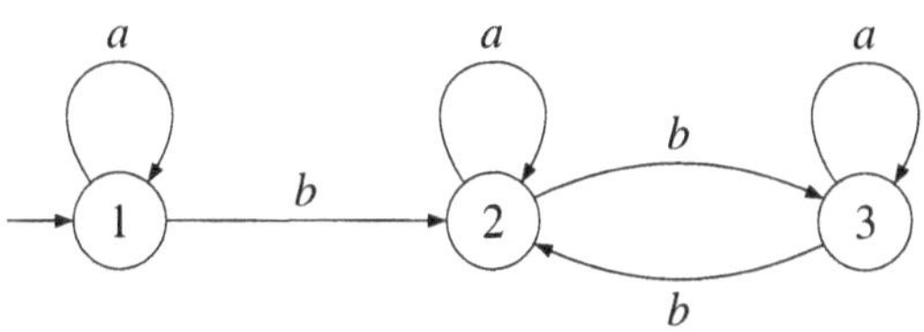

Figure 9.11. The Rabin automaton obtained by Safra's algorithm.

Example 9.2. The Büchi automaton represented in Figure 9.12 recognizes the set of words with a finite number of b. The deterministic automaton obtained by Safra's algorithm is represented in Figure 9.13. We recognize the automaton computing the parity of the number of b. The acceptance conditions are, in Rabin's form, the pairs $\{(\{1\},\{2\}),(\{2\},\{1\})\}$, which gives the table $\{\{1\},\{2\}\}$.

Example 9.3. Consider the set $X = (\{b,c\}^*a \cup b)^\omega$. A Büchi automaton recognizing X is given in Figure 9.15: The application of Safra's algorithm gives the deterministic automaton of Figure 9.16, in which the set of states is $\{I, II, III, IV, V\}$:

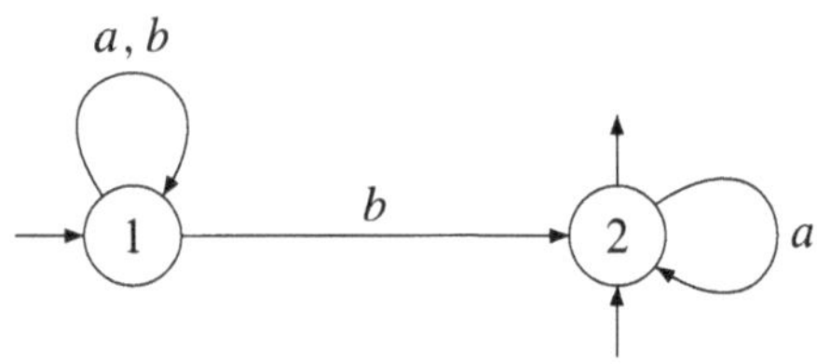

Figure 9.12. A Büchi automaton for the set A^*a^ω.

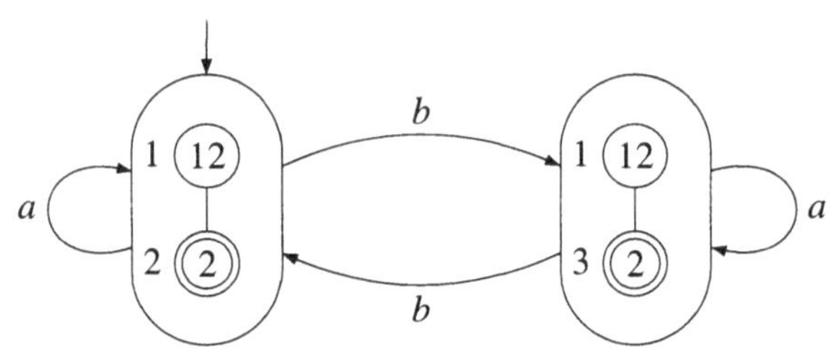

Figure 9.13. The deterministic automaton obtained by Safra's algorithm.

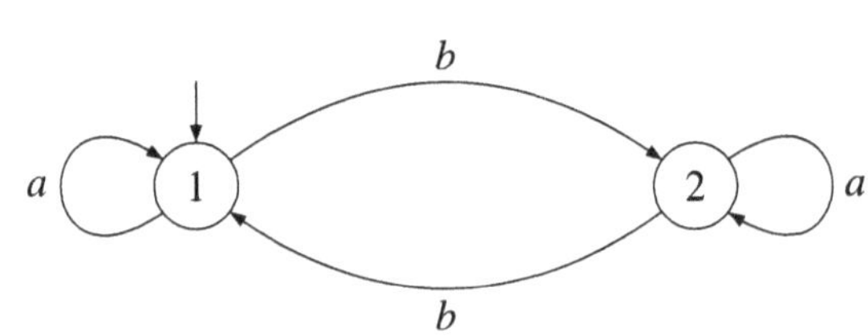

Figure 9.14. The same automaton after renaming the states.

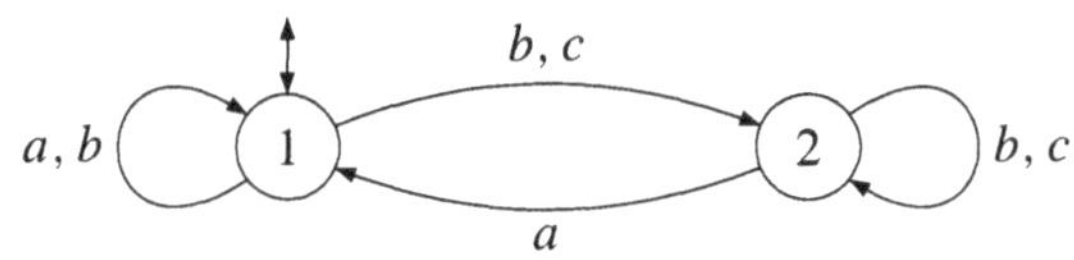

Figure 9.15. A Büchi automaton for $(\{b, c\}^* a \cup b)^\omega$.

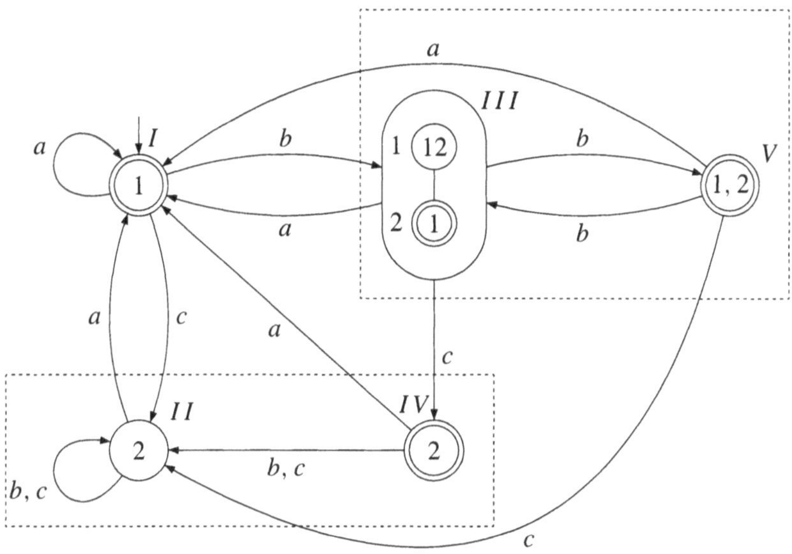

Figure 9.16. A deterministic Muller automaton for $(\{b, c\}^* a \cup b)^\omega$.

The Rabin pairs are $(\emptyset, \{I, IV, V\})$ and $(\{I, II, IV, V\}, \{III\})$. Therefore the table of the corresponding Muller automaton is

$$\mathcal{T} = \{T \subset Q \mid T \text{ contains either } I, IV \text{ or } V\} \cup \{\{III\}\}$$

This table is full. Indeed, if a set T contains I, IV or V, any superset of T has the same property. If $T = \{III\}$, any superset of T contains I, IV or V, or is equal to the set $\{II, III\}$. But this latter set is not admissible.

One can in fact obtain a smaller automaton by merging the states grouped inside each dashed rectangle. After renaming the states, the resulting automaton is represented in Figure 9.17. The table is $\mathcal{T} = \{\{1\}, \{2\}, \{1, 2\}, \{1, 3\}, \{1, 2, 3\}\}$, which is the comple-

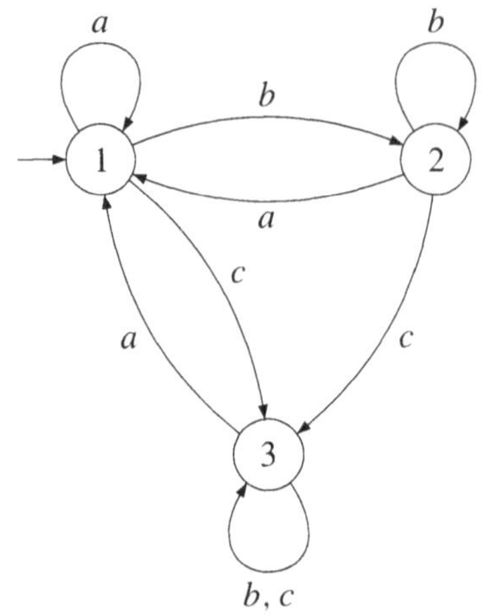

Figure 9.17. Applying Safra's construction.

ment of the table $\{\{3\}\}$. Since the table is full, X is deterministic. We have actually $X = \overrightarrow{\{a, b, c\}^* ab^* \cup b^+}$.

We are now going to prove that the deterministic automaton $\mathcal{D}$ is equivalent to the automaton $\mathcal{A}$ we started from. We shall need a lemma which makes more precise the behavior of $\mathcal{D}$. Let $u = a_1 \cdots a_n$ be a finite word and let R_0 be a state of $\mathcal{D}$ containing a marked node v labeled S_0. We suppose that, for $1 \leq i \leq n$, the states $R_i = \Delta(R_0, a_1 \cdots a_i)$ also contain the node v, with a label S_i, but that this node is marked only for $i = n$. The hypotheses are represented in Figure 9.18.

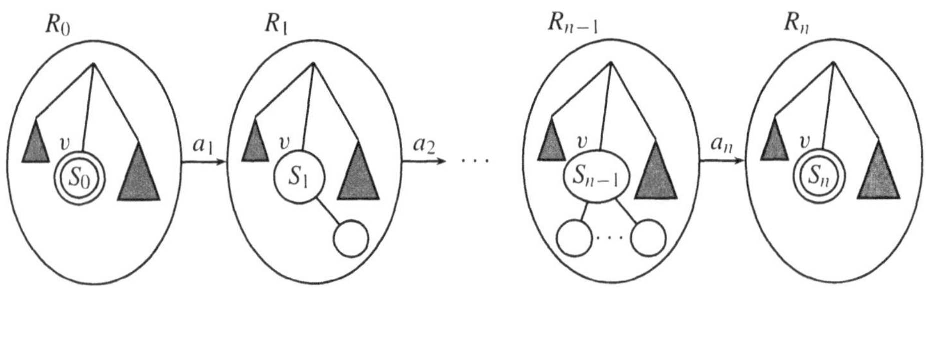

Figure 9.18. The states R_i.

Lemma 9.3. *For $0 \leq i \leq n-1$, S_{i+1} is contained in $\delta(S_i, a_{i+1})$. Moreover, for every $q \in S_n$, there is a path in $\mathcal{A}$ starting in S_0, ending at q, labeled u and visiting at least one final state after its origin.*

Proof. We follow the construction step by step. We first compute at step 1 the set $S_{i+1} = \delta(S_i, a_{i+1})$, then we suppress some states of S_{i+1} during step 3. The first part of the lemma follows.

Let us show by induction on i that, for $0 \leq i \leq n-1$ and for every state q_i appearing in the label of a descendant of v in R_i, there exists a path in $\mathcal{A}$ starting in S_0, ending with q_i, labeled $a_1 \cdots a_i$ and passing at least once more by a final state. The result holds for $i = 0$, since v, which is marked has to be a leaf of R_0 and has no descendants. On the other hand, if q_{i+1} appears in the label of a descendant of v in R_{i+1}, either $q_{i+1} \in \delta(q_i, a_{i+1})$ for some q_i appearing in the label of a strict descendant of v in R_i, and we conclude by induction, or q_{i+1} appears in a label created at step 2 and thus $q_{i+1} \in F$, which also allows one to conclude.

Finally, since v is marked in R_n, it received its mark at step 5. Thus if $q \in S_n$, either $q \in \delta(S_{n-1}, a_n) \cap F$, or q belongs to the union of the $\delta(q_{n-1}, a_n)$ where q_{n-1} appears in the label of a descendant of v in R_{n-1}. In the first case, there exists a path labeled u, starting in S_0 and ending with q, which is a final state. In the second case, we use the conclusion of the above induction: there exists a path in $\mathcal{A}$ starting in S_0, ending with q_{n-1}, labeled $a_1 \cdots a_{n-1}$ and passing at least once more by a final state. The lemma follows immediately. □

Consider now a successful path c in $\mathcal{D}$ and let $u \in A^\omega$ be the label of c. There exists a $v \in V$ such that, ultimately, the path visits only states in which v is a node and infinitely often states in which v is a marked node. Setting $S_0 = I$, there exists by Lemma 9.3 a factorization $u = u_0 u_1 u_2 \cdots$ and subsets S_n of Q, such that

(a) for every $n \geq 0$, $S_{n+1} \subset \delta(S_n, u_n)$,

(b) for every $n > 0$ and for every $q \in S_{n+1}$, there exists a path in $\mathcal{A}$ starting in S_n, ending with q, labeled u_n and visiting at least one final state after its origin.

In order to apply König's lemma, we build a tree (N, r, p) as follows. The set of nodes is

$$N = \{r\} \cup \{(q, n) \mid q \in S_n, n \in \mathbb{N}\}$$

The parent of each node of the form $(q, 0)$ is r and, for $n > 0$, the parent of each node of the form $(q, n+1)$ is chosen among the states (q', n) such that there is a path in $\mathcal{A}$ starting in q', ending in q, labeled u_n and visiting at least one final state after its origin. Conditions (a) and (b) guarantee the possibility of such a construction. Since the tree thus obtained is infinite and since each child has only a finite number of children, it contains an infinite path by König's lemma. This implies the existence of an infinite path in $\mathcal{A}$, labeled u, starting in I and passing infinitely often through a final state. Thus u is accepted by $\mathcal{A}$.

Conversely, let us consider a successful path c of the automaton $\mathcal{A}$

$$c : q_0 \xrightarrow{a_0} q_1 \xrightarrow{a_1} q_2 \ \cdots$$

There is a unique initial path in $\mathcal{D}$ with the same label u

$$d : I = R_0 \xrightarrow{a_0} R_1 \xrightarrow{a_1} R_2 \ \cdots$$

Each of the states q_i belongs to the label of the root of the trees R_i. This root is never suppressed and it is thus a fixed element v_0 of V. If v_0 is marked infinitely often in the R_i's, the path is successful in $\mathcal{D}$ and the word u is accepted. Otherwise, there is a largest integer n such that v_0 is marked in R_n. Let n_0 be this integer and let us consider the smallest integer $m > n_0$ such that q_m is an infinitely repeated final state. Since q_m is final, it appears in a child of the root, and from some time $n_1 \geq m$ on, each q_n with $n \geq n_1$ appears in a fixed child v_1 of the root of R_n. Indeed, if q_n occurs in the label of given node v, then q_{n+1} occurs again in the label of v at the next step, unless it occurs on the left of v (step 3). But such a left shift can occur only a finite number of times. If v_1 is marked infinitely often, the path is successful in $\mathcal{D}$. Otherwise, we repeat the same process, replacing v_0 by v_1. Since the tree has a finite height, we always find some node which is marked infinitely often. □

We shall see later other proofs of McNaughton's theorem which cast a different light upon it (see Section II.9). Among its numerous consequences, we begin with the most important one, known as Büchi's theorem.

Theorem 9.4. *The class of recognizable subsets of* A^ω *is closed under complement.*

Proof. By McNaughton's theorem, any recognizable set can be recognized by a Muller automaton. This automaton can be supposed to be complete by Proposition 7.3. Conversely, by Theorem 7.1 and Theorem 5.4 any set recognized by a Muller automaton is recognizable. The result follows from the fact that, by Proposition 7.6, the class of sets recognized by Muller automata is closed under all boolean operations. □

Büchi's theorem can also be proved directly using congruences (see Chapter II). But the size of the automaton for the complement given by Safra's algorithm is asymptotically optimal, as will be shown in Section 10.6 using the following result.

Theorem 9.5. *For each* $n > 0$*, there exists a set* L_n *of infinite words recognized by a Büchi automaton with* $n + 2$ *states, such that any Büchi automaton recognizing the complement of* L_n *has at least* $n!$ *states.*

Proof. Let $A_n = \{0, 1, \ldots, n\}$ and let $\mathcal{A}_n$ be the automaton on the alphabet A_n represented in Figure 9.19 and let $L_n = L^\omega(\mathcal{A}_n)$. One could of course describe precisely

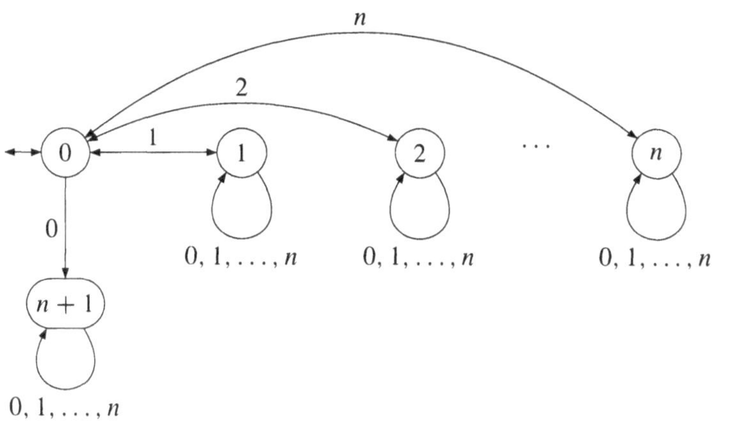

Figure 9.19. A Büchi automaton recognizing L_n.

L_n, but two weaker lemmas will be sufficient for our purpose. We start with a sufficient condition for a word to be in L_n.

Lemma 9.6. *Let $\{i_1, i_2, \dots, i_k\}$ be a subset of $\{1, 2, \dots, n\}$. If an infinite word u contains infinitely many occurrences of each of the factors $i_1 i_2$, $i_2 i_3$, ..., $i_k i_1$, and if in $\mathcal{A}_n$, there is a finite path from 1 to i_1 labeled by a prefix of u, then $u \in L_n$.*

Proof. It suffices to describe a successful path of label u in $\mathcal{A}_n$. By hypothesis, there is a path from 1 to i_1 labeled by a prefix of u. We then stay in state i_1 until the next occurrence of $i_1 i_2$, that is used to produce the transitions $i_1 \xrightarrow{i_1} 0 \xrightarrow{i_2} i_2$. Then we stay in state i_2 until the next occurrence of $i_2 i_3$, that is used to produce the transitions $i_2 \xrightarrow{i_2} 0 \xrightarrow{i_3} i_3$, etc. This process, repeated infinitely often on the cycle $(i_1 i_2, i_2 i_3, \dots, i_k i_1)$, produces the desired successful path. □

With each permutation σ of $\{1, 2, \dots, n\}$, associate the infinite word $u_\sigma = (\sigma(1) \cdots \sigma(n)0)^\omega$.

Lemma 9.7. *For any permutation σ of $\{1, \dots, n\}$, the infinite word u_σ is not in L_n.*

Proof. Clearly, $L_n \subset K^\omega$, where $K = \bigcup_{1 \le i \le n} i A_n^* i$. Therefore, if $u_\sigma \in L_n$, $u = u_1 u_2 \dots$, where each u_i is in K. It follows that $\sigma(1)$ is the first and last letter of u_1, $\sigma(2)$ is the first and last letter of u_2, and $\sigma(n)$ is the first and last letter of u_n. Consequently, the first letter of u_{n+1} is 0, a contradiction, since $u_{n+1} \in K$. □

Let now $\mathcal{B}$ be a Büchi automaton accepting the complement of L_n. By Lemma 9.7, each word u_σ is accepted by $\mathcal{B}$. Therefore, there is in $\mathcal{B}$ a successful path p_σ of label u_σ.

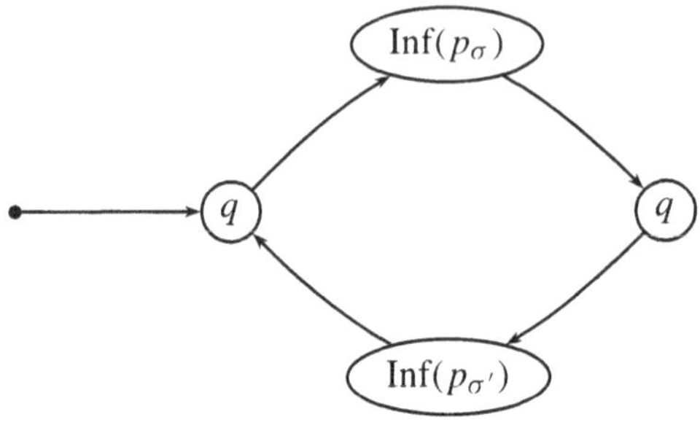

Figure 9.20. The path p.

We claim that if $\sigma \neq \sigma'$, then $\mathrm{Inf}(p_\sigma) \cap \mathrm{Inf}(p_{\sigma'}) = \emptyset$. Assume by contradiction that some state q belongs to both $\mathrm{Inf}(p_\sigma)$ and $\mathrm{Inf}(p_{\sigma'})$. Using the two paths, we build a new path p in $\mathcal{B}$ which, at the beginning, follows a prefix of p_σ of length at least $n(n+1)$ until it reaches q. Then p enters a loop which is repeated infinitely often. This loop consists of two parts that we also take of length at least $n+1$: in the first part, p follows a portion of p_σ to go from q to q after visiting at least once all states of $\mathrm{Inf}(p_\sigma)$ and in the second part, p follows a portion of $p_{\sigma'}$ to go from q to q after visiting at least once all states of $\mathrm{Inf}(p_{\sigma'})$ (see Figure 9.20). Then $\mathrm{Inf}(p)$ contains $\mathrm{Inf}(p_\sigma)$ (and $\mathrm{Inf}(p_{\sigma'})$) and in particular contains a final state, since p_σ is successful. It follows that p is successful and thus its label u is not in L_n. We shall arrive to a contradiction by showing that u satisfies the conditions of Lemma 9.6, and therefore belongs to L_n.

We first verify the existence of a cycle of infinitely repeated factors of length two. Let k be the smallest integer such that $\sigma(k) \neq \sigma'(k)$. Then $\sigma'(k) = \sigma(l)$ for some $l > k$ and $\sigma(k) = \sigma'(m)$ for some $m > k$. Since u is a concatenation of factors of length at least $n+1$ of u_σ and $u_{\sigma'}$, each of the factors $\sigma(k)\sigma(k+1)$, $\sigma(k+1)\sigma(k+2)$, $\ldots$, $\sigma(\ell-1)\sigma(\ell)(=\sigma(\ell-1)\sigma'(k))$, $\sigma'(k)\sigma'(k+1), \ldots, \sigma'(m-1)\sigma'(m)(=\sigma'(m-1)\sigma(k))$ occur infinitely often in u.

It suffices now to verify that the state $\sigma(k)$ is reachable in $\mathcal{A}_n$ by a path labelled by a prefix of u. By construction, the word $(\sigma(1)\cdots\sigma(n)0)^n$ is a prefix of u. Therefore, the path

$$0 \xrightarrow{\sigma(1)} \sigma(1) \xrightarrow{\sigma(2)\cdots\sigma(n)0} \sigma(1) \xrightarrow{\sigma(1)} 0 \xrightarrow{\sigma(2)} \sigma(2) \cdots$$
$$\sigma(k-1) \xrightarrow{\sigma(k-1)} 0 \xrightarrow{\sigma(k)} \sigma(k)$$

is suitable for our purpose.

This proves the claim, and since there are $n!$ permutations on $\{1, \ldots, n\}$, there are at least $n!$ disjoint sets of the form $\mathrm{Inf}(p_\sigma)$, which clearly implies that $\mathcal{B}$ has at least $n!$ states. □

As announced above, recognizable sets are determined by the ultimately periodic words they contain.

Corollary 9.8. *Let X and Y be two recognizable subsets of* A^ω*. Let* $U \subset A^\omega$ *be the set of ultimately periodic words. If* $X \cap U \subset Y$*, then* $X \subset Y$*. In particular* $X = Y$ *if and only if X and Y contain the same ultimately periodic words, i.e. if* $X \cap U = Y \cap U$*.*

Proof. In fact, if X is not contained in Y, the set $X \setminus Y$ is, by Büchi's theorem, a nonempty recognizable subset of A^ω. By Lemma 5.1, there exists an ultimately periodic word which is in X but not in Y. □

We now turn to another consequence of McNaughton's theorem, which solves a subtle point raised in Section 6.

Theorem 9.9. *A subset of* A^ω *is recognizable by a finite deterministic Büchi automaton if and only if it is both deterministic and recognizable.*

Proof. Any set recognized by a finite deterministic Büchi automaton satisfies certainly these two conditions. Conversely, let X be a subset of A^ω satisfying the two conditions. By McNaughton's theorem, the set X is recognized by a Muller automaton $\mathcal{A} = (Q, A, E, i, \mathcal{T})$. But since X is deterministic, the table $\mathcal{T}$ is full by Proposition 7.10. Finally, Proposition 7.9 shows that X can be recognized by a finite deterministic Büchi automaton. □

Corollary 9.10. *It is decidable whether a given recognizable subset of* A^ω *is deterministic or not.*

Proof. Let X be a recognizable subset of A^ω. We may build, using the previously described algorithms, a Muller automaton recognizing X. Proposition 7.10 allows one to conclude. □

Example 9.4. The set $X = (\{b, c\}^*a \cup b)^\omega$ of Example 9.3 is deterministic. On the contrary, the set $Y = (a\{b, c\}^* \cup b)^\omega$ is not deterministic. In fact Y is recognized by the Büchi automaton represented in Figure 9.21. The deterministic automaton obtained

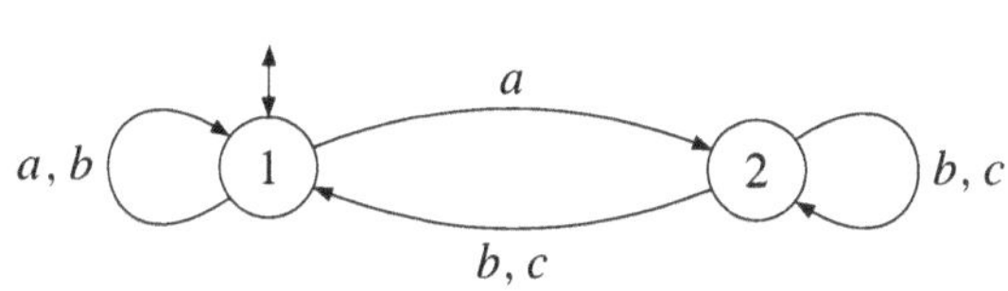

Figure 9.21. A co-deterministic but non deterministic automaton.

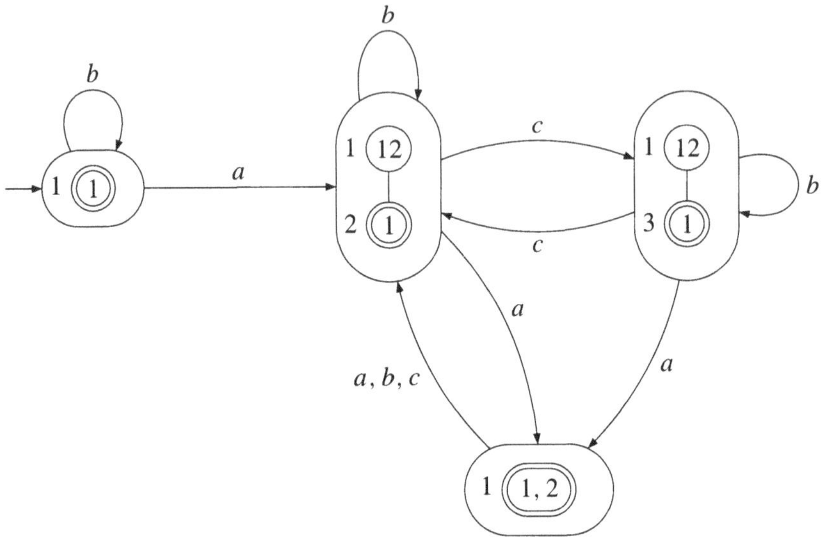

Figure 9.22. The resulting automaton.

by Safra's algorithm is represented in Figure 9.22 One can check directly that Y is not deterministic, by imitating the construction used in Example 6.2. Let us indeed suppose that $Y = \overrightarrow{L}$. Since $acb^\omega \in Y$, there is an integer n_1 such that $acb^{n_1} \in L$. Again, since $acb^{n_1}cb^\omega \in Y$, there is an integer n_2 such that $acb^{n_1}cb^{n_2} \in L$, etc. and the infinite word $u = acb^{n_1}cb^{n_2}cb^{n_3}\cdots$ has an infinite number of prefixes in L. This implies that $u \in Y$, which is impossible since u contains infinitely many c's but a finite number of a's.

10 Computational complexity issues

In this section, we address the problem of the computational complexity of the various transformations introduced in this chapter. The results are summarized in Figure 10.1. The nodes of this graph illustrate various representations of sets of infinite words, such as ω-rational expression, Büchi automaton, etc. An arrow between two nodes indicates an algorithm to convert one representation into another one. The label of the arrow indicates the complexity of the corresponding algorithm. The label P stands for a polynomial time algorithm and *Exp* for an exponential one.

The size of the various objects is defined according to the following conventions. As a general rule, we consider the cardinality of the alphabet as being a constant.

The *size* of an ω-rational expression is the number of symbols that it involves, without the parenthesis but taking the dot into account for the product. Thus size(ε) =

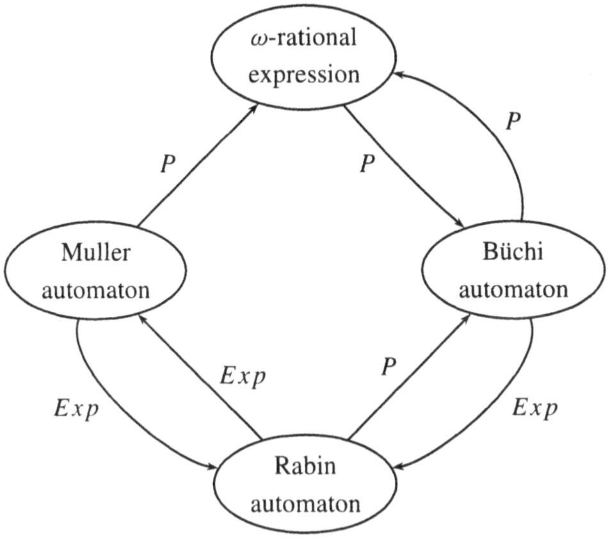

Figure 10.1. Summary of transformations.

size($\{a\}$) = 1 and size($X + Y$) = size(XY) = size(X) + size(Y) + 1, size(X^*) = size(X) + 1.

The size of a Büchi automaton $\mathcal{A} = (Q, E, I, T)$ is max(Card(Q), Card(E)). It is thus at most equal to Card$(Q)^2 \times$ Card(A).

The size of a Muller automaton $\mathcal{A} = (Q, E, i, \mathcal{T})$ is

$$\max(\mathrm{Card}(Q), \mathrm{Card}(E), \mathrm{Card}(\mathcal{T}))$$

It is thus bounded by $\max(\mathrm{Card}(Q) \times \mathrm{Card}(A), 2^{\mathrm{Card}(Q)})$. Note that it may be exponential in Card(Q). We describe separately the transformation associated with each arrow in Figure 10.1. The algorithms associated with these transformations have the same complexity as the size of the resulting objects.

10.1 From ω-rational expressions to Büchi automata and back. The following result shows that ω-rational expressions and Büchi automata are objects of essentially equivalent computational complexity.

Proposition 10.1. *For any ω-rational expression of size n describing a set $X \subset A^\omega$, there is a Büchi automaton of size $O(n)$ recognizing X and conversely.*

Proof. Given an ω-rational expression of size s, the size of a Büchi automaton built by the method described in Section 5 is bounded by $2s$.

Conversely, given a Büchi automaton of size s, one can compute a corresponding ω-rational expression of size bounded by some linear function of s. □

10.2 From Muller automata to ω-rational expressions. The following result shows that the size of an ω-rational expression for the set recognized by a Muller automaton is polynomially bounded by the size of the automaton.

Proposition 10.2. *Let $\mathcal{A}$ be a Muller automaton of size n. The set $L^{\omega}(\mathcal{A})$ can be described by an ω-rational expression of size $O(n^3)$.*

Proof. The ω-rational expression obtained from a Muller automaton

$$\mathcal{A} = (Q, E, i, \mathcal{T})$$

by the method of Theorem 7.1 is a union of $\mathrm{Card}(\mathcal{T})$ terms of the form

$$X(Y_0 Y_1 \dots Y_k)^{\omega}$$

with $k \leq \mathrm{Card}(Q)$ and where X and the Y_i's are recognized by Büchi automata of size lesser than or equal to the size of $\mathcal{A}$.

Its size is bounded by $\mathrm{Card}(\mathcal{T}) \times \mathrm{Card}(Q) \times \mathrm{Card}(E)$ and hence by a polynomial of degree 3 in the size of $\mathcal{A}$. □

Note that the size of the ω-rational expression can be exponential in the number of states (if the number of accepting sets is exponential).

10.3 From Büchi automata to Rabin automata. The following result implies that, given a Büchi automaton of size n, there is an equivalent Rabin automaton of size $2^{O(n \log n)}$.

Proposition 10.3. *For any Büchi automaton of size n, there exists an equivalent Rabin automaton with $2^{O(n \log n)}$ states and n pairs.*

Given a Büchi automaton with n states, the algorithm of Section 9 builds a Rabin automaton on a set of states $\mathcal{T}_n$ with $O(n)$ pairs. Thus the result follows from the following proposition.

Proposition 10.4. *The set $\mathcal{T}_n$ satisfies* $\ln(\mathrm{Card}(\mathcal{T}_n)) = O(n \ln n)$.

Proof. Let, as in the proof of Proposition 9.2, $r(v)$ be the set of states which appear in the label of v but in none of its children. The relation $s = r^{-1}$, is a function from Q onto T which completely determines e, since we have

$$e(v) = \bigcup_{w \text{ ancestor of } v} s^{-1}(w).$$

An element of $\mathcal{T}_n$ is described by the tuple (T, f, s, M) where $(T, 1, f)$ is an oriented tree with at most n nodes, s is a function from Q onto T and M is a subset of T. By Proposition 2.1, the number of planar trees with k nodes is

$$C_k = \frac{(2k-2)!}{k!(k-1)!}.$$

To obtain the number of oriented trees with k nodes chosen in V, we multiply by the number of surjective functions from a k element set to V, which is

$$I_k = \frac{(2n)!}{(2n-k)!}.$$

We can then bound the number of surjective functions S_k from Q onto a k element set by n^{k+1}. Finally, the number of subsets of a k element set is 2^k. We obtain

$$\text{Card}(\mathcal{T}_n) \leq \sum_{1 \leq k \leq n} C_k I_k S_k 2^k \leq n C_n I_n S_n 2^n \leq n \frac{(2n-2)!}{n!(n-1)!} \frac{(2n)!}{n!} (n)^{n+1} 2^n$$

and, observing that $\frac{(2n)!}{(n)!^2} \leq 4^n$, we conclude that

$$\ln(\text{Card}(\mathcal{T}_n)) \leq (2n-1)\ln 4 + \ln((n-1)!) + (n+2)\ln n + n \ln 2$$

whence the result since $\ln(n!) = O(n \ln n)$. □

10.4 From Rabin automata to Muller automata. The following result shows that the transformation from a Rabin automaton into a Muller automaton involves an exponential blow-up in the size of the automaton.

Proposition 10.5. *For any Rabin automaton with n states, there is an equivalent Muller automaton with n states and $O(2^n)$ accepting sets.*

Proof. A Rabin automaton with n states and m pairs can be transformed into a Muller automaton with n states and 2^n accepting sets of states, as shown in the proof of Proposition 7.7. □

The method indicated in Exercise 17 shows that, conversely, any Muller automaton with n states and m accepting sets can be converted into an equivalent Rabin automaton with $2^{O(n)}$ states and m pairs. Thus, the size of the Rabin automaton is exponential in the size of the original Muller automaton.

10.5 From Rabin automata to Büchi automata. The following result shows that there is a polynomial algorithm transforming a Rabin automaton into an equivalent Büchi automaton.

Proposition 10.6. *For any Rabin automaton $\mathcal{A}$ with n states and m accepting pairs, there is a Büchi automaton $\mathcal{B}$ of size $O(nm)$ such that $L^\omega(\mathcal{A}) = L^\omega(\mathcal{B})$.*

Proof. Let $\mathcal{A} = (Q, i, \mathcal{R})$ be a Rabin automaton with $\mathcal{R} = \{(L_i, U_i) \mid 1 \le i \le m\}$. The set of states of the automaton $\mathcal{B}$ is the union of Q and of the set

$$\{(q, i) \in Q \times \{1, 2, \dots, m\} \mid q \notin L_i\}$$

The initial state is the initial state of $\mathcal{A}$. Each transition $p \xrightarrow{a} q$ of $\mathcal{A}$ gives rise to the additional transitions $p \xrightarrow{a} (q, i)$ for each i such that $q \notin L_i$ and $(p, i) \xrightarrow{a} (q, i)$ for each i such that $p \notin L_i$ and $q \notin L_i$. The terminal states are the (p, i)'s such that $p \in U_i$. Thus a successful path in $\mathcal{B}$ begins with a path of $\mathcal{A}$ and chooses nondeterministically to avoid some L_i, while checking that it meets U_i infinitely often. It is clear that $\mathcal{A}$ and $\mathcal{B}$ are equivalent and that $\mathcal{B}$ has at most $n + nm = O(nm)$ states. □

10.6 From Streett automata to Büchi automata. We describe a construction which will be used below to compute a Büchi automaton recognizing the complement.

Proposition 10.7. *Any Streett automaton with n states and m accepting pairs is equivalent to a Büchi automaton of size $n2^{O(m)}$.*

Proof. Let $\mathcal{A} = (Q, i, \mathcal{P})$ be a Streett automaton with a set $\mathcal{P} = \{(L_i, U_i) \mid 1 \le i \le m\}$ of accepting pairs. We construct an equivalent Büchi automaton $\mathcal{B} = (U, j, T)$ with U as set of states, j as initial state and T as set of terminal states as follows. The states have the form (q, S_1, S_2) where $q \in Q$ is a state of $\mathcal{A}$, and S_1, S_2 are finite sets of integers either reduced to $\{0\}$ or equal to a subset of $\{1, 2, \dots, m\}$. The initial state is $j = (i, 0, 0)$. The transitions are of two kinds. There is a first set of initial transitions formed of all transitions $(p, 0, 0) \xrightarrow{a} (q, 0, 0)$ or $(q, \emptyset, \emptyset)$ where $p \xrightarrow{a} q$ is a transition of $\mathcal{A}$. The second set consists of transitions of the form $(p, S_1, S_2) \xrightarrow{a} (q, S_1', S_2')$ where $p \xrightarrow{a} q$ is a transition of $\mathcal{A}$ and

$$S_1' = \begin{cases} S_1 \cup \{i\} & \text{if } q \in U_i \\ S_1 & \text{otherwise} \end{cases} \quad \text{and} \quad S_2' = \begin{cases} \emptyset & \text{if } S_1 \subset S_2 \\ S_1 \cup \{i\} & \text{if } q \in L_i \text{ and } S_1 \not\subset S_2 \\ S_1 & \text{otherwise} \end{cases}$$

Thus the automaton $\mathcal{B}$ first behaves as $\mathcal{A}$ during some initial period. In the rest of the time, it uses the sets S_1, S_2 to remember which of the sets L_i, U_i were visited. The set T of terminal states is formed of the states (q, S_1, S_2) such that $S_2 = \emptyset$. A path β of the Büchi automaton $\mathcal{B}$ is of the form

$$\beta : (i, 0, 0) \xrightarrow{a_0} \dots (p_n, S_n, T_n) \xrightarrow{a_n} (p_{n+1}, S_{n+1}, T_{n+1}) \xrightarrow{a_{n+1}} \dots$$

where S_n is ultimately equal to some subset S of $\{1, 2, \ldots, m\}$ (unless, of course, $S_n = \{0\}$ for all n). It corresponds to a path

$$\gamma : i \xrightarrow{a_0} \ldots p_n \xrightarrow{a_n} p_{n+1} \xrightarrow{a_{n+1}} \ldots$$

of $\mathcal{A}$. The path β is successful if and only if $T_n = \emptyset$ for infinitely many n. This happens if and only if $S \subset T_n$ for infinitely many n. This last condition is equivalent to $p_n \in L_i$ for infinitely many n and for each $i \in S$. This is again equivalent to the fact that for each $i \in \{1, 2, \ldots, m\}$, either $i \notin S$, i.e. $\mathrm{Inf}(\gamma) \cap U_i = \emptyset$, or $i \in S$ and $\mathrm{Inf}(\gamma) \cap L_i \neq \emptyset$. This shows that β is successful if and only if γ is successful. Thus $\mathcal{A}$ and $\mathcal{B}$ are equivalent. □

In the case where $m = O(\log n)$, we obtain as a corollary.

Corollary 10.8. *Any Streett automaton with n states and $O(\log n)$ accepting pairs is equivalent to a Büchi automaton of size $O(n)$.*

Finally, we obtain the following result.

Theorem 10.9. *For any set $X \subset A^\omega$ recognized by an n-state Büchi automaton, there is an automaton of size $2^{O(n \log n)}$ recognizing $A^\omega \setminus X$.*

Proof. Indeed, let $\mathcal{A}$ be a Büchi automaton of size n recognizing X. By Proposition 10.3, we can first build a Rabin automaton $\mathcal{B}$ with $2^{O(n \log n)}$ states and n accepting pairs equivalent to $\mathcal{A}$. Then, considered as a Streett automaton, $\mathcal{A}$ recognizes $A^\omega \setminus X$. By Corollary 10.8, there is an equivalent Büchi automaton of size $2^{O(n \log n)}$. □

Let us formulate two comments about Theorem 10.9. First of all, the complexity of computing the complement of an ω-rational set is not considerably higher than for a rational set of finite words. Indeed, we obtain $2^{O(n \log n)}$ instead of 2^n. Second, this bound is essentially optimal. Indeed, Theorem 9.5 shows that the size of a Büchi automaton for the complement can be $n! = 2^{\Theta(n \log n)}$.

10.7 Complexity of algorithms on automata. We consider now the algorithmic complexity of some algorithms on ω-rational sets. For a recognizable set $X \subset A^\omega$, let $\zeta(X)$ be the minimal size of a Büchi automaton recognizing X. The following result gives an evaluation of the complexity of some of the basic operations on ω-rational sets.

Proposition 10.10. *Let X_1 and X_2 be recognizable subsets of A^ω and let $m_1 = \zeta(X_1)$, $m_2 = \zeta(X_2)$. Then*

(1) $\zeta(X_1 \cup X_2) = O(m_1 + m_2)$.

(2) $\zeta(X_1 \cap X_2) = O(m_1 m_2)$.

(3) $\zeta(A^\omega \setminus X_1) = m_1^{O(m_1)}$.

Proof. The construction of an automaton recognizing the union of two recognizable sets has been described in the proof of Kleene's theorem (Theorem 5.4). The number m of edges of the new automaton satisfies $m \leq (m_1 + 2) + (m_2 + 2)$.

The construction of a Büchi automaton recognizing $X_1 \cap X_2$ is given in Exercise 7. The number m of edges satisfies $m \leq 4m_1m_2$.

The last statement is Theorem 10.9. □

In the following statement, we put together the consequences of several algorithms seen before in terms of decidability. The complexity of these algorithms will be considered afterwards.

Proposition 10.11. *One may effectively decide*

(1) *the emptiness of a recognizable subset of* A^ω,
(2) *the inclusion of two recognizable subsets of* A^ω,
(3) *the equality of two recognizable subsets of* A^ω.

Proof. (1) Let $\mathcal{A} = (Q, A, E, I, F)$ be a finite Büchi automaton. By Proposition 5.2, we may suppose that $\mathcal{A}$ is trim. In this case, $L^\omega(\mathcal{A})$ is empty if and only if Q is empty.

(2) The inclusion $X \subset Y$ is equivalent to $X \setminus Y = \emptyset$. We may thus apply (1) and Proposition 10.10.

(3) The equality of two subsets X and Y of A^ω is equivalent to the double inclusion $X \subset Y$ and $Y \subset X$ and is thus decidable. □

We now consider the algorithmic complexity of the basic problems on Büchi automata. We will use here some notation and results from the theory of algorithms (see the Notes Section for references). The *nonemptiness problem* for a Büchi automaton is to decide if $L^\omega(\mathcal{A}) \neq \emptyset$. A problem is in the class *NL* if it can be solved by a nondeterministic algorithm operating with a logarithmic amount of space.

Proposition 10.12. *The nonemptiness problem for Büchi automata is decidable in linear time. It is NL-complete.*

Proof. The set $L^\omega(\mathcal{A})$ is nonempty if and only if there is a cycle in the underlying graph of the automaton which is accessible from the initial state. A depth-first search in this graph can verify this in linear time.

To show that the nonemptiness problem is *NL*-complete, we first have to show that it is in the class *NL* and then that it is *NL*-hard, i.e. that any problem in *NL* can be reduced to it. We shall prove the first part only. See the Notes Section for a reference to a proof of the second part.

The nondeterministic algorithm to verify nonemptiness consists in choosing nondeterministically at each step a state of the automaton $\mathcal{A} = (Q, i, T)$ of size n. If, from state p, one chooses q, we check that there is an edge connecting p to q. If this is the case, we can forget p and continue. Otherwise, we choose another state q'. When we reach a terminal state $t \in T$, we remember it. We then continue until we find t again. We thus verify nondeterministically the existence of a successful path in $\mathcal{A}$. The space needed consists in memorizing two states (the current state on the path and the terminal state when reached). Since $\log n$ bits are enough to represent an element of Q, the proof is complete. □

The *nonuniversality problem* for a Büchi automaton $\mathcal{A}$ is to decide whether $L^\omega(\mathcal{A}) \neq A^\omega$. A problem is in the class *PSPACE* if it can be solved by an algorithm using a polynomial amount of space.

Proposition 10.13. *The nonuniversality problem for automata is decidable in exponential time. It is PSPACE-complete.*

Proof. The nonuniversality problem for $\mathcal{A}$ is the same as the nonemptiness problem for the complement of $L^\omega(\mathcal{A})$. Since the size of an automaton $\bar{\mathcal{A}}$ for the complement is exponential, the first assertion follows easily. The second assertion requires more attention. First, by Proposition 10.12, we can verify the nonemptiness of $\bar{\mathcal{A}}$ with a nondeterministic polynomial space algorithm. Second, by a result known as Savitch theorem (see the Notes Section for references), any problem solvable by a nondeterministic polynomial space algorithm is in *PSPACE*. We again leave the proof of the *PSPACE*-hardness to be found in the references of the Notes Section. □

11 Exercises

11.1 Words and trees.

Exercise 1. Let A be a finite alphabet. Show that every word $x \in A^\omega$ can be factorized as $x = yz$, where $y \in A^*$, $z \in A^\omega$, and each letter of z occurs infinitely often in x.

11.2 Rational sets.

Exercise 2. Let X, X_1 and X_2 be subsets of A^∞. Verify the following equalities.

(1) $\mathit{Pref}(\mathit{Pref}(X)) = \mathit{Pref}(X)$,

(2) $\mathit{Pref}(X_1 \cup X_2) = \mathit{Pref}(X_1) \cup \mathit{Pref}(X_2)$,

(3) $\mathit{Pref}(X_1 X_2) = \mathit{Pref}(X_1) \cup (X_1 \cap A^*)\mathit{Pref}(X_2)$,

(4) $\mathit{Pref}(X^\omega) = \mathit{Pref}(X^+) = (X \cap A^+)^*\mathit{Pref}(X)$.

Conclude that if X is an ω-rational subset of A^∞, $\mathit{Pref}(X)$ is a rational subset of A^*.

Exercise 3. Verify the following identities, where K and L are subsets of A^*:

$$(K + L)^* = (K^*L)^*K^*$$
$$(KL)^* = 1 + K(LK)^*L$$

Exercise 4. Prove that any subset of A^+ can be recognized by a (possibly infinite) deterministic automaton.

11.3 Büchi automata.

Exercise 5. Show that any subset of A^ω can be recognized by a (possibly infinite) Büchi automaton in which all the states are final.

Exercise 6. Prove that the following conditions are equivalent for a subset X of A^ω.

(1) X is stable by shift, i.e. $\sigma(X) \subset X$.

(2) If X is recognized by a Büchi automaton $\mathcal{A} = (Q, A, E, I, F)$, X is also recognized by the automaton (Q, A, E, P, F) where P is the set of accessible states of $\mathcal{A}$.

(3) X is recognized by a Büchi automaton in which all the states are initial.

Exercise 7. Let $\mathcal{A}_1 = (Q_1, A, E_1, i_1, F_1)$ and $\mathcal{A}_2 = (Q_2, A, E_2, i_2, F_2)$ be two Büchi automata with a single initial state. The aim of this problem is to build a Büchi automaton recognizing $L^\omega(\mathcal{A}_1) \cap L^\omega(\mathcal{A}_2)$. Let

$$Q = Q_1 \times Q_2 \times \{0, 1\}$$

and consider the automaton $\mathcal{A} = (Q, A, E, (i_1, i_2, 1), F)$ where

$$E = \big\{((p_1, p_2, s), a, (q_1, q_2, t)) \mid (p_1, a, q_1) \in E_1, (p_2, a, q_2) \in E_2 \text{ and}$$
$$t = 0 \text{ if and only if } (s = 1, p_2 \in F_2 \text{ and } q_1 \notin F_1) \text{ or } (s = 0 \text{ and } q_1 \notin F_1\big\}$$

and

$$F = Q_1 \times F_2 \times \{1\}.$$

Show that $L^\omega(\mathcal{A}) = L^\omega(\mathcal{A}_1) \cap L^\omega(\mathcal{A}_2)$.

11.4 Deterministic Büchi automata.

Exercise 8. Let $\mathcal{A}$ be a deterministic Büchi automaton. The set of infinite words recognized *in the weak sense* by $\mathcal{A}$ is the set, denoted $L^\omega_w(\mathcal{A})$ of all words which are the label of an infinite path starting in the initial state and passing at least once by a final state.

Show that a set of infinite words can be recognized in the weak sense by a complete deterministic Büchi automaton if and only if it is of the form XA^ω for some $X \subset A^*$.

Exercise 9. Show that if a subset X of A^+ is a finite union of prefix-free sets (and thus in particular either if X is finite or if X is prefix), then $\overrightarrow{X} = \emptyset$.

Exercise 10. Let X be a rational subset of A^+. Show that $\overrightarrow{X} = \emptyset$ if and only if X is a finite union of prefix-free sets.

Show that the above result is not always true if X is not rational (consider the set $X = \{a^n b^m \mid 0 < m \leq n\}$).

Exercise 11. Let X be a subset of A^ω. Show that the following conditions are equivalent:

(1) the set X can be recognized by a deterministic Büchi automaton in which all states are final,

(2) there exists a subset P of A^* such that X is the set of infinite words having all its prefixes in P,

(3) there exists a prefix-closed subset P of A^* such that X is the set of infinite words having all its prefixes in P,

(4) there exists a subset R of A^* such that X is the set of infinite words having no prefix in R,

(5) there exists a prefix-closed subset P of A^* such that $X = \overrightarrow{P}$,

(6) the equality $X = \overrightarrow{Pref(X)}$ holds.

These are the closed sets of the natural topology on A^ω (see Chapter III).

Exercise 12. A subset of A^ω is called *open* if its complement satisfies the equivalent conditions of Exercise 11. The link with topology will, here again, be covered in Chapter III.

Show that any recognizable open subset of A^ω is deterministic.

It is also possible to give a characterization of open subsets in terms of automata (see Exercise 8).

Exercise 13. Let X be a subset of A^ω. Prove that the following conditions are equivalent.

(1) X is closed and shift invariant.

(2) X is recognized by a Büchi automaton with deterministic transitions with all its states initial and final.

(3) There exists a subset P of A^* such that X is the set of infinite words having all its factors in P.

11.5 Muller automata.

Exercise 14. Let $\mathcal{A}_1 = (Q_1, A, E_1, i_1, \mathcal{T}_1)$ and $\mathcal{A}_2 = (Q_2, A, E_2, i_2, \mathcal{T}_2)$ be two Muller automata. Build a Muller automaton recognizing $L^\omega(\mathcal{A}_1) \cup L^\omega(\mathcal{A}_2)$.

Exercise 15. A *non deterministic Muller automaton* is a tuple

$$\mathcal{A} = (Q, A, E, I, \mathcal{T})$$

where (Q, A, E) is a finite automaton, I is a subset of Q whose elements are the initial states and $\mathcal{T}$ is a family of subsets of Q, called the *table* of the automaton.

An infinite path p in $\mathcal{A}$ is *initial* if it starts in the initial state and *final* if $\mathrm{Inf}(p) \in \mathcal{T}$. It is *successful* if it is both initial and final. The set of infinite words recognized by $\mathcal{A}$ is the set of labels of infinite successful paths in $\mathcal{A}$.

Show that any subset of A^ω recognized by a non deterministic Muller automaton can be also recognized by a deterministic one. (Hint: the proof of Theorem 7.1 does not use the hypothesis that the automata are deterministic.)

Exercise 16. Let $\mathcal{A} = (Q, A, E, i, F)$ and $\mathcal{A}' = (Q', A, E', i', F')$ be two deterministic Büchi automata. Our aim is to build directly a Rabin $\mathcal{B}$ automaton recognizing $L^*(\mathcal{A})L^\omega(\mathcal{A}')$. The idea is to use the automaton $\mathcal{A}$ starting a new computation of $\mathcal{A}'$ each time $\mathcal{A}$ visits a final state. To make the total size finite, we keep only the earliest copy of $\mathcal{A}'$ when two of them are in the same state. As an acceptance condition, we require a final state of a fixed copy of $\mathcal{A}'$ be visited infinitely often.

For this, we first define the set of *simple* words on Q' as the elements of Q'^* in which every state $q \in Q'$ appears at most once. For each $v \in Q'^*$, we denote by $[v]$ the simple word obtained by keeping only the leftmost occurrence of each state.

For a simple word $v = q_1 q_2 \cdots q_n \in Q'^*$ and a letter $a \in A$, we define the action of a on v by

$$v \cdot a = (q_1 \cdot a)(q_2 \cdot a) \cdots (q_n \cdot a)$$

The word $v \cdot a$ may perhaps not be simple. Let k be the largest integer at most equal to n such that the states $q_1 \cdot a, \ \ldots\ , \ q_k \cdot a$ are distinct. The integer k will be denoted $\ell_{v,a}$. If v is the empty word, we let $\ell_{v,a} = 0$.

We define now a Rabin automaton $\mathcal{B}$ by choosing the state set as being the set T of triples

$$(q, v, \ell)$$

with $q \in Q$, v a simple word on Q' and ℓ an integer such that $0 \leq \ell \leq |v|$. The states are thus formed of a state of $\mathcal{A}$ and a sequence of distinct states of $\mathcal{A}'$ marked at a position ℓ which is used to indicate the position of the last deleted state. The initial state is $(i, 1, 0)$. The transitions are defined by

$$(q, v, \ell) = \begin{cases} (q \cdot a, [v \cdot a], \ell_{v,a}) & \text{if } q \cdot a \notin F, \\ (q \cdot a, [(v \cdot a)i'], \ell_{v,a}) & \text{if } q \cdot a \in F. \end{cases}$$

The acceptance is defined by the family $\mathcal{R} = \{(L_k, U_k) \mid 1 \leq k \leq \mathrm{Card}(Q)\}$, with

$$L_k = \{(q, v, \ell) \in T \mid \ell > 0 \text{ and the } k\text{-th symbol of } v \text{ is in } F'\},$$
$$U_k = \{(q, v, \ell) \in T \mid k \leq \ell\}.$$

Show that the Rabin automaton $\mathcal{B}$ built in this way recognizes $L^*(\mathcal{A})L^\omega(\mathcal{A}')$.

11.6 McNaughton's theorem.

Exercise 17. Let $\mathcal{A} = (Q, E, i, \mathcal{T})$ be a Muller automaton, with

$$\mathcal{T} = \{T_0, \ldots, T_{k-1}\}$$

Let

$$\mathcal{A}' = (Q', E', i', \mathcal{R}')$$

be the Rabin automaton defined by $Q' = \mathcal{P}(T_0) \times \ldots \times \mathcal{P}(T_{k-1}) \times Q$, $i' = (\emptyset, \ldots, \emptyset, i)$ with transitions defined by $(U_0, \ldots, U_{k-1}, q) \cdot a = (U'_0, \ldots, U'_{k-1}, q')$ where $q' = q \cdot a$,

$$U'_i = \begin{cases} \emptyset & \text{if } U_i = T_i, \\ T_i \cap (U_i \cup \{q'\}) & \text{otherwise.} \end{cases}$$

and $\mathcal{R} = \{(L_i, R_i) \mid 0 \le i \le k-1\}$ with $L_i = \{(U_0, \ldots, U_{k-1}, q) \in Q' \mid q \notin T_i\}$ and $R_i = \{(U_0, \ldots, U_{k-1}, q) \in Q' \mid U_i = T_i\}$. Show that $L^\omega(\mathcal{A}) = L^\omega(\mathcal{A}')$.

Exercise 18. Verify Example 9.4 by developing in detail Safra's algorithm.

Exercise 19. Consider a Büchi automaton in which all states are final. Show that the algorithm described in Section 9 coincides with the usual determinization algorithm.

12 Notes

The use of finite automata to recognize infinite words appears for the first time in the work of Büchi [45] and not much later in a paper of Muller [210]. Kleene's seminal paper on automata [160] already considers the case of infinite words, but on the left side. The motivations of Büchi came from the investigation of decidable logical theories. Those of Muller arose from the study of asynchronous circuits.

Looking for earlier origins of this kind of question leads one to the foundations of topology and measure theory. This dates back to the beginning of the twentieth century, with the work of Borel and Lebesgue and, after them, the Polish school, in particular Suslin, who developed the foundations of analysis (see Bourbaki [37]). We shall come back to this point in the chapter on topology (Chapter III). Even closer to our subject, infinite sequences have been studied as sequences of events in classical probability theory. And, for example, the lemma of Borel-Cantelli makes use of the favorite predicates used in this chapter: "there is an infinity of occurrences of some event" (cf. Feller [110]). This aspect of the study of infinite sequences leads to ergodic theory and symbolic dynamics.

Earlier introductory presentations of automata on infinite words include chapters in books on automata theory (e.g. Chapter XIV of [97]) or in multi-author Handbooks (like the surveys of Thomas in [330] or [332] and the survey of Staiger in [301]). The recent multi-author monograph [127] gives a consolidated overview of the recent research results achieved in the theory.

The notation used in this chapter is, in general, the one in use elsewhere in automata theory. In some cases, however, we follow the terminology introduced by Eilenberg [97]. Thus, we use systematically the terms "recognizable" and "rational" instead of "regular". The use of the term "rational" deserves a comment. One may, as first shown by Schützenberger, generalize the theory of Kleene to formal series in noncommuting variables. In this framework, rational series appear as the natural generalization of the classical notion of rational series in one variable.

Besides Kleene's theorem, for which we have followed Conway's presentation [79], the main results of this chapter are due to Büchi and to McNaughton. The chronology places first the work of Büchi. He introduced ω-rational sets and automata in connection with the logical monadic second-order theory of $(\mathbb{N}, <)$. We shall come to this theory in detail in Chapter VIII. His main result says that the class of recognizable sets is closed under complement. His original proof differs from the one given in this chapter. It uses semigroup congruences which are introduced in Chapter II. Example 9.4 is due to Gire (personal communication). See also Staiger [298, Example 2, p. 489]. Theorem 7.8 is due to Landweber [169]. It is the essential part of Theorem 9.9.

The notation $\overrightarrow{L}$ is from Eilenberg [97], whereas the classical notation X^ω is borrowed from the theory of ordinals.

McNaughton's theorem was conjectured by Muller [210], who first introduced the automata that bear his name. McNaughton's original proof [189] uses a construction which has been the object of several further studies. Variants of it were described by Rabin [253], Choueka [68] and independently by Eilenberg and Schützenberger [97] in two successive steps:

(1) any recognizable subset of A^ω can be written as a finite union of sets of the form $K\overrightarrow{L}$, where K and L are recognizable subsets of A^+,

(2) if K and L are recognizable subsets of A^+, it is possible to recognize $K\overrightarrow{L}$ by a deterministic Muller automaton.

Part (1) will be developed in Chapter II with the syntactic analysis of recognizable subsets. Part (2) is the subject of Exercise 16.

The proof presented here is due to Safra [269] (see also the tutorial in [265]). It represents a genuinely new proof both because of its reduced computational complexity and because it is a direct construction. Theorem 10.9 is an unpublished result of Michel, presented in the survey of Thomas [332] (see also [109]).

For a general reference concerning the complexity of algorithms, see [118] or [4] or the volume on algorithms of the Handbook of Theoretical Computer Science [339]. Proposition 10.7 is from [269], where it is credited to Vardi. Theorem 10.9 is also from [269]. Safra has obtained in [270] a construction which generalizes Proposition 10.3. It shows that for any nondeterministic Streett automaton with n states and m pairs, there is an equivalent (deterministic) Rabin automaton with $2^{nm \cdot \log(nm)}$ states and nm pairs (see the tutorial in [276]).

Two earlier references had obtained a construction of an automaton of exponential size for the complement (Proposition 10.7). Both constructions, by Sistla, Vardi and Wolper [290] and by Pécuchet [226] make use of semigroups (see Proposition VI.4.5). Other constructions are possible, either working directly on Büchi automata, or using alternating automata (see [155]). Proposition 10.12 is from [342]. The *NL*-hardness relies on the analogous result for automata on finite words, first obtained in [148]. Proposition 10.13 is from [290]. It is proved in the same reference that the equivalence problem for Büchi automata is also *PSPACE*-complete (for logspace reductions, as the universality problem). The proof of *PSPACE*-hardness relies on the proof of the corresponding result for finite words which is due to [195]. The theorem of Savitch [272] used in the proof of Proposition 10.13 states that if a problem can be solved by a nondeterministic algorithm in space $f(n) \geq \log n$, it can be solved by a deterministic one in space $f(n)^2$ (see for instance [4]). Thus, it implies that *NPSPACE* = *PSPACE*. Proposition 10.6 is due to Choueka [68]. Streett automata were introduced by Streett in [316]. Exercise 17 is from Rabin [255].

Although the topic is not treated in this book, let us mention that beyond finite automata and rational expressions, the notions of a context-free grammar, a pushdown automaton or even a Turing machine have also been extended to the case of infinite words (see the survey by Staiger in [301]). In particular, the class of context-free ω-languages can be defined in several equivalent ways including the recognition by a pushdown automaton or the description as finite unions of sets of the form XY^ω with X, Y being ordinary context-free languages (Linna [181], Cohen and Gold [75, 76]).

The theory of infinite words has a natural extension to words indexed by ordinals (see Section V.2 or Rosenstein [267] for a general introduction to the theory of ordinals). A first advantage of this generalization is to unify the presentation of finite and infinite words. Indeed, finite words correspond to words indexed by finite ordinals and infinite words to words indexed by the ordinal ω. The extension to higher ordinals was first considered by Büchi [47, 48]. He extended his proof of the decidability of the monadic second-order theory of $(\mathbb{N}, <)$ to countable ordinals. The theory was later developed by Choueka [69] and Wojciechowski [361, 362, 363] who proved an extension of Kleene's theorem. Recent developments include an algebraic approach (see the notes of Chapter II) and an extension to words indexed by linear orders (see the notes of Chapter IX).

II

AUTOMATA AND SEMIGROUPS

1 Introduction

Given a finite automaton $\mathcal{A}$, it is natural to associate with each word x a binary relation $\varphi(x)$ defined on the set of states Q of $\mathcal{A}$ in the following way: two states p and q are related by $\varphi(x)$ if there exists a path from p to q labeled by x. This map satisfies the relation $\varphi(xx') = \varphi(x)\varphi(x')$ since the concatenation of two words is associated with the composition of the corresponding relations. In other words, φ is a morphism from the semigroup A^+ of nonempty words on A into the semigroup S of binary relations on Q. In this way, the behavior of an automaton can be simulated by a morphism from A^+ into a finite semigroup S. Such a morphism can be used to verify whether a given finite word is recognized by the automaton. Indeed, the image under φ of the set of words accepted by $\mathcal{A}$ is a special subset of S, consisting of the elements of S containing at least one pair (i, f) made up of an initial state and a final state.

It turns out that, to study automata on finite words, it is convenient to use morphisms onto finite semigroups, without any reference to the underlying automaton. As a matter

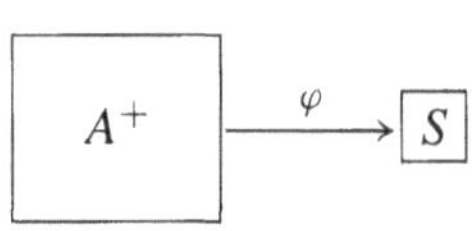

Figure 1.1. A semigroup morphism.

of fact, a number of constructions are more natural on semigroups than on automata. This is the case for instance for the study of "star-free" sets, which are related to first-order logic on words, a topic that will be treated in Chapters VI and VIII.

This situation contrasts even more for automata recognizing infinite words. However, instead of semigroups, we shall use semigroups equipped with infinite products, or ω-semigroups. We shall see that a few simple conditions lead to the notion of recognition of a finite or infinite word by a morphism of ω-semigroups, in such a way that a set and its complement are simultaneously recognized. This is, by the way, the technique used from the very beginning by Büchi for proving that recognizable subsets of infinite words are closed under complement.

However, some natural classes of recognizable sets are closed under finite union and intersection, but not under complement. A solution in this case is to use *ordered* semigroups. This definition is really a generalization of the standard one since a semigroup can always be ordered by the equality relation. After a few natural definitions, what was true for semigroups can be extended to ordered semigroups.

For infinite words, there is the corresponding notion of ordered ω-semigroups. This is the case in particular for the open (resp. closed) sets that will be studied in Chapters III and VI.

We have postponed to Annex A a reminder of properties of finite semigroups which are used in this book. This includes Green relations and basic decomposition results.

The key for the constructions of this chapter is a special case of a theorem which is proved and discussed in Section 2. The ω-semigroups are defined in Section 4 and the results concerning finite ω-semigroups are gathered in Section 5. Recognition by morphism of ω-semigroups is introduced in Section 3 and refined in Section 8, in which the syntactic ω-semigroup of a set X of finite or infinite words is defined. As for finite words, this is the smallest ω-semigroup recognizing X. However, it does not necessarily exist for an arbitrary, not necessarily recognizable set of infinite words — a major difference with the case of finite words. But for recognizable sets, the syntactic ω-semigroup is always defined, and it is an invariant which reflects many important properties, as we shall see in subsequent chapters, in particular in Chapter VI.

In Section 7, we give a method for converting a finite Büchi automaton into an ω-semigroup morphism recognizing the same set of infinite words.

In Section 9, we show how the notions introduced in this chapter lead to a new proof of the theorems of McNaughton and Landweber stated in Chapter I. These new proofs have the advantage of being easily generalized to other situations, for instance to bi-infinite words treated in Chapter IX or for particular classes of recognizable sets, that form the subject of Chapter VI.

Section 10 presents an important result (Theorem 10.3) on a class of automata called prophetic automata. It states that any Büchi automaton is equivalent to an unambiguous one, in the sense that it is co-deterministic and co-complete.

2 Ramseyan factorizations and linked pairs

2.1 Ramseyan factorizations. Let $u \in A^\omega$ be an infinite word. A *factorization* of u is a sequence $(u_n)_{n\geq 0}$ of words of A^+ such that $u = u_0u_1\cdots$. A factorization $(u'_n)_{n\geq 0}$ of u is a *superfactorization* of $(u_n)_{n\geq 0}$ if there exists a strictly increasing sequence of strictly positive integers $(k_n)_{n\geq 0}$ such that $u'_0 = u_0u_1\cdots u_{k_0-1}$ and $u'_{n+1} = u_{k_n}u_{k_n+1}\cdots u_{k_{n+1}-1}$ for each $n \geq 0$. More intuitively, $u = u'_0u'_1\cdots$ is a superfactorization of $u = u_0u_1\cdots$ if the sequence u'_n is obtained by grouping the u_n:

$$u = \underbrace{(u_0u_1\cdots u_{k_0-1})}_{u'_0}\underbrace{(u_{k_0}u_{k_0+1}\cdots u_{k_1-1})}_{u'_1}\cdots$$

Let

$$\varphi : A^+ \to E$$

be a map from A^+ into a set E. A factorization of u

$$u = xy_0y_1y_2\cdots$$

is said to be *Ramseyan for* φ if there exists an element e of E such that, for each i, j satisfying $i \leq j$,

$$\varphi(y_iy_{i+1}\cdots y_j) = e.$$

Traditionally, the elements of E are called *colors* and the previous property amounts to saying that the set of blocks defined by the factorization is monochromatic.

Theorem 2.1. *Let $\varphi : A^+ \to E$ be a map from A^+ into a finite set E. Every infinite word of A^ω admits a Ramseyan factorization for φ.*

Proof. Define a sequence of pairs (n_i, U_i) consisting of an index $n_i \geq 0$ and an infinite subset U_i of $\mathbb{N}$ as follows. First choose $U_0 = \mathbb{N}$. Suppose that U_i has been chosen, and let $n_i = \min U_i$. Since E is finite, there exists at least one element e_i of E such that the set

$$T = \{n \in U_i \mid \varphi(u[n_i, n-1]) = e_i\}$$

is infinite. Then define $U_{i+1} = T$ (see Figure 2.1). The resulting sequence $n_0, n_1, n_2, \ldots$ is illustrated in Figure 2.1. By construction, $\varphi(u[n_i, n_{i+j} - 1]) = e_i$ for all $i \geq 0$ and $j > 0$. Again, as E is finite, there exists an element e of E such that $e_i = e$ for infinitely many integers i. These integers define a subsequence $n_{i_0}, n_{i_1}, \ldots$, giving the required factorization $y_j = u[n_{i_j}, n_{i_{j+1}} - 1]$. □

Theorem 2.1 is actually a special case of the Ramsey theorem (cf. the notes at the end of the chapter). Theorem 2.1 will be systematically used in the case where the coloring is given by a morphism into a finite semigroup. In this case, a more precise result can be stated.

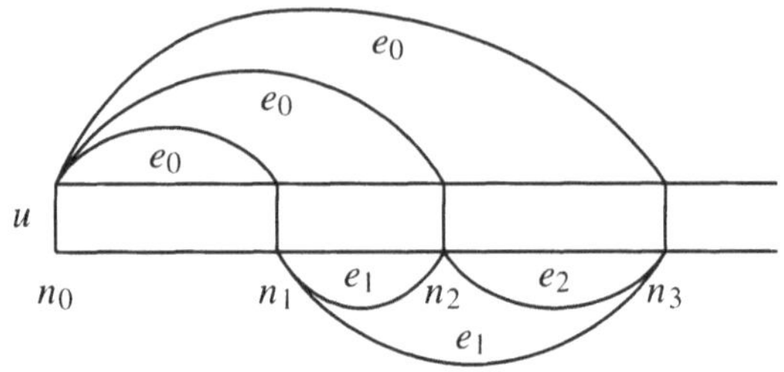

Figure 2.1. The sequence of indices.

Theorem 2.2. *Let $\varphi : A^+ \to S$ be a morphism from A^+ into a finite semigroup S. For each infinite word $u \in A^\omega$, there exist a pair (s, e) of elements of S satisfying $se = s$, $e^2 = e$, and a factorization $u = u_0u_1 \cdots$ of u as a product of words of A^+ such that $\varphi(u_0) = s$ and $\varphi(u_n) = e$ for all $n > 0$.*

Proof. By Theorem 2.1, there exist elements r and t of S such that $u \in \varphi^{-1}(r)(\varphi^{-1}(t))^\omega$. Since S is finite, Proposition A.1.9 shows that t has an idempotent power e. By grouping the successive factors of u of image t, we get a factorization $u = x_0x_1 \cdots$ such that $\varphi(x_0) = r$ and $\varphi(x_n) = e$ for each $n > 0$. Finally, setting $s = re$, we obtain $se = ree = re = s$, and the required factorization is obtained by putting $u_0 = x_0x_1$, and, for each $n > 0$, $u_n = x_{n+1}$. □

We shall frequently use Theorem 2.2 in a slightly different form, which relies on the fact that, in the statement of Theorem 2.1, the alphabet is not necessarily finite.

Proposition 2.3. *Let A be an alphabet, S a finite semigroup and $\varphi : A^+ \to S$ a semigroup morphism. Let u be an infinite word of A^ω, and let $(u_n)_{n\geq 0}$ be a factorization of u in words of A^+. Then there exist $s \in S$, $e \in E(S)$ and a superfactorization $(u'_n)_{n\geq 0}$ of $(u_n)_{n\geq 0}$ such that $se = s$, $\varphi(u'_0) = s$ and $\varphi(u'_n) = e$ for each $n > 0$.*

Proof. Let $B = \{b_0, b_1, \cdots\}$ be a countable alphabet and let $\alpha : B^+ \to A^+$ be the semigroup morphism defined by $\alpha(b_i) = u_i$. Let us apply Theorem 2.2 to the infinite word $b_0b_1 \cdots$ and the morphism $\psi = \varphi \circ \alpha : B^+ \to S$. There exist $s \in S$, $e \in E(S)$ and a strictly increasing sequence of integers $(k_n)_{n\geq 0}$ such that $se = s$, $\psi(b_0b_1 \cdots b_{k_0-1}) = s$ and $\varphi(b_{k_n}b_{k_n+1} \cdots b_{k_{n+1}-1}) = e$ for each $n \geq 0$. The required result follows immediately, since $\alpha(b_i) = u_i$. □

Corollary 2.4. *Let S be a finite semigroup and let $(s_n)_{n\in\mathbb{N}}$ be a sequence of elements of S. Then there exist $s \in S$, $e \in E(S)$ and a strictly increasing sequence of integers $(k_n)_{n\geq 0}$ such that $se = s$, $s_0s_1 \cdots s_{k_0-1} = s$ and $s_{k_n}s_{k_n+1} \cdots s_{k_{n+1}-1} = e$ for each $n \geq 0$.*

Proof. It suffices to apply Proposition 2.3 with $A = \{a_n \mid n \in \mathbb{N}\}$, $\varphi(a_n) = s_n$ and $u = a_0 a_1 \cdots$. □

2.2 Linked pairs. An element s of a semigroup S is *linked* to an idempotent $e \in E(S)$ if $se = s$. A pair $(s, e) \in S \times S$ is said to be *linked on the right*, or simply *linked*, if $e \in E(S)$ and $se = s$. Given a semigroup morphism $\varphi : A^+ \to S$ from A^+ into a finite semigroup S, a factorization $(u_n)_{n \geq 0}$ of a word u is *associated with a linked pair* (s, e) of S if $\varphi(u_0) = s$ and $\varphi(u_n) = e$ for each $n > 0$. By extension, (s, e) is said to be a linked pair of u if there exists a factorization of u associated with (s, e). The sets of the form $\varphi^{-1}(s)(\varphi^{-1}(e))^\omega$ play an important role in the sequel. They are called *φ-simple subsets* or *simple subsets* when the reference to φ is understood.

When an infinite word u is suffix of an infinite word v, one can easily pass from a linked pair of v to a linked pair of u.

Proposition 2.5. *Let u be an infinite word, x a finite word and let $r = \varphi(x)$. Then (s, e) is a linked pair of xu if and only if there exists an element $t \in S$ satisfying $s = rt$ such that (t, e) is a linked pair of u.*

Proof. Let (t, e) be a linked pair of u and let $s = rt$. Then (s, e) is a linked pair of xu since $se = rte = rt = s$ and since xu is in $\varphi^{-1}(r)\varphi^{-1}(t)(\varphi^{-1}(e))^\omega$, which is a subset of $\varphi^{-1}(rt)(\varphi^{-1}(e))^\omega = \varphi^{-1}(s)(\varphi^{-1}(e))^\omega$.

Conversely, let (s, e) be a linked pair of $v = xu$. Then there exists a factorization $v = v_0 v_1 v_2 \cdots$ such that $\varphi(v_0) = s$ and $\varphi(v_n) = e$ for each $n > 0$. Let n be an integer such that x is a prefix of $v_0 v_1 \cdots v_{n-1}$. Set $v_0 v_1 \cdots v_{n-1} = xv'$ and $t = \varphi(v' v_n) = \varphi(v')e$. The pair (t, e) is linked since $te = \varphi(v')ee = \varphi(v')e = t$ and associated with u, since $u = v' v_n v_{n+1} \cdots$, and thus $u \in \varphi^{-1}(t)(\varphi^{-1}(e))^\omega$. □

Note that a given infinite word may have several linked pairs. These pairs are nevertheless related by the so-called *conjugacy relation*, defined as follows:

Two linked pairs (s, e) and (s', e') of $S \times E(S)$ are said to be *conjugate* if there exist $x, y \in S^1$ such that $e = xy$, $e' = yx$ and $s' = sx$. These equalities also imply $s = s'y$ (since $s'y = sxy = se = s$), showing the symmetry of the definition. Note that if (s, e) and (s', e') are conjugate, then $s \mathrel{\mathcal{R}} s'$ and $e \mathrel{\mathcal{D}} e'$, where $\mathcal{R}$ and $\mathcal{D}$ denote the Green relations defined in Annex A.

Proposition 2.6. *The conjugacy relation is an equivalence relation on the linked pairs.*

Proof. Reflexivity is trivial. Symmetry follows from the remark above. Finally, if (s, e) and (s', e') are conjugate and if (s', e') and (s'', e'') are conjugate, there exist elements x, y, x', y' of S such that

$$s' = sx,\ e = xy,\ e' = yx,\ s'' = s'x',\ e' = x'y',\ e'' = y'x'.$$

Therefore $x'y' = yx$, and hence $(xx')(y'y) = x(x'y')y = x(yx)y = (xy)(xy) = e$, and similarly, $(y'y)(xx') = e''$ and $s'' = s(xx')$. Therefore (s, e) and (s'', e'') are conjugate. □

Here is a more technical characterization of conjugate pairs.

Proposition 2.7. *Let S be a finite semigroup. Two linked pairs (s, e) and (s', e') of S are conjugate if and only if there exist $x \in S^1$ such that $s' = sx$, $xe' \mathcal{R}\ e$ and $e\ \mathcal{D}\ e'$.*

Proof. If (s, e) and (s', e') are conjugate, there exists $x, y \in S^1$ such that $e = xy$, $e' = yx$ and $s' = sx$. Therefore $xe' = xyx\ \mathcal{R}\ xy$ since $(xyx)y = xy$. Moreover, the equalities $y(xy)x = yx$ and $x(yx)y = xy$ show that e and e' generate the same ideal. Consequently $e\ \mathcal{D}\ e'$ by Proposition A.2.5.

Conversely, suppose there exists $x \in S^1$ such that $s' = sx$, $xe' \mathcal{R}\ e$ and $e\ \mathcal{D}\ e'$. Let $v \in S^1$ be such that $xe'v = e$. Then we have $xe' \mathcal{R}\ e = xe'v\ \mathcal{L}\ e'v\ \mathcal{R}\ e'$ and thus, by Proposition A.2.4, $h = (e'v)(xe')\ \mathcal{H}\ e'$. By the same proposition, the $\mathcal{H}$-class of e' is a group with identity e', and therefore, there exists an integer n such that $h^n = e'$. Setting $y = h^{n-1}e'v = (e'vxe')^{n-1}e'v$, we obtain

$$(xe')y = xe'(e'vxe')^{n-1}e'v = (xe'v)^n = e^n = e,$$
$$y(xe') = h^{n-1}e'vxe' = h^n = e'$$
$$sxe' = s'e' = s'$$

and thus (s, e) and (s', e') are conjugate. □

There is also a notion of conjugacy for the factorizations of a word associated with two linked pairs. Let $(u_n)_{n\geq 0}$ and $(v_n)_{n\geq 0}$ be two factorizations of u associated with linked pairs. Then $(u_n)_{n\geq 0}$ is said to be *conjugate* of $(v_n)_{n\geq 0}$ if there exist elements $x, y \in S$, superfactorizations $(\bar{u}_n)_{n\geq 0}$ of $(u_n)_{n\geq 0}$ and $(\bar{v}_n)_{n\geq 0}$ of $(v_n)_{n\geq 0}$ and, for each $n > 0$, a factorization $\bar{u}_n = \bar{u}'_n\bar{u}''_n$ such that $\varphi(\bar{u}'_n) = x$, $\varphi(\bar{u}''_n) = y$ $\bar{v}_0 = \bar{u}_0\bar{u}'_1$ and $\bar{v}_n = \bar{u}''_n\bar{u}'_{n+1}$:

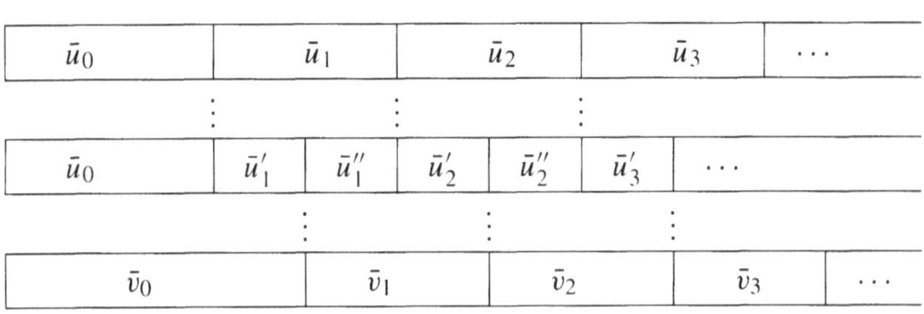

Note that two conjugate factorizations are necessarily associated with conjugate pairs. Indeed, setting, with the previous notations, $\varphi(u_0) = s$ (resp. $\varphi(v_0) = t$) and $\varphi(u_n) = e$

(resp. $\varphi(v_n) = f$) for each $n > 0$, we have $sx = t$, $ty = s$, $xy = e$ and $yx = f$. We can now state the main result about conjugate factorizations.

Proposition 2.8. *Any two factorizations of u associated with linked pairs are conjugate.*

Proof. Let $(u_n)_{n \geq 0}$ (resp. $(v_n)_{n \geq 0}$) be a factorization of u associated with the linked pair (s, e) (resp. (t, f)). Up to taking superfactorizations, we may assume that $u_0 \leq v_0 \leq u_0 u_1 \leq v_0 v_1 \leq u_0 u_1 u_2 \leq v_0 v_1 v_2 \leq \cdots$

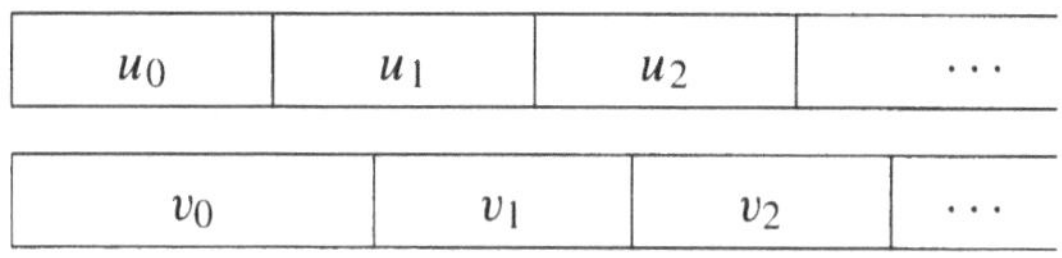

Therefore, for each $n > 0$, $u_n = u'_n u''_n$, with $v_0 = u_0 u'_1$ and $v_n = u''_n u'_{n+1}$. Since S is finite, there are only finitely many pairs $(\varphi(u'_n), \varphi(u''_n))$. Thus one can find elements $x, y \in S$ and an increasing sequence $(k_n)_{n \geq 0}$ such that $k_n \geq k_{n_1} + 2$, $\varphi(u'_{k_n}) = x$ and $\varphi(u''_{k_n}) = y$ for each $n \geq 0$. Put $\bar{u}_0 = u_0 \cdots u_{k_0 - 1}$ (resp. $\bar{v}_0 = v_0 \cdots v_{k_0 - 1}$) and, for each $n > 0$, $\bar{u}_n = u_{k_{n-1}} \cdots u_{k_n - 1}$ (resp. $\bar{v}_n = v_{k_{n-1}} \cdots v_{k_n - 1}$). Finally, put, for each $n > 0$, $\bar{u}'_n = u'_{k_{n-1}}$ and $\bar{u}''_n = u''_{k_{n-1}} u_{k_{n-1}+1} \cdots u_{k_n - 1}$. We have $\bar{u}_n = \bar{u}'_n \bar{u}''_n$, $\varphi(\bar{u}'_n) = x$ and $\varphi(\bar{u}''_n) = ye$ for all $n > 0$. Moreover, $\bar{v}_0 = \bar{u}_0 \bar{u}'_1$ and $\bar{v}_n = \bar{u}''_n \bar{u}'_n$. Therefore the factorizations $(u_n)_{n \geq 0}$ and $(v_n)_{n \geq 0}$ are conjugate. □

Corollary 2.9. *Let S be a finite semigroup, let (s, e) and (s', e') be two linked pairs of S and let $\varphi : A^+ \to S$ be a surjective morphism. The pairs (s, e) and (s', e') are conjugate if and only if the intersection $\varphi^{-1}(s)(\varphi^{-1}(e))^\omega \cap \varphi^{-1}(s')(\varphi^{-1}(e'))^\omega$ is nonempty.*

Proof. Suppose that (s, e) and (s', e') are conjugate. Then there exist elements $x, y \in S^1$ such that $s' = sx$, $e = xy$, and $e' = yx$. Since φ is surjective, there exist words $\bar{s} \in \varphi^{-1}(s)$, $\bar{x} \in \varphi^{-1}(x)$, and $\bar{y} \in \varphi^{-1}(y)$. Consider the word $u = \bar{s}(\bar{x}\bar{y})^\omega$. This word can be written on the one hand

$$u = \bar{s}(\bar{x}\bar{y})(\bar{x}\bar{y}) \cdots, \quad \text{whence } u \in \varphi^{-1}(s)(\varphi^{-1}(e))^\omega,$$

and on the other hand,

$$u = \bar{s}\bar{x}(\bar{y}\bar{x})(\bar{y}\bar{x}) \cdots, \quad \text{whence } u \in \varphi^{-1}(s')(\varphi^{-1}(e'))^\omega.$$

Consequently, the sets $\varphi^{-1}(s)(\varphi^{-1}(e))^\omega$ and $\varphi^{-1}(s')(\varphi^{-1}(e'))^\omega$ have a nonempty intersection.

Conversely, let (s, e) and (s', e') be two linked pairs of a word u. Then u admits two factorizations associated with (s, e) and (s', e'), respectively, which are conjugate by Proposition 2.8. It follows that the pairs (s, e) and (s', e') are conjugate. □

2.3 Ultimate classes. Let u be an infinite word of A^ω. Corollary 2.9 shows that the various linked pairs of u are strongly related. In particular, for any choice of such a pair (s, e), the $\mathcal{R}$-class of s and the $\mathcal{D}$-class of e only depend on u. They are called the *ultimate $\mathcal{R}$-class* and the *ultimate $\mathcal{D}$-class* of u, and are respectively denoted $R(u)$ and $D(u)$. They can also be defined directly as follows.

Proposition 2.10. *Let u be an infinite word of A^ω and let $\varphi : A^+ \to S$ be a morphism from A^+ into a finite semigroup S.*

(1) *Every element of S which is the image under φ of infinitely many prefixes of u is in $R(u)$.*

(2) *Every element of S which is in $R(u[i, \infty[)$ for infinitely many integers i is in $D(u)$.*

Proof. (1) Let (s, e) be a linked pair of u. Then we have a factorization $u = u_0u_1u_2\cdots$ with $\varphi(u_0) = s$, and, for each $n > 0$, $\varphi(u_n) = e$. Let s' be an element of S which is the image under φ of infinitely many prefixes of u. In particular, $s' = \varphi(p)$ for some prefix p of u longer than u_0. This prefix can be written

$$p = u_0u_1\cdots u_r u' \text{ with } r \geq 0 \text{ and } u_{r+1} = u'u''.$$

It follows on the one hand

$$s' = \varphi(p) = \varphi(u_0)\varphi(u_1\cdots u_r u') \in sS^1,$$

and on the other hand,

$$s = se = \varphi(u_0u_1\cdots u_r u_{r+1}) = \varphi(pu'') = \varphi(p)\varphi(u'') \in s'S^1,$$

and hence $s \mathrel{\mathcal{R}} s'$ and $s' \in R(u)$.

(2) Let t be an element of S which is in $R(u[i, \infty[)$ for infinitely many integers i. In particular, there exists an integer $k > 0$ and a factorization $u'_k u''_k$ of u_k such that, setting $v = u''_k u_{k+1}u_{k+2}\cdots$, we have $t \in R(v)$. Let s' be an element of S which is the image under φ of infinitely many prefixes of v, and which is therefore, by (1), an element of $R(v)$. In particular, $s' = \varphi(p)$ for some prefix p of u longer than $u''_k u_{k+1}$. This prefix can be written

$$p = u''_k u_{k+1}\cdots u_{k+j}u'_{k+j+1} \text{ with } j > 0 \text{ and } u_{k+j+1} = u'_{k+j+1}u''_{k+j+1}.$$

It follows on the one hand, since $\varphi(u_{k+1}) = e$,

$$s' = \varphi(p) = \varphi(u''_k)\varphi(u_{k+1})\varphi(u_{k+2}\cdots u_{k+j}u'_{k+j+1}) \in S^1eS^1,$$

and on the other hand, since $\varphi(p) = s'$,

$$e = \varphi(u_k\cdots u_{k+j+1}) = \varphi(u'_k)\varphi(p)\varphi(u''_{k+j+1}) \in S^1s'S^1,$$

and hence $s' \mathrel{\mathcal{D}} e$ by Proposition A.2.5. Now, since $t \in R(v)$ and $s' \in R(v)$, we have $t \mathrel{\mathcal{R}} s'$ and thus $t \mathrel{\mathcal{D}} s' \mathrel{\mathcal{D}} e$. Therefore $t \in D(u)$. □

In the case of an ultimately periodic word $u = xy^\omega$, the $\mathcal{R}$-class and the ultimate $\mathcal{D}$-class are easy to identify. Let $s = \varphi(x)$ and let e be an idempotent power of $\varphi(y)$. We have $R(u) = R(se)$ and $D(u) = D(e)$. The next result elaborates these remarks.

Proposition 2.11. *Let S be a finite semigroup, $\varphi : A^+ \to S$ a semigroup morphism, and (s, e) a linked pair of S. Then every ultimately periodic word of $\varphi^{-1}(s)(\varphi^{-1}(e))^\omega$ can be written as xy^ω, with $\varphi(x) = s$ and $\varphi(y) = e$.*

Proof. Let $u = xy^\omega$ be an ultimately periodic word of $\varphi^{-1}(s)(\varphi^{-1}(e))^\omega$. Thus $u = u_0u_1\cdots$, where $\varphi(u_0) = s$ and $\varphi(u_n) = e$ for each $n > 0$. Since $se = s$ and $e^2 = e$, one may assume, up to grouping the consecutive u_i, that x is prefix of u_0 and that $|u_i| \geq |y|$ for each $i > 0$. Therefore $u_0 = xy^{k_0}y'_1$, and for each $n > 0$, $u_n = y''_n y^{k_n} y'_{n+1}$, with $y'_n y''_n = y$. Since y has only finitely many prefixes, one can find two indices n_0 and n_1 such that $y'_{n_0} = y'_{n_1} = y'$ and $y''_{n_0} = y''_{n_1} = y''$. Put $f = u_0u_1\cdots u_{n_0-1}$ and $g = u_{n_0}u_{n_0+1}\cdots u_{n_1}$. We have $\varphi(f) = s$, $\varphi(g) = e$ and there exists two integers p and q such that $f = xy^py'$ and $g = (y''y')^q$. Consequently, $u = fg^\omega$, concluding the proof. □

The factorization of an infinite word associated with a linked pair is related to an equivalence relation defined as follows. Let $u \in A^\omega$ be an infinite word and let $\varphi : A^+ \to S$ be a semigroup morphism. We write $x \sim_\varphi y$ for $\varphi(x) = \varphi(y)$ (see Annex A). Define a relation on $\mathbb{N}$, denoted by $\sim$, by setting $i \sim j$ if and only if there exist k greater than or equal to i and j such that

$$u[i, k] \sim_\varphi u[j, k] \tag{2.1}$$

Proposition 2.12. *The relation $\sim$ is an equivalence relation of index bounded by* Card(S).

Proof. First of all, it can be observed that if the equality 2.1 is true for an integer k, it is also true for each $k' > k$. This implies that the relation $\sim$ is an equivalence relation: indeed, if $i \sim j$ and $j \sim k$, we have $u[i, \ell] \sim_\varphi u[j, \ell]$ and $u[j, \ell] \sim_\varphi u[k, \ell]$ for some integer ℓ, whence $u[i, \ell] \sim_\varphi u[k, \ell]$ and thus $i \sim k$.

Let us show that the index of $\sim$ is bounded by Card(S). Let $i_1, i_2, \ldots, i_k$ be distinct integers with $k >$ Card(S) and let n be an integer greater than $i_1, i_2, \ldots, i_k$. By the pigeon hole principle, there exist two indices p and q such that $\varphi(u[i_p, n]) = \varphi(u[i_q, n])$, that is, such that $i_p \sim i_q$. □

The link between this equivalence and the Ramsey theorem is the following.

Proposition 2.13. *Let $\varphi : A^+ \to S$ be a morphism from A^+ into a finite semigroup S and let $u \in A^\omega$ be an infinite word. Then, for each integer $i \geq 0$, the $\sim$-class of i is infinite if and only if there exists an idempotent $e \in S$ such that $u[i, \infty[\in (\varphi^{-1}(e))^\omega$.*

Proof. Up to a replacement of u by $u[i, \infty[$, we may assume that $i = 0$. The condition is sufficient. Indeed, if there exists an idempotent e of S and a factorization $u = u_0 u_1 \cdots$ such that $\varphi(u_j) = e$ for each $j \geq 0$, we have $\varphi(u_0 u_1) = \varphi(u_0)\varphi(u_1) = ee = e = \varphi(u_1)$, and similarly $\varphi(u_1 u_2) = \varphi(u_2)$, ...and hence $0 \sim |u_0| \sim |u_0 u_1| \sim \ldots$.

Conversely, if the class of 0 is infinite, there exists a factorization of u as $u = u_0 u_1 \cdots$ with

$$0 \sim |u_0| \sim |u_0 u_1| \sim \ldots \tag{2.2}$$

Then we build by induction a sequence of integers $(k_i)_{i \geq 0}$ such that the words $v_i = u_{k_{i-1}+1} \cdots u_{k_i}$ satisfy

$$0 \sim |v_1| \sim |v_1 v_2| \sim \ldots \quad \text{and} \quad \varphi(v_i v_{i+1}) = \varphi(v_{i+1}) \text{ for each } i > 0 \tag{2.3}$$

For this, we first put $k_0 = 0$ and $k_1 = 1$. Then, if $k_0, \ldots, k_i$ are already defined, we have $|v_1 \cdots v_{i-1}| \sim |v_1 \cdots v_i|$ and k_{i+1} can be chosen such that

$$\varphi(v_i u_{k_i+1} \cdots u_{k_{i+1}}) = \varphi(u_{k_i+1} \cdots u_{k_{i+1}})$$

that is, such that $\varphi(v_i v_{i+1}) = \varphi(v_{i+1})$. Then we also have $0 \sim |v_1| \sim |v_1 v_2| \sim \ldots$ by (2.2). Now, since the map $k \to \varphi(v_1 \cdots v_{k-1})$ takes only finitely many values, there exists a sequence $(r_n)_{n>0}$ such that

$$\varphi(v_1 \cdots v_{r_1 - 1}) = \varphi(v_1 \cdots v_{r_2 - 1}) = \ldots = e$$

Therefore, setting $w_0 = v_1 \cdots v_{r_1-1}$ and $w_i = v_{r_i} \cdots v_{r_{i+1}-1}$, we have

$$\varphi(w_0) = \varphi(w_0 w_1) = \varphi(w_1) = \varphi(w_0 w_1 w_2) = \varphi(w_2) = \ldots = e$$

giving the required factorization. □

Proposition 2.13 gives a new proof of Theorem 2.2. It also permits one to prove the following statement.

An infinite word $u \in A^\omega$ is said to be *recurrent* if each factor x of u has infinitely many occurrences in u, i.e. if $u \in A^* x A^\omega$ implies $u \in (A^* x)^\omega$.

Theorem 2.14. *Let $\varphi : A^+ \to S$ be a morphism from A^+ into a finite semigroup S and let $u \in A^\omega$ be an infinite word. If u is recurrent, there exists an idempotent e of S such that $u \in (\varphi^{-1}(e))^\omega$.*

Proof. Let y be a prefix of u such that the set

$$S(y) = \{\varphi(s) \mid s \text{ is a suffix of } y\}$$

is maximal. Since u is recurrent, there exists a factorization $u = xyv$ with $x \neq 1$. By the choice of y, and since every suffix of y is also suffix of xy, $S(y)$ is equal to $S(xy)$.

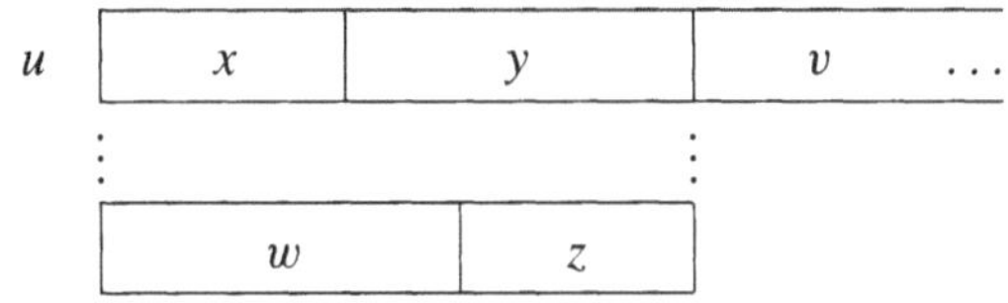

Figure 2.2. Two factorizations of u.

In particular, there exists a suffix z of y such that $\varphi(xy) = \varphi(z)$. Putting $xy = wz$, the situation is summarized in Figure 2.2. Since $\varphi(wz) = \varphi(z)$, we have $0 \sim |w|$. As v is also a recurrent word, this construction can be iterated, thereby showing that the $\sim$-class of 0 is infinite. □

Theorem 2.14 cannot be extended to an arbitrary map from A^+ into a finite set (see Exercise 4).

2.4 Idempotent linked pairs. A pair of the form (e, e), where e is idempotent, is called an *idempotent linked pair.*

Proposition 2.15. *A linked pair (s, e) is conjugate with an idempotent linked pair (f, f) if and only if $s \mathcal{D} e$. In this case, s and f are $\mathcal{R}$-equivalent.*

Proof. If (s, e) and (f, f) are conjugate, then $s \mathcal{R} f$ and $e \mathcal{D} f$. Thus $s \mathcal{D} e$. Conversely, suppose that $s \mathcal{D} e$. Then $s = se$ since (s, e) is a linked pair and by Corollary A.2.6, $s \mathcal{L} e$. Therefore, there exists $t \in S^1$ such that $ts = e$. Now $stst = set = st$, and thus $f = st$ is idempotent. It follows that the pairs (s, e) and (f, f) are conjugate. □

Corollary 2.16. *Two idempotent linked pairs (e, e) and (f, f) are conjugate if and only if $e \mathcal{R} f$.*

Proposition 2.17. *Let (s, e) and (s', e') be two linked pairs such that $s \mathcal{D} s' \mathcal{D} e \mathcal{D} e'$. Then the two pairs are conjugate if and only if $s \mathcal{R} s'$.*

Proof. If the pairs are conjugate, then $s \mathcal{R} s'$. Conversely, assume that $s \mathcal{R} s'$. By Proposition 2.15, (s, e) (resp. (s', e')) is conjugate with a pair (f, f) (resp. (f', f')) such that $s \mathcal{R} f$ (resp. $s' \mathcal{R} f'$). Since $s \mathcal{R} s'$, it follows $f \mathcal{R} f'$ and by Corollary 2.16, the pairs (f, f) and (f', f') are conjugate. Thus, by Proposition 2.6, the pairs (s, e) and (s', e') are conjugate. □

Proposition 2.18. *Let u be an infinite word having an idempotent linked pair. Then every suffix of u has a linked pair conjugate with an idempotent linked pair.*

Proof. Let $u = u_0 u_1 \cdots$ be a factorization associated with some idempotent linked pair (e, e). If v is a suffix of u, v can be factorized as $w = u'_i u_{i+1} u_{i+2} \cdots$ for some $i \geq 0$ and some suffix u'_i of u_i. Setting $t = \varphi(u'_i u_{i+1})$, we have $t \mathcal{L} e$ since $e = \varphi(u_i u_{i+1}) \leq_{\mathcal{L}} \varphi(u'_i u_{i+1}) \leq_{\mathcal{L}} \varphi(u_{i+1}) = e$. Now, (t, e) is a linked pair of w and by Proposition 2.15, (t, e) is conjugate with an idempotent linked pair. □

3 Recognition by morphism

We give in this section two algebraic definitions of recognizable sets. The first definition makes use of semigroup morphisms. According to this definition, a subset of a semigroup and its complement are recognized by the same semigroup morphisms. The second one, which is more general, makes use of morphisms of ordered semigroups. It breaks the symmetry between a subset and its complement, an essential feature for some applications.

3.1 Semigroup morphisms. Let S and T be two semigroups. A surjective morphism of semigroups $\varphi : S \to T$ *recognizes* a subset I of S if there exists a subset J of T such that

$$I = \varphi^{-1}(J).$$

By extension, a semigroup T *recognizes* a subset I of S if there exists a surjective morphism of semigroups $\varphi : S \to T$ that recognizes I. Finally, a subset of a semigroup is said to be *recognizable* if it is recognized by a finite semigroup.

Example 3.1. Let T be the additive semigroup defined on $\{0, 1, 2\}$ by $x + y = \max\{x, y\}$ and let φ be the surjective morphism from $(\mathbb{N}, +)$ onto T defined by $\varphi(0) = 0$, $\varphi(1) = 1$ and $\varphi(n) = 2$ for each $n \geq 2$. The subsets of $\mathbb{N}$ recognized by φ are $\varphi^{-1}(\emptyset) = \emptyset$, $\varphi^{-1}(0) = \{0\}$, $\varphi^{-1}(1) = \{1\}$, $\varphi^{-1}(2) = \{2, 3, \ldots\}$, $\varphi^{-1}(\{0, 1\}) = \{0, 1\}$, $\varphi^{-1}(\{0, 2\}) = \{0, 2, 3, 4, \ldots\}$, $\varphi^{-1}(\{1, 2\}) = \{1, 2, 3, 4, \ldots\}$ and $\varphi^{-1}(\{0, 1, 2\}) = \mathbb{N}$.

Example 3.2. Let S be the multiplicative semigroup of matrices with integer entries generated by the two matrices $a = \begin{pmatrix} 0 & 1 \\ 0 & 0 \end{pmatrix}$ and $b = \begin{pmatrix} 0 & 0 \\ 1 & 0 \end{pmatrix}$. This semigroup contains five elements: a, b and the matrices $ab = \begin{pmatrix} 1 & 0 \\ 0 & 0 \end{pmatrix}$, $ba = \begin{pmatrix} 0 & 0 \\ 0 & 1 \end{pmatrix}$, and $0 = \begin{pmatrix} 0 & 0 \\ 0 & 0 \end{pmatrix}$ and satisfies the following relations: $aa = bb = 0$, $aba = a$ and $bab = b$. Let $A = \{a, b\}$ and let $\varphi : A^+ \to S$ be the morphism defined by $\varphi(a) = a$ and $\varphi(b) = b$. We have $\varphi^{-1}(a) = (ab)^* a$, $\varphi^{-1}(b) = (ba)^* b$, $\varphi^{-1}(ab) = (ab)^+$, $\varphi^{-1}(ba) = (ba)^+$ and $\varphi^{-1}(0) = A^* aa A^* \cup A^* bb A^*$.

This definition is actually a generalization of the definition of a subset of the free semigroup recognized by an automaton. To see this we shall show that one can easily pass from an automaton recognizing a subset L of A^+ to a semigroup recognizing L, and conversely. Let $\mathcal{A} = (Q, A, E, I, F)$ be a Büchi automaton. Denote by $\varphi_{\mathcal{A}}$:

$A^+ \to \mathcal{R}(Q)$ the semigroup morphism from A^+ into the semigroup of binary relations on the set Q (see Annex A) defined, for each $a \in A$, by

$$\varphi_{\mathcal{A}}(a) = \{(q, q') \in Q \times Q \mid (q, a, q') \in E\}.$$

Equivalently, we have, for each $u \in A^+$,

$$\varphi_{\mathcal{A}}(u) = \{(q, q') \in Q \times Q \mid \text{ there exists in } \mathcal{A} \text{ a path from } q \text{ to } q' \text{ with label } u\}$$

The semigroup $S(\mathcal{A}) = \varphi_{\mathcal{A}}(A^+)$ is called the *transition semigroup of* $\mathcal{A}$.

Conversely, if $\varphi : A^+ \to S$ is a surjective morphism of semigroups recognizing a subset L of A^+, there exists a subset F of S such that $L = \varphi^{-1}(F)$. We construct a deterministic automaton $\mathcal{A}_\varphi$ by taking the monoid S^1 as set of states, and, as transition function, the right action of A on S^1, defined, for $q \in S^1$ and $a \in A$, by

$$q \cdot a = q\varphi(a).$$

Finally, we take 1, the identity of S^1, as initial state, and F as set of final states.

Proposition 3.1. *If $\mathcal{A}$ is an automaton, the morphism $\varphi_{\mathcal{A}} : A^+ \to S(\mathcal{A})$ recognizes $L^+(\mathcal{A})$. Conversely, if $\varphi : A^+ \to S$ is a surjective morphism of semigroups recognizing a subset L of A^+, the deterministic automaton $\mathcal{A}_\varphi$ recognizes L.*

Proof. Let $\mathcal{A} = (Q, A, E, I, F)$. The set $L^+(\mathcal{A})$ is recognized by the morphism $\varphi_{\mathcal{A}}$ since we have the equality

$$L^+(\mathcal{A}) = \varphi_{\mathcal{A}}^{-1}(\{r \in S(\mathcal{A}) \mid r \cap I \times F \neq \emptyset\}).$$

We have on the other hand, if $\varphi : A^+ \to S$ is a surjective morphism of semigroups recognizing a subset L of A^+,

$$\begin{aligned} L^+(\mathcal{A}_\varphi) &= \{u \in A^+ \mid 1.u \in F\} \\ &= \{u \in A^+ \mid \varphi(u) \in F\} \\ &= \varphi^{-1}(F), \end{aligned}$$

showing that $\mathcal{A}_\varphi$ recognizes L. □

Note that the previous constructions do not require any finiteness assumption on the semigroup or on the automaton.

Example 3.3. The semigroup of Example 3.2 is in fact the transition semigroup of the automaton represented in Figure 3.1, which recognizes $(ab)^+$.

Example 3.4. Conversely, let $\varphi : A^+ \to S$ be the morphism of Example 3.2. Applying the construction given above, one gets the automaton shown in Figure 3.2, which also recognizes $(ab)^+$.

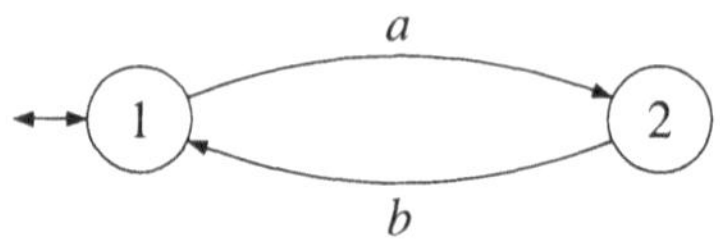

Figure 3.1. A finite automaton recognizing $(ab)^+$.

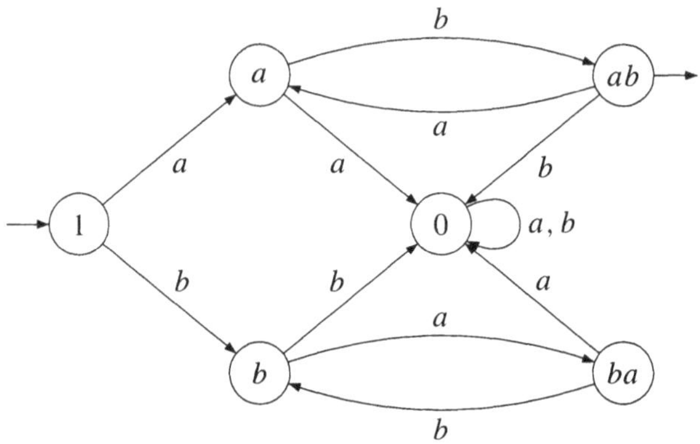

Figure 3.2. The automaton associated with the semigroup S.

3.2 Morphisms of ordered semigroups. Let S be an ordered semigroup (see Annex A). A surjective morphism of ordered semigroups $\varphi : S \to T$ *recognizes an order ideal* I of S if there exists an order ideal J of T such that

$$I = \varphi^{-1}(J).$$

By extension, an ordered semigroup T *recognizes an order ideal* I of S if there exists a surjective morphism of ordered semigroups $\varphi : S \to T$ that recognizes I. Finally, an order ideal I of S is said to be *recognizable* if it is recognized by a finite ordered semigroup.

In the case when the order relation on S (resp. on T) is the equality relation, every subset of S (resp. T) is an order ideal. Thus we recover in this case the definition of a subset of a semigroup recognized by a morphism given in Section 3.1.

Example 3.5. The semigroup $T = \{0, 1, 2\}$ of Example 3.1, ordered by the relation $0 < 1 < 2$ is an ordered semigroup. The order ideals of T are $\{0\}$, $\{0, 1\}$ and $\{0, 1, 2\}$. The map φ from $\mathbb{N}$ on T defined by $\varphi(0) = 0$, $\varphi(1) = 1$ and $\varphi(n) = 2$ for each $n \geq 3$ defines a surjective morphism of ordered semigroups of $(\mathbb{N}, +, \leq)$ on T, which recognizes the ideals $\{0\}$, $\{0, 1\}$ and $\mathbb{N}$ of $\mathbb{N}$.

But φ also defines a surjective morphism of ordered semigroups from $(\mathbb{N}, +, =)$ on the dual ordered semigroup of T, denoted by $\check{T}$. The order ideals of $\check{T}$ are $\{2\}$, $\{1, 2\}$ and

$\{0, 1, 2\}$. Therefore φ recognizes in this case the ideals $\{2, 3, 4, 5, \dots\}$, $\{1, 2, 3, 4, \dots\}$ and $\mathbb{N}$ of $\mathbb{N}$.

Proposition 3.1 can be extended to the case of ordered semigroups. An *ordered automaton* is a Büchi automaton $\mathcal{A} = (Q, A, E, I, F)$ in which the set of states Q is equipped with a partial order $\leq$ satisfying the two following conditions, for each $p, q, q' \in Q$ and $a \in A$:

(1) if $p \leq q$ and $(q, a, q') \in E$, there exists a state $p' \leq q'$ such that $(p, a, p') \in E$,

(2) the set of final states is an order ideal: if $q \in F$ and $p \leq q$, then $p \in F$.
Condition (1) can be extended by transitivity as follows:

(3) if $p \leq q$ and if there exists a path from q to q' with label u, there exists a state $p' \leq q'$ and a path from p to p' with label u.

Note that in the case of a deterministic automaton, condition (1) simplifies to

(1′) if $p \leq q$ and if $q \cdot a$ are defined, then $p \cdot a$ is defined and satisfies $p \cdot a \leq q \cdot a$.

If $\mathcal{A}$ is an ordered automaton, one defines a relation $\preceq$ on A^+ by putting $u' \preceq u$ if and only if, for each $(p, q) \in u$, there exists $q' \leq q$ such that $(p, q') \in u'$.

Let us give immediately an equivalent formulation of this definition:

Lemma 3.2. *We have $u' \preceq u$ if and only if, for each $(p, q) \in u$ and for each $p' \leq p$, there exists $q' \leq q$ such that $(p', q') \in u$.*

Proof. Suppose that $u' \preceq u$ and let $p, p', q \in Q$ be such that $(p, q) \in u$ and let $p' \leq p$. By definition of an ordered automaton, there exists $q' \leq q$ such that $(p', q') \in u$. Now, since $u' \preceq u$, there exists $q'' \leq q'$ such that $(p', q'') \in u$. Finally, by transitivity, we have $q'' \leq q$ and the condition of the statement is fulfilled. □

Lemma 3.3. *The relation $\preceq$ is a congruence of ordered semigroup on A^+.*

Proof. The relation $\preceq$ is obviously reflexive. It is also transitive. Indeed, if $u'' \preceq u'$ and $u' \preceq u$ and if $(p, q) \in u$, there exist $q' \leq q$ such that $(p, q') \in u'$ and thus there exists $q'' \leq q'$ such that $(p, q'') \in u''$. Since $q'' \leq q'$ and $q' \leq q$, we have $q'' \leq q$ and hence $u'' \preceq u$.

Let us show that the relation is stable. Let $u, u', v \in A^+$ with $u' \preceq u$. If $(p, q) \in vu$, there exists $r \in Q$ such that $(p, r) \in v$ and $(r, q) \in u$. Thus there exists $q' \leq q$ such that $(r, q') \in u'$ and hence $(p, q') \in vu'$. Therefore $vu' \preceq vu$. Furthermore, if $(p, q) \in uv$, there exists $r \in S$ such that $(p, r) \in u$ and $(r, q) \in v$. Since $u' \preceq u$, there exists $r' \leq r$ such that $(p, r') \in u'$. By condition (3) of the definition of an ordered automaton, there exists $q' \leq q$ such that $(r', q') \in v$. Therefore $(p, q') \in u'v$ and $u'v \preceq uv$. □

The relation $\preceq$ might not be antisymmetric, as is shown in Example 3.6.

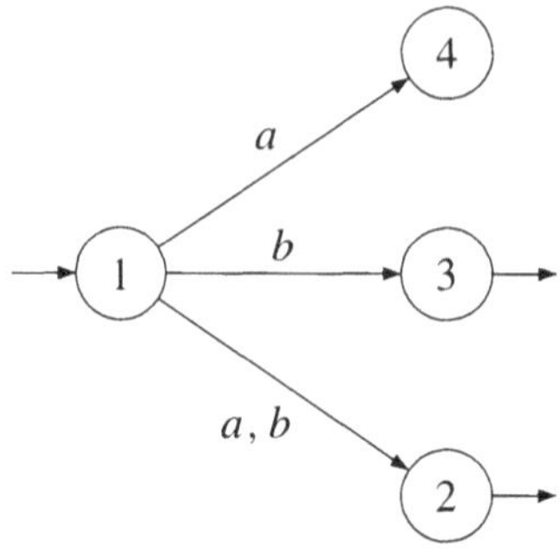

Figure 3.3. An ordered automaton.

Example 3.6. Consider the automaton of Figure 3.3, equipped with the order $2 \leq 3 \leq 4$. We have, in the transition semigroup, $a \preceq b$ and $b \preceq a$, although a is different from b.

The ordered semigroup $SO(\mathcal{A}) = (A^+/\preceq)$ is the *ordered transition semigroup* of $\mathcal{A}$.

Conversely, if S is an ordered semigroup and $\varphi : A^+ \to S$ a surjective morphism, the order on S defines an order on the deterministic automaton $\mathcal{A}_\varphi$. Indeed, if $s \leq t$, then $s \cdot a = s\varphi(a) \leq t\varphi(a) = t \cdot a$. Then the analogue of Proposition 3.1 holds, establishing the equivalence between ordered semigroups and ordered automata.

Proposition 3.4. *If $\mathcal{A} = (Q, \leq, A, E, I, F)$ is an ordered automaton, the morphism $\varphi_\mathcal{A} : A^+ \to SO(\mathcal{A})$ recognizes $L^+(\mathcal{A})$. Conversely, if $\varphi : A^+ \to S$ is a surjective morphism of ordered semigroups recognizing a subset L of A^+, the deterministic ordered automaton $\mathcal{A}_\varphi$ recognizes L.*

Proof. Put $P = \{u \in S(\mathcal{A}) \mid u \cap I \times F \neq \emptyset\}$ and let us show that if $u \in P$ and $u' \preceq u$, we still have $u' \in P$. Indeed, if there exist $p \in I$ and $q \in F$ such that $(p, q) \in u$, there exists, by definition of the relation $\preceq$, a state $q' \leq q$ such that $(p, q') \in u'$. Since F is an order ideal, we have $q' \in F$ and hence $u' \in P$. It follows that $P/\sim$ is an order ideal of $SO(\mathcal{A})$. Furthermore $\varphi_\mathcal{A}^{-1}(P) = L^+(\mathcal{A})$, showing that $\varphi_\mathcal{A}$ recognizes $L^+(\mathcal{A})$.

The second part of the statement follows from the fact, already established in the proof of Proposition 3.1, that $L^+(\mathcal{A}_\varphi) = \varphi^{-1}(F)$. □

3.3 Congruence and syntactic order. Let I be a subset of a semigroup S and let s be an element of S. The *set of contexts* of s in I is defined as the set of pairs

$$C_I(s) = \{(u, v) \in S^1 \times S^1 \mid usv \in I\}.$$

The equivalence relation $\sim_I$ defined on S by

$$s \sim_I t \text{ if and only if } C_I(s) = C_I(t)$$

is a congruence, called the *syntactic congruence* of I. The quotient semigroup $S(I) = S/{\sim_I}$ is called the *syntactic semigroup* of I.

Example 3.7. Let $A = \{a, b\}$, $S = (A^+, =)$ and $I = (ab)^+$. A computation shows that the syntactic congruence of I has five classes: $a(ba)^*$, $b(ab)^*$, $(ab)^+$, $(ba)^+$ and $A^*aaA^* \cup A^*bbA^*$. The syntactic semigroup of I is thus the semigroup $T = \{a, b, ab, ba, 0\}$, defined by the relations $aa = 0$, $bb = 0$, $aba = a$, $bab = b$ and $a0 = 0a = b0 = 0b = 0$, which is also the semigroup of Example 3.2.

One can verify that the syntactic congruence of a subset I of a semigroup S is the coarsest congruence on S recognizing I. It follows that a subset of a semigroup is recognizable if and only if its syntactic semigroup is finite.

A similar definition can be given for ordered semigroups. Let I be an order ideal of an ordered semigroup S. The relation $\preceq_I$ defined by

$$s \preceq_I t \text{ if and only if } C_I(t) \subset C_I(s)$$

is a congruence of ordered semigroup and the quotient of the relation $\preceq_I$ by $\sim_I$ is a stable order on $S/{\sim_I}$, called the *syntactic order* of I, and denoted by $\leq_I$. The quotient ordered semigroup $(S/{\sim_I}, \leq_I)$ is called the *syntactic ordered semigroup*. It is also denoted by $S(I)$. The natural morphism from S onto $S(I)$ is called the *syntactic morphism*.

Example 3.8. Let $A = \{a, b\}$, $S = (A^+, =)$ and $I = (ab)^+$. It can be verified that the syntactic ordered semigroup of I is T, where T is the semigroup defined in the example 3.7 and the order $\leq$ is defined by $a < 0$, $b < 0$, $ab < 0$, $ba < 0$.

Once again, the syntactic congruence of an order ideal I of an ordered semigroup S is the coarsest congruence on S recognizing I. An order ideal of an ordered semigroup is thus recognizable if and only if its syntactic ordered semigroup is finite.

3.4 Residuals. Let s be an element of S^1 and P a subset of S. Put

$$s^{-1}P = \{t \in S \mid st \in P\} \qquad \text{and} \qquad Ps^{-1} = \{t \in S \mid ts \in P\}$$

The set $s^{-1}P$ (resp. Ps^{-1}) is the left (resp. right) *residual* of P by s.

Example 3.9. Let $A = \{a, b\}$, $S = A^+$ and $L = A^*aba$. We have $a^{-1}L = L \cup ba$, $b^{-1}L = L$, $La^{-1} = A^*ab$ and $Lb^{-1} = \emptyset$.

Proposition 3.5. *Let I be an order ideal of an ordered semigroup S. Then for each $s \in S^1$, $s^{-1}I$ and Is^{-1} are order ideals of S.*

Proof. If $y \leq x$ and $x \in s^{-1}I$, we have $sx \in I$ by definition. Since $sy \leq sx$, it comes $sy \in I$, whence $y \in s^{-1}I$. Therefore $s^{-1}I$ is an order ideal. A dual proof works for Is^{-1}. □

4 Semigroups and infinite products

Semigroups are perfectly adapted to the study of sets of finite words. However, they are much less convenient to deal with infinite words. Indeed, the analog of the infinite product of words is lacking in the algebraic structure of a semigroup. The solution simply consists in introducing a new structure, richer than semigroups, in which infinite products are allowed.

4.1 ω-semigroups. An *ω-semigroup* is an algebra consisting of two components, $S = (S_+, S_\omega)$ equipped with the following operations:

(a) A binary operation defined on S_+ and denoted multiplicatively.

(b) A mapping $S_+ \times S_\omega \to S_\omega$, called *mixed product*, that associates with each pair $(s, t) \in S_+ \times S_\omega$ an element of S_ω denoted st.

(c) A surjective mapping $\pi : S_+^\omega \to S_\omega$, called *infinite product*.

These three operations satisfy the following properties:

(1) S_+, equipped with the binary operation, is a semigroup,

(2) for every $s, t \in S_+$ and for every $u \in S_\omega$, $s(tu) = (st)u$,

(3) for every increasing sequence $(k_n)_{n>0}$ and for each sequence $(s_n)_{n\in\mathbb{N}} \in S_+^\omega$,

$$\pi(s_0 s_1 \cdots s_{k_1-1}, s_{k_1} s_{k_1+1} \cdots s_{k_2-1}, \ldots) = \pi(s_0, s_1, s_2, \ldots),$$

(4) for every $s \in S_+$ and for every $(s_n)_{n\in\mathbb{N}} \in S_+^\omega$

$$s\pi(s_0, s_1, s_2, \ldots) = \pi(s, s_0, s_1, s_2, \ldots).$$

Conditions (1) and (2) can be thought of as an extension of associativity. Conditions (3) and (4) show that one can replace the notation $\pi(s_0, s_1, s_2, \ldots)$ by the notation $s_0 s_1 s_2 \cdots$ without ambiguity. We shall use this simplified notation in the sequel. Intuitively, an ω-semigroup is a sort of semigroup in which infinite products are defined.

Example 4.1. We denote by A^∞ the ω-semigroup (A^+, A^ω) equipped with the usual concatenation product.

Example 4.2. The *trivial* ω-semigroup is the ω-semigroup $1 = (\{1\}, \{a\})$, obtained by equipping the trivial semigroup $\{1\}$ with an infinite product: the unique way is to declare that every infinite product is equal to a.

Let us warn the reader immediately about a possible misunderstanding: the sets S_+ and S^ω are disjoint and the infinite product

$$\pi(s_0, s_1, s_2, \ldots)$$

is not defined, in any sense, as the "limit" of the finite products $s_0 s_1 \cdots s_n$, even if the sequence $s_0, s_0 s_1, s_0 s_1 s_2, \ldots$ is constant (for instance when $s_0 = s_1 = s_2 = \ldots = e$, an idempotent of S). This is illustrated in the next example.

Example 4.3. Consider the ω-semigroup $S = (\{0, 1\}, \{a\})$ defined as follows: every infinite product is equal to a and every finite product $s_0 s_1 \ldots s_n$ is equal to 0 except if all the s_i's are equal to 1. In particular, the elements 0 and 1 are idempotents and thus, for all $n > 0$, $1^n \neq 0^n$. Nevertheless $1^\omega = 0^\omega = a$.

An *ordered ω-semigroup* is an ordered algebra with two components $(S, \leq) = ((S_+, \leq), (S_\omega, \leq))$ such that (S_+, S_ω) is an ω-semigroup and such that the partial orders on S_+ and S_ω are stable for the operations of ω-semigroup:

(1) for each $s, s', u \in S_+$, $s \leq s'$ implies $us \leq us'$ and $su \leq s'u$,

(2) for each $s, s' \in S_+$ and for each $u \in S_\omega$, $s \leq s'$ implies $su \leq s'u$,

(3) if $(s_n)_{n\in\mathbb{N}}$ and $(s'_n)_{n\in\mathbb{N}}$ are two sequences of elements of S_+^ω such that $s_n \leq s'_n$ for each n, then $s_0 s_1 s_2 \cdots \leq s'_0 s'_1 s'_2 \cdots$.

If S is an ordered ω-semigroup, the *dual* of S is the ordered ω-semigroup $\check{S}$ obtained by replacing the order on S by the dual order.

An *order ideal* of an ordered ω-semigroup $(S\ \leq) = ((S_+, \leq), (S_\omega, \leq))$ is a pair $I = (I_+, I_\omega)$, where I_+ is an order ideal of $(S_+, \leq)$ and I_ω an order ideal of $(S_\omega, \leq)$, i.e. a subset of S_ω such that $x \in S_\omega$ and $y \leq x$ implies $y \in S_\omega$.

4.2 Morphisms of ω-semigroups. Given two ω-semigroups $S = (S_+, S_\omega)$ and $T = (T_+, T_\omega)$, a *morphism of ω-semigroups* S is a pair $\varphi = (\varphi_+, \varphi_\omega)$ consisting of a semigroup morphism $\varphi_+ : S_+ \to T_+$ and of a mapping $\varphi_\omega : S_\omega \to T_\omega$ preserving the infinite product: for every sequence $(s_n)_{n\in\mathbb{N}}$ of elements of S_+,

$$\varphi_\omega(s_0 s_1 s_2 \cdots) = \varphi_+(s_0)\varphi_+(s_1)\varphi_+(s_2)\cdots$$

We claim that φ also preserves the mixed product, that is, for all $s \in S_+$, and for each $t \in S_\omega$,

$$\varphi_+(s)\varphi_\omega(t) = \varphi_\omega(st)$$

To prove the claim, it suffices to write t as an infinite product $t_0 t_1 t_2 \cdots$ and to multiply on the left by $\varphi_+(s)$:

$$\begin{aligned}\varphi_+(s)\varphi_\omega(t) &= \varphi_+(s)\varphi_\omega(t_0 t_1 \cdots) = \varphi_+(s)\varphi_+(t_0)\varphi_+(t_1)\cdots \\ &= \varphi_\omega(s t_0 t_1 \cdots) = \varphi_\omega(st)\end{aligned}$$

In the sequel, we shall often omit the subscripts, and use the simplified notation φ instead of φ_+ and φ_ω. A *morphism of ordered ω-semigroups* is a morphism of ω-semigroups preserving the order.

A morphism of ω-semigroups (resp. of ordered ω-semigroups) $\varphi : S \to T$ is an *isomorphism* if there exists a morphism $\psi : T \to S$ such that $\varphi \circ \psi = Id_T$ and $\psi \circ \varphi = Id_S$.

Algebraic concepts like ω-subsemigroup, quotient and division are easily adapted to ω-semigroups. For instance, an *ω-subsemigroup* of an ω-semigroup $S = (S_+, S_\omega)$ is a pair $T = (T_+, T_\omega)$ where T_+ is a subset of S_+ and T_ω a subset of S_ω such that the operations on S induce a structure of ω-semigroup on T. In other words,

(1) the infinite product of any sequence of elements of T_+ belongs to T_ω and every element of T_ω can be written as an infinite product of elements of T_+,

(2) if $s \in T_+$ and $t \in T_\omega$, then $st \in T_\omega$.

If P is a subset of S_+, the *ω-subsemigroup of S generated by P* is the ω-semigroup $\langle P \rangle = (T_+, T_\omega)$ where T_+ is the subsemigroup of S_+ generated by P and T_ω is the set of elements of S_ω which can be written as an infinite product of elements of P.

An ω-semigroup T is *quotient* of an ω-semigroup S if there exists a surjective morphism of ω-semigroups from S onto T. A *congruence* on an ω-semigroup $S = (S_+, S_\omega)$ is a pair $(\sim_+, \sim_\omega)$, where $\sim_+$ is a congruence of semigroup on S_+ and $\sim_\omega$ is an equivalence relation on S_ω, these relations being stable for the infinite product and the mixed product: if $(s_0, s_1, s_2, \dots)$ and $(s'_0, s'_1, s'_2, \dots)$ are sequences of elements of S such that $s_n \sim_+ s'_n$ for each $n \geq 0$, we have $s_0 s_1 s_2 \cdots \sim_\omega s'_0 s'_1 s'_2 \cdots$ and if $s, s' \in S_+$ and $t, t' \in S_\omega$ are such that $s \sim_+ s'$ and $t \sim_\omega t'$, we have $st \sim_\omega s't'$. In this case, the quotient set $S/{\sim} = (S_+/{\sim_+}, S_\omega/{\sim_\omega})$ is naturally equipped with a structure of ω-semigroup.

A *congruence* on an ordered ω-semigroup $S = ((S_+, \leq), (S_\omega, \leq))$ is a pair $(\preceq_+, \preceq_\omega)$ where $\preceq_+$ is a congruence of ordered semigroup on S and $\preceq_\omega$ is a preorder on S_ω, coarser than $\leq$ and stable for the infinite product and the mixed product. If $\sim = (\sim_+, \sim_\omega)$ denotes the equivalence relation associated with $\preceq$, the quotient set $S/{\sim}$ is naturally equipped with a structure of ordered ω-semigroup.

Example 4.4. If $\varphi : S \to T$ is a morphism of ω-semigroups, the *nuclear congruence of* φ is the congruence $\sim_\varphi$ defined on S_+ (resp. S_ω) by $s \sim_\varphi t$ if and only if $\varphi_+(s) = \varphi_+(t)$ (resp. $\varphi_\omega(s) = \varphi_\omega(t)$).

The following results are proved exactly as in the case of semigroups.

Proposition 4.1. *Let $\varphi : S \to T$ be a morphism of ω-semigroups and let $\pi : S \to S/{\sim_\varphi}$ be the quotient morphism. Then there exists a unique morphism of ω-semigroups $\bar\varphi : S/{\sim_\varphi} \to T$ such that $\varphi = \bar\varphi \circ \pi$. Moreover, $\bar\varphi$ is a isomorphism from $S/{\sim_\varphi}$ onto $\varphi(S)$.*

Proposition 4.2. *Let $\sim_1$ and $\sim_2$ be two congruences defined on an ω-semigroup S and let π_1 (resp. π_2) the canonical morphism from S on $S/{\sim_1}$ (resp. $S/{\sim_2}$). Then $\sim_1$ is thinner than $\sim_2$ if and only if there exists a morphism $\pi : S/{\sim_1} \to S/{\sim_2}$ such that $\pi \circ \pi_1 = \pi_2$.*

Proposition 4.3. *Let $\preceq_1$ and $\preceq_2$ be two congruences defined on an ordered ω-semigroup S and let π_1 (resp. π_2) be the canonical morphism from S onto $S/{\preceq_1}$ (resp. $S/{\preceq_2}$). Then $\preceq_1$ is thinner than $\preceq_2$ if and only if there exists a surjective morphism $\pi : S/{\preceq_1} \to S/{\preceq_2}$ such that $\pi \circ \pi_1 = \pi_2$.*

Our last result states that division is stable under product.

Proposition 4.4. *Let $(S_i)_{i \in I}$ and $(T_i)_{i \in I}$ two families of ordered ω-semigroups such that S_i divides T_i. Then $\prod_{i \in I} S_i$ divides $\prod_{i \in I} T_i$.*

Proof. By assumption, each S_i is quotient of an ordered ω-subsemigroup R_i of T_i. It follows that $\prod_{i \in I} S_i$ is a quotient of $\prod_{i \in I} R_i$ and that $\prod_{i \in I} R_i$ is an ordered ω-subsemigroup of $\prod_{i \in I} T_i$. Therefore $\prod_{i \in I} S_i$ divides $\prod_{i \in I} T_i$. □

4.3 Free structures. Just as the semigroup A^+ of finite words on the alphabet A is the free semigroup on A, the ω-semigroup A^∞ is the free ω-semigroup on A. Indeed, it satisfies the universal property characterizing free structures.

Proposition 4.5. *Let A be an alphabet, $S = (S_+, S_\omega)$ an ω-semigroup and φ a map from A into S_+. Then there exists a unique morphism $\bar{\varphi} : A^\infty \to S$ such that, for each $a \in A$, $\bar{\varphi}(a) = \varphi(a)$. Moreover, $\bar{\varphi}$ is surjective if and only if the set $\varphi(A)$ generates S.*

Proof. Define a mapping $\bar{\varphi}$ from A^∞ into S by setting, for each finite word $a_0a_1 \cdots a_n$,

$$\bar{\varphi}(a_0a_1 \cdots a_n) = \varphi(a_0)\varphi(a_1) \cdots \varphi(a_n)$$

and, for each infinite word $a_0a_1a_2 \cdots$,

$$\bar{\varphi}(a_0a_1a_2 \cdots) = \varphi(a_0)\varphi(a_1)\varphi(a_2) \cdots$$

One can easily verify that $\bar{\varphi}$ is the required morphism. On the other hand, any morphism $\bar{\varphi}$ such that $\bar{\varphi}(a) = \varphi(a)$ for each $a \in A$ must satisfy these two equalities, which shows it is unique.

By construction, the set $\varphi(A)$ generates $\bar{\varphi}(A)$. Consequently, the morphism $\bar{\varphi}$ is surjective if and only if the set $\varphi(A)$ generates S. □

Similarly, the ordered ω-semigroup A^∞ (ordered by the equality relation) is the free ordered ω-semigroup on A. Indeed, it suffices to imitate the proof of Proposition 4.5 to obtain the following statement.

Proposition 4.6. *Let A be an alphabet, S an ordered ω-semigroup and φ a map from A in S_+. Then there exists a unique morphism of ordered ω-semigroups $\bar{\varphi} : A^\infty \to S$ such that, for all $a \in A$, $\bar{\varphi}(a) = \varphi(a)$. Moreover, $\bar{\varphi}$ is surjective if and only if the set $\varphi(A)$ generates S.*

We now state without proofs the ω-semigroup versions of Corollaries A.1.6 and A.1.7.

Corollary 4.7. *Let S be an [ordered] ω-semigroup and let A be a subset of S_+ generating S. The identity map from A into S_+ induces a surjective morphism of [ordered] ω-semigroups from A^∞ onto S.*

This morphism is called the *natural* morphism from A^∞ onto S.

Corollary 4.8. *Let $\eta : A^\infty \to S$ and $\beta : T \to S$ be surjective morphisms of [ordered] ω-semigroups. Then there exists a morphism of [ordered] ω-semigroups $\varphi : A^\infty \to T$ such that $\eta = \beta \circ \varphi$.*

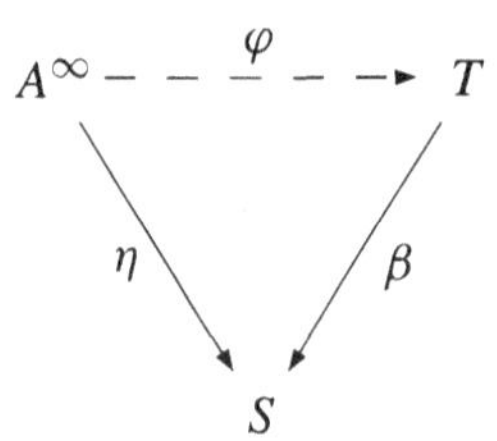

Let $(S_i)_{i\in I}$ be any family of ω-semigroups. The *product* of the family $(S_i)_{i\in I}$ is the ω-semigroup defined on the product set $\prod_{i\in I} S_i$ by the operations

$$(s_i)_{i\in I}(t_i)_{i\in I} = (s_i t_i)_{i\in I}$$
$$(s_{0,i})_{i\in I}(s_{1,i})_{i\in I}(s_{2,i})_{i\in I}\cdots = (s_{0,i}s_{1,i}s_{2,i}\cdots)_{i\in I}$$

In the case of a finite number of ω-semigroups $S_1, S_2, \dots, S_k$, we also use the notation $S_1 \times S_2 \times \cdots \times S_k$. Since the trivial ω-semigroup 1 is an identity for the product, it is convenient to put

$$\prod_{i\in\emptyset} S_i = 1$$

Let $(\sim_i)_{i\in I}$ be a family of congruences on an ω-semigroup S. The congruence $\sim$ defined by

$$s \sim t \text{ if and only if, for each } i \in I,\ s \sim_i t$$

is the lower bound of the family $(\sim_i)_{i\in I}$, denoted by $\bigwedge_{i\in I} \sim_i$.

Proposition 4.9. *Let $(\sim_i)_{i \in I}$ be a family of congruences on an ω-semigroup S, and let $\sim$ be the lower bound of the family $(\sim_i)_{i \in I}$. Then $S/{\sim}$ is an ω-subsemigroup of $\prod_{i \in I} S/{\sim_i}$.*

Proof. Denote by $\pi_i : S \to S/{\sim_i}$ the quotient morphisms and let $\pi : S \to \prod_{i \in I} S/{\sim_i}$ be the morphism defined, for each $s \in S$, by $\pi(s) = (\pi_i(s))_{i \in I}$. The nuclear congruence of π is nothing else than $\sim$ and by Proposition 4.1, $S/{\sim}$ is isomorphic to $\pi(S)$. □

Symmetrically, one can define the upper bound of the family $(\sim_i)_{i \in I}$, denoted by $\bigvee_{i \in I} \sim_i$, as the lower bound of the congruences $\sim$ that are coarser than all the congruences $(\sim_i)_{i \in I}$. But contrary to the case of the lower bound, it is not possible in general to give a constructive definition of the upper bound.

5 Wilke Algebras

We describe in this section the structure of finite ω-semigroups, which is reminiscent of the structure of finite semigroups. The key result is that, in a finite ω-semigroup, the infinite product is entirely determined by the value of the infinite products of the form

$$s^\omega = sss\cdots$$

Theorem 5.1. *Let S_+ be a finite semigroup and let S_ω be a finite set. Suppose there exists a mixed product $S_+ \times S_\omega \to S_\omega$ and a map from S_+ in S_ω, denoted by $s \to s^\omega$, satisfying, for each $s, t \in S_+$ and $u \in S_\omega$,*

$$\begin{aligned} s(tu) &= (st)u \\ s(ts)^\omega &= (st)^\omega \\ (s^n)^\omega &= s^\omega \quad \text{for each } n > 0 \end{aligned}$$

and such that every element of S_ω can be written under the form st^ω with $s, t \in S_+$. Then the pair $S = (S_+, S_\omega)$ can be equipped, in a unique way, with a structure of ω-semigroup that inherits the given mixed product and such that, for each $s \in S_+$, the infinite product $sss\cdots$ is equal to s^ω.

Proof. Let $(s_n)_{n \in \mathbb{N}}$ be a sequence of elements of S_+. By Corollary 2.4, there exists $s \in S_+$, $e \in E(S_+)$ and a strictly increasing sequence of integers $(k_n)_{n \in \mathbb{N}}$ such that

$$s_0 s_1 \cdots s_{k_0 - 1} = s \quad \text{and} \quad s_{k_n} s_{k_n + 1} \cdots s_{k_{n+1} - 1} = e \text{ for each } n \geq 0 \qquad (5.1)$$

It follows that if S can be equipped with a structure of ω-semigroup satisfying the conditions of the statement, the infinite product $s_0 s_1 s_2 \cdots$ is necessarily equal to se^ω.

Let now S be a finite semigroup equipped with an external unary operation $s \to s^\omega$ satisfying the conditions of the statement. If $(s_n)_{n\in\mathbb{N}}$ is a sequence of elements of S, we put

$$\pi(s_0, s_1, s_2, \cdots) = se^\omega$$

where (s, e) is a linked pair of S chosen, according to Proposition 2.3, such that there exists a strictly increasing sequence of integers $(k_n)_{n\in\mathbb{N}}$ satisfying (5.1). To warrant a consistent definition, we have to verify that the result is independent of the choice of the pair (s, e). But by Corollary 2.9, if (s', e') is another choice of linked pair, (s, e) and (s', e') are conjugate. Therefore, there exist $x, y \in S^1$ such that $e = xy$, $e' = yx$ and $s' = sx$. A simple computation, based on the identities $s(tu) = (st)u$ and $s(ts)^\omega = (st)^\omega$, now gives

$$s'e'^\omega = sx(yx)^\omega = s(xy)^\omega = se^\omega$$

Let us now verify that the map π has the properties required for an infinite product. Let $(s_n)_{n\in\mathbb{N}}$ be a sequence of elements of S and let t be an element of S. Let (s, e) be a linked pair of S and let $(k_n)_{n\in\mathbb{N}}$ be a strictly increasing sequence of integers satisfying (5.1). It comes $\pi(s_0, s_1, s_2, \dots) = se^\omega$, and since $ts_0s_1 \cdots s_{k_0-1} = ts$, we also have $\pi(t, s_0, s_1, s_2, \dots) = (ts)e^\omega$. Therefore $t\pi(s_0, s_1, s_2, \dots) = \pi(t, s_0, s_1, s_2, \dots)$. Finally, let $(r_n)_{n>0}$ be a strictly increasing sequence of integers. It follows immediately from the definition of π that

$$\pi(s_0s_1 \dots s_{r_1-1}, s_{r_1}s_{r_1+1} \dots s_{r_2-1}, \dots) = \pi(s_0, s_1, s_2, \dots)$$

It remains to verify that for each $s \in S$, $\pi(s, s, s, \dots) = s^\omega$. By Proposition A.1.9, there exists an integer $n > 0$ such that s^n is idempotent. It follows, by the definition of π, $\pi(s, s, s, \dots) = (s^n)^\omega$ and hence $\pi(s, s, s, \dots) = s^\omega$ since $(s^n)^\omega = s^\omega$. □

Theorem 5.1 can be reformulated by saying that a finite ω-semigroup is fully defined by specifying the value of se^ω for each conjugacy class $[s, e]$ of linked pairs.

Theorem 5.1 justifies the introduction of a new algebraic structure. A *Wilke algebra* is an algebra with two components $S = (S_+, S_\omega)$, equipped with the following operations:

(1) an associative product on S_+,

(2) a mixed product, which maps each pair $(s, t) \in S_+ \times S_\omega$ onto an element of S_ω denoted by st, such that, for every $s, t \in S_+$ and for every $u \in S_\omega$, $s(tu) = (st)u$,

(3) a map from S_+ in S_ω, denoted by $s \to s^\omega$ satisfying, for each $s, t \in S_+$,

$$s(ts)^\omega = (st)^\omega$$
$$(s^n)^\omega = s^\omega \quad \text{for each } n > 0$$

and such that every element of S_ω can be written as st^ω with $s, t \in S_+$.

Theorem 5.1 states that every finite ω-semigroup is equivalent to a finite Wilke algebra. This result has an interesting algorithmic consequence. It is clear that a finite semigroup can be entirely described by a finite amount of data (for instance by giving its multiplication table). But the description of a finite ω-semigroup seems to require an infinite amount of data to define properly all infinite products. Theorem 5.1, by reducing this description to that of Wilke algebras, overcomes this difficulty. Indeed, Wilke algebras are clearly finite objects, that can be entirely described by the multiplication table of the product and the mixed product and by the operation $s \to s^{\omega}$. This fact will play an important role in Chapter VI.

Similarly, an *ordered Wilke algebra* is an algebra $S = ((S_+, \leq), (S_\omega, \leq))$ such that (S_+, S_ω) is a Wilke algebra and the order relation on S_+ and S_ω are stable under the operations of Wilke algebra:

(1) for each $s, s', u \in S_+$, $s \leq s'$ implies $us \leq us'$ and $su \leq s'u$,

(2) for each $s, s' \in S_+$ and for each $u \in S_\omega$, $s \leq s'$ implies $su \leq s'u$,

(3) for each $s, s' \in S_+$, $s \leq s'$ implies $s^{\omega} \leq s'^{\omega}$.

We now show how a finite Wilke algebra $\bar{S} = (S, S_\omega)$ can be naturally associated with each finite semigroup S. Denote by π the exponent of S and by $[s, e]$ the conjugacy class of a linked pair (s, e). We take for S_ω the set of conjugacy classes of the linked pairs of S. The set $\bar{S}$ is equipped with a structure of Wilke algebra by setting, for each $[s, e] \in S_\omega$ and $t \in S$,

$$t[s, e] = [ts, e] \quad \text{and} \quad t^{\omega} = [t^{\pi}, t^{\pi}]$$

The definition is consistent since if (s', e') is a conjugate pair of (s, e), then (ts', e') is a conjugate pair of (ts, e). There remains to verify the identities of a Wilke algebra. Let us show that $u(vu)^{\omega} = (uv)^{\omega}$. If $u, v \in S$, we have

$$u(vu)^{\omega} = u[(vu)^{\pi}, (vu)^{\pi}] = [u(vu)^{\pi}, (vu)^{\pi}] \text{ and } (uv)^{\omega} = [(uv)^{\pi}, (uv)^{\pi}]$$

Now

$$[u(vu)^{\pi}, (vu)^{\pi}] = [(uv)^{\pi}, (uv)^{\pi}]$$

since the pairs $\big(u(vu)^{\pi}, (vu)^{\pi}\big)$ and $\big((uv)^{\pi}, (uv)^{\pi}\big)$ are conjugate. Moreover, we have for each $n > 0$, $(u^n)^{\omega} = [(u^n)^{\pi}, (u^n)^{\pi}] = [u^{\pi}, u^{\pi}] = u^{\omega}$. Finally, $\bar{S}$ is generated by S since $[s, e] = s[e, e] = s[e^{\pi}, e^{\pi}] = se^{\omega}$. Moreover, every semigroup morphism $\varphi : A^+ \to S$ can be extended as a morphism of ω-semigroups $\varphi : A^{\infty} \to \bar{S}$ defined by $\bar{\varphi}_+ = \varphi$ and $\bar{\varphi}_\omega(u) = [s, e]$ where (s, e) is a linked pair of u. The ω-semigroup $\bar{S}$ has the following universal property.

Proposition 5.2. *Let $T = (T_+, T_\omega)$ be a finite ω-semigroup and let S be a finite semigroup. For every semigroup morphism φ from S into T_+, there exists a unique morphism of ω-semigroups $\bar{\varphi} : \bar{S} \to T$ extending φ. Furthermore, if $\varphi(S)$ generates T_+ as a semigroup, then $\bar{\varphi}$ is onto.*

Proof. A morphism of ω-semigroups $\bar{\varphi} : \bar{S} \to T$ satisfying the conditions of the statement is necessarily defined by

$$\bar{\varphi}(s) = \varphi(s) \quad \text{if } s \in S$$
$$\bar{\varphi}([s, e]) = \varphi(s)\varphi(e)^{\omega} \quad \text{if } [s, e] \in I$$

This definition is consistent, since it does not depend on the choice of the linked pair (s, e). Indeed, if (s', e') is a conjugate pair of (s, e), there exist $x, y \in S^1$ such that $s' = sx$, $e = xy$ and $e' = yx$, and hence

$$\begin{aligned}\varphi(s')\varphi(e')^{\omega} &= \varphi(sx)\big(\varphi(yx)\big)^{\omega} = \varphi(s)\varphi(x)\big(\varphi(y)\varphi(x)\big)^{\omega} \\ &= \varphi(s)\big(\varphi(x)\varphi(y)\big)^{\omega} = \varphi(s)\varphi(e)^{\omega}\end{aligned}$$

It is not difficult to verify that $\bar{\varphi}$ is a morphism of ω-semigroups. Finally, if $\varphi(S)$ generates T_+ as a semigroup, then $\bar{\varphi}$ is surjective by construction. □

Corollary 5.3. *Let S be a finite semigroup and let $T = (T_+, T_\omega)$ be an ω-semigroup such that T_+ is a quotient of S. Then T is a quotient of $\bar{S}$.*

Proof. Let φ be a surjective semigroup morphism from S onto T_+. Since φ is onto, we have $\varphi_+(S) = T_+$ and hence $\varphi(S)$ generates T as a semigroup. By Proposition 5.2, $\bar{\varphi}$ is surjective and hence T is a quotient of $\bar{S}$. □

Corollary 5.4. *Every finite ω-semigroup $T = (T_+, T_\omega)$ is a quotient of $\bar{T}_+$.*

Proof. It suffices to apply Corollary 5.3 with $S = T_+$. □

We now extend to ω-semigroups the definitions of residuals of a subset of a semigroup by an element. Let $S = (S_+, S_\omega)$ be an ω-semigroup, $P = (P_+, P_\omega)$ a subset of S and s an element of S_+^1. We define

$$s^{-1}P = \{u \in S \mid su \in P\}$$
$$Ps^{-\omega} = \{u \in S_+ \mid (us)^{\omega} \in P_\omega\}$$

Similarly, if s is an element of S_ω, we put

$$Ps^{-1} = \{u \in S_+ \mid us \in P_\omega\}$$

The next proposition summarizes some basic properties of residuals.

Proposition 5.5. *For each $s, t \in S_+$, the formula $(st)^{-1}P = t^{-1}(s^{-1}P)$ holds.*

Proof. We have indeed

$$t^{-1}(s^{-1}P) = \{u \in S_\omega \mid tu \in s^{-1}P\} = \{u \in S_\omega \mid stu \in P\} = (st)^{-1}P. \qquad \square$$

However, we have in general $(s^{-1}P)t^{-\omega} \neq s^{-1}(Pt^{-\omega})$.

Proposition 5.6. *Let $\varphi : S \to T$ be a morphism of ω-semigroups, let P be a subset of T and let $Q = \varphi^{-1}(P)$.*

(1) *If $s \in S_+^1$ and $t = \varphi(s)$, then $s^{-1}P = \varphi^{-1}(t^{-1}Q)$ and $Ps^{-\omega} = \varphi^{-1}(Qt^{-\omega})$.*

(2) *If $s \in S_\omega$ and $t = \varphi(s)$, $Ps^{-1} = \varphi^{-1}(Qt^{-1})$.*

Proof. (1) Let $s \in S_+^1$. We have

$$\begin{aligned} s^{-1}P = \{v \in S \mid sv \in P\} &= \{v \in S \mid \varphi(sv) \in Q\} \\ &= \{v \in S \mid t\varphi(v) \in Q\} = \varphi^{-1}(t^{-1}Q) \end{aligned}$$

Similarly,

$$\begin{aligned} Ps^{-\omega} = \{v \in S_+ \mid (vs)^\omega \in P\} &= \{v \in S_+ \mid \varphi(vs)^\omega \in Q\} \\ &= \{v \in S_+ \mid \big(\varphi(v)t\big)^\omega \in Q\} \\ &= \{v \in S_+ \mid \varphi(v) \in Qt^{-\omega}\} = \varphi^{-1}(Qt^{-\omega}) \end{aligned}$$

Let $s \in S_\omega$ and $t = \varphi(s)$. We have

$$\begin{aligned} Ps^{-1} = \{v \in S_+ \mid vs \in P\} &= \{v \in S_+ \mid \varphi(vs) \in Q\} \\ &= \{v \in S_+ \mid \varphi(v)t \in Q\} \\ &= \{v \in S_+ \mid \varphi(v) \in Qt^{-1}\} = \varphi^{-1}(Qt^{-1}) \end{aligned}$$

which concludes the proof. $\square$

6 Recognition by morphism of ω-semigroups

Let $S = (S_+, S_\omega)$ and $T = (T_+, T_\omega)$ be two ω-semigroups. A surjective morphism of ω-semigroups $\varphi : S \to T$ *recognizes* a subset $I = (I_+, I_\omega)$ of S if there exists a subset $J = (J_+, J_\omega)$ of T such that $I = \varphi^{-1}(J)$, that is if $I_+ = \varphi^{-1}(J_+)$ and $I_\omega = \varphi^{-1}(J_\omega)$, which amounts to saying that, for each $u \in I$, the set $\varphi^{-1}\varphi(u)$ is contained in I, or, equivalently, that $\varphi^{-1}\varphi(I) = I$. The set $J = \varphi(I)$ is called *the image of I in T*.

By extension, we say that an ω-semigroup T *recognizes* a subset I of S if there exists a surjective morphism of semigroups $\varphi : S \to T$ which recognizes I. A congruence (of ω-semigroup) $\sim$ defined on an ω-semigroup S *recognizes* a subset I of S if the natural morphism $\varphi : S \to S/{\sim}$ recognizes I. Finally, a subset of an ω-semigroup is said to be *recognizable* if it is recognized by a finite ω-semigroup.

This definition can be extended to ordered ω-semigroups as follows. A surjective morphism of ordered ω-semigroups $\varphi : S \to T$ *recognizes an order ideal* $I = (I_+, I_\omega)$ of S if there exists an order ideal $J = (J_+, J_\omega)$ of T such that $I = \varphi^{-1}(J)$, i.e. $I_+ = \varphi^{-1}(J_+)$ and $I_\omega = \varphi^{-1}(J_\omega)$. In this case, we have $J = \varphi(I)$ and J is called the image of I in T. By extension, we say that an ordered ω-semigroup T *recognizes an order ideal* I of S if there exists a morphism of ordered ω-semigroups $\varphi : S \to T$ which recognizes I.

Example 6.1. Let $A = \{a, b\}$, and consider the ω-semigroup

$$S = (\{s, t\}, \{u, v\})$$

equipped with the operations $ss = s$, $st = ts = tt = t$, $s^\omega = u$, $t^\omega = v$, $su = u$, $sv = tv = tu = v$. Let $\varphi : A^\infty \to S$ be the morphism of ω-semigroups defined by $\varphi(a) = s$ and $\varphi(b) = t$. We have

$$\begin{aligned}
\varphi^{-1}(s) &= a^+ && \text{(finite words containing no occurrence of } b\text{),}\\
\varphi^{-1}(t) &= A^*bA^* && \text{(finite words containing at least one occurrence of } b\text{),}\\
\varphi^{-1}(u) &= a^\omega && \text{(infinite words containing no occurrence of } b\text{),}\\
\varphi^{-1}(v) &= A^\omega \setminus a^\omega && \text{(infinite words containing at least one occurrence of } b\text{),}
\end{aligned}$$

The morphism φ recognizes each of these sets (cf. Figure 6.1), as well as any union of these sets.

Example 6.2. Let us take the same ω-semigroup S and consider the morphism of ω-semigroups $\varphi : A^\infty \to S$ defined by $\varphi(a) = s$ for each $a \in A$. We have $\varphi^{-1}(s) = A^+$, $\varphi^{-1}(t) = \emptyset$ and $\varphi^{-1}(u) = A^\omega$. Thus the morphism φ recognizes the empty set and the sets A^+, A^ω and A^∞.

Example 6.3. Let $A = \{a, b\}$, and consider the ω-semigroup

$$S = (\{a, b\}, \{c, d\})$$

equipped with the following operations:

$$\begin{array}{llll}
aa = a & ab = a & ac = c & ad = c\\
ba = b & bb = b & bc = d & bd = d\\
a^\omega = c & b^\omega = d & &
\end{array}$$

Finite words without any b	Finite words containing at least one b
Infinite words without any b	Infinite words containing at least one b

Table 6.1. The subsets recognized by φ.

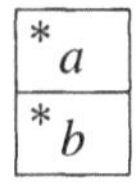

Figure 6.1. The $\mathcal{D}$-class structure of S_+.

The $\mathcal{D}$-class structure of S_+ is represented in Figure 6.1.The morphism of ω-semigroups $\varphi : A^\infty \to S$ defined by $\varphi(a) = a$ and $\varphi(b) = b$ recognizes aA^ω since we have $\varphi^{-1}(c) = aA^\omega$.

Example 6.4. Let S be the ordered ω-semigroup $(\{1, 0\}, \{u, v\})$, equipped with the following operations:

$$\begin{array}{cccc} 1 \cdot 1 = 1 & 1 \cdot 0 = 0 & 1u = u & 1v = v \\ 0 \cdot 1 = 0 & 0 \cdot 0 = 0 & 0u = u & 0v = v \\ 0^\omega = u & 1^\omega = v & & \end{array}$$

and the order $0 \leq 1$ and $u \leq v$. Let $A = \{a, b\}$ and let $\varphi : A^\infty \to S$ be the morphism of ω-semigroups defined by $\varphi(a) = 0$ and $\varphi(b) = 1$. The set $I = (\emptyset, \{u\})$ is an order ideal of S, and $\varphi^{-1}(I) = (A^*a)^\omega$.

The next proposition is an immediate consequence of Corollary 4.8

Proposition 6.1. *Let S and T be (ordered) ω-semigroups, with S quotient of T. Every subset of A^ω recognized by S is also recognized by an (ordered) ω-semigroup of T.*

Boolean operations can be easily translated in terms of morphisms. Let us start with a result which allows us to treat separately, in an ω-semigroup S, the subsets of S_+ and those of S_ω.

Proposition 6.2. *Let $\varphi : S \to T$ be a morphism of ordered ω-semigroups recognizing an order ideal $I = (I_+, I_\omega)$ of S. Then the subsets $(\emptyset, \emptyset)$, $(I_+, \emptyset)$, (I_+, S_ω), $(\emptyset, I_\omega)$, (S_+, I_ω) and (S_+, S_ω) are order ideals of S recognized by φ.*

Proof. It is clear that the subsets considered in the statement are order ideals. Since φ recognizes I, there exists an order ideal J of T such that $I = \varphi^{-1}(J)$ i.e. $I_+ = \varphi^{-1}(J_+)$ and $I_\omega = \varphi^{-1}(J_\omega)$. On the other hand, since φ is surjective, we have the equalities $\varphi^{-1}(T_+) = S_+$ and $\varphi^{-1}(T_\omega) = S_\omega$. It follows that

$$\begin{array}{lll} (\emptyset, \emptyset) = \varphi^{-1}(\emptyset, \emptyset) & (I_+, \emptyset) = \varphi^{-1}(J_+, \emptyset) & (I_+, S_\omega) = \varphi^{-1}(J_+, T_\omega) \\ (\emptyset, I_\omega) = \varphi^{-1}(\emptyset, J_\omega) & (S_+, I_\omega) = \varphi^{-1}(T_+, J_\omega) & (S_+, S_\omega) = \varphi^{-1}(T_+, T_\omega) \end{array}$$

thereby proving the result. $\square$

We now consider the complement.

Proposition 6.3. *Let T be an ordered ω-semigroup recognizing an order ideal $I = (I_+, I_\omega)$ of S. Then the subset $S \setminus I = (S_+ \setminus I_+, S_\omega \setminus I_\omega)$ is recognized by the dual of T.*

Proof. Let $\varphi : S \to T$ be a morphism of ordered ω-semigroups recognizing I and let J be an order ideal of S such that $I = \varphi^{-1}(J)$. Then $S \setminus I = \varphi^{-1}(T \setminus J)$ and since J is an order ideal of T, $T \setminus J$ is an order ideal of the dual of T. □

In particular, if T is simply an ω-semigroup (and therefore equal to its dual), Proposition 6.3 shows that the subsets of S_+ (resp. S_ω) recognized by T are closed under complement.

For union and intersection, we have the following results.

Proposition 6.4. *Let S be an ordered ω-semigroup and let $(\varphi_i)_{i\in F} : S \to S_i$ be a family of surjective morphisms recognizing an order ideal I_i of S. Then the order ideals $\bigcup_{i\in F} I_i$ and $\bigcap_{i\in F} I_i$ are recognized by an ordered ω-subsemigroup of the product $\prod_{i\in F} S_i$.*

Proof. Put, for $i \in F$, $J_i = \varphi(I_i)$. Since φ_i recognizes I_i, we have $I_i = \varphi_i^{-1}(J_i)$. Let $\varphi : S \to \prod_{i\in F} S_i$ the morphism defined by $\varphi(u) = (\varphi_i(u))_{i\in F}$. Put

$$S' = \varphi(S),\ \ J = S' \cap \prod_{i\in F} J_i \text{ and } J' = S' \cap \bigcup_{j\in I} \prod_{i\in I} J'_{i,j},\ \text{ where } J'_{i,j} = \begin{cases} J_i & \text{if } i = j \\ S_i & \text{if } i \neq j \end{cases}$$

By construction, S' is an ordered ω-subsemigroup of $\prod_{i\in F} S_i$. Now, if $y \leq x$ with $y \in S'$ and $x \in J$ (resp. J'), then for each (resp. for at least one) $i \in F$, $x_i \in I_i$. Therefore $y_i \in I_i$ and $y \in J$ (resp. J'). Consequently, J and J' are order ideals of S'. Finally, $\bigcap_{i\in F} I_i = \varphi^{-1}(J)$ and $\bigcup_{i\in F} I_i = \varphi^{-1}(J')$ and hence S' recognizes $\bigcup_{i\in F} I_i$ and $\bigcap_{i\in F} I_i$. □

If the set I is finite, it is possible to characterize all the subsets recognized by an ordered ω-subsemigroup of a product $\prod_{i\in F} S_i$.

Proposition 6.5. *Let S be an ordered ω-semigroup, $(S_i)_{i\in F}$ a finite family of ordered ω-semigroups and T an ordered ω-subsemigroup of $\prod_{i\in F} S_i$. Let φ be a morphism from S onto T. Then every subset of S recognized by φ is a finite union of finite intersections of subsets recognized by one of the S_i's.*

Proof. Denote by $\pi_i : \prod_{i \in F} S_i \to S_i$ the i-th projection and put $\varphi_i = \pi_i \circ \varphi$. Let P be an order ideal of T. Then we have

$$\varphi^{-1}(P) = \bigcup_{s \in P} \varphi^{-1}(\downarrow s)$$

where $\downarrow s$ denotes the order ideal generated by s (see Annex A), and if $s = (s_i)_{i \in F}$,

$$\varphi^{-1}(\downarrow s) = \bigcap_{1 \leq i \leq n} \varphi_i^{-1}(\downarrow s_i)$$

concluding the proof of the proposition. □

In the same spirit, the following properties hold:

Proposition 6.6. *Let $\varphi : S \to T$ be a surjective morphism of ordered ω-semigroups recognizing an order ideal I of S and let $\alpha : R \to S$ be a morphism of ordered ω-semigroups. Then $\alpha^{-1}(I)$ is an order ideal of R recognized by the morphism $\varphi \circ \alpha : R \to T'$, where $T' = \varphi(\alpha(R))$.*

Proof. Indeed, if $I = \varphi^{-1}(J)$, $\alpha^{-1}(I) = \alpha^{-1}\varphi^{-1}(I) = (\varphi \circ \alpha)^{-1}(J)$. □

Proposition 6.7. *Let $\varphi : S \to T$ be a morphism of ordered ω-semigroups recognizing an order ideal I of S and let $s \in S_+^1$. Then the sets $s^{-1}I$ and $Is^{-\omega}$ are also recognized by φ. If $s \in S_\omega$, Is^{-1} is recognized by φ.*

Proof. It is an immediate consequence of Proposition 5.6.

7 The two modes of recognition

Let X be a recognizable subset of A^ω. Our aim in this section is to give a construction to pass from a finite Büchi automaton recognizing X to a finite ω-semigroup recognizing X. We shall then show that every set of infinite words recognized by a finite ω-semigroup is ω-rational, thereby proving the equivalence between the recognition by a finite Büchi automaton and by a finite ω-semigroup.

We seek to associate with each finite path in an automaton a *multiplicity* expressing the following possibilities:

(1) the path does not exist,

(2) the path exists but does not visit any final state,

(3) the path exists and visits a final state.

These three possibilities enrich the zero-one choice corresponding to the existence or the nonexistence of a path. Our construction makes use of the semiring $k = \{-\infty, 0, 1\}$

	$-\infty$	0	1
$-\infty$	$-\infty$	$-\infty$	$-\infty$
0	$-\infty$	0	1
1	$-\infty$	1	1

Table 7.1. The multiplication table.

in which addition is the maximum for the ordering $-\infty < 0 < 1$ and multiplication, which extends the boolean addition, is given in Table 7.1.

Let $\mathcal{A} = (Q, A, E, I, F)$ be a Büchi automaton. With each letter $a \in A$ is associated a matrix $\mu(a)$ with entries in k defined by

$$\mu(a)_{p,q} = \begin{cases} -\infty & \text{if } (p, a, q) \notin E \\ 0 & \text{if } (p, a, q) \in E,\ p \notin F \text{ and } q \notin F \\ 1 & \text{if } (p, a, q) \in E,\ p \in F \text{ or } q \in F \end{cases}$$

We have already used a similar technique to encode automata, but now we want to keep track of the visits to final states. The idea is to extend μ into a morphism of ω-semigroups. We know already that the set $k^{Q \times Q}$ of $Q \times Q$-matrices over k, equipped with the usual product of matrices, is a semiring. It is easy to extend μ into a semigroup morphism from A^+ into $k^{Q \times Q}$. If u is a finite word, one gets

$$\mu(u)_{p,q} = \begin{cases} -\infty & \text{if there exists no path of label } u \text{ from } p \text{ to } q, \\ 1 & \text{if there exists a path from } p \text{ to } q \text{ with label } u \\ & \text{visiting a final state,} \\ 0 & \text{if there exists a path from } p \text{ to } q \text{ with label } u \\ & \text{but no such path visits a final state} \end{cases}$$

However, trouble arises when one tries to equip $k^{Q \times Q}$ with a structure of ω-semigroup. Indeed, the coefficients $\mu(u)_{p,q}$ encode the existence or nonexistence of a path from p to q labeled by u. When trying to extend this definition to infinite words, we encounter the problem of the interpretation of the state q, which designated the end of the path in the finite case and is of course meaningless in the infinite case. The solution consists in coding infinite paths not by square matrices, but by column matrices, in such a way that each coefficient $\mu(u)_p$ codes the existence of an infinite path of label u starting at p.

To this end, consider the set $S = (S_+, S_\omega)$ where $S_+ = k^{Q \times Q}$ is the set of square matrices of size Card Q with entries in $\{-\infty, 0, 1\}$ and $S_\omega = k^Q$ is the set of column matrices of size Card Q with entries in $\{-\infty, 1\}$.

There remains to equip S with a structure of ω-semigroup. In order to define the operation ω on square matrices, we need the following definition. If s is a matrix of S_+, we call *infinite s-path starting at p* a sequence $p = p_0, p_1, \ldots$ of elements of Q such that, for $0 \leq i \leq n - 1$, $s_{p_i, p_{i+1}} \neq -\infty$.

The s-path is said to be *successful* if $s_{p_i,p_{i+1}} = 1$ for an infinite number of coefficients. Then we define s^ω as the element of S_ω such that, for every $p \in Q$,

$$s^\omega_p = \begin{cases} 1 & \text{if there exists a successful } s\text{-path of origin } p, \\ -\infty & \text{otherwise} \end{cases}$$

Note that the coefficients of this matrix can be effectively computed. Indeed, computing s^ω_p amounts to checking the existence of circuits containing a given edge in a finite graph. There remains to verify that S, equipped with these operations, is a Wilke algebra.

Proposition 7.1. *The operations defined above provide S with a structure of a Wilke algebra.*

Proof. Let $s, t \in S_+$. For proving the equality $s(ts)^\omega = (st)^\omega$, let us compare, for each state $p \in Q$, the elements $s(ts)^\omega_p$ and $(st)^\omega_p$. We have

$$s(ts)^\omega_p = \sum_{q \in Q} s_{p,q}(ts)^\omega_q$$

and consequently, if $s(ts)^\omega_p = 1$, there exists a state q such that $s_{p,q} \neq -\infty$ and an infinite sequence $q_0, q_1, \ldots$ of states such that $q_0 = q$, $(ts)_{q_i,q_{i+1}} \neq -\infty$ for each $i \geq 0$ and $(ts)_{q_i,q_{i+1}} = 1$ for infinitely many values of i. Now $(ts)_{q_i,q_{i+1}} \neq -\infty$ if and only if there exists a state p_{i+1} such that $t_{q_i,p_{i+1}} \neq -\infty$ and $s_{p_{i+1},q_{i+1}} \neq -\infty$. Setting $p_0 = p$, it follows $(st)_{p_0,p_1} \neq -\infty$, since $s_{p_0,q_0} \neq -\infty$ and $t_{q_0,p_1} \neq -\infty$ and more generally, $(st)_{p_i,p_{i+1}} \neq -\infty$, since $s_{p_i,q_i} \neq -\infty$ and $t_{q_i,p_{i+1}} \neq -\infty$. Moreover, if $s(ts)^\omega_p = 1$, one can choose the q_i's such that $(ts)_{q_i,q_{i+1}} = 1$ for infinitely many i's. Therefore $t_{q_i,p_{i+1}} = 1$ for infinitely many i's or $s_{p_{i+1},q_{i+1}} = 1$ for infinitely many i's, and in both cases, $(st)_{p_i,p_{i+1}} = 1$ for infinitely many i's. Then there exists an infinite st-path starting at p, showing that $(st)^\omega_p = 1$.

Conversely, if $(st)^\omega_p = 1$, there exists a successful infinite st-path starting at p, and by a similar argument, there exist a state q such that $s_{p,q} \neq -\infty$ and a successful infinite ts-path starting at q. Consequently $s(ts)^\omega_p = 1$, concluding the proof of the identity $s(ts)^\omega = (st)^\omega$.

There remains to verify the equality $s^\omega = (s^n)^\omega$. Suppose that $s^\omega_p = 1$. Thus there exists a sequence $p_0, p_1 \ldots$ such that $p = p_0$, $s_{p_i,p_{i+1}} \neq -\infty$ for each $i \geq 0$ and $s_{p_i,p_{i+1}} = 1$ for infinitely many i's. Therefore, for every $i \geq 0$, $s^n_{p_{ni},p_{n(i+1)-1}} \neq -\infty$ and $s^n_{p_{ni},p_{n(i+1)-1}} = 1$ for infinitely many i's, showing that $(s^n)^\omega_p = 1$.

Conversely, if $(s^n)^\omega_p = 1$, there exists a sequence $p_0, p_1 \ldots$ such that $p = p_0$, $(s^n)_{p_i,p_{i+1}} \neq -\infty$ for each $i \geq 0$ and $(s^n)_{p_i,p_{i+1}} = 1$ for infinitely many i's. Then there exist states $p_{i_0}, p_{i_1}, \ldots, p_{i_{n-1}}$ such that $p_{i_0} = p_i$ and $s_{p_{i_0},p_{i_1}} \neq -\infty$, $s_{p_{i_1},p_{i_2}} \neq -\infty$, $\ldots$, $s_{p_{i_{n-1}},p_{(i+1)_0}} \neq -\infty$. Furthermore, if $(s^n)_{p_i,p_{i+1}} = 1$, one of the entries $s_{p_{i_0}}, \ldots, s_{p_{i_{n-1}}}$ is equal to 1. Consequently, there exists a successful infinite s-path starting at p and $s^\omega_p = 1$. □

The morphism μ can now be extended in a unique way as a morphism of ω-semigroups from A^∞ into S. Furthermore, we have the following result.

Proposition 7.2. *The morphism of ω-semigroups $\mu : A^\infty \to S$ recognizes the set $L^\omega(\mathcal{A})$.*

Proof. Set

$$P = \{s \in S_\omega \mid \text{there exists } i \in I \text{ such that } s_i = 1\}$$

Then we have by construction $\mu(u) \in P$ if and only if there exists an infinite successful path labeled by u. Therefore μ recognizes $L^\omega(\mathcal{A})$. □

Example 7.1. Let $\mathcal{A}$ be the Büchi automaton represented in Figure 7.1. The morphism $\mu : A^\infty \to S(\mathcal{A})$ is defined by the formula

$$\mu(a) = \begin{pmatrix} 0 & -\infty \\ -\infty & 1 \end{pmatrix} \quad \text{and} \quad \mu(b) = \begin{pmatrix} 0 & 1 \\ -\infty & -\infty \end{pmatrix}$$

The ω-semigroup generated by these matrices contains five elements:

$$a = \begin{pmatrix} 0 & -\infty \\ -\infty & 1 \end{pmatrix} \qquad b = \begin{pmatrix} 0 & 1 \\ -\infty & -\infty \end{pmatrix}$$

$$a^\omega = \begin{pmatrix} -\infty \\ 1 \end{pmatrix} \qquad b^\omega = \begin{pmatrix} -\infty \\ -\infty \end{pmatrix} \qquad ba^\omega = \begin{pmatrix} 1 \\ -\infty \end{pmatrix}$$

It is defined by the relations:

$$a^2 = a \quad ab = b \quad ba = b \quad b^2 = b \quad aa^\omega = a^\omega \quad ab^\omega = b^\omega \quad bb^\omega = b^\omega$$

Example 7.2. Let $X = (a\{b, c\}^* \cup \{b\})^\omega$. This set was already considered in Example I.9.4 in which we gave a Büchi automaton recognizing X, represented in Figure 7.2. For this automaton, the previous computation provides an ω-semigroup with eleven el-

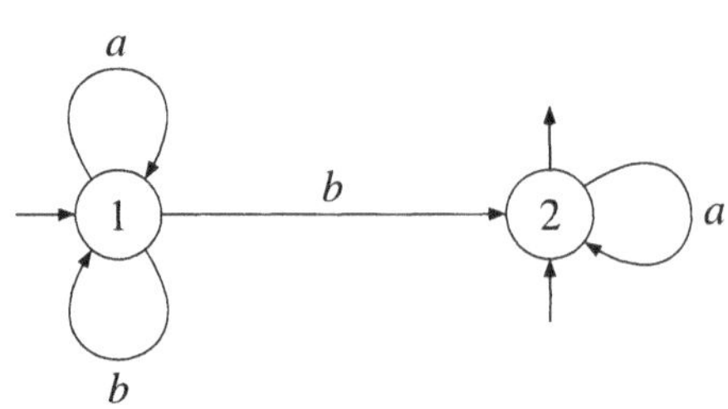

Figure 7.1. A Büchi automaton.

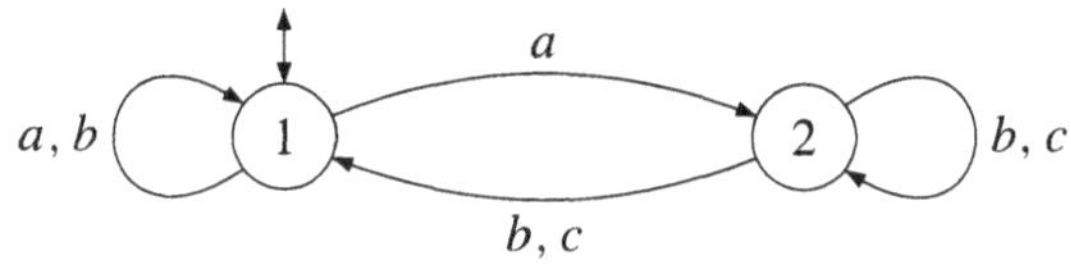

Figure 7.2. An automaton.

ements

$$a = \begin{pmatrix} 1 & 1 \\ -\infty & -\infty \end{pmatrix} \quad b = \begin{pmatrix} 1 & -\infty \\ 1 & 0 \end{pmatrix} \quad c = \begin{pmatrix} -\infty & -\infty \\ 1 & 0 \end{pmatrix} \quad ba = \begin{pmatrix} 1 & 1 \\ 1 & 1 \end{pmatrix}$$

$$ca = \begin{pmatrix} -\infty & -\infty \\ 1 & 1 \end{pmatrix} \quad a^\omega = \begin{pmatrix} 1 \\ -\infty \end{pmatrix} \quad b^\omega = \begin{pmatrix} 1 \\ 1 \end{pmatrix} \quad c^\omega = \begin{pmatrix} -\infty \\ -\infty \end{pmatrix}$$

$$(ca)^\omega = \begin{pmatrix} -\infty \\ 1 \end{pmatrix}$$

It is defined by the following relations:

$$\begin{array}{ccccc} a^2 = a & ab = a & ac = a & b^2 = b & bc = c \\ cb = c & c^2 = c & c^\omega = 0 & (ba)^\omega = b^\omega & aa^\omega = a^\omega \\ ab^\omega = a^\omega & a(ca)^\omega = a^\omega & ba^\omega = b^\omega & bb^\omega = b^\omega & b(ca)^\omega = (ca)^\omega \\ ca^\omega = (ca)^\omega & cb^\omega = (ca)^\omega & c(ca)^\omega = (ca)^\omega & & \end{array}$$

Proposition 7.2 shows that every recognizable subset of A^ω is recognized by a finite ω-semigroup. We now establish the reciprocal property.

Proposition 7.3. *Every subset of A^ω recognized by a finite ordered ω-semigroup is ω-rational.*

Proof. Let $\varphi : A^\infty \to S$ be a morphism of ω-semigroups. Set, for each $s \in S_+$, $X_{\downarrow s} = \varphi^{-1}(\downarrow s)$. Let X be a subset of A^ω recognized by a morphism of ordered ω-semigroups $\varphi : A^\infty \to S$ and let I be an order ideal of S_ω such that $X = \varphi^{-1}(I)$. We claim that

$$X = \bigcup_{(s,e)\in F} X_{\downarrow s} X_{\downarrow e}^\omega \tag{7.1}$$

where F is the set of linked pairs (s, e) of elements of S_+ such that $se^\omega \in I$. Let Y be right hand side of (7.1). If $u \in X$, there exists by Theorem 2.2 a linked pair (s, e) of S_+ such that $u \in X_s X_e^\omega$. Since $X_s X_e^\omega \subset X_{\downarrow s} X_{\downarrow e}^\omega$, we also have $u \in X_{\downarrow s} X_{\downarrow e}^\omega$. On the other hand, $\varphi(u) = se^\omega \in I$ and hence $(s, e) \in F$. Consequently, $u \in Y$.

Conversely, if $u \in Y$, there exists a linked pair $(s, e) \in F$ and a factorization $u = u_0u_1u_2\cdots$ with $\varphi(u_0) \leq s$ and $\varphi(u_n) \leq e$ for each $n > 0$. It follows $\varphi(u) \leq se^\omega$ and hence $\varphi(u) \in I$ since I is an order ideal. Therefore $u \in \varphi^{-1}(I) = X$, which establishes the formula (7.1). It follows that X is an ω-rational subset. □

Corollary 7.4. *Every subset of A^ω recognized by a finite ω-semigroup is ω-rational.*

Proof. Indeed, if S an ω-semigroup recognizing X, $(S, =)$ is an ordered ω-semigroup recognizing X. Note that (7.1) can be slightly simplified: setting, for each $s \in S_+$, $X_s = \varphi^{-1}(s)$, we have

$$X = \bigcup_{(s,e)\in F} X_s X_e^\omega \tag{7.2}$$

where F is the set of linked pairs (s, e) of elements of S_+ such that $se^\omega \in P$. □

The conjunction of Theorem I.3.2 and Propositions 7.2 and 7.4 shows that the notions of recognition by ω-semigroup, by ordered ω-semigroup and by a finite Büchi automaton are equivalent:

Theorem 7.5. *Let X be a subset of A^ω. The following conditions are equivalent:*

(1) *X is ω-rational,*

(2) *X is recognizable,*

(3) *X is recognized by a finite ω-semigroup,*

(4) *X is recognized by a finite ordered ω-semigroup.*

Here is another important consequence of formula (7.2) and (7.1).

Proposition 7.6. *Let S be a finite ordered ω-semigroup and let $\varphi : A^\infty \to S$ be a morphism of ordered ω-semigroups. Then there are only finitely many subsets of A^ω recognized by φ.*

Proof. Since S_+ is finite, there are finitely many subsets of the form $X_{\downarrow s} X_{\downarrow e}^\omega$. Now (7.1) shows that every subset recognized by φ is a finite union of subsets of the form $X_{\downarrow s} X_{\downarrow e}^\omega$. Therefore, there are only finitely many such subsets. □

Corollary 7.7. *Let X be a recognizable subset of A^ω, $s \in A^*$ and $t \in A^\omega$. There are only finitely many subsets of the form $s^{-1}X$, $Xs^{-\omega}$ and Xt^{-1}.*

Proof. Let S be a finite ω-semigroup and let $\varphi : A^\infty \to S$ be a morphism of ω-semigroups recognizing X. By Proposition 6.7, the subsets of the form $s^{-1}X$, $Xs^{-\omega}$ and Xt^{-1} are also recognized by φ. By Proposition 7.6, there are only finitely many such subsets. □

8 Syntactic congruence

It is known that for each order ideal X of an ordered semigroup S, there exists a coarsest congruence of ordered semigroup among the congruences recognizing X. This result does not, however, extend to ordered ω-semigroups because the upper bound of a family of congruences recognizing a given subset does not necessarily recognize this subset. Nevertheless, this result still holds in the case of a finite family.

An order ideal X of an ordered ω-semigroup S is said to have a *syntactic congruence* if the upper bound of the congruences recognizing X still recognizes X. In this case, this upper bound is called the syntactic congruence of X and denoted by $\sim_X$. The quotient $S(X) = S/{\sim_X}$ is the *ordered ω-semigroup syntactic* of X, the morphism of ω-semigroups $\eta : S \to S(X)$ is the *syntactic morphism* and the subset $\eta(X)$ of $S(X)$ is called *the syntactic image* of X.

Unlike the semigroup case, there is no simple constructive definition of the syntactic congruence, except for some particular cases that we now consider. Let us start with the case where X is an order ideal of S_+.

Proposition 8.1. *If X is an order ideal of S_+, the syntactic congruence $\sim_X$ exists and is defined on S_+ by $u \sim_X v$ if and only if, for each $x, y \in S_+^1$,*

$$xuy \in X \iff xvy \in X$$

and it is the universal congruence on S_ω.

Proof. Denote by $\sim$ the equivalence relation defined in the statement. It is a semigroup congruence. It is also stable for the infinite product and the mixed product, since any two infinite products are equivalent. Consequently, $\sim$ is a congruence of ω-semigroup.

This congruence recognizes X: indeed, if $u \in X$ and $u \sim v$, we have $1 \cdot u \in X$ and hence $1 \cdot v = v \in X$. Furthermore, if $\approx$ is a congruence recognizing X, $\sim$ is coarser than $\approx$. Indeed, if u and v are two finite words such that $u \approx v$, we have for each $x \in S_+^1$ and $y \in S_+^1$, $xuy \approx xvy$ and hence

$$xuy \in X \iff xvy \in X$$

since $\approx$ recognizes X. Therefore $u \sim v$, showing that $\sim$ is the syntactic congruence. □

Proposition 8.1 allows a simple computation of the syntactic ω-semigroup of a subset X of A^+, given its syntactic semigroup (in the usual sense of finite words).

Corollary 8.2. *Let S be the syntactic ordered semigroup of a subset X of A^+. The syntactic ordered ω-semigroup of X is equal to $(S, \{0\})$, where all the infinite products are equal to 0 and the mixed product is given by $s0 = 0$ for each $s \in S$.*

Here is another important special case:

Proposition 8.3. *If X is a deterministic subset of A^ω whose complement is deterministic, the syntactic congruence $\sim_X$ exists and is defined on A^+ by $u \sim_X v$ if and only if, for each $x \in A^*$ and for each $y \in A^\omega$,*

$$xuy \in X \Longleftrightarrow xvy \in X$$

and on A^ω by $u \sim_X v$ if and only if, for each $x \in A^*$,

$$xu \in X \Longleftrightarrow xv \in X$$

Proof. Since X and $A^\omega \setminus X$ are deterministic subsets, there exist subsets K and L of A^+ such that $X = \overrightarrow{K}$ and $A^\omega \setminus X = \overrightarrow{L}$. Denote by $\sim$ the equivalence relation defined in the statement. We claim that $\sim$ is a congruence of ω-semigroup. If it is not the case, one can find two sequences $(u_n)_{n\geq 0}$ and $(v_n)_{n\geq 0}$ of elements of A^∞ such that for each $n \geq 0$, $u_n \sim v_n$, but such that

$$u_0u_1u_2\cdots \not\sim v_0v_1v_2\cdots$$

Then there exist $x \in A^*$ such that $xu_0u_1u_2\cdots \in X$ and $xv_0v_1v_2\cdots \notin X$. Now, since $xu_0u_1u_2\cdots \in X = \overrightarrow{K}$, there exists an integer n_0 such that the word $w_0 = xu_0u_1\cdots u_{n_0}$ has a prefix in K. Since on the other hand $u_i \sim v_i$ for $0 \leq i \leq n_0$, we have $w_0v_{n_0+1}v_{n_0+2}\cdots \in A^\omega \setminus X = \overrightarrow{L}$ and thus there exists an integer n_1 such that $w_1 = w_0v_{n_0+1}\cdots v_{n_1}$ has a prefix in L. The same argument allows us to find an integer n_2 such that $w_1u_{n_1+1}\cdots u_{n_2}$ has a prefix in K of length greater than $|w_1|$, next an integer n_3 such that $w_3 = w_2v_{n_2+1}\cdots v_{n_3}$ has a prefix in L of length greater than $|w_2|$, etc.

The sequence of words $(w_n)_{n\geq 0}$ built in this way satisfies $w_0 < w_1 < w_2 < \cdots$ for the prefix order and the word $w = \lim_{n\to\infty} w_n$ has infinitely many prefixes in K and in L. Therefore $w \in \overrightarrow{K} \cap \overrightarrow{L}$, which is impossible. Therefore $\sim$ is a congruence of ω-semigroup.

This congruence recognizes X: indeed, if $u \in X$ and $u \sim v$, we have $1 \cdot u \cdot 1 \in X$ and hence $1 \cdot v \cdot 1 = v \in X$. Furthermore, if $\approx$ is a congruence recognizing X, $\sim$ is coarser than $\approx$. Indeed, if u and v are finite words such that $u \approx v$, we have for each $x \in A^*$ and $y \in A^* \cup A^\omega$, $xuy \approx xvy$ and thus

$$xuy \in X \Longleftrightarrow xvy \in X$$

since $\approx$ recognizes X. Similarly, if u and v are infinite words such that $u \approx v$, we have for each $x \in A^*$, $xu \approx xv$. Thus $u \sim v$, showing that $\sim$ is the syntactic congruence. □

We end by the more important special case. An order ideal of an ordered ω-semigroup is said to be *recognizable* if it is recognized by a finite ordered ω-semigroup. We shall see that every recognizable ideal has a syntactic congruence.

Let $S = ((S_+, \leq), (S_\omega, \leq))$ be an ordered ω-semigroup and let P be an order ideal of S. For $s \in S_+$, the *sets of finite and infinite contexts of s in P* are respectively defined by

$$CF_P(s) = \{(x, y) \in S_+^1 \times S_\omega \mid xsy \in P\}$$
$$CI_P(s) = \{(x, y) \in S_+^1 \times S_+^1 \mid x(sy)^\omega \in P\}$$

Finally, for $s \in S_\omega$, the set of left contexts of s in P is defined as the set

$$CG_P(s) = \{x \in S_+^1 \mid xs \in P\}$$

Define a relation $\leq_P$ on S by setting, for each $s, t \in S_+$, $s \leq_P t$ if and only if,

$$CF_P(t) \subset CF_P(s) \quad \text{and} \quad CI_P(t) \subset CI_P(s)$$

and for each $s, t \in S_\omega$, $s \leq_P t$ if and only if

$$CG_P(t) \subset CG_P(s)$$

Proposition 8.4. *If P is recognizable, the relation $\leq_P$ is a congruence of ordered ω-semigroup.*

Proof. The relation $\leq_P$ defines by construction a preorder on S_+ (resp. on S_ω) coarser than the relation $\leq$. Indeed, let s and t be two elements of S_+ such that $s \leq t$. If $(x, y) \in CF_P(t)$, then $xty \in P$ and as $xsy \leq xty$ and since P is an ideal, we have $xsy \in P$ and hence $(x, y) \in CF_P(s)$. Therefore $CF_P(t) \subset CF_P(s)$ and, by a similar argument, $CI_P(t) \subset CI_P(s)$. The proof for S_ω is analogous.

Let u and v be two elements of S_+ such that $u \leq_P v$ and let $x \in S_+$. Let us show that $xu \leq_P xv$ and $ux \leq_P vx$. Indeed, let $r, s \in S_+^1$ and $t \in S_+$ such that $r(xv)st^\omega \in P$. Since $u \leq_P v$, we have

$$(rx)vst^\omega \in P \Rightarrow (rx)ust^\omega \in P$$

and hence $r(xu)st^\omega \in P$. Suppose now that $r(xvs)^\omega \in P$. Since $r(xvs)^\omega = (rx)(vsx)^\omega$ and $u \leq_P v$, we have $rx(usx)^\omega \in P$. Since $(rx)(usx)^\omega = r(xus)^\omega$, it follows $r(xus)^\omega \in P$. Therefore $xu \leq_P xv$. A similar proof would show that $ux \leq_P vx$.

Likewise if s and s' are two elements of S_+ such that $s \leq_P s'$ and if $x \in S_\omega$, then $sx \leq_P s'x$.

Finally, let $(s_n)_{n \geq 0}$ and $(s'_n)_{n \geq 0}$ be two sequences of elements of S_+ such that, for each $n \geq 0$, $s_n \leq_P s'_n$. It remains to showing that $s_0 s_1 \cdots \leq_P s'_0 s'_1 \cdots$ and we now use our key hypothesis that P is recognizable. Let indeed φ be a surjective morphism from S onto a finite ordered ω-semigroup T recognizing P and let $I = \varphi(P)$. Let $u \in S_+^1$ be such that $us'_0s'_1 \cdots \in P$. One first observes that if $s \leq_P s'$ and if $s' \in P$, then $s \in P$. It follows that if $\varphi(s') \in I$ then $\varphi(s) \in I$. By Proposition 2.3 applied to the morphism $\psi \times \varphi : (S \times S)^+ \to T \times T$ defined by $\psi(s, s') = (\varphi(s), \varphi(s'))$, there exists a linked pair $((s, s'), (e, e'))$ of $T \times T$ and a strictly increasing sequence of integers $(k_n)_{n \in \mathbb{N}}$ such that

$$\begin{aligned} \varphi(s_0s_1 \cdots s_{k_0-1}) &= s & \varphi(s_{k_n}s_{k_n+1} \cdots s_{k_{n+1}-1}) &= e \text{ for each } n \geq 0 \\ \varphi(s'_0s'_1 \cdots s'_{k_0-1}) &= s' & \varphi(s'_{k_n}s'_{k_n+1} \cdots s'_{k_{n+1}-1}) &= e' \text{ for each } n \geq 0 \end{aligned}$$

In particular, if $us'_0s'_1 \cdots \in P$, we have

$$\varphi(us'_0s'_1 \cdots) = \varphi(u)s'e'^{\omega} = \varphi(us'_0s'_1 \cdots s'_{k_0-1}(s'_{k_0}s'_{k_0+1} \cdots s'_{k_1-1})^{\omega}) \in I$$

and hence $us'_0s'_1 \cdots s'_{k_0-1}(s'_{k_0}s'_{k_0+1} \cdots s'_{k_1-1})^{\omega} \in P$. Now, by the first part of the proof, we have

$$s_{k_0}s_{k_0+1} \cdots s_{k_1-1} \leq_P s'_{k_0}s'_{k_0+1} \cdots s'_{k_1-1}$$

and consequently $us'_0s'_1 \cdots s'_{k_0-1}(s_{k_0}s_{k_0+1} \cdots s_{k_1-1})^{\omega} \in P$. Similarly,

$$s_0s_1 \cdots s_{k_0-1} \leq_P s'_0s'_1 \cdots s'_{k_0-1}$$

and hence $us_0s_1 \cdots s_{k_0-1}(s_{k_0}s_{k_0+1} \cdots s_{k_1-1})^{\omega} \in P$. Finally, as

$$\varphi(us_0s_1 \cdots s_{k_0-1}(s_{k_0}s_{k_0+1} \cdots s_{k_1-1})^{\omega}) = \varphi(us_0s_1 \cdots) = \varphi(u)se^{\omega}$$

we have $us_0s_1 \cdots \in P$. Therefore $s_0s_1 \cdots \leq_P s'_0s'_1 \cdots$ and $\leq_P$ is a congruence of ordered ω-semigroup. □

Denote by $\sim_P$ the equivalence relation on S associated with $\leq_P$. This relation is defined on S_+ by $u \sim_P v$ if and only if, for every $x, y, z \in S_+$

$$\begin{aligned} xuyz^{\omega} \in P &\iff xvyz^{\omega} \in P \\ x(uy)^{\omega} \in P &\iff x(vy)^{\omega} \in P \end{aligned} \tag{8.1}$$

and on S^{ω} by $u \sim_P v$ if and only if, for each $x \in S_+$,

$$xu \in P \iff xv \in P \tag{8.2}$$

Corollary 8.5. *The relation $\sim_P$ is a congruence of ω-semigroups.*

Proposition 8.6. *If P is a recognizable order ideal of S, the congruence $\sim_P$ is the syntactic congruence of P.*

Proof. Since P is recognizable, there exists a surjective morphism φ from S onto a finite ordered ω-semigroup T recognizing P. Setting $I = \varphi(P)$, we claim that if $u, v \in S_+$ satisfy $\varphi(u) = \varphi(v)$, then $u \sim_P v$. We have indeed, for all $x, y \in S_+^1$ and for each $z \in S_+$, $\varphi(xuyz^\omega) = \varphi(x)\varphi(u)\varphi(y)\varphi(z)^\omega = \varphi(x)\varphi(v)\varphi(y)\varphi(z)^\omega = \varphi(xvyz^\omega)$, whence

$$xuyz^\omega \in P \iff \varphi(xuyz^\omega) \in I \iff \varphi(xvyz^\omega) \in I \iff xvyz^\omega \in P$$

Similarly, $\varphi(x(uy)^\omega) = \varphi(x)(\varphi(u)\varphi(y))^\omega = \varphi(x)(\varphi(v)\varphi(y))^\omega = \varphi(x(vy)^\omega)$ and hence

$$x(uy)^\omega \in P \iff \varphi(x(uy)^\omega) \in I \iff \varphi(x(vy)^\omega) \in I \iff x(vy)^\omega \in P$$

proving the claim. One would prove in the same way that if u and v are two infinite words such that $\varphi(u) = \varphi(v)$, then $u \sim_P v$. This leads to an equivalent definition of $\sim_P$.

Lemma 8.7. *We have $u \sim_P v$ if and only if $\varphi(u) \sim_I \varphi(v)$.*

Proof. We claim that if $\varphi(u) \sim_I \varphi(v)$, then $u \sim_P v$. Indeed, if $u, v \in S_+$, then for each $x, y \in S_+^1$ and for each $z \in S_+$,

$$\begin{aligned} xuyz^\omega \in P &\iff \varphi(xuyz^\omega) \in I \iff \varphi(x)\varphi(u)\varphi(y)\varphi(z)^\omega \in I \\ &\iff \varphi(x)\varphi(v)\varphi(y)\varphi(z)^\omega \in I \iff \varphi(xvyz^\omega) \in I \iff xvyz^\omega \in P \end{aligned}$$

and the two other equivalences of 8.1 are proved in a similar way by computing the image under φ.

In the opposite direction, let $u, v \in S_+$ be such that $u \sim_P v$ and let $r, s \in T_+^1$ and $t \in T_+$ be such that $r\varphi(u)st^\omega \in I$. Since φ is surjective, there exist elements $x, y \in S_+^1$ and $z \in S_+$ such that $\varphi(x) = r$, $s = \varphi(y)$ and $t = \varphi(z)$. Now, by the first equivalence of (8.1), we have $xuyz^\omega \in P$ if and only if $xvyz^\omega \in P$, whence, by taking the images under φ, $r\varphi(u)st^\omega \in I$ if and only if $r\varphi(v)st^\omega \in I$. A similar argument would give the second equivalence of the definition of $\sim_I$. Therefore $\varphi(u) \sim_I \varphi(v)$. The proof is analogous in the case where $u, v \in S_\omega$. □

We come back to the proof of Proposition 8.6. Lemma 8.7 shows that $\sim_P$ is a congruence of ordered ω-semigroup coarser than all the congruences recognizing P. Furthermore, it recognizes P itself, since if $u \sim_P v$ and $u \in P$, we get $v \in P$ by taking $x = 1$ in the equivalence 8.2. Therefore $\sim_P$ is the syntactic congruence of P. □

In particular, the syntactic congruence of a recognizable subset of A^ω is defined on A^+ by $u \sim_X v$ if and only if, for each $x, y \in A^*$ and for each $z \in A^+$,

$$\begin{aligned} xuyz^\omega \in X &\iff xvyz^\omega \in X \\ x(uy)^\omega \in X &\iff x(vy)^\omega \in X \end{aligned} \tag{8.3}$$

and on A^ω by $u \sim_X v$ if and only if, for each $x \in A^*$,

$$xu \in X \Longleftrightarrow xv \in X \tag{8.4}$$

Example 8.1. Let $A = \{a, b\}$ and $X = \{a^\omega\}$. The syntactic congruence of X divides A^+ into two classes: a^+ and A^*bA^* and A^ω into two classes also: A^*bA^ω and a^ω.

Example 8.2. When X is not recognizable, the equivalence relation $\sim$ defined on A^+ by (8.3) and on A^ω by (8.4) is not in general a congruence. For instance, let $A = \{a, b\}$ and $X = \{ba^1ba^2ba^3b\cdots\}$. We have, for each $n > 0$, $b \sim_X ba^n$, but nevertheless $ba^1ba^2ba^3b\cdots$ is not equivalent to b^ω since $ba^1ba^2ba^3b\cdots \in X$ but $b^\omega \notin X$.

Proposition 8.6 allows one to compute the syntactic ω-semigroup of a recognizable subset X of A^ω (resp. A^+) given an ω-semigroup recognizing X.

Proposition 8.8. *Let X be a subset of A^ω (resp. A^+) recognized by a finite ω-semigroup S and let P be the image of X in S. The syntactic ω-semigroup of X is equal to $S/{\sim_P}$.*

Proof. It is an immediate consequence of Lemma 8.7. □

Example 8.3. Let $A = \{a, b\}$ and let $X = a^\omega$. The syntactic ω-semigroup of X is the four element ω-semigroup of Example 6.1.

Example 8.4. Let $A = \{a, b\}$ and let $X = aA^\omega$. The syntactic ω-semigroup of X is the ω-semigroup of Example 6.3.

Example 8.5. Let $A = \{a, b\}$ and let $X = (A^*a)^\omega$. The syntactic ordered ω-semigroup of X is the ordered ω-semigroup given in Example 6.4.

Example 8.6. Let $X = (a\{b, c\}^* \cup \{b\})^\omega$. We have computed in Example 7.2 a ω-semigroup S recognizing this set. We have

$$S = \{a, b, c, ba, ca, a^\omega, b^\omega, (ca)^\omega, 0\}$$

with the following relations:

$$\begin{array}{llllll} a^2 = a & ab = a & ac = a & b^2 = b & bc = c & cb = c \\ c^2 = c & c^\omega = 0 & (ba)^\omega = b^\omega & aa^\omega = a^\omega & ab^\omega = a^\omega & a(ca)^\omega = a^\omega \end{array}$$
$$ba^\omega = b^\omega bb^\omega = b^\omega b(ca)^\omega = (ca)^\omega ca^\omega = (ca)^\omega cb^\omega = (ca)^\omega c(ca)^\omega = (ca)^\omega$$

Since $P = \{a^\omega, b^\omega\}$, the congruence $\sim_P$ is defined by

$$a \sim_P ba \qquad \text{and} \qquad a^\omega \sim_P b^\omega$$

Therefore $S(X) = \{a, b, c, ca, a^\omega, (ca)^\omega, 0\}$, with the relations

$$\begin{array}{llllll} a^2 = a & ab = a & ac = a & ba = a & b^2 = b & bc = c \\ cb = c & c^2 = c & b^\omega = a^\omega & c^\omega = 0 & \multicolumn{2}{l}{aa^\omega = a^\omega a(ca)^\omega = a^\omega} \end{array}$$
$$ba^\omega = a^\omega b(ca)^\omega = (ca)^\omega ca^\omega = (ca)^\omega c(ca)^\omega = (ca)^\omega$$

If an order ideal I of S admits a syntactic congruence, its syntactic ordered ω-semigroup is the least ordered ω-semigroup recognizing I. More precisely, we have the following statement:

Proposition 8.9. *Let I be an order ideal of S having a syntactic congruence and let $\eta : S \to S(I)$ its syntactic morphism. Then a surjective morphism $\alpha : S \to T$ recognizes I if and only if there exists a surjective morphism $\beta : T \to S(I)$ such that $\eta = \beta \circ \alpha$.*

Proof. Suppose there exist morphisms $\alpha : S \to T$ and $\beta : T \to S(I)$ such that $\eta = \beta \circ \alpha$ and let $J = \eta(I)$. Since η recognizes I, it comes $I = \varphi^{-1}(J)$ and hence $I = \alpha^{-1}(\beta^{-1}(J))$. Therefore α recognizes I.

Conversely, suppose that α recognizes I. By definition of the syntactic congruence, the nuclear congruence of α is thinner than $\sim_I$ and by Proposition 4.2, there exists a surjective morphism $\beta : T \to S(I)$ such that $\eta = \beta \circ \alpha$. □

Corollary 8.10. *Let X be a recognizable subset of A^∞. An ordered ω-semigroup S recognizes X if and only if $S(X)$ is a quotient of S.*

Proof. Let $\eta : A^\infty \to S(X)$ be the syntactic morphism of X. If there exists a surjective morphism of ordered ω-semigroups $\beta : S \to S(X)$, there exists by Proposition 8.9 a surjective morphism of ordered ω-semigroups $\alpha : A^\infty \to S$ such that $\eta = \beta \circ \alpha$ and hence S recognizes X by Proposition 8.9. Conversely, if S recognizes X, the same proposition shows that $S(X)$ is a quotient of S. □

9 Back to McNaughton's theorem

We give in this section another proof of McNaughton's theorem which makes use of ω-semigroups. This proof relies on a few useful formulas of independent interest on deterministic sets. Let S be a finite ω-semigroup and let $\varphi : A^\infty \to S$ be a surjective morphism recognizing a subset X of A^ω. As above, we set, for each $s \in S_+$, $X_s = \varphi^{-1}(s)$ and we denote by P the image of X in S. For each $s \in S_+$, the set $P_s = X_s \setminus X_s A^+$ is prefix-free, since a word of P_s cannot be, by definition, prefix of another word of P_s. Put

$$E_s = \{e \in E(S_+) \mid se = s\},$$

and denote by $\leq$ the relation on E_s defined by

$$f \leq e \text{ if and only if } ef = f.$$

It is the restriction to the set E_s of the preorder $\leq_{\mathcal{R}}$, since, if $f = ex$ then $ef = eex = ex = f$. We shall use the notation $e < f$ if $e \leq f$ and if $f \nleq e$. To simplify notations, we shall suppose implicitly that for every expression of the form $X_s X_f^{\omega}$ or $X_s P_f$, the pair (s, f) is a linked pair of S_+.

Proposition 9.1. *For each linked pair (s, e) of S_+, the following formula holds*

$$X_s X_e^{\omega} \subset \overrightarrow{X_s P_e} \subset \bigcup_{f \leq e} X_s X_f^{\omega}. \tag{9.1}$$

Proof. Let $u \in X_s X_e^{\omega}$. Then $u = u_0 u_1 \cdots$, where $u_0 \in X_s$ and $u_n \in X_e$ for each $n > 0$. Each word u_n, for $n > 0$, has a unique prefix p_n in P_e. Since $\varphi(u_0 u_1 \cdots u_{n-1}) = se = s$, we have $u_0 u_1 \cdots u_{n-1} p_n \in X_s P_e$, and u has infinitely many prefixes in $X_s P_e$. Consequently $u \in \overrightarrow{X_s P_e}$, thereby proving the first inclusion.

Let $u \in \overrightarrow{X_s P_e}$. Then u has infinitely many prefixes of the form $x_n p_n$ with $x_n \in X_s$ and $p_n \in P_e$. If the length of the words x_n is bounded, one can find two indices $i < j$ such that $x_i = x_j$, and hence one of the words p_i or p_j is prefix of the other one, a contradiction with the fact that P_e is prefix-free. Consequently, up to taking a subsequence, we may assume that, for each $n \geq 0$, the word $x_n p_n$ is a strict prefix of x_{n+1}. Setting, for all $n \geq 0$, $x_{n+1} = x_n v_n$, we have $u = x_0 v_0 v_1 \cdots$, as is shown in Figure 9.1. By Proposition 2.3, there exists a linked pair (t, f) and a factorization

$$u = (x_0 v_0 \cdots v_{n_0 - 1})(v_{n_0} \cdots v_{n_1 - 1})(v_{n_1} \cdots v_{n_2 - 1}) \cdots$$

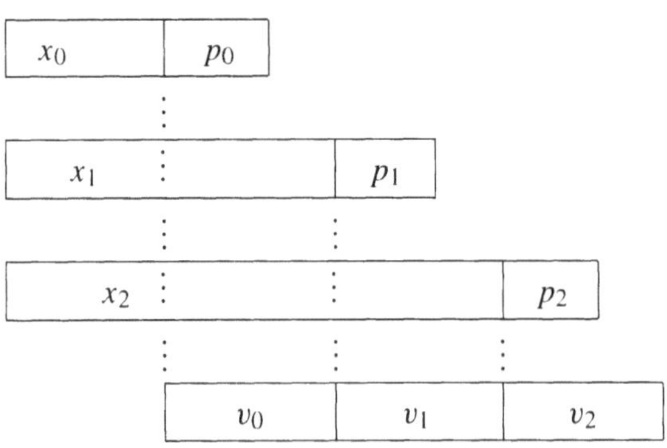

Figure 9.1. $u = x_0 v_0 v_1 \cdots$.

such that $\varphi(x_0 v_0 \cdots v_{n_0-1}) = t$, $\varphi(v_{n_0} \cdots v_{n_1-1}) = \varphi(v_{n_1} \cdots v_{n_2-1}) = \cdots = f$. It follows that

$$t = \varphi(x_0 v_0 \cdots v_{n_0-1}) = \varphi(x_{n_0}) = s$$

whence $s = t$. In particular, (s, f) is a linked pair and $u \in X_s X_f^\omega$. Moreover, since p_{n_0} is a prefix of v_{n_0}, we have $f = \varphi(v_{n_0} \cdots v_{n_1-1}) \in \varphi(p_{n_0})S = eS$. Consequently $f \leq e$, which proves the second inclusion. □

Corollary 9.2.

(1) *For every idempotent e of S_+, the following formula holds*

$$X_e^\omega = \overrightarrow{X_e P_e}. \tag{9.2}$$

(2) *For every linked pair (s, e) of S_+, we have*

$$\bigcup_{f \leq e} X_s X_f^\omega = \bigcup_{f \leq e} \overrightarrow{X_s P_f}. \tag{9.3}$$

Proof. (1) Applying formula (9.1) with $s = e$ gives the inclusion

$$X_e^\omega \subset \overrightarrow{X_e P_e} \subset \bigcup_{f \leq e} X_e X_f^\omega.$$

Now we have $ef = e$ since (e, f) is a linked pair and $ef = f$ since $f \leq e$. Therefore $f = e$ and the formula reduces to $X_e^\omega = \overrightarrow{X_e P_e}$.

(2) For every linked pair (s, f), formula (9.1) gives the following inclusion

$$X_s X_f^\omega \subset \overrightarrow{X_s P_f} \subset \bigcup_{g \leq f} X_s X_g^\omega.$$

It follows that

$$\bigcup_{f \leq e} X_s X_f^\omega \subset \bigcup_{f \leq e} \overrightarrow{X_s P_f} \subset \bigcup_{f \leq e} \Big(\bigcup_{g \leq f} X_s X_g^\omega \Big).$$

But if $g \leq f$ and $f \leq e$, then $g \leq e$, and therefore

$$\bigcup_{f \leq e} \bigcup_{g \leq f} X_s X_g^\omega \subset \bigcup_{g \leq e} X_s X_g^\omega$$

thereby proving formula (9.3). □

The previous statement shows that a set of the form X_e^ω, with e idempotent, is always deterministic. This may lead the reader to the conjecture that every subset of the form X^ω, where X is a recognizable subset of A^+, is deterministic. However, this conjecture is ruined by the next example.

Example 9.1. Let $X = (a\{b,c\}^* \cup \{b\})^\omega$. This set was already considered in Example I.9.4 in which we gave a Büchi automaton and a Muller automaton recognizing X. We have also seen that this set is not deterministic. The syntactic ω-semigroup of Y has been computed in Example 8.6. In this ω-semigroup, b is the identity, and all the elements are idempotent. The set X can be split into simple elements as follows:

$$\begin{aligned} X &= \varphi^{-1}(a)\varphi^{-1}(b)^\omega \cup \varphi^{-1}(a)^\omega \\ &= b^*a\{a,b,c\}^*b^\omega \cup (b^*a\{a,b,c\}^*)^\omega. \end{aligned}$$

We shall now deduce from the previous formula a new proof of McNaughton theorem.

Theorem 9.3. *Let S be a finite ω-semigroup and let $\varphi : A^\infty \to S$ be a morphism of ω-semigroups. Let X be a subset of A^ω recognized by φ and let P be the image of X in S. Finally, let $F(P)$ be the set of linked pairs (s,e) of S_+ such that $se^\omega \in P$. The following formula hold*

$$X = \bigcup_{(s,e)\in F(P)} \bigcup_{f \mathcal{R} e} X_s X_f^\omega \tag{9.4}$$

and, for each $(s,e) \in F(P)$,

$$\bigcup_{f \mathcal{R} e} X_s X_f^\omega = (\overrightarrow{U}_{s,e} \setminus \overrightarrow{V}_{s,e}) \tag{9.5}$$

where $U_{s,e}$ and $V_{s,e}$ are the subsets of A^+ defined by:

$$U_{s,e} = \bigcup_{f \leq e} X_s P_f \quad \text{and} \quad V_{s,e} = \bigcup_{f < e} X_s P_f$$

In particular, X is a boolean combination of deterministic recognizable subsets.

Proof. Let X' be the right hand side of (9.4). If $x \in X$, there exists by Theorem 2.2 a linked pair (s,e) of S_+ such that $x \in X_s X_e^\omega$. Then we have $\varphi(x) = se^\omega$ and hence $se^\omega \in P$ since $P = \varphi(X)$. This proves the inclusion of X into X'. Conversely, if $x \in X'$, there exists a linked pair (s,e) of elements of S_+ such that $se^\omega \in P$ and an idempotent $f \in E_s$ such that $e \mathcal{R} f$ and $x \in X_s X_f^\omega$. Since $e \mathcal{R} f$, we have $ef = f$ and $fe = e$, whence $sf^\omega = s(ef)^\omega = se(fe)^\omega = see^\omega = se^\omega$. It follows $\varphi(x) = sf^\omega \in P$ and hence $x \in X$, proving (9.4).

To establish (9.5), we observe that $f \mathcal{R} e$ if and only if $f \leq e$ and $f \not< e$. Furthermore, Corollary 2.9 shows that the condition $f < e$ implies $X_s X_e^\omega \cap X_s X_f^\omega = \emptyset$, since the pairs (s,e) and (s,f) cannot be conjugate in this case. We obtain

$$\bigcup_{f \mathcal{R} e} X_s X_f^\omega = \bigcup_{f \leq e} X_s X_f^\omega \setminus \bigcup_{f < e} \bigcup_{g \leq f} X_s X_g^\omega$$

It follows, by Corollary 9.2,

$$\begin{aligned}\bigcup_{f \mathcal{R} e} X_s X_f^\omega &= \bigcup_{f \le e} \overrightarrow{X_s P_f} \setminus \bigcup_{f<e} \bigcup_{g \le f} \overrightarrow{X_s P_g} \\ &= \bigcup_{f \le e} \overrightarrow{X_s P_f} \setminus \bigcup_{f<e} \overrightarrow{X_s P_f} \\ &= \overrightarrow{U}_{s,e} \setminus \overrightarrow{V}_{s,e}\end{aligned}$$

whence the conclusion. □

Formula (9.5) can also be restated in some more convenient ways. One of these variations is the following formula

$$X = \bigcup_{(s,e) \in F(P)} (\overrightarrow{U}_{s,e} \setminus \overrightarrow{W}_e) \tag{9.6}$$

where $W_e = A^* K_e$, with $K_e = J_e \setminus J_e A^+$ and $J_e = \{k \in A^+ \mid \varphi(k) \le_{\mathcal{D}} e\}$. Indeed, we have on one hand $X_s X_e^\omega \cap \overrightarrow{W}_e = \emptyset$ and on the other hand, for each $g < e$, $P_g \subset K_e$.

We shall now use these formula to obtain a characterization of the deterministic subsets. This will lead to a new proof of the Landweber theorem (Theorem I.9.9), which states that, if X is simultaneously recognizable and deterministic, then $X = \overrightarrow{L}$ for some recognizable set L. This follows directly from Formula (9.7) and avoids considering automata with a full state table.

Theorem 9.4. *Let S be a finite ω-semigroup and let $\varphi : A^\infty \to S$ a morphism of ω-semigroups. Let X be a subset of A^ω recognized by φ and let P be the image of X in S.*

(1) *The set X is deterministic if and only if, for each linked pairs (s, e) and (s, f) of S_+ such that $f \le e$ and $se^\omega \in P$, we have $sf^\omega \in P$.*

(2) *The complement of the set X is deterministic if and only if, for each linked pair (s, e) and (s, f) of S_+ such that $e \le f$ and $se^\omega \in P$, we have $sf^\omega \in P$.*

Proof. We have $X = \bigcup_{(s,e) \in F(P)} X_s X_f^\omega$ by (9.4) and, by (9.1),

$$X = \bigcup_{(s,e) \in F(P)} X_s X_e^\omega \subset \bigcup_{(s,e) \in F(P)} \overrightarrow{X_s P_e} \subset \bigcup_{(s,e) \in F(P)} \bigcup_{f \le e} X_s X_f^\omega.$$

If P satisfies the condition of the statement, we have

$$\bigcup_{(s,e) \in F(P)} \bigcup_{f \le e} X_s X_f^\omega = \bigcup_{(s,e) \in F(P)} X_s X_e^\omega$$

and consequently

$$X = \bigcup_{(s,e)\in F(P)} \overrightarrow{X_s P_e} \tag{9.7}$$

Thus X is deterministic.

Conversely, if X is deterministic, $X = \overrightarrow{L}$ for some subset L of A^+. Let (s, e) and (s, f) be two linked pairs such that $f \leq e$ and $se^\omega \in P$, and let $x \in X_s$, $y \in X_e$ and $z \in X_f$. Then we have $xy^\omega \in X_s X_e^\omega \subset X$, and as $X = \overrightarrow{L}$, there exists an integer n_1 and a prefix y_1 of y such that $xy^{n_1} y_1 \in L$. Furthermore, since $f \leq_{\mathcal{R}} e$, we have $ef = f$, and hence $\varphi(xy^{n_1+1}z) = sef = sf = s$. Consequently $xy^{n_1+1}zy^\omega \in X$, thus there exists an integer n_2 and a prefix y_2 of y such that $xy^{n_1+1}zy^{n_2}y_2 \in L$. Therefore $\varphi(xy^{n_1+1}zy^{n^2+1}z) = sef = sf = s$. By induction, we obtain an infinite sequence $n_1, n_2, \cdots$ and a sequence of prefixes $y_1, y_2, \cdots$ of y such that, for each k, $xy^{n_1+1}z \cdots y^{n_{k-1}+1}zy^{n_k}y_k \in L$. The infinite word $u = xy^{n_1+1}zy^{n_2+1}z \cdots$ thus belongs to $\overrightarrow{L}$, and hence to X. On the other hand, u can be factorized as $u = (xy^{n_1+1}z)(y^{n_2+1}z)(y^{n_3+1}z)\cdots$, showing that $u \in X_s X_f^\omega$. Therefore $sf^\omega \in P$. The second part of the statement is easily proved by taking the complement. □

Example 9.2. We return to Example 9.1. The image of X in its syntactic ω-semigroup is the set $P = \{a^\omega\}$. Now, the pairs (a, b) and (a, c) are linked pairs of S_+ since $ab = ac = a$ and we have $c \leq b$ since $bc = c$. But $ab^\omega = a^\omega \in P$, and $ac^\omega \notin P$. Therefore X is not deterministic.

10 Prophetic automata

In this section, we introduce a new type of automata, called prophetic, because in some sense, all the information concerning the future is encoded in the initial state. By definition a *prophetic Büchi automaton* is a co-deterministic, co-complete Büchi automaton. Similarly, a *prophetic transition automaton* is co-deterministic, co-complete transition Büchi automaton.

The main result of this section states that any recognizable set of A^ω can be recognized by a prophetic automaton. This also gives yet another proof of the fact that recognizable sets are closed under complement. We start with a simple characterization.

Proposition 10.1. *Let $\mathcal{A} = (Q, A, E, I, F)$ be a Büchi (resp. transition Büchi) automaton and let, for each $q \in Q$, $L_q = L^\omega(E, q, F)$.*

(1) *$\mathcal{A}$ is co-deterministic if and only if the L_q's are pairwise disjoint.*

(2) *$\mathcal{A}$ is co-complete if and only if $\bigcup_{q\in Q} L_q = A^\omega$.*

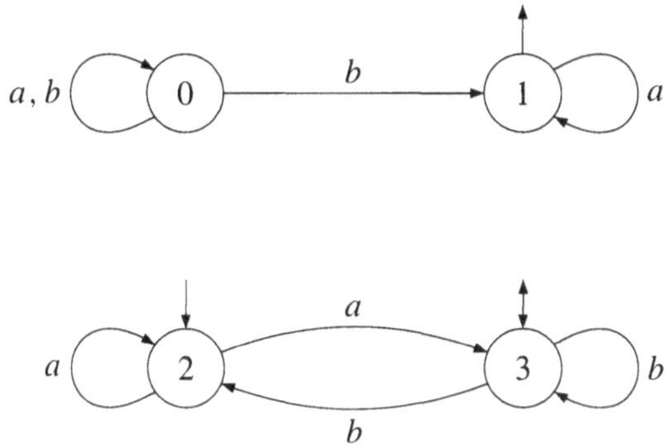

Figure 10.1. A prophetic Büchi automaton.

Proof. (1) If $\mathcal{A}$ is co-deterministic, the L_q's are clearly pairwise disjoint. Suppose that the L_q's are pairwise disjoint and let

$$p_0 \xrightarrow{a_0} p_1 \xrightarrow{a_1} p_2 \cdots \quad \text{and} \quad q_0 \xrightarrow{a_0} q_1 \xrightarrow{a_1} q_2 \cdots$$

be two infinite paths with the same label $u = a_0a_1\cdots$. Then, for each $i \geq 0$, $a_ia_{i+1}\cdots \in L(p_i) \cap L(q_i)$, and thus $p_i = q_i$. Thus $\mathcal{A}$ is co-deterministic.

(2) follows immediately from the definition of co-complete automata. □

Complementation becomes easy with prophetic Büchi automata.

Proposition 10.2. *Let $\mathcal{A} = (Q, A, E, I, F)$ be a prophetic Büchi automaton recognizing a subset X of A^ω. Then the Büchi automaton $(Q, A, E, Q \setminus I, F)$ recognizes the complement of X.*

Theorem 10.3. *Any recognizable set of A^ω can be recognized by a prophetic transition automaton.*

Proof. Let X be a recognizable set of A^ω, let $\varphi : A^\infty \to S$ be the syntactic morphism of X and let $P = \varphi(X)$. Our construction strongly relies on the properties of $\geq_\mathcal{R}$-chains of the semigroup S_+ and requires a few preliminaries.

We shall denote by R the set of all nonempty $>_\mathcal{R}$-chains of S_+:

$$R = \{(s_0, s_1, \ldots, s_n) \mid n \geq 0, s_0, s_1, \ldots, s_n \in S \text{ and } s_0 >_\mathcal{R} s_1 >_\mathcal{R} \cdots >_\mathcal{R} s_n\}$$

In order to convert a $\geq_{\mathcal{R}}$-chain into a strict $>_{\mathcal{R}}$-chain, we introduce the reduction ρ, defined inductively as follows

$$\rho(s) = (s)$$

$$\rho(s_1, \ldots, s_n) = \begin{cases} \rho(s_1, \ldots, s_{n-1}) & \text{if } s_n \mathrel{\mathcal{R}} s_{n-1} \\ (\rho(s_1, \ldots, s_{n-1}), s_n) & \text{if } s_{n-1} >_{\mathcal{R}} s_n \end{cases}$$

In particular, for each finite word $u = a_0 a_1 \cdots a_n$ (where the a_i's are letters), let $\hat{\varphi}(u)$ be the $>_{\mathcal{R}}$-chain $\rho(s_0, s_1, \ldots, s_n)$, where $s_i = \varphi(a_0 a_1 \cdots a_i)$ for $0 \leq i \leq n$. The definition of $\hat{\varphi}$ can be extended to infinite words. Indeed, if $u = a_0 a_1 \cdots$ is an infinite word,

$$s_0 \geq_{\mathcal{R}} s_1 \geq_{\mathcal{R}} s_2 \ldots$$

and since S_+ is finite, there exists an integer n, such that, for all $i, j \geq n$, $s_i \mathrel{\mathcal{R}} s_j$. Then we set $\hat{\varphi}(u) = \hat{\varphi}(a_0 \ldots a_n)$.

Define a left action of A on S_+^1 by setting, for each $a \in A$ and $s \in S_+^1$,

$$a \cdot s = \varphi(a)s$$

We extend this action to a left action of A on R by setting, for each $a \in A$ and $(s_1, \ldots, s_n) \in R$,

$$a \cdot (s_1, \ldots, s_n) = \rho(a \cdot 1, a \cdot s_1, \ldots, a \cdot s_n)$$

To extend this action to A^+, it suffices to apply the following induction rule, where $u \in A^+$ and $a \in A$

$$(ua) \cdot (s_1, \ldots, s_n) = u \cdot (a \cdot (s_1, \ldots, s_n))$$

The connections between this action, φ and $\hat{\varphi}$ are summarized in the next lemma.

Lemma 10.4. *The following formulas hold:*

(1) *For each $u \in A^+$ and $v \in A^\omega$, $u \cdot \varphi(v) = \varphi(uv)$.*

(2) *For each $u, v \in A^+$, $u \cdot \hat{\varphi}(v) = \hat{\varphi}(uv)$.*

Proof. (1) follows directly from the definition of the action and it suffices to establish (2) when u reduces to a single letter a. Let $v = a_0 a_1 \ldots a_n$, where the a_i's are letters and let, for $0 \leq i \leq n$, $s_i = \varphi(a_0 a_1 \ldots a_i)$. Then, by definition, $\hat{\varphi}(v) = \rho(s_0, \ldots, s_n)$ and since, by Proposition A.2.1, the relation $\geq_{\mathcal{R}}$ is stable on the left,

$$a \cdot \hat{\varphi}(v) = \rho(a \cdot 1, a \cdot s_0, a \cdot s_1, \ldots, a \cdot s_n) = \hat{\varphi}(av)$$

which gives (2). □

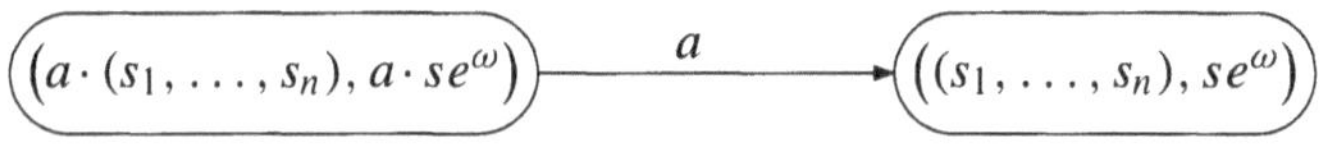

Figure 10.2. A transition of $\mathcal{A}$.

We now define a transition Büchi automaton $\mathcal{A} = (Q, A, E, I, F)$ by setting

$$\begin{aligned}
Q &= \big\{\big((s_1, \ldots, s_n), se^\omega\big) \mid (s_1, \ldots, s_n) \in R, \\
&\qquad (s, e) \text{ is a linked pair of } S_+ \text{ and } s_n \ \mathcal{R}\ s\big\} \\
I &= \big\{\big((s_1, \ldots, s_n), se^\omega\big) \in Q \mid se^\omega \in P\big\} \\
E &= \big\{\big(\big(a \cdot (s_1, \ldots, s_n), a \cdot se^\omega\big), a, \big((s_1, \ldots, s_n), se^\omega\big)\big) \\
&\qquad \big| \ a \in A \text{ and } \big((s_1, \ldots, s_n), se^\omega\big) \in Q\big\}
\end{aligned}$$

A transition $((a \cdot (s_1, \ldots, s_n), a \cdot se^\omega), a, ((s_1, \ldots, s_n), se^\omega))$ is said to be *cutting* if the last two elements of the $\geq_{\mathcal{R}}$-chain $(a \cdot 1, a \cdot s_1, \ldots, a \cdot s_n)$ are $\mathcal{R}$-equivalent.

We choose for F the set of cutting transitions of the form

$$\big(\big(a \cdot (s_1, \ldots, s_n), a \cdot e^\omega\big), a, \big((s_1, \ldots, s_n), e^\omega\big)\big)$$

where e is an idempotent of S_+ such that $s_n \ \mathcal{R}\ e$.

Note that $\mathcal{A}$ has co-deterministic transitions. A typical transition is shown in Figure 10.2. The first part of the proof consists in proving that every infinite word is the label of a final path. Let $u = a_0 a_1 \cdots$ be an infinite word, and let, for each $i \geq 0$, $x_i = a_i a_{i+1} \cdots$ and $q_i = (\hat{\varphi}(x_i), \varphi(x_i))$. Each q_i is a state of Q, and Lemma 10.4 shows that

$$p = q_0 \xrightarrow{a_0} q_1 \xrightarrow{a_1} q_2 \ \cdots$$

is a path of $\mathcal{A}$.

Lemma 10.5. *The path p is final.*

Proof. Let $(u_i)_{i \geq 0}$ be a factorization of u associated with the linked pair (s, e). Then for each $i > 0$, $\varphi(u_i u_{i+1} \cdots) = e^\omega$. Fix some $i > 0$ and let $n_i = |u_0 u_1 \cdots u_i|$. Then $q_{n_i} = ((s_1, \ldots, s_n), e^\omega)$ with $(s_1, \ldots, s_n) = \hat{\varphi}(u_{i+1} u_{i+2} \cdots)$. In particular, $s_n \ \mathcal{R}\ e$ and hence $es_n = s_n$ by Proposition A.2.2. Suppose first that $n \geq 2$. Then $\varphi(u_i)s_{n-1} = es_{n-1} \leq_{\mathcal{R}} e$ and $\varphi(u_i)s_n = es_n = s_n \ \mathcal{R}\ e$. Therefore the relation $\varphi(u_i)s_{n-1} >_{\mathcal{R}} \varphi(u_i)s_n$ does not hold. If $n = 1$, the same argument works by replacing s_{n-1} by 1. It follows that in the path of label u_i from $q_{n_{i-1}}$ to q_{n_i}, at least one of the transitions is cutting. Thus p contains infinitely many cutting transitions and one can select one, say (q, a, q'), that occurs infinitely often. This gives a factorization of the form

$$p = q_0 \xrightarrow{x_0} q \xrightarrow{a} q' \xrightarrow{x_1} q \xrightarrow{a} q' \xrightarrow{x_2} \cdots$$

Up to taking a superfactorization, we can assume, by Proposition 2.3, that for some idempotent f, $\varphi(x_i a) = f$ for every $i > 0$. It follows that the second component of q' is $\varphi(x_i a x_{i+1} a \cdots) = f^\omega$ and thus the transition (q, a, q') is final, which proves the lemma. □

Furthermore, p is successful if and only if $\varphi(u) \in P$, or, equivalently, if $u \in X$. Thus $\mathcal{A}$ recognizes X and is co-complete. It just remains to prove that $\mathcal{A}$ is co-deterministic, which, by Proposition 10.1, will be a consequence of the following lemma.

Lemma 10.6. *Any final path of label u starts at state $(\hat{\varphi}(u), \varphi(u))$.*

Proof. Let p be a final path of label u. Then some final transition, say (q, a, q'), occurs infinitely often in p. Highlighting this transition yields a factorization of p

$$q_0 \xrightarrow{v_0} q \xrightarrow{ⓐ} q' \xrightarrow{v_1} q \xrightarrow{ⓐ} q' \xrightarrow{v_2} \cdots$$

Let $q' = ((s_1, \dots, s_n), e^\omega)$, and consider a factor of the path p labelled by a word of the form $v = v_i a v_{i+1} a \cdots v_j a$, with $i > 0$ and $j - i \geq n$. By the choice of v, $q' = v \cdot q'$, and the first component of q' is obtained by reducing the $\geq_{\mathcal{R}}$-chain

$$\big(\varphi(v[0,0]), \varphi(v[0,1]), \dots, \varphi(v), \varphi(v)s_1, \dots, \varphi(v)s_n\big)$$

Now, since the cutting transition (q, a, q') occurs $n+1$ times in this factor, the last $n+1$ elements of this chain are $\mathcal{R}$-equivalent. It follows that the first component of q' is simply equal to $\hat{\varphi}(v)$.

Consider now a superfactorization $u = w_0 w_1 w_2 \cdots$ obtained by grouping the factors $v_i a$

$$u = \underbrace{(v_0 a \cdots v_{i_0 - 1} a)}_{w_0} \underbrace{(v_{i_0} a \cdots v_{i_1 - 1} a)}_{w_1} \underbrace{(v_{i_1} a \cdots v_{i_2 - 1} a)}_{w_2}$$

in such a way that, for some idempotent f, $\varphi(w_1) = \varphi(w_2) = \cdots = f$. We may also assume that $i_0 > 0$ and $i_1 - i_0 \geq n + 1$. Thus $q' = w_1 \cdot q' = w_1 w_2 \cdot q' = \cdots$, and

$$(s_1, \dots, s_n) = \hat{\varphi}(w_1) = \hat{\varphi}(w_1 w_2) = \cdots = \hat{\varphi}(w_1 w_2 \cdots)$$

It follows in particular $s_n \;\mathcal{R}\; \varphi(w_1) = f$. Furthermore, $s_n \;\mathcal{R}\; e$ since (q, a, q') is a final transition and thus $e \;\mathcal{R}\; f$. Therefore $e^\omega = f^\omega = \varphi(w_1 w_2 \cdots)$. Thus $q' = (\hat{\varphi}(w_1 w_2 \cdots), \varphi(w_1 w_2 \cdots))$ and it follows from Lemma 10.4 that $q_0 = w_0 \cdot q' = (\hat{\varphi}(u), \varphi(u))$. □

As an immediate consequence of Theorem 10.3 and Proposition I.8.1 be obtain the result announced at the beginning of this section.

Corollary 10.7. *Any recognizable set of A^ω can be recognized by a prophetic Büchi automaton.*

The construction given in the proof of Theorem 10.3 is illustrated in the following examples.

Example 10.1. Let $A = \{a, b\}$ and let $X = aA^\omega$. The syntactic ω-semigroup S of X, already computed in Example 8.4 is $S = (S_+, S_\infty)$ where $S_+ = \{a, b\}$, $S_\omega = \{a^\omega, b^\omega\}$, submitted to the following relations

$$aa = a \qquad ab = a \qquad aa^\omega = a^\omega \qquad ab^\omega = a^\omega$$
$$ba = b \qquad bb = b \qquad ba^\omega = b^\omega \qquad bb^\omega = b^\omega$$

The syntactic morphism φ of X is defined by $\varphi(a) = a$ and $\varphi(b) = b$. The transition Büchi automaton associated with φ is shown in Figure 10.3. The final transitions are circled.

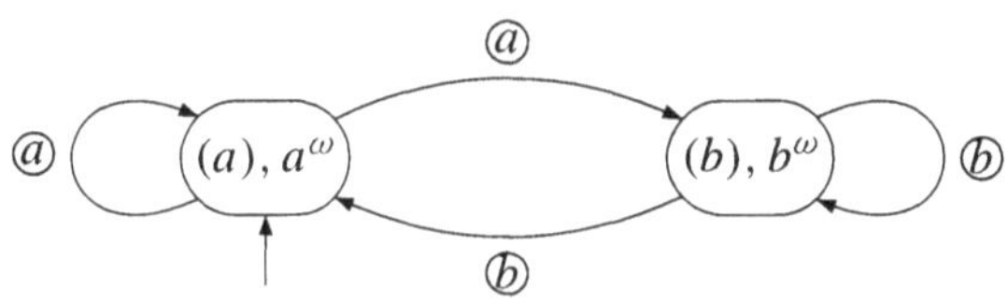

Figure 10.3. The transition Büchi automaton associated with φ.

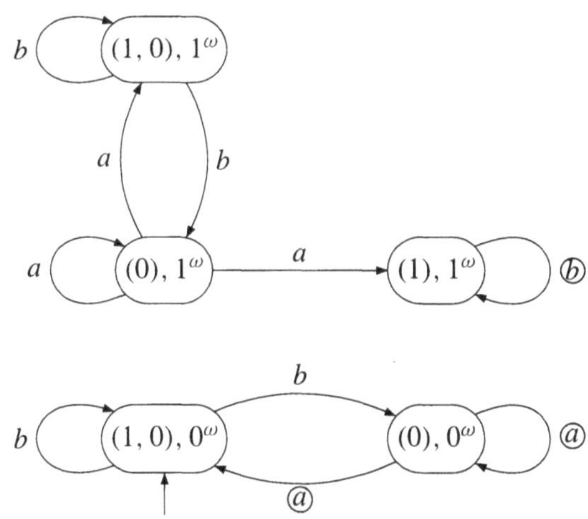

Figure 10.4. The transition Büchi automaton associated with φ.

Example 10.2. Let $A = \{a, b\}$ and let $X = (A^*a)^\omega$. The syntactic ω-semigroup S of X, already computed in Example 8.5 is $S = (S_+, S_\infty)$ where $S_+ = \{1, 0\}$, $S_\omega = \{0^\omega, 1^\omega\}$, submitted to the following relations

$$\begin{array}{llll} 1 \cdot 1 = 1 & 1 \cdot 0 = 0 & 10^\omega = 0^\omega & 11^\omega = 1^\omega \\ 0 \cdot 1 = 0 & 0 \cdot 0 = 0 & 00^\omega = 0^\omega & 01^\omega = 1^\omega \end{array}$$

The syntactic morphism φ of X is defined by $\varphi(a) = 0$ and $\varphi(b) = 1$. The transition Büchi automaton associated with φ is shown in Figure 10.4.

11 Exercises

11.1 Ramseyan factorizations.

Exercise 1. Let S be a finite semigroup and let $\varphi : A^+ \to S$ be a semigroup morphism. Show that for every infinite word $x \in A^\omega$, one of the following two cases arises:

(a) the ultimate $\mathcal{R}$-class of x contains a left zero which is also an element of the ultimate $\mathcal{D}$-class of x,

(b) the word x has a factorization $x = u_0v_0v_1 \cdots$ such that $\varphi(u_0v_0 \cdots v_n) = \varphi(u_0)$ for each $n \geq 0$, and all the elements $\varphi(v_i)$ lie in a subsemigroup strictly included in S.

Deduce by induction on Card(S) that x admits a Ramseyan factorization.

Exercise 2. Let $\varphi : A^+ \to E$ be a map from A^+ into a finite set E. A factorization of a finite word $u \in A^+$ as

$$u = xy_1y_2 \cdots y_nz$$

is said to be *n-Ramseyan factorization* if $\varphi(y_iy_{i+1} \cdots y_{i+j})$ is independent of i and of j. Show that there exists an integer p depending of n and Card(E) such that every word of length greater than or equal to p has an n-Ramseyan factorization.

Exercise 3. Let S be a finite semigroup with k elements and let $\varphi : A^+ \to S$ be a semigroup morphism. Show that every word of length at least $(nk)^k$ has an n-Ramseyan factorization with respect to φ.

Exercise 4. Let x be an infinite recurrent square-free word and on an alphabet A containing at least three letters (the existence of such a word is proved for instance in Lothaire [183]. Let $E = \{0, 1\}$ and let $\varphi : A^+ \to E$ be the map defined by

$$\varphi(w) = \begin{cases} 0 \text{ if } w \text{ is a prefix of } x \\ 1 \text{ otherwise.} \end{cases}$$

Show that there is no element e of E such that $x \in \varphi^{-1}(e)^\omega$.

Exercise 5. Two elements a and b of a semigroup S are said to be *conjugate* if there exist $u, v \in S^1$ such that $a = uv$ and $b = vu$. Show that two idempotents of a finite semigroup are conjugate if and only if they belong to the same $\mathcal{D}$-class.

11.2 Recognition by morphism.

Exercise 6. Let $A = \{a, b\}$ and $L = A^*abA^*$. Build an automaton recognizing L, and then a semigroup recognizing L. Compute the syntactic semigroup S of L and verify that the relation $\mathcal{D}$ in S is the equality. Finally compute the syntactic ordered semigroup S of L and verify that, for each idempotent e of S and for each $s \in S$, $ese \leq e$.

Exercise 7. Let $A = \{a, b, c\}$ and $L = A^*abA^*$. Do the previous exercise again, but verify this time that the syntactic semigroup of L has a nontrivial $\mathcal{D}$-class.

Exercise 8. Let M be a finite monoid and let e be an idempotent of M. Show that $a \leq_{\mathcal{R}} e$, if and only if $ea = a$ and dually, that $a \leq_{\mathcal{L}} e$, if and only if $ae = a$.

11.3 Wilke algebras.

Exercise 9. Let $S = (S_+, S_\omega)$ be a Wilke algebra. Show that for each $a \in S_+$, $aa^\omega = a^\omega$.

11.4 The two modes of recognition.

Exercise 10. Let $A = \{a, b\}$ and $X = A^*abA^\omega$. Build a Büchi automaton recognizing X, and then compute an ω-semigroup recognizing X.

Exercise 11. Write a program that takes as input a finite Büchi automaton $\mathcal{A}$ and produces the finite ω-semigroup associated with $\mathcal{A}$.

11.5 Syntactic congruence.

Exercise 12. Let $A = \mathbb{N}$ and let X be the subset of A^ω consisting of the bounded sequences of integers.

(1) Show that every equivalence on A all of which classes are finite generates a congruence recognizing X.

(2) Deduce from (1) that the upper bound of all the congruences recognizing X does not recognize X.

Exercise 13. Let X be the subset of A^ω consisting of the infinite periodic words.

(1) Show that the only congruence recognizing X is the equality.

(2) Show that for every word $u, v \in A^+$, $u \sim_X v$.

(3) Deduce that the congruence $\sim_X$ is not the upper bound of all the congruences recognizing X.

Exercise 14. Let S be a finite ω-semigroup and let $\varphi : A^\infty \to S$ be a morphism. Let X be a recognizable subset of A^∞ recognized by φ and let P be the image of X in S. Thus, by (7.2),

$$X = \bigcup_{(s,e)\in F} X_s X_e^\omega$$

where F is the set of linked pairs (s, e) of elements of S_+ such that $se^\omega \in P$.

(1) Show that X is countable if and only if each set X_e occurring in the formula is a singleton.

(2) Show that X is finite if and only if each set X_e is a singleton and each set X_s is finite.

(3) Deduce that it is decidable whether a given ω-rational set of infinite words is finite (resp. countable).

11.6 Back to McNaughton's theorem.

Exercise 15. Let S be a finite semigroup and let $\varphi : A^+ \to S$ be a semigroup morphism. Let X be a subset of A^ω recognized by φ and let P be the image of X in S. Show that X is shift invariant if and only if the following condition is satisfied: for each linked pairs (s, e) and (s', e') such that $s \leq_{\mathcal{L}} s'$ and $(s, e) \in P$, we have $(s', e) \in P$.

Exercise 16 (Difficult). A transformation semigroup (P, S) is *fixpoint-free* if, for every $p \in P$, and every $s \in S$, the condition $p \cdot s = p$ implies that s is idempotent. Show that every finite transformation semigroup divides a finite fixpoint-free transformation semigroup.

11.7 Prophetic automata.

Exercise 17. Let A be an alphabet and let $\mathcal{A} = (\{q\}, A, \cdot, q, \{q\})$ be the deterministic automaton whose transition function is defined by $q \cdot a = q$ for every $a \in A$. Show that $\mathcal{A}$ is the unique deterministic and prophetic automaton on the alphabet A.

12 Notes

The historical reference for Ramsey's theorem is [259]. The complete statement is the following: "if the k-subsets of an infinite set E are colored with c colors, there exists an infinite subset F of E whose k-subsets have the same color". The original statement is even more precise: "there exists a function $R(k, c, n)$ such that if the k-subsets of an n-set E are colored with c colors, there exists a $R(k, c, n)$-subset F of E whose k-subsets have the same color". A compactness argument makes it possible to recover the finite case from the infinite case. The key reference for the so-called Ramsey theory is Graham, Rothschild, Spencer [128]. The Ramsey theorem is applied to words with $E = \mathbb{N}$ and $k = 2$. Indeed, one can identify the block $u[i, j]$ of a word u with a two-element subset $\{i, j\}$ of the set of integers. The Ramsey theory for words has numerous applications in automata theory besides the one used in this chapter. For a general exposition, see Lothaire [183], Simon [286]. The reformulation for finite words discussed in Exercises 2 and 3 is from Simon [286]. There are a few variations of the result on finite words, including a lemma of Brown later generalized by Justin, stating that when the coloring is given by a morphism into a semigroup, every infinite word contains an arbitrary large number of consecutive blocks of bounded length of the same color (see Lothaire [183]). A unified presentation and some extensions of these results can be found in Justin, Pirillo [149], and in [287] where Simon has settled a theory of Ramseyan factorization forests.

The equivalence relation introduced in Proposition 2.12 goes back at least to the original article of McNaughton [189]. Theorem 2.14 appeared in Perrin, Schupp [238].

The definition of recognition by semigroup morphisms occurred for the first time in Rabin and Scott [257], where the authors give credit for the idea to Myhill. The introduction of ordered semigroups, due to Pin [244], is much more recent.

The definition of the syntactic congruence of a recognizable subset of A^ω is due to Arnold [11]. Proposition 8.3 is due to Maler and Staiger [185]. The notion of ω-semigroup first appeared in [237]. The notion of Wilke algebra is an original idea of Wilke [355]. It was introduced to study the semigroup varieties associated with sets of infinite words and, in particular, to extend to infinite words Eilenberg's variety theorem which gives a correspondence between sets and syntactic semigroups (see Chapter VI). The notions introduced in Section 6 are a reformulation in terms of ω-semigroup of the original construction of Büchi to recognize the complement [46]. The formulation in terms of ordered ω-semigroup is new.

An extension of the notion of ω-semigroup to countable ordinals was given by Bedon and Carton in [25, 26, 27]. A further extension to countable linear orders, due to Bruyère and Carton, is given in [41].

The proof of McNaughton's theorem given in Section 9 is due to Schützenberger [273]. It is presented here in a simplified form. Another algebraic proof of McNaughton theorem was given by Le Saec, Pin and Weil [175, 176].

The results presented in Section 10 are due to Carton and Michel [61, 62]. The main result (Theorem 10.3) was originally formulated by Michel in the eighties and remained unpublished for a long time. The clarification brought by the published version [61, 62] leads to two distinct proofs, one based on ω-semigroups and the other one based directly on the automaton. Our presentation follows the first proof.

The counterexample given in Exercise 12 is due to Pouzet (personal communication).

III

AUTOMATA AND TOPOLOGY

1 Introduction

The notions of limits, neighborhood and continuity occur naturally in the study of infinite words. We have so far avoided getting into these topological considerations, sometimes at the price of a few complications, notably in the discussion around determinism in Chapter I.

In this chapter, we study in a systematic way the topological aspects of infinite words. This leads us to a presentation of a domain that nowadays belongs jointly to analysis and mathematical logic. This domain is known as *descriptive set theory*.

We give here a presentation of some basic results of this theory, but we limit ourselves to those results that shed light on our main topic, the theory of automata. This leads us to present a theory of *countable automata*, which shares many properties with the theory of finite automata. Countable automata are subject to the restriction that their alphabet as well as their set of states are countable. This restriction is useless for sets of finite words since any set of finite words on a countable alphabet is itself countable and thus can be recognized by a countable automaton. However, in the case of infinite words, it is the case that this restriction leads to a class of sets traditionally known as *analytic* or *Suslin* sets. Unlike the class of recognizable sets, the class of Suslin sets is not closed under complement. This phenomenon is, as might be expected, linked to the inherent ambiguity of the associated automata.

It is remarkable that the class of sets recognized by countable ω-unambiguous automata has another, more constructive, definition. This is indeed the class of *Borel* sets, which can be directly defined as the class obtained from the open sets by applying iteratively all the countable boolean operations. This result constitutes a sort of equivalent of Kleene's theorem for countable automata. This is a consequence of another fundamental result, the *separation theorem*, which states that a set X and its complement are both Suslin if and only if X is Borel. This statement is of course reminiscent of the situation encountered in the theory of *recursivity*. Indeed, *effective* descriptive set theory allows to develop both theories in the same framework.

As we shall see, the topological point of view sheds light on the theory developed in the previous chapters, by emphasizing the role played by the boolean operations in general and the complement in particular. It will also be an opportunity to take advantage of another view-point which consists of considering an automaton as the representation of a continuous map.

The plan of this chapter is the following. In the introduction, we give a brief, but self-contained, overview of the elements of topology used in the sequel. This presentation is not intended to be an introduction to topology suitable for a reader discovering open and closed sets for the first time. It can, however, serve as a reminder to the reader whose knowledge of the domain is somewhat rusty. In particular, Section 2.5 contains the definition and basic properties of Borel sets.

In Section 3, we give a characterization, due to H. Landweber, of deterministic sets as countable intersections of open sets, a class that forms the second level of the finite Borel hierarchy (Proposition 3.11). We also introduce the Borel automata and the Suslin sets. The closure properties of the class of Suslin sets are established (to sum up: all but complementation). We also establish the precise link between automata and continuous functions (Corollary 6.6). This allows us to give an alternative definition of the Suslin sets: they are the continuous images of open sets.

In Section 7, we prove the two main results of this chapter: the separation theorem (Theorem 7.1) and the characterization of Borel sets as those recognized by countable ω-unambiguous automata or, equivalently, as the image of a closed set under a continuous bijection (Theorem 7.4 and Corollary 7.6).

2 Topological spaces

2.1 Countable sets. A set E is said to be *countable* if there exists an injective map from E into $\mathbb{N}$. In other words, a set is countable if one can associate with each of its elements an integer key that suffices to identify them. Such a restriction seems to be rather weak but the next proposition, that makes use of a so-called *diagonal argument*, shows this is not the case and that there exist very simple sets that are not countable.

Proposition 2.1. *The set of infinite words on an alphabet A with at least two letters is not countable.*

Proof. Indeed, suppose the contrary and let $f : A^{\omega} \to \mathbb{N}$ an injective map. Consider an infinite word x such that for each $n \geq 0$,

$$x_n \neq \left(f^{-1}(n)\right)_n$$

Here we use our hypothesis that A contains at least two elements. Let $m = \varphi(x)$. We have

$$x_m \neq \left(f^{-1}(m)\right)_m = x_m$$

whence the contradiction. □

Two basic operations preserve the countable sets: the *finite product* and the *countable union*. Proposition 2.1 shows that, on the contrary, a countable product of countable sets is not countable in general.

Proposition 2.2. *Every finite product of countable sets is countable.*

Proof. By induction, it suffices to treat the case of the product of two countable sets E and F. Let $f : E \to \mathbb{N}$ and $g : F \to \mathbb{N}$ be two injective functions. The function $f \times g : E \times F \to \mathbb{N} \times \mathbb{N}$ that maps (x, y) onto $(f(x), g(y))$ is injective. Furthermore, there exist bijections from $\mathbb{N} \times \mathbb{N}$ onto $\mathbb{N}$. One of them is given by the "crab" walk on $\mathbb{N} \times \mathbb{N}$ drawn in Figure 2.1. Thus $E \times F$ is countable. □

Proposition 2.3. *The union of any countable family of countable sets is countable.*

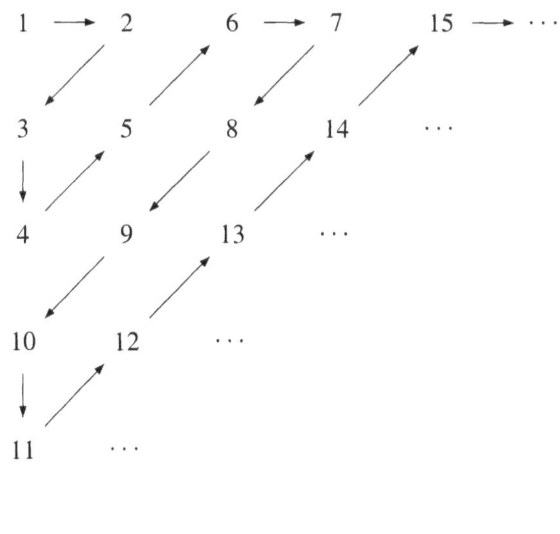

Figure 2.1. A bijection from $\mathbb{N} \times \mathbb{N}$ onto $\mathbb{N}$.

Proof. Let $(E_n)_{n\geq 0}$ be a family of countable sets and let E be their union. We may assume that the E_n's are pairwise disjoint. Then there exists, for each $n \geq 0$, an injective map $f_n : E_n \to \mathbb{N}$. The function $f : E \to \mathbb{N} \times \mathbb{N}$ defined by $f(x) = (n, f_n(x))$ if $x \in E_n$ is an injection from E into $\mathbb{N} \times \mathbb{N}$. Since $\mathbb{N} \times \mathbb{N}$ is countable, so is E. □

2.2 General topology. Recall that a *topology* on a set E is a set $\mathcal{T}$ of subsets of E satisfying the following conditions:

(i) $\emptyset$ and E are in $\mathcal{T}$,

(ii) $\mathcal{T}$ is closed under arbitrary union: if $(X_i)_{i\in I}$ is a family of elements of $\mathcal{T}$, then $\bigcup_{i\in I} X_i \in \mathcal{T}$,

(iii) F is closed under finite intersection: if $(X_i)_{1\leq i\leq n}$ is a finite sequence of elements of $\mathcal{T}$ then $\bigcap_{1\leq i\leq n} X_i \in \mathcal{T}$.

Condition (i) is actually superfluous, since we have by definition

$$\bigcup_{i\in\emptyset} X_i = \emptyset \quad \text{and} \quad \bigcap_{i\in\emptyset} X_i = E$$

The elements of $\mathcal{T}$ are called the *open sets*. The complement of an open set is called a *closed set*.

A *topological space* is a set E together with a topology on E. A topology $\mathcal{T}_1$ on a set is a *refinement* of a topology $\mathcal{T}_2$ on the same set, if the inclusion $\mathcal{T}_1 \subset \mathcal{T}_2$ holds, that is, if each open set for $\mathcal{T}_1$ is also an open set for $\mathcal{T}_2$. One also says that $\mathcal{T}_1$ is *coarser* than $\mathcal{T}_2$ or that $\mathcal{T}_2$ is *stronger* than $\mathcal{T}_1$.

The coarsest topology on E is the *trivial* topology, which reduces to the empty set and E. The strongest topology is the *discrete* topology defined by $\mathcal{T} = \mathcal{P}(E)$.

If F is a subset of a topological space $(E, \mathcal{T})$, then the traces $F \cap X$ for $X \in \mathcal{T}$, define a topology on F, called the *relative topology*.

It is sometimes convenient to give a *basis* for a topology on E. This is a collection $\mathcal{B}$ of open subsets of E such that every open set is the union of elements of $\mathcal{B}$. An equivalent condition is that $\mathcal{B}$ satisfies the two following conditions:

(i) E is the union of all the elements of $\mathcal{B}$,

(ii) every finite intersection of elements of $\mathcal{B}$ is a union of elements of $\mathcal{B}$.

The open sets of the topology generated by $\mathcal{B}$ are by definition the arbitrary unions of elements of $\mathcal{B}$.

A map from a topological space into another one is *continuous* if the inverse image of each open set is an open set. It is an *homeomorphism* if it is a continuous bijection and the inverse bijection is also continuous. Two topological spaces are *homeomorphic* is there is an homeomorphism between them.

Let $(E_i, \mathcal{T}_i)_{i\in I}$ be a family of topological spaces, and let $E = \prod_{i\in I} E_i$ be the Cartesian product of the E_i's. Denote by π_i the natural projection from E onto E_i, defined by $\pi_i((e_j)_{j\in I}) = e_i$. The *product topology* on E is the topology generated by the basis consisting of the finite intersections of sets of the form $\pi_i^{-1}(X_i)$ where X_i is an open set of E_i. These sets are nothing else than the products $\prod_{i \in I} X_i$ where the X_i's are open sets of E_i and where $X_i = E_i$ except for a finite number of indices. The natural projections $\pi_i : E \to E_i$ are then continuous and a mapping $\varphi : F \to E$ is continuous if and only if the mappings $\pi_i \circ \varphi : F \to E_i$, for $i \in I$, are all continuous.

A topological space $(E, \mathcal{T})$ is *Hausdorff* if for each $u, v \in E$ with $u \neq v$, there exist *disjoint* open sets U and V such that $u \in U$ and $v \in V$. If f is a continuous map from a topological space X into an Hausdorff space Y, then the graph of f is closed in $X \times Y$.

An important notion is that of *compact* set. A family of open sets $(U_i)_{i \in I}$ is said to *cover* a topological space $(E, \mathcal{T})$ if $E = \bigcup_{i \in I} U_i$. A topological space $(E, \mathcal{T})$ is said to be *compact* if it is Hausdorff and if, for each family of open sets covering E, there exists a finite subfamily that still covers E.

If X is a closed set of a compact space E, then X, equipped with the relative topology, is also a compact space. We shall use freely the well-known result that every product of compact spaces is compact (Tychonov's theorem).

We now recall the standard notion of *converging sequence*: a sequence $(x_n)_{n \geq 0}$ of elements of a topological space $(E, \mathcal{T})$ is converging to a *limit* x if every open set containing x contains all but a finite number of the x_n's. If the space is Hausdorff, the limit is unique and denoted by $\lim_{n \to \infty} x_n$. A sequence is said to be *converging* if it admits a limit.

A weaker notion is that of *adherence point*. An element x of E is an adherence point of the sequence $(x_n)_{n \geq 0}$ if, for each open set U containing x, there exists infinitely many n such that $x_n \in U$. More generally, an element x of E is an *adherence point* to a subset F of E if and only if each open set containing x meets F. The set of adherence points to F is called the *closure* of F, denoted by $\overline{F}$. It is also the intersection of the closed sets containing F. A subset of E is *dense* if its closure is equal to E.

One can show that if $f : E \to F$ is a continuous map from a topological space E into a topological Hausdorff space F, the image of a compact under f is still compact. However the image of an open (resp. closed) set under a continuous map is not necessarily open (resp. closed), even if the map is injective. A map $f : E \to F$ is said to be *open* (resp. *closed*) if, for every open (resp. closed) set X of E, the image $f(X)$ of X is an open (resp. closed) set of F. Compact sets admit the following characterization in terms of closed maps: a topological space E is compact if and only if, for every topological space F, the projection $\pi : E \times F \to F$ is closed.

2.3 Metric spaces. A *metric* d on a set E is a map $d : E \to \mathbb{R}_+$ from E into the set of nonnegative real numbers satisfying the three following conditions, for every $x, y, z \in E$:

(i) $d(x, y) = 0$ if and only if $x = y$,

(ii) $d(y, x) = d(x, y)$,

(iii) $d(x, z) \leq d(x, y) + d(y, z)$.

A *metric space* is a set E together with a metric d on E. The topology defined by d is obtained by taking as a basis the *open ε-balls* defined for $x \in E$ and $\varepsilon > 0$ by

$$B(x, \varepsilon) = \{y \in E \mid d(x, y) < \varepsilon\}$$

A topological space is said to be *metrizable* if its topology can be defined by a metric. Such a metric is said to be *compatible* with the topology. Every metrizable space is Hausdorff. Indeed, given two distinct elements x and y, the open balls $B(x, \varepsilon)$ and $B(y, \varepsilon)$ are disjoint if ε is set to be half the distance between x and y.

A *Cauchy sequence* in a metric space (E, d) is a sequence $(x_n)_{n \geq 0}$ of elements of E such that for each $\varepsilon > 0$, there exists a integer k such that for each $n \geq k$ and $m \geq k$, $d(x_n, x_m) < \varepsilon$. Every converging sequence is a Cauchy sequence, but the converse does not hold in general. A metric space in which each Cauchy sequence is convergent is said to be *complete*. Every closed subset of a complete space is complete. A topological space is said to be *completely metrizable* if its topology can be defined by a metric for which it is complete. Such a metric is said to be *completely compatible* with the topology. Note that a metric space (E, d) might be completely metrizable even if it is not complete.

The product of any family of metrizable spaces is not in general metrizable. However, the following result holds.

Proposition 2.4. *The product of any countable family of metrizable (resp. completely metrizable) spaces is a metrizable (resp. completely metrizable) space.*

Proof. Let $(E_n)_{n \geq 0}$ be a family of metrizable spaces and let E be the product of the spaces E_n. Let d_n be a compatible metric for E_n. We may assume that d_n is bounded by 1, for otherwise, we may replace d_n by the metric d'_n given by $d'_n(x, y) = \min(1, d_n(x, y))$, which defines the same topology. Then the function $d : E \to \mathbb{R}_+$ given by $d((x_n), (y_n)) = \sum_{n=0}^{\infty} 2^{-n} d_n(x_n, y_n)$ is a metric that defines the product topology.

In the completely metrizable case, we may assume that (E_n, d_n) is complete, and in that case, (E_n, d'_n) is also complete. Let $(x^{(m)})_{m \geq 0}$ be a Cauchy sequence in E, with $x^{(m)} = (x_n^{(m)})_{n \geq 0}$. Then, for each $n \geq 0$, the sequence $(x_n^{(m)})_{m \geq 0}$ is Cauchy and hence converges to some $x_n \in E_n$. It follows that $(x^{(m)})$ converges to $x = (x_n)_{n \geq 0}$. □

If X is subset of a metrizable space E, the two following conditions are equivalent:

(i) X is compact,

(ii) every infinite sequence of elements of X admits an adherence point.

A subset X of a metric space E is said to be *relatively compact* if it is contained in a compact subset of E. It is equivalent to saying that the closure of X is compact, or yet that each sequence of elements of X, admits a converging subsequence.

A subset X of a metric space is said to be *bounded* if there exists a number $\delta > 0$ such that for each $x, y \in X$, $d(x, y) < \delta$. Every compact subset of a metric space is closed and bounded, but the converse is not true in general.

Let (E, d) and (E', d') be two metric spaces. A function φ from E into E' is said to be *uniformly continuous* if for each $\varepsilon > 0$, there exists $\delta > 0$ such that the relation $d(x, y) < \delta$ implies $d'(\varphi(x), \varphi(y)) < \varepsilon$. If φ is uniformly continuous, the image under φ of a Cauchy sequence of E is a Cauchy sequence of E'. We say that φ is a *uniform homeomorphism* if it is a uniformly continuous bijection and φ^{-1} is also uniformly continuous. Two metric spaces are *uniformly homeomorphic* if there is a uniform homeomorphism between them.

We say that φ is an *isometry* if it is a bijection from E onto E' such that, for each $x, y \in E$,

$$d(x, y) = d'(\varphi(x), \varphi(y))$$

Isometries are particular cases of Lipschitz functions. A function φ from E into E' is said to be *Lipschitz* of modulus k if, for each $x, y \in E$,

$$d'(\varphi(x), \varphi(y)) \leq kd(x, y)$$

A function is said to be *Lipschitz* if it is Lipschitz of modulus k for some $k > 0$. Every Lipschitz function is uniformly continuous, but the converse is not true in general.

The *completion* of a metric space E is a complete metric space $\hat{E}$ together with an isometric embedding of E as a dense subspace of $\hat{E}$. One can prove that every metric space admits a completion, which is unique up to isometric equivalence: if $\hat{E}_1$ and $\hat{E}_2$ are two completions of E, there exists an isometry of $\hat{E}_1$ onto $\hat{E}_2$, whose restriction to E is the identity. If φ is a uniformly continuous function from (E, d) in a metric space (E', d'), φ admits a uniformly continuous extension $\hat{\varphi} : \hat{E} \to \hat{E}'$ and this extension is unique.

We conclude this section with a property connecting the open sets and the closed sets of a metric space.

Proposition 2.5. *In a metric space, each closed set is a countable intersection of open sets and each open set is a countable union of closed sets.*

Proof. Let F be a closed set of the metric space (E, d). Set, for each $n > 0$,

$$\omega_n = \left\{ y \in E \mid \text{there exists } x \in F \text{ such that } d(x, y) < \tfrac{1}{n} \right\}$$

Since $\omega_n = \bigcup_{x \in F} B(x, 1/n)$, each ω_n is an open set. Moreover, if $x \in \bigcap_{n>0} \omega_n$, there exists a sequence of elements of F such that $d(x, x_n) < 1/n$. It follows that the sequence x_n converges to x and, since F is closed, we have $x \in F$. Therefore $F = \bigcap_{n>0} \omega_n$, which proves the first part of the statement. The second part can be obtained by arguing on the complement. □

2.4 Polish spaces. A *Polish space* is a completely metrizable space which admits a countable basis of open sets. For instance, the set of real numbers, equipped with the usual topology, is a Polish space, since the open balls of the form $B(x, r)$, where x and r are rational numbers, form a countable basis of the topology. We shall see later (Proposition 3.4) that if A is countable, then A^ω is a Polish space. The metrizable spaces with a countable basis of open sets admit a simple characterization.

Proposition 2.6. *A metrizable space has a countable basis of open sets if and only if it contains a countable dense subset.*

Proof. If $(U_n)_{n\in\mathbb{N}}$ is a countable basis of open sets of a metrizable space E, and if u_n is an element of U_n, then the set of the u_n forms a dense subset of E.

Conversely, let D be a countable dense subset of a metric space E. Then the finite intersections of balls of the form $B(x, 1/n)$, with $x \in D$ and $n > 0$, form a countable basis of open sets. □

The notion of Polish space is quite robust, in the sense that it is preserved under several natural operations. We start with countable product and countable intersection.

Proposition 2.7.

(1) *The product of any countable family of Polish spaces is Polish.*

(2) *Any countable intersection of open subsets of a Polish space is Polish.*

Proof.

(1) Let E be the product of a countable family $(E_n)_{n\geq 0}$ of Polish spaces. By Proposition 2.4, E is completely metrizable. Furthermore, if $\mathcal{B}_n$ is a countable basis for the topology of E_n, the topology of E is generated by the countable basis consisting of the finite intersections of open sets of the form $\Pi_{n\geq 0} X_n$, where $X_n = E_n$ except for a finite number of indices, for which $X_n \in \mathcal{B}_n$. Therefore E is Polish.

(2) Let E be a Polish space and let d be a completely compatible metric for E. Let $(X_n)_{n\geq 0}$ be a sequence of open subsets of E and let $X = \bigcap_{n\geq 0} X_n$. Let F_n be the complement of X_n. Define a new metric on X, by letting

$$d'(x, y) = d(x, y) + \sum_{n\geq 0} \min\left\{2^{-n-1}, \left|\frac{1}{d(x, F_n)} - \frac{1}{d(y, F_n)}\right|\right\}$$

We let the reader verify that d' is a metric compatible with the topology of X. Let $(x_n)_{n\geq 0}$ be a Cauchy sequence in (X, d'). Then it is Cauchy in (X, d), and hence converges to some $x \in X$. Furthermore, for any fixed $n \geq 0$, the sequence $(\frac{1}{d(x_m, F_n)})_{m\geq 0}$, is Cauchy and hence converges in $\mathbb{R}$. Therefore, there is a real number $r_n > 0$ such that, for m large enough, $d(x_m, F_n) > r_n$. Since $\lim_{m\to\infty} d(x_m, F_n) = d(x, F_n)$, we have $d(x, F_n) > 0$ for all n, whence $x \in X$. It follows that $(x_n)_{n\geq 0}$ converges to x in (X, d').

Finally, by the definition of an induced topology, it is clear that any subspace of a Polish space admits a countable basis of open sets. Thus X is Polish. □

Other properties will be proved later on in this chapter.

2.5 Borel sets. Let E be a topological space. The class of *Borel sets* of E is the smallest class of sets containing the open sets and closed under countable union and complementation.

Denote by X^c the complement of a subset X of E. The De Morgan law

$$\Big(\bigcap_{n\geq 0} X_n\Big)^c = \bigcup_{n\geq 0} X_n^c$$

shows that the class of Borel sets of E is also closed under countable intersection. If E is a metric space, it is possible to give a definition of Borel sets without any reference to complementation.

Proposition 2.8. *Let (E, d) be a metric space. The class of Borel sets of E is the smallest class of subsets of E containing the open sets (resp. the closed sets) which is closed under countable union and countable intersection.*

Proof. Denote by $\mathcal{F}$ the smallest class of subsets of E containing the open sets which is closed under countable union and countable intersection. It follows from the observation preceding the proposition that every element of $\mathcal{F}$ is a Borel set.

Denote by $\mathcal{C}$ the class of subsets X of E such that $X \in \mathcal{F}$ and $X^c \in \mathcal{F}$. Then $\mathcal{C}$ is contained in $\mathcal{F}$ and is closed under complementation. Let us show that $\mathcal{C}$ is closed under countable union: if $X_n \in \mathcal{C}$ for each $n \geq 0$, we have by definition $X_n \in \mathcal{F}$ and $X_n^c \in \mathcal{F}$. It follows

$$\bigcup_{n\geq 0} X_n \in \mathcal{F} \quad \text{and} \quad \Big(\bigcup_{n\geq 0} X_n\Big)^c = \bigcap_{n\geq 0} X_n^c \in \mathcal{F}$$

since $\mathcal{F}$ is closed under countable union and countable intersection. Finally $\mathcal{C}$ contains the open sets. Indeed, if X is open, we have $X \in \mathcal{F}$ by hypothesis and $X^c \in \mathcal{F}$ by Proposition 2.5. Therefore $\mathcal{C}$ contains the Borel sets of E.

A similar proof would work with closed sets instead of open sets. □

By alternating the use of the complementation and of the union, one can define a hierarchy, called the *finite Borel hierarchy*, in the following way. Let Σ_1 be the set of open sets of E. Define by induction for $n \geq 1$, the *Borel classes of finite order* of E by

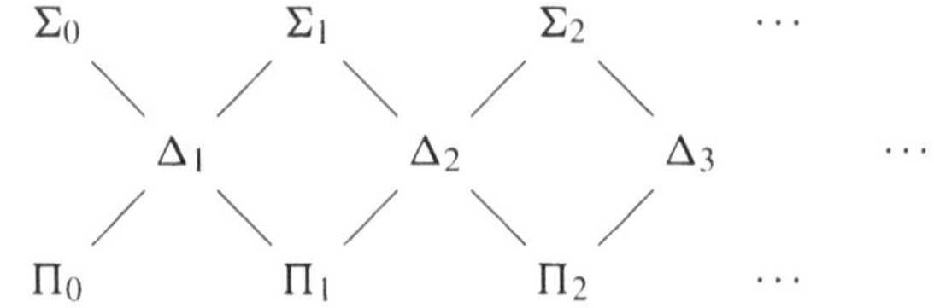

Figure 2.2. The finite Borel hierarchy.

setting

$$\Pi_n = \{X^c \mid X \in \Sigma_n\}$$
$$\Sigma_{n+1} = \Big\{\bigcup_{k\geq 0} X_k \mid X_k \in \Pi_n\Big\}$$
$$\Delta_n = \Sigma_n \cap \Pi_n$$

Thus Π_1 is the class of closed sets, Δ_1 consists of the sets which are simultaneously open and closed (sometimes called the *clopen* sets), Σ_2 is the set of countable unions of closed sets and Π_2 is the set of countable intersections of open sets. More generally, Π_{n+1} is the closure of Σ_n under countable intersection.

It is convenient to use the following quantifier notation to manipulate these classes. For a class Γ of subsets of E, we set

$$\neg\Gamma = \{X^c \mid X \in \Gamma\}$$
$$\exists^\omega\Gamma = \Big\{\bigcup_{k\geq 0} X_k \mid X_k \in \Gamma\Big\}$$
$$\forall^\omega\Gamma = \Big\{\bigcap_{k\geq 0} X_k \mid X_k \in \Gamma\Big\}.$$

The class $\neg\Gamma$ is called the *dual class* of Γ. A class is called *self-dual* if it is closed under complement, i.e. if $\Gamma = \neg\Gamma$.
With these notations, the Borel classes are defined by the condensed formulas

$$\Pi_n = \neg\Sigma_n, \qquad \Sigma_{n+1} = \exists^\omega\Pi_n$$

If we need to make explicit the underlying topological space E, we use instead the notations $\Pi_n(E)$, $\Sigma_n(E)$, $\Delta_n(E)$.

Proposition 2.9. *In a metric space, the following inclusions hold:*

$$\Delta_1 \subset \Sigma_1 \subset \Delta_2 \subset \Sigma_2 \subset \dots$$
$$\Delta_1 \subset \Pi_1 \subset \Delta_2 \subset \Pi_2 \subset \dots$$

Proof. The inclusion $\Sigma_n \subset \Pi_{n+1}$ is almost trivial since Π_{n+1} is the closure of Σ_n under countable intersection. Proposition 2.5 shows that $\Sigma_1 \subset \Sigma_2$. Supposing by induction that $\Sigma_n \subset \Sigma_{n+1}$, we obtain

$$\Sigma_{n+1} = \exists^\omega \neg \Sigma_n \subset \exists^\omega \neg \Sigma_{n+1} = \Sigma_{n+2}$$

thereby proving that $\Sigma_n \subset \Sigma_{n+1}$ for each $n > 0$. □

The inclusions stated in Proposition 2.9 may fail in an arbitrary topological space (Exercise 1).

Borel classes are stable under inverse image by a continuous map.

Proposition 2.10. *Let $\varphi : E \to F$ be a continuous map. If X is a Borel subset of F (resp. in Σ_n, Π_n, Δ_n) then so is $\varphi^{-1}(X)$.*

Proof. If Γ is a class of subsets of F, denote by $\varphi^{-1}(\Gamma)$ the set $\{\varphi^{-1}(X) \mid X \in \Gamma\}$. The proposition holds for Σ_1 by definition of a continuous function. The formulas

$$\begin{aligned} \varphi^{-1}(\neg\Gamma) &= \neg\varphi^{-1}(\Gamma) \\ \varphi^{-1}(\exists^\omega\Gamma) &= \exists^\omega\varphi^{-1}(\Gamma) \end{aligned} \tag{2.1}$$

give a proof for the Borel classes of finite order.

Finally, let Γ be the class of Borel subsets X of F such that $\varphi^{-1}(X)$ is Borel. It was shown previously that Γ contains Σ_1 and (2.1) shows that Γ is closed under countable union and complementation. Therefore Γ contains the Borel sets and thus, if X is a Borel set, so is $\varphi^{-1}(X)$. □

However, only the Π_k classes are stable under countable product.

Proposition 2.11. *Every countable product of Borel (resp. Π_k) sets is Borel (resp. Π_k).*

Proof. Let $(E_n)_{n \geq 0}$ be a family of nonempty metric spaces and let E be the product of the E_n's. Let, for each $n \geq 0$, X_n be a set of E_n and let X be the product of the X_n's. Since the class of Borel sets (resp. Π_k) is closed under countable intersection, the formula

$$\prod_{n \geq 0} X_n = \bigcap_{n \geq 0} E_0 \times E_1 \times \cdots \times E_{n-1} \times X_n \times E_{n+1} \times \cdots$$

reduces the problem to the case where only one of the X_n's is different from E_n. But now since each projection

$$\pi_n : \prod_{n \geq 0} E_n \to E_n$$

is continuous by definition of the product topology, if X_n is Borel (resp. Π_k), so are the sets

$$E_0 \times E_1 \times \cdots \times E_{n-1} \times X_n \times E_{n+1} \times \cdots = \pi^{-1}(X_n) \qquad \square$$

Proposition 2.11 does not generalize to the classes Σ_k, and not even to open sets. Consider for instance the discrete sets $E_n = \{a, b\}$. Each set $X_n = \{a\}$ is an open set of E_n but the product of these sets, which can be identified with the subset $\{a^\omega\}$ of $\{a, b\}^\omega$, is not open.

3 The space of infinite words

Let A be an alphabet. There are several equivalent ways to define a natural topology on A^ω. The first method consists in considering A as a discrete space, that is a topological space in which every subset is open — and hence also closed. Then A^ω is equipped with the product topology. We shall refer implicitly to this topology by speaking of *the topological space* A^ω. We shall give two other equivalent definitions of the same topology, one obtained by describing directly the open sets, the other one defined by a metric. Depending on the context, it can be more appropriate to use one or the other of these three equivalent definitions.

3.1 The topology of A^ω. Let us first describe the open sets.

Proposition 3.1. *The open sets of A^ω are the sets of the form XA^ω with $X \subset A^*$. They also are the sets of the form XA^ω where X is a prefix subset of A^*.*

Proof. By definition, the sets of the form

$$A^{n_1}a_1A^{n_2}a_2\cdots A^{n_k}a_kA^\omega \qquad \text{with } k \geq 0, n_1, \ldots, n_k \geq 0$$

form a basis of open sets for the product topology. These sets are indeed of the form XA^ω and, conversely, each set of the form XA^ω is union of such sets since

$$XA^\omega = \bigcup_{a_1 \ldots a_k \in X} a_1 \ldots a_k A^\omega$$

Therefore, the sets of the form XA^ω form a basis of open sets, but since this base is closed under arbitrary union, it contains all the open sets.

The second part of the statement follows from the formula

$$XA^\omega = (X \setminus XA^+)A^\omega,$$

since a set of the form $X \setminus XA^+$ is prefix by construction. $\square$

Since A is a discrete space, A is metrizable: it suffices to define a metric δ on A by setting

$$\delta(a,b) = \begin{cases} 0 & \text{if } a = b \\ 1 & \text{if } a \neq b \end{cases}$$

Proposition 2.4 now shows that A^ω is metrizable. The next proposition provides a suitable metric for this metrizable space.

Proposition 3.2. *The topology on A^ω is given by the metric d defined by*

$$d(u,v) = 2^{-r} \quad \textit{where} \quad r = \min\{n \mid u(n) \neq v(n)\}$$

with the usual conventions $\min \emptyset = +\infty$ *and* $2^{-\infty} = 0$.

Proof. On one hand, the formula

$$B(u, 2^{-k}) = \{v \in A^\omega \mid d(u,v) < 2^{-k}\} = u[0,k]A^\omega$$

shows that, for each $u \in A^\omega$ and each $k \geq 0$, every open ball of d is open in the product topology. On the other hand, the formulas

$$XA^\omega = \bigcup_{x \in X} xA^\omega \qquad \text{and} \qquad xA^\omega = B(x, 2^{-|x|+1})$$

show that every open set of the product topology is union of open balls for d. Therefore the two topologies coincide. □

Note that $d(u,v) \leq 2^{-k}$ if and only if the k first letters of u and v coincide. So, intuitively, the closer are two words for this metric, the longest is their common prefix. For instance we obtain, by underlying the first letter for which the words differ

$$d(aba\underline{a}b^\omega, aba\underline{b}ba^\omega) = 2^{-3},$$
$$d(\underline{a}ba^\omega, (\underline{b}a)^\omega) = 2^{-0},$$
$$d(aba^\omega, aba^\omega) = 2^{-\infty} = 0.$$

Proposition 3.3. *For each alphabet A, the metric space (A^ω, d) is complete.*

Proof. It follows directly from Proposition 2.4, since (A, δ) is a complete metric space. □

Proposition 3.4. *For each countable alphabet A, A^ω is a Polish space.*

Proof. Since A is countable, the open balls of the form $B(a, 1/2)$, with $a \in A$, form a countable basis of the topology of (A, δ). Therefore A is Polish, and A^ω is also Polish by Proposition 2.7. □

3.2 Closed sets. Proposition 3.1 provides a combinatorial description of the open sets of A^ω. In the same spirit, one can give a characterization of the closure of a subset of A^ω.

Proposition 3.5. *Let X be a subset of A^ω and let $u \in A^\omega$. Then $u \in \overline{X}$ if and only if every prefix of u is a prefix of a word of X.*

Proof. We have $u \in \overline{X}$ if and only if, for each $n \geq 0$, there exists $v \in X$ such that $d(u, v) < 2^{-n}$. This amounts to saying that, for each $n \geq 0$, $u[0, n]$ is prefix of a word of X. □

Example 3.1.

(1) If B is a subset of A, B^ω is a closed subset of A^ω.

(2) If X is a set of words of length n of A^+, the set XA^ω is closed. Indeed, we have $A^\omega \setminus XA^\omega = (A^n \setminus X)A^\omega$, showing that XA^ω is the complement of an open set.

(3) If B and C are two subsets of A, the set $E = B^+C^\omega \cup B^\omega$ is a closed subset of A^ω. Indeed, let u be an infinite word each prefix of which is a prefix of a word of E, which amounts to saying that all the prefixes of u are elements of B^+C^*. Let $M = \sup \{n \mid u(n) \in B\}$. If $M = \infty$, then $u \in B^\omega$. If $M < \infty$, then necessarily $u(n) \in C$ for $n > M$, and hence $u \in B^+C^\omega$. In both cases, $u \in E$.

As was announced in Chapter I, it is possible to characterize the closed subsets of A^ω in terms of automata. We first need a lemma

Lemma 3.6. *Let X be a subset of A^ω. The following conditions are equivalent:*

(1) *there exists a subset P of A^* such that X is the set of infinite words all of which prefixes are in P,*

(2) *there exists a prefix-closed subset P of A^* such that X is the set of infinite words all of which prefixes are in P,*

(3) *there exists a subset R of A^* such that X is the set of infinite words having no prefixes in R,*

(4) *there exists a prefix-closed subset P of A^* such that $X = \overrightarrow{P}$,*

(5) $X = \overrightarrow{Pref(X)}$.

Proof. (1) implies (2). Setting $P' = A^* \setminus (A^* \setminus P)A^*$, we see that P' is a prefix-closed subset of P. Furthermore, if all the prefixes of an infinite word u are in P, all its prefixes are in P'. Suppose indeed that u has a prefix $x \notin P'$. Then $x \in (A^* \setminus P)A^*$ and thus x has a factor in $A^* \setminus P$, a contradiction.

(2) implies (3). Put $R = A^* \setminus P$. By (1), all the prefixes of a word of X are in P and therefore, none of these prefixes belong to R. Conversely, a word with no prefix in R has all its prefixes in P. Thus X is the set of infinite words having no prefixes in R.

(3) implies (1). A similar proof would show that if X is the set of infinite words having no prefixes in R, then X is the set of infinite words all of which prefixes are in P, where $P = A^* \setminus RA^*$.

(2) and (4) are equivalent. Indeed, it amounts to showing that if P is a prefix-closed subset of A^*, then every infinite word u having infinitely many prefixes in P has all its prefixes in P. Now, by hypothesis, the length of the prefixes of u that are in P has no upper bound. Thus if v is a prefix of u, it is prefix of another prefix p of u which is in P. As P is prefix-closed, v is itself in P.

(5) implies (4) since $\mathit{Pref}(X)$ is a prefix-closed subset of A^*.

(4) implies (5). Suppose that $X = \overrightarrow{P}$, where P is prefix-closed. Then if $u \in \mathit{Pref}(X)$, there exists v such that $uv \in X$, and thus u is prefix of some word in P. It follows that $u \in P$, since P is prefix-closed. Thus $\mathit{Pref}(X) \subset P$ whence $X \subset \overrightarrow{\mathit{Pref}(X)} \subset \overrightarrow{P} = X$ and therefore $X = \overrightarrow{\mathit{Pref}(X)}$. □

Proposition 3.7. *Let X be a subset of A^ω. The following conditions are equivalent:*

(1) *X is closed,*

(2) *X is recognized by a deterministic Büchi automaton in which each state is final,*

(3) *every infinite word, each of which prefix is a prefix of a word of X, belongs to X,*

(4) *$X = \overrightarrow{\mathit{Pref}(X)}$.*

Proof. The equivalence of (1) and (3) follows from Proposition 3.5.

(3) implies (4). The inclusion $X \subset \overrightarrow{\mathit{Pref}(X)}$ always holds. Assume that X satisfies (3) and let $u \in \overrightarrow{\mathit{Pref}(X)}$. Then each prefix of u is a prefix of some word of X and thus $u \in X$. This proves (4).

(4) implies (1). By Lemma 3.6, there exists a subset R of A^* such that X is the set of infinite words with no prefix in R. This amounts to saying that the complement of X in A^ω is of the form RA^ω, that is, an open set. Thus X is closed.

(2) implies (4). Let $\mathcal{A} = (Q, A, \cdot, i, Q)$ be a deterministic Büchi automaton in which each state is final recognizing X. Then $L^*(\mathcal{A}) = \mathit{Pref}(X)$ and thus $X = \overrightarrow{\mathit{Pref}(X)}$ by Proposition I.6.1.

(4) implies (2). Let $P = \mathit{Pref}(X)$ and let $\mathcal{A} = (Q, A, \cdot, i, F)$ be the minimal (deterministic) automaton of P. Since P is prefix-closed, $F = Q$ and $X = L^\omega(\mathcal{A})$ by Proposition I.6.1. □

Example 3.2.

(1) For each word $u \in A^\omega$, the set $\{u\}$ consisting of the unique infinite word u is closed. It is recognizable if and only if u is ultimately periodic.

(2) Let $x, y \in A^*$. The set of infinite words in which x is a prefix and y is not a factor is a recognizable closed set.

The hypothesis "deterministic" in condition (2) is crucial. However, it can be omitted in the case of recognizable sets. As a preparation to this result, let us state a proposition of independent interest, which is a special case of a more general result (cf. Proposition 5.1).

Proposition 3.8. *Let $\mathcal{A} = (Q, A, E, I, F)$ be a finite Büchi automaton. The set of infinite initial paths of $\mathcal{A}$ is a closed subset of E^ω. The set of labels of these paths is a closed subset of A^ω.*

Proof. Let $I(\mathcal{A})$ be the set of infinite initial paths of $\mathcal{A}$ and let $L(I)$ be the set of their labels. The two sets are connected by the relation $L(I) = \lambda_{\mathcal{A}}(I(\mathcal{A}))$, where $\lambda_{\mathcal{A}}$ is the function from $I(\mathcal{A})$ into A^ω which maps an infinite path onto its label. Observe that

$$I(\mathcal{A}) = SE^\omega \setminus E^*FE^\omega$$

where

$$S = \{(q, a, q') \in E \mid q \in I\}$$
$$F = \{ee' \mid e \text{ and } e' \text{ are two nonconsecutive transitions of } E\}.$$

It follows immediately that $I(\mathcal{A})$ is a closed subset of E^ω. Since E is finite, E^ω is compact and hence $I(\mathcal{A})$ is also compact. Finally, $\lambda_{\mathcal{A}}$ is continuous. Indeed, if two infinite paths have the same r first transitions, their labels have the same prefixes of length r. Therefore, $L(I)$ is the continuous image of a compact set in an Hausdorff space, and hence is also compact. Therefore, $L(I)$ is a closed subset of A^ω. □

Proposition 3.9. *Let X be a recognizable subset of A^ω. The following conditions are equivalent:*

(1) *X is closed,*

(2) *if X is recognized by a finite Büchi automaton $\mathcal{A} = (Q, A, E, I, F)$, X is also recognized by the automaton (Q, A, E, I, C) where C is the set of coaccessible states of $\mathcal{A}$,*

(3) *X is recognized by a finite Büchi automaton in which each state is final,*

(4) *X is recognized by a finite deterministic Büchi automaton in which each state is final.*

Proof. (1) implies (2). Suppose that X is recognized by a finite Büchi automaton $\mathcal{A} = (Q, A, E, I, F)$. If $u \in X$, u is the label of an initial path visiting infinitely often some final state q. Then q is coaccessible and hence $u \in L^\omega(E, I, C)$.

Conversely, let $u = a_0 a_1 \cdots \in L^\omega(E, I, C)$. Then there exists an infinite path

$$(q_0, a_0, q_1)(q_1, a_1, q_2) \cdots$$

such that $q_0 \in I$ and an infinite sequence $n_0 < n_1 < n_2 < \ldots$ such that $q_{n_0} = q_{n_1} = \ldots = q \in C$. Since q is coaccessible, there exists an infinite path starting from q and visiting infinitely often a final state. Let v be the label of this path. Then the words

$$u_k = a_0 a_1 \cdots a_{n_k - 1}$$

are prefixes of u such that $u_k v \in X$. But the sequence $(u_k v)_{k \geq 0}$ converges to u, and as X is closed, we have $u \in X$. Consequently $X = L^\omega(E, I, C)$.

(2) implies (3). By Proposition I.5.2, X is recognized by a finite Büchi automaton $\mathcal{A} = (Q, A, E, I, F)$ in which each state is coaccessible. By (2), X is also recognized by the automaton (Q, A, E, I, Q) in which each state is final.

(3) implies (1). Let $\mathcal{A}$ be a finite Büchi automaton recognizing X, in which each state is final. Then X is the set of labels of the initial paths of $\mathcal{A}$, and hence is closed by Proposition 3.8.

It follows that conditions (1), (2) and (3) are equivalent.

(4) implies (1) follows from Proposition 3.7.

(3) implies (4). By hypothesis, X is recognized by a finite Büchi automaton

$$\mathcal{A} = (Q, A, E, I, Q)$$

in which each state is final. Let C be the set of coaccessible states of $\mathcal{A}$. Let $T = L^+(E, I, C)$. We claim that T is the set of finite prefixes of words of X. Indeed, if $u \in T$, there exists an initial path ending in $q \in C$ and labeled by u. Since q is coaccessible, there exists an infinite path starting from q. If v is the label of this path, we have $uv \in X$, and hence u is a prefix of a word of X. Conversely, if $v = a_0 a_1 \cdots \in X$, there exists an initial path $p = (q_0, a_0, q_1)(q_1, a_1, q_2) \cdots$ with label v. Therefore, for each $n > 0$, $v_n = a_0 a_1 \cdots a_{n-1}$ is the label of an initial path ending in q_n. Since q_n is coaccessible by construction, $v_n \in T$ and each prefix of v is in T.

Therefore T is a recognizable subset of A^*. Thus there exists a deterministic Büchi automaton $\mathcal{A}' = (Q', A, E', I', F')$ such that $T = L^*(\mathcal{A}')$. Since T is closed under prefixes, we may assume that $F' = Q'$. Furthermore, since X is closed by (1), Proposition 3.7 shows that $X = \overrightarrow{T}$. Finally, we have $L^\omega(\mathcal{A}') = \overrightarrow{(L^*(\mathcal{A}'))} = \overrightarrow{T} = X$. This establishes (4). □

Proposition 3.9 gives an algorithm for deciding whether a given recognizable subset of A^ω is closed.

3.3 Clopen sets. We have described the open sets and the closed sets of A^ω. We shall now describe the clopen sets.

Proposition 3.10. *If A is a finite alphabet, a subset of A^ω is clopen if and only if it is of the form XA^ω, where X is a finite subset of A^*.*

Proof. The subsets of A^ω of the form XA^ω, where X is a finite subset of A^*, are open by Proposition 3.1 and are also closed, for if n is an upper bound of the length of the words of X, then

$$A^\omega \setminus XA^\omega = (A^n \setminus XA^*)A^\omega$$

Conversely, let Y be a clopen subset of A^ω. By Proposition 3.1, there exists a prefix subset X of A^* such that $Y = XA^\omega$. Let us show that X is finite. This is obvious if A is empty. Otherwise, let us choose a letter a of A. If X is infinite, it contains, by compactness of A^ω, an infinite sequence of words x_n of X of increasing length such that the sequence $x_n a^\omega$ converges to an element x of A^ω. But since $x_n a^\omega \in XA^\omega = Y$, and Y is closed by hypothesis, we have $x \in Y$. Then there exists a word $p \in X$ which is a prefix of x and hence of all but a finite number of the x_n's, contradicting the fact that X is prefix. Therefore X is finite. □

When A is infinite, the subsets of A^ω of the form XA^ω, where X is a set of words of bounded length of A^*, are clopen. Indeed, let X be a set of words of A^* of length bounded by an integer n. Then XA^ω is an open set by Proposition 3.1 and closed by the formula

$$A^\omega \setminus YA^\omega = (A^n \setminus YA^*)A^\omega$$

There exist however some clopen sets that are not of this form. For instance, if $A = \{a_n \mid n \geq 0\}$, the set XA^ω, where $X = \{a_n^n \mid n > 0\}$ is clopen.

3.4 The second level of the Borel hierarchy. We now consider the second level of the Borel hierarchy. Let us first give some examples.

Example 3.3. The set $X = (a^*b)^\omega$, consisting of the infinite words on $\{a, b\}$ with an infinite number of b is in Π_2. Indeed, we have

$$X = \bigcap_{n \geq 0} X_n$$

where X_n is the open set of words having at least n occurrences of b.

Example 3.4. The set

$$X = \{xy^\omega \mid x, y \in A^+\}$$

of all ultimately periodic words is in Σ_2 since

$$X = \bigcup_{x \in A^+} \bigcup_{y \in A^+} \{xy^\omega\}$$

Example 3.5. The set $X = (a+b)^*a^\omega$ of the words with a finite number of b is in Σ_2 but not in Π_2. Indeed, suppose it is in Π_2, that is, can be written in the form $X = \bigcap_{n\geq 0} X_n A^\omega$. Since $a^\omega \in X$, there exists an integer n_0 such that $a^{n_0} \in X_0$. Next, as $a^{n_0}ba^\omega \in X$, there exists an integer n_1 such that $a^{n_0}ba^{n_1} \in X_1$. We obtain in this way an infinite word $a^{n_0}ba^{n_1}b\cdots$ which is in X although it contains infinitely many b's, a contradiction. This example, that was already considered in Example I.6.2, shows that Δ_2 is strictly included into Σ_2 and hence also into Π_2.

The class Π_2 of countable intersections of open sets can be also defined by using the operator $\overrightarrow{L}$ introduced in Chapter I. Recall that for a subset L of A^*, $\overrightarrow{L}$ is the set of infinite words having infinitely many prefixes in L. The following result extends Theorem I.6.2.

Theorem 3.11. *Let X a be subset of A^ω. Then the following conditions are equivalent:*

(1) *There exists a subset L of A^+ such that $X = \overrightarrow{L}$,*

(2) *X is a countable intersection of open sets,*

(3) *X is recognizable by a deterministic Büchi automaton.*

Proof. (1) implies (2). Put, for each $n \geq 0$, $L_n = \{u \in L \mid |u| \geq n\}$. Since the sets $L_n A^\omega$ are open, it suffices to establish the formula $\overrightarrow{L} = \bigcap_{n\geq 0} L_n A^\omega$. First of all, if $u \in \overrightarrow{L}$, then for each $n \geq 0$, u has a prefix of length greater than or equal to n in L, and $u \in L_n A^\omega$. Conversely, if $u \in L_n A^\omega$, u has a prefix in L_n. Consequently, if $u \in \bigcap_{n\geq 0} L_n A^\omega$, u has infinitely many prefixes in L and $u \in \overrightarrow{L}$.

(2) implies (1). Suppose that $X = \bigcap_{n\geq 0} U_n$, where the U_n's are open sets of A^ω. Put $V_n = \bigcap_{0\leq i\leq n} U_i$. The V_n's are still open sets and we have $X = \bigcap_{n\geq 0} V_n$ and $V_{n+1} \subset V_n$ for each $n \geq 0$. By Proposition 3.1, we have $V_n = L_n A^\omega$ with L_n a prefix subset of A^+. Since $L_n A^\omega = L_n A^n A^\omega$, we also have $X = \bigcap_{n\geq 0} L_n A^n A^\omega$. We claim that $X = \overrightarrow{L}$, where $L = \bigcup_{n\geq 0} L_n A^n$. First of all, if $x \in X$, then $x \in L_n A^n A^\omega$ for each $n \geq 0$, and hence x has a prefix p_n in $L_n A^n$. Since $|p_n| \geq n$, x has infinitely many prefixes in L and hence $x \in \overrightarrow{L}$.

Conversely, if $x \in \overrightarrow{L}$, x has infinitely many prefixes in L, say $p_0, p_1, \ldots$ with $p_0 < p_1 < p_2 < \ldots$. Then there exists for each i an index n_i such that $p_i \in L_{n_i} A^{n_i}$. But since $L_n A^n$ is a prefix set for all $n \geq 0$, one cannot have $p_i, p_j \in L_{n_i} A^{n_i}$ if $i < j$. It follows that $x \in L_n A^n A^\omega = V_n$ for infinitely many values of n. Since the V_n are decreasing, we have $x \in \bigcap_{n\geq 0} V_n = X$.

The equivalence of (1) and (3) was already established in Theorem I.6.2. □

Let us add two remarks on this result. First, it implies that the class of sets recognized by a deterministic Büchi automaton is closed under countable intersection. The direct construction of an automaton to recognize the intersection leads to another proof of this theorem (see Exercise 20). Next, since every recognizable subset of A^ω is, by McNaughton's theorem, a boolean combination of sets of the form $\overrightarrow{X}$, the class of recognizable sets is contained in Δ_3. It is a remarkable fact that the level of the recognizable sets in the Borel hierarchy is bounded.

3.5 Compactness. We now analyze the properties related to compactness. Recall that π_n denotes the projection from A^ω onto A defined by $\pi_n(u) = u(n)$.

Proposition 3.12. *A subset X of A^ω is relatively compact if and only if, for each $n \geq 0$, $\pi_n(X)$ is finite.*

Proof. If for each $k \geq 0$, $\pi_k(X)$ is finite, hence compact, then X is contained in the compact space $\prod_k \pi_k(X)$. On the contrary, if there exists an index k such that $\pi_k(X)$ is infinite, there exists an infinite sequence a_n of pairwise distinct letters of A, and a sequence u_n of words of X such that $\pi_k(u_n) = a_n$. It follows, for each $p, q \geq 0$, $d(u_p, u_q) \geq 2^{-n}$, and hence the sequence u_n has no adherence point. □

Corollary 3.13. *The space A^ω is compact if and only if A is finite.*

Another corollary of Proposition 3.12 is König's lemma stated in Chapter I (cf. Exercise 6).

In the case of a binary alphabet $A = \{0, 1\}$, the space A^ω is often called the *Cantor set*. This denomination refers to an identification of $\{0, 1\}^\omega$ with a subset of the interval $[0, 1]$ of real numbers obtained as follows. With each word $u = u_1 u_2 \cdots \in \{0, 1\}^\omega$ is associated the number x whose development in base 3 is u:

$$x = \sum_{n \geq 1} u_n 3^{-n}$$

or in ternary notation

$$x = 0.u_1 u_2 \cdots$$

This defines an injective mapping from $\{0, 1\}^\omega$ onto a subset C of the interval $[0, 1]$. It is easy to see that this mapping is also continuous and thus it identifies the space $\{0, 1\}^\omega$ with C. One can visualize the Cantor set by removing successively the median subsets in an infinite succession of trichotomies as is shown in Figure 3.1. This figure emphasizes the unusual aspect of the spaces A^ω from a topological point of view. In fact, every space of the form A^ω, with A finite, is uniformly homeomorphic to the Cantor space.

Proposition 3.14.

(1) *If A and B are two finite alphabets with at least two letters, the metric spaces A^ω and B^ω are uniformly homeomorphic.*

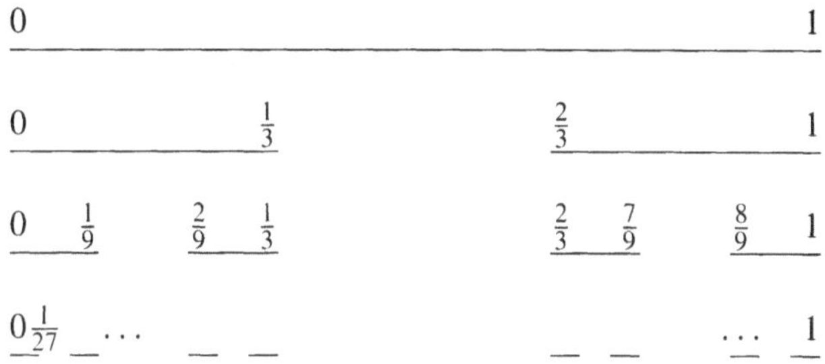

Figure 3.1. The Cantor set.

(2) *If A is a countable alphabet, there is surjective, uniformly continuous map from $\mathbb{N}^\omega$ onto A^ω, but A^ω and $\mathbb{N}^\omega$ are not homeomorphic if A is finite.*

(3) *For each alphabet A, the topological spaces A^ω and $A^\omega \times A^\omega$ are homeomorphic.*

(4) *For each countable alphabet A, the topological spaces N^ω and $A^\omega \times N^\omega$ are homeomorphic.*

Proof. (1) We may clearly assume that $B = \{0, 1\}$ and $A = \{a_1, \ldots, a_k\}$. Let r be the unique integer such that $2^r \leq k < 2^{r+1}$. We build a subset P of B^* such that

(i) $|P| = |A|$,

(ii) P is prefix,

(iii) every word of B^ω has a prefix in P,

(iv) every word of P has length either r or $r + 1$.

To define P, select a partition $S \cup T$ of B^r such that $|S| = 2^{r+1} - k$ and $|T| = k - 2^r$, and put $P = S\{0, 1\} \cup T$. Figure 3.2 illustrates this construction for $k = 6$. Let now f be a bijection from A onto P, extended to a map from A^ω into B^ω by setting $f(u_0u_1 \cdots) = f(u_0)f(u_1) \cdots$. Then f is injective since P is prefix and is onto by

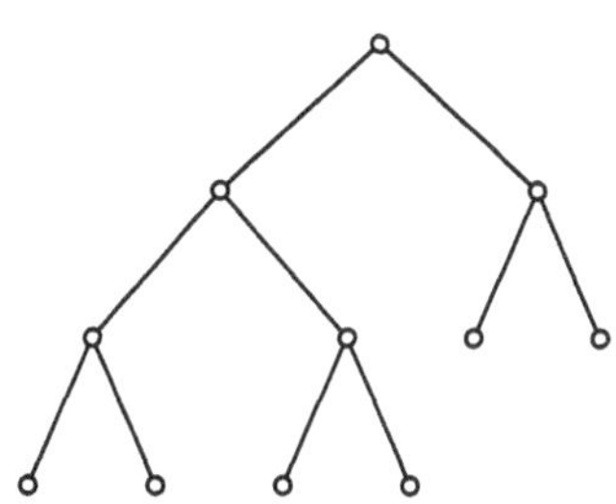

Figure 3.2. The prefix set P, for $|A| = 6$.

(iii). Furthermore, f and f^{-1} are uniformly continuous, since

$$(d(u,v))^r \le d(f(u), f(v)) \le 2^{r-1}(d(u,v))^r$$

(2) The result is trivial if $A = \mathbb{N}$. Assume that $A = \{a_1, \ldots, a_k\}$, and let f be the map from $\mathbb{N}$ onto A defined by $f(n) = \min\{n, k\}$. Then f extends in a natural way to a surjective, uniformly continuous map from $\mathbb{N}^\omega$ onto A^ω. However, if A is finite, Corollary 3.13 shows that A^ω is compact but $\mathbb{N}^\omega$ is not. Therefore, A^ω and $\mathbb{N}^\omega$ are not homeomorphic in that case.

(3) Let $g : A^\omega \times A^\omega \to A^\omega$ be the bijection defined by $g(u, v) = u_0 v_0 u_1 v_1 \cdots$. Then the relations

$$d^2((u,v),(u',v')) \le d(g(u,v), g(u',v')) \le 2d^2((u,v),(u',v'))$$

show that g is a uniform homeomorphism.

(4) By (2), there is a surjective, uniformly continuous map f from $\mathbb{N}^\omega$ onto A^ω. It follows that the map $x \to (x, f(x))$ is an homeomorphism from $\mathbb{N}^\omega$ onto $A^\omega \times N^\omega$. □

3.6 The Borel hierarchy of A^ω. The main result of this section states that, if A contains at least two letters, the Borel hierarchy of A^ω is strict. The proof of this result requires a few auxiliary tools, namely the Polish classes and the universal sets, that are introduced below. We first fix some terminology and notation concerning binary relations.

If $\mathcal{R}$ is a binary relation over E and F, that is, if $\mathcal{R}$ is a subset of $E \times F$, we adopt the following notations, where $x \in E$, $y \in F$, $X \subset E$ and $Y \subset F$:

$$\mathcal{R}(x) = \{y \in F \mid (x,y) \in \mathcal{R}\}$$

$$\mathcal{R}(X) = \bigcup_{x \in X} \mathcal{R}(x) = \{y \in F \mid (x,y) \in \mathcal{R} \text{ for some } x \in X\}$$

$$\mathcal{R}^{-1}(y) = \{x \in E \mid (x,y) \in \mathcal{R}\}$$

$$\mathcal{R}^{-1}(Y) = \bigcup_{y \in Y} \mathcal{R}^{-1}(y) = \{x \in E \mid \mathcal{R}(x) \cap Y \neq \emptyset\}$$

We let the reader verify the following elementary formulas, in which, as usual, X^c denotes the complement of X (we assume that the set in which the complement is taken is implicitly given by the context).

(1) $\mathcal{R}^c(x) = (\mathcal{R}(x))^c$
(2) $\mathcal{R}(\bigcup_{i \in I} X_i) = \bigcup_{i \in I} \mathcal{R}(X_i)$
(3) $\mathcal{R}^{-1}(\bigcup_{i \in I} X_i) = \bigcup_{i \in I} \mathcal{R}^{-1}(X_i)$
(4) $\mathcal{R}^{-1}(\bigcap_{i \in I} X_i) = \bigcap_{i \in I} \mathcal{R}^{-1}(X_i)$
(5) $\mathcal{R}^{-1}(X^c) = (\mathcal{R}^{-1}(X))^c$

A *Polish class* is a correspondence Γ which associates with each Polish space E a collection $\Gamma(E)$ of subsets of E such that, if $\varphi : E \to F$ is a continuous map, $X \in \Gamma(F)$ implies $\varphi^{-1}(X) \in \Gamma(E)$. Proposition 2.10 shows that the class of Borel sets and the classes Σ_n, Π_n, Δ_n are Polish classes.

Let Γ be a Polish class, let E be a Polish space and let A be an alphabet. A subset $\mathcal{U}$ of $A^\omega \times E$ is said to be A-*universal* for $\Gamma(E)$ if $\mathcal{U} \in \Gamma(A^\omega \times E)$ and

$$\Gamma(E) = \{\mathcal{U}(y) \mid y \in A^\omega\}$$

The next theorem shows in particular that if there is a universal set for the class of open sets, then there is also one for each of the classes Σ_n and Π_n of the Borel hierarchy.

Theorem 3.15. *Let Γ be a Polish class. If there is an A-universal set for a class $\Gamma(E)$, there is also one for the classes $\neg\Gamma(E)$ and $\exists^\omega\Gamma(E)$.*

Proof. Let $\mathcal{U}$ be an A-universal class for $\Gamma(E)$. Then $\mathcal{U}^c$ is an A-universal class for $\neg\Gamma(E)$, since

$$\neg\Gamma(E) = \{(\mathcal{U}(y))^c \mid y \in A^\omega\} = \{\mathcal{U}^c(y) \mid y \in A^\omega\}$$

Let $(n, m) \to \langle n, m\rangle$ be a fixed bijection from $\mathbb{N} \times \mathbb{N}$ onto $\mathbb{N}$. For each $y \in A^\omega$, define $\bar{y}_n \in A^\omega$ by $\bar{y}_n(m) = y(\langle n, m\rangle)$. We claim that the map $y \to (\bar{y}_n)_{n\in\mathbb{N}}$ defines a uniformly continuous bijection from A^ω onto $(A^\omega)^\omega$. First, the map is bijective, since for any sequence $(y_n)_{n\in\mathbb{N}}$ of elements of A^ω, there is $y \in A^\omega$ such that $\bar{y}_n = y_n$ for all $n \in \mathbb{N}$. Furthermore, given $k > 0$, let

$$K = \max\{\langle i, j\rangle \mid i \leq k, j \leq k\}$$

Let now x and y be two elements of A^ω. If $d(x, y) \leq 2^{-K+1}$, that is, if $x(0) = y(0), \ldots, x(K) = y(K)$, then $x(\langle i, j\rangle) = y(\langle i, j\rangle)$ for $0 \leq i, j \leq k$. It follows that $d(\bar{x}_0, \bar{y}_0) \leq 2^{-(k+1)}, \ldots, d(\bar{x}_k, \bar{y}_k) \leq 2^{-(k+1)}$, whence

$$d((\bar{x}_n)_{n\in\mathbb{N}}, (\bar{y}_n)_{n\in\mathbb{N}}) = \sum_{n=0}^{\infty} 2^{-n} d_n(\bar{x}_n, \bar{y}_n) \leq 2^{-k}$$

which proves uniform continuity.

We now show that the set

$$\exists^\omega\mathcal{U} = \{(y, x) \in A^\omega \times E \mid \exists n \in \mathbb{N}\ (\bar{y}_n, x) \in \mathcal{U}\}$$

is A-universal for $\exists^\omega\Gamma(E)$. First

$$\exists^\omega\mathcal{U} = \bigcup_{n\in\mathbb{N}} \{(y, x) \in A^\omega \times E \mid (\bar{y}_n, x) \in \mathcal{U}\}$$

and since each map $y \to \bar{y}_n$ is continuous, each set

$$(y, x) \in A^\omega \times E \mid (\bar{y}_n, x) \in \mathcal{U}\}$$

belongs to $\Gamma(A^\omega \times E)$. Therefore $\exists^\omega \mathcal{U}$ belongs to $\exists^\omega \Gamma(E)$. Furthermore, we have

$$\begin{aligned}
\exists^\omega \Gamma(E) &= \Big\{ \bigcup_{n\in\mathbb{N}} X_n \mid X_n \in \Gamma(E) \Big\} = \Big\{ \bigcup_{n\in\mathbb{N}} \mathcal{U}(y_n) \mid (y_n)_{n\in\mathbb{N}} \in (A^\omega)^\omega \Big\} \\
&= \Big\{ \bigcup_{n\in\mathbb{N}} \mathcal{U}(\bar{y}_n) \mid y \in A^\omega \Big\} = \Big\{ \bigcup_{n\in\mathbb{N}} \{x \in E \mid (\bar{y}_n, x) \in \mathcal{U}\} \mid y \in A^\omega \Big\} \\
&= \big\{ \{x \in E \mid \exists\, n \in \mathbb{N}\ (\bar{y}_n, x) \in \mathcal{U}\} \mid y \in A^\omega \big\} = \big\{ \exists^\omega \mathcal{U}(y) \mid y \in A^\omega \big\}
\end{aligned}$$

showing that $\exists^\omega \mathcal{U}$ is universal for $\exists^\omega \Gamma(E)$. □

We first establish a generic result on universal sets.

Proposition 3.16. *Let A and B be two countable alphabets, and assume that B has at least two letters. Then, for each $n > 0$, there is a B-universal set for the classes $\Sigma_n(A^\omega)$ and $\Pi_n(A^\omega)$.*

Proof. By Theorem 3.15, it suffices to establish the result for $\Sigma_1(A^\omega)$. Let 0 and 1 be two distinct letters of B and let $(u_n)_{n\geq 0}$ be an enumeration of A^*. Then the set

$$\mathcal{U} = \{(y, x) \in B^\omega \times A^\omega \mid \exists n\ x \in u_n A^\omega \text{ and } y(n) = 1\}$$

is open and universal for $\Sigma_1(A^\omega)$, since if $X \subset A^\omega$, then $XA^\omega = \mathcal{U}(y)$, with

$$y(n) = \begin{cases} 1 & \text{if } u_n \in X \\ 0 & \text{otherwise.} \end{cases} \quad □$$

The next proposition shows that the complement of an universal set for a given topological class never belongs to the same class. It is based on a remarkable diagonal argument.

Proposition 3.17. *If $\mathcal{U}$ is A-universal for $\Gamma(A^\omega)$, then $\mathcal{U}^c \notin \Gamma(A^\omega \times A^\omega)$.*

Proof. Assume that $\mathcal{U}^c \in \Gamma(A^\omega \times A^\omega)$. Setting

$$X = \{x \in A^\omega \mid (x, x) \notin \mathcal{U}\}$$

we have $X = \varphi^{-1}(\mathcal{U}^c)$, where φ is the continuous map from A^ω into $A^\omega \times A^\omega$ defined by $\varphi(x) = (x, x)$. Therefore $X \in \Gamma(A^\omega)$ and thus $X = \mathcal{U}(y)$ for some $y \in A^\omega$, that is,

$$X = \{x \in A^\omega \mid (y, x) \in \mathcal{U}\}$$

and the sequence of equivalences

$$(y, y) \in \mathcal{U} \iff y \in X \iff (y, y) \notin \mathcal{U}$$

leads to a contradiction. □

Corollary 3.18. *Let Γ be a self-dual class. Then there cannot be an A-universal set for $\Gamma(A^\omega)$.*

We now show that the Borel hierarchy of A^ω is strict.

Theorem 3.19. *If A is a countable alphabet with at least two letters, $\Sigma_n(A^\omega)$ is not equal to $\Pi_n(A^\omega)$. Therefore, $\Delta_n(A^\omega) \subsetneq \Sigma_n(A^\omega) \subsetneq \Delta_{n+1}(A^\omega)$ and $\Delta_n(A^\omega) \subsetneq \Pi_n(A^\omega) \subsetneq \Delta_{n+1}(A^\omega)$.*

Proof. By Proposition 3.16, there exists an A-universal set $\mathcal{U}$ for $\Sigma_n(A^\omega)$. Thus $\mathcal{U}^c \in \Pi_n(A^\omega \times A^\omega)$, but, by Proposition 3.17, $\mathcal{U}^c \notin \Sigma_n(A^\omega \times A^\omega)$. The inequality $\Sigma_n(A^\omega) \neq \Pi_n(A^\omega)$ follows, since, by Proposition 3.14, the spaces A^ω and $A^\omega \times A^\omega$ are homeomorphic.

In particular, $\Delta_n(A^\omega) = \Sigma_n(A^\omega) \cap \Pi_n(A^\omega)$ is strictly included in $\Sigma_n(A^\omega)$. Finally, if $\Sigma_n(A^\omega) = \Delta_{n+1}(A^\omega)$, then $\Sigma_n(A^\omega)$ is closed under complement, and $\Sigma_n(A^\omega) = \Pi_n(A^\omega)$. □

4 The space of finite or infinite words

We return in this section to the set of finite or infinite words defined in Chapter I. It is indeed sometimes convenient to extend to this space the algebraic properties of finite words and the topological properties of infinite words. It is in particular the right framework to present the computation of the closure of a rational set. To avoid any problem with the empty word when dealing with infinite products, we shall slightly depart from our original notation in this section and use A^∞ for the set $A^+ \cup A^\omega$ of nonempty finite or infinite words.

From an algebraic point of view, we simply extend the concatenation product defined in Chapter I by setting $uv = u$ if u is an infinite word. We shall see later on that this definition is well suited to the topological properties of A^∞. One can next define in a natural way the "rational" operations (union, product, plus and omega) on A^∞.

From a topological point of view, we equip A^∞ with a metric that extends the metric already defined on the space of infinite words. This makes A^∞ a complete metric space, which turns out to be the completion of the space A^+ of finite words. This property allows one to extend by continuity to infinite words every uniformly continuous function defined on finite words. It justifies in particular the definition of the concatenation product given above. We now proceed to the formal definitions.

We extend to A^∞ the metric defined on A^ω by identifying each finite word u with $u\$^\omega$ where \$ is a new symbol. This defines a topology on A^∞, whose open sets are the sets of the form

$$X \cup YA^\omega$$

for $X \subset A^+$ and $Y \subset A^*$. In this way, we keep the topology previously defined on A^ω while A^+ is equipped with the discrete topology. Every infinite word now appears naturally as the limit of its prefixes. More precisely:

Proposition 4.1. *A sequence $(u_n)_{n\geq 0}$ of finite words converges to an infinite word u if and only if each prefix of u is prefix of all but a finite number of the u_n's. In this case, we have $\lim_{n\to\infty} |u_n| = +\infty$.*

Proof. Suppose that $\lim_{n\to\infty} u_n = u \in A^\omega$. Then for each $k > 0$, there exists n_0 such that for each $n \geq n_0$, $d(u, u_n) \leq 2^{-k}$, which implies that u and u_n have a common prefix of length k. The condition of the statement is thus fulfilled.

Conversely, if this condition is satisfied, there exists a sequence of integers $(n_k)_{k\geq 0}$ such that, for each $k \geq 0$, for each $n \geq n_k$, $u[0, k]$ is prefix of u_n. It follows that $|u_n| > k$ and $d(u, u_n) < 2^{-k}$. Consequently, $\lim_{n\to\infty} u_n = u$ and $\lim_{n\to\infty} |u_n| = +\infty$. □

Proposition 4.1 shows in particular that each infinite word u is the limit of the sequence of its prefixes $u[0, n]$, which is in agreement with our intuition. There are however other converging sequences. For example, we have

$$\lim_{n\to\infty} a^n b^\omega = a^\omega$$

although none of the words $a^n b$ is a prefix of a^ω.

The space A^∞ is actually the completion of the metric space (A^+, d).

Proposition 4.2. *For each alphabet A, (A^∞, d) is a complete metric space isomorphic to the completion of the metric space (A^+, d).*

Proof. By Proposition 4.1, every infinite word is the limit of the sequence of its prefixes, showing that A^+ is dense in A^∞. There remains to prove that A^∞ is complete. Let $(u_n)_{n\geq 0}$ be a Cauchy sequence. By definition, there exists for each $k > 0$ an integer n_k such that, for each $p, q \geq n_k$, $d(u_p, u_q) < 2^{-k}$, which means that u_p and u_q have a common prefix of length k, the word $v_k = u_{n_k}[0, k-1]$. If n_k is chosen to be minimal, the sequence n_k is increasing and each v_k is a prefix of v_{k+1}. It follows that the infinite word u defined by $u[0, k-1] = v_k$ for each $k \geq 0$ is an adherence point of the sequence. □

Let us mention a first consequence of Proposition 4.2.

Proposition 4.3. *Let X be a nonempty closed subset of A^ω. Then there exists a uniformly continuous surjection from A^ω onto X whose restriction to X is the identity.*

Proof. By Lemma 3.6, there exists a prefix-closed subset P of A^* such that X is the set of infinite words all of which prefixes are in P. Define by induction a map $\varphi : A^* \to A^*$ as follows:

$$\varphi(1) = 1$$

$$\varphi(ua) = \begin{cases} ua & \text{if } ua \in P \\ \varphi(u)b & \text{if } ua \notin P, \text{ where } b \text{ is such that } \varphi(u)b \in P \end{cases}$$

Then $d(\varphi(u), \varphi(v)) \leq d(u, v)$ by construction and therefore φ is uniformly continuous. Therefore, φ extends to a unique uniformly continuous map φ from A^ω into itself. By construction, $\varphi(x) = x$ for $x \in X$ and the range of φ is X. □

In the opposite direction, A^ω is the continuous image of A^∞.

Proposition 4.4. *For each nonempty alphabet A, there is a continuous surjection from A^∞ onto A^ω.*

Proof. Let $a \in A$. Then the map f from A^∞ onto A^ω defined by

$$f(u) = \begin{cases} u & \text{if } u \in A^\omega \\ ua^\omega & \text{otherwise} \end{cases}$$

is uniformly continuous and surjective. □

Proposition 4.2 can also be used to extend by continuity the concatenation product, considered as a function from $A^+ \times A^+$ into A^+. Of course, we first need to choose a metric for defining the topology on $A^\infty \times A^\infty$. We set, for each $u, u', v, v' \in A^\infty$,

$$d((u, v), (u', v')) = \max\{d(u, u'), d(v, v')\}$$

Proposition 4.5. *The function $\varphi : A^+ \times A^+ \to A^+$ defined by $\varphi(u, v) = uv$ is Lipschitz of modulus 1, and hence is uniformly continuous. It admits a unique uniformly continuous extension from $A^\infty \times A^\infty$ into A^∞ given by*

$$\varphi(u, v) = \begin{cases} u & \textit{if } u \textit{ is an infinite word,} \\ uv & \textit{if } u \textit{ is a finite word.} \end{cases}$$

Proof. If $u \neq u'$, we have $d(uv, u'v') \leq d(u, u')$ and if $u = u'$, we have $d(uv, u'v') \leq d(v, v')$. Consequently, $d(uv, u'v') \leq d((u, v), (u', v'))$ in every case, proving the first part of the proposition. Let us verify the second part. If u is an infinite word, then by continuity $\varphi(u, v) = \lim_{n\to\infty} u[0, n]v[0, n] = u$. If u is finite, but if v is infinite, we have $\varphi(u, v) = \lim_{n\to\infty} uv[0, n]$. □

The product and the plus operation can be extended in a natural way to subsets of A^∞ by setting

$$XY = \{xy \mid x \in X, y \in Y\} \qquad X^+ = \bigcup_{n>0} X^n$$

It is also possible to define an infinite product on A^∞. If $(u_n)_{n\in\mathbb{N}}$ is a sequence of words of A^∞, the product $u_0u_1\cdots$ is equal to the usual infinite product if all the u_i's are finite and to $u_0u_1\cdots u_k$ if u_k is the first infinite word of the sequence. More generally, if $(X_n)_{n\geq 0}$ is a sequence of subsets of A^∞, the infinite product $X_0X_1\cdots$ is defined by setting

$$X_0X_1\cdots = \{u_0u_1\ldots \mid u_0 \in X_0, u_1 \in X_1, \cdots\}$$

For example, if $X_n = \{a^nb\} \cup b^\omega$,

$$X_0X_1\cdots = \{baba^2ba^3b\cdots\} \cup \{baba^2\cdots ba^nb^\omega \mid n \geq 0\}$$

In particular, we put, for each subset X of A^∞

$$X^\omega = \{x_0x_1x_2\cdots \mid x_0, x_1, x_2 \in X\} \quad \text{and} \quad X^\infty = X^+ \cup X^\omega$$

Note that if X_f and X_i denote resp. the set of finite and infinite words of X, we have $X^\omega = X_f^\omega \cup X_f^* X_i$.

Theorem 4.6. *Every finite or infinite product of closed sets of A^∞ is closed.*

Proof. We shall only treat the case of an infinite product of closed sets, but the other case is similar. Let $(F_n)_{n\geq 0}$ be a sequence of closed sets, and put $F = F_0F_1\cdots$. Let $(u_n)_{n\geq 0}$ be a sequence of elements of F converging to an element u of A^∞. Each u_n can be decomposed as a product $u_n = u_{n,0}u_{n,1}\cdots$ where, for each $r \geq 0$, $u_{n,r} \in F_r$. In particular, one has necessarily $u_n \in A^\omega$ and hence also $u \in A^\omega$.

Suppose there exists an integer r such that the sequence $n \to |u_{n,r}|$ is not bounded, but the sequences $|u_{n,0}|, \ldots, |u_{n,r-1}|$ are bounded by an integer K. In this case, Proposition 4.1 shows that, for some large enough n, $u[0, K]$ is a prefix of u_n. But as $u_{n,0} \leq K$, $u_{n,0}$ is a prefix of $u[0, K]$ and hence of u. Therefore, there exists a subsequence of the sequence $u_{n,0}$ which is constant and equal to a finite word v_0. By induction, one can find a subsequence $(u_{n_k})_{k\geq 0}$ such that the sequences $u_{n_k,0}, \ldots, u_{n_k,r-1}$ are constant and equal to finite words $v_0, \ldots, v_{r-1}$ resp., and such that $|u_{n_k,r}| > k$ for each $k > 0$. Then we have

$$d(u_{n_k,r}, u_{n_k,r}u_{n_k,r+1}\ldots) \leq 2^{-k} \text{ and hence } \lim_{k\to\infty} u_{n_k,r} = (v_0\cdots v_{r-1})^{-1}u \in F_r$$

It follows $u = v_0\cdots v_{r-1}((v_0\cdots v_{r-1})^{-1}u) \in F_0F_1\cdots F_r$. But since u is an infinite word, we also have for instance $u = uu_{0,r+1}u_{0,r+2}\cdots \in F$.

There remains to study the case where each of the sequences $n \to |u_{n,r}|$ is bounded by a constant K_r. In this case, one can find an infinite subset E_0 of $\mathbb{N}$ and a finite word $v_0 \in F_0$, such that, for each $n \in E_0$, $u_{n,0} = v_0$; then an infinite subset E_1 of E_0 and a finite word $v_1 \in F_1$, such that, for each $n \in E_1$, $u_{n,1} = v_1$, etc. For each $r \geq 0$, the word $v_0 v_1 \cdots v_r$ is then a prefix of u_n for $n \in E_r$ and hence also a prefix of u. It follows that u is equal to the infinite product $v_0 v_1 \cdots$ and consequently $u \in F$. □

Corollary 4.7. *If X is a closed set of A^∞, the subsets X^ω and X^∞ are also closed.*

Proof. Let X be a closed subset of A^∞. Then X^ω is closed by Theorem 4.6. Put $B = A \cup \{\$\}$. Since $X \cup \{\$\}$ is a closed subset of B^∞, $(X \cup \{\$\})^\omega$ is also a closed subset of B^∞. But A^∞ can be identified to the closed set $E = A^+\$^\omega \cup A^\omega$ of B^ω, and X^∞ can be identified to the set $(X \cup \{\$\})^\omega \cap E$, which is a closed subset of B^ω. Therefore X^∞ is closed in A^∞. □

Theorem 4.6 leads to an algorithm for computing the closure of a rational set, and gives a characterization of the rational closed subsets of A^∞.

Proposition 4.8. *For each $X, X_1, X_2 \subset A^\infty$, the following formulas hold*

(1) $\overline{X_1 \cup X_2} = \overline{X}_1 \cup \overline{X}_2$,

(2) $\overline{X_1 X_2} = \overline{X}_1 \overline{X}_2$,

(3) $\overline{X^+} = \overline{X^\infty} = (\overline{X})^\infty$,

(4) $\overline{X^\omega} = (\overline{X})^\omega$.

Finally, if X is a rational subset of A^∞, $\overline{X}$ is also rational.

Proof. (1) follows from general topological arguments.

(2) The inclusion $\overline{X}_1 \overline{X}_2 \subset \overline{X_1 X_2}$ follows from the continuity of the product. But on the other hand, we have $X_1 X_2 \subset \overline{X}_1 \overline{X}_2$, and since $\overline{X}_1 \overline{X}_2$ is closed by Theorem 4.6, it comes $\overline{X_1 X_2} \subset \overline{X}_1 \overline{X}_2$, and hence $\overline{X_1 X_2} = \overline{X}_1 \overline{X}_2$.

(3) Since $X \subset \overline{X}$, we have $X^+ \subset X^\infty \subset (\overline{X})^\infty$. Since $(\overline{X})^\infty$ is closed by Corollary 4.7, we deduce that $\overline{X^+} \subset \overline{X^\infty} \subset (\overline{X})^\infty$. Conversely, we claim that the following formula holds

$$(\overline{X})^\infty \subset \overline{\overline{X}^+} \subset \overline{X^+} \tag{4.1}$$

To establish the first inclusion, we observe that if $u \in (\overline{X})^\omega$, u can be written as $u = u_0 u_1 \cdots$ where, for each $n \geq 0$, $u_n \in \overline{X}$. Therefore $u = \lim_{n \to \infty} (u_0 u_1 \cdots u_n)$ and $(u_0 u_1 \cdots u_n) \in (\overline{X})^+$. To establish the second inclusion, it suffices to verify that $(\overline{X})^+ \subset \overline{X^+}$, which follows from the continuity of the product. Therefore $\overline{X^+} = \overline{X}^\infty$.

(4) The inclusion $(\overline{X})^\omega \subset \overline{X^\omega}$ follows from the continuity of π, the function which maps an infinite sequence of words onto its infinite product. Furthermore, we have

$X \subset \overline{X}$ and hence $X^\omega \subset (\overline{X})^\omega$. Since $(\overline{X})^\omega$ is closed by Corollary 4.7, it comes

$$\overline{X^\omega} \subset (\overline{X})^\omega$$

and hence $\overline{X^\omega} = (\overline{X})^\omega$.

Finally, let $\mathcal{F}$ be the class of subsets X of A^∞ such that $\overline{X}$ is rational. The set $\mathcal{F}$ contains the finite sets, since the finite sets are closed and is stable by finite union, product, plus and omega as was shown above. Therefore $\mathcal{F}$ contains all the rational subsets of A^∞, thereby proving the last part of the proposition. □

Corollary 4.9. *The class of closed rational subsets of A^∞ is the smallest class $\mathcal{F}$ of subsets of A^∞ such that*

(a) *$\emptyset \in \mathcal{F}$, and for each $a \in A$, $\{a\} \in \mathcal{F}$,*

(b) *$\mathcal{F}$ is stable by finite union, finite product and by the operations $L \to L^\omega$ and $L \to L^\infty$.*

Proof. By Theorem 4.6 and Corollary 4.7, every subset $X \in \mathcal{F}$ is rational and closed. Conversely, let $\mathcal{R}$ be the set of subsets X of A^∞ such that $\overline{X} \in \mathcal{F}$. Then $\mathcal{R}$ contains the empty set and the singletons, and Proposition 4.8 shows that $\mathcal{R}$ is stable by union, product and for the operations $X \to X^+$ and $X \to X^\omega$. Therefore $\mathcal{R}$ contains the rational subsets of A^∞. In particular, every closed rational subset is element of $\mathcal{F}$. □

We now survey the topological properties of the morphisms between two free semigroups.

Proposition 4.10. *Let A and B be two alphabets. Every semigroup morphism $\varphi : A^+ \to B^+$ is Lipschitz of modulus 1, hence is uniformly continuous. Therefore, it admits a unique uniformly continuous extension from A^∞ in B^∞.*

Proof. Let $u, v \in A^+$. If x is a common prefix of u and of v, $\varphi(x)$ is a common prefix of $\varphi(u)$ and $\varphi(v)$, of length greater than or equal to $|x|$. Consequently, $d(\varphi(u), \varphi(v)) \leq d(u, v)$. □

When A is a one letter alphabet, (A^+, d) can be identified to the space $\mathbb{N}_+ = \{1, 2, \ldots\}$, equipped with the metric $d(n, m) = 2^{-\min(n,m)}$, and the product becomes the addition on integers. In particular, we derive the following result.

Corollary 4.11. *The function from (A^+, d) into $(\mathbb{N}_+, d)$, which maps a word onto its length, is a continuous semigroup morphism.*

The surjective morphisms share an additional property. We let the reader verify that a morphism semigroup $\varphi : A^+ \to B^+$ is surjective if and only if $\varphi(A)$ contains B.

Proposition 4.12. *Let $\varphi : A^+ \to B^+$ be a surjective morphism semigroup. Then*
(1) *the function $\varphi : A^\omega \to B^\omega$ induced by φ is open,*
(2) *the function $\varphi : A^\infty \to B^\infty$ induced by φ is open.*

Proof. If $U = XA^\omega$ is an open subset of A^ω (with $X \subset A^+$), it comes $\varphi(XA^\omega) = \varphi(X)B^\omega$ and hence $\varphi(U)$ is an open set of B^ω. The proof is analogous for (2). □

Proposition 4.13. *Let A and B be two alphabets, with A finite, and let $\varphi : A^+ \to B^+$ be a semigroup morphism. Then*
(1) *the function $\varphi : A^\omega \to B^\omega$ induced by φ is closed,*
(2) *the function $\varphi : A^\infty \to B^\infty$ induced by φ is closed.*

Proof. Since A is finite, A^ω and A^∞ are compact. Therefore if F is closed in A^ω (resp. A^∞), F is compact, and so is $\varphi(F)$ since φ is continuous and B^ω (resp. B^∞) is Hausdorff. Consequently $\varphi(F)$ is closed. □

Proposition 4.13 does not hold if A is infinite, even in the case of a length preserving morphism. Indeed, let $A = \{a_{m,n} \mid 0 \leq m \leq n\}$, $B = \{a_n \mid n \geq 0\}$ and let $\varphi : A^+ \to B^+$ be the semigroup morphism defined by $\varphi(a_{m,n}) = a_{n-m}$. Put $F = \{a_{m,m}a_{m,m+1} \cdots a_{m,2m} \mid m \geq 0\}$. Then F is closed, since the only sequences of elements of F that converge in A^∞ are the ultimately constant ones. However, $\varphi(F) = \{a_0a_1 \ldots a_m \mid m \geq 0\}$ is not closed, since the infinite word $a_0a_1 \cdots$ is not in $\varphi(F)$.

It is nevertheless possible to get a result that holds for every alphabet A, but only for a more restricted class of morphisms. A length preserving morphism $\varphi : A^+ \to B^+$ is said to be *fiber bounded* if there exists a constant K such that, for each $b \in B$, $\mathrm{Card}(\varphi^{-1}(b)) \leq K$. It is clear in particular that if A is finite, every morphism from A^+ into B^+ is fiber bounded.

Proposition 4.14. *Let $\varphi : A^+ \to B^+$ a fiber bounded, length preserving morphism. Then φ induces a closed function from A^ω into B^ω (resp. from A^∞ into B^∞).*

Proof. We shall give the proof in the case of A^ω. Let K be an integer such that, for each $b \in B$, $\mathrm{Card}(\varphi^{-1}(b)) \leq K$. Set $C = \{1, \ldots, K\}$ and let $\pi : B \times C \to B$ be the projection defined, for each $(b, c) \in B \times C$, by $\pi(b, c) = b$. Then there exists an injection $\iota : A \to B \times C$ such that, for each $a \in A$, $\pi\iota(a) = \varphi(a)$. The natural extensions $\varphi : A^\omega \to B^\omega$, $\pi : (B \times C)^\omega \to B^\omega$ and $\iota : A^\omega \to (B \times C)^\omega$ also satisfy the equality $\varphi = \pi \circ \iota$. Since ι is an isometry from A^ω onto a closed subset of $(B \times C)^\omega$, ι is a closed function. Therefore, proving that φ is closed reduces to verify that π is a closed function.

The function π is continuous by Proposition 4.10. On the other hand, one can identify $(B \times C)^\omega$ to $B^\omega \times C^\omega$. Let F be a closed subset of $(B \times C)^\omega$ and let $(u_n)_{n \geq 0}$ be a sequence of elements of $\pi(F)$ converging to an element $u \in B^\omega$. Then there exists for each $n \geq 0$ a pair of the form (u_n, v_n) in F. Since C is finite, C^ω is compact, and the sequence $(v_n)_{n \geq 0}$ has an adherence point v in C^ω. The pair (u, v) is thus an adherence point of the sequence $(u_n, v_n)_{n \geq 0}$ and hence $(u, v) \in F$. Consequently $u = \pi(u, v) \in \pi(F)$ and F is closed. □

5 Borel automata

Recall that an automaton $\mathcal{A} = (Q, A, E)$ is countable if the set E is countable. We shall denote by $C(\mathcal{A})$ the set of infinite paths of $\mathcal{A}$ and by $\lambda_\mathcal{A} : C(\mathcal{A}) \to A^\omega$ the function that maps an infinite path onto its label.

Proposition 5.1. *The set $C(\mathcal{A})$ is a closed subset of E^ω and $\lambda_\mathcal{A}$ is a Lipschitz map of modulus* 1, *hence a uniformly continuous map, from $C(\mathcal{A})$ into A^ω.*

Proof. The formula

$$C(\mathcal{A}) = E^\omega \setminus E^* F E^\omega$$

with

$$F = \{ee' \mid e \text{ and } e' \text{ are two nonconsecutive transitions of } E\}$$

shows that $C(\mathcal{A})$ is a closed subset of E^ω.

Furthermore, if two infinite paths have the same r first transitions, their labels have the same prefixes of length r. It follows that, for each $p, p' \in C(\mathcal{A})$,

$$d\big(\lambda_\mathcal{A}(p), \lambda_\mathcal{A}(p')\big) \leq d(p, p')$$

and hence $\lambda_\mathcal{A}$ is Lipschitz of modulus 1. □

An *Borel automaton* is a 4-tuple

$$\mathcal{A} = (Q, A, E, R)$$

where (Q, A, E) is a countable automaton and R a Borel subset of E^ω, called the *acceptance condition*. A path is *successful* if it is an element of R. The set of successful paths is, by Proposition 5.1, the intersection of R with a closed, hence Borel, subset of E^ω. An infinite word $x \in A^\omega$ is recognized by $\mathcal{A}$ if it is the label of a successful path. The set of words recognized by $\mathcal{A}$ is denoted by $L^\omega(\mathcal{A})$. The following proposition is a direct consequence of the definition.

Proposition 5.2. *Let $\mathcal{A} = (Q, A, E, R)$ be a Borel automaton. Then $L^\omega(\mathcal{A}) = \lambda_\mathcal{A}(R \cap C(\mathcal{A}))$.*

A Borel automaton $\mathcal{A}$ is *ω-unambiguous* if two infinite successful paths with the same label are equal. In particular, each word of $L^\omega(\mathcal{A})$ is the label of a unique successful path.

A Borel automaton $\mathcal{A} = (Q, A, E, R)$ is *deterministic* if, on the one hand, the automaton (Q, A, E) is deterministic and if, on the other hand, there exists a state which is the origin of all the successful paths. When $\mathcal{A}$ is a deterministic automaton, one can improve Proposition 5.1 as follows.

Proposition 5.3. *If $\mathcal{A}$ is a deterministic Borel automaton, $\lambda_\mathcal{A}$ defines an isometry from the set of successful paths onto $L^\omega(\mathcal{A})$.*

Proof. Since $\mathcal{A}$ is deterministic, each word of $L^\omega(\mathcal{A})$ is the label of a unique successful path. Moreover, if two words u and u' of $L^\omega(\mathcal{A})$ have the same prefix of length r, the successful paths p and p' they define have the same prefix of length r. It follows

$$d(p, p') \leq d(u, u') = d(\lambda_\mathcal{A}(p), \lambda_\mathcal{A}(p'))$$

and, by Proposition 5.1, $\lambda_\mathcal{A}$ defines an isometry from the set of successful paths onto $L^\omega(\mathcal{A})$. □

Countable Büchi and Rabin automata are particular cases of Borel automata. A Büchi automaton corresponds to an acceptance condition of the form

$$R = (E^* F)^\omega$$

where F is the set of transitions whose end is a final state. A Rabin automaton $\mathcal{A} = (Q, A, E, i, \mathcal{R})$ with $\mathcal{R} = \{(L_i, U_i) \mid i \in I\}$ is defined by the acceptance condition

$$\bigcup_{i \in I} ((E^* L_i)^\omega \setminus (E^* U_i)^\omega)$$

We have seen that, in the case of finite automata, Rabin automata are equivalent to Muller automata. The latter of the two classes does not have any natural equivalent in the countable case.

Proposition 5.4. *Every Borel subset of A^ω can be recognized by a Borel automaton.*

Proof. If $X \subset A^\omega$ is Borel, it suffices to take the automaton $\mathcal{A} = (Q, A, E, R)$ with $Q = A^*$, $E = \{(u, a, ua) \mid u \in A^*, a \in A\}$ and to take for R the set of paths whose label is in X. □

If $\mathcal{A} = (Q, A, E, R)$ is a Borel automaton, the function which maps a path onto its label is a continuous map and hence $L(\mathcal{A})$ is the continuous image of a Borel subset of the set of paths. The next proposition will allow us to speak almost indifferently of Borel automata or continuous functions in the sequel.

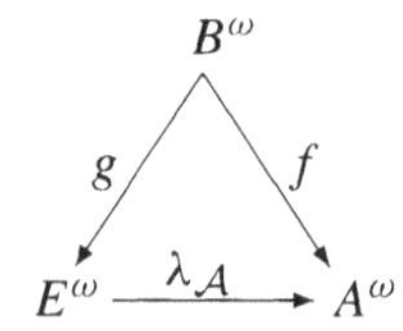

Figure 5.1. A commutative diagram.

Proposition 5.5. *Let A and B be countable alphabets, let $f : B^\omega \to A^\omega$ be a continuous map and let Y be a Borel subset of B^ω. There exists a Borel automaton $\mathcal{A} = (Q, A, E, R)$ recognizing $f(Y)$ and an injective bicontinuous map $g : B^\omega \to E^\omega$ such that $R = g(Y)$ and $\lambda_{\mathcal{A}} \circ g = f$. Finally, if the restriction of f to Y is injective, the automaton is ω-unambiguous.*

Proof. A word $v \in B^*$ is said to be *suitable* for a word $u \in A^*$ if

(i) $f(vB^\omega) \subset uA^\omega$,

(ii) $|u| \leq |v|$,

(iii) v is of minimal length among the words verifying (i) and (ii).

We take as set of states the set Q of pairs $(u, v) \in A^* \times B^*$ such that v is suitable for u. The transitions are the triples (p, a, q) with $p = (u, v)$ and $q = (ua, w)$. All the paths of the automaton start in $(1, 1)$ and thus are of the form

$$c : (1, 1) \xrightarrow{a_0} (a_0, v_0) \xrightarrow{a_1} (a_0a_1, v_0v_1) \xrightarrow{a_2} \cdots \tag{5.1}$$

and we have $a_0a_1a_2 \cdots = f(v_0v_1v_2 \cdots)$ as shown in Figure 5.2. The set R consists of the paths of this form such that $v_0v_1v_2 \cdots \in Y$. We thus have $L^\omega(\mathcal{A}) = f(Y)$. The function g from B^ω into E^ω maps each $x \in B^\omega$ onto a path of the form (5.1) such that $a_0a_1 \cdots = f(x)$ and $v_0v_1 \cdots = x$. The computation of the v_n's can be done by induction as follows. If $v_0, v_1, \ldots, v_{n-1}$ are already known, v_n is the shortest word such that

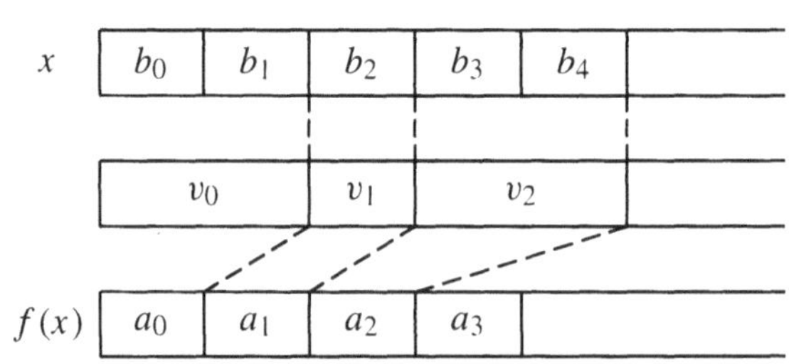

Figure 5.2. Turning a continuous map into an automaton.

(a) $f(v_0 \cdots v_n B^\omega) \subset a_0 \cdots a_n A^\omega$,

(b) $n + 1 \leq |v_0 \cdots v_n|$,

(c) $v_0 \cdots v_n$ is prefix of x.

It follows in particular that $v_0 \cdots v_n$ only depends on the prefix $a_0 \cdots a_n$ of $f(x)$. Let $k > 0$. Since f is continuous by hypothesis, there exists an integer n_k such that, if two words x and x' of B^ω have a common prefix of length n_k, $f(x)$ and $f(x')$ have a common prefix of length k. It follows that $g(x)$ and $g(x')$ have a common prefix of length k, showing that g is continuous.

The function g is injective since $x = v_0 v_1 v_2 \cdots$. Moreover, $g(Y)$ is by definition the set of paths of the form (5.1) such that $v_0 v_1 v_2 \cdots \in Y$, that is, R. We also have, for each

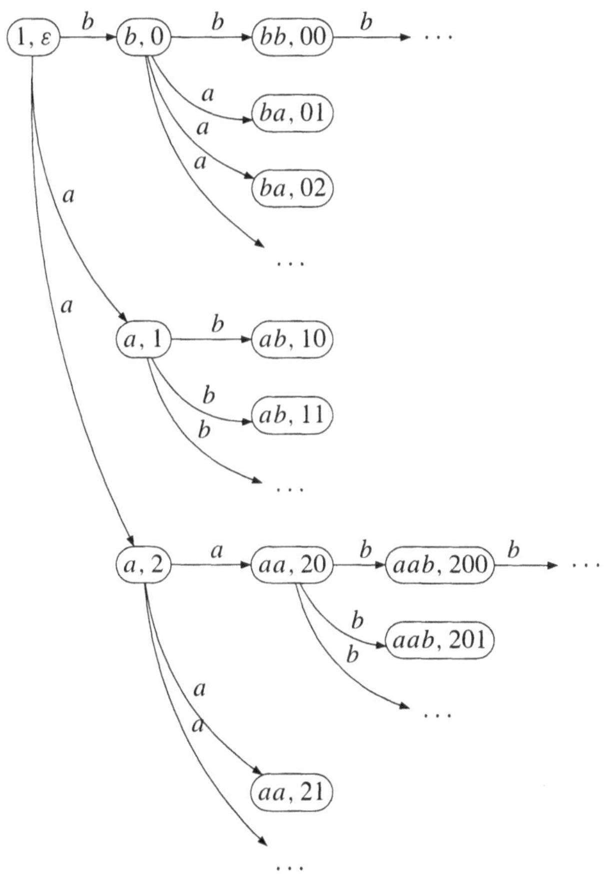

Figure 5.3. The run-length coding.

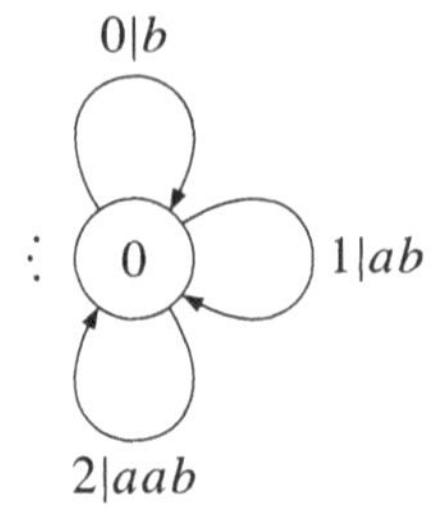

Figure 5.4. A transducer.

$x, x' \in B^\omega$,

$$d(x, x') \leq d(g(x), g(x'))$$

since if $g(x)$ and $g(x')$ have the path

$$(1, 1) \xrightarrow{a_0} (a_0, v_0) \xrightarrow{a_1} (a_0a_1, v_0v_1) \cdots \xrightarrow{a_{k-1}} (a_0a_1 \cdots a_{k-1}, v_0v_1 \cdots v_{k-1})$$

as a common prefix of length k, then $v_0v_1v_2 \cdots v_{k-1}$ is a common prefix of x and x' of length greater than or equal to k. Consequently, the function $g^{-1} : R \to B^\omega$ is continuous.

Finally, since $f = \lambda_\mathcal{A} \circ g$, we have $\lambda_\mathcal{A} = g^{-1} \circ f$. Therefore, if the restriction of f to Y is injective, $\lambda_\mathcal{A}$ is injective and the automaton $\mathcal{A}$ is ω-unambiguous. □

Example 5.1. The *run-length coding* is the function from $\mathbb{N}^\omega$ into $\{a, b\}^\omega$ that maps the sequence $(n_0, n_1, \ldots)$ onto the word $a^{n_0}ba^{n_1}b \cdots$ The Borel automaton constructed in the proof of Proposition 5.5 is represented in Figure 5.3.

Another simpler realization, obtained by putting the entries on the arrows is given in Figure 5.4. Such an automaton with an input and an output alphabet is called a *transducer*.

6 Suslin sets

As we shall see later, the image of a Borel subset of a Polish space under a continuous map is not always Borel. Historically, this negative result led to the introduction of the class of Suslin sets. The abstract definition of a Suslin set will be given at the end of this section, but we shall first present a definition restricted to the subsets of A^ω, which is intimately related to Borel automata.

Let A be a countable alphabet. A subset X of A^ω is a *Suslin set* if there is a countable alphabet B, a Borel subset Y of B^ω and a continuous map $f : B^\omega \to A^\omega$ such that $X = f(Y)$. Equivalent characterizations will be given later on (Theorem 6.11). We first make precise the link between Suslin sets and Borel automata.

Theorem 6.1. *Let A be a countable alphabet. A subset of A^ω is Suslin if and only if it is recognized by a Borel automaton.*

Proof. If X is a subset of A^ω recognized by a Borel automaton

$$\mathcal{A} = (Q, A, E, R)$$

then $X = \lambda_{\mathcal{A}}(R \cap C(\mathcal{A}))$ and, since $C(\mathcal{A})$ is closed, $R \cap C(\mathcal{A})$ is a Borel subset of E^ω. Now, by Proposition 4.3, there exists a continuous surjection $\varphi : E^\omega \to C(\mathcal{A})$ whose restriction to $C(\mathcal{A})$ is the identity, and by Proposition 5.1, $\lambda_{\mathcal{A}}$ is a continuous map from $C(\mathcal{A})$ into A^ω. Therefore, $f = \lambda_{\mathcal{A}} \circ \varphi$ is a continuous map from E^ω into A^ω and

$$X = f(R \cap C(\mathcal{A})) \tag{6.1}$$

Since E is countable, X is Suslin.

Conversely, let $f : B^\omega \to A^\omega$ be a continuous map and let Y be a Borel subset of B^ω such that $f(Y) = X$. Proposition 5.5 shows that X is recognized by a Borel automaton. □

An immediate consequence of the definition is the fact that the class of Suslin sets, contrary to that of Borel sets, is closed under continuous functions.

Proposition 6.2. *Let A and B be countable alphabets and $g : A^\omega \to B^\omega$ be a continuous function. If X is a Suslin subset of A^ω, $g(X)$ is a Suslin subset of B^ω.*

Proof. By definition, there is a countable alphabet C, a Borel subset Y of C^ω and a continuous map $f : C^\omega \to A^\omega$ such that $X = f(Y)$. It follows that $g(X) = (g \circ f)(X)$ and hence $g(X)$ is Suslin. □

We shall next prove successively that, like the class of Borel subsets of A^ω, the class of Suslin subsets of A^ω is closed under countable union and intersection. However, it is not closed under complement, as we shall see in Section 7. Closure under countable union can be proved exactly in the same way as for finite union.

Proposition 6.3. *The class of Suslin subsets of A^ω is closed under countable union.*

Proof. Let $(\mathcal{A}_i)_{i \in I}$ be a countable family of Borel automata, with

$$\mathcal{A}_i = (Q_i, A, E_i, R_i).$$

We may assume that the different sets Q_i are pairwise disjoint. Let $\mathcal{A} = (Q, A, E, R)$ be the automaton defined by

$$Q = \bigcup_{i \in I} Q_i, \quad E = \bigcup_{i \in I} E_i \text{ and } R = \bigcup_{i \in I} R_i \tag{6.2}$$

Then $\mathcal{A}$ is a Borel automaton by Proposition 2.3 and $L(\mathcal{A}) = \bigcup_{i \in I} L(\mathcal{A}_i)$. □

The case of a countable intersection is more involved. Indeed, we cannot use the product $\prod_{i \in I} Q_i$ as a set of states, since it is not countable. Therefore, a more economical procedure is in order. The idea is the same as the one we used to show that $\mathbb{N} \times \mathbb{N}$ is countable. We put successively to work the various automata but keep in memory the input word. In this way, the last automaton to be activated reads the same word as its predecessors, but with some delay.

Let $(\mathcal{A}_n)_{n \geq 0}$ be an infinite sequence of Borel automata with

$$\mathcal{A}_n = (Q_n, A, E_n, R_n)$$

Define the *diagonal product* of the automata $\mathcal{A}_n$, denoted by $\mathcal{A} = (Q, A, E, R)$, as follows. The set of states Q is

$$Q = A^* \times \Big(\bigcup_{n \geq 0} \prod_{0 \leq i \leq n} Q_i \Big)$$

The transitions are the triples of the form (p, a_n, q) where

$$p = \big(a_0 a_1 \cdots a_{n-1}, (p_0, p_1, \ldots, p_n)\big), \quad q = \big(a_0 a_1 \cdots a_{n-1} a_n, (q_0, q_1, \ldots, q_{n+1})\big)$$

with, for $0 \leq i \leq n$, $(p_i, a_{n-i}, q_i) \in E_i$ (see Figure 6.1) The definition of the acceptance condition R requires some developments. We first identify the set of paths of $\mathcal{A}$ with a product of paths in the automata $\mathcal{A}_i$ as follows. Let

$$c : p_0 \xrightarrow{a_0} p_1 \xrightarrow{a_1} p_2 \xrightarrow{a_2} \cdots$$

be a path of $\mathcal{A}$, with

$$p_n = \big(b_0 b_1 \cdots b_{k-1} a_0 a_1 \cdots a_{n-1}, (p_0^n, p_1^n, \cdots, p_{k+n}^n)\big)$$

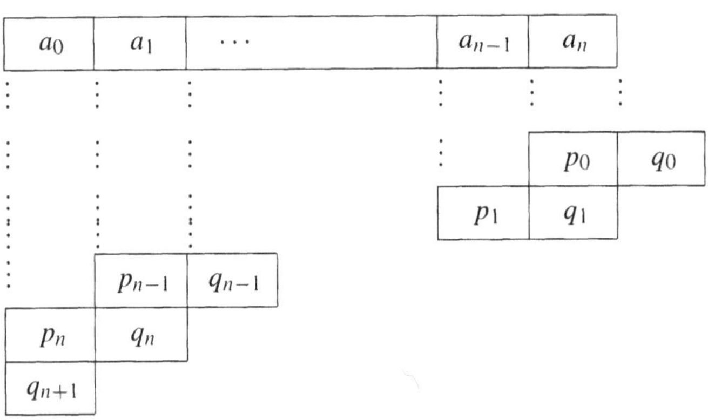

Figure 6.1. The transitions of the diagonal product.

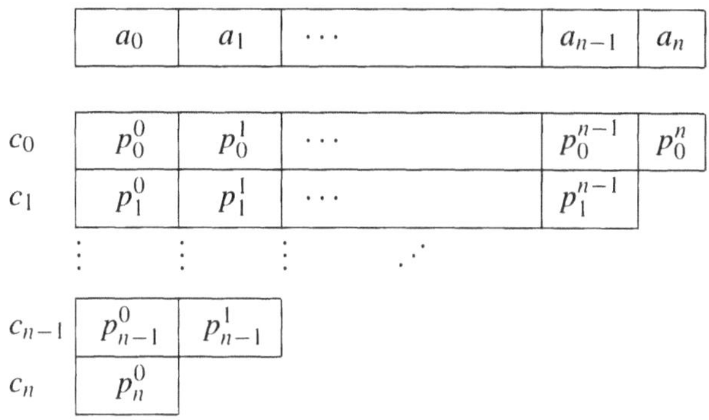

Figure 6.2. The paths of the diagonal product.

with $p_i^n \in Q_i$ for $0 \leq i \leq n$. We associate with the path c a sequence $\varphi(c) = (c_n)_{n \geq 0}$ of paths in the automaton $\mathcal{A}_n$ by setting

$$c_n : p_n^n \xrightarrow{a_0} p_n^{n+1} \xrightarrow{a_1} p_n^{n+2} \xrightarrow{a_2} \cdots$$

(see Figure 6.2). Finally, we choose for R the set of paths p starting in a state of the form $(1, p_0)$ (with $p_0 \in Q_0$) and such that $\varphi(c)_n \in R_n$ for each $n \geq 0$.

The set of paths starting in $(1, p_0)$ is a closed subset F of E^ω. Furthermore, as each of the R_n's is a Borel subset of E_n^ω, their product $\prod_{n \geq 0} R_n$ is a Borel subset of the space $\prod_{n \geq 0} E_n^\omega$. Since we have the formula

$$R = F \cap \varphi^{-1}\Big(\prod_{n \geq 0} R_n\Big)$$

it will suffice to prove that φ is continuous to verify that R is Borel.

Proposition 6.4. *The function $\varphi : C(\mathcal{A}) \to \prod_{n \geq 0} C(\mathcal{A}_n)$ is continuous.*

Proof. According to the characterization of continuous functions whose range is a product space recalled in Section 2.2, it suffices to prove that, for each n, the function $\varphi_n : C(\mathcal{A}) \to C(\mathcal{A}_n)$ defined by $\varphi_n(c) = c_n$ is continuous. Now, if two paths c and c' have the same prefixes of length $n + r$, the paths $\varphi_n(c)$ and $\varphi_n(c')$ have the same prefixes of length r. Therefore φ_n is continuous for each n. □

The properties of the diagonal product are summarized into the following statement:

Proposition 6.5. *Let $(\mathcal{A}_n)_{n \geq 0}$ be a sequence of Borel automata, equipped with the acceptance condition R_n and let $\mathcal{A} = (Q, A, E, R)$ be their diagonal product.*

(1) *Let k be an positive integer. If each of the sets R_n is in Π_k, R is in Π_k.*
(2) *If each of the automata $\mathcal{A}_n$ is deterministic, the automaton $\mathcal{A}$ is deterministic.*
(3) *The set $L^\omega(\mathcal{A})$ is equal to the intersection of the sets $L^\omega(\mathcal{A}_n)$ for $n \geq 0$.*

Proof. (1) If each of the sets R_n are in Π_k, their product

$$V = \prod_{n \geq 0} R_n$$

is in Π_k by Proposition 2.11 and the set $\varphi^{-1}(V)$ is also in Π_k by Proposition 2.10. Since the set F of paths starting in $(1, p_0)$ is closed (that is, in Π_1), the set $R = F \cap \varphi^{-1}(V)$ is in Π_k.

(2) If each of the automata $\mathcal{A}_n$ is deterministic, the automaton (Q, A, E) is deterministic by construction. Furthermore, all the successful paths of E start in $(1, p_0)$ by hypothesis, where p_0 is the initial state of $\mathcal{A}_0$.

(3) By construction, a path c is successful if and only if the paths $\varphi_n(c)$ are successful. Therefore $u \in L^\omega(\mathcal{A})$ if and only if $u \in L^\omega(\mathcal{A}_n)$ for each $n \geq 0$. □

We deduce from the previous statement the result announced above:

Corollary 6.6. *The class of Suslin subsets of A^ω is closed under countable intersection.*

The following statement gives an equivalent definition of the Suslin sets: they are the continuous images of closed sets. Note that a stronger version of this result will be proved in Theorem 6.11 below.

Proposition 6.7. *A subset X of A^ω is Suslin if and only if there exists a closed set Y of $\mathbb{N}^\omega$ and a continuous map $f : \mathbb{N}^\omega \to A^\omega$ such that $X = f(Y)$.*

Proof. If X is the continuous image of a closed subset of $\mathbb{N}^\omega$, it is Suslin by definition.

Conversely, if X is Suslin, there is a countable alphabet B, a Borel subset Z of B^ω and a continuous map $g : B^\omega \to E$ such that $X = g(Z)$. Therefore, it suffices to show that Z is the continuous image of a closed set of $\mathbb{N}^\omega$.

Consider the class $\mathcal{F}$ of subsets of A^ω recognized by a Borel automaton in which the acceptance condition is closed. We claim that every element of $\mathcal{F}$ is the continuous image of a closed subset of $\mathbb{N}^\omega$. Indeed, if X is recognized by a Borel automaton (Q, A, E, R) with R closed, then $F = R \cap C(\mathcal{A})$ is also closed by Proposition 5.1 and $X = \lambda_\mathcal{A}(F)$ by Proposition 5.2. Finally, since E is countable, there is a continuous surjection $\pi : \mathbb{N}^\omega \to E^\omega$ and $X = (\lambda_\mathcal{A} \circ \pi)(\pi^{-1}(F))$, which proves the claim, since $\pi^{-1}(F)$ is closed and $\lambda_\mathcal{A} \circ \pi$ is continuous. It suffices now to show that $\mathcal{F}$ contains the Borel subsets of A^ω.

Clearly $\mathcal{F}$ contains the closed sets, and, by Proposition 6.3, is closed under countable union (more precisely, the set R constructed in the proof of Proposition 6.3 is a closed subset of E). The class $\mathcal{F}$ is also closed under countable intersection by Proposition 6.5. Therefore, by Proposition 2.8, it contains the Borel sets of A^ω. □

One can obtain an equivalent formulation in terms of automata. The next result states that each Borel automaton is equivalent to a Borel automaton in which the set of successful paths is closed (such an automaton is said to be a *closed automaton*). This is a major difference with the case of finite automata.

Proposition 6.8. *Every Suslin subset of A^ω can be recognized by a Borel automaton in which the acceptance condition is closed.*

Proof. Let X be a Suslin subset of A^ω. By Proposition 6.7, there exist a closed set Y of $\mathbb{N}^\omega$ and a continuous map $f : \mathbb{N}^\omega \to A^\omega$ such that $X = f(Y)$. By Proposition 5.5, there exists a Borel automaton $\mathcal{A} = (Q, A, E, R)$ such that $X = L(\mathcal{A})$ and an injective bicontinuous map $g : \mathbb{N}^\omega \to E^\omega$ such that $R = g(Y)$. Since Y is closed, R is also closed. □

Exercise 21 provides a direct construction of a closed automaton equivalent to a given Büchi automaton.

Example 6.1. The set $X = (a^*b)^\omega$ of words with an infinite number of occurrences of b's is recognized by the closed automaton represented in Figure 6.3.

The next proposition shows that, in the deterministic case, the complexity of the acceptance condition cannot be chosen so freely.

Proposition 6.9. *Let $\mathcal{A} = (Q, A, E, R)$ be a deterministic Borel automaton. For each integer $n \geq 1$ and each Borel class Σ_n, Π_n or Δ_n, the sets R and $L(\mathcal{A})$ are in the same Borel class.*

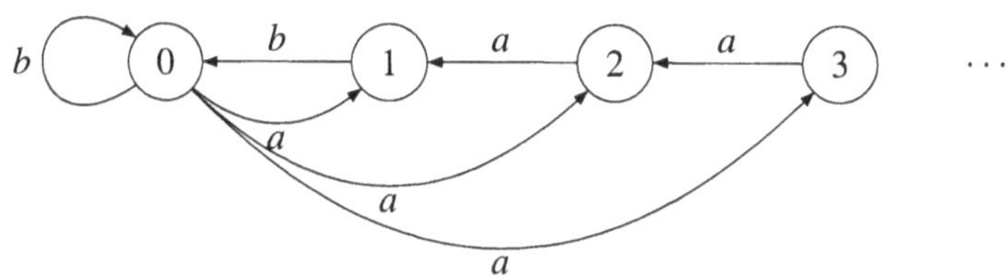

Figure 6.3. A closed automaton.

Proof. By Proposition 5.3, the function which maps any path onto its label is an isometry. Therefore, it preserves the Borel classes. □

As a special case of the previous statement, we have the correspondence between the deterministic Büchi automata and the Π_2-sets (see Theorems I.6.2 and 3.11). See also the correspondence between the Rabin automata and the Σ_3 sets specified in Exercise 22.

Another construction related to the same ideas is the simulation of a Muller automaton with a full table by a deterministic Büchi automaton (Proposition I.7.9). The set of successful paths of a Muller automaton with a full table is indeed in Π_2 and hence, equivalently, is recognizable by a deterministic Büchi automaton. The constructed automaton corresponds to the composition of functions (it is actually a wreath product of the two automata).

Before stating our characterizations of Suslin subsets of A^ω, we need an auxiliary result on Polish spaces.

Proposition 6.10. *For every Polish space X, there exists a continuous surjection from $\mathbb{N}^\omega$ onto X.*

Proof. Since X is Polish, it contains by Proposition 2.6 a countable dense subset, say

$$D = \{r_0, r_1, \ldots\}$$

Assign to each sequence $u \in \mathbb{N}^\omega$ the sequence x_n^u defined by the induction

$$x_0^u = r_{u(0)}$$
$$x_{n+1}^u = \begin{cases} r_{u(n+1)} & \text{if } d(x_n^u, r_{u(n+1)}) < 2^{-n} \\ x_n^u & \text{otherwise} \end{cases}$$

Now, for each n,

$$d(x_n^u, x_{n+1}^u) < 2^{-n}$$

Therefore, x_n is Cauchy and we can set

$$\pi(u) = \lim x_n^u$$

Then π is continuous, since if $u[0, n] = v[0, n]$, then $x_0^u = x_0^v, \ldots, x_n^u = x_n^v$, whence

$$d(\pi(u), \pi(v)) \leq d(\pi(u), x_n^u) + d(x_n^v, \pi(v)) \leq 2^{-n+1} + 2^{-n+1} = 2^{-n+2}$$

Furthermore, for each $x \in X$, we have $x = \pi(u)$, where $u(n)$ is the least k such that $d(x, r_k) < 2^{-n-1}$. Thus π is a continuous surjection. □

We are now ready to state our main characterization of Suslin subsets of A^ω. We have followed the terse kind of statement traditionally used in descriptive set theory, but this terminology deserves a comment. The fourth condition, for instance, should be interpreted as: "there is a continuous function $\varphi : \mathbb{N}^\omega \to A^\omega$ and a closed subset Y of $\mathbb{N}^\omega$ such that $\varphi(Y) = X$". And in the third condition, the projection is the map $(x, y) \to y$ from $\{a, b\}^\omega \times A^\omega$ onto A^ω.

Theorem 6.11. *Let X be a nonempty subset of A^ω. The following conditions are equivalent:*

(1) *X is Suslin,*

(2) *X is the continuous image of some Polish space,*

(3) *X is the projection of some Π_2-subset of $\{a, b\}^\omega \times A^\omega$,*

(4) *X is the continuous image of some closed subset of $\mathbb{N}^\omega$,*

(5) *X is the continuous image of $\mathbb{N}^\omega$.*

Proof. We prove the implications (1) $\Rightarrow$ (4) $\Rightarrow$ (2) $\Rightarrow$ (5) $\Rightarrow$ (3) $\Rightarrow$ (1) in this order.

(1) implies (4) by Proposition 6.7.

(4) implies (2) since every closed subset of $\mathbb{N}^\omega$ is Π_2 by Proposition 2.5 and every Π_2-subset of $\mathbb{N}^\omega$ is Polish by Propositions 2.7 and 3.4.

(2) implies (5) is a consequence of Proposition 6.10.

(5) implies (3). By assumption, there is a continuous map f from N^ω onto X. Consider the continuous map $g : (a^*b)^\omega \to \mathbb{N}^\omega$ defined by

$$g(a^{n_0}ba^{n_1}ba^{n_k}\cdots) = (n_0, n_1, \ldots, n_k, \ldots)$$

and let

$$Y = \{(x, y) \in \{a, b\}^\omega \times A^\omega \mid y = f(g(x)) \text{ and } x \in (a^*b)^\omega\}$$

Since $g \circ f$ is continuous, its graph is a closed subset of $(a^*b)^\omega \times A^\omega$. Now since $(a^*b)^\omega$ is a countable intersection of open sets of $\{a, b\}^\omega$, it follows that Y itself is a countable intersection of open sets of $\{a, b\}^\omega \times A^\omega$. Finally, X is the projection of Y, since g maps $(a^*b)^\omega$ onto $\mathbb{N}^\omega$.

(3) implies (1). Suppose that X is the projection of some Π_2-subset of $\{a, b\}^\omega \times A^\omega$. By Proposition 3.14, $\{a, b\}^\omega \times A^\omega$ is homeomorphic to $\{a, b\}^\omega$ or to $\mathbb{N}^\omega$. Since the projection from $\{a, b\}^\omega \times A^\omega$ onto A^ω is continuous, X is the continuous image of some Borel subset of either $\{a, b\}^\omega$ or $\mathbb{N}^\omega$. Therefore, X is Suslin. □

We now come to the general abstract definition of a Suslin set. A topological space X is said to be *Suslin* if it is Hausdorff and if there exists a Polish space E and a continuous map from E onto X. In particular, every Polish set is Suslin. Condition (2) of Theorem 6.11 shows that our new definition is compatible with the one given for the subsets of A^ω. Suslin sets are closed under continuous images and pre-images.

Proposition 6.12. *Let X be a Suslin space, Y an Hausdorff space and $f : X \to Y$ be a continuous map. Then $f(X)$ is Suslin, and for every Suslin subset Z of Y, $f^{-1}(Z)$ is a Suslin subset of X.*

Proof. If X is Suslin, there exists a Polish space E and a continuous map g from E onto X. It follows that $g \circ f$ is a continuous map from E onto $f(X)$, and thus $f(X)$ is Suslin.

If Z is a Suslin subset of Y, there exists a Polish space F and a continuous map h from F onto Z. Let R be the set of elements $(x, y) \in E \times F$ such that $f(g(x)) = h(y)$. Then R is closed in $E \times F$, and is a Polish space by Proposition 2.7. Let r be the restriction to R of the projection from $E \times F$ onto E. Then $f^{-1}(Z) = (g \circ r)(R)$, showing that $f^{-1}(Z)$ is Suslin. □

It follows in particular that the Suslin subsets of a Polish space form a Polish class.

Theorem 6.13. *Let A be a countable alphabet with at least two letters. There is an A-universal set for the class of Suslin subsets of A^ω.*

Proof. We shall use freely the fact, established in Proposition 3.14, that the spaces A^ω, $A^\omega \times A^\omega$ and $\{a, b\}^\omega \times A^\omega$ are homeomorphic. By Proposition 3.16, there is an A-universal set $\mathcal{F}$ for $\Pi_2(A^\omega \times A^\omega)$. Let

$$\mathcal{U} = \{(y, x) \in A^\omega \times A^\omega \mid (y, x, z) \in \mathcal{F} \text{ for some } z \in A^\omega\}$$

We claim that $\mathcal{U}$ is an A-universal set for the class of Suslin subsets of A^ω. By construction, $\mathcal{U}$ is the projection of a Π_2-subset of $A^\omega \times A^\omega$ and therefore is Suslin. Furthermore, if X is a Suslin subset of A^ω, there is by Theorem 6.11 a Π_2-subset S of A^ω and a surjective continuous map from A^ω onto A^ω such that $X = f(S)$. Let

$$R = \{(y, x) \in A^\omega \times A^\omega \mid y = f(x) \text{ and } x \in S\}$$

Then $R \in \Pi_2(A^\omega \times A^\omega)$ and $x \in X$ if and only if there exists $y \in A^\omega$ such that $(y, x) \in R$. Let $y \in A^\omega$ be such that $R = \mathcal{F}(y)$. Then $X = \mathcal{U}(y)$, proving the claim. □

Corollary 6.14. *Let A be a countable alphabet with at least two letters. There is a Suslin subset of A^ω which is not Borel.*

Proof. Let $\mathcal{U}$ be a universal subset for the class of Suslin subsets of A^ω. By Proposition 3.17, $\mathcal{U}^c$ is not Suslin. It follows in particular that $\mathcal{U}^c$ is not Borel, and since Borel sets are closed under complement, $\mathcal{U}$ is not Borel. □

7 The separation theorem

Two subsets X and Y of A^ω are said to be *separable* by a Borel set if there exists a Borel set Z containing X and disjoint from Y (see Figure 4.1). We also say in this case that Z separates X and Y. Note that this definition is symmetrical in X and Y since the complement of Z is a Borel set containing Y and disjoint from X (see Figure 7.1). The next result is known as the separation theorem.

Theorem 7.1. *Two disjoint Suslin sets can be separated by a Borel set.*

The proof relies on the following combinatorial lemma.

Lemma 7.2. *If, for each index $n, m \geq 0$, the subsets S_n and T_m are separated by the Borel set $R_{n,m}$, then $\bigcup_{n\geq 0} S_n$ and $\bigcup_{m\geq 0} T_m$ are separated by the Borel set $R = \bigcup_{n\geq 0}\bigcap_{m\geq 0} R_{n,m}$.*

Proof. Set $R_n = \bigcap_{m\geq 0} R_{n,m}$, so that $R = \bigcup_{n\geq 0} R_n$. Each S_n is contained in each $R_{n,m}$ for $m \geq 0$ and hence S_n is contained in R_n. Thus $\bigcup_{n\geq 0} S_n$ is contained in $R = \bigcup_{n\geq 0} R_n$. Furthermore, each T_m is disjoint from $R_{n,m}$ for each $n \geq 0$, and therefore, is also disjoint from R_n. Consequently $\bigcup_{m\geq 0} T_m$ is disjoint from R_n for each $n \geq 0$ and finally $\bigcup_{m\geq 0} T_m$ is disjoint from R. □

Proof of Theorem 7.1. Let X and Y be two disjoint Suslin sets of A^ω. By Proposition 6.7, there exist two continuous functions $f : B^\omega \to A^\omega$ and $g : C^\omega \to A^\omega$ and two closed subsets $Z \subset B^\omega$, $T \subset C^\omega$ such that $X = f(Z)$ and $Y = g(T)$. Considering the product $B \times C$ and the projections on each component, we may reduce the problem to the case where $B = C$ and $Z = T$. Suppose now that X and Y are not separable by a Borel set. Setting, for $u \in B^*$, $Z_u = Z \cap uB^\omega$, we obtain

$$X = \bigcup_{b\in B} f(Z_b), \quad Y = \bigcup_{c\in B} g(Z_c)$$

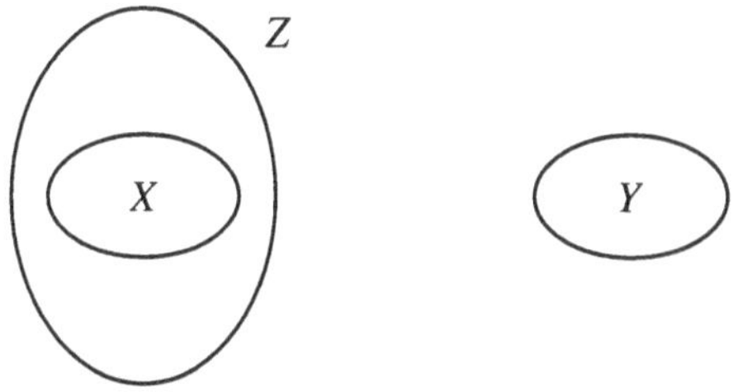

Figure 7.1. Z separates X and Y.

If there was a Borel set separating each $f(Z_b)$ from each $g(Z_c)$, then, by Lemma 7.2, it would be possible to separate X and Y by a Borel set. Therefore there exists $b_0, c_0 \in B$ such that no Borel set can separate $f(Z_{b_0})$ and $g(Z_{c_0})$. Iterating this argument, we obtain two words $x = b_0 b_1 \dots$ and $y = c_0 c_1 \dots$ such that, for each $n \geq 0$, $f(Z_{b_0 \dots b_n})$ and $g(Z_{c_0 c_1 \dots c_n})$ cannot be separated by a Borel set. Since Z is closed, we have $x, y \in Z$ and $f(x) \in X$, $g(y) \in Y$. Since X and Y are disjoint, we have $f(x) \neq g(y)$. Then there exist two distinct words $u, v \in A^*$ such that $f(x) \in uA^\omega$, $g(y) \in vA^\omega$. One can choose u and v of the same length so that both sets $U = uA^\omega$ and $V = vA^\omega$ are disjoint open sets. Since f and g are continuous, there exists an integer n such that $f(Z_{b_0 \cdots b_n}) \subset U$ and $g(Z_{c_0 \cdots c_n}) \subset V$, a contradiction with the fact that $f(Z_{b_0 \cdots b_n})$ and $g(Z_{c_0 \cdots c_n})$ are not separable. □

The first corollary of the separation theorem is the following:

Theorem 7.3. *A subset of A^ω is Borel if and only if it is Suslin and its complement is Suslin.*

Proof. The condition is obviously necessary. Furthermore, if a subset and its complement are Suslin, they are separable by a Borel set by the separation theorem and hence are also Borel. □

The next result is another consequence of the separation theorem. It is reminiscent of the characterization of the Suslin sets as continuous images of closed sets.

Theorem 7.4. *A subset X of A^ω is Borel if and only if there exists a continuous map $f : B^\omega \to A^\omega$ and a closed set $Y \subset B^\omega$ such that the restriction of f to Y is a bijection from Y onto X.*

Proof. The condition is necessary. Consider indeed the class $\mathcal{F}$ of the subsets of A^ω satisfying this condition, or equivalently, recognizable by a countable ω-unambiguous automaton. The set $\mathcal{F}$ contains obviously the closed sets. Next, let $(X_n)_{n \geq 0}$ be a countable family of elements of $\mathcal{F}$. Put $Y_n = X_n \setminus \bigcup_{m<n} X_m$. Then the Y_n's are pairwise disjoint and $\bigcup_{n \geq 0} Y_n = \bigcup_{n \geq 0} X_n$. Let $\mathcal{A}_n = (Q_n, A, E_n, R_n)$ be a countable ω-unambiguous automaton recognizing Y_n. Then the union of the $\mathcal{A}_n$'s defined by Formula (6.2) is still ω-unambiguous and $R = \bigcup_{n \geq 0} R_n$ is closed. For countable intersection, we observe that the diagonal product of the ω-unambiguous automata recognizing the X_n's is ω-unambiguous. This shows that $\mathcal{F}$ contains the class of Borel sets.

For the opposite implication, we make use of the separation theorem under the following form: if $(X_n)_{n \geq 0}$ is a family of pairwise disjoint Suslin sets, there exists a sequence $(Y_n)_{n \geq 0}$ of pairwise disjoint Borel sets such that $X_n \subset Y_n$ for every $n \geq 0$. Let $\mathcal{A} = (Q, A, E, R)$ be an ω-unambiguous automaton with R closed. For each finite path u in $\mathcal{A}$, denote by X_u the set of labels of successful paths starting by the path u. For each integer n, the family $(X_u)_{|u|=n}$ is a family of pairwise disjoint Suslin sets. Then there exist pairwise disjoint Borel sets $(Y_u)_{|u|=n}$ such that $X_u \subset Y_u$. We now define by induction a new family Z_u of Borel sets realizing a thinner separation, by setting for $e \in E$

$$Z_e = Y_e \cap \overline{X}_e$$

where $\overline{X}_e$ denotes the topological closure of X_e. Next we define for a path u and a consecutive transition e

$$Z_{ue} = Y_{ue} \cap \overline{X}_{ue} \cap Y_u$$

This gives, for each u of a given length, a family of pairwise disjoint Borel sets such that $Z_u \subset Z_v$ if v is a prefix of u. Furthermore

$$X_u \subset Z_u \subset \overline{X}_u$$

We shall now verify that the set X recognized by the automaton $\mathcal{A}$ can be written

$$X = \bigcap_{n \geq 0} \bigcup_{|u|=n} Z_u \qquad (7.1)$$

which implies of course that X is a Borel set. The inclusion from left to right is obvious. For the opposite inclusion, observe that if a word w belongs to the set defined by the right hand side of (7.1), it is the label of a path which is in R and thus $w \in X$. □

Corollary 7.5. *Let $f : A^\omega \to B^\omega$ be a continuous injective map. The image under f of every Borel subset of A^ω is a Borel subset of B^ω.*

Proof. Let X be a Borel subset of A^ω. By Theorem 7.4, there exists a continuous map $g : C^\omega \to A^\omega$ and a closed subset Y of C^ω such that the restriction of g to Y is a bijection from Y onto X. The map $g \circ f : C^\omega \to B^\omega$ is thus continuous and induces a bijection from Y onto $f(X)$. It suffices to apply once again Theorem 7.4 to conclude that $f(X)$ is a Borel set. □

Theorem 7.4 can be reformulated in terms of automata as follows:

Corollary 7.6. *Let X be a subset of A^ω. The following conditions are equivalent:*

(1) *X is a Borel set,*

(2) *X can be recognized by a closed ω-unambiguous automaton,*

(3) *X can be recognized by an ω-unambiguous Borel automaton.*

Proof. (1) implies (2). By Theorem 7.4, there exists a continuous map $f : B^\omega \to A^\omega$ and a closed subset Y of B^ω such that the restriction of f to Y is a bijection from Y onto X. By Proposition 5.5, there exists a Borel automaton $\mathcal{A} = (Q, A, E, R)$ such that $X = L(\mathcal{A})$ and an injective bicontinuous map $g : B^\omega \to E^\omega$ such that $R = g(Y)$. Furthermore, $\mathcal{A}$ is ω-unambiguous since the restriction of f to Y is injective.

(2) implies (3) is trivial.

(3) implies (1). If X is recognized by an ω-unambiguous Borel automaton $\mathcal{A}$, the map $\lambda_\mathcal{A} : C(\mathcal{A}) \to A^\omega$ is injective. Furthermore, by Proposition 5.1, $\lambda_\mathcal{A}$ is continuous and $C(\mathcal{A})$ is closed. Therefore, X is a Borel set by Corollary 7.5. □

This statement can be considered as a version of Kleene's theorem for the countable automata. It is, of course, to be compared with the analogous result for the finite automata (Theorem I.3.2).

8 Exercises

8.1 Topological spaces.

Exercise 1. Consider the topology on an infinite noncountable set whose open sets are the subsets of finite complement. Show that the closed sets are the finite subsets, that Σ_2 is the class of countable subsets and Π_2 the class of subsets with a countable complement. Deduce that Δ_2 is empty and consequently that Σ_1 is not contained in Δ_2.

Exercise 2. Proposition 2.6 shows that, for a metric space E, the following conditions are equivalent:

(1) E admits a countable basis of open sets.

(2) E contains a countable dense subset.

Show that, in an arbitrary topological space, condition (2) does not necessarily imply (1), even if E is compact.

8.2 The space of infinite words.

Exercise 3. Let X and Y be two arbitrary nonempty subsets of the metric space (E, d). Define the *distance between the sets X and Y* as the number

$$d(X, Y) = \inf_{x \in X, y \in Y} d(x, y)$$

In particular, we denote by $d(u, X)$ the distance between the set $\{u\}$ and the set X (also called the *distance from u to X*).

(a) Show that $d(u, X) = 0$ if and only if u belongs to the closure of X.

(b) Show that the function $u \to d(u, X)$ is uniformly continuous on E.

Exercise 4. Show that the metric d defined on A^ω is an *ultrametric*, that is, for each $x, y, z \in A^\omega$,

$$d(x, z) \leq \max\{d(x, y), d(y, z)\}$$

Figure 8.1. An infinite tree.

Exercise 5. Let d be an ultrametric on a set E. Let $x, y, z \in E$. Show that

(a) if $d(x, y) \neq d(y, z)$ then $d(x, z) = \max((d(x, y), d(y, z))$,

(b) every open (resp. closed) ball B is also closed (resp. open), and every point of B is the center of the ball,

(c) if two balls of E have a common point, then one of them is contained in the other,

(d) a sequence $(x_n)_{n \in \mathbb{N}}$ is Cauchy if and only if $\lim_{n \to \infty} d(x_n, x_{n+1}) = 0$.

Exercise 6. The aim of this exercise is to relate König's lemma ("every infinite tree in which each node has a finite arity contains an infinite branch") with topology.

First identify a tree with the set T of words of $(A \cup \{\$\})^{\mathbb{N}}$ of the form u, where u is an infinite branch, or $u\$^{\omega}$, where u is a finite branch. For example, the tree drawn in Figure 8.1 is represented by the set $a^{\omega} \cup a^{+}\$^{\omega} \cup a^{*}b\$^{\omega}$.

Show that if each node has a finite arity, each projection $\pi_n(T)$ is finite. Deduce from this fact that T is relatively compact and that if T is infinite, there exists a sequence of pairwise distinct words of T converging to a word $u \in (A \cup \{\$\})^{\mathbb{N}}$. Show that necessarily $u \in A^{\omega}$, and that every prefix $u[0, n]$ of u is prefix of a word of T. Conclude that there must exist an infinite branch.

8.3 The space of finite or infinite words.

Exercise 7. Recall that if x is a finite word and u a finite or infinite word, $x^{-1}u$ denotes the unique word v, if it exists, such that $xv = u$. Show that if x a word of A^{+}, the function φ from xA^{∞} into A^{∞} defined by $\varphi(u) = x^{-1}u$ is Lipschitz and hence uniformly continuous.

Exercise 8. Let F be a closed (resp. compact) subset of A^{∞}. Show that the set

$$C(F) = \{u \in A^{\infty} \mid \text{there exists } v \in F \text{ such that } u \leq v\}$$

is a closed (resp. compact) subset of A^{∞}.

Exercise 9. (Redziejowski) This problem makes use of a few notions of general topology. For each word $u, v, w \in A^\infty$, set

$$r_w(u) = \begin{cases} 0 & \text{if } u = w, \\ 2^{-|u|} & \text{if } u < w, \\ 1 & \text{in the other cases.} \end{cases}$$

and

$$e_w(u, v) = |r_w(u) - r_w(v)|.$$

(a) Show that for each $w \in A^\infty$, e_w is an écart. Let us call *strong topology* on A^∞ the topology defined on A^∞ by the family of écarts $\{e_w \mid w \in A^\infty\}$.

(b) Show that the open balls associated with this family of écarts are of the form $B_u(n) = \{x \in A^\infty \mid x \leq u \text{ and } |x| \geq n\}$ (where $u \in A^\infty$ and $n \geq 0$). Thus these balls constitute a basis of open sets for the strong topology.

(c) Show that the strong topology is the strongest topology such that, for each infinite word u, the sequence of prefixes $u[0, n]$ of u converges to u.

(d) Denote by $\overline{X}^F$ the closure of a subset X of A^∞ in the strong topology. Show that, for every subset L of A^+, we have $\overline{X}^F = L \cup \overrightarrow{L}$ and $\overrightarrow{L} = \overline{X}^F \cap A^\omega$. Deduce that $\overrightarrow{L}$ is the set of accumulation points of L (sometimes called "derived set").

(e) Show that, if $\mathrm{Card}(A) \geq 2$, the multiplication in A^∞ is not continuous for the strong topology.

(f) Show that A^∞, equipped with the strong topology, is a *completely regular* topological space (i.e. if F is closed and if $x \notin F$, there exist two disjoint open sets U and V such that $F \subset U$ and $x \in V$).

(g) Show that if $\mathrm{Card}(A) \geq 2$, A^∞ is not a *normal* space (a space is normal if, whenever F_1 and F_2 are two disjoint closed sets, there exist two disjoint open sets U_1 and U_2 such that $F_1 \subset U_1$ and $F_2 \subset U_2$).

(h) Deduce that the strong topology cannot be defined by any metric.

Exercise 10. This problem makes use of a few notions of general topology and of semigroup theory (see Annex B and Chapter VI).

A semigroup *separates* two words u and v of A^+ if there exists a semigroup morphism $\varphi : A^+ \to S$ such that $\varphi(u) \neq \varphi(v)$. Let $\mathbf{V}$ be a variety of finite semigroups. Put, for $u, v \in A^+$

$$d_{\mathbf{V}}(u, v) = 2^{-\min\{\mathrm{Card}(S) \mid S \in \mathbf{V} \text{ and } S \text{ separates } u \text{ and } v\}}$$

(with the usual conventions $\min \emptyset = +\infty$ and $2^{-\infty} = 0$).

(a) Show that $d_{\mathbf{V}}$ is an ultrametric *écart*, that is, satisfies, for $u, v, w \in A^+$,

$$d_{\mathbf{V}}(u, u) = 0$$
$$d_{\mathbf{V}}(u, v) = d_{\mathbf{V}}(v, u)$$
$$d_{\mathbf{V}}(u, v) \leq \max\{d_{\mathbf{V}}(u, w), d_{\mathbf{V}}(w, v)\}.$$

(b) Denote by $\mathcal{T}_{\mathbf{V}}$ the topology defined by this écart. Show that $\mathcal{T}_{\mathbf{V}}$ is the coarsest topology such that every semigroup morphism $\varphi : A^+ \to S$ (where S is a finite semigroup of $\mathbf{V}$, equipped with the discrete topology), is continuous. Show that the subsets of A^+ recognized by a semigroup of $\mathbf{V}$ are clopen sets, and form a basis of $\mathcal{T}_{\mathbf{V}}$.

(c) Show that, if $\mathbf{V}$ is the variety of finite groups, $d_{\mathbf{V}}$ is an *ultrametric*, and that the completion of A^+ for this metric is a group, which is compact if A is finite. (Hint. Show that if $u, v \in A^+$, then $\lim_{n\to\infty} vu^{n!} = \lim_{n\to\infty} u^{n!}v = v$).

(d) Let $\mathbf{V}$ be the variety of finite semigroups S such that, for each $n \geq \text{Card}(S)$, and for every finite sequence $s_0, s_1, \ldots, s_n$ of elements of S, one has $s_1 s_2 \cdots s_n s_0 = s_1 s_2 \cdots s_n$. Show that $d_{\mathbf{V}}$ is an ultrametric uniformly equivalent to d. Deduce from this that the completion of $(A^+, d_{\mathbf{V}})$ is (A^∞, d).

Exercise 11. Let L, L_0 and L_1 be subsets of A^+. Establish the following formulas:

(a) $\overrightarrow{L_0 \cup L_1} = \overrightarrow{L_0} \cup \overrightarrow{L_1}$,

(b) $L_0\overrightarrow{L_1} \subset \overrightarrow{L_0 L_1} \subset \overrightarrow{L_0} \cup L_0\overrightarrow{L_1}$,

(c) if $L_0 \subset L_0 L_1$, then $\overrightarrow{L_0 L_1} = \overrightarrow{L_0} \cup L_0\overrightarrow{L_1}$,

(d) if $\overrightarrow{L_0} = \emptyset$, then $\overrightarrow{L_0 L_1} = L_0\overrightarrow{L_1}$,

(e) $\overrightarrow{L^+} = L^*\overrightarrow{L} \cup L^\omega$,

(f) if $\overrightarrow{L} = \emptyset$, then $\overrightarrow{L^+} = L^\omega$.

Show that if $L = ba^*$, then $L^+ = bA^*$, $\overrightarrow{L^+} = bA^\omega$. Show that $L^\omega = (ba^*)^\omega$ is the set of infinite words whose first letter is a b and containing infinitely many occurrences of b's. Deduce that the formula $\overrightarrow{L^+} = L^\omega$, which is true if $\overrightarrow{L} = \emptyset$, does not hold in general.

Exercise 12. Let L be a subset of A^+ and let $P = L \setminus LA^+$. Show that

(a) if $A^*L \subset L$ then $\overrightarrow{P^+} = L^\omega$,

(b) if $A^*LA^* \subset L$ then $\overrightarrow{A^*P} = L^\omega$.

Exercise 13. Let L be a subset of A^+ and let φ be a semigroup morphism from A^+ into B^+. Show that $\varphi(\overrightarrow{L}) = \overrightarrow{\varphi(L)}$.

Exercise 14. Let A be an alphabet. Set, for each $n \geq 0$, $A_n = A$ and equip the space $\prod_{n \geq 0} A_n^\infty$ with the metric d defined by

$$d((x_n)_{n \geq 0}, (x'_n)_{n \geq 0}) = \sum_{n \geq 0} 2^{-n} d(x_n, x'_n)$$

Show that the map $\pi : \prod_{n \geq 0} A_n^+ \to A^\infty$ defined by $\pi((x_n)_{n \geq 0}) = x_0 x_1 \cdots$ is uniformly continuous. It follows that this map admits a unique uniformly continuous extension from $\prod_{n \geq 0} A_n^\infty$ to A^∞.

Exercise 15. Show that every finite or infinite product of compact sets of A^∞ is compact (use the Exercise 14). Derive from this result another proof of Theorem 4.6 when A is a finite alphabet.

Exercise 16. Show that, if A is an infinite alphabet, the map $\pi : A^\infty \times A^\infty \to A^\infty$ defined by $\pi(u, v) = uv$ *is not* a closed map. (Hint: if $a_0, a_1, \ldots$ is an infinite sequence of distinct letters, consider the closed subset F of $A^\infty \times A^\infty$ defined by $F = \{(a_0^n, a_n) \mid n \geq 0\}$).

8.4 Borel automata.

Exercise 17. Show that for each finite trim Büchi automaton $\mathcal{A}$, the automaton obtained by making all the states final recognizes the topological closure of $L(\mathcal{A})$.

Exercise 18. Show that, for each set $X, Y \subset A^+$, the set XY^ω is a Suslin set.

Exercise 19. Show that, for each set $X \subset A^*$ and each Suslin set $Z \subset A^\omega$, the set XZ is Suslin.

Exercise 20. Show that if $\mathcal{A}_n = (Q_n, \delta_n, F_n)$, with $n \in \mathbb{N}$, is a countable family of Büchi deterministic automata, one can build a deterministic Büchi automaton recognizing $\bigcap_{n \geq 0} L(\mathcal{A}_n)$. (Hint: take $A^* \times \mathbb{N} \times \mathbb{N}$ as set of states and memorize in the state (u, i, j) whether the automaton $\mathcal{A}_i$ has visited j times a final state. The transitions lead to a "crab" walk on $\mathbb{N} \times \mathbb{N}$. The final states are those of the form (u, i, i).)

Exercise 21. Given a countable Büchi automaton $\mathcal{A} = (Q, A, E, I, F)$, construct directly an equivalent closed automaton. (Hint: take $Q \times \mathbb{N}$ as set of states, where the state (q, i) indicates that the distance to the next final state is i. Every transition decrements the index i except for those starting from 0.)

Exercise 22. A countable Rabin automaton is a deterministic automaton

$$(Q, A, E, i, \mathcal{R})$$

where $\mathcal{R}$ is a countable family of pairs $(L_m, U_m)_{m\geq 0}$. An initial path is successful if there exists an index n such that the path visits L_n infinitely many times and R_n finitely often.

Show that the sets recognized by a countable Rabin automaton form the class Σ_3.

9 Notes

Unlike most of the notions presented in this book, those presented in this chapter have a long history. It goes back at least to the work of Borel, Baire and Lebesgue on the abstract notion of function in calculus and measure theory. The introduction of Suslin sets is due to Suslin, who is also the author of the separation theorem, which is nowadays considered a fundamental result of the theory. The systematic development of this topic is due to the mathematicians Lusin and Suslin and also to Sierpinski.

In this chapter, we have only presented a very small part of the descriptive set theory. One of the important issues of this theory is the introduction of the so-called *projective sets*, also called Lusin sets, after their discoverer Lusin. It is the starting point of a hierarchy obtained from the Suslin sets by alternating the closure under complementation and projection. The main results bring this theory closer to mathematical logic and, in particular, closer to the axioms of set theory and the continuum hypothesis. For an extensive survey, we refer the reader to Moschovakis [204] or Kechris [153] and, for a shorter presentation, to Bourbaki [37] or to the chapter of the Handbook of Mathematical Logic devoted to this topic [17]. Infinite games form one of the very interesting developments of this theory. They will be treated in Chapter IV, because they are also linked to automata.

The notation used for the finite Borel hierarchy has fluctuated over the years. We denote by Σ_n, Π_n, Δ_n what Moschovakis [204] denotes Σ^0_n, Π^0_n, Δ^0_n. The top index 0 corresponds to the level 0 of a hierarchy whose level 1 is formed by the *projective* or *Lusin* sets. We shall encounter Lusin sets in Chapter X when dealing with automata on trees. A somewhat older notation for the Borel hierarchy consists of using the notation $\mathcal{F}$ for the class Π_1 (the closed sets), $\mathcal{G}$ for Σ_1 (the open sets), and then $\mathcal{F}_\sigma$ for Σ_2, $\mathcal{G}_\delta$, $\mathcal{F}_{\sigma\delta}$ for Π_3 and so on. The notations $\mathcal{F}_\sigma$ and $\mathcal{G}_\delta$ are still commonly used.

Theorem 3.11 is due to Landweber [169], who was the first to consider the link between recognition by automata and topological properties.

Proposition 4.14 is a particular case of a more general statement: if X is a topological compact space, and Y a topological space, the projection $\pi : X \times Y \to Y$ is closed (cf. Bourbaki [37, Chapter 1, p. 77]).

Theorem 7.1 is due to Suslin. Theorem 4.4 is, according to Moschovakis [204], due to Lusin and Suslin. The formulation in terms of ω-unambiguous automata is due to Arnold [8]. Exercise 9 is due to Redziejowski [261]. Exercise 10 describes a special case of profinite topology (see [7] for more details).

IV

GAMES AND STRATEGIES

1 Introduction

Games can be used as a mathematical technique to prove some results that are not formulated in the vocabulary of games. We shall see examples of this with Wadge games and later on, in Chapter VIII, with Ehrenfeucht games used to study the quantifier complexity of first-order formulas.

In this chapter, we systematically study games. These games are abstract two-player games in which each player chooses in turn a symbol from an alphabet. The infinite word obtained is the result of the play. There is a fixed set of plays which makes the first player the winner. The second player wins on the complement of this set. This is the link with the subject of this book: each set of infinite words defines a particular game. Intuitively, a large set favours the first player, who moreover plays first.

Games can also be studied as a subject interesting in its own right. Moreover, games are often used in practical situations as a model for the control of discrete systems. The infinite games which are studied here correspond to systems which are supposed to run indefinitely, for example operating systems.

We are concerned here with the description of winning strategies. It turns out that, for reasonable winning sets, the game is determined, i.e. one of the two players has a winning strategy. In fact, by Martin's theorem, every Borel game is determined. The complexity of this strategy depends however of the complexity of the winning set. The main result of this chapter is the Büchi-Landweber theorem: if the winning set is rational, one of the two players has a winning strategy which requires only a bounded memory of the past.

The chapter is organized as follows. We first introduce the terminology of games. We then study games of increasing Borel complexity: closed games, Π_2-games, Borel games. We prove Martin's theorem (Theorem 3.3), asserting that all Borel games are determined, in the case of Borel classes of finite index. We then turn to rational games which, through the automaton recognizing the winning set, can be reduced to games on finite graphs. We prove that, in a game with a winning chain condition, one of the players has a memoryless strategy (Theorem 4.6). We deduce the Büchi-Landweber theorem (Theorem 4.9) from this result. The last section is devoted to a particular class of games called Wadge games. Its results will be used in Chapter V to provide a classification of recognizable sets of infinite words.

2 Infinite games

A *game* $G(X)$ on a given alphabet A is given by a set $X \subset A^\omega$ of infinite words on A. We shall often denote the game by X instead of $G(X)$. A *play* in the game is a word $x = a_0 a_1 \cdots \in A^\omega$. We consider that there are actually two *players* called Player I and Player II. The players are often given children names: Player I is usually a girl called Alice and Player II a boy called Bob. We shall keep the names Player I and Player II but use also "she" for Player I and "he" for Player II. Each of them plays in turn, and Player I begins. Thus Player I plays a_0 and Player II then chooses a_1, and so on. We say that Player I *wins* the play if $x \in X$, and otherwise that Player II wins the play.

A word $u \in A^*$ is often called a *position* of the game. If the length of u is even, it is a position of Player I, otherwise it is a position of Player II.

For a word $u \in A^*$, we denote by $G_u(X)$ the game which is obtained by starting the game $G(X)$ at u. Thus $G_u(X) = G(u^{-1}X)$ with the convention that when the length of u is odd, Player II plays first in $G_u(X)$.

A *strategy* for Player I is a function $f : (A^2)^* \to A$ from the set of words of even length into A. We say that I has *followed* the strategy f in the play x if for all $n \geq 0$

$$a_{2n} = f(a_0 a_1 \cdots a_{2n-1})$$

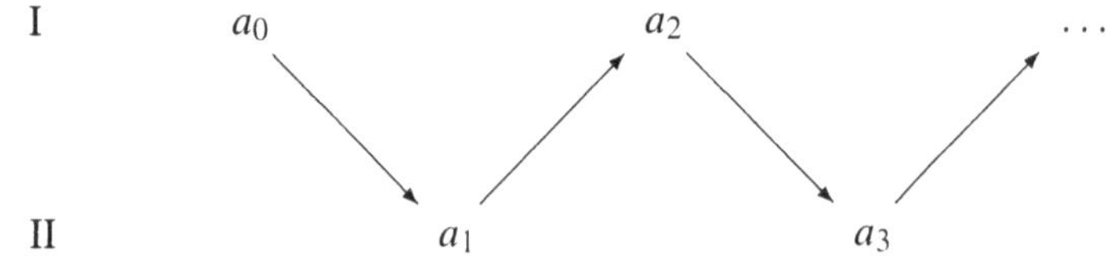

Figure 2.1. A play.

A strategy g for Player II is, in a dual way, a function $g : (A^2)^*A \to A$ from the set of words of odd length into A.

A strategy f of Player I is a *winning strategy* if she wins all plays in which she follows f. In other words, f is a winning strategy of Player I if any infinite word $x \in A^\omega$ such that for all $n \geq 0$, $a_{2n} = f(a_0 \cdots a_{2n-1})$ belongs to X. The notion of a winning strategy for Player II is similar. When Player I has a winning strategy, we also say that she *wins the game*. This means of course that she is able to win any play, whatever her opponent plays. A word of $u \in A^*$ of even length is called a *winning position* for Player I if Player I wins the game $G_u(X)$.

A game is *determined* if one of the players has a winning strategy. It is intuitively clear that "simple" games should be determined. However, we will see that a precise statement and a proof of this assertion requires some machinery.

Example 2.1. Let $A = \{a, b\}$ and let $X = (a^*b)^\omega$. Then Player I has a winning strategy in $G(X)$ consisting in playing always b. There are other possible winning strategies for Player I. For example, she can play systematically a when Player II plays b and conversely. This example shows that the winning strategy is not unique in general.

In the games considered in this chapter, all plays are infinite. In the games of everyday life, the plays are finite (and, in some cases, the infinite plays are considered draws). We can always consider that the play goes on at infinity with a winning set $X = WA^\omega \cup U$ where W is the set of (finite) winning plays for Player I and U is the set of infinite plays, in which none of the players wins (if we credit Player I with the draws).

Also, in most games in use, there are specific *rules* of the game. Let us indicate how one can introduce rules of the game in our setting. Let us consider a game $G(X)$ in which both players are required to remain inside a closed set $R \subset A^\omega$, with the rule that a player leaving the set R has lost. Such a game is actually identical with a game $G(Y)$ defined as follows: denoting by P the set of prefixes of the words of R, let Y be the set of infinite words $y = y_0y_1 \cdots$ such that either $y \in X \cap R$ (i.e. Player I has won $G(X)$ and both players have played consistently with the rules) or the smallest index n such that $y_0y_1 \cdots y_n \notin P$ is odd (i.e. Player II has made an illegal move).

To close this preliminary section, let us consider how a game $G(X)$ behaves with respect to boolean operations on the set X. We begin with complementation. If Player I wins the game $G(X)$, then Player II wins the game $G(AX^c)$. Indeed, whatever be the first symbol a played by I, Player II can use the strategy of Player I in $G(X)$ resulting in a word $ax \in AX$. Conversely, if Player II wins $G(X)$, then Player I wins every game $G(a^{-1}X^c)$ for $a \in A$.

Thus, games behave in a simple way with respect to complementation. In particular, since AX and $a^{-1}X$ belong to the same Borel class as X, we will be able to replace sets by their complement in the following discussion. It is important to realize that this is not true for other boolean operations. For instance, if Player II wins $G(X)$ and $G(Y)$, nothing can be said in general about $G(X \cup Y)$.

3 Borel games

In this section, we shall consider games of increasing topological complexity. In each case, we shall prove that the games are determined and we shall consider the complexity of the construction of a winning strategy. We begin with open games.

3.1 Open games. Open games are games for which the winning set has the form $X = UA^\omega$. Therefore, in these games, the outcome of the play only depends on a finite, initial part of the play.

An even simpler situation occurs when A is finite and the game is clopen, i.e. if $X = UA^\omega$ for some finite set U. In this case, the outcome of a play only depends on a bounded number of rounds and the game can be really considered as a finite game. In this simple case, it is obvious that one of the players has a winning strategy: a simple induction on the maximal length of the words of U suffices to establish this fact.

We prove the following result, known as the Gale-Stewart theorem.

Proposition 3.1. *Any open (or closed) game is determined.*

Proof. We suppose that X is open and thus that $X = UA^\omega$ for some $U \subset A^*$. We assume that Player I does not have a winning strategy and we prove that Player II does. Let $P \subset A(A^2)^*$ be the set of words p of odd length such that Player I does not have a winning strategy in the game $G_p(X)$ (since $|p|$ is odd, Player II plays first in $G_p(X)$). By our hypothesis, whatever be the first move of Player I, she does not have a winning strategy afterwards. Thus $A \subset P$. For the same reason, for each $p \in P$, there is a symbol $a \in A$ such that for all $b \in A$, $pab \in P$, since otherwise, Player I wins the game $G_p(X)$, contrary to the definition of P. We define a function $f : P \to A$ by choosing $f(p) = a$ with a as above.

Let us verify that f is a winning strategy for Player II. Let $x \in A^\omega$ be a play in which Player II has followed the strategy f. Then x has no prefix in U since for every prefix p of odd length, Player I does not win $G_p(X)$ and in particular p has no prefix in U. This shows that $x \notin X$, i.e. that Player II wins. □

The proof of the above result is by no means constructive. We describe a second proof based on rank functions.

We suppose that the game is open. Let $U \subset A^*$ be a set of words such that $X = UA^\omega$. We define a sequence W_i of subsets of A^* by $W_0 = U$ and for $i \geq 0$

$$\begin{aligned} W_{i+1} = W_i &\cup \{w \in (A^2)^* \mid wa \in W_i \text{ for some } a \in A\} \\ &\cup \{w \in (A^2)^*A \mid wa \in W_i \text{ for every } a \in A\} \end{aligned}$$

We then define a function $\rho : A^* \to \mathbb{N} \cup \{\infty\}$ called the *rank function* of the game by

$$\rho(x) = \min\{i \geq 0 \mid x \in W_i\}$$

with the convention that $\min \emptyset = +\infty$.

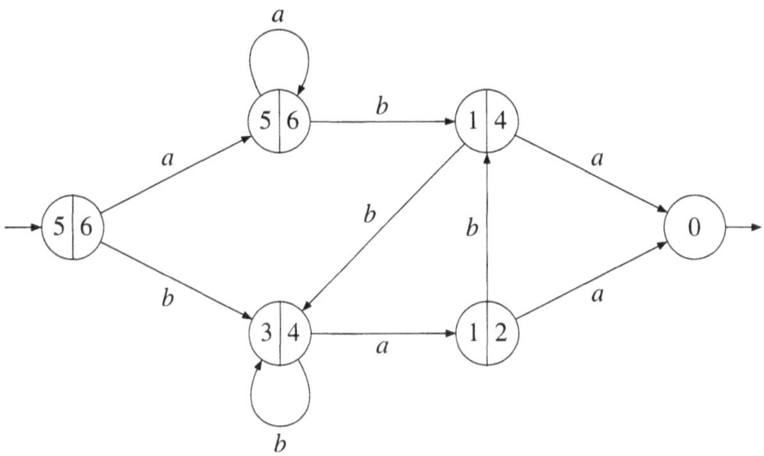

Figure 3.1. The game of words, for $P = \{aba, baa\}$.

Knowing the rank function makes it easy to determine a winning strategy: if $\rho(1)$ is finite, then I has a winning strategy which consists in decreasing the rank. If $\rho(1) = +\infty$, then Player II has a winning strategy: stay off the positions of finite rank.

Example 3.1. Let $U = b^*a$. Then $W_0 = b^*a$, $W_1 = W_0 \cup (b^2)^*$ and $W_2 = W_1 \cup (b^2)^*b$.

Let P be a finite set of words. We say that Player I wins the *game of words* on P if some word of P appears as a factor of the play. This game is therefore $G(A^*PA^\omega)$.

Example 3.2. For instance, if $P = \{aba, baa\}$, Player I wins. The values of the rank function in this case are represented in Figure 3.1. For each state of the automaton, two values are indicated, corresponding to the value of $\rho(x)$ according to the parity of the length of x.

Example 3.3. If $P = \{aaa, baa\}$, then Player II wins. Indeed, Player II has a winning strategy consisting in playing b all the time.

Example 3.4. Let $X = a(ab + ba)^\omega$. Player I wins this game. Indeed, she plays initially a and whenever Player II plays a (resp. b), then she plays b (resp. a). This game is a closed game.

3.2 Π_2-games. We now consider games in the class Π_2, consisting of countable intersections of open sets. Theorem III.3.11 shows that $X \in \Pi_2$ if and only if $X = \overrightarrow{U}$ for some set $U \subset A^*$ (recall that $x \in \overrightarrow{U}$ if x has an infinite number of prefixes in U). We shall prove the following result, known as Wolfe's theorem.

Proposition 3.2. *Any Π_2-game is determined.*

Proof. Suppose that $X = \overrightarrow{U}$. Let D be the set of winning positions for Player II. We claim that for every $x \notin D$, Player I has a winning strategy f_x to reach $U \setminus D$ from x. Indeed, otherwise, Player II can either

(1) avoid U and win $G(X)$, or

(2) reach D and subsequently win $G(X)$.

In both cases, Player II wins the game $G(X)$, a contradiction.

Player I has thus a winning strategy from each $x_0 \notin D$: play according to f_{x_0} to reach some $x_1 \in U \setminus D$. Then play according to f_{x_1} to reach some $x_2 \in U \setminus D$, and so on. The resulting play will have an infinite number of prefixes in U, and thus Player I wins. □

Rank functions can also be used to prove Proposition 3.2, as we did for Proposition 3.1, but in general, it requires ordinals instead of ordinary integers (see problem 1). We shall see later that in the case of ω-rational Π_2-sets, there is a rank function with values in $\mathbb{N} \cup \{\infty\}$ just as for open sets.

Example 3.5. Given a set P of words, let us consider the game $G((A^*P)^\omega)$, in which Player I wins if the play contains infinitely many factors from P. The winner of this game is the winner of the game of words $G(A^*PA^\omega)$ introduced in the preceding section. Indeed, Player I is able to force one occurrence of a word of P if and only if she is able to force an infinity of occurrences.

3.3 Martin's theorem. The following result is known as Martin's theorem. It is a generalization of Propositions 3.1 and 3.2.

Theorem 3.3 (Martin). *Any Borel game is determined.*

We shall prove the theorem for the Borel class Σ_3. Recall that the Borel classes of finite order are the dual classes Σ_n and Π_n defined for each $n \geq 1$ by

$$\Sigma_{n+1} = \exists^\omega \Pi_n, \qquad \Pi_{n+1} = \forall^\omega \Sigma_n$$

where Σ_1 (resp. Π_1) is the class of open (resp. closed) sets. Any ω-rational set is a finite boolean combination of Π_2-sets and thus in Σ_3. The proof given below extends easily to all Borel classes of finite order. The general proof makes use of ordinals (see the Notes section).

We shall use the following terminology. Given a word $x \in A^*$, we say that a set U of words of A^* beginning by x is a I-*imposed subgame* on x if

(a) $U \subset xA^*$,

(b) $ua \in U$ for each $a \in A$ and for each word $u \in U$ of odd length. More formally, $(U \cap A(A^2)^*)A \subset U$.

Thus a I-imposed subgame is a set of restrictions on the moves of Player I but not of Player II.

Proof. We consider a set $X \in \Sigma_3$ given by

$$X = \bigcup_{i \geq 0} \bigcap_{j \geq 0} X_{i,j}$$

where each $X_{i,j}$ is open. Let $(i, j) \to \langle i, j \rangle$ be a fixed bijection from $\mathbb{N} \times \mathbb{N}$ onto $\mathbb{N}$ and let $Y_{\langle i,j \rangle} = X_{i,j}$. For each $n \geq 0$, let also $P_n \subset A^*$ be a set such that $Y_n = P_n A^\omega$.

We shall build a Σ_2-game $G' = G(X')$ such that a player wins $G(X)$ if and only if he or she wins $G(X')$. The game G' includes *auxiliary moves* as indicated in Figure 3.2. The game G' is played in turn on the sets A, A again, $\mathcal{P}(A^*)$ and $\{0, 1\} \times A^*$. Thus, a play of G' has the form

$$(a_0, a_1, U_0, (t_0, u_0), a_2, a_3, \ldots, a_{2n}, a_{2n+1}, U_n, (t_n, u_n), \ldots)$$

where for each $n \geq 0$, $a_{2n}, a_{2n+1} \in A$, $U_n \subset A^*$ and $(t_n, u_n) \in \{0, 1\} \times A^*$. The following conditions are imposed on the auxiliary moves.

X_{00}	X_{01}	X_{02}	X_{03}	$\cdots$
X_{10}	X_{11}	X_{12}	$\cdots$	
X_{20}	X_{21}			
X_{30}				
$\vdots$				

Y_0	Y_1	Y_3	Y_6	$\cdots$
Y_2	Y_4	Y_7	$\cdots$	
Y_5	Y_8			
Y_9				
$\vdots$				

Table 3.1. The correspondence between the X_i's and the Y_i's.

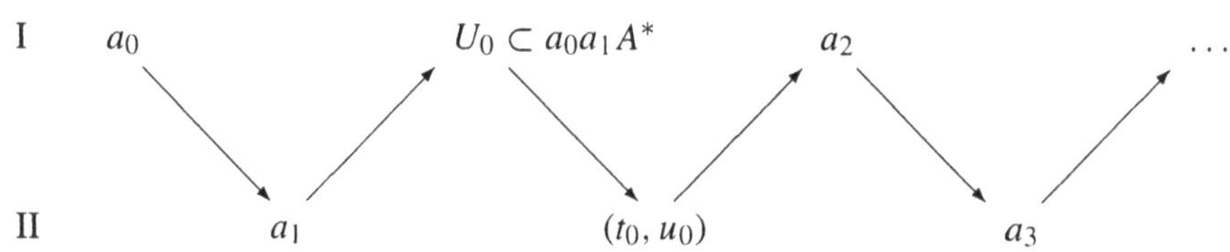

Figure 3.2. The game G'.

(1) Each U_n is a I-imposed subgame on $x_n = a_0a_1 \cdots a_{2n+1}$.

(2) Each t_n is equal to 0 or 1 and each u_n is a word of odd length such that $u_n \in x_nA^*$.

There is one more restriction on the moves to define completely the game G'. It will be given below after some informal explanation.

Intuitively, when Player I plays U_n, she offers her opponent to restrict her moves within U_n. Now Player II can answer in two ways:

(a) Either play $t_n = 0$ to express that he accepts the proposal. In return, he promises to avoid Y_n, i.e. to consider Player I as a winner if she reaches P_n. In this case, the choice of u_n is irrelevant.

(b) Play $t_n = 1$ to mean that he rejects the offer of Player I and at the same time chooses a word $u_n \in P_nA^*$. Both players will be required to play consistently with u_n.

To describe the rules formally, we introduce a sequence S_n defined by $S_{-1} = A^*$ and for $n \geq 0$,

(a) if $t_n = 0$, then $S_n = S_{n-1} \cap U_n$.

(b) if $t_n = 1$, then $S_n = S_{n-1} \cap C(u_n)$ where $C(u_n)$ denotes the set of finite words x which are compatible with u_n, i.e. such that x is a prefix of u_n or conversely.

The last rule of the game G' now reads

(3) For each n, $a_0 \cdots a_{2n}$ and $a_0 \cdots a_{2n+1}$ are in S_{n-1} and if $t_n = 1$, then $u_n \in P_nA^*$.

Finally, the winning set X' of G' is the set of sequences $(a_0, a_1, U_0, (t_0, u_0), \ldots)$ such that either

(i) for some $n \geq 0$, $t_n = 0$ and $x = a_0a_1a_2 \ldots \in Y_n$ (i.e. Player II did not keep his promise),

(ii) there is an $i \geq 0$ such that $t_{\langle i,j \rangle} = 1$ for all integers $j \geq 0$ (and in particular, by rule (3), $x \in X$).

It is clear that X' is in Σ_2. Indeed let T_n be the set of all words of G' such that $t_n = 0$. Then T_n is closed, and

$$X' = \bigcup_{n \geq 0} (Y_n \cap T_n) \cup \Big(\bigcup_{i \geq 0} \bigcap_{j \geq 0} T^c_{\langle i,j \rangle}\Big)$$

Thus G' is determined. We shall see that the winner of G' also wins G. Let us say that Player II has *kept all his commitments* in a play $x' = (a_0, a_1, U_0, (t_0, u_0), \ldots)$ of G' if for every n

$$x = (a_0, a_1, \ldots) \in Y_n \text{ if and only if } t_n = 1.$$

In this case, Player I wins the play in G' if and only if she wins the corresponding play x in G. Indeed, $x' \in X'$ if and only if $x \in X$.

Let us first consider the easier of the two possible cases, that is when Player I has a winning strategy σ' in G'. We shall build a winning strategy σ on G by following the strategy σ' on a simulated play of G'. Player I will have to guess (or simulate) the moves of Player II in G' in order to obtain a play in which Player II keeps all his commitments. She chooses systematically $t_i = 0$ unless $a_0 a_1 \cdots a_{2n} \in P_i A^*$ for some $i < n$. In this case, she changes (t_i, u_i) to $(1, a_0 \cdots a_{2n})$ and updates the following moves in G' to insure they always follow the strategy σ'.

More precisely, suppose that the moves $(a_0, a_1, \ldots, (t_{n-1}, u_{n-1}), a_{2n})$ have already been played with $t_{n-1} = 0$. If it happens that, for some $i < n$, $t_i = 0$ and $a_0 \cdots a_{2n} \in P_i A^*$, she chooses the smallest of such indices. Then she changes (t_i, u_i) to $(1, u_i')$ with $u_i' = a_0 \cdots a_{2n}$ and updates the moves between i and n to always follow the strategy σ'. This cannot change the sequence $(a_0, a_1, \ldots, a_{2n})$ because the rules of G' force both players to play consistently with u_i'.

In this way, no move U_n or (t_n, u_n) in G' is changed more than a finite number of times. And no move of G is ever changed since the choice of u_n forces the players in G' to play in the same way in G as before. In this way, Player I wins G.

Let us now consider the case where Player II wins G'. Let τ' be a winning strategy for Player II in G'. We shall again play in G by simulating the additional moves in G'. This time, it is the turn of Player II to simulate the auxiliary moves U_n or (t_n, u_n). The key idea is to choose the set U_0 as

$$\begin{aligned} U_0 &= \{u \in a_0 a_1 A^* \mid \text{for every I-imposed subgame } U \text{ on } a_0 a_1, \\ &\qquad (1, u) \neq \tau'(a_0, a_1, U)\} \end{aligned}$$

Since U_0 contains all words of $a_0 a_1 A^*$ of even length, it is a I-imposed subgame on $a_0 a_1$ and thus a legitimate move for Player I in G'. We also remark that $\tau'(a_0, a_1, U_0)$ is of the form $(0, u_0)$. Indeed, if we had $t_0 = 1$, then u_0 would be in U_0 by the rules of G', a contradiction. Suppose that we have played until stage n with

$$\begin{aligned} U_i &= \{u \in a_0 \cdots a_{2i+1} A^* \mid \text{for every I-imposed subgame } U \text{ on } a_0 \cdots a_{2i+1}, \\ &\qquad (1, u) \neq \tau'(a_0, a_1, U_0, \ldots, a_{2i+1}, U)\} \end{aligned}$$

Suppose that for some $i < n$, $a_0 a_1 \cdots a_{2n} \notin U_i$ and $t_i = 0$, contradicting rule (3) of G'. This means that there is some U_i' such that

$$\tau'(a_0, a_1, \ldots, a_{2i+1}, U_i') = (1, a_0 a_1 \cdots a_{2n}).$$

We can then change U_i for U_i'. This will change some further moves in G' as we follow the strategy τ', but none of the moves $(a_0, a_1, \ldots, a_{2n})$ by the definition of G'. Playing in this way, Player II will keep all his commitments because his strategy is a winning one and win G', thus winning G. □

One may note that the proof above is not effective in any sense, because of the uncountably many choices required by the simulation on the game G'. Consequently, the previous proof gives a general existence result on determinacy but it does not address the problem of the computational complexity of a solution.

4 Games on graphs

We now consider games played on graphs, in which each player chooses in turn a vertex adjacent to the current vertex. The abstract games $G(X)$ considered until now can be considered as games on the Cayley graph of A^*. But actually, a game $G(X)$ such that X is an ω-rational set can also be considered as a game on a finite graph by playing the game on an automaton recognizing X. Thus the notion of a game played on a graph will give us more flexibility by allowing us to choose the more appropriate graph to play the game. As a counterpart, the properties of the game graph obtained by playing the game on an automaton will the depend on the automaton chosen to recognize the winning set.

Let us define formally a game on a graph. A graph $G = (V, E)$ on a countable set of vertices V is called an *arena* or *game graph* if

(1) it is bipartite, i.e. its vertex set V is partitioned into V_1 and V_2 and the edges connect vertices belonging to different subsets,

(2) there is at least one edge starting from every vertex (i.e. there are no dead ends).

If G is an arena, a game on G is given by a set X of winning paths. We shall always suppose that the set of winning paths does not depend on the starting vertex, i.e. that the set X is suffix-closed. All the particular winning sets considered later have this property.

We consider that Player I plays on vertices in V_1 and Player II on vertices in V_2. A play is thus an infinite path in G. Thus, if the first vertex is in V_1, then Player I plays first and otherwise, Player II plays first. Player I wins the play if it is a winning path, i.e. belongs to X. Otherwise, Player II wins the play.

Thus a game on a graph is essentially a particular case of the notion of game defined in Section 2, the alphabet being the set of vertices of the graph. The only difference is that the first player is not always Player I. This defines the notion of a strategy and of a winning strategy for each player as a function from the set of paths of even (or odd length) into the set V of vertices.

Example 4.1. Let G be the graph of Figure 4.1. We use simple circles for the positions of Player I and double ones for those of Player II. Thus $V_1 = \{1, 3\}$ and $V_2 = \{2\}$. If X is the set of paths passing infinitely often by 1, Player II wins the game by always choosing vertex 3.

Let G be a game graph. A *memoryless strategy*, say for Player I, is a strategy which depends only on the last vertex of the path. When it is moreover a winning strategy,

Figure 4.1. A game graph.

we will speak of a *memoryless winning strategy*. Actually, a memoryless strategy can be considered as a subgraph since it consists in selecting one edge for each vertex on which Player I makes a move.

Formally, we say that a pair (P, F) consisting of a set $P \subset V$ of vertices and a subset $F \subset E \cap (P \times P)$ is a *winning policy* for Player I if

(a) for each $q \in P \cap V_1$, there is exactly one edge in F starting at q,

(b) for each $q \in P \cap V_2$, all edges starting at q are in F,

(c) all paths in (P, F) are winning for Player I.

The corresponding notion for Player II is symmetrical. It is clear that each player has a memoryless winning strategy from vertex p if and only if he or she has a winning policy (P, F) such that $p \in P$ (we use here the hypothesis that the set of winning paths is suffix-closed).

We say that a player has a *winning policy on a set* W if he or she has a winning policy of the form (W, F).

Example 4.2. In the game of Example 4.1, Player II has a memoryless winning strategy from every vertex.

The following auxiliary result allows one to merge different memoryless winning strategies into one winning policy. It shows that there is a maximal set on which a player has a winning policy.

Proposition 4.1. *Each player has a winning policy on the set of all vertices from which he or she has a memoryless winning strategy.*

Proof. Let W be the set of all vertices from which Player I has a memoryless winning strategy. Thus, for each $p \in W$ we can choose a winning policy (P_p, F_p) for Player I such that $p \in P_p$ (note that this requires the axiom of choice). Since the set W, as a subset of V, is countable, we may index the set of these strategies by integers. For each vertex $p \in W \cap V_1$, we select the pair (P_x, F_x) such that $p \in P_x$ which has minimal index. This defines a unique edge going out of p. Let F be the set formed by all edges of this type and by those in $(W \cap V_2) \times W$. Then the pair (W, F) is a winning policy on W for Player I.

The same result is of course true for Player II. □

For a vertex set U, we define the *attractor* of U for player I, denoted $A_1(G, U)$, or simply $A_1(U)$, as the set of vertices from which Player I can force a visit in U. The complement W of $A_1(U)$ is a set which is a *trap* for Player I: Player II can force Player I to remain inside W. This implies that for each vertex of W, there is at least one edge leading to a vertex in W. Thus the subgraph restricted to W is again a game graph, sometimes called a *subgame*.

The attractor of U for Player II, denoted by $A_2(U)$, is defined in the same way. In the same way, a trap for Player II is a set X of vertices such that Player I can force Player II to remain inside X. The complement of $A_2(U)$ is a trap for Player II.

4.1 Simple games. It is interesting to come back with these new definitions to the simple games defined by open or Π_2-sets that we have treated before.

Let us first consider open games. Such a game can always be obtained as a game on a graph G in which the set of winning paths is formed by the paths which pass through a given set $F \subset V$ of vertices. Let us denote by (G, F) such a game.

Proposition 4.2. *In an open game (G, F), Player I has a memoryless winning strategy on the attractor $A_1(F)$ and Player II on the complement.*

Proof. The winning strategies of Players I and II can be computed using rank functions defined directly using definitions similar to those of Section 3.1. Indeed, the rank of a vertex q can be defined as the smallest integer i such that $q \in W_i$ where W_i is an increasing sequence of subsets of Q defined by $W_0 = F$ and inductively

$$\begin{aligned} W_{i+1} = W_i &\cup \{p \in Q_1 \mid q \in W_i \text{ for some } (p, q) \in E\} \\ &\cup \{p \in Q_2 \mid q \in W_i \text{ for every } (p, q) \in E\}. \end{aligned}$$

The attractor of F is $\bigcup_{k \geq 0} W_k$.

The strategy of Player I on this set consists in decreasing the rank. The strategy of Player II on the complement consists in keeping off the positions of $A_1(F)$. Both strategies are memoryless strategies. □

The case of Π_2-games corresponds to games on a graph G in which the winning paths are those which pass infinitely often through F. Let us consider here the case where the graph G is finite. For a set W of vertices, we denote by $R(W)$ the set of vertices from which Player I can force a visit to W after a path of length ≥ 1. The set $R(W)$ is close to the attractor $A_1(W)$ and can be computed in a similar way. We then consider the decreasing sequence of sets defined by $W_1 = F$ and

$$W_{i+1} = R(F \cap W_i)$$

Since the set Q of vertices is finite, the sequence W_i is stationary. Let k be such that $W_k = W_{k+1}$. Then Player I has a memoryless winning strategy from the set $U = W_k$ consisting in reaching a vertex of $F \cap U$. Player II has a memoryless winning strategy on the other vertices. It consists in avoiding U.

Example 4.3. Let G be the graph represented in Figure 4.2 where 2, 4 are positions of Player I and 1, 3, 5 are positions of Player II. Player I wins if vertex 1 is visited infinitely often. Player I can force an infinity of visits of 1 from 1 and 2 but not from 3, 4 or 5. Thus Player I has a memoryless strategy on $\{1, 2\}$ and Player II on $\{3, 4, 5\}$.

Figure 4.2. A game graph.

4.2 Winning conditions. We shall consider games on graphs in which the winning set is defined through a finite set of colors in the following way. Let $G = (V, E)$ be a game graph and let $c : V \to Q$ be a map from the set of vertices into a finite set Q of colors. If x is an infinite path on G, we denote by $\mathrm{Inf}_c(x) = \mathrm{Inf}(c(x))$ the set of infinitely repeated colors in x.

We choose a particular collection $\mathcal{F}$ of subsets of Q and we define the set X of winning paths as those paths in G such that $\mathrm{Inf}_c(x)$ belongs to $\mathcal{F}$. In this section, we study particular classes $\mathcal{F}$ of sets of states used to define the winning paths. This game will be denoted by $(G, \mathcal{F})$.

We shall denote by $\mathcal{F}^c$ the complement of $\mathcal{F}$ in $\mathcal{P}(Q)$, that is, the set of subsets of Q which are not in $\mathcal{F}$. We define the *split tree* of $\mathcal{F}$ as follows. It is a tree T whose vertices are pairs $(1, X)$ for $X \in \mathcal{F}$ or $(2, X)$ for $X \notin \mathcal{F}$. The root of T is (σ, Q) with $\sigma = 1$ or 2 according to $Q \in \mathcal{F}$ or not. Inductively, if $x = (1, X)$ is a vertex of T, then

(1) if X contains subsets which do not belong to $\mathcal{F}$, then the children of x are all the $(2, Y)$'s where Y is a maximal subset of X which do not belong to $\mathcal{F}$,

(2) otherwise, x is a leaf of T.

A symmetrical condition holds if $x = (2, X)$

(1) if X contains subsets which belong to $\mathcal{F}$, then the children of x are all the $(1, Y)$ where Y is a maximal subset of X which is in $\mathcal{F}$,

(2) otherwise, x is a leaf of T.

Since the collection $\mathcal{F}$ is finite, the split tree of $\mathcal{F}$ has a finite height.

Example 4.4. Let G be the graph of Figure 4.1 with $\mathcal{F} = \{\{1, 2, 3\}\}$. The corresponding split tree is represented in Figure 4.3. If $\mathcal{F} = \{\{1, 2, 3\}, \{1\}\}$, the split tree is represented in Figure 4.4.

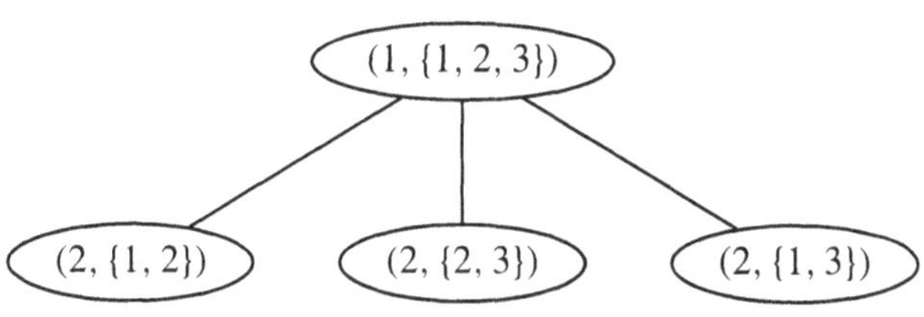

Figure 4.3. A split tree.

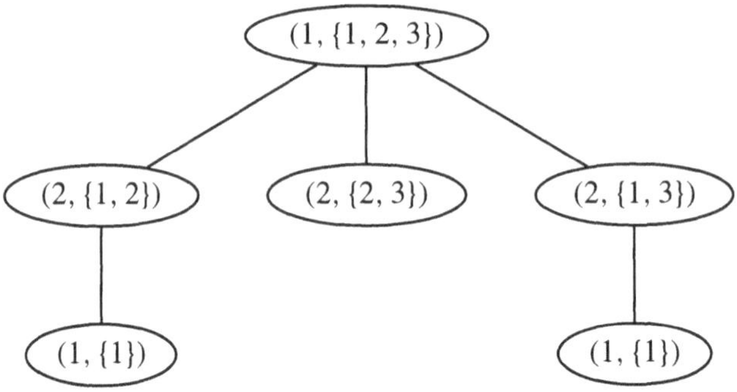

Figure 4.4. Another split tree.

The following proposition shows that the collection $\mathcal{F}$ can be computed from its split tree T.

Proposition 4.3. *A set X is in $\mathcal{F}$ if and only if there is a vertex $x = (1, Y)$ such that $X \subset Y$ and $X \not\subset Z$ for every child $(2, Z)$ of x.*

Proof. If X satisfies the condition, then X has to be in $\mathcal{F}$ since otherwise there would be a child $(2, Z)$ of x with $X \subset Z$. Thus $X \in \mathcal{F}$.

Conversely, let $x = (1, Y)$ be a node of T as low as possible such that $X \subset Y$. Such a node exists since the root satisfies this condition. No child $z = (2, Z)$ of x can satisfy $X \subset Z$ since otherwise z would have a child $(1, W)$ with $X \subset W$, a contradiction with the choice of x. Thus the property holds for x. □

Example 4.5. Let $F \subset Q$ and let $\mathcal{F} = \{X \subset Q \mid X \cap F \neq \emptyset\}$. The split tree of $\mathcal{F}$ has two vertices: the root $(1, Q)$ with one child $(2, Q \setminus F)$.

Let $\mathcal{P} = (L_i, R_i)_{i \in I}$ be a family of pairs of subsets of a set Q. A subset X of a set Q is said to satisfy *Streett condition* $S(\mathcal{P})$ if for every $i \in I$, $L_i \cap X \neq \emptyset$ or $R_i \cap X = \emptyset$. Thus a Streett condition is just the negation of a Rabin condition (see Chapter I). A collection $\mathcal{F}$ of subsets of Q is said to be expressible by a Streett condition if there is a set $\mathcal{P}$ of pairs such that $X \in \mathcal{F}$ if and only if X satisfies $S(\mathcal{P})$.

Proposition 4.4. *A collection $\mathcal{F}$ of nonempty subsets of a finite set Q is expressible by a Streett condition if and only if it is closed under union.*

Proof. If X and Y satisfy the Streett condition $S(\mathcal{P})$, then so does $X \cup Y$. Indeed, if $\mathcal{P} = (L_i, R_i)_{i \in I}$, then for every i, either one of X, Y meets L_i and so does $X \cup Y$, or none of them meets R_i and neither does $X \cup Y$.

Conversely, let $\mathcal{F}$ be a collection of nonempty subsets of Q closed under union and let T be the split tree of $\mathcal{F}^c$. Let I be the set of all subsets U of Q such that $(1, U)$ is a node of T (thus none of the U's are in $\mathcal{F}$). Each node $(1, U)$ can have at most one child since $\mathcal{F}$ is closed under union. Let V_U be the label of this child if there is one and let V_U be the empty set otherwise.

Then, by Proposition 4.3, $X \in \mathcal{F}^c$ if and only if, for some U, $X \subset U$ and $X \not\subset V_U$. Let $L_U = U^c$ and $R_U = V_U^c$. Then $X \in \mathcal{F}^c$ if and only if for some U, one has $X \cap R_U \neq \emptyset$ and $X \cap L_U = \emptyset$. Thus $X \in \mathcal{F}$ if and only if for each $U \in I$, one has $X \cap L_U \neq \emptyset$ or $X \cap R_U = \emptyset$. It follows that X is defined by the Streett condition $S(\mathcal{P})$ with $\mathcal{P} = (L_U, R_U)_{U \in I}$. □

Let $\mathcal{C}$ be an increasing sequence of subsets of Q

$$\mathcal{C} : E_1 \subset F_1 \subset E_2 \subset F_2 \subset \ldots \subset E_n \subset F_n$$

A subset P of Q is said to satisfy the *Rabin chain condition* $\mathcal{C}$ if there is an index k such that $P \cap E_k = \emptyset$ and $P \cap F_k \neq \emptyset$.

There is an alternative formulation of Rabin chain condition using a *parity condition*. Given a function $\mu : Q \to \mathbb{N}$, we say that X *satisfies the parity condition* μ if and only if

$$\min \{ \mu(q) \mid q \in X \} \text{ is odd}$$

This formulation of the chain condition makes it extremely easy to use since it entails a very compact representation.

We say that a collection $\mathcal{F}$ of subsets of Q is *expressible* by a Rabin chain condition, (resp. by a parity condition μ) if there exists a chain $\mathcal{C}$ such that $X \in \mathcal{F}$ if and only if X satisfies $\mathcal{C}$ (resp. μ).

Proposition 4.5. *Let $\mathcal{F}$ be a collection of finite nonempty subsets of a set Q. The following conditions are equivalent:*

(1) *$\mathcal{F}$ and $\mathcal{F}^c$ are closed under union.*

(2) *$\mathcal{F}$ can be defined by a Rabin chain condition.*

(3) *$\mathcal{F}$ can be defined by a parity condition.*

Proof. (1) implies (2). Let T be the split tree of the collection $\mathcal{F}$. Since $\mathcal{F}$ and $\mathcal{F}^c$ are closed under union, each vertex of T has at most one child. It follows that T has exactly one leaf. We may suppose that $Q \in \mathcal{F}^c$, so that the root of T is $(2, Q)$. Let $((2, V_0), (1, U_1), (2, V_1), \ldots)$ be the unique path from the root to the leaf. For $i \geq 0$, let $F_i = U_i^c$ and $E_i = V_i^c$. Then the sequence $\mathcal{C}$ formed by $E_1 \subset F_1 \subset \cdots$ is increasing and, by Proposition 4.3, one has $X \in \mathcal{F}$ if and only if X satisfies $\mathcal{C}$. Thus $\mathcal{F}$ can be defined by a Rabin chain condition.

(2) implies (3). Let

$$E_1 \subset F_1 \subset E_2 \subset F_2 \subset \ldots \subset E_n \subset F_n$$

be an increasing sequence of subsets of Q defining a Rabin chain condition. Let us define a function $\mu : Q \to \mathbb{N}$ by setting, for $1 \leq k \leq n$,

$$\mu(q) = \begin{cases} 0 & \text{if } q \in E_1 \\ 2k-2 & \text{if } q \in E_k \setminus F_{k-1} \text{ and } k > 1 \\ 2k-1 & \text{if } q \in F_k \setminus E_k \\ 2n & \text{if } q \notin F_n \end{cases}$$

Then a set X satisfies $\mathcal{C}$ if and only if $\mu(X)$ is odd. Thus X is defined by a parity condition.

(3) implies (1). If $\mathcal{F}$ is defined by a parity condition, then $\mathcal{F}$ and $\mathcal{F}^c$ are clearly closed under union. □

4.3 Parity games. We now consider games on graphs, called *parity games*, in which the winning set is defined by a parity or Rabin chain condition defined by a chain $\mathcal{C}$

$$E_1 \subset F_1 \subset \ldots \subset E_n \subset F_n.$$

More precisely, let $G = (V, E)$ be a game graph and let $c : V \to Q$ be a coloring. The *parity game* defined by $\mathcal{C}$ is the game $(G, \mathcal{F})$, where $\mathcal{F}$ is the collection of subsets of Q defined by the Rabin chain condition $\mathcal{C}$. Therefore, the set X of winning paths consists of the paths x such that $\text{Inf}_c(x) \cap E_k = \emptyset$ and $\text{Inf}_c(x) \cap F_k \neq \emptyset$ for some $k \in \{1, \ldots, n\}$.

Observe that any play visiting infinitely often E_1-colored vertices is winning for Player II. Indeed, let x be a path in G and let $P = \text{Inf}_c(x)$. If $P \cap E_1 \neq \emptyset$, any of the sets E_k is met infinitely often and thus there can be no index k satisfying the condition $P \cap E_k = \emptyset$.

Theorem 4.6. *In a parity game, one of the players has a memoryless winning strategy from each vertex.*

Proof. We shall prove by induction on the length of the chain $\mathcal{C}$ that there is a partition $Q = W_1 \cup W_2$ on the set of vertices such that Player I has a memoryless winning strategy on W_1 and Player II has one on W_2. We make the assumption that $E_1 \neq \emptyset$. Otherwise, we would exchange the roles of Players I and II in the forthcoming discussion. Thus, whenever we find a game with a chain of length n and $E_1 = \emptyset$, we can invoke the induction hypothesis.

If $n = 0$, then Player II wins anyway.

Let W be the set of vertices from which Player I has a memoryless winning strategy. By Proposition 4.1, Player I has a winning policy on the set W. We want to prove that Player II has a memoryless winning strategy from every vertex in $L = W^c$. We first notice that, for Player I, W is its own attractor. Thus L is a trap for Player I. This implies that the graph induced by G on L is a game graph G'.

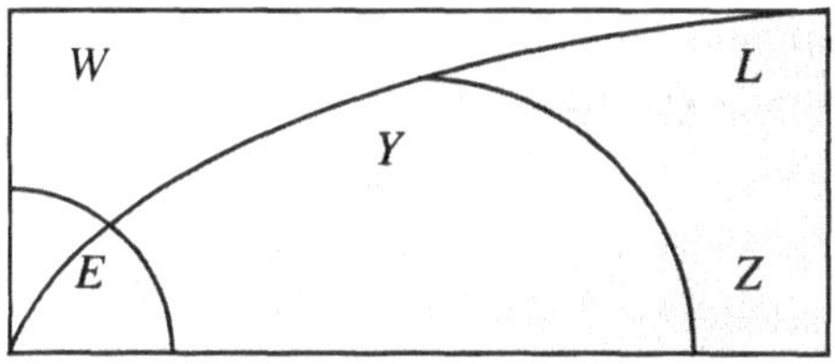

Figure 4.5. The memoryless strategies.

Let Y be the attractor for Player II of the set E_1 inside the game G'.

$$Y = A_2(G', L \cap E_1)$$

Let finally Z be the complement of Y in L. Since $Z \cap E_1 = \emptyset$, we may apply the induction hypothesis to the game restricted to Z. There can be no positions in Z on which Player I has a winning strategy because Z is disjoint from W. Thus Player II has a memoryless winning strategy on Z (provided the game remains within Z). Let us consider the strategy for Player II on L consisting in following the winning strategy on Z and to reach E_1 on the vertices of Y. This is clearly a memoryless strategy. It is actually winning because either the play passes infinitely often through E_1 or it stays out of Y from some moment on and then it stays within Z and is thus winning for Player II. □

4.4 Parity automata. An *m-parity automaton* is an automaton $\mathcal{A} = (Q, i, \mu)$ where μ is a function from Q into $\{0, 1, \ldots, m\}$. For a path c in $\mathcal{A}$, we define

$$\mu(c) = \max\{\mu(q) \mid q \text{ occurs infinitely often in } c\}.$$

By definition, a path c in $\mathcal{A}$ is successful if it starts at i and the integer $\mu(c)$ is odd. As for parity games, an equivalent definition is obtained by considering an increasing sequence $\mathcal{C} = E_1 \subset F_1 \subset \ldots \subset E_n \subset F_n$. A *Rabin chain automaton* is an automaton $\mathcal{A} = (Q, i, \mathcal{C})$, with $\mathcal{C}$ as above. A path in $\mathcal{A}$ is final if the set of infinitely repeated states satisfies the Rabin chain condition $\mathcal{C}$. As a consequence of Proposition 4.5, any parity automaton can be viewed as a Rabin automaton and vice versa.

We shall use here a construction that allows one to build a parity automaton from a Muller automaton. We shall meet this construction later in Chapter V. It is based on the notion of a *memory extension* of a finite automaton.

Let $\mathcal{A} = (Q, i, \mathcal{F})$ be a Muller automaton. We build a deterministic automaton $\mathcal{B}$ as follows. Let $\mathrm{Arr}(Q)$ denote the set of sequences of elements of Q, each appearing at most once (sometimes called arrangements). The set of states of $\mathcal{B}$ is $S = \{(u, v) \mid uv \in \mathrm{Arr}(Q)\}$. An element of S can be called *last appearance record* since the transitions are defined in such a way that the arrangement uv gives the order of last occurrence of each state. The division of uv into a pair (u, v) marks the previous position of the last state. The initial state of $\mathcal{B}$ is the pair (ε, i) where i is the initial state of $\mathcal{A}$ and where ε denotes the empty sequence. The transitions are defined as follows. Let $(u, v) \in S$ and $a \in A$. Let p be the last element of uv and let $q = p \cdot a$. Then

$$(u, v) \cdot a = \begin{cases} (x, yq) & \text{if } uv = xqy \\ (uv, q) & \text{if } q \notin uv \end{cases}$$

The automaton $\mathcal{B}$ is called the *memory extension* of $\mathcal{A}$.

Example 4.6. The memory extension of the automaton $\mathcal{A}_{\text{l}}$ of Figure 4.6 is pictured in Figure 4.7.

The fundamental property of the memory extension is the following one. For a path c in an automaton, we denote by $\text{Inf}(c)$ the set of states occurring infinitely often in c. In the following proposition, we use the notation $\underline{v}$ to denote the set of elements appearing in a sequence v.

Proposition 4.7. *Let $\mathcal{A}$ be an automaton and let $\mathcal{B}$ be its memory extension. Let c be an initial path in $\mathcal{A}$ and let c' be the corresponding path in $\mathcal{B}$. Then $T = \text{Inf}(c)$ if and only if all states $(u, v) \in \text{Inf}(c')$ satisfy $\underline{v} \subset T$ and at least one satisfies $\underline{v} = T$.*

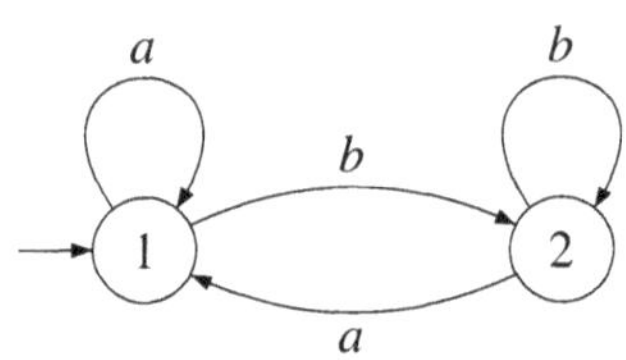

Figure 4.6. A Muller automaton.

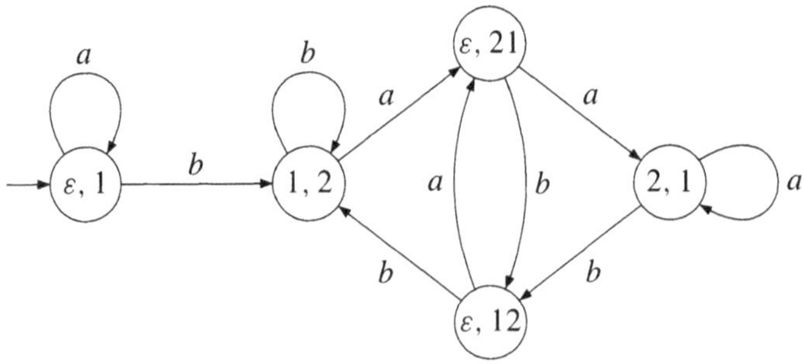

Figure 4.7. The memory extension of the automaton $\mathcal{A}_{\text{l}}$.

Proof. Let $(q_0, q_1, \ldots)$ be the sequence of states appearing along c. We first observe that all states of c' are ultimately of the form (uv'_n, v''_n) with $\underline{u} = S$ and $\underline{v_n} = T$, where $v_n = v'_n v''_n$ and where S is the set of states appearing finitely often along c. Next, for each state of this form with $v'_n \neq \varepsilon$, there is later on the path a state of the form (u, v_m). Let indeed $v'_n = qw'$ with $q \in Q$. Since $q \in T$, there is an occurrence of q on c later on. For the first index $m > n$ such that $q_m = q$, we have $v'_m = \varepsilon$.

This shows that the condition is necessary and sufficient. □

We define a chain $E_0 \subset F_0 \subset \cdots \subset E_n \subset F_n \subset \cdots$ as follows. For $i \geq 0$, let E_i be the set of states (u, v) of $\mathcal{B}$ such that either $|u| < i$ or $|u| = i$ and $\underline{v} \notin \mathcal{F}$. And let F_i be the union of E_i and the set of states (u, v) such that $|u| = i$ and $\underline{v} \in \mathcal{F}$. This defines a chain automaton which is clearly equivalent to $\mathcal{A}$.

We have thus proved the following result.

Theorem 4.8. *For any Muller automaton, there exists an equivalent parity automaton.*

It would not change anything to use as set of states the pairs (u, v) where uv is a permutation of Q. In this case, the initial state can be chosen as any of the states of the form (u, vi), where i is the initial state. This can be used to reduce the number of states of the resulting automaton, as in the following example.

Example 4.7. Let $\mathcal{A} = (Q, i, \mathcal{F})$ be the Muller automaton represented in Figure 4.8 with $i = 2$ and $\mathcal{F} = \{1, 2, 3\}$. It recognizes the set of infinite words in $(ab + ba)^\omega$ with both an infinite number of occurrences of ab and ba.

The memory extension $\mathcal{B}$ of $\mathcal{A}$ is represented in Figure 4.9. Actually, we have represented only the states which are permutations of Q. Both states $(3, 12)$ or $(1, 32)$ can be used as initial state. The chain reduces to $F_0 = \{(\varepsilon, 123), (\varepsilon, 321)\}$ since $E_0 = \emptyset$ and $F_1 = F_0$. Thus $\mathcal{B}$ is actually a Büchi deterministic automaton.

4.5 Rational winning strategies. Let G be a game graph in which the winning condition is given in Muller form, i.e. by a collection $\mathcal{F}$ of subsets of Q such that Player I wins the play if the set of infinitely repeated vertices belongs to $\mathcal{F}$. The following example shows that, in general, there is no memoryless winning strategy.

Example 4.8. Let G be the game graph of Figure 4.10 with $\mathcal{F} = \{\{1, 2, 3\}\}$. Player I has a winning strategy from each vertex consisting in choosing alternately 1 and 3 from

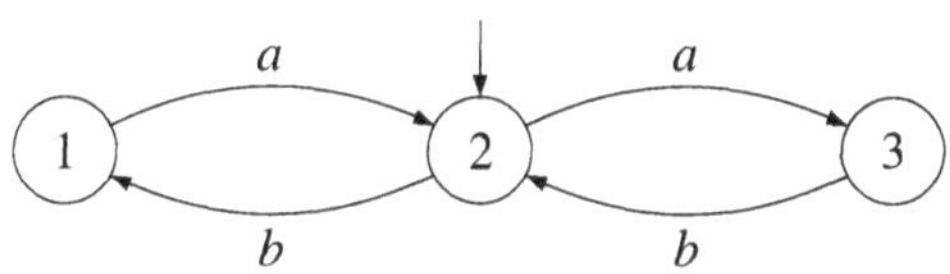

Figure 4.8. A Muler automaton.

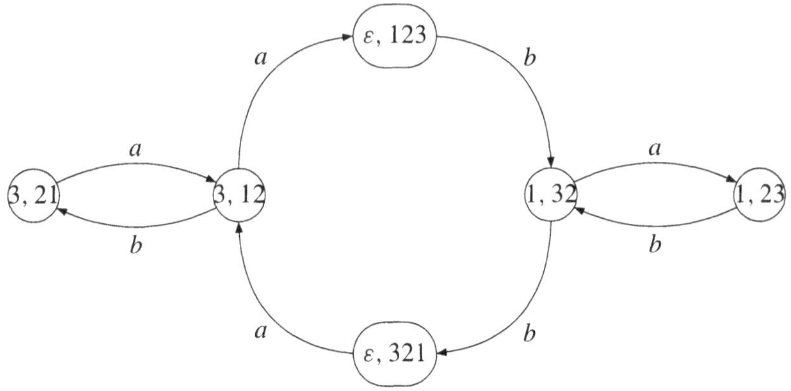

Figure 4.9. The memory extension.

Figure 4.10. Player I has no memoryless strategy.

vertex 2. However, there is no memoryless strategy since it would force Player I to always choose either 1 or 3 after 2, resulting in a loop either on $\{1, 2\}$ or on $\{2, 3\}$.

We now come back to abstract games given by the winning set $X \subset A^\omega$. A *rational* or *finite memory* strategy for Player I is given by a finite deterministic automaton $\mathcal{S} = (M, i, \delta)$ and a function

$$f : M \to A.$$

We say that Player I follows the strategy $(\mathcal{S}, f)$ in the play $a_0 a_1 \cdots$ if for every $n \geq 0$, $a_{2n} = f(m)$ where $m = \delta(i, a_0 \cdots a_{2n-1})$.

We prove the following result, known as the Büchi-Landweber theorem.

Theorem 4.9. *In a rational game, one of the players has a rational winning strategy.*

Proof. By Theorem 4.8, there is a parity automaton $\mathcal{A}$ recognizing X. We may suppose, by duplicating the states that the set of states Q is partitioned into $Q = Q_1 \cup Q_2$ in such a way that the initial state is in Q_1 and that the graph of $\mathcal{A}$ is bipartite. The game $G(X)$ defines a parity game on the graph of $\mathcal{A}$ and this game is equivalent to the original one. By Theorem 4.6, one of the players, say Player I, has a memoryless winning strategy in this game. This player has a rational winning strategy in $G(X)$. It uses the automaton $\mathcal{A}$ and the function $f : Q_1 \to A$ defined as follows (the value of f on Q_2 is irrelevant).

For $p \in Q_1$, there is a state $q \in Q_2$ given by the memoryless strategy of Player I. Let a be a symbol such that (p, a, q) is a transition of $\mathcal{A}$. Then we define $f(p) = a$. This is clearly a rational winning strategy for Player I. □

Example 4.9. Let X be the set recognized by the Muller automaton of Example 4.7. The graph of the automaton coincides with the graph of Example 4.8 and the winning condition is the same. Accordingly, Player I wins $G(X)$ by choosing alternately the states 1 and 3, i.e. by playing alternately a and b (or any other strategy ensuring to play infinitely often a and b).

A parity automaton recognizing X is represented in Figure 4.9. This time, we have a memoryless strategy on the graph of the automaton. It consists in playing b in $(1, 32)$ and a in $(3, 12)$. It happens to be the same strategy as above, resulting in one of the two possible plays $(abba)^\omega$ or $(baab)^\omega$.

5 Wadge games

Let $X \subset A^\omega$ and $Y \subset B^\omega$. The Wadge game $G(X, Y)$ is a game on $A \cup B$ defined as follows. Player I first chooses $a_0 \in A$. Then Player II chooses $b_0 \in B$. Player I chooses $a_1 \in A$, and so on. Thus a play in this game is a sequence $a_0 b_0 a_1 b_1 \cdots \in (A \cup B)^\omega$ which is the interleaving of the two sequences $x = a_0 a_1 \cdots \in A^\omega$ and $y = b_0 b_1 \cdots \in B^\omega$ played by each player. Player II wins if either ($x \in X$ and $y \in Y$) or ($x \notin X$ and $y \notin Y$).

Such a game can be viewed as a game on the alphabet $A \cup B$ with a rule forcing Player I to choose a symbol from A and Player II a symbol from B. Observe that, if X and Y are Borel sets, then so is the winning set $Z \subset (A \cup B)^\omega$.

These games are strongly related to the following notion. We say that $X \subset A^\omega$ *Wadge reduces* or simply *reduces* to $Y \subset B^\omega$, denoted $(X, A^\omega) \leq_W (Y, B^\omega)$ or simply $X \leq_W Y$ if there exists a continuous function $f : A^\omega \to B^\omega$ such that

$$X = f^{-1}(Y)$$

The function f is called a *reduction*. It is important that the definition of the reduction is relative to the embedding of X in A^ω (see Example 5.2 below).

Obviously, $X \leq_W Y$ if and only if $X^c \leq_W Y^c$. It is possible to have simultaneously $X \leq_W Y$ and $Y \leq_W X$, in which case X and Y are called *Wadge equivalent*, denoted $X \equiv_W Y$. The $\equiv_W$-class of X is called the *Wadge class* of X.

The class of X^c and the class of X are called *dual*. The class of X is called self-dual if $X \equiv_W X^c$.

Example 5.1. Let A be a nonempty alphabet and let $X = A^\omega$. We claim that $X \leq_W Y$ for any nonempty set $Y \subset B^\omega$. Indeed, given a fixed element $y \in Y$, the constant function defined for all $x \in A^\omega$ by $f(x) = y$ is a reduction from X to Y. The Wadge class of A^ω is formed by all B^ω where B is nonempty.

Example 5.2. Let $X \subset A^\omega$ be a closed set with $A = \{a, b\}$. Then $(X, A^\omega) \leq_W (a^\omega, A^\omega)$, i.e. is reducible to a^ω (as a subset of A^ω). By Proposition III.3.7, X is recognized by a deterministic Büchi automaton in which all states are final. Therefore, X is recognized by a complete deterministic Büchi automaton $\mathcal{A} = (Q, i, E)$ in which all states, except the sink s, are final. We define a continuous function $f : A^\omega \to A^\omega$ by $f(x) = g(c)$ where c is the initial path of label x in $\mathcal{A}$ and g is obtained by sending all states to a except s which is sent to b. Then $f^{-1}(a^\omega) = X$.

Conversely, if X is reducible to a^ω, then X is closed since a^ω is closed and f is continuous.

5.1 Wadge lemma. The following result is known as Wadge's lemma. We state it for Borel sets although we only need it for Σ_3. Thus, although the proof relies on the determinacy of Borel games, our proof of Martin's theorem for Σ_3-games is sufficient.

Theorem 5.1. *Let $X \subset A^\omega$ and $Y \subset B^\omega$ be Borel sets. Then either $X \leq_W Y$ or $Y \leq_W X^c$.*

Proof. Since X, Y are Borel sets, by Martin's theorem the game $G(X, Y)$ is determined. Let us first suppose that Player II has a winning strategy. Thus there is a function $\varphi : (A \times B)^* A \to B$ which makes Player II a winner when he follows this strategy. Let $f : A^\omega \to B^\omega$ be the function defined by $f(a_0 a_1 \cdots) = b_0 b_1 \cdots$ if for each $i \geq 0$, $b_i = \varphi(a_0 b_0 \cdots a_{i-1} b_{i-1} a_i)$. Then f is a continuous function from A^ω into B^ω. Let, for $x \in A^\omega$, $y = f(x)$. If $y \in Y$, then $x \in X$ since otherwise Player I wins the play. Thus $f^{-1}(Y) = X$ since any $x \in X$ can be played by Player I. This shows that $X \leq_W Y$.

If we now suppose that Player I has a winning strategy, we have a function $\varphi : (A \times B)^* \to A$ giving this strategy. We then define $f : B^\omega \to A^\omega$ by $f(b_0 b_1 \cdots) = a_0 a_1 \cdots$ where for each $i \geq 0$, $a_i = \varphi(a_0 b_0 \cdots a_{i-1} b_{i-1})$. The function f is continuous and $f^{-1}(X) = Y^c$. This shows that $Y \leq_W X^c$. □

There are several consequences of Wadge's Lemma. Let us first consider the order defined on pairs (X, X^c) by

$$(X, X^c) \leq (Y, Y^c) \text{ if and only if } X \leq_W Y \text{ or } X \leq_W Y^c.$$

Wadge's lemma expresses that, for Borel sets, this is a total preorder.

Another formulation is the following. The pairs formed by the classes of X and X^c when $X \not\equiv_W X^c$ are the only incomparable pairs of Wadge classes.

5.2 Rational reduction. Let $\mathcal{A}$ be an automaton with an initial state i and let $f : Q \times A \to B^*$ be a function. We assume that for each loop in $\mathcal{A}$, there is at least one edge with a nonempty output label. We denote by $f_{\mathcal{A}}$ the function from A^ω to B^ω

defined by

$$f_{\mathcal{A}}(a_0a_1a_2\ldots) = f(i, a_0)f(i \cdot a_0, a_1)\ldots$$

The condition on the labels of loops guarantees that the image of an infinite word under $f_{\mathcal{A}}$ is again infinite. A function $f : A^\omega \to B^\omega$ is called *sequential* if $f = f_{\mathcal{A}}$ for some automaton $\mathcal{A}$. It can be easily verified that a sequential function is continuous. When f is a function from $Q \times A$ to B, the function $f_{\mathcal{A}}$ is said to be *synchronous*. The preorder $\leq_W$ can be relativized to sequential and to synchronous functions as follows. For $X \subset A^\omega$ and $Y \subset B^\omega$, we set $X \leq_{DA} Y$ (resp. $X \leq_{DS} Y$) if there exists a sequential (resp. synchronous sequential) function f such that $X = f^{-1}(Y)$.

Example 5.3. Let $\mathcal{A} = (Q, i, F)$ be a deterministic Büchi automaton and let X be the ω-rational set recognized by this automaton. Let f be the function from $Q \times A$ to $\{0, 1\}$ defined by $f(q, a) = 1$ if $q \in F$ and $f(q, a) = 0$ otherwise. Then $X = f^{-1}((0^*1)^\omega)$. Thus, since f is synchronous, we have $X \leq_{DS} (0^*1)^\omega$.

Since a sequential mapping is continuous, $X \leq_{DA} Y$ implies $X \leq_W Y$. The converse is not true in general. However, we have the following important result.

Theorem 5.2. *The relation $\leq_{DA}$ coincides with the Wadge reduction $\leq_W$ on the class of ω-rational sets.*

Proof. Let $X \subset A^\omega$ and $Y \subset B^\omega$ be such that $X \leq_W Y$. Let $f : A^\omega \to B^\omega$ be a continuous function such that $X = f^{-1}(Y)$. We consider a modified Wadge game, denoted $G^+(X, Y)$, in which Player II has the possibility of passing his turn, playing ε. He is allowed to pass only during a finite number of consecutive turns. This may clearly be considered as a Wadge game (X, Z) using an additional symbol $\# \notin B$ and the appropriate set Z. The function f gives a winning strategy to Player II. Indeed, Player II leaves Player I play $x_0 = a_0a_1\ldots a_{n_0}$ until $f(x_0A^\omega) \subset b_0B^\omega$. Then, he plays b_0. He then passes his turn until Player I has played a word $x_0x_1 = a_0\ldots a_{n_0}\ldots a_{n_1}$ such that $f(x_0x_1A^\omega) \subset b_0b_1B^\omega$. He then plays b_1, and so on. Clearly, this defines a winning strategy for Player II. According to Theorem 4.9, Player II has a rational winning strategy for this game. This gives a sequential reduction form X to Y. □

The relations $\leq_{DS}$ and $\leq_{DA}$ do not coincide in general as shown by the following example.

Example 5.4. Let $A = \{a, b\}$, $X = aA^\omega$ and $Y = aaA^\omega$. Then $X \leq_{DS} Y$, since $X = f^{-1}(Y)$, where f is the synchronous sequential function realized by the automaton represented in Figure 5.1. In fact, $X \equiv_W Y$, since $Y = g^{-1}(X)$, where g is the sequential function realized by the automaton represented in Figure 5.2. However, it is not true that $Y \leq_{DS} X$. Indeed, if f is a synchronous sequential function from A^ω

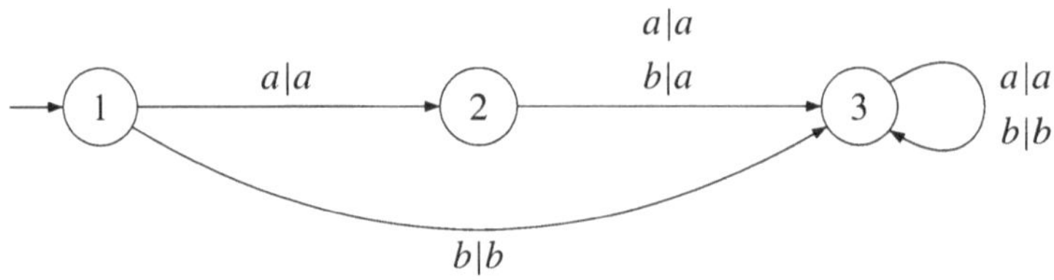

Figure 5.1. A synchronous sequential automaton.

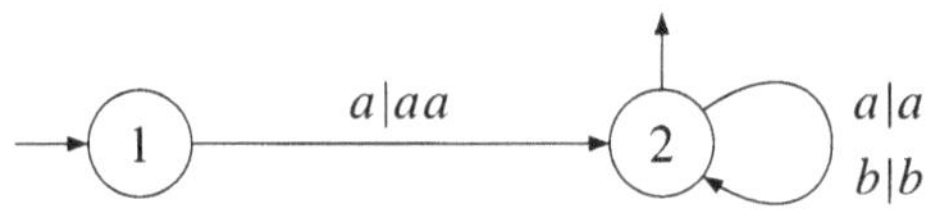

Figure 5.2. An asynchronous sequential automaton.

into itself, then $f^{-1}(aA^{\omega})$ contains $f^{-1}(a)A^{\omega}$. But since f is synchronous, $f^{-1}(a)$ is a subset B of A. Therefore, $f^{-1}(a)A^{\omega}$ cannot be a subset of aaA^{ω} unless $f^{-1}(a)$ is empty. But in this case, $f^{-1}(aA^{\omega})$ would also be empty.

6 Exercises

6.1 Borel games.

Exercise 1. Let $X = \bigcap U_n A^{\omega}$ and let $U = \bigcup_n U_n$. For a set $P \subset A^*$, denote by $\rho_n(P, x)$ the rank function of the set $P \cap U_n$, as in Section 3.1.

Define by induction on a countable ordinal α a sequence of sets P_α as follows.

(1) $P_0 = A^*$,

(2) $P_{\alpha+1}$ is the set of words such that $\rho_n(P_\alpha, x) < \infty$ for all $n \geq 0$,

(3) If α is a limit ordinal, then $P_\alpha = \bigcap_{\beta<\alpha} P_\beta$.

The sequence of sets P_α is decreasing and there is an ordinal α such that $P_\alpha = P_{\alpha+1}$. Let $P = P_\alpha$ for such an ordinal α. If $x \in P$, then $\rho_n(P, x) < \infty$ for all $n \geq 0$.

Show that Player I wins $G(X)$ if and only if $x \in P$.

6.2 Infinite games.

Exercise 2. Let G be an acyclic connected directed graph on a set Q of vertices. Let i be a particular vertex of G and let T be the set of leaves in G. We define a game on G by letting Player I start at vertex i and letting each player choose at his turn a vertex q

adjacent to the current vertex p, i.e. a vertex q such that (p, q) belongs to the set E of vertices of G. A player who cannot move, i.e. who is placed on a vertex in T has lost.

We define a function $f : Q \to \mathbb{N}$ by the following conditions. For $t \in T$ we have $f(t) = 0$. For other vertices, we have

$$f(p) = \text{mex}\{f(q) \mid (p, q) \in E\}$$

where $\text{mex}(E)$ is the *minimum excluded* of a set E of natural numbers, i.e. the smallest $n \geq 0$ such that $n \notin E$.

Show that a vertex p is a winning position if and only if $f(p) \neq 0$. The function f is called the *Grundy function* of the game.

Exercise 3. The game of Nim is played with n pebbles arranged in heaps of sizes n_1, $n_2, \ldots, n_k$. At each step, the player can take an arbitrary number of pebbles from one of the heaps (but from only one). The player who cannot play (because there are no more pebbles) has lost.

Show that the graph of the game of Nim is acyclic and that its Grundy function is

$$f(n_1, n_2, \ldots, n_k) = n_1 +_2 n_2 +_2 \ldots +_2 n_k$$

where $+_2$ is the binary addition consisting in adding componentwise modulo 2 the binary expansions of the numbers. For example, $f(1, 3, 5, 7) = 0$, showing that Player I loses the game.

Exercise 4. We consider the following game, called the *game of partitions*. It is played with n pebbles. At each turn, there is a partition $(n_1, n_2, \ldots, n_k)$ of the n pebbles into k parts. The player can either

(1) divide one of the parts into equal parts (of arbitrary size),

(2) join two parts of different size into one part.

The player reaching the partition $(1, 1, \ldots, 1)$, in which no move is possible, has lost. Find, for $n \leq 5$ the winning positions. Show that, for $n \geq 7$, there are cycles in the game.

6.3 Games on graphs.

Exercise 5. Let $G = (Q, E)$ be a parity game on a finite graph with a winning condition given by a chain $\mathcal{S} = (E_1, F_1, E_2, F_2, \ldots, E_m, F_m)$. The aim of this exercise is to prove Theorem 4.6 by induction on $Card(Q)$ instead of m as we did.

If Q has at most two elements, the problem is trivial. The induction starts with the hypothesis $E_1 = \emptyset$ (compare with the other proof). Let U be the complement of the attractor set $A_1(q)$. The induction hypothesis can be applied to U, giving a partition $U = U_1 + U_2$ into sets from which each player has a memoryless winning strategy in the subgame played on U. We distinguish two complementary cases: either Player I can force a direct transition from q to $A_1(q) + U_1$ (case 1) or Player II can force a direct transition from q to U_2 (case 2).

Prove that, in case 1, Player I has a memoryless winning strategy on the set $A_1(q) + U_1$ and Player II on U_2.

In case 2, let $V = A_2(U_2)$ be the attractor set of U_2 for Player II and let W be the complement of V. The induction hypothesis gives a partition $W = W_1 + W_2$ into sets from which each player has a memoryless winning strategy in the subgame played on W.

Prove that player I has a memoryless winning strategy on W_1 and player 2 on $V + W_2$.

7 Notes

The games introduced here are known as *games of perfect information*. This means that each player is informed at each step of the moves of the opponent and both share the same information. This excludes card games in which the cards in each player's hand are unknown to the other player. The reader interested in the mathematical theory of games will find many interesting results and examples in the book of Berlekamp, Conway and Guy [29, 30]. Standard references also include the book of J. von Neumann and O. Morgenstern [343] and the Handbook edited by Aumann and Hart [15].

The proof of Martin's theorem (Theorem 3.3) is the culmination of a long series of results proving the determinacy of games of increasing complexity. The original reference is [187]. The case of open (or closed) games (Proposition 3.1) is due to Gale and Stewart [117]. The case of Π_2-games (Proposition 3.2) is due to Wolfe [364]. The proof of Martin's theorem presented here is essentially reproduced from Moschovakis [204]. It is given here for the case of Σ_3 but the argument can be readily generalized to any Borel class of finite order Σ_n. A proof for Borel classes of any order can be found in Kechris [153].

The notion of a split tree appears in Zielonka [370]. The Rabin chain condition for automata on infinite words was first considered by Mostowski [206] and independently by Carton [58].

The results of Section 4.1 are the result of several contributions from Gurevich and Harrington [131], McNaughton [191] and Zielonka [370]. The presentation follows closely Thomas [331]. Several chapters of the recent monograph [127] present material connected with ours. This includes a general presentation of infinite games [188], the memory less determinacy of parity games [165] and algorithms for parity games [159]. It has been proved by N. Klarlund [157] that Player I has a memoryless winning strategy on her winning set if the game is defined by a Rabin pair condition. This is clearly a generalization of Theorem 4.6 since, by Proposition 4.5, the game is a parity game if and only if the winning set and its complement are both definable by a Rabin pair condition. Klarlund's condition is actually necessary and sufficient, provided that the winning condition is defined through a partial coloring of the vertices (see [370]). A proof of Theorem 4.6 by induction on the number of vertices (provided it is finite) has been given by McNaughton [191] (see Exercise 5).

The theorem of Büchi and Landweber (Theorem 4.9) appeared in Büchi and Landweber [54]. It can be considered as the solution of a specification problem, raised by Church [70]: given a condition relating pairs $(x, y) \in A^\omega \times B^\omega$ of sequences which can be defined by a finite automaton, is there a finite automaton definable function $f : A^\omega \to B^\omega$ such that all pairs $(x, f(x))$ satisfy the given condition?

The results of Wadge are from his Ph.D. thesis [344]. Our presentation follows Kechris [153]. It can be shown that the order defined by $X \leq_W Y$ or $X \leq_W Y^c$ is actually a well ordering on pairs (X, X^c) of Borel sets (see Kechris [153]). We shall come back to the Wadge hierarchy in Chapter V.

The function defined in Exercise 2 is called the *Sprague-Grundy function* of the game (see Conway [80]). Exercise 1 is from Gurevich [130]. The game of partitions (Exercise 4) is from Berlekamp et al. [29, 30] where it is called "fair shares, unequal pairs". It is recommended to play with almonds (around ten almonds shared in small parts of size 2 or 3 at the beginning). The game is not guaranteed to end in a bounded number of steps. Actually, it turns out that for $n \leq 10$, one of the players can always win in a finite number of steps but for $n \geq 11$, there are positions from which each player can force a draw.

V

WAGNER HIERARCHY

1 Introduction

In this chapter, we describe a hierarchy on ω-rational sets, called the Wagner hierarchy, which refines the Borel hierarchy introduced in Chapter III. It completely characterizes the equivalence classes of ω-rational sets for the *Wadge preorder*, already introduced in Section IV.5.

Recall that given two topological spaces E, F and two subsets $X \subset E$, $Y \subset F$, X reduces to Y in the Wadge preorder, denoted $X <_W Y$, if there is a continuous function f from E into F such that $X = f^{-1}(Y)$. This preorder is actually very close to be a total order on Borel sets, in the sense that its antichains have at most two elements. Furthermore, since we are only interested in ω-rational sets, Theorem IV.5.2 allows us to replace continuous functions by asynchronous sequential functions in the definition of the Wadge preorder.

The complete description of the Wagner hierarchy is rather involved, but can be obtained as the result of three successive refinements.

The first refinement relies on the concept of chains, which can be defined either on Muller automata or on ω-semigroups. Using the length of chains as a measure of complexity, this leads to the chain hierarchy, in which classes of increasing complexity form a sequence which exhausts the class of ω-rational sets.

The second refinement makes use of a similar concept, the superchains. They can also be defined either on Muller automata or on ω-semigroups. A superchain is formed by a sequence of chains.

The Wagner hierarchy relies on the notion of derivation of a Muller automaton and is indexed by ordinals. More precisely, the classification uses the concepts of chains, superchains and derivation to associate with each ω-rational set X an ordinal $\gamma(X)$ in such a way that, up to a three valued variable called the sign, one has $X <_W Y$ if and only if $\gamma(X) < \gamma(Y)$. We shall see that several concepts and results introduced in previous chapters appear as particular cases of this result. The chain hierarchy is the subhierarchy indexed by ordinals of the form ω^m and the superchain hierarchy is the subhierarchy indexed by ordinals of the form $\omega^m n$.

The chapter is organized as follows. It begins with a reminder on ordinals. Section 3 presents some definitions relating to classes of sets. It contains a result due to Hausdorff describing the boolean closure of a family of sets by alternating sums (Theorem 3.1). Chains are introduced in Section 4. The section begins with the definition of chains in Muller automata and in ω-semigroups. The main result of the section (Theorem 4.10) is a characterization in topological terms of the classes of the chain hierarchy. Section 5 presents the superchain hierarchy. It begins with a description of superchains, either defined in ω-semigroups or in Muller automata. The main result (Theorem 5.14) is a topological characterization of the classes of the hierarchy. Section 6 contains a description of the final hierarchy. The notion of the derived automaton is then presented. The main result is Wagner's Theorem (Theorem 6.2) asserting that the classes of the hierarchy are the Wadge classes of ω-regular sets.

It is possible to give an alternative presentation of the full Wagner hierarchy, avoiding the use of derived automata and extending the parallelism between automata and ω-semigroups. See the notes section.

2 Ordinals

It is convenient to use ordinals as indices to the Wagner hierarchy. We start this chapter by a brief reminder on ordinals.

A *linear order* on a set E is an order relation $\leq$ such that, for every pair $(x, y) \in E \times E$, either $x \leq y$ or $y \leq x$. A *linear ordering* is a pair $(E, \leq)$ where $\leq$ is a linear order on E. Given two linear orderings $(E, \leq)$ and $(F, \leq)$, an *embedding* from $(E, \leq)$ into $(F, \leq)$ is a map $\varphi : E \to F$ such that

$$x < y \text{ if and only if } \varphi(x) < \varphi(y)$$

A surjective embedding is called an *isomorphism*. We say that $(E, \leq)$ and $(F, \leq)$ have the same *order type* if there is an isomorphism from $(E, \leq)$ onto $(F, \leq)$.

If two linear orderings $(E, \leq)$ and $(F, \leq)$ have the same order type τ, then $(E, \geq)$ and $(F, \geq)$ have the same order type, denoted $\tilde{\tau}$.

Example 2.1. Let $\leq$ denote the usual order on integers. The order type of $(\mathbb{N}, \leq)$ is denoted ω. Thus the order type of $(\mathbb{N}, \geq)$ is $\tilde{\omega}$. The order type of $(\mathbb{Z}, \leq)$ is denoted ξ. It is easy to see that $\xi = \tilde{\tilde{\xi}}$. The order type of $(E, \leq)$, where E has n elements, is denoted n. This is a way to identify natural numbers with finite order types.

Let $(E, \leq)$ and $(F, \leq)$ be two linear orderings and assume that E and F are disjoint. We define the sum $(E, \leq_E) + (F, \leq_F)$ to be the linear ordering $(E \cup F, \leq)$ where the new order $\leq$ is defined by

$$x < y \text{ if and only if } \begin{cases} x \in E \text{ and } y \in E \text{ and } x <_E y \\ x \in E \text{ and } y \in F \\ x \in F \text{ and } y \in F \text{ and } x <_F y. \end{cases}$$

One can show that if $(E, \leq)$ and $(E', \leq)$ have the same order type then $(E, \leq) + (F, \leq)$ and $(E', \leq)+(F, \leq)$ have the same order type. Similarly, $(F, \leq)+(E, \leq)$ and $(F, \leq)+(E', \leq)$ have the same order type. Therefore addition can be defined without ambiguity on order types.

Example 2.2. Since $(\mathbb{Z}, \leq)$ is isomorphic to $(\mathbb{N}, \geq) + (\mathbb{N}, \leq)$, the relation $\xi = \tilde{\omega} + \omega$ holds. One can also verify that addition on finite order types can be identified with the usual addition on natural numbers.

Addition of order types is associative but not commutative. For instance $\tilde{\omega}+\omega \neq \omega+\tilde{\omega}$, and $1 + \omega = \omega \neq \omega + 1$.

Let $(I, \leq)$ be a linear ordering and let $(E_i, \leq_i)_{i \in I}$ be a family of pairwise disjoint linear orderings. The generalized sum

$$\sum_{i \in (I, \leq)} (E_i, \leq_i)$$

is the linear ordering $(E, \leq)$ where $E = \sum_{i \in I} E_i$ and

$$x < y \text{ if and only if } \begin{cases} \text{for some } i \in I,\, x, y \in E_i \text{ and } x <_i y \quad \text{or} \\ \text{for some } i < j,\, x \in E_i \text{ and } y \in E_j. \end{cases}$$

If all the linear orderings $(E_i, \leq_i)$ are isomorphic to the same linear ordering $(E, \leq)$, the generalized sum $\sum_{i \in (I, \leq)} (E_i, \leq_i)$ is called the *product* of $(I, \leq)$ by $(E, \leq)$ and is simply denoted $(E, \leq)(I, \leq)$.

Again, the generalized sum as well as the product can be extended to order types. One can show that multiplication is an associative but not commutative operation on order types. Distributivity holds only on one side: if τ, τ_1 and τ_2 are order types,

$$\tau(\tau_1 + \tau_2) = \tau\tau_1 + \tau\tau_2$$

but $(\tau_1 + \tau_2)\tau$ is in general different from $\tau_1\tau + \tau_2\tau$.

A linear ordering $(E, \leq)$ is called a *well-ordering* if every nonempty subset of E has a least element. It can be shown that a linear ordering $(E, \leq)$ is a well-ordering if and only if there is no embedding from $(\mathbb{N}, \geq)$ into $(E, \leq)$. The order type of a well-ordering is called an *ordinal*.

An ordinal α is called a *successor ordinal* if there is an ordinal β such that $\alpha = \beta + 1$. Any other ordinal, except 0, is called a *limit ordinal*. It can be shown that every ordinal can be written uniquely in the form $\alpha + n$, where α is a limit ordinal and n is a natural number.

Given two well-orderings $(E, \leq)$ and $(F, \leq)$, we write $(E, \leq) \leq (F, \leq)$ if $(E, \leq)$ is isomorphic to an initial segment of $(F, \leq)$. This relation depends only on the order type of $(E, \leq)$ and $(F, \leq)$ and can therefore can be extended to ordinals in a natural way. The following proposition summarizes some basic properties of ordinals.

Proposition 2.1. *If α, β and γ are ordinals, the following properties hold:*

(1) *if $\alpha \leq \beta$ then $\alpha + \gamma \leq \beta + \gamma$, $\gamma + \alpha \leq \gamma + \beta$, $\alpha\gamma \leq \beta\gamma$ and $\gamma\alpha \leq \gamma\beta$.*

(2) *if $\alpha < \beta$, there is a unique ordinal δ such that $\beta = \alpha + \delta$.*

(3) *if $\alpha + \beta = \alpha + \gamma$, then $\beta = \gamma$.*

Note, however, that the condition $\beta + \alpha = \gamma + \alpha$ does not implies that $\beta = \gamma$. If α is an ordinal, the set

$$E_\alpha = \{\beta \mid \beta \text{ is an ordinal smaller than } \alpha\}$$

is linearly ordered by $<$, and has order type α. Let α be a limit ordinal, and let φ be a function from E_α into E_β such that

$$\gamma < \delta \text{ implies } \varphi(\gamma) < \varphi(\delta)$$

We define $\lim_{\gamma<\alpha} \varphi(\gamma)$ to be the smallest ordinal larger than $\varphi(\gamma)$ for each $\gamma < \alpha$. Given two ordinals α and β, the ordinal α^β is defined by induction as follows:

(1) $\alpha^0 = 1$,

(2) $\alpha^{\beta+1} = \alpha^\beta\alpha$

(3) if β is a limit ordinal, $\alpha^\beta = \lim_{\gamma<\beta} \alpha^\gamma$.

The ordinals we shall use are all smaller than ω^ω. One can show that any ordinal smaller than ω^ω admits a unique *Cantor normal form*

$$\alpha = \omega^{m_1} n_1 + \cdots + \omega^{m_k} n_k$$

where k is a nonnegative integer, $n_1, \ldots, n_k$ are positive integers and $m_1, \ldots, m_k$ are nonnegative integers such that $m_1 > \ldots > m_k$.

3 Classes of sets

In this section, we adopt a simplified notation to handle more easily the usual set operations. Let E be a set. The union and intersection of two subsets X and Y of E are denoted by $X + Y$ and XY, respectively. The complement of X is denoted by X^c. Furthermore, if Y is contained in X, the set difference $X \setminus Y$ will be simply denoted by $X - Y$.[1]

A set $\mathcal{F}$ of subsets of E containing $\emptyset$ and E and closed under finite union and complement is called a *positive boolean algebra*. It is a *boolean algebra* if, furthermore, it is closed under complement. For instance, the classes of open (resp. closed) sets form a positive boolean algebra.

3.1 Wadge classes. A *class* of sets is a correspondence that associates with each finite alphabet A, a set $\Gamma(A^\omega)$ of subsets of A^ω. We use here the terms class and correspondence instead of set and function to avoid any paradox of set theory, since it is known, for instance, that the finite sets do not form a set.

We say, slightly improperly, that a class Γ satisfies a property $\mathcal{P}$ if each class $\Gamma(A^\omega)$ has the property $\mathcal{P}$. For instance, the class of open sets is closed under finite intersections and arbitrary union. When the alphabet A is understood, we often say that a set of infinite words of A^ω belongs to Γ instead of saying that it belongs to $\Gamma(A^\omega)$.

If Γ is a class, its *dual class* is the class $\neg\Gamma$ formed of the complements X^c of the elements of Γ. We say that Γ is *self-dual* if $\Gamma = \neg\Gamma$.

A class of sets Γ is *a Wadge class* if $X \in \Gamma$ and $Y \leq_W X$ implies $Y \in \Gamma$. This notion is a slight restriction of the notion of a Polish class introduced in Chapter III: the difference is that our classes are defined on spaces of the form A^ω, with A finite, and not on arbitrary Polish spaces.

A class of sets Γ is *a Wadge degree* if it is an equivalence class for $\equiv_W$.

A set X is Γ*-complete* if, for each alphabet A

$$\Gamma(A^\omega) = \{Y \subseteq A^\omega \mid Y \leq_W X\}$$

It is clear from the definition that a class which admits a Γ-complete set is closed under Wadge reduction.

3.2 The boolean hierarchy. Let $\mathcal{F}$ be a set of subsets of E. For each integer $n \geq 1$, we denote by $D_n(\mathcal{F})$ the class of all sets of the form

$$X = X_1 - X_2 + \ldots \pm X_n \tag{3.1}$$

where the sets X_i satisfy $X_i \in \mathcal{F}$ and $X_1 \supset X_2 \supset X_3 \supset \cdots \supset X_n$. By convention, the expression on the right hand side of 3.1 should be evaluated from left to right, but given

[1] Outside of this section, the notation XY will denote concatenation, but we shall use the notations $X + Y$ and $X - Y$ episodically in the other sections of this chapter.

the conditions on the X_i's, it can also be evaluated as $(X_1 - X_2) + (X_3 - X_4) + (X_5 - X_6) \ldots$. We also set $D_0 = \{\emptyset\}$. An equivalent definition of D_n is given in Exercise 1.

By extension, if Γ is a class of sets, the classes $D_n(\Gamma)$ are defined in the natural way by setting $D_n(\Gamma)(A^\omega) = D_n(\Gamma(A^\omega))$.

The classes $D_n(\mathcal{F})$ define a hierarchy within the boolean closure of $\mathcal{F}$. Actually, the union of all $D_n(\mathcal{F})$ is the boolean closure of $\mathcal{F}$ as soon as $\mathcal{F}$ is a positive boolean algebra.

Theorem 3.1. *Let $\mathcal{F}$ be a positive boolean algebra of subsets of E. The union of the classes $D_n(\mathcal{F})$ for $n \geq 1$ is the boolean closure of $\mathcal{F}$.*

Proof. Let $D(\mathcal{F}) = \bigcup_{n\geq 1} D_n(\mathcal{F})$. We have $D_1(\mathcal{F}) = \mathcal{F}$ and thus $\mathcal{F} \subset D$. It is therefore enough to prove that $D(\mathcal{F})$ is closed under complement and intersection. For $X = X_1 - X_2 + \ldots \pm X_n$, we have

$$U - X = U - X_1 + X_2 - \ldots \mp X_n$$

and thus $X \in D(\mathcal{F})$ implies $U - X \in D(\mathcal{F})$. Let $X = X_1 - X_2 + \ldots \pm X_n$ and $Y = Y_1 - Y_2 + \ldots \pm Y_m$ be two elements of D. We claim that

$$XY = Z_1 - Z_2 + \ldots \pm Z_{n+m} \tag{3.2}$$

with

$$Z_k = \sum_{\substack{i+j=k+1 \\ i \text{ and } j \text{ not both even}}} X_i Y_j$$

Therefore

$$\begin{aligned}
Z_1 &= X_1Y_1, \\
Z_2 &= X_1Y_2 + X_2Y_1, \\
Z_3 &= X_1Y_3 + X_3Y_1, \\
Z_4 &= X_1Y_4 + X_2Y_3 + X_3Y_2 + X_4Y_1, \\
Z_5 &= X_1Y_5 + X_3Y_3 + X_5Y_1, \\
&\vdots \\
Z_{n+m-1} &= \begin{cases} X_nY_m & \text{if } m \text{ and } n \text{ are not both even} \\ \emptyset & \text{otherwise.} \end{cases}
\end{aligned}$$

Let $Z = Z_1 - Z_2 + \ldots \pm Z_{n+m}$. To prove the claim, consider for each set $X = X_1 - X_2 + \ldots \pm X_n$ associated with the decreasing sequence $X_1, \ldots, X_n$ of subsets of E, the function μ_X defined on E by

$$\mu_X(x) = \max\{i \geq 1 \mid x \in X_i\}$$

with the convention that $\mu_X(x) = 0$ if $x \in U - X_1$. Then $x \in X$ if and only if $\mu_X(x)$ is odd. We now evaluate $\mu_Z(x)$, as a function of $i = \mu_X(x)$ and $j = \mu_Y(x)$. We first observe that if $k \geq i + j$, then $x \notin Z_k$. Next, if i and j are not both even, then $x \in X_i Y_j$ and $X_i Y_j \subset Z_{i+j-1}$, whence $\mu_Z(x) = i + j - 1$. Finally, if i and j are both even, then $x \notin Z_{i+j-1}$ and thus $\mu_Z(x)$ is either equal to 0 or to $i + j - 2$. Summarizing the various cases, we observe that $\mu_X(x)$ and $\mu_Y(x)$ are both odd if and only if $\mu_Z(s)$ is odd, which proves the claim. It follows that $D(\mathcal{F})$ is closed under complement. □

In general, $D_n(\mathcal{F})$ is neither closed under intersection nor union (see Exercises 3 and 4). However, the following property holds

Proposition 3.2. *Let $\mathcal{F}$ be a positive boolean algebra of subsets of E. Let X and Y be subsets of E such that X and X^c belong to $\mathcal{F}$ and Y belongs to $D_n(\mathcal{F})$. Then $X + Y$ and XY belong to $D_n(\mathcal{F})$.*

Proof. The result follows from the following elementary formula

$$X(Y_1 - Y_2 + \ldots \pm Y_n) = XY_1 - XY_2 + \ldots \pm XY_n$$
$$X + (Y_1 - Y_2 + \ldots \pm Y_n) = (X + Y_1) - Y_2X^c + \ldots \pm Y_nX^c$$

The first formula follows from (3.2) and the second one is illustrated in Figure 3.1, in which the elements of $X + (Y_1 - Y_2 + \ldots \pm Y_n)$ appear in white. □

3.3 Separated and biseparated union. We now introduce the notions of separated and biseparated union. Let X_1, X_2, X be three subsets of A^ω. We say that Z is the union of X_1 and X_2 *separated* by X if $Z = X^cX_1 + XX_2$. If $X_1 \subset X^c$ and $X_2 \subset X$ (see Figure 3.2), Z is simply the union of X_1 and X_2. We also define, for two classes Γ,

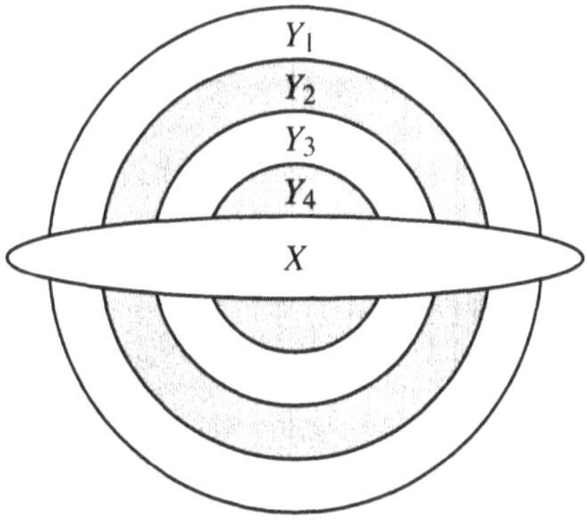

Figure 3.1. A representation of $X + (Y_1 - Y_2 + \ldots \pm Y_n)$.

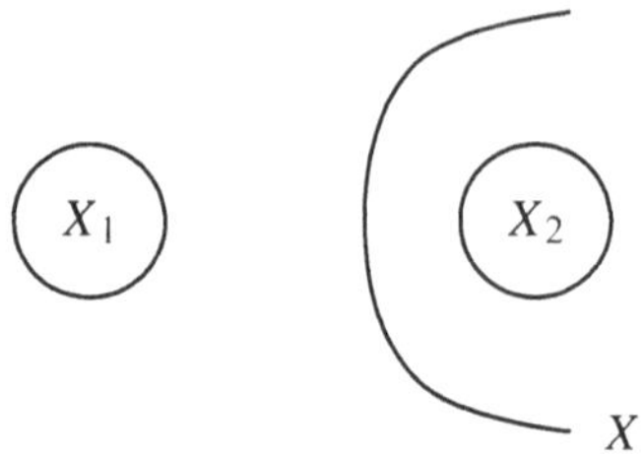

Figure 3.2. Separated union of X_1 and X_2.

Δ of subsets of A^ω, a new class $\text{Sep}(\Gamma, \Delta)$ as the class of all sets which are the union of some $X_1 \in \Delta$ and $X_2 \in \neg\Delta$ separated by some $X \in \Gamma$. Note that, if Γ is the empty class, $\text{Sep}(\Gamma, \Delta)$ reduces to Δ.

Let X_1, X_2, Y_1, Y_2 and X be five subsets of A^ω such that $X_1 X_2 = \emptyset$. The *biseparated union* of X_1, X_2 and X protected by Y_1 and Y_2 is the set

$$Z = X_1 Y_1 + X_2 Y_2 + X_1^c X_2^c X.$$

If $X_1 \subset Y_1$, $X_2 \subset Y_2$, $Y_1 Y_2 = \emptyset$, $XY_1 = \emptyset$ and $XY_2 = \emptyset$, then

$$Z = X_1 + X_2 + X$$

(see Figure 3.3). If Φ, Γ, Δ are three classes of subsets of A^ω, we denote by $\text{Bisep}(\Phi, \Gamma, \Delta)$ the class of sets Z which are the biseparated union of X_1, X_2 and X protected by Y_1 and Y_2 with $Y_1, Y_2 \in \Phi$, $X_1 \in \Gamma$, $X_2 \in \neg\Gamma$ and $Z + Y_2 \in \Delta$. Actually, if Δ is closed under boolean operations with elements of Φ, as it will be the case in general, we can assume that $Z \in \Delta$ instead of $Z + Y_2 \in \Delta$.

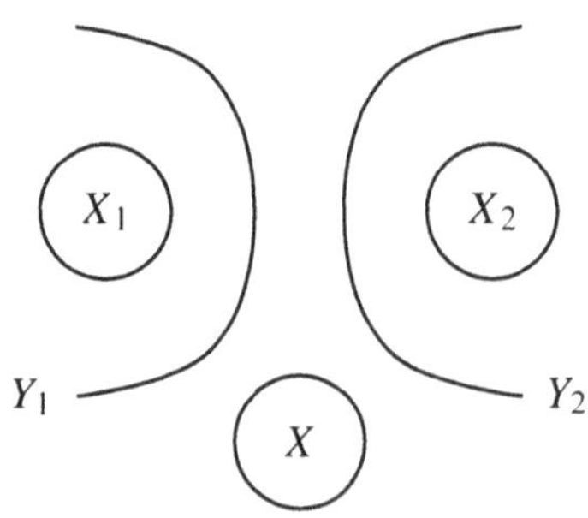

Figure 3.3. Biseparated union of X_1, X_2 and X.

4 Chains

In this section, we introduce the concept of *chains*, first in Muller automata and subsequently in ω-semigroups. We show that these two concepts are in natural correspondence and study their basic properties. Next we define the chain hierarchy, which can be considered as a simplified version of the Wagner hierarchy.

4.1 Chains in automata. Let $\mathcal{A} = (Q, E, i, \mathcal{T})$ be a complete Muller automaton. Recall from Chapter I that a subset T of Q is admissible if there exists a path p starting in i such that $\mathrm{Inf}(p) = T$. An $\mathcal{A}$-*chain* of length m is an increasing sequence

$$R_0 \subset R_1 \subset \cdots \subset R_{m-1}$$

of m admissible subsets of Q such that, for $0 \leq i \leq m-1$, the R_i are alternately accepting (i.e. in $\mathcal{T}$) and rejecting (i.e. outside $\mathcal{T}$).

We say that the chain is *positive* if $R_0 \in \mathcal{T}$ and *negative* if $R_0 \notin \mathcal{T}$. We denote by $m^+(\mathcal{A})$ (resp. $m^-(\mathcal{A})$) the maximal length of positive (resp. negative) $\mathcal{A}$-chains and we set

$$m(\mathcal{A}) = \max(m^+(\mathcal{A}), m^-(\mathcal{A}))$$

It is obvious by the definition that $m(\mathcal{A})$ is finite for any finite Muller automaton $\mathcal{A}$. One indeed has the relation $m(\mathcal{A}) \leq \mathrm{Card}(Q)$. Note that $m^+(\mathcal{A}) = 1$ if and only if the table is full. It was shown in Theorem I.7.8 that full table Muller automata are equivalent to deterministic Büchi automata.

Example 4.1. The automaton $\mathcal{A}_1$ represented in Figure 4.1, with table $\mathcal{T} = \{\{1\}, \{1,2\}\}$, accepts the set X of Example 4.2. The only chain of length 2 is $\{2\} \subset \{1,2\}$ and it is negative. Hence we have $m^+(\mathcal{A}_1) = 1$ and $m(\mathcal{A}_1) = m^-(\mathcal{A}_1) = 2$.

We conclude this section by establishing a few basic properties of the integers $m^+(X)$, and $m^-(X)$ introduced above.

Proposition 4.1. *For any complete Muller automaton $\mathcal{A}$, the following properties hold:*

(1) $m(\mathcal{A}) \geq 1$,

(2) $|m^+(\mathcal{A}) - m^-(\mathcal{A})| \leq 1$.

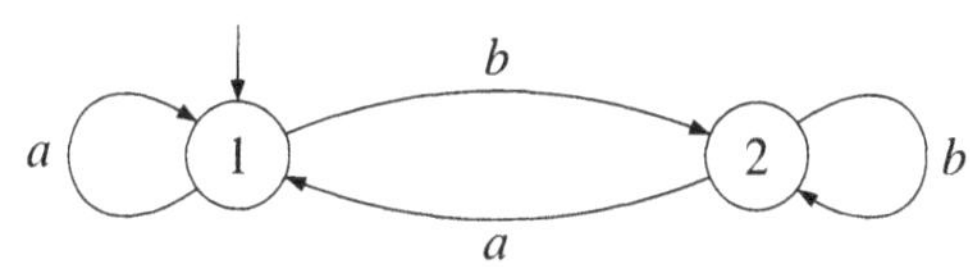

Figure 4.1. The automaton $\mathcal{A}_1$, with $\mathcal{T} = \{\{1\}, \{1,2\}\}$.

Proof. (1) is obvious.

(2) If $R_0, \ldots, R_{m-1}$ is an $\mathcal{A}$-chain then $R_1, \ldots, R_{m-1}$ is an $\mathcal{A}$-chain of opposite sign. This shows that $|m^+(X) - m^-(X)| \leq 1$. □

4.2 Chains in ω-semigroups. We now come to the definition of chains in an ω-semigroup. Although the definition might appear somehow artificial at first sight, it is designed to be equivalent to the definition on automata given in the previous section. It is worth mentioning that our definition can be applied in particular to free ω-semigroups and to finite ω-semigroups. We shall see later that, in the case of finite ω-semigroups, it is possible to give the definition in a different form (see Theorem 4.4).

Let $S = (S_+, S_\omega)$ be an ω-semigroup and let X be a subset of S_ω. Let Y be a nonempty subset of S_+ and let $z_0, z_1, \ldots, z_{m-1}$ be a sequence of m elements of S_+. Set $Z = \{z_0, z_1, \ldots, z_{m-1}\}$ and

$$\text{for } -1 \leq i \leq m-1, \qquad Z_i = \{z_0, z_1, \ldots, z_i\}$$
$$\text{for } 0 \leq i \leq m-1, \qquad W_i = YZ^*(z_i Z_i^*)^\omega \tag{4.1}$$

Observe that W_i can also be written as

$$W_i = YZ^*(Z_{i-1}^* z_i)^\omega \tag{4.2}$$

We say that $C = (Y, z_0, \ldots, z_{m-1})$ is an X-*chain* of S if the sets W_i are alternately included in X and in X^c. The chain is said to be *positive* (resp. *negative*) if W_0 is a subset of X (resp. X^c). Two positive (resp. negative) chains are said to be of the *same sign*.

The number m is called the *length* of the chain C. We denote by $m^+(X)$ (resp. $m^-(X)$) the maximal length of the positive (resp. negative) X-chains. We set $m^+(X) = 0$ (resp. $m^-(X) = 0$) if the set of positive (resp. negative) chains is empty, and $m^+(X) = m^-(X) = \infty$ if the lengths of X-chains are unbounded. Finally, we set

$$m(X) = \max(m^+(X), m^-(X))$$

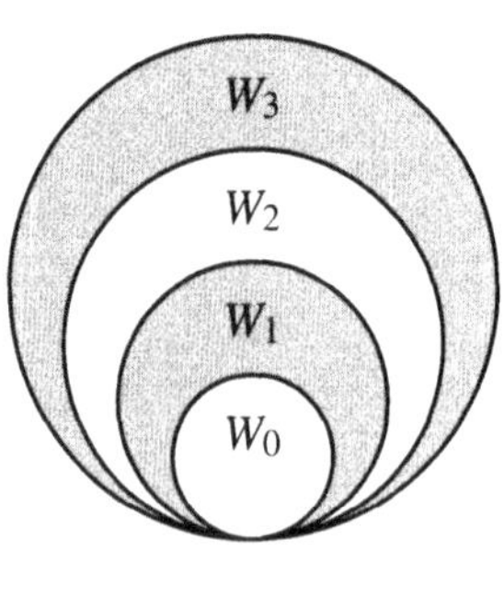

Figure 4.2. A chain of length 4.

Example 4.2. Let $A = \{a, b\}$ and $X = (b^*a)^\omega$ be the set of words with an infinite number of occurrences of a. Then $(\{a\}, b, a)$ is a negative X-chain of length 2. Indeed, $Z = A$, $W_0 = aA^*(bb^*)^\omega = aA^*b^\omega$ is disjoint from X and $W_1 = aA^*(aA^*)^\omega = (aA^*)^\omega$ is contained in X. We claim there is no positive X-chain of length 2. Indeed, if (Y, z_0, z_1) is such a chain, then $W_0 = YZ^*z_0^\omega$ is contained in X, and therefore the letter a occurs in z_0. It follows that $W_1 = YZ^*(z_1Z_1^*)^\omega$ contains $YZ^*(z_1z_0)^\omega$, and therefore meets X, proving the claim. Thus $m^+(X) = 1$ and $m(X) = m^-(X) = 2$.

Proposition 4.2. $m^+(X) = m^-(X^c)$.

Proof. An X-chain is also an X^c-chain of opposite sign. Hence $m^+(X) = m^-(X^c)$ and $m^-(X) = m^+(X^c)$. □

We shall see that $m(X)$ is always finite if X is a recognizable set, but $m(X)$ might be infinite if X is not recognizable, as shown in the next example.

Example 4.3. Let X be the set of words over the alphabet $A = \{a, b\}$ such that the maximal length of a-runs appearing infinitely often is bounded and even (where an a-run is a factor of type a^k with $k > 0$). It may be easily checked that $b, ab, a^2b, a^3b, \dots, a^nb$ is a positive X-chain for any integer n. Thus, we have $m(X) = \infty$.

Since we are mainly interested in the maximal length of chains, it is sometimes convenient to modify a chain without changing its length. We introduce for this purpose the notion of *derivation* of a chain. A chain $C' = (Y', z'_0, \dots, z'_{m-1})$ is derived from a chain $C = (Y, z_0, \dots, z_{m-1})$ if $Y' \subset YZ^*$ and, for $0 \leq i \leq m-1$, $z'_i \in Z_i^*z_iZ_i^*$. The next proposition shows that derivation preserves X-chains.

Proposition 4.3. *Let $S = (S_+, S_\omega)$ be an ω-semigroup and let X be a subset of S_ω. Let C be an X-chain. Then every chain derived from C is an X-chain of the same length and sign as C.*

Proof. Let $C = (Y, z_0, \dots, z_{m-1})$ be a chain and let $C' = (Y', z'_0, \dots, z'_{m-1})$ be a chain derived from C. First, since $z'_i \in Z_i^*z_iZ_i^*$, $Z'_i \subset Z_i^*Z_iZ_i^*$, whence $z'_iZ'^*_i \subset Z_i^*z_iZ_i^*$. The inclusions $W'_i \subset W_i$ follow, since $W'_i = Y'Z'^*(z'_iZ'^*_i)^\omega \subset YZ^*(Z_i^*z_iZ_i^*)^\omega = W_i$. It follows that the set W'_i are alternatively included in X and X^c and thus C' is an X-chain. □

4.3 Chains in finite ω-semigroups. In this section, we show that in a finite ω-semigroup, the chains can be put in a special form related to the ideal structure of the semigroup S_+. The set of idempotents of a finite semigroup can be ordered by the relation $\leq$ defined by

$$e \leq f \iff ef = fe = e$$

Theorem 4.4. *Let S be a finite ω-semigroup and let X be a subset of S_ω. From each X-chain $C = (Y, z_0, \dots, z_m)$, one can derive an X-chain $C' = (\{s\}, e_0, \dots, e_{m-1})$ such that*

(1) *for $0 \le i \le m-1$, the pair (s, e_i) is linked,*

(2) $e_0 > e_1 > \dots > e_{m-1}$,

(3) *for $1 \le i \le m-1$, $S_+ e_i S_+ \cap \{e_1, \dots, e_{i-1}\}^+ = \emptyset$.*

Proof. Let π be the exponent of the semigroup S_+ (see Annex A) and let y be any element of Y. We define the elements $s, e_0, \dots, e_{m-1}$ by setting

$$\begin{aligned}
e_0 &= z_0^\pi \\
e_1 &= (e_0 z_1 e_0)^\pi \\
e_2 &= (e_1 z_2 e_1)^\pi \\
&\vdots \\
e_{m-1} &= (e_{m-2} z_{m-1} e_{m-2})^\pi \\
s &= y e_{m-1}
\end{aligned}$$

By construction, we have $s \in YZ^*$ and $e_i \in Z_i^* z_i Z_i^*$ for $0 \le i \le m-1$ and thus the X-chain C' is derived from C. Since $e_{i+1} = (e_i z_{i+1} e_i)^\pi$, each e_i is an idempotent of S_+ and the relation $e_{i+1} e_i = e_i e_{i+1} = e_{i+1}$ holds. This ensures that

$$e_i e_j = e_j e_i = e_j \tag{4.3}$$

for any integers i and j such that $0 \le i \le j \le m-1$. Furthermore, $se_i = ye_{m-1}e_i = ye_{m-1} = s$. Thus the pairs (s, e_i) are linked and Property (1) is satisfied. Formula 4.3 also implies $e_0 \ge e_1 \ge \dots \ge e_{m-1}$. Furthermore, if $e_i = e_{i+1}$, we obtain a contradiction since se_i^ω and se_{i+1}^ω respectively belong to the disjoint subsets W_i and W_{i+1} of S_ω. This proves Property (2).

Property (2) also implies that $\{e_1, \dots, e_i\}^+ = \{e_1, \dots, e_i\}$ and that $e_0 >_{\mathcal{J}} e_1 >_{\mathcal{J}} \dots >_{\mathcal{J}} e_{m-1}$. Condition (3) follows immediately. □

Example 4.4. Let A be the alphabet $\{a, b\}$. Let S be the ω-semigroup (S_+, S_ω) where S_+ is the semigroup $\{a, b\}$ with the relations $b^2 = b$ and $a^2 = ab = ba = a$ and S_ω is the set $\{b^\omega, ab^\omega, a^\omega\}$ with the natural action of S_+ on S_ω. The $\mathcal{J}$-classes of S_+ are represented in Figure 4.3. The symbol $*$ indicates an $\mathcal{H}$-class containing an idempotent. In this example, each $\mathcal{H}$-class is reduced to an idempotent. The morphism $\varphi : A^\infty \to S$ defined by $\varphi(a) = a$ and $\varphi(b) = b$ recognizes the set $X = (b^*a)^\omega$ of Example 4.2. The image of X in S_ω is $P = \{a^\omega\}$. If the proof of Theorem 4.4 is applied to the P-chain $(\{a, b\}, b, a)$, the P-chain $(\{a\}, b, a)$ is obtained. This chain is in fact the only chain of length 1 satisfying the conditions of Theorem 4.4.

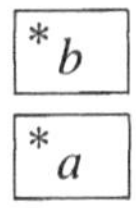

Figure 4.3. The $\mathcal{J}$-classes of $S_+ = \{a, b\}$.

We now show that integers $m^+(X)$, $m^-(X)$ and $m(X)$ can be computed in any ω-semigroup recognizing X. It follows that if two recognizable sets have the same syntactic ω-semigroup and the same image in their syntactic ω-semigroup, then they have the same parameters m^+, m^- and m.

Let $\varphi : S \to S'$ be a morphism of ω-semigroups. If $C = (Y, z_0, \ldots, z_{m-1})$ is a chain of S_ω, we denote by $\varphi(C)$ the chain $(\varphi(Y), \varphi(z_0), \ldots, \varphi(z_{m-1}))$.

Proposition 4.5. *Let $\varphi : S \to S'$ be a surjective morphism of ω-semigroups. Let X' be a subset of S'_ω and let $X = \varphi^{-1}(X')$.*

(1) *If C is an X-chain, then $\varphi(C)$ is an X'-chain of the same length and sign.*

(2) *Each X'-chain is the image under φ of some X-chain.*

Proof. (1) Let $C = (Y, z_0, \ldots, z_{m-1})$ be an X-chain and let C' be its image under φ. Let W_i and W'_i be the sets canonically associated with the chains C and C'. The equalities $W'_i = \varphi(W_i)$ obviously hold. Since $X = \varphi^{-1}(X')$, the sets W'_i are alternately included in X' and X'^c and thus C' is an X'-chain of the same length and sign as C.

(2) Let $C' = (Y', z'_1, \ldots, z'_{m-1})$ be an X'-chain. Since φ is onto, $Y = \varphi^{-1}(Y')$ is nonempty and we can choose elements $z_1, \ldots, z_{m-1}$ such that $\varphi(z_i) = z'_i$ for $0 \le i \le m-1$. Let $C = (Y, z_1, \ldots, z_{m-1})$. The sets W_i associated with C satisfy the inclusions $W_i \subset \varphi^{-1}(W'_i)$. Since $X = \varphi^{-1}(X')$, the condition $W'_i \subset X'$ implies

$$W_i \subset \varphi^{-1}(W'_i) \subset \varphi^{-1}(X') = X$$

and similarly, $W'_i \subset X'^c$ implies $W_i \subset X^c$. Therefore, the sets W_i are alternately included in X and X^c and C is an X-chain of the same length and sign as C'. □

The assumption that the morphism φ is onto is necessary since otherwise, X' might be of greater complexity than X.

The following corollary states that the integers $m^+(X)$ and $m^-(X)$ can be computed in any ω-semigroup recognizing the set X. In particular, this can be done in the syntactic ω-semigroup of X.

Corollary 4.6. *Let $\varphi : S \to S'$ be a surjective morphism of ω-semigroups. Let X' be a subset of S'_ω and let $X = \varphi^{-1}(X')$. Then $m^+(X) = m^+(X')$ and $m^-(X) = m^-(X')$.*

4.4 Equivalence of the definitions of chains. We now prove that the definitions of a chain in automata and in A^∞ are in correspondence. This result has two main consequences: first it shows that the integers $m(X)$ are finite and computable for any ω-rational set X. Second, it shows that the integers $m(\mathcal{A})$ do not depend on the automaton but only on the set X.

Theorem 4.7. *Let $\mathcal{A} = (Q, E, i, \mathcal{T})$ be a complete Muller automaton recognizing a subset X of A^ω. The following equalities hold:*

$$m^+(X) = m^+(\mathcal{A}) \quad \textit{and} \quad m^-(X) = m^-(\mathcal{A})$$

Before proceeding to the proof, we illustrate this result by combining the data concerning the set X considered in Examples 4.2 and 4.1.

Example 4.5. The set $X = (b^*a)^\omega$ is recognized by the automaton $\mathcal{A}_1$ of Figure 4.1. We have the following values:

$$m^+(X) = m^+(\mathcal{A}) = 0$$
$$m^-(X) = m^-(\mathcal{A}) = 1$$

To prove the theorem, with each X-chain we shall associate an $\mathcal{A}$-chain of the same length and sign. Conversely, with each $\mathcal{A}$-chain we shall associate an X-chain of the same length and sign. We actually prove a more precise result, for further use in Section 5.4.

Proposition 4.8. *Let $\mathcal{A} = (Q, E, i, \mathcal{T})$ be a complete Muller automaton recognizing the set $X \subset A^\omega$. For each X-chain $C = (Y, z_0, z_1, \ldots, z_{m-1})$, there exists an $\mathcal{A}$-chain $C' = (R_0, \ldots, R_{m-1})$ of the same length and sign as C such that $i \cdot y = q$ for some state q in R_0 and some word $y \in YZ^*$.*

Proof. Let $Q' = i \cdot YZ^*$ be the set of states accessible from the initial state i by a path labeled by some word in YZ^*. The inclusion $Z_j \subset Z$ implies that Q' is stable by the action of each Z_j, that is, $Q' \cdot Z_j = Q'$. Define a preorder $\prec_j$ and the associated equivalence relation $\equiv_j$ on Q' by

$$q' \prec_j q \quad \Longleftrightarrow \quad \text{there exists } w \in Z_j^* \text{ such that } q \xrightarrow{w} q'$$

Each preorder $\prec_j$ induces an order on the set of $\equiv_j$-classes. If $j \leq k$, Z_j is a subset of Z_k, and thus the relation $\prec_j$ (resp. $\equiv_j$) is thinner than $\prec_k$ (resp. $\equiv_k$).

Define Q_{m-1} as an $\equiv_{m-1}$-class which is minimal for $\prec_{m-1}$. Since $\prec_{m-2}$ refines $\prec_{m-1}$, the set Q_{m-1} is a union of $\equiv_{m-2}$-classes. Then select for Q_{m-2} an $\equiv_{m-2}$-class included in Q_{m-1} which is minimal for $\prec_{m-2}$. Iterate this process to obtain a sequence

$$Q_0 \subset Q_1 \subset \cdots \subset Q_{m-1}$$

where, for $0 \le j < m-1$, each Q_j is an $\equiv_j$-class included in Q_{j+1} which is minimal for $\prec_j$. Choose a state q in Q_0. Since $q \in Q' = i \cdot YZ^*$, we may choose a word $y \in YZ^*$ such that $i \cdot y = q$. For each $0 \le j \le m-1$, set $q_j = q \cdot z_j$ (this state always exists because the automaton is complete). Since the sets Q_j are minimal classes, we have $q \equiv_j q_j$, and there exist words $z'_j \in Z_j^*$ such that $q_j \cdot z'_j = q$. Then define words w_j by

$$\begin{aligned} w_0 &= z_0 z'_0 \\ w_1 &= w_0 z_1 z'_1 \\ &\vdots \\ w_{m-1} &= w_{m-2} z_{m-1} z'_{m-1} \end{aligned}$$

The words w_j satisfy the following properties:

(a) w_j stabilizes the state q, that is $q \cdot w_j = q$,

(b) $yw_j^\omega \in W_j = YZ^*(z_j Z_j^*)^\omega$.

Properties (a) and (b) are clear by construction of w_j. Let $R_0, \ldots, R_{m-1}$ be the respective contents of the paths $q \xrightarrow{w_j} q$ for $j = 0, \ldots, m-1$. This sequence of admissible subsets forms an $\mathcal{A}$-chain since the set of infinitely repeated states in the path labeled by yw_j^ω is exactly R_j. Furthermore the state q belongs to R_0 and is accessible from the initial state by a path labeled by a word $y \in YZ^*$. □

It is worth mentioning that Proposition 4.8 no longer holds if the automaton is not complete. For example, the set $X = a^\omega$ can be recognized by a Muller automaton $\mathcal{A}$ with just one state. Then $m^-(\mathcal{A}) = 0$ whereas $m^-(X) = 1$ since $(\{b\}, b)$ is a negative chain of length 1.

We now transform $\mathcal{A}$-chains into X-chains.

Proposition 4.9. *Let $\mathcal{A} = (Q, E, i, T)$ be a complete Muller automaton recognizing a subset X of A^ω. For each $\mathcal{A}$-chain there exists an X-chain of the same length and sign.*

Proof. Let $R_0 \subset \cdots \subset R_{m-1}$ be an $\mathcal{A}$-chain. We choose a state $q \in R_0$ and a word y such that $i.y = q$. We also choose words $z_0, \ldots, z_{m-1} \in A^*$ such that R_j is the content of the path $q \xrightarrow{z_j} q$. A straightforward verification shows that $(\{y\}, z_0, \ldots, z_{m-1})$ is an X-chain of the same length and sign as the chain $R_0 \subset \cdots \subset R_{m-1}$. □

Theorem 4.7 now follows easily from Propositions 4.8 and 4.9. Proposition 4.8 shows that $m^+(X) \le m^+(\mathcal{A})$ and $m^-(X) \le m^-(\mathcal{A})$ since any X-chain can be transformed into a $\mathcal{A}$-chain. Conversely, Proposition 4.9 shows that $m^+(X) \ge m^+(\mathcal{A})$ and $m^-(X) \ge m^-(\mathcal{A})$.

4.5 The chain hierarchy. We now build a hierarchy among the sets in the class Rec of recognizable sets based on the lengths of maximal chains, called the chain hierarchy. This hierarchy will be refined in the next sections. The final stage is the Wagner hierarchy, which is indexed by ordinals. To keep a coherent notation, we already use ordinals instead of integers for the chain hierarchy. For reasons which will become clear later, the indices are ordinals of the form ω^m.

For $m \geq 0$, we define classes $\mathcal{C}_{\omega^m}$, $\mathcal{D}_{\omega^m}$, $\mathcal{E}_{\omega^m}$ as follows.

$$\begin{aligned}
\mathcal{C}_{\omega^m} &= \{X \in \text{Rec} \mid m^+(X) \leq m\},\\
\mathcal{D}_{\omega^m} &= \{X \in \text{Rec} \mid m^-(X) \leq m\},\\
\mathcal{E}_{\omega^m} &= \mathcal{C}_{\omega^m} \cap \mathcal{D}_{\omega^m}
\end{aligned}$$

Thus $\mathcal{C}_1(A^\omega) = \mathcal{C}_{\omega^0}(A^\omega) = \{\emptyset\}$, $\mathcal{D}_1(A^\omega) = \mathcal{D}_{\omega^0}(A^\omega) = \{A^\omega\}$, $\mathcal{E}_1(A^\omega) = \mathcal{E}_{\omega^0}(A^\omega) = \emptyset$, and $\mathcal{C}_\omega$ is the class $\text{Rec} \cap \Pi_2$ of sets recognized by deterministic Büchi automata (see Landweber's theorem III.3.11). Similarly, $\mathcal{D}_\omega = \text{Rec} \cap \Sigma_2$, and $\mathcal{E}_\omega$ is the class $\text{Rec} \cap \Delta_2$ formed of boolean combinations of rational clopen sets (see Theorem VI.3.7).

For each $m \geq 0$, the class $\mathcal{C}_{\omega^m}$ and $\mathcal{D}_{\omega^m}$ are dual. The inclusion diagram of the classes $\mathcal{C}_{\omega^m}$, $\mathcal{D}_{\omega^m}$, $\mathcal{E}_{\omega^m}$ is represented in Figure 4.4. The fact that the hierarchy is strict follows easily from the existence of Muller automata with arbitrary positive value of $m(\mathcal{A})$. We now state the main result of this section, which gives a characterization of the class $\mathcal{C}_{\omega^m}$.

The statement uses m-parity automata introduced in Chapter IV. Recall that an *m-parity automaton* is a deterministic automaton $\mathcal{A} = (Q, i, \mu)$ where μ is a function from Q into $\{0, 1, \ldots, m\}$. For a path c in $\mathcal{A}$, we define

$$\mu(c) = \max\{\mu(q) \mid q \in \text{Inf}(c)\}.$$

By definition, a path c in $\mathcal{A}$ is successful if it starts at i and the integer $\mu(c)$ is odd. When $m = 1$, the notion of a parity automaton coincides with the notion of a deterministic Büchi automaton. A path is then successful if it meets infinitely often the set of final states since $F = \{q \mid \mu(q) = 1\}$.

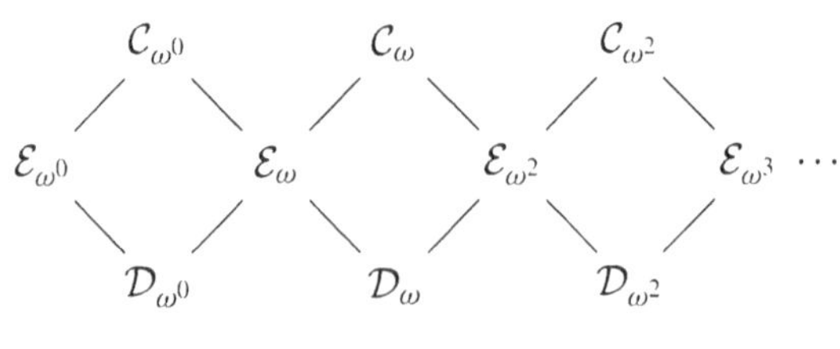

Figure 4.4. The chain hierarchy.

Theorem 4.10. *Let X be an ω-rational set and $m \geq 1$. The following conditions are equivalent:*

(1) $X \in \mathcal{C}_{\omega^m}$, *i.e.* $m^+(X) \leq m$,

(2) X *is recognized by an* m*-parity automaton,*

(3) $X \in D_m(\mathcal{C}_\omega)$,

(4) $X \in D_m(\Pi_2)$.

The proof of the theorem relies on the following lemma, which holds independently of any hypothesis of rationality. The case $m = 1$ follows from Proposition I.7.9, but we shall give a direct proof of it.

Lemma 4.11. *For $m \geq 1$, a set X in $D_m(\Pi_2)$ satisfies $m^+(X) \leq m$.*

Proof. The proof is by induction on m. The result is trivial if $m = 0$.

For $m = 1$, we show that for any finite words $y, z_0, z_1 \in A^+$, we have

$$y(z_0 + z_1)^* z_0^\omega \subset X \Rightarrow y(z_0^* z_1)^\omega \cap X \neq \emptyset.$$

This proves that X has no chain of length 1. Since X is a Π_2-set, it can be written $X = \overrightarrow{U}$ for some $U \subset A^*$. Since $yz_0^\omega \in X$, there are finite words u_1 and v_1 such that $yu_1 \in U$ and $w_1 = u_1 v_1 \in z_0^*$. Again, since $yw_1 z_1 z_0^\omega \in X$, there are finite words u_2 and v_2 such that $yw_1 z_1 u_2 \in U$ and $w_2 = u_2 v_2 \in z_0^*$. Continuing in this way, we build a sequence $(w_i)_{i \geq 1}$ such that $yw_1 z_1 w_2 z_1 \cdots \in X$. This gives the conclusion.

For the induction step, we need to argue on the parity of m. Let us first suppose that m is even. We claim that for any sequence of finite words $y, z_0, z_1, \ldots, z_{m-1} \in A^+$, with $Z_i = z_0 + \cdots + z_i$, if

$$\begin{aligned} yZ^* z_0^\omega &\subset X \\ yZ^*(Z_0^* z_1)^\omega &\subset X^c \\ yZ^*(Z_1^* z_2)^\omega &\subset X \\ &\vdots \\ yZ^*(Z_{m-3}^* z_{m-2})^\omega &\subset X^c \end{aligned}$$

then

$$y(Z^* z_{m-1})^\omega \not\subset X.$$

Since m is even, we have $X = Y - \overrightarrow{U}$ for some $U \subset A^*$ such that $Y \in D_{m-1}(\Pi_2)$ and $\overrightarrow{U} \subset Y$. In particular $X \subset Y$, and thus $yZ^*(Z_{i-1}^* z_i)^\omega \subset Y$ for each even i. Now, by the induction hypothesis, the set Y has no positive chains of length $m - 1$. This implies that some of the sets $yZ^*(Z_{i-1}^* z_i)^\omega$ for some odd i meets Y and therefore also $\overrightarrow{U}$. Thus there are words u_1 and v_1 such that $yu_1 \in U$ and $w_1 = u_1 v_1 \in Z^*$. By prefixing all sets above by $yw_1 z_{m-1}$, we have again

$$\begin{aligned}
yw_1 z_{m-1} Z^* z_0^\omega &\subset X \\
yw_1 z_{m-1} Z^*(Z_0^* z_1)^\omega &\subset X^c \\
yw_1 z_{m-1} Z^*(Z_1^* z_2)^\omega &\subset X \\
&\vdots \\
yw_1 z_{m-1} Z^*(Z_{m-2}^* z_{m-2})^\omega &\subset X^c
\end{aligned}$$

For the same reason, there is some odd $i \leq m-2$ such that $yw_1 z_{m-1} Z^*(Z_i^* z_i)^\omega \cap \overrightarrow{U} \neq \emptyset$. Thus, there are again finite words u_2 and v_2 such that $yw_1 z_m u_2 \in U$ and $w_2 = u_2 v_2 \in Z^*$. Continuing in this way, we obtain a word

$$yw_1 z_{m-1} w_2 z_{m-1} \cdots \in \overrightarrow{U}$$

and therefore a word in $y(Z^* z_{m-1})^\omega - X$, proving the claim.

Let us now suppose that m is odd. We have this time to prove that for any sequence of finite words $y, z_0, z_1, \ldots, z_{m-1}$ in A^+, with $Z_i = z_0 + \cdots + z_i$, if

$$\begin{aligned}
yZ^* z_0^\omega &\subset X \\
yZ^*(Z_0^* z_1)^\omega &\subset X^c \\
yZ^*(Z_1^* z_2)^\omega &\subset X \\
&\vdots \\
yZ^*(Z_{m-3}^* z_{m-2})^\omega &\subset X
\end{aligned}$$

then

$$y(Z^* z_{m-1})^\omega \not\subset X^c.$$

Since m is odd, we have $X = Y + \overrightarrow{U}$ with $Y \in D_{m-1}(\Pi_2)$ and $\overrightarrow{U} \subset Y^c$. By the induction hypothesis, Y has no positive chain of length $m-1$. Thus some of the inclusions $yZ^*(Z_i^* z_i)^\omega \subset Y$ is false for some even i. This implies that $yZ^*(Z_i^* z_i)^\omega \cap \overrightarrow{U} \neq \emptyset$. The conclusion follows as in the case of an even m. □

We can now conclude the proof of Theorem 4.10. We use the notation introduced in Section IV.4.4.

(1) implies (2). Let $\mathcal{A} = (Q, i, \mathcal{T})$ be a Muller automaton recognizing X and let $\mathcal{B}$ be the memory extension of $\mathcal{A}$ with S as set of states. We shall turn $\mathcal{B}$ into an m-parity

automaton by defining a function $\mu : S \to \{0, \dots, m\}$. For a set $T \subset Q$ we denote by $m^+(T)$ the maximal length of positive chains $R_0 \subset R_1 \subset \cdots \subset R_k$ such that $R_k = T$. We define for $(u, v) \in S$,

$$\mu(u, v) = m^+(\underline{v}) + 1$$

The automaton $\mathcal{B}$ is equivalent to $\mathcal{A}$. Let indeed c be a path in $\mathcal{A}$ with $T = \text{Inf}(c)$ and let c' be the corresponding path in $\mathcal{B}$. By Proposition IV.4.7, the maximal value of the $\mu(p)$ for $p \in \text{Inf}(c')$ is reached for a state (u, v) such that $\underline{v} = T$. But $T \in \mathcal{T}$ if and only if $m^+(T)$ is even and thus T is accepting if and only if $\mu(u, v)$ is odd.

(2) implies (3). Let $\mathcal{A} = (Q, i, \mu)$ be an m-parity automaton recognizing X. Let $R_k = \{q \in Q \mid \mu(q) \geq k\}$ and let X_k be the set recognized by the deterministic Büchi automaton (Q, i, R_k). Then it follows from the definitions that $X_1 \supset X_2 \supset \dots \supset X_m$ and that

$$X = X_1 - X_2 + \cdots \pm X_m$$

whence the conclusion.

(3) implies (4) is obvious.

(4) implies (1) is exactly Lemma 4.11. □

As a special case of Theorem 4.10, we obtain for $m = 1$ a characterization of the class of sets recognized by deterministic Büchi automata (see Exercise 13). According to Theorem 4.10, this class is formed of the ω-rational sets X such that $m^+(X) \leq 0$. This result is originally due to H. Landweber (see Chapter III). The next corollary is another important consequence of Theorem 4.10.

Corollary 4.12. *The classes $\mathcal{C}_{\omega^m}$, $\mathcal{D}_{\omega^m}$ and $\mathcal{E}_{\omega^m}$ are Wadge classes.*

Proof. For $m = 1$, it was already shown that the class $\mathcal{C}_\omega$ coincide with the topological class Π_2. Therefore, it is a Wadge class. Let now $X \in \mathcal{C}_{\omega^m}(A^\omega)$ and let $\varphi : B^\omega \to A^\omega$ be a sequential function. Then by Theorem 4.10, X can be written as $X_1 - X_2 + X_3 \dots \pm X_m$, where $X_1 \supset X_2 \supset \cdots \supset X_m$ is a decreasing sequence of Π_2-sets. Since φ is continuous, each set $\varphi^{-1}(X_i)$ is also a Π_2 set and furthermore

$$\varphi^{-1}(X) = \varphi^{-1}(X_1) - \varphi^{-1}(X_2) + \varphi^{-1}(X_3) \dots \pm \varphi^{-1}(X_m)$$

Therefore $\varphi^{-1}(X) \in \mathcal{C}_{\omega^m}(B^\omega)$. Therefore, the class $\mathcal{C}_{\omega^m}$ is a Wadge class. The proof is dual for the classes $\mathcal{D}_{\omega^m}$. Finally, the classes $\mathcal{E}_{\omega^m}$, which are defined as the intersection of $\mathcal{C}_{\omega^m}$ and $\mathcal{D}_{\omega^m}$, are also Wadge classes. □

We shall now exhibit $\mathcal{C}_{\omega^m}$-complete sets for each nonnegative integer n. Setting

$$A_m = \{a_0, a_1, \dots, a_m\}$$

define, for each word $x \in A_m^\omega$, a function $\mu : A_m^\omega \to \mathbb{N}$ by

$$\mu(x) = \max\{k \leq m \mid a_k \text{ occurs infinitely often in } x\}$$

Finally, set

$$X_m = \{x \in A_m^\omega \mid \mu(x) \text{ is odd}\}$$

Thus $X_0 = \emptyset$, $X_1 = (a_0^* a_1)^\omega$, $X_2 = A_2^* X_1$, $X_3 = A_3^* X_1 + (A_2^* a_3)^\omega$, etc.

Theorem 4.13. *For each m, X_m is $\mathcal{C}_{\omega^m}$-complete and X_m^c is $\mathcal{D}_{\omega^m}$-complete.*

Proof. It suffices, by duality, to prove that X_m is $\mathcal{C}_{\omega^m}$-complete. We first prove that $X_m \in \mathcal{C}_{\omega^m}$. Let $\mathcal{A}_m = (Q, 0, E, \mu)$ be the m-parity automaton defined by $Q = \{0, 1, \ldots, m\}$, $E = \{(i, a_k, k) \mid i, k \in Q\}$ and $\mu(k) = k$ for each $k \in Q$. Then $\mathcal{A}_m$ recognizes X_m since

$$\begin{aligned} L^\omega(\mathcal{A}) &= \{x \in A_m^\omega \mid \max\{k \mid k \in \mathrm{Inf}(x)\} \text{ is odd}\} \\ &= \{x \in A_m^\omega \mid \mu(x) \text{ is odd}\} \\ &= X_m \end{aligned}$$

Next we show that if $X \in \mathcal{C}_{\omega^m}(A^\omega)$, then $X \leq_W X_m$. Since $X \in \mathcal{C}_{\omega^m}$, Theorem 4.10 shows that X is recognized by an m-parity automaton $\mathcal{A} = (Q, i, E, \mu)$. Now each infinite word $a_0 a_1 a_2 \cdots$ defines a unique path starting in i:

$$i = p_0 \xrightarrow{a_0} p_1 \xrightarrow{a_0} p_2 \cdots$$

Let φ be the sequential function from A^ω into A_m^ω defined by

$$\varphi(a_0 a_1 a_2 \cdots) = a_{\mu(p_0)} a_{\mu(p_1)} a_{\mu(p_2)} \cdots$$

Then $x \in X$ if and only if $\varphi(x) \in X_m$ and thus $X \leq_W X_m$. □

5 Superchains

In this section, we introduce a refinement of the chain hierarchy called the *superchain hierarchy*.

We follow the presentation adopted in the previous section. Superchains are first defined in Muller automata and subsequently in ω-semigroups. Then we prove the equivalence between the two definitions and study their basic properties. The superchain hierarchy is introduced in the last subsection.

5.1 Superchains in automata. Let $\mathcal{A} = (Q, E, i, \mathcal{T})$ be a complete Muller automaton. Given two subsets C and C' of Q, we say that C' is *accessible* from C if there exists a path from some state in C to some state in C'. By extension, an $\mathcal{A}$-chain $R' = (R'_0, R'_1, \dots, R'_{m-1})$ is *accessible* from an $\mathcal{A}$-chain $R = (R_0, R_1, \dots, R_{m-1})$ if the set R'_{m-1} is accessible from the set R_{m-1}.

An $\mathcal{A}$-*superchain* of length n is a sequence

$$(C_0, C_1, \dots, C_{n-1})$$

of n $\mathcal{A}$-chains of length $m(\mathcal{A})$ such that:

(a) for $1 \le i \le n-1$, C_i is accessible from C_{i-1},

(b) the $\mathcal{A}$-chains C_i are alternately positive and negative.

We say that the superchain is *positive* if C_0 is positive and *negative* otherwise. We denote by $n^+(\mathcal{A})$ (resp. $n^-(\mathcal{A})$) the maximal length of positive (resp. negative) superchains. We let $n^+(\mathcal{A}) = 0$ (resp. $n^-(\mathcal{A}) = 0$) if the set of positive (resp. negative) superchains is empty. Finally, we set

$$n(\mathcal{A}) = \max(n^+(\mathcal{A}), n^-(\mathcal{A}))$$

Note that $n(\mathcal{A})$ is finite for any finite Muller automaton $\mathcal{A}$, since $n(\mathcal{A}) \le \mathrm{Card}(Q)$.

Example 5.1. Consider the Muller automaton $\mathcal{A}_2$ of Figure 5.1 with table

$$\mathcal{T} = \{\{1, 2\}\}$$

All chains are of length 1 and $m(\mathcal{A}_2) = m^+(\mathcal{A}_2) = m^-(\mathcal{A}_2) = 1$. The sequence $(\{1\}, \{2\}, \{3\})$ is a negative superchain of length 3 and we have $n(\mathcal{A}_2) = n^-(\mathcal{A}_2) = 3$ and $n^+(\mathcal{A}_2) = 2$.

Example 5.2. The automaton $\mathcal{A}$ of Figure 5.2 with table

$$\mathcal{T} = \{\{1\}, \{1, 2\}, \{4\}\}$$

recognizes the set X of Example 5.4. There are two chains of length 1 and no chain of length greater than 1. The $\mathcal{A}$-chain $C_0 = (\{2\}, \{1, 2\})$ is negative and the $\mathcal{A}$-chain

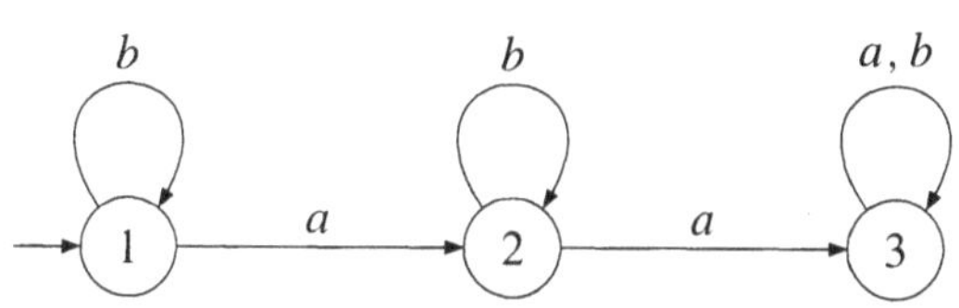

Figure 5.1. The automaton $\mathcal{A}_2$, with $\mathcal{T} = \{\{1, 2\}\}$.

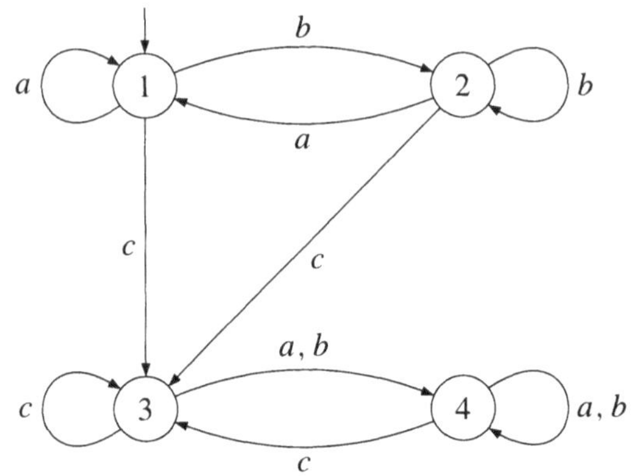

Figure 5.2. The automaton $\mathcal{A}$, with $\mathcal{T} = \{\{1\}, \{1, 2\}, \{4\}\}$.

$C_1 = (\{4\} \subset \{3, 4\})$ is positive. We have $m(\mathcal{A}) = m^+(\mathcal{A}) = m^-(\mathcal{A}) = 2$. This gives a negative $\mathcal{A}$-superchain (C_0, C_1) and we have $n^+(\mathcal{A}) = 1$ and $n(\mathcal{A}) = n^-(\mathcal{A}) = 2$.

The next proposition essentially states that accessibility in a Muller automaton is an order for chains of maximal length.

Proposition 5.1. *Let $\mathcal{A}$ be a complete Muller automaton. If two $\mathcal{A}$-chains of length $m(\mathcal{A})$ are accessible from each other, then they have the same sign.*

Proof. Let $\mathcal{A} = (Q, E, i, \mathcal{T})$ and let $m = m(\mathcal{A})$. Let $C = (R_0, \ldots, R_{m-1})$ and $C' = (R'_0, \ldots, R'_{m-1})$ be two $\mathcal{A}$-chains. If there are paths from R_{m-1} to R'_{m-1} and back from R'_{m-1} to R_{m-1}, then $R_{m-1} \cup R'_{m-1}$ is an admissible subset of Q. If, for instance, $R_{m-1} \cup R'_{m-1}$ is accepting, then both R_{m-1} and R'_{m-1} must be also accepting since otherwise, if, say, R_{m-1} was rejecting, then $(R_0, \ldots, R_{m-1}, R_{m-1} \cup R'_{m-1})$ would be a chain of length $m + 1$. $\square$

5.2 Superchains in ω-semigroups. We now come to the definition of a superchain in an ω-semigroup. As in the case of chain, it is worth mentioning that our definition is valid for the free ω-semigroup $A^\infty = (A^+, A^\omega)$ as well for finite ω-semigroups. We shall see later that, in the case of finite ω-semigroups, it is possible to give the definition in a different form (see Theorem 5.3).

Let $S = (S_+, S_\omega)$ be an ω-semigroup and let X be a subset of S_ω. A chain $C' = (Y', z'_0, z'_1, \ldots, z'_{m-1})$ is said to be *accessible* from a chain $C = (Y, z_0, z_1, \ldots, z_{m-1})$ if there is an element $u \in S_+$ such that $YZ^*u \subset Y'$. Note that accessibility is a transitive relation on the set of chains. Indeed, if $Y_0Z_0^*u_0 \subset Y_1$ and $Y_1Z_1^*u_1 \subset Y_2$, then $Y_0Z_0^*u_0u_1 \subset Y_1u_1 \subset Y_2$.

An X-*superchain* of length n is a sequence

$$(C_0, C_1, \dots, C_{n-1})$$

of n X-chains $C_i = (Y_i, z_{i,0}, z_{i,1}, \dots, z_{i,m-1})$, all of maximal length $m = m(X)$, such that, for $1 \le i \le n-1$,

(a) each C_i is accessible from C_{i-1},

(b) the chains C_i are alternately positive and negative.

To simplify notation, we set, for $0 \le i, j \le m-1$,

$$Z_{i,j} = \{z_{i,0}, z_{i,1}, \dots, z_{i,j}\}$$

and $Z_i = Z_{i,m-1}$, so that $C_i = (Y_i, Z_i)$.

We say that the superchain is *positive* if C_0 is positive and *negative* otherwise. We denote by $n^+(X)$ (resp. $n^-(X)$) the maximal length of positive (resp. negative) superchains. We let $n^+(X) = 0$ (resp. $n^-(X) = 0$) if the set of positive (resp. negative) superchains is empty. We shall see that $n(X)$ is always finite if X is ω-rational. Finally, we set

$$n(X) = \max(n^+(X), n^-(X))$$

Example 5.3. Let $A = \{a, b\}$ and $X = a^*b^\omega$. We have $m(X) = m^+(X) = m^-(X) = 1$ as one may check easily. Then $C_0 = (\{a\}, a)$, $C_1 = (a^+b, b)$, and $C_2 = (a^+b^+a, a)$ form a sequence of X-chains of maximal length. The words $u_1 = b$ and $u_2 = a$ ensure that C_1 and C_2 are respectively accessible from C_0 and C_1. The sequence (C_0, C_1, C_2) forms a negative X-superchain of length 3 since C_0 and C_2 are negative X-chains and C_1 is a positive X-chain. In fact we have $n^+(X) = 2$ and $n(X) = n^-(X) = 3$.

Example 5.4. Let $A = \{a, b, c\}$ and let $X = (b^*a)^\omega + A^*c(a+b)^\omega$ be the set of ω-words having either:

(1) no occurrence of c and infinitely many occurrences of a, or

(2) a finite non zero number of occurrences of c.

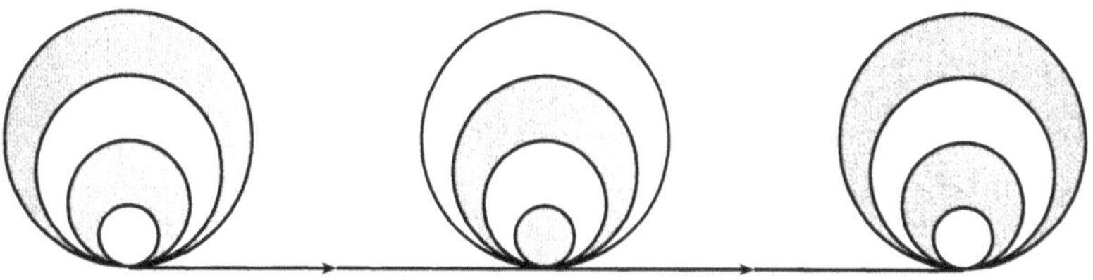

Figure 5.3. A superchain of length 2 with chains of length 3.

The set X has two chains of length 2: the negative chain $C_0 = (\{b\}, b, a)$ and the positive chain $C_1 = ((a+b)^*c, a, c)$. The word $u_1 = c$ ensures that the chain C_1 is accessible from C_0 and therefore the sequence (C_0, C_1) is a negative X-superchain of length 2. We shall see later that actually $n^+(X) = 1$ and $n(X) = n^-(X) = 2$.

We now come, as we did for chains, to some elementary properties of the integers $n^+(X)$, $n^-(X)$ and $n(X)$.

Proposition 5.2. *For any ω-rational set $X \subset A^\omega$ one has*

(1) $1 \leq n(X) < \infty$
(2) $|n^+(X) - n^-(X)| \leq 1$
(3) $n^+(X) = n^-(X^c)$
(4) $m^+(X) = m^-(X)$ *as soon as* $n^+(X) \geq 1$ *and* $n^-(X) \geq 1$.

Proof. (1) Any ω-rational set has at least one chain since $m(X) \geq 1$ and it has some chain of maximal length since $m(X) < \infty$. It therefore has at least one superchain of length 1. Hence $n(X) \geq 1$. Since maximal chains of opposite sign cannot be accessible from one another by Proposition 5.1, the length of superchains in automata is bounded by the number of states. Thus $n(X) < \infty$.

(2) As with chains, if the first element of a superchain is removed, a superchain of opposite sign is obtained and therefore $|n^+(X) - n^-(X)| \leq 1$.

(3) As with chains, an X-superchain constitutes an X^c-superchain of opposite sign. Hence $n^+(X) = n^-(X^c)$ and $n^-(X) = n^+(X^c)$.

(4) A superchain of length at least equal to 1 contains both a positive and a negative chain of length $m(X)$. This implies $m^+(X) = m^-(X) = m(X)$. □

5.3 Superchains in finite ω-semigroups. We now prove the superchain counterpart of Theorem 4.4. It states that superchains of a finite ω-semigroup can be transformed into superchains of a special form.

Theorem 5.3. *Let S be a finite ω-semigroup and X be a subset of S_ω. Let $(C_0, \ldots, C_{n-1})$ be an X-superchain. There is an X-superchain $(C'_0, \ldots, C'_{n-1})$ where each $C'_i = (s_i, e_{i,0}, e_{i,1}, \ldots, e_{i,m-1})$ is an X-chain derived from C_i such that, for $0 \leq i \leq n-1$ and $0 \leq j \leq m-1$,*

(1) *the pair $(s_i, e_{i,j})$ is linked.*
(2) $e_{i,0} > e_{i,1} > \ldots > e_{i,m-1}$.
(3) $s_0 >_{\mathcal{R}} s_1 >_{\mathcal{R}} \cdots >_{\mathcal{R}} s_{m-1}$.

Proof. Let $C_i = (Y_i, z_{i,0}, z_{i,1}, \ldots, z_{i,m-1})$. Since each X-chain C_i is accessible from C_{i-1}, there are elements $u_i \in S_+$ for $1 \leq i \leq n-1$ such that $Y_{i-1}Z^*_{i-1}u_i \subset Y_i$. Let y be an element of Y_0 and π is the exponent of the semigroup S_+. We define the elements

s_i and $e_{i,j}$ for $0 \le i \le n-1$ and $0 \le j \le m-1$ of S_+ by the following relations.

$$\begin{aligned}
e_{i,0} &= z_{i,0}^{\pi}\\
e_{i,1} &= (e_{i,0}z_{i,1}e_{i,0})^{\pi}\\
&\vdots\\
e_{i,m-1} &= (e_{i,m-2}z_{i,m-1}e_{i,m-2})^{\pi}\\
s_0 &= ye_{0,m-1}\\
s_1 &= s_0u_1e_{1,m-1}\\
s_2 &= s_1u_2e_{2,m-1}\\
&\vdots\\
s_{n-1} &= s_{n-2}u_{n-1}e_{n-1,m-1}
\end{aligned}$$

By definition of the elements s_i and $e_{i,j}$, each X-chain $C'_i = (s_i, e_{i,0}, \ldots, e_{i,m-1})$ is derived from the X-chain C_i. The proof of properties (1) and (2) is the same as in the proof of Theorem 4.4. By construction, $s_0 \ge_{\mathcal{R}} s_1 \ge_{\mathcal{R}} \cdots \ge_{\mathcal{R}} s_{m-1}$. Furthermore, if $s_i \mathrel{\mathcal{R}} s_{i+1}$, there are elements u and v of S_+ such that $s_iu = s_{i+1}$ and $s_{i+1}v = s_i$. Setting $f_1 = (e_{i,m-1}ue_{i+1,m-1}v)^{\pi}$ and $f_2 = (e_{i+1,m-1}ve_{i,m-1}u)^{\pi}$, a simple computation shows that $s_if_1^{\omega} = s_{i+1}f_2^{\omega}$. Indeed,

$$\begin{aligned}
s_if_1^{\omega} &= s_i(e_{i,m-1}ue_{i+1,m-1}v)^{\omega}\\
&= s_ie_{i,m-1}u(e_{i+1,m-1}ve_{i,m-1}u)^{\omega}\\
&= s_{i+1}(e_{i+1,m-1}ve_{i,m-1}u)^{\omega}\\
&= s_{i+1}f_2^{\omega}
\end{aligned}$$

Now, one of the sequences

$$(\{s_i\}, e_{i,0}, \ldots, e_{i,m-1}, f_1) \text{ or } (\{s_{i+1}\}, e_{i+1,0}, \ldots, e_{i+1,m-1}, f_2)$$

is an X-chain of length $m+1$. This is a contradiction since the chains of a superchain must be of maximal length. This proves property (3). □

Example 5.5. Let A be the alphabet $A = \{a, b, c\}$. Let S be the ω-semigroup (S_+, S_ω) where S_+ is the semigroup $\{a, b, c\}$ with the relations $b = 1$, $a^2 = a$ and $c = 0$ and

$$S_\omega = \{b^{\omega}, ab^{\omega}, cb^{\omega}, a^{\omega}, cb^{\omega}, ca^{\omega}, c^{\omega}\}$$

with the natural action of S_+ on S_ω. The $\mathcal{J}$-classes of S_+ are represented in Figure 5.4. The morphism $\varphi : A^{\infty} \to S$ defined by $\varphi(a) = a$, $\varphi(b) = b$ and $\varphi(c) = c$ recognizes the set $X = (b^*a)^{\omega} + A^*c(a+b)^{\omega}$ of Example 5.4. The image of X in S_ω is $P =$

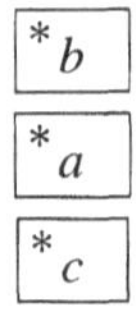

Figure 5.4. The $\mathcal{J}$-classes of S_+.

$\{a^\omega, cb^\omega, ca^\omega\}$. The chains $C_0 = (\{a\}, b, a)$ and $C_1 = (\{c\}, a, c)$ are the only P-chains of length 3 satisfying the conditions of Theorem 4.4. The chain C_0 is negative and the chain C_1 is positive. The sequence (C_0, C_1) is a negative P-superchain of length 2 satisfying the conditions of Theorem 5.3 since $a \geq_{\mathcal{R}} c$.

We now show that the superchains of ω-semigroups are syntactic invariants. Let $\varphi : S \to S'$ be a morphism of ω-semigroups. If $C = (C_0, \ldots, C_n)$ is a superchain of S_ω, we denote by $\varphi(C)$ the superchain $(\varphi(C_0), \varphi(C_1), \ldots, \varphi(C_n))$.

Proposition 5.4. *Let $\varphi : S \to S'$ be a surjective morphism of ω-semigroups. Let X' be a subset of S'_ω and let $X = \varphi^{-1}(X')$.*

(1) *If C is an X-superchain, then $\varphi(C)$ is an X'-superchain of the same length and sign.*
(2) *Each X'-superchain is the image under φ of some X-superchain.*

Proof. (1) Let $C = (C_0, \ldots, C_{n-1})$ be an X-superchain and let C' be its image under φ. By Proposition 4.5, every chain $\varphi(C_i)$ is an X'-chain of length $m(X)$ of the same sign as C_i. Since C_i is accessible from C_{i-1}, there is an element $u_i \in S_+$ such that $Y_{i-1}Z^*_{i-1}u_i \subset Y_i$. It follows that $\varphi(Y_{i-1}Z^*_{i-1})\varphi(u_i) = \varphi(Y_i)$ and thus C' is an X-superchain.

(2) Let $C' = (C'_0, \ldots, C'_{m-1})$ be an X'-chain with

$$C'_i = (Y'_i, z'_{i,0}, \ldots, z'_{i,m-1})$$

Since C'_i is accessible from C'_{i-1}, there exists $u'_i \in S_+$ such that

$$Y'_{i-1}Z'^*_{i-1}u'_i \subset Y'_i$$

Since φ is onto, we can select for each i, j, an element $u_i \in S_+$ such that $\varphi(u_i) = u'_i$ and elements $z_{i,j}$ such that $\varphi(z_{i,j}) = z'_{i,j}$. Finally, let $Y_i = \varphi^{-1}(Y'_i)$. Since φ is onto, Y_i is nonempty and, according to the proof of Proposition 4.5, $C_i = (Y_i, z_{i,0}, \ldots, z_{i,m-1})$ is an X-chain of the same sign as C'_i. Furthermore,

$$\varphi(Y_{i-1}Z^*_{i-1}u_i) = Y'_{i-1}Z'^*_i u'_i \subset Y'_i$$

Therefore,

$$Y_{i-1}Z_{i-1}^{*}u_i \subset \varphi^{-1}(Y_i') = Y_i$$

and thus $C = (C_0, \ldots, C_{m-1})$ is an X-superchain. □

The following corollary states that the integers $m(X)$, $n^+(X)$ and $n^-(X)$ can be computed in any ω-semigroup recognizing the set X. In particular, this can be done in the syntactic ω-semigroup of X.

Corollary 5.5. *Let $\varphi : S \to S'$ be a surjective morphism of ω-semigroups. Let X' be a subset of S'_ω and let $X = \varphi^{-1}(X')$. Then $m(X) = m(X')$, $n^+(X) = n^+(X')$ and $n^-(X) = n^-(X')$.*

5.4 Equivalence of the definitions of superchains. We now prove that the definitions of a superchain in automata and in ω-semigroups are in correspondence. As in the case of chains, this result has two main consequences: first it shows that the integers $n(X)$ are finite and computable for any ω-rational set X. Second, it implies that the integers $n(X)$ do not depend on the automaton recognizing X.

Theorem 5.6. *Let $\mathcal{A} = (Q, E, i, \mathcal{T})$ be a complete Muller automaton recognizing an ω-rational set X. The following equalities hold:*

$$n^+(X) = n^+(\mathcal{A}) \quad \textit{and} \quad n^-(X) = n^-(\mathcal{A})$$

Before proceeding to the proof, we put together the information concerning the set X considered in Examples 5.4 and 5.2.

Example 5.6. The set $X = (b^*a)^\omega + A^*c(a+b)^\omega$ on the alphabet $A = \{a, b, c\}$ of Example 5.4 is recognized by the automaton $\mathcal{A}$ of Example 5.2 represented in Figure 5.2. We have the following values:

$$m^+(X) = m^+(\mathcal{A}) = 2$$
$$m^-(X) = m^-(\mathcal{A}) = 2$$
$$n^+(X) = n^+(\mathcal{A}) = 1$$
$$n^-(X) = n^-(\mathcal{A}) = 2$$

To prove Theorem 5.6, with each X-superchain we first associate an $\mathcal{A}$-superchain of the same length and sign. Conversely, with each $\mathcal{A}$-superchain we shall associate an X-superchain of the same length and sign.

Proposition 5.7. *Let $\mathcal{A}$ be a complete Muller automaton recognizing a subset X of A^ω. For each X-superchain, there is an $\mathcal{A}$-superchain of the same length and sign.*

Proof. Let, $C = (C_0, C_1, \dots, C_m)$ be an X-superchain and let, for $0 \le j \le n-1$, $C_j = (Y_j, z_{i,0}, z_{i,1}, \dots, z_{i,m-1})$. Since, for $1 \le j \le n-1$, each X-chain C_j is accessible from C_{j-1}, there are words u_j such that $Y_{j-1}Z^*_{j-1}u_j \subset Y_j$.

We first apply Proposition 4.8 to the X-chain C_0. We obtain an $\mathcal{A}$-chain C'_0 with a state q_0 accessible from the initial state by a path labeled by a word y_0 in $Y_0Z_0^*$. Since $y_0u_0 \in Y_1$, we can replace the X-chain (Y_1, Z_1) by $(\{y_0u_0\}, Z_1)$. We then apply Proposition 4.8 to this new X-chain to obtain an $\mathcal{A}$-chain C'_1 with a state q_1 accessible from the initial state by a path labeled by a word y_1 in $y_0u_0Z_1^*$. The word y_0u_0 is then a prefix of y_1 and the $\mathcal{A}$-chain C'_1 is accessible from C'_0. Since $y_0u_0u_1 \in Y_2$ we can now replace the X-chain (Y_2, Z_2) by $(\{y_0u_0u_1\}, Z_2)$ and we continue the process. Since each $\mathcal{A}$-chain C'_j has the same length and sign as C_j, the resulting sequence $(C'_0, \dots, C'_{n-1})$ is an $\mathcal{A}$-superchain of the same length and sign as $(C_0, \dots, C_{n-1})$. □

Proposition 5.8. *Let $\mathcal{A}$ be a complete Muller automaton recognizing a subset X of A^ω. For each $\mathcal{A}$-superchain, there is an X-superchain of the same length and sign.*

Proof. Let $(C_0, \dots, C_{m-1})$ be an $\mathcal{A}$-superchain, with

$$C_j = (R_{j,0}, R_{j,1}, \dots, R_{j,m-1})$$

Select a state q_j in each subset $R_{j,0}$. There is a word y labeling a path from the initial state to q_0 and since each $\mathcal{A}$-chain C_{j+1} is accessible from C_j, there is a word u_j for $1 \le j \le n-1$ labeling a path from q_{j-1} to q_j. We also choose words $z_{j,k}$ such that the admissible sets $R_{j,k}$ are the respective content of the paths $q_j \xrightarrow{z_{j,k}} q_j$ for $0 \le j \le n$ and $0 \le k \le m-1$. Let $Y_j = yZ_0^*u_1Z_1^*u_2 \dots Z^*_{j-1}u_j$. A straightforward verification shows that each $C'_j = (Y_j, Z_j)$ is an X-chain and that the sequence $(C'_0, \dots, C'_{n-1})$ is an X-superchain of the same length and sign as $(C_0, \dots, C_{n-1})$. □

Putting Propositions 5.7 and 5.8 together, we now obtain Theorem 5.6.

5.5 The superchain hierarchy. We introduce a new hierarchy of sets, called the *superchain hierarchy*. It is indexed by ordinals of the form $\omega^m n$, where $m \ge 1$ and $n \ge 1$.

$$\begin{aligned}
\mathcal{C}_{\omega^m n} &= \{X \in \text{Rec} \mid m(X) \le m \text{ or } m(X) = m+1 \text{ and } n^+(X) \le n-1\} \\
\mathcal{D}_{\omega^m n} &= \{X \in \text{Rec} \mid m(X) \le m \text{ or } m(X) = m+1 \text{ and } n^-(X) \le n-1\} \\
\mathcal{E}_{\omega^m n} &= \mathcal{C}_{\omega^m n} \cap \mathcal{D}_{\omega^m n}
\end{aligned}$$

Observe that our notation is consistent with the definition of the chain hierarchy, which corresponds to the case $n = 1$. It suffices indeed to observe that, for a recognizable subset of A^ω, the condition

$$m(X) \le m \text{ or } m(X) = m+1 \text{ and } n^+(X) \le 0$$

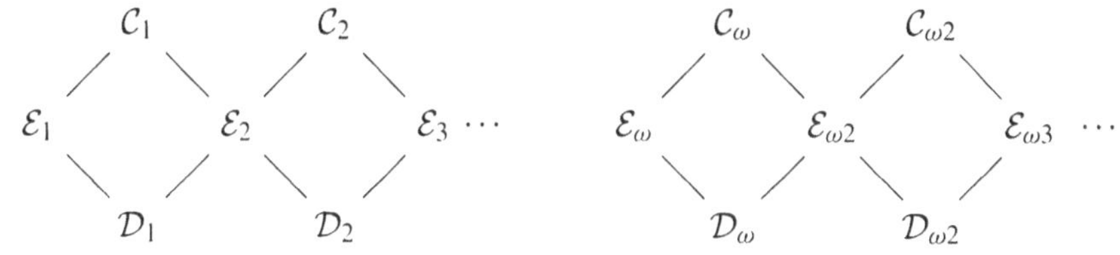

Figure 5.5. The superchain hierarchy.

reduces to $m^+(X) \leq m$, which follows easily from Proposition 4.1. Thus the superchain hierarchy refines the chain hierarchy.

In particular, for each alphabet A, we have $\mathcal{C}_1(A^\omega) = \{\emptyset\}$, $\mathcal{D}_1(A^\omega) = \{A^\omega\}$, $\mathcal{C}_\omega = \text{Rec} \cap \Pi_2$, $\mathcal{D}_\omega = \text{Rec} \cap \Sigma_2$ and $\mathcal{E}_\omega = \text{Rec} \cap \Delta_2$. Also, $\mathcal{C}_2$ is the class of recognizable open sets, $\mathcal{D}_2$ is the class of recognizable closed sets (this follows easily from the definitions, but it will appear as a particular case of the next theorem). We shall see later (Corollary 5.17) that each $\mathcal{C}_{\omega^m n}$ is a Wadge class.

Proposition 5.9. *For each integer $m \geq 0$, the following equalities hold:*

$$\bigcup_{n \geq 1} \mathcal{E}_{\omega^m n} = \mathcal{E}_{\omega^{m+1}}$$

Proof. By definition,

$$\mathcal{E}_{\omega^m n} = \{X \in \text{Rec} \mid m(X) \leq m \text{ or}$$
$$m(X) = m+1, n^+(X) \leq n-1 \text{ and } n^-(X) \leq n-1\}$$

Therefore, $X \in \bigcup_{n \geq 1} \mathcal{E}_{\omega^m n}$ if and only if $X \in \text{Rec}$ and $m(X) = m+1$. □

The inclusions of the classes thus defined is summarized in Figure 5.5. We now proceed to the topological characterization of the classes $\mathcal{C}_{\omega^m n}$. We treat first the case $m = 0$.

Let $n \geq 0$ be an integer. A *weak parity automaton* of length n is an automaton $\mathcal{A} = (Q, i, \nu)$ where ν is a function from Q into $\{0, 1, \ldots, n\}$. For a path c in $\mathcal{A}$, we define

$$\nu(c) = \max\{\nu(q) \mid q \text{ occurs in } c\}.$$

By definition, a path c in $\mathcal{A}$ is successful if it starts at i and the integer $\nu(c)$ is odd.

Let us consider the particular cases $n = 0$ and $n = 1$. For $n = 0$, no path of the automaton is successful, and the set recognized by such an automaton is the empty set.

For $n = 1$, the notion of a weak parity automaton corresponds to what can be called a weak Büchi automaton (see Exercise 12). Indeed, a path of a weak parity automaton of length 1 is successful if it meets at least one state of the set $F = \{q \mid \nu(q) = 1\}$.

We are now ready to prove the following result.

Theorem 5.10. *The following conditions are equivalent for any ω-rational set X and any integer $n \geq 0$.*

(1) $X \in \mathcal{C}_{n+1}$.

(2) *X is recognizable by a weak parity automaton of length n.*

(3) $X \in D_n(\mathcal{C}_2)$.

(4) $X \in D_n(\Sigma_1)$.

The proof relies on the following lemma.

Lemma 5.11. *Let $n > 0$. A set $X \in D_n(\Sigma_1)$ satisfies $m(X) = 1$ and $n^+(X) \leq n$.*

Proof. We shall prove that X has no positive superchain of length n of chains of length 1. This will prove both that $m(X) = 1$ and $n^+(X) \leq n$. Indeed, if $C = (Y, Z)$ where $Z = z_0, z_1$ is an X-chain of length 2, both $C_0 = (YZ^*, z_0)$ and $C_1 = (YZ^*, z_1)$ are X-chains of length 1 and one is positive while the other is negative. The sequence $C_0, C_1, C_0, \ldots$ then forms an X-superchain of arbitrary length since C_0 and C_1 are accessible from one another.

We prove the claim by induction on n. For $n = 1$, the set X is open and thus $X = UA^\omega$ for some $U \subset A^*$. Suppose that C_0, C_1 is a positive X-superchain where $C_0 = (Y_0, z_0)$ and $C_1 = (Y_1, z_1)$ are two X-chains of length 1 and that $u \in A^*$ is such that $Y_0 z_0^* u \subset Y_1$. Since C_1 is a negative chain, we have $Y_1 z_1^\omega \cap X = \emptyset$. Let y_0 be a word of Y_0. Since $y_0 z_0^\omega \in X$, there is an integer k such that $y_0 z_0^k \in UA^+$. Thus $y_0 z_0^k u z_1^\omega \in X$ and this is a contradiction since $Y_1 z_1^\omega \cap X = \emptyset$.

Let us consider $n \geq 1$ and suppose first that n is even. The set X is then equal to $X = Y - UA^\omega$ for some $U \subset A^*$ with $Y \in D_{n-1}(\mathcal{C}_2)$ and $UA^\omega \subset Y$. Suppose that $C_0, \ldots, C_n$ is an X-superchain where each $C_k = (Y_k, z_k)$ is an X-chain of length 1. Let $y_0 \in Y_0$ and $u_k \in A^*$ such that $Y_{k-1} z_{k-1}^* u_k \subset Y_i$ for $1 \leq k \leq n$. We then have

$$y_0 z_0^\omega \subset X, \quad y_0 z_0^* u_1 z_1^\omega \cap X = \emptyset, \quad \ldots, \quad y_0 z_0^* u_1 z_1^* \cdots u_n z_n^\omega \subset X$$

since n is even. By the induction hypothesis, the set Y has no positive superchains of length n of chains of length 1. There is then an integer $k \geq 1$ such that $y_0 z_0^* u_1 \cdots u_k z_k^\omega \cap UA^\omega \neq \emptyset$. This contradicts the fact that $y_0 z_0^* \cdots u_n z_n^\omega \subset X$. The case of n odd is quite similar. □

The converse of Lemma 5.11 is false since the set X of ultimately periodic words satisfies $m(X) = 0$ and $n(X) = 0$ although $X \notin \mathcal{C}_2$ (actually $X \in \mathcal{C}_3$).

We now conclude the proof of Theorem 5.10.

(1) implies (2). Let $\mathcal{A} = (Q, i, \mathcal{T})$ be a Muller automaton recognizing X. Since $m(\mathcal{A}) = 1$ and $n^+(\mathcal{A}) \leq n$, the positive superchains of $\mathcal{A}$ are sequences

$(R_0, R_1, \dots, R_k)$ of subsets of Q alternately accepting or not, with $k \le n - 1$. For $q \in Q$, we define $\nu(q) = k + 1$ where k is the maximal length of a positive superchain from which q is accessible, i.e., q is accessible from a state of R_k. The weak parity automaton of length n (Q, i, ν) recognizes X.

(2) implies (3). Let $\mathcal{A} = (Q, i, \nu)$ be a weak parity automaton recognizing X. Let U_j be the set of finite words labeling a path from the initial state i to a state q such that $\nu(q) \ge j$. Let $X_j = U_j A^\omega$. Then

$$X = X_1 - X_2 + \cdots \pm X_n$$

so that $X \in D_n(\mathcal{C}_2)$.

(3) implies (4) is obvious.

(4) implies (1) is exactly Lemma 5.11. □

When $n = 1$, we obtain the well-known characterization of rational open sets as those recognizable by a weak Büchi automaton (see Exercise 12). As an immediate consequence of Theorem 5.10, we obtain the following characterization of the class $\mathcal{E}_\omega$.

Corollary 5.12. *One has*

$$\mathcal{E}_\omega = \bigcup_{n \ge 0} D_n(\mathcal{C}_2) = \bigcup_{n \ge 0} D_n(\Pi_1)$$

and this class coincides with the boolean closure of the family of ω-rational open sets.

The characterization of the self-dual classes $\mathcal{E}_n$ is the following. The class $\mathcal{E}_2(A^\omega)$ is formed of the clopen sets, i.e., sets equal to UA^ω for some finite set $U \subset A^*$. More generally, we have the following proposition which will be used in Section 6.4.

Proposition 5.13. *Let $n \ge 1$ be an integer and let X be an ω-rational set. The following conditions are equivalent.*

(1) $X \in \mathcal{E}_{n+1}$.

(2) *There exists a clopen set C such that $X \cap C \in \mathcal{D}_n$ and $X \cap C^c \in \mathcal{C}_n$.*

Proof. (1) implies (2). Let $\mathcal{A} = (Q, i, \mathcal{T})$ be a complete Muller automaton recognizing X. Let $P \subset Q$ be the set of states q which belong to a cycle, which means that there is a word $w \in A^+$ such that $q \xrightarrow{w} q$. Let P_+ (resp. P_-) be the set of states which belong to a positive (resp. negative) cycle. Since $m(\mathcal{A}) = 0$, the pair (P_+, P_-) is a partition of P. Let U_+ (resp. U_-) be the set of words u such that there is a path $i \xrightarrow{u} q$ with $q \in P_+$ (resp. $q \in P_-$) which do not meet P before q. Since $\mathcal{A}$ is complete, every infinite word is the label of a path and this path contains a cycle. Therefore, $A^\omega = U_+A^\omega + U_-A^\omega$ is a partition of A^ω into clopen sets. Let $C = U_+A^\omega$. Then $X \cap C \in \mathcal{D}_n$ and $X \cap C^c \in \mathcal{C}_n$.

(2) implies (1). Since C is clopen, a superchain of X is either a superchain of $X \cap C$ or a superchain of $X \cap C^c$. Thus $n(X) \le \max(n(X \cap C), n(X \cap C^c))$. Thus $n(X) \ge n$. □

We now come to the description of the classes $\mathcal{C}_{\omega^m n}$. It makes use of a new kind of automata which generalizes both parity automata and of weak Büchi automata. Let $m, n \geq 0$ be two integers. An *(m, n)-superparity automaton* is an automaton $\mathcal{A} = (Q, i, \mu, \nu)$ where μ is a function from Q into $\{0, \ldots, m\}$ and ν is a function from Q into $\{0, \ldots, n\}$. For a path c in $\mathcal{A}$, we define

$$\mu(c) = \max\{\mu(q) \mid q \text{ occurs infinitely often in } c\}$$
$$\nu(c) = \max\{\nu(q) \mid q \text{ occurs in } c\}.$$

By definition, the path c is successful if it starts at i and if $\mu(c) + \nu(c)$ is odd i.e. if the integers $\mu(c)$ and $\nu(c)$ do not have the same parity.

The case $n = 0$ corresponds to a parity automaton and the case $m = 0$ to a weak parity automaton.

Theorem 5.14. *Let X be an ω-rational set and let $m, n \geq 0$ be two positive integers. The following conditions are equivalent.*

(1) $X \in \mathcal{C}_{\omega^m (n+1)}$.
(2) *X can be recognized by an (m, n)-superparity automaton.*
(3) $X \in \mathrm{Sep}(D_n(\mathcal{C}_2), \mathcal{C}_{\omega^m})$.
(4) $X \in \mathrm{Sep}(D_n(\Sigma_1), D_m(\Pi_2))$.

The proof relies on the following lemma.

Lemma 5.15. *Let X be the union of X_1 and X_2 separated by an open set O. The following relations hold:*

(1) $m(X) \leq \max(m(X_1), m(X_2))$.
(2) $n^+(X) \leq k + \ell$, *where*

$$k = \begin{cases} n^+(X_1) & \text{if } m(X_1) = m(X) \\ 0 & \text{otherwise} \end{cases} \quad \text{and } \ell = \begin{cases} n(X_2) & \text{if } m(X_2) = m(X) \\ 0 & \text{otherwise.} \end{cases}$$

Proof. Since the set O is open, it is equal to UA^ω for some $U \subset A^*$.

(1) It suffices to show that for any X-chain, one can find either an X_1-chain or an X_2-chain of the same length. Let

$$C = (Y, z_0, z_1, \ldots, z_{m-1})$$

be an X-chain of length m. Since $X = (O^c \cap X_1) + (O \cap X_2)$, if none of the sets W_i intersects O, then C is an X_1-chain and thus $m \leq m(X_1)$.

If one set W_i intersects O, then there is a word y of Y and a word $y' \in yZ^*$ which has a prefix in U. Consider now the chain

$$C' = (y', z_0, z_1, \ldots, z_{m-1})$$

The corresponding W_i's are all contained in O and it follows that C' is an X_2-chain of length m and thus $m \leq m(X_2)$ in this case. This proves that $m(X) \leq \max(m(X_1), m(X_2))$.

(2) We come to the second inequality. Let $(C_0, C_1, \ldots, C_{n-1})$ be a positive X-superchain, with $C_i = (Y_i, Z_i)$. We claim that $(C_0, \ldots, C_k)$ cannot be an X_1-superchain. Indeed, if $m(X_1) = m(X)$, then $k = n^+(X_1)$ and the claim follows from the definition of n^+. If $m(X_1) \neq m(X)$ and $k = 0$, C_0 is a chain of length $m(X)$ and hence cannot be an X_1-chain of length $m(X_1)$. Therefore (C_0) is not an X_1-superchain, which proves the claim.

Our next claim is that $Y_k Z_k^\omega$ meets O. Suppose, on the contrary, that $Y_k Z_k^\omega$ and O are disjoint. Then, for $0 \leq i \leq k$, $Y_i Z_i^\omega$ and O are also disjoint. Otherwise, there is, for some i, a word u_i in $Y_i Z_i^* \cap UA^*$. But since C_k is accessible from C_i, there exists a word v such that $Y_i Z_i^* v \subset Y_k$, and now $u_i v \in Y_k \cap UA^*$, a contradiction. We also observe that, in a chain $C = (Y, Z)$, all the sets W_i are subsets of YZ^ω. In particular, if YZ^ω is disjoint from O, each W_i is disjoint from O, and, as was already observed, C is an X_1-chain. It follows that if $Y_k Z_k^\omega$ and O are disjoint, $(C_0, \ldots, C_k)$ is actually a positive X_1-superchain, in contradiction with the definition of k. This proves our second claim.

It is now easy to conclude. Since $Y_k Z_k^\omega$ meets O, some word $y \in Y_k Z_k^*$ has a prefix in U. For $k \leq i \leq n-1$, let $C'_i = (Y'_i, Z_i)$ with $Y'_i = yA^* \cap Y_i Z_i^*$. Then $S' = (C'_k, \ldots, C'_{n-1})$ is an X_2-superchain of length $n-k$ and thus $n \leq k + n(X_2)$. It follows that $n \leq k + \ell$ if $m(X_2) = m(X)$, If now $m(X_2)$ is different from $m(X)$, $\ell = 0$, but S' has to be empty: otherwise, the chains C'_{n-1} – of length $m(X_2)$ – and C_n – of length $m(X)$ – would be of different length. It follows that $n - k = 0$ and the relation $n \leq k + \ell$ still holds in this case. □

Iterating the assertion of the previous lemma, we prove the following one.

Lemma 5.16. *Let $m, n \geq 0$ and let $X \in \mathrm{Sep}(D_n(\Sigma_1), D_m(\Pi_2))$. Then $m(X) \leq m + 1$ and if $m(X) = m + 1$, then $n^+(X) \leq n$.*

Proof. By induction on n. For $n = 0$, the set X is in $D_m(\Pi_2)$ and by Theorem 4.10 and Proposition 5.9, $D_m(\Pi_2) = \mathcal{C}_{\omega^m}$. Therefore $m(X) \leq m$ or $m(X) = m + 1$ and $n^+(X) = 0$.

Suppose that the result holds for $n - 1$ and let X be the union of X_1 and X_2 separated by Y, with $X_1, X_2^c \in D_m(\Pi_2)$ and $Y \in D_n(\Sigma_1)$. Thus $X = (Y^c \cap X_1) + (Y \cap X_2)$.

We first suppose that n is odd. It follows from the definition of $D_n(\Sigma_1)$ that $Y = Z + T$ for some disjoint sets $Z \in D_{n-1}(\Sigma_1)$ and $T \in \Sigma_1$. Let $V = (Z^c \cap X_1) + (Z \cap X_2)$. We claim that X is the union of V and X_2 separated by the open set T. Indeed

$$\begin{aligned}(T^c \cap V) + (T \cap X_2) &= (T^c \cap Z^c \cap X_1) + (T^c \cap Z \cap X_2) + (T \cap X_2)\\ &= ((T \cup Z)^c \cap X_1) + ((T^c \cap Z) + T) \cap X_2)\\ &= (Y^c \cap X_1) + (Y \cap X_2) = X\end{aligned}$$

Note that $V \in \mathrm{Sep}(D_{n-1}(\Sigma_1), D_m(\Pi_2))$, since V is the union of X_1 and X_2 separated by Z. By the induction hypothesis, $m(V) \leq m+1$ and if $m(V) = m+1$, then $n^+(V) \leq n-1$.

Furthermore, since $X_2^c \in D_m(\Pi_2) = \mathcal{C}_{\omega^m}$, X_2 belongs to the dual class $\mathcal{D}_{\omega^m}$. It follows that $m(X_2) \leq m$ or $m(X_2) = m+1$ and $n^-(X_2) = 0$. Let us apply Lemma 5.15 to V and X_2. First

$$m(X) \leq \max(m(V), m(X_2)) \leq m+1$$

Suppose now that $m(X) = m+1$. Then $\max(m(V), m(X_2)) = m+1$. Furthermore, with the notations of Lemma 5.15, we have $k = 0$ if $m(V) \leq m$ and $k \leq n-1$ otherwise. Similarly, $\ell = 0$ if $m(X_2) \leq m$ and $\ell = n(X_2) \leq 1$ otherwise. It follows that $n^+(X) \leq k + \ell \leq n$.

If n is even, we have $Y = Z - T$ with $Z \in D_{n-1}(\Sigma_1)$, $T \in \Sigma_1$ and $T \subset Z$. One verifies that X is the union of V and X_1 separated by T and the rest of the proof is quite similar. □

We can now conclude the proof of Theorem 5.14.

Proof. (1) implies (2). Let $\mathcal{A} = (Q, i, \mathcal{T})$ be a Muller automaton recognizing X and let $\mathcal{B}$ be the memory extension of $\mathcal{A}$. Let S be the set of states of $\mathcal{B}$. We define two functions $\mu : S \to \{0, \ldots, m\}$ and $\nu : S \to \{0, \ldots, n\}$ to turn $\mathcal{B}$ into a superparity automaton. For a state q of $\mathcal{A}$, we define $n^+(q)$ as the maximal length of a positive superchain $(C_0, \ldots, C_k)$ such that q is accessible from C_k but C_k is not accessible from q. If no such superchain exists, we set $n^+(q) = -1$. We then define $\nu(u, v) = n^+(q) + 1$ where q is the last element of uv. We finally define $\mu(u, v)$ as follows.

$$\mu(u, v) = \begin{cases} m^+(v) + 1 & \text{if } \nu(u, v) \text{ is even} \\ m^-(v) + 1 & \text{otherwise} \end{cases}$$

One may verify that the superparity automaton $\mathcal{B}$ is equivalent to $\mathcal{A}$.

(2) implies (3). Let $\mathcal{B} = (S, i, \mu, \nu)$ be a $(m, n-1)$-superparity automaton. Let $Y \in \mathcal{C}_n$ be the set recognized by the weak parity automaton (Q, i, ν). Let $\mu', \mu'' : Q \to \{0, \ldots, m\}$ be defined by

$$\mu'(q) = \begin{cases} \mu(q) & \text{if } \nu(q) \text{ is even} \\ 0 & \text{otherwise} \end{cases}$$

and $\mu''(q) = \mu(q) - \mu'(q)$. Let $X_1 \in \mathcal{C}_{\omega^m}$ be the set recognized by the parity automaton (Q, i, μ') and let $X_2 \in \Pi_{\omega^m}$ be the complement of the set recognized by the parity automaton (Q, i, μ''). One may verify that X is the union of X_1 and X_2 separated by Y.

(3) implies (4) is obvious.

(4) implies (1) follows directly from Lemma 5.16. □

Corollary 5.17. *The classes $\mathcal{C}_{\omega^m n}$, $\mathcal{D}_{\omega^m n}$ and $\mathcal{E}_{\omega^m n}$ are Wagner classes.*

Proof. We give the proof for the classes $\mathcal{C}_{\omega^m n}$. The proof is dual for the classes $\mathcal{D}_{\omega^m n}$ and the result for the classes $\mathcal{E}_{\omega^m n}$ follows easily.

For $n = 1$, the result follows from Corollary 4.12. Let now $X \in \mathcal{C}_{\omega^m (n+1)}(A^\omega)$ and let $\varphi : B^\omega \to A^\omega$ be a sequential function. Then by Theorem 5.14, $X \in \mathrm{Sep}(D_n(\Sigma_1), D_m(\Pi_2))$. Applying φ^{-1}, it is not difficult to see that $\varphi^{-1}(X)$ is also in $\mathrm{Sep}(D_n(\Sigma_1), D_m(\Pi_2))$, and thus in $\mathcal{C}_{\omega^m (n+1)}(B^\omega)$. □

We shall now exhibit $\mathcal{C}_{\omega^m n}$-complete sets for each pair of nonnegative integers (n, m). Setting

$$A_m = \{a_0, a_1, \ldots, a_m\} \quad B_n = \{b_0, b_1, \ldots, b_n\} \quad C_{m,n} = A_m \cup B_n$$

define, two functions μ and ν from $(A_m \cup B_n)^\omega$ into $\mathbb{N}$ by setting, for each word $x \in A_m^\omega$,

$$\mu(x) = \max\{\, k \leq m \mid a_k \text{ occurs infinitely often in } x \,\}$$
$$\nu(x) = \max\{\, k \leq n \mid b_k \text{ occurs in } x \,\}$$

with the convention that $\mu(x) = 0$ if no a_k occurs infinitely often in x and $\nu(x) = 0$ if no b_k occurs in x. Finally, set

$$X_{m,n} = \{x \in (A_m \cup B_n)^\omega \mid \mu(x) + \nu(x) \text{ is odd}\}$$

Thus for $m = n = 0$, $C_{0,0} = \{a_0, b_0\}$ and $X_{0,0} = \emptyset$, for $m = 0$ and $n = 1$, $C_{0,1} = \{a_0, b_0, b_1\}$ and $X_{0,1} = C_{0,1}^* b_1 C_{0,1}^\omega$, for $m = 0$ and $n = 2$, $C_{0,2} = \{a_0, b_0, b_1, b_2\}$ and $X_{0,2} = \{a_0, b_0\}^\omega \cup C_{0,2}^* b_2 C_{0,2}^\omega$, and for $m = 1$ and $n = 0$, $C_{1,0} = \{a_0, a_1, b_0\}$ and $X_{1,0} = (C_{1,0} a_1)^\omega$.

Theorem 5.18. *For each $m, n \geq 0$, $X_{m,n}$ is $\mathcal{C}_{\omega^m (n+1)}$-complete and $X_{m,n}^c$ is $\mathcal{D}_{\omega^m (n+1)}$-complete.*

Proof. We first prove that $X_{m,n} \in \mathcal{C}_{\omega^m (n+1)}$. Let $\mathcal{A}_{m,n} = (Q, (0,0), E, \mu, \nu)$ be the (m, n)-superparity automaton defined by

$$Q = \{(k, \ell) \mid 0 \leq k \leq m, 0 \leq \ell \leq n\}$$
$$E = \big\{\big((k, \ell), a_j, (j, \ell)\big) \mid 0 \leq j \leq m, 0 \leq \ell \leq n\big\}$$
$$\cup \big\{\big((k, \ell), b_j, (k, \max(\ell, j))\big) \mid 0 \leq j \leq n\big\}$$

and $\mu(k, \ell) = k$, $\nu(k, \ell) = \ell$ for each $(k, \ell) \in Q$. Then, by construction, $\mathcal{A}_{m,n}$ recognizes $X_{m,n}$.

Next we show that if $X \in \mathcal{C}_{\omega^m(n+1)}(A^\omega)$, then $X \leq_W X_{m,n}$. Since $X \in \mathcal{C}_{\omega^m(n+1)}$, Theorem 5.14 shows that X is recognized by an (m,n)-superparity automaton $\mathcal{A} = (Q, i, E, \mu, \nu)$. Now each infinite word $a_0a_1a_2\cdots$ defines a unique path starting in i:

$$i = p_0 \xrightarrow{a_0} p_1 \xrightarrow{a_0} p_2 \cdots$$

Let φ be the sequential function from A^ω into A_m^ω defined by

$$\varphi(a_0a_1a_2\cdots) = a_{\mu(p_0)}b_{\nu(p_0)}a_{\mu(p_1)}b_{\nu(p_1)}\cdots$$

Then $x \in X$ if and only if $\varphi(x) \in X_{m,n}$ and thus $X \leq_W X_{m,n}$. □

6 The Wagner hierarchy

We shall now define the final hierarchy, called the Wagner hierarchy.

6.1 The derived automaton. For a Muller automaton $\mathcal{A} = (Q, i, \mathcal{T})$, we define another Muller automaton called the *derived* automaton and denoted $\partial\mathcal{A}$. It is nonempty only when $m(\mathcal{A}) > 1$ and $n^+(\mathcal{A}) = n^-(\mathcal{A})$. It is then obtained from $\mathcal{A}$ by the following transformation. Let Q_+ (resp. Q_-) be the set of states from which a maximal positive (resp. negative) superchain can be reached. The set of states of $\partial\mathcal{A}$ is $\partial Q = (Q_+ \cap Q_-) \cup \{q_+, q_-\}$ where q_+ and q_- are two new states. The transitions of $\partial\mathcal{A}$ are the following. For $q \in \partial Q$ and $a \in A$,

$$\delta(q,a) = \begin{cases} q\cdot a & \text{if } q\cdot a \in Q_+ \cap Q_- \\ q_+ & \text{if } q\cdot a \in Q_+ - Q_- \\ q_- & \text{otherwise} \end{cases}$$

and moreover, $\delta(q_+, a) = q_+$ and $\delta(q_-, a) = q_-$. The family of accepting sets is $\{S \subset Q_+ \cap Q_- \mid S \in \mathcal{T}\} \cup \{\{q_+\}\}$.

Thus, in the automaton $\partial\mathcal{A}$, all states of $\mathcal{A}$ that belong to a maximal positive superchain are collapsed into a single state q_+ (and the set $\{q_+\}$ is positive) and all states that belong to a maximal negative superchain are collapsed into a single state q_- (and the set $\{q_-\}$ is negative). The definition is not completely symmetrical. States of $\mathcal{A}$ from which no maximal superchain can be reached and that do not belong to a maximal superchain are merged with q_-. Note that for any Muller automaton $\mathcal{A}$ such that $m(\mathcal{A}) \geq 2$, one has $m(\partial\mathcal{A}) < m(\mathcal{A})$.

Example 6.1. Consider the automaton $\mathcal{A}$ pictured in Figure 6.1, with the table $\mathcal{T} = \{\{1\}, \{1,2\}, \{3\}\}$. For $\mathcal{A}$, $(\{2\} \subset \{1,2\})$ is a negative chain and $(\{3\} \subset \{3,4\})$ is a positive chain. There are no superchain of length 2. We have $m(\mathcal{A}) = m^+(\mathcal{A}) = m^-(\mathcal{A}) = 2$, $n(\mathcal{A}) = n^+(\mathcal{A}) = n^-(\mathcal{A}) = 1$, $Q_- = \{0,1,2\}$ and $Q_+ = \{0,3,4\}$.

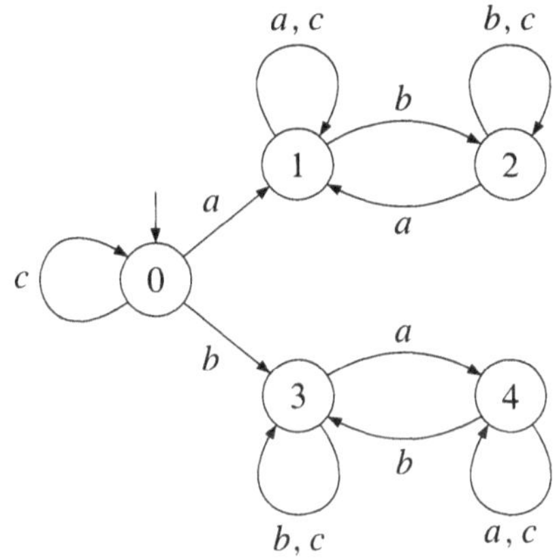

Figure 6.1. The automaton $\mathcal{A}$.

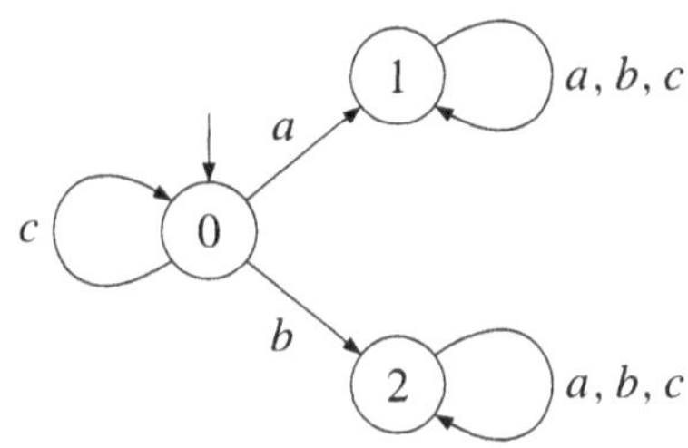

Figure 6.2. The derived automaton $\partial\mathcal{A}$.

The derived automaton $\partial\mathcal{A}$ is represented in Figure 6.2, in which q_- is the state 1 and q_+ the state 2. Its table is $\{\{2\}\}$. We then have $m(\partial\mathcal{A}) = m^+(\partial\mathcal{A}) = m^-(\partial\mathcal{A}) = 1$, $n^+(\partial\mathcal{A}) = 1$ and $n(\partial\mathcal{A}) = n^-(\partial\mathcal{A}) = 2$. Since $m(\partial\mathcal{A}) = 0$, we have $\partial\partial\mathcal{A} = \emptyset$.

It is possible to define directly the set recognized by the derived automaton as follows. Let $X \subset A^\omega$ be an ω-rational set and let $m = m(X)$ and $n = n(X)$. We first set, for a finite word u,

$$m_u(X) = m(uA^\omega \cap X)$$
$$n_u^+(X) = n^+(uA^\omega \cap X)$$
$$n_u^-(X) = n^-(uA^\omega \cap X)$$

Next we set

$$U_X = \{u \in A^* \mid m_u(X) = m \text{ and } n_u^+(X) = n \text{ and } n_u^-(X) = n - 1\}$$
$$V_X = \{v \in A^* \mid m_v(X) < m \text{ or } (m_v(X) = m \text{ and } n_v^+(X) < n)\}$$

Finally, we define the *derivative* of $X \subset A^\omega$, as

$$\partial X = (X - V_X A^\omega) + U_X A^\omega$$

We first verify the following statement.

Proposition 6.1. *Let $\mathcal{A}$ be a Muller automaton recognizing a set $X \subset A^\omega$. Then the automaton $\partial\mathcal{A}$ recognizes ∂X.*

Proof. Let x be a word accepted by $\partial\mathcal{A}$. Then either $\mathrm{Inf}(x) = \{q_+\}$ or $\mathrm{Inf}(x) \subset Q_+ \cap Q_-$ and $\mathrm{Inf}(x) \in \mathcal{T}$.

In the first case, let u be the prefix of x such that u reaches q_+ for the first time. This means that in $\mathcal{A}$, $i \cdot u \in Q_+ - Q_-$, that is, one can reach a positive chain of length n but no negative chain of length n from $i \cdot u$. In other words, $m_u(X) = m$, $n_u^+(X) = n$ and $n_u^-(X) = n - 1$, that is, $u \in U_X$. Thus $x \in U_X A^\omega$.

In the second case, $x \in X$, since $\mathrm{Inf}(x) \in \mathcal{T}$. Furthermore, the condition $\mathrm{Inf}(x) \subset Q_+ \cap Q_-$ implies that x never visits q_- in $\partial\mathcal{A}$, which amounts to saying that x never visits $Q - Q_+$ in $\mathcal{A}$, or, equivalently, that x has no prefix in V_X. Thus $x \in X - V_X A^\omega$.

Conversely, if $x \in U_X A^\omega$, x has a first prefix in U_X and thus $i \cdot u \in Q_+ - Q_-$ in $\mathcal{A}$, so that $i \cdot u = q_+$ in $\partial\mathcal{A}$. Since q_+ is fixed by any letter, x is recognized by $\partial\mathcal{A}$.

Finally, if $x \in X - (U_X A^\omega + V_X A^\omega)$, then in particular $x \in X$, whence $\mathrm{Inf}(x) \in \mathcal{T}$. Furthermore, since $x \notin U_X A^\omega + V_X A^\omega$, x has no prefix in u_X nor V_X, that is, never visits q_- nor q_+. It follows that $\mathrm{Inf}(x) \subset Q_+ \cap Q_-$ and x is accepted by $\partial\mathcal{A}$. □

We associate with each ω-rational set X two ordinals denoted $\gamma(X)$ and $\mu(X)$. The ordinal $\mu(X)$ is defined as follows

$$\mu(X) = \omega^{m(X)-1} n(X)$$

For example, for the sets X_1 and X_2 recognized by the automata $\mathcal{A}_1$ of Figure 4.1 and $\mathcal{A}_2$ of Figure 5.1, we have $m(X_1) = 2$, $n(X_1) = 1$, $m(X_2) = 1$ and $n(X_2) = 3$. Therefore

$$\mu(X_1) = \omega \quad \text{and} \quad \mu(X_2) = 3.$$

The ordinal $\gamma(X)$ is then defined by

$$\gamma(X) = \begin{cases} 0 & \text{if } X = \emptyset \\ \mu(X) + \gamma(\partial X) & \text{otherwise} \end{cases}$$

The ordinal $\gamma(X)$ is an arbitrary ordinal $< \omega^\omega$ and moreover, since $m(\partial X) < m(X)$ as soon as $m(X) \geq 1$, the decomposition given by the definition of $\gamma(X)$ produces the Cantor normal form of the ordinal $\gamma(X)$ which can be written as follows.

$$\gamma(X) = \mu(X) + \mu(\partial X) + \cdots + \mu(\partial^k X) = \sum_{n \geq 0} \mu(\partial^n(X)) \tag{6.1}$$

where k is the least integer such that either $\partial^{k+1} X = \emptyset$.

Both ordinals $\gamma(X)$ and $\mu(X)$ can be computed from any Muller automaton $\mathcal{A}$ recognizing the ω-rational set X since the integers $m(\mathcal{A})$ and $n(\mathcal{A})$ depend only on the set recognized by $\mathcal{A}$ (see [64]).

Example 6.2. For the set X recognized by the automaton $\mathcal{A}$ of Figure 6.1 $\gamma(X) = \omega + 2$.

We finally associate with an ω-rational set X a three valued variable called its *sign* and denoted $s(X)$. It is an element of the three element set $\{\sigma, \delta, \pi\}$ defined as follows.

$$s(X) = \begin{cases} \sigma & \text{if } n^- > n^+ \\ \pi & \text{if } n^- < n^+ \\ \delta & \text{if } n^- = n^+ \text{ and } m = 0 \\ s(\partial X) & \text{otherwise} \end{cases}$$

Thus $s(X)$ is given by the last term of the right hand side of Equation (6.1).

It is clear that $s(X) = \sigma$ if and only if $s(\overline{X}) = \pi$ and that $s(X) = \delta$ if and only if $s(\overline{X}) = \delta$. Also, if $s(X) = \delta$, then $\gamma(X)$ is a successor ordinal, i.e. $\gamma(X) = \beta + 1$. Indeed the last nonempty derivative $Y = \partial^k(X)$ of X is such that $\mu(Y)$ is a positive integer.

For an ω-rational set X, we call *degree* of X the pair

$$\varphi(X) = (\gamma(X), s(X)).$$

6.2 The Wagner hierarchy. We introduce a preorder on the class Rec defined by lexicographically ordering the pairs $\varphi(X)$ with the convention that $\delta < \sigma$ and $\delta < \pi$ (σ and π being incomparable). We set for each ordinal $\alpha < \omega^\omega$

$$\mathcal{C}_\alpha = \{X \in \text{Rec} \mid \varphi(X) \leq (\alpha, \sigma)\}$$
$$\mathcal{D}_\alpha = \{X \in \text{Rec} \mid \varphi(X) \leq (\alpha, \pi)\}$$
$$\mathcal{E}_\alpha = \{X \in \text{Rec} \mid \varphi(X) \leq (\alpha, \delta)\}$$

We shall see later that each of these classes is a Wagner class. For any ordinal $\alpha < \omega^\omega$, we have

$$\mathcal{E}_\alpha = \mathcal{C}_\alpha \cap \mathcal{D}_\alpha$$

For any ordinal $\alpha < \omega^\omega$, the classes $\mathcal{C}_\alpha$ and $\mathcal{D}_\alpha$ are dual to one another while the class $\mathcal{E}_\alpha$ is self-dual. The classes $\mathcal{C}_\alpha$, $\mathcal{D}_\alpha$ and $\mathcal{E}_\alpha$ are usually called "downward" classes since $\mathcal{C}_\alpha \subset \mathcal{C}_{\alpha'}$ for $\alpha < \alpha'$. The class of ω-rational sets of degree (α, σ) is $\mathcal{C}_\alpha - \mathcal{D}_\alpha$ and the class of ω-rational sets of degree (α, π) is $\mathcal{D}_\alpha - \mathcal{C}_\alpha$. The class of ω-rational sets of degree (α, δ) is empty if α is a limit ordinal and is equal to $\mathcal{E}_\alpha - (\mathcal{C}_\beta \cup \Pi_\beta)$ if $\alpha = \beta + 1$ is a successor ordinal.

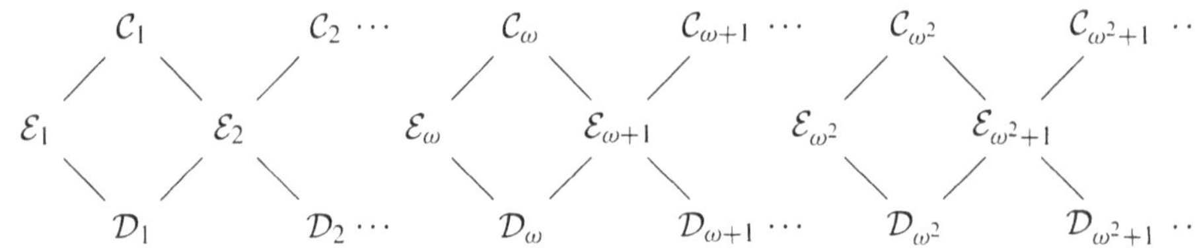

Figure 6.3. The Wagner hierarchy.

With this notation, the definition of the sets U_X and V_X used in the definition of the derivative of X has the following form.

$$U_X = \{u \in A^* \mid uA^\omega \cap X \in \Pi_{\omega^m \cdot n} - \mathcal{C}_{\omega^m \cdot n}\}$$
$$V_X = \{v \in A^* \mid vA^\omega \cap X \in \mathcal{C}_{\omega^m \cdot n}\}.$$

The diagram of the order defined by inclusion on the above defined classes has the familiar shape given by Figure 6.3. The main result is the following theorem showing that $\varphi(X)$ completely characterizes the Wadge preorder on ω-rational sets X.

Theorem 6.2 (Wagner). *For any ω-rational sets X and Y, $X \le_W Y$ if and only if $\varphi(X) \le \varphi(Y)$.*

We postpone the proof of this result to Section 6.4. It shows in particular that for two ω-rational sets X, Y one can decide whether $X \le_\omega Y$.

6.3 Biseparated differences. We now relate the definition of the set ∂X with the topological structure of X. We shall need the following results.

Lemma 6.3. *Let X be the biseparated union of X_1, X_2 and Z protected by O_1 and O_2, where O_1 and O_2 are open sets and let m and n be two integers. Suppose that both $\omega^{m(X_1)} \cdot n(X_1) \le \omega^m \cdot n$ and $\omega^{m(X_2)} \cdot n(X_2) \le \omega^m \cdot n$ and that $m(Z) < m$. Then, we have $\omega^{m(X)} \cdot n(X) \le \omega^m \cdot n$.*

Proof. Let $O_1 = U_1 A^\omega$ for some $U_1 \subset A^*$.

By Lemma 5.15, we have $m(X) \le m(X_1)$ and thus $m(X) \le m$. If this inequality is strict, the result is true.

We now suppose that $m(X) = m$. Let now $S = (C_0, C_1, \ldots, C_n)$ be an X-superchain where $C_i = (Y_i, Z_i)$. The set $Y_0 Z_0^\omega$ cannot be disjoint from $O_1 + O_2$ since otherwise C_0 would be a chain of Z. We may suppose that $Y_0 Z_0^\omega \cap O_1 \ne \emptyset$. Then some word $y \in Y_0 Z_0^*$ has a prefix in U_1. Let $C_i' = (Y_i', Z_i)$ with $Y_i' = yA^* \cap Y_i Z_i^*$ for $0 \le i \le n$. Then $(C_0', \ldots, C_n')$ is an X_1-superchain of length n. This implies that $m(X_1) = m$ and $n(X_1) \le n$. Since $n(X) \le n(X_1)$, we have $n(X) \le n$. □

Lemma 6.4. *Let $\alpha = \omega^m \cdot n + \beta$ be an ordinal in Cantor normal form and let $X \in \mathcal{S}_\alpha$. Then X satisfies $\omega^{m(X)} \cdot n(X) \le \omega^m \cdot n$.*

Proof. We use an induction on α.

If $\alpha = \omega^m \cdot n$, the result follows directly from Lemma 5.16. Otherwise, X is the biseparated union of X_1, X_2 and Z protected by O_1 and O_2 with $O_1, O_2 \in \Sigma_1$, $X_1, X_2^c \in \mathcal{S}_{\omega^m \cdot n}$ and $Z + O_2 \in \mathcal{S}_\beta$. By the induction hypothesis and by Lemma 6.3, the result is true. □

The following theorem gives a topological characterization of the classes $\mathcal{C}_\alpha$ for $\alpha < \omega^\omega$.

For an ordinal $\alpha < \omega^\omega$, we define a class $\mathcal{S}_\alpha$ of sets as follows. For $n \ge 0$ and $m \ge 1$ and $\beta < \omega^m$, let

$$\begin{aligned} \mathcal{S}_n &= D_n(\Sigma_1) \\ \mathcal{S}_{\omega^m \cdot n} &= \mathrm{Sep}(D_{n-1}(\Sigma_1), D_m(\Pi_2)) \\ \mathcal{S}_{\omega^m \cdot n + \beta} &= \mathrm{Bisep}(\Sigma_1, \mathcal{S}_{\omega^m \cdot n}, \mathcal{S}_\beta). \end{aligned}$$

Theorem 6.5. *Let X be an ω-rational set and let $\alpha = \omega^m \cdot n + \beta$ be an ordinal in Cantor normal form. The following conditions are equivalent.*

(1) $X \in \mathcal{C}_\alpha$.

(2) $X \in \mathrm{Bisep}(\Sigma_1, \mathcal{C}_{\omega^m n}, \mathcal{C}_\beta)$.

(3) $X \in \mathcal{S}_\alpha$.

Proof. We prove the equivalence of the three conditions by induction on α.

(1) implies (2). Let $\mathcal{A} = (Q, i, \mathcal{T})$ be a Muller automaton recognizing X and let $\mathcal{B} = \partial\mathcal{A}$ be the derived automaton. Let q_+ and q_- the new states of $\mathcal{B}$ defined as in Section 6. Let O_1 be the set of labels of paths in $\mathcal{B}$ from i to q_- and let O_2 be the set of labels of paths in $\mathcal{B}$ from i to q_-. O_1 and O_2 are then disjoint open sets. Let $X_1 = X \cap O_1$, $X_2 = X \cap O_2$ and $Z = X - X_1 - X_2$. Then X is the biseparated union of X_1, X_2 and Z protected by O_1 and O_2. Then it follows from the definitions that $\partial X = Z + O_2$. Since both X_1 and X_2^c belong to $\mathcal{C}_{\omega^m \cdot n}$ and $\partial X \in \mathcal{C}_\beta$, the result follows.

(2) implies (3) follows from the induction hypothesis.

(3) implies (1). The case $\alpha = \omega^m \cdot n$ results from Theorems 4.10 and 5.14. We assume the result true for β. Let X be the biseparated union of X_1, X_2 and Z protected by O_1 and O_2 with $O_1, O_2 \in \Sigma_1$, $X_1, X_2^c \in \mathcal{S}_{\omega^m \cdot n}$ and $Z + O_2 \in \mathcal{S}_\beta$. By Lemma 6.4, we have $\omega^{m(X)} \cdot n(X) \le \omega^m \cdot n$. If this inequality is strict, the result is true. There remains to show that $\partial X \in \mathcal{C}_\beta$ when equality holds. We have, with the notation of Section 6, $\partial X = Z + O_2 - V_X$. When $\beta \ge \omega$, it is clear that $\partial X \in \mathcal{C}_\beta$. When $\beta = k$ is an integer, it can be checked that $n(\partial X) \le k$ (in fact, if the last negative chain in a superchain of

∂X would be included in V_X, we could change it to a chain included in X_2 and thus in O_2. The result would be a chain of equal length in $Z + O_2$). □

We deduce from Theorem 6.5 the following corollary which shows that the class $\mathcal{C}_\alpha$ restricted to ω-rational sets is closed under the inverse of continuous functions.

Corollary 6.6. *Let $\alpha < \omega^\omega$ be an ordinal. Then $\mathcal{C}_\alpha$ is a Wadge class.*

Proof. If $Y \in \mathcal{C}_\alpha$, then by Theorem 6.5, we have $Y \in \mathcal{S}_\alpha$. Since $\mathcal{S}_\alpha$ is closed under inverse continuous functions, we have $X \in \mathcal{S}_\alpha$ whence $X \in \mathcal{C}_\alpha$, again by Theorem 6.5. □

The same statement obviously holds with $\mathcal{D}$ or $\mathcal{E}$ instead of $\mathcal{C}$.

6.4 The standard representatives. In this section, we give the proof of Theorem 6.2. We first introduce sets which are representatives of the classes $\mathcal{C}_\alpha$.

For an ordinal $\alpha < \omega^\omega$, we define an alphabet A_α and a set $X_\alpha \subset (A_\alpha)^\omega$. We first define A_α and X_α when $\alpha = \omega^m \cdot n$ for some integers m and n.

If $\alpha = n < \omega$ be an integer, let $A_n = \{a_0, b_0, \ldots, b_n\}$. For $\alpha = \omega^m \cdot n$ with $m \geq 1$, we set $A_\alpha = \{a_0, \ldots, a_m, b_0, \ldots, b_{n-1}\}$. For $x \in A_\alpha^* \cup A_\alpha^\omega$, we denote

$$\mu(x) = \max\{k \leq m \mid a_k \text{ occurs infinitely often in } x\}$$
$$\nu(x) = \max\{k \leq n \mid b_k \text{ occurs in } x\}$$

with the convention $\mu(x) = 0$ if none of the a_i occurs infinitely often in x and $\nu(x) = 0$ if none of the b_i occurs in x.

Let X_α be the set of words x of A_α^ω such that $\mu(x) + \nu(x)$ is odd.

Example 6.3. For instance, we have

$$\begin{aligned}
A_0 &= \{a_0, b_0\} & &\text{and} & X_0 &= \emptyset \\
A_1 &= \{a_0, b_0, b_1\} & &\text{and} & X_1 &= A_1^* b_1 A_1^\omega \\
A_2 &= \{a_0, b_0, b_1, b_2\} & &\text{and} & X_2 &= (a_0 + b_0)^\omega + A_2^* b_2 A_2^\omega \\
A_\omega &= \{a_0, a_1, b_0\} & &\text{and} & X_\omega &= (A_\omega^* a_1)^\omega.
\end{aligned}$$

Let now $\alpha = \omega^m \cdot n + \beta$ be an ordinal in Cantor normal form. We now define A_α and X_α by induction on α. Let $B = \{0, 1, 2\}$ and let $A_\alpha = A_\beta \times A_{\omega^m \cdot n} \times A_{\omega^m \cdot n} \times B$ and X_α be the set defined as follows

$$\begin{aligned}
X_\alpha = \{(x, y, z, t) \mid\ & (x \in X_\beta \text{ and } t = 0^\omega) \\
& \text{or } (y \in X_{\omega^m \cdot n} \text{ and } t \in 0^* 1 B^\omega) \\
& \text{or } (z \in X^c_{\omega^m \cdot n} \text{ and } t \in 0^* 2 B^\omega)\}
\end{aligned}$$

The following proposition shows that the sets X_α form a system of representatives of the exact classes $\mathcal{C}_\alpha \setminus \mathcal{D}_\alpha$.

Proposition 6.7. *For any ordinal* $\alpha < \omega^\omega$, $\varphi(X_\alpha) = (\alpha, \sigma)$. *Consequently,* $X_\alpha \in \mathcal{C}_\alpha - \mathcal{D}_\alpha$.

Proof. Let us first suppose that $\alpha = \omega^m \cdot n$. The set X_α is recognized by the (m, n)-superparity automaton $\mathcal{A}_\alpha = (Q, q_{0,0}, \mathcal{T})$ where $\mathcal{A}_\alpha$ is defined as follows. The set of states is $Q = \{q_{k,l} \mid k \leq m \text{ and } l \leq n\}$ and the transitions are given by

$$E = \{(q_{k,l}, a_j, q_{j,l}) \mid a_j \in A_\alpha\} \cup \{(q_{k,l}, b_j, q_{k,\max(l,j)}) \mid b_j \in A_\alpha\}.$$

The table $\mathcal{T}$ is given by

$$\mathcal{T} = \bigl\{P \subset Q \bigm| \max\{k \mid \exists l\ q_{k,l} \in P\} + \max\{l \mid \exists k\ q_{k,l} \in P\} \text{ is odd}\bigr\}.$$

It is straightforward to verify that $\gamma(\mathcal{A}_\alpha) = \alpha$ and thus $\varphi(X_\alpha) = (\alpha, \sigma)$.

Let now suppose that $\alpha = \omega^m \cdot n + \beta$. We first define the automaton $\mathcal{B} = (\{q_0, q_1, q_2\}, q_0)$ where the transitions of $\mathcal{B}$ are given by

$$E = \{(q_0, k, q_k) \mid k = 0, 1, 2\} \cup \{(q_k, l, q_k) \mid k = 1, 2 \text{ and } l = 0, 1, 2\}.$$

This automaton just remembers if a 1 or a 2 appears in the input and, in this case, which appears first. We recall that the alphabet A_α is equal to $A_\beta \times A_{\omega^m \cdot n} \times A_{\omega^m \cdot n} \times B$. The automaton $\mathcal{A}_\alpha$ is then equal to the product $\mathcal{A}_\beta \times \mathcal{A}_{\omega^m \cdot n} \times \bar{\mathcal{A}}_{\omega^m \cdot n} \times \mathcal{B}$ where each automaton of the product reads the corresponding component of the letter. The table of the automaton is

$$\begin{aligned}\mathcal{T} = \bigl\{\{(p_1, q_1, r_1, l_1), \ldots, (p_k, q_k, r_k, l_k)\} \mid \\ l_1 = \cdots = l_k = 0 \text{ and } \{p_1, \ldots, p_k\} \in \mathcal{T}_0 \\ \text{or } l_1 = \cdots = l_k = 1 \text{ and } \{q_1, \ldots, q_k\} \in \mathcal{T}_1 \\ \text{or } l_1 = \cdots = l_k = 2 \text{ and } \{r_1, \ldots, r_k\} \in \mathcal{T}_2\bigr\}\end{aligned}$$

where $\mathcal{T}_0$, $\mathcal{T}_1$ and $\mathcal{T}_2$ are the respective tables of $\mathcal{A}_\beta$, $\mathcal{A}_{\omega^m \cdot n}$ and $\bar{\mathcal{A}}_{\omega^m \cdot n}$. It is again straightforward to verify that $\gamma(\mathcal{A}_\alpha) = \alpha$ and thus $\varphi(X_\alpha) = (\alpha, \sigma)$. □

The following proposition, together with Proposition 6.7, shows that each X_α is $\mathcal{C}_\alpha$-complete.

Proposition 6.8. *Let X be an ω-rational subset of A^ω and let $\alpha < \omega^\omega$ be an ordinal. Then $X \in \mathcal{C}_\alpha$ if and only if $X \leq_W X_\alpha$.*

Proof. The right to left implication is an immediate consequence of Corollary 6.6 and Proposition 6.7.

To prove the converse, we introduce a notation. When f is a function from $A^* \times A$ to B^+, we denote by $\hat{f}$ the function from A^ω to B^ω defined by

$$\hat{f}(x_0x_1x_2x_3\dots) = f(\varepsilon, x_0)f(x_0, x_1)f(x_0x_1, x_2)f(x_0x_1x_2, x_3)\dots$$

where $x_i \in A$. The function $\hat{f}$ is obviously continuous.

We first suppose that $\alpha = \omega^m \cdot n$. The set X is then recognized by an (m, n)-superparity automaton $\mathcal{A} = (Q, i, \mu, \nu)$. We then define a function f from $A^* \times A$ to A^+_α by $f(w, a) = a_{\mu(q)}b_{\nu(q)}$ where $q = i \cdot wa$ is the state reached in $\mathcal{A}$ by the word wa. We have $X = \hat{f}^{-1}(X_\alpha)$.

We now prove the general case by induction on α. Suppose that $\alpha = \omega^m \cdot n + \beta$. The set X is the biseparated union of X_1, X_2 and Z protected by Y_1 and Y_2 with $Y_1, Y_2 \in \Sigma_1$, $X_1 \in \mathcal{C}_{\omega^m n}$, $\overline{X}_2 \in \mathcal{C}_{\omega^m \cdot n}$ and $Z \in \mathcal{C}_\beta$. Since Y_1 and Y_2 are disjoint open sets, there is a continuous function h from A^ω to B^ω which outputs a word in 0^*1B^ω (resp. in 0^*2B^ω and in 0^ω if the input belongs to Y_1 (resp. Y_1 and $A^\omega - (Y_1 + Y_2)$).

By induction hypothesis, the set Z reduces to X_β. We then have a continuous rational function g from A^ω to A^ω_β such that $Z = f^{-1}(X_\beta)$. By the previous result, the sets X_1, $\overline{X}_2$ reduce to $X_{\omega^m \cdot n}$. We then have two functions f and $\bar{f}$ from A^ω to $A^\omega_{\omega^m \cdot n}$ such that $X_1 = f^{-1}(X_{\omega^m \cdot n})$ and $X_2 = \bar{f}^{-1}(\overline{X}_{\omega^m \cdot n})$. We finally consider the function $(g, f, \bar{f}, h)$ form A^ω to A^ω_α. □

If X and Y are two subsets of A^ω, we define $X \oplus Y = 1X + 2Y$ where 1 and 2 are two different letters of the alphabet A.

The next proposition is a complement to Proposition 6.7.

Proposition 6.9. *For any ordinal $\alpha < \omega^\omega$, we have*

$$\varphi(X_\alpha \oplus \overline{X}_\alpha) = (\alpha + 1, \delta)$$

Consequently, $X_\alpha \oplus \overline{X}_\alpha \in \mathcal{E}_{\alpha+1} \setminus (\mathcal{C}_\alpha + \mathcal{D}_\alpha)$.

We now prove the following result showing with the previous proposition that $X_\alpha \oplus \overline{X}_\alpha$ is $\Delta_{\alpha+1}$-complete.

Proposition 6.10. *Let X be an ω-rational subset of A^ω and let $\alpha < \omega^\omega$ be an ordinal. Then $X \in \mathcal{E}_{\alpha+1}$ if and only if $X \leq_W X_\alpha \oplus \overline{X}_\alpha$.*

Proof. The right to left implication is again an immediate consequence of Corollary 6.6 with $\mathcal{E}$ instead of $\mathcal{C}$ and Proposition 6.9.

Since $X \in \mathcal{E}_{\alpha+1}$, there is an integer k such that $Y = \partial^k X \in \mathcal{E}_{n+1}$ for some integer $n \geq 0$. By Proposition 5.13, there exists a clopen set $U \subset A^\omega$ such that $Y \cap U \in \mathcal{D}_n$ and $Y \cap U^c \in \mathcal{C}_n$. Then $X \cap U \in \mathcal{D}_\alpha$ and $X \cap U^c \in \mathcal{C}_\alpha$, as one may verify. Since U is a clopen, we have $X \leq_W X \cap U \oplus X \cap U^c$. By Proposition 6.8, we have $X \cap U \leq_W X^c_\alpha$ and $X \cap U^c \leq_W X_\alpha$. Thus $X \leq_W X_\alpha \oplus X^c_\alpha$. □

Putting together the previous results, we obtain the following statement.

Theorem 6.11. *Any ω-rational set X is $\equiv_W$-equivalent to one and only one of X_α, $\bar{X}_\alpha$, or $X_\alpha \oplus \overline{X}_\alpha$.*

We finally give a proof of Theorem 6.2.

Proof. Let us first suppose that $X \leq_W Y$. If $s(Y) = \sigma$, then by Corollary 6.6, we have $X \in \mathcal{C}_{\gamma(Y)}$. Thus $\varphi(X) \leq \varphi(Y)$. The cases $s(Y) = \pi$ or $s(Y) = \delta$ are similar.

Conversely, let us first suppose that $s(Y) = \sigma$. By Proposition 6.8, we have $Y \leq_W X_{\gamma(Y)}$. Let us suppose that $X_{\gamma(Y)} \not\leq_W Y$. Then, by Theorem IV.5.1, we would have $Y \leq_W X^c_{\gamma(Y)}$ and thus $Y \in \Pi_{\gamma(Y)}$, a contradiction. Thus $X_{\gamma(Y)} \leq_W Y$. Moreover, since $\varphi(X) \leq \varphi(Y)$, by Proposition 6.8, $X \leq_W X_{\gamma(Y)}$. Finally, we obtain $X \leq_W X_{\gamma(Y)} \leq_W Y$ and thus $X \leq_W Y$.

The case $s(Y) = \pi$ is similar. Finally, let us suppose that $s(Y) = \delta$. The ordinal $\gamma(Y)$ is a successor and we let $\gamma(Y) = \alpha + 1$. By Proposition 6.10, we have $Y \leq_W X_\alpha \oplus X^c_\alpha$. Using again Theorem IV.5.1, we have $X_\alpha, X^c_\alpha \leq_W Y$. Thus $X_\alpha \oplus X^c_\alpha \leq_W Y$. If $\varphi(X) = \varphi(Y)$, then by the same argument, $X \leq_W X_\alpha \oplus X^c_\alpha$ and the conclusion follows. If $\varphi(X) \leq_W \varphi(Y)$, then either $X \in \mathcal{C}_\alpha$ and $X \leq_W X_\alpha$ or $X \in \mathcal{D}_\alpha$ and $X \leq_W X^c_\alpha$. In both cases we obtain $X \leq_W Y$. □

7 Exercises

7.1 Classes of sets.

Exercise 1. Let E be a set. Recall that the *symmetrical difference* of two subsets X and Y of E is the set $X \,\Delta\, Y$ consisting of the elements which belong to X or to Y but not to both of them. Thus, formally, $X \,\Delta\, Y = (X + Y) - XY$. The symmetrical difference is a commutative and associative operation.

Let $\mathcal{F}$ be a positive algebra of subsets of E. Show that a set X belongs to $D_n(\mathcal{F})$ if and only if it can be written as

$$X = X_1 \,\Delta\, X_2 \,\Delta \cdots \Delta\, X_n$$

for some $X_1, X_2, \ldots, X_n \in \mathcal{F}$.

Exercise 2. Let $\mathcal{F}$ be a positive algebra of subsets of E and let $X_1, X_2, \ldots, X_n \in \mathcal{F}$. Show that

$$(X_1 \,\Delta\, X_2 \,\Delta \cdots \Delta\, X_n)^c = \begin{cases} X_1^c \,\Delta\, X_2^c \,\Delta \cdots \Delta\, X_n^c & \text{if } n \text{ is even} \\ (X_1^c \,\Delta\, X_2^c \,\Delta \cdots \Delta\, X_n^c)^c & \text{if } n \text{ is odd} \end{cases}$$

Exercise 3. Let $\mathcal{F}$ be a positive algebra of subsets of E. Prove that if $X \in D_n(\mathcal{F})$ and $Y \in D_m(\mathcal{F})$ then $X \cap Y \in D_p(\mathcal{F})$ with

$$p = \begin{cases} m+n-2 & \text{if } m \text{ and } n \text{ are even} \\ m+n-1 & \text{otherwise} \end{cases}$$

Hint: use Formula (3.2).

Exercise 4. Let $X = X_1 - X_2 + \ldots \pm X_n$ and $Y = Y_1 - Y_2 + \ldots \pm Y_m$ be subsets of a set E, with $X_1 \supset X_2 \supset \ldots X_n$ and $Y_1 \supset Y_2 \supset \ldots Y_m$. Show that

$$X + Y = Z_1 - Z_2 + \ldots \pm Z_p$$

with $X_0 = Y_0 = E$,

$$p = \begin{cases} m+n-1 & \text{if } m \text{ and } n \text{ are odd} \\ m+n & \text{otherwise} \end{cases}$$

and

$$Z_{2i} = \sum_{\substack{k+\ell=2i \\ k \text{ even}}} X_k Y_\ell, \quad Z_{2i+1} = \sum_{k+\ell=2i+1} X_k Y_\ell.$$

Exercise 5. Let $\mathcal{F}$, X, Y be as in Exercise 4. Prove that $XY \in D_p(\mathcal{F})$ with

$$p = \begin{cases} m+n-1 & \text{if } m \text{ and } n \text{ are odd} \\ m+n & \text{otherwise.} \end{cases}$$

Exercise 6. Let $\mathcal{F}$ be as above. Prove that $D_n(\mathcal{F})$ is equal to the set of differences of length n

$$X = (X_1 - X_2) + (X_3 - X_4) + \ldots + (X_{n-1} - X_n)$$

7.2 Chains.

Exercise 7. Let $\mathcal{A} = (Q, i, \mathcal{F})$ be a Muller automaton. Let $\mathcal{P}$ be the set of admissible subsets of $\mathcal{F}$ and $\mathcal{N}$ be the set of admissible subsets which are not in $\mathcal{F}$. Let also $\mathcal{M}$ be the set of maximal admissible subsets of Q. For $P \in \mathcal{P}$, we denote by $c(P)$ the maximal length of a chain in $\mathcal{A}$ starting with P.

(1) Show that if P is maximal in $\mathcal{P}$, then $c(P) = 2$ if $P \notin \mathcal{M}$ and $c(P) = 1$ otherwise.

(2) If P is not maximal in $\mathcal{P}$, let $k = \max\{c(P') \mid P' \in \mathcal{P}, P \subset P'\}$ and let $\mathcal{P}' = \{P' \in \mathcal{P} \mid P \subset P', c(P') = k\}$. Show that

$$c(P) = \begin{cases} k+2 & \text{if there are } N \in \mathcal{N} \text{ and } P' \in \mathcal{P} \text{ such that } P \subset N \subset P' \\ k & \text{otherwise.} \end{cases}$$

Exercise 8. A chain $C = (Y, z_0, \ldots, z_m)$ is said to be *strong* if, for $1 \leq i \leq m$, it satisfies the condition

$$S_+ z_i S_+ \cap Z_{i-1}^+ = \emptyset \tag{7.1}$$

Let, for $0 \leq i \leq m$,

$$U_i = \bigcup_{k=0}^{i} W_k$$

with $W_i = YZ^*(z_i Z_i^*)^\omega$.

(1) Show that, for $0 \leq i \leq m$,

$$U_i = YZ^* Z_i^\omega \tag{7.2}$$

(2) Show that if C is a strong chain, then the sets $W_i = YZ^*(z_i Z_i^*)^\omega$, for $0 \leq i \leq m$, are pairwise disjoint.

(3) Show that if X is an ω-rational set, any X-chain of A^ω can be replaced by a strong one.

Exercise 9. The following exercise gives an example of a chain which is not strong. Let $Y = \{a\}$ and $Z = \{z_0, z_1, z_2)$ with $z_0 = abb$, $z_1 = babba$, $z_2 = b$. Show that (Y, z_0, z_1, z_2) is a negative chain for $X = YZ^*(z_1 z_0^*)^\omega$ which is not strong. (Hint: show that the automaton of Figure 7.1 recognizes Z^*.)

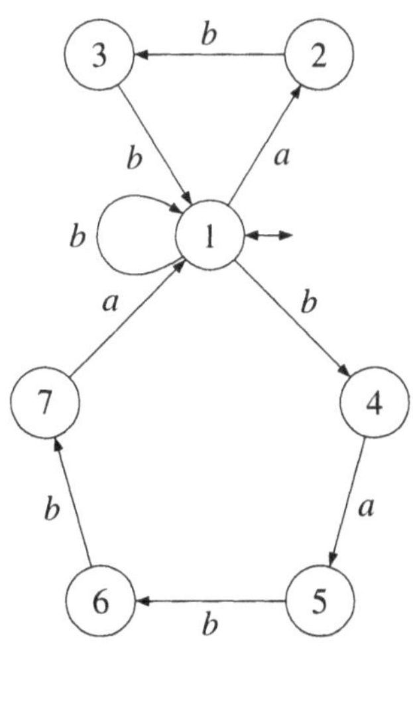

Figure 7.1. The automaton of Z^*.

Exercise 10. Let X be the set of words over the alphabet $A = \{a, b\}$ such that the maximal length of a-runs appearing infinitely often is bounded and even (where an a-run is a factor of type a^k with $k > 0$). Show that $m(X) = \infty$.

7.3 Wagner hierarchy.

Exercise 11. The *Rabin index* of a recognizable set $X \subset A^\omega$, denoted $i(X)$, is the least integer n such that X can be recognized by a Rabin automaton with n pairs. Show that

$$i(X) = \left\lfloor \frac{m^+(X) + 1}{2} \right\rfloor.$$

(Hint: use Exercise 6).

Exercise 12. Prove directly that the following equivalent conditions characterize recognizable open sets in A^ω (it is a particular case of Theorem 5.10).

(1) $n^+(X) = 0$.

(2) For all $x, y, z, t \in A^*$,

$$xy^\omega \in X \Rightarrow xy^*zt^\omega \cap X \neq \emptyset$$

(3) If $\varphi : A^\infty \to S$ is a morphism on a finite ω-semigroup recognizing X

$$se^\omega \in \varphi(X) \Rightarrow stf^\omega \in \varphi(X)$$

for all $s, t \in S_+$ and all idempotents e, f such that $s = se$ and $st = stf$.

See also Theorem VI.3.1.

Exercise 13. Prove directly that the following equivalent conditions characterize the Π_2 recognizable sets in A^ω (it is a particular case of Theorem 4.10).

(1) $m^+(X) \leq 0$.

(2) For all $x, y, z \in A^*$,

$$x(y + z)^*y^\omega \subset X \Rightarrow x(y^*z)^\omega \cap X \neq \emptyset$$

(3) If $\varphi : A^\infty \to S$ is a morphism on a finite ω-semigroup recognizing X then, for all $s \in S_+$ and all idempotents $e, f \in S_+$ such that $s = se = sf$ and $e \geq f$

$$se^\omega \in \varphi(X) \Rightarrow sf^\omega \in \varphi(X)$$

See also Theorem VI.3.4.

8 Notes

The Wagner hierarchy was introduced by Wagner in [349]. Our presentation follows the two papers [64] and [65] whose content is essentially reproduced here.

Wagner's theory has been rediscovered several times in a weaker form (see [16, 150]). As we have seen, it happens to coincide with the restriction on ω-rational sets of the hierarchy defined by Wadge (see [289, 184]). Recently, Selivanov has given a presentation of Wagner's hierarchy which uses, as we do here, concepts of descriptive set theory [279, 278]. A recent reference to Wagner's hierarchy is Staiger's chapter in the Handbook of Formal Languages [301]. We refer the reader to [267] for an introduction to the theory of ordinals.

The concepts of chain and superchain are defined in Wagner's paper only for Muller automata and it is not obvious from the definitions that the length of maximal chains and superchains depends only on the set recognized. The concept of chains in ω-semigroups can be extracted from Wagner's original paper. The proof of Lemma 4.9 is actually similar to a lemma in Wagner's original paper (Lemma 14 of [349]). We would like to warn the reader that our notation slightly differs from Wagner's. The correspondence for the superchain hierarchy is given in the table below

Our notation	Wagner's notation
$\mathcal{C}_{\omega^m(n+1)}$	$\hat{\mathcal{C}}^{n+1}_{m+1}$
$\mathcal{D}_{\omega^m(n+1)}$	$\hat{\mathcal{D}}^{n+1}_{m+1}$
$\mathcal{E}_{\omega^m(n+1)}$	$\hat{\mathcal{E}}^{n+1}_{m+1}$

Theorem 4.7 is not really stated in Wagner's paper [349]. The fact that $m(X) = m(\mathcal{A})$ follows in fact from the characterization of the classes of ω-sets X defined in an equivalent way using the integer $m(x)$.

The notations Sep and Bisep are borrowed from [184].

The hierarchy which is studied here is based on the preorder $\leq_W$ on ω-rational sets defined by continuous functions. According to Theorem IV.5.2, the same hierarchy could be defined using the preorder $\leq_{DA}$. The preorder $\leq_{DS}$ defines a slightly different hierarchy. It has also been studied by Wagner [349] but we shall not describe it here.

Corollary 5.12 is due to Staiger and Wagner [303]. It is a counterpart for ω-rational sets of a theorem of Hausdorff according to which, in a topological space such as A^ω,

$$\Pi_2 \cap \Sigma_2 = \bigcup_{\alpha < \omega_1} D_\alpha(\Sigma_1)$$

with a natural definition of D_α generalizing the usual definition of D_n (see [153, Thm. 22.27]). The construction of the memory extension is virtually folklore and has been used in various forms (see [131] or [332] for example). Theorem 4.10 essentially appears in [349] (quoting a result from [348]). Theorem 5.14 is analogous to a statement given in [349] (Theorem 15, proved in [348]). Theorem 6.11 appears almost verbatim in [279].

Theorem 3.1 is due to Hausdorff (see [153] or [135]).

Wilke and Yoo [358, 359] have shown that, given a Muller automaton $\mathcal{A}$, one can compute $m^+(\mathcal{A})$ in polynomial time. More precisely, if $\mathcal{A}$ has n states and m accepting sets the time is $O(m^2nc)$ where c is the cardinality of the alphabet. The algorithm is based on a lemma reproduced in Exercise 7. The computation of $m^+(\mathcal{A})$ gives also the Rabin index of $\mathcal{A}$ (see Exercise 11).

Exercises 3, 4, 5, 6 are from [56]. Exercise 11 is already in [349].

Recently, Duparc and Riss [93] have completed the algebraic treatment initiated in [64] and [65]. Their result relies on an extension of the notion of superchain in ω-semigroups and in Muller automata which allows a full description of the Wagner hierarchy.

VI

VARIETIES

1 Introduction

We have seen in Chapter II that, in order to study recognizable sets of finite words, it was possible to make use of semigroup morphisms in an abstract way. This point of view leads to a classification of the recognizable sets of finite words according to the algebraic properties of their syntactic semigroups (see Annex B). The first result in this approach is Kleene's theorem, which states that a set of finite words is rational if and only if it is recognized by a finite semigroup. Next, chronologically but also by order of importance, comes the result of Schützenberger on "star-free" subsets, which are the sets of finite words obtained from the finite subsets by using the boolean operations and the concatenation product, but not the star operation. Schützenberger's theorem states that a set of finite words is star-free if and only if it is recognized by a finite aperiodic semigroup. There exist numerous other results of this type, such as Simon's theorem which characterizes the so-called "piecewise testable" subsets as the sets of words recognized by a finite $\mathcal{J}$-trivial semigroup.

The natural framework in which to formulate this type of result is the theory of varieties developed by Eilenberg. A variety of finite semigroups is a class of finite semigroups stable under taking subsemigroup, quotients and finite direct products. Varieties can also be characterized in terms of identities — this justifies the terminology "variety": for instance, the variety of commutative semigroups is defined by the identity $xy = yx$, that of idempotent semigroups by the identity $x^2 = x$, etc. One then associates with each variety of semigroups **V** the class $\mathcal{V}$ of the sets of finite words recognized by a semigroup of **V**. Such a class $\mathcal{V}$ is called a +-variety (the prefix + refers to the free semigroup A^+, to indicate that we are dealing with finite words) and admits an abstract characterization: it is a class of recognizable subsets closed under boolean operations, inverse morphisms between free semigroups, and right and left quotients. Eilenberg's variety theorem asserts that the correspondence $\mathbf{V} \to \mathcal{V}$ between varieties of finite semigroups and +-varieties is bijective. For instance, the variety of rational subsets corresponds to the variety of finite semigroups, the variety of star-free subsets corresponds to the variety of aperiodic semigroups, and that of piecewise testable subsets corresponds to the variety of $\mathcal{J}$-trivial semigroups.

Varieties of ordered semigroups can be defined in the same way. This notion leads to a formulation of the variety theorem which takes into account classes of recognizable subsets that are not necessarily closed under complement.

One can also obtain a variety theorem for sets of infinite words at the cost of a few modifications. From an algebraic point of view, we replace finite (ordered) semigroups by (ordered) finite ω-semigroups, which are more suitably considered here as (ordered) Wilke algebras. We recover in this way a standard setting of universal algebra, which allows us to extend the characterization of varieties in terms of identities. Of course, these identities involve the three operations of a Wilke algebra, and in particular the ω-power.

From a combinatorial point of view, one needs to work simultaneously with finite or infinite words. There are good reasons to do so, and this was recently confirmed by the extension of the theory of automata to words over ordinals. For instance, the free ω-semigroup on the alphabet A is A^∞ and not A^ω, and various attempts to restrict the theory to infinite words have led to much less elegant results.

Once the right framework has been found, the variety theorem can be extended without much trouble and gives a one-to-one correspondence between "varieties of ordered ω-semigroups" — the algebraic aspect — and "∞-varieties" — the combinatorial aspect. There remains to give examples of such a correspondence. We first give some examples of topological nature. For instance, we show that the recognizable sets of the topological classes Σ_1 (open sets), Π_1 (closed sets), Δ_1 (clopen) and Σ_2, Π_2 and Δ_2 introduced in Chapter III form ∞-varieties for which we describe the corresponding varieties of ordered ω-semigroups.

To obtain other examples, a natural idea is to extend the known results on varieties of finite semigroups. However, there are at least three natural ways to associate with a variety of finite semigroups **V** a class of recognizable subsets of infinite words. First, one can take the subsets recognized by a finite ω-semigroup S such that $S_+ \in \mathbf{V}$. Secondly, by analogy with Kleene's theorem, one can also consider the subsets recognized by a Büchi automaton whose transition semigroup belongs to **V**. Finally, inspired by McNaughton's theorem, one may introduce the boolean combinations of subsets of the form $\overrightarrow{L}$ where L is recognized by a semigroup of **V** (class three). Unfortunately, these three classes are in general distinct, and the last two do not define, in general, an ∞-variety.

The presentation of the above material is organised as follows. First of all, the presentation of the theory of varieties on finite words has been postponed to Appendix B. The notions of varieties of ω-semigroups and the corresponding notion of ∞-variety of sets are introduced in Section 2. The varieties theorem (Theorem 2.8) is proved in this framework.

Section 3 presents the ∞-varieties defined by topological properties and the corresponding varieties of ω-semigroups. In the next two sections, we study the three types of varieties mentioned earlier.

Section 4 is devoted to the second type (varieties defined in terms of the transition semigroup of a Büchi automaton). We introduce for this purpose the notion of weak recognition. The key result is Theorem 4.5 which gives, in the most general setting, three equivalent formulations of this notion. In order to compare the first two classes, we associate with each finite ordered semigroup S an ordered ω-semigroup $\hat{S}$ with the following property: every set of infinite words weakly recognized by S is recognized by $\hat{S}$. The detailed study of this construction leads to a characterization of the varieties $\mathbf{V}$ for which the two first classes are equal. Unfortunately, this characterization is only workable when identities defining $\mathbf{V}$ are known. For the third class (boolean combinations of deterministic sets), the few available general results are presented in Section 5.

With these tools in hand, we broach in Section 6 the study of varieties closed under concatenation product. In this case, the three classes are equal, giving in turn a generalization of McNaughton's theorem. This is the case in particular for the variety of star-free subsets, which corresponds to the variety of aperiodic semigroups. The fine study of this variety leads to concatenation hierarchies, which form the topic of Section 7.

Some further examples of varieties are presented in the next chapter (Chapter VII).

2 Varieties of finite or infinite words

We give in this section an extension of the variety theorem for finite or infinite words. In this section, every ω-semigroup is supposed to be finite (with the obvious exception of the free ω-semigroup).

2.1 Varieties of ω-semigroups. A *variety of ω-semigroups* is a class of ω-semigroups, closed under division and finite product. In other words, it is a class $\mathbf{V}$ of ω-semigroups satisfying the three following conditions:

(1) if $S \in \mathbf{V}$ and if T is a ω-subsemigroup of S, then $T \in \mathbf{V}$,

(2) if $S \in \mathbf{V}$ and if T is a quotient of S, then $T \in \mathbf{V}$,

(3) if $(S_i)_{i\in I}$ is a finite family of ω-semigroups of $\mathbf{V}$, then $\prod_{i\in I} S_i \in \mathbf{V}$.

A *variety of ordered ω-semigroups* is defined in the same way as a class of ordered ω-semigroups closed under division and finite product.

The intersection of an arbitrary family of varieties of ordered ω-semigroups is still a variety of ordered ω-semigroups. In particular, if $\mathcal{C}$ is a class of ordered ω-semigroups, the intersection of all varieties of ordered ω-semigroups containing $\mathcal{C}$ is a variety of ordered ω-semigroups, called the variety generated by $\mathcal{C}$, and denoted by $\langle \mathcal{C} \rangle$. The next result and its corollary provide a more constructive definition. They are only stated for varieties of semigroups, but similar results hold for varieties of ordered semigroups.

Proposition 2.1. *Let $\mathcal{C}$ be a class of ω-semigroups. An ω-semigroup belongs to $\langle \mathcal{C} \rangle$ if and only if it divides a finite product of ω-semigroups of $\mathcal{C}$.*

Proof. Let $\mathbf{V}$ be the class of ω-semigroups that divide a finite product of ω-semigroups $\mathcal{C}$. We shall show that $\mathbf{V} = \langle \mathcal{C} \rangle$. By construction, $\mathbf{V}$ contains $\mathcal{C}$ and is closed under division. Furthermore, by Proposition II.4.4, $\mathbf{V}$ is closed under finite product. Therefore $\mathbf{V}$ is a variety and thus $\langle \mathcal{C} \rangle$ is contained in $\mathbf{V}$. Let now be $S \in \mathbf{V}$. Then S divides a product T of ω-semigroups T_i of $\mathcal{C}$. If $\mathbf{W}$ is a variety of ω-semigroups containing $\mathcal{C}$, we have $T_i \in \mathbf{W}$ by definition, and hence $T \in \mathbf{W}$ and $S \in \mathbf{W}$ since a variety of ω-semigroups is closed under finite product and division. Therefore S belongs to all varieties containing $\mathcal{C}$ and thus $S \in \langle \mathcal{C} \rangle$. □

Corollary 2.2. *Let $\mathcal{C}$ be a class of ω-semigroups closed under product. An ω-semigroup belongs to $\langle \mathcal{C} \rangle$ if and only if it divides an ω-semigroup of $\mathcal{C}$.*

2.2 Identities. One can develop a theory of identities for ω-semigroups which is quite similar to the corresponding theory for semigroups, as described in Annex B. In particular, Theorem B.1.13, which was stated for varieties of ordered semigroups, carries over to varieties of ω-semigroups. Rewriting the whole theory would be quite tedious, so we just indicate the two major changes when dealing with ω-semigroups. First, since ω-semigroups have two components, there are two types of identities. Second, identities may involve infinite products. An example will suffice to clarify these differences: an ordered ω-semigroup $S = (S_+, S_\omega)$ satisfies the identities $xy^\pi = y^\pi$ on S_+ and $(x^\pi yz)^\pi x^\omega \leq (yx)^\omega$ if, for every $x, y, z \in S_+$, $xy^\pi = y^\pi$ and $(x^\pi yz)^\pi x^\omega \leq (yx)^\omega$.

In the sequel, we shall often describe varieties of ω-semigroups by a set of identities.

2.3 The varieties theorem. If $\mathbf{V}$ is a variety of ordered ω-semigroups, we denote by $\mathcal{V}(A^\infty)$ the set of recognizable subsets of A^∞ whose syntactic ordered ω-semigroup belongs to $\mathbf{V}$. It is also, by Proposition II.8.9, the set of all subsets of A^∞ recognized by an ordered ω-semigroup of $\mathbf{V}$.

As for finite words, we call *class* of recognizable subsets a correspondence that associates with each finite alphabet A, a set $\mathcal{C}(A^\infty)$ of recognizable subsets of A^∞. The correspondence $\mathbf{V} \to \mathcal{V}$ associates with each variety of ordered ω-semigroups a class of recognizable subsets. As shown in the next theorem, this correspondence is injective.

Theorem 2.3. *Let* $\mathbf{V}$ *and* $\mathbf{W}$ *be two varieties of ordered* ω*-semigroups and suppose that* $\mathbf{V} \to \mathcal{V}$ *and* $\mathbf{W} \to \mathcal{W}$*. Then* $\mathbf{V} \subset \mathbf{W}$ *if and only if, for each alphabet* A, $\mathcal{V}(A^\infty) \subset \mathcal{W}(A^\infty)$*. In particular,* $\mathbf{V} = \mathbf{W}$ *if and only if* $\mathcal{V} = \mathcal{W}$*.*

Proof. If $\mathbf{V} \subset \mathbf{W}$, it follows immediately from the definitions that, for each alphabet A, $\mathcal{V}(A^\infty) \subset \mathcal{W}(A^\infty)$. The proof of the opposite inclusion is based upon the following proposition.

Proposition 2.4. *Let* $\mathbf{V}$ *be a variety of ordered* ω*-semigroups and let* $S \in \mathbf{V}$*. There exists an integer* k *and subsets* $X_1, X_2, \ldots, X_k \in \mathcal{V}(S_+^\infty)$ *such that* S *is an ordered* ω*-subsemigroup of* $S(X_1) \times S(X_2) \times \cdots \times S(X_k)$*.*

Proof. By Corollary II.4.7, there exists a surjective morphism of ω-semigroups $\varphi : S_+^\infty \to S$. For each $s \in S$, the set $X_s = \varphi^{-1}(\downarrow s)$ is recognized by S and hence $X_s \in \mathcal{V}(A^\infty)$. Since S recognizes X_s, there exists, by Proposition II.8.9, a surjective morphism $\pi_s : S \to S(X_s)$. Let $\pi : S \to \prod_{s \in S} S(X_s)$ be the morphism defined by $\pi(u) = (\pi_s(u))_{s \in S}$. Let $x, y \in S$. Since π is a morphism of ordered ω-semigroups, the condition $x \leq y$ implies $\pi(x) \leq \pi(y)$. Conversely, if $\pi(x) \leq \pi(y)$, then in particular $\pi_y(x) \leq \pi_y(y)$. Suppose first that $x \in S_+$. Then, by definition of the order on $S(X_y)$, $syt \leq y$ implies $sxt \leq y$ for each $s, t \in S_+^1$. In particular, we have, for $s = t = 1$, $1y1 \leq y$ and hence $x = 1x1 \leq y$. If now $x \in S_\omega$, then for each $s \in S_+^1$, $sy \leq y$ implies $sx \leq y$, whence $x \leq y$ by taking $s = 1$. Therefore $x \leq y$ if and only if $\pi(x) \leq \pi(y)$, and thus π is one-to-one. Therefore S is isomorphic to an ordered ω-subsemigroup of $S(X_1) \times \cdots \times S(X_k)$. □

End of the proof of Theorem 2.3. Suppose that for each alphabet A, $\mathcal{V}(A^\infty) \subset \mathcal{W}(A^\infty)$ and let $S \in \mathbf{V}$. Then by Proposition 2.4, S is an ordered ω-subsemigroup of a direct product of the form $S(X_1) \times S(X_2) \times \cdots \times S(X_k)$ with $X_1, X_2, \ldots, X_k \in \mathcal{V}(S_+^\infty)$. Since $\mathcal{V}(S_+^\infty) \subset \mathcal{W}(S_+^\infty)$, $X_1, X_2, \ldots X_k \in \mathcal{W}(S_+^\infty)$, and hence $S(X_1), \ldots, S(X_n)$ are in $\mathbf{W}$ by definition of $\mathcal{W}$. Since a variety is closed under division and finite product, it follows that $S \in \mathbf{W}$. □

The definition of an ∞-variety is close to that of a $+$-variety. The main difference lies in the use of the additional operator $Xu^{-\omega}$. Recall that we identify a subset X of A^∞ with the pair $(X \cap A^+, X \cap A^\omega)$. According to the general definition of Chapter II, the residuals of a subset X of A^∞ are defined as follows. If $u \in A^*$,

$$u^{-1}X = \{x \in A^\infty \mid ux \in X\}$$
$$Xu^{-\omega} = \{x \in A^+ \mid (xu)^\omega \in X\}$$

Similarly, if u is an element of A^ω, we set

$$Xu^{-1} = \{x \in A^+ \mid xu \in X\}$$

An *∞-positive variety* is a class of recognizable subsets such that

(1) for each alphabet A, $\mathcal{V}(A^\infty)$ is closed under finite union and finite intersection,

(2) if $\varphi : A \to B^+$ is a map, defining a morphism of ω-semigroups $\varphi : A^\infty \to B^\infty$, then $X \in \mathcal{V}(B^\infty)$ implies $\varphi^{-1}(X) \in \mathcal{V}(A^\infty)$,

(3) if $X \in \mathcal{V}(A^\infty)$, then, for each $u \in A^*$, $u^{-1}X, Xu^{-\omega} \in \mathcal{V}(A^\infty)$ and for each $u \in A^\omega$, $Xu^{-1} \in \mathcal{V}(A^\infty)$.

An *∞-variety* is a positive ∞-variety closed under complementation, which amounts to replacing Condition (1) by Condition (1′) in the previous definition

(1′) for each alphabet A, $\mathcal{V}(A^\infty)$ is a boolean algebra (for union and complementation).

If $\mathcal{V}$ is a positive ∞-variety, denote by $\mathcal{V}(A^+)$ (resp. $\mathcal{V}(A^\omega)$) the set of subsets X of A^+ (resp. A^ω) such that $X \in \mathcal{V}(A^\infty)$. Note that the correspondence $A \to \mathcal{V}(A^+)$ defines a $+$-positive variety, called the *finite component* of $\mathcal{V}$. Similarly, the class defined by the correspondence $A \to \mathcal{V}(A^\omega)$ is called the *ω-component* of $\mathcal{V}$.

Proposition 2.5. *Let* $\mathbf{V}$ *be a variety of ordered ω-semigroup. If* $\mathbf{V} \to \mathcal{V}$, *then* $\mathcal{V}$ *is a positive ∞-variety.*

Proof. The empty set is recognized by the trivial ordered ω-semigroup which belongs to $\mathbf{V}$. Therefore $\emptyset \in \mathcal{V}(A^\infty)$. Let $(X_i)_{i \in I}$ be a family of subsets of $\mathcal{V}(A^\infty)$. By definition, their ordered syntactic ω-subsemigroup $S(X_i)$ belongs to $\mathbf{V}$, as well as the product $S = \prod_{i \in I} S(X_i)$. Now Proposition II.6.4 shows that an ordered ω-subsemigroup of S recognizes simultaneously the union and the intersection of the X_i's. Therefore $\mathcal{V}(A^\infty)$ is closed under finite union and intersection.

One could prove in a similar way that $\mathcal{V}$ satisfies the two other properties of an ∞-positive variety by using Propositions II.6.6 and II.6.7, respectively. □

Theorem 2.6. *For each ∞-positive variety $\mathcal{V}$, there exists a variety of ordered ω-semigroups* $\mathbf{V}$ *such that* $\mathbf{V} \to \mathcal{V}$.

Proof. Let $\mathbf{V}$ be the variety of ordered ω-semigroups generated by all the ordered ω-semigroups of the form $S(X)$ with $X \in \mathcal{V}(A^\infty)$ for some alphabet A. Suppose that $\mathbf{V} \to \mathcal{W}$. We shall prove that actually, $\mathcal{V} = \mathcal{W}$. First of all, if $X \in \mathcal{V}(A^\infty)$, then $S(X) \in \mathbf{V}$ by definition and hence $X \in \mathcal{W}(A^\infty)$. Consequently, for each alphabet A, $\mathcal{V}(A^\infty) \subset \mathcal{W}(A^\infty)$.

The proof of the inclusion $\mathcal{W}(A^\infty) \subset \mathcal{V}(A^\infty)$ is more tricky. Let $X \in \mathcal{W}(A^\infty)$. Then $S(X) \in \mathbf{V}$ and by the definition of $\mathbf{V}$, there exists an integer $n > 0$, alphabets A_i $(1 \leq i \leq n)$ and subsets $X_i \in \mathcal{V}(A_i^\infty)$ $(1 \leq i \leq n)$ such that $S(X)$ divides $S(X_1) \times \cdots \times S(X_n) = S$. Denote by $\pi_i : S \to S(X_i)$ the morphism of ordered ω-semigroups defined by $\pi_i(s_1, \ldots, s_n) = s_i$. Since $S(X)$ divides S, $S(X)$ is quotient of an ordered ω-subsemigroup T of S. Therefore T recognizes X by Corollary II.8.10

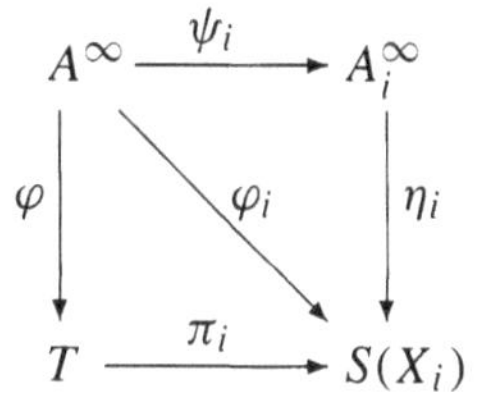

Figure 2.1.

and there exists a morphism of ordered ω-semigroups $\varphi : A^\infty \to S$ and an order ideal P of S such that $X = \varphi^{-1}(P)$. Set $\varphi_i = \pi_i \circ \varphi$, and let, for $1 \leq i \leq n$, $\eta_i : A_i^\infty \to S(X_i)$ be the syntactic morphisms of the X_i's. Since η_i is surjective, there exists by Proposition II.4.8, a morphism of ω-semigroups $\psi_i : A^\infty \to A_i^\infty$ such that $\varphi_i = \eta_i \circ \psi_i$. The situation is summarized in the following diagram: We prove that X belongs to $\mathcal{V}(A^\infty)$, by a sequence of reductions. Since

$$X = \varphi^{-1}(P) = \bigcup_{s \in P} \varphi^{-1}(\downarrow s)$$

and as $\mathcal{V}(A^\infty)$ is closed under finite union, it suffices to establish that, for each $s \in S$, $\varphi^{-1}(\downarrow s) \in \mathcal{V}(A^\infty)$. Next, setting $s = (s_1, \ldots, s_n)$, we have

$$\varphi^{-1}(\downarrow s) = \bigcap_{1 \leq i \leq n} \varphi_i^{-1}(\downarrow s_i)$$

and since $\mathcal{V}(A^\infty)$ is closed under finite intersection, it suffices to prove that, for $1 \leq i \leq n$, $\varphi_i^{-1}(\downarrow s_i) \in \mathcal{V}(A^\infty)$. Since $\varphi_i = \eta_i \circ \psi_i$, one gets $\varphi_i^{-1}(\downarrow s_i) = \psi_i^{-1}(\eta_i^{-1}(s_i))$. Since $\mathcal{V}$ is a ∞-variety, it remains to prove that $\eta_i^{-1}(\downarrow s_i) \in \mathcal{V}(A_i^\infty)$.

Lemma 2.7. *Let $\mathcal{V}$ be an ∞-variety and let A be a finite alphabet. Let $X \in \mathcal{V}(A^\infty)$ and let $\eta : A^\infty \to S(X)$ be the syntactic morphism of X. Then for each $s \in S$, $\eta^{-1}(\downarrow s) \in \mathcal{V}(A^\infty)$.*

Proof. Let $P = \eta(X)$. Then P is an order ideal of $S(X)$ such that $X = \eta^{-1}(P)$. Let us recall the notations introduced on page 113. For each $s \in S_+$, we set

$$CF_P(s) = \{(x, y) \in S_+^1 \times S_\omega \mid xsy \in P\}$$
$$CI_P(s) = \{(x, y) \in S_+^1 \times S_+^1 \mid x(sy)^\omega \in P\}$$

We claim that if $s \in S_+$,

$$\downarrow s = \bigcap_{(x,y) \in CF_P(s)} x^{-1}Py^{-1} \cap \bigcap_{(x,y) \in CI_P(s)} x^{-1}Py^{-\omega} \tag{2.1}$$

Let R be the right hand side of 2.1 and let $(x, y) \in CF_P(s)$. Then $x^{-1}Py^{-1}$ is an order ideal by Proposition II.3.5, which contains s by the definition of $CF_P(s)$. Similarly, if $(x, y) \in CI_P(s)$, it is easy to see that $x^{-1}Py^{-\omega}$ is an order ideal. Therefore $\downarrow s$ is contained in R. Conversely, let $t \in R$. If $(x, y) \in CF_P(s)$ (resp. $CI_P(s)$), then $t \in x^{-1}Py^{-1}$ (resp. $t \in x^{-1}Py^{-\omega}$. In other words, $xsy \in P$ (resp. $x(sy)^\omega \in P$) implies $xty \in P$ (resp. $x(ty)^\omega \in P$). It follows that $t \leq s$ and hence $t \in \downarrow s$, proving (2.1).

Now, each set $\eta^{-1}(x^{-1}Py^{-1})$ (resp. $\eta^{-1}(x^{-1}Py^{-\omega})$) is a residual of X and hence belongs to $\mathcal{V}(A^\infty)$. Furthermore, since $\mathcal{V}(A^\infty)$ is closed under finite intersection, $\eta^{-1}(\downarrow s) \in \mathcal{V}(A^\infty)$. □

Grouping Theorems 2.3 and 2.6, we obtain the varieties theorem.

Theorem 2.8. *The correspondence* $\mathbf{V} \rightarrow \mathcal{V}$ *defines a bijection between varieties of ordered* ω*-semigroups and* ∞*-positive varieties.*

Corollary 2.9. *The correspondence* $\mathbf{V} \rightarrow \mathcal{V}$ *defines a bijection between varieties of* ω*-semigroups and* ∞*-varieties.*

3 Varieties and topology

We shall see in this section that the recognizable subsets of various topological classes form positive varieties. Thus they correspond to varieties of ordered ω-semigroups for which precise algebraic descriptions can be given. These results give an alternative characterization of some of the levels of the Wagner hierarchy seen in Chapter V. In the forthcoming statements, $S(X)$ denotes, as usual, the syntactic ordered ω-semigroup of a subset X of A^ω and our algebraic characterizations are of the form "$S(X)$ satisfies the identities of E", where E is a set of identities. Proposition II.8.9 shows that this condition can be rephrased by saying that "X is recognized by an ordered ω-semigroup satisfying the identities of E". We shall use this equivalence in the subsequent proofs without any further warning.

Let us start with the open, closed and clopen sets.

Theorem 3.1. *Let* X *a recognizable subset of* A^ω. *The following conditions are equivalent:*

(1) *X is an open set,*

(2) *$S(X)$ satisfies the identity* $x^\pi yz^\omega \leq x^\omega$.

Proof. (1) implies (2). Let X be an open set of A^ω, let $\varphi : A^\infty \rightarrow S(X)$ be the syntactic morphism of X and let $P = \varphi(X)$. Since X is open, there exists a subset U of A^* such that $X = UA^\omega$. Up to replacing U by UA^*, we may assume that $UA^* = U$. In order to prove the identity $x^\pi yz^\omega \leq x^\omega$, it suffices to verify that, for each $s \in A^*$ and for every

$x, y, z \in A^+$, the condition $sx^\omega \in X$ implies $sx^\pi yz^\omega \in X$, where π is the exponent of $S(X)$. If $sx^\omega \in X$, sx^ω has a prefix in U of the form $s(x^\pi)^k$ with $k > 0$. Therefore $s(x^\pi)^k yz^\omega \in X$ and as $\varphi((x^\pi)^k) = \varphi(x^\pi)$, it comes $\varphi(s(x^\pi)^k yz^\omega) = \varphi(sx^\pi yz^\omega)$, whence $sx^\pi yz^\omega \in X$.

(2) implies (1). Suppose that X is recognized by a morphism $\varphi : A^\infty \to S$, where S is an ordered ω-semigroup satisfying the identity $x^\pi yz^\omega \leq x^\omega$. Let P be an order ideal of S such that $X = \varphi^{-1}(P)$. Set

$$T = \{s \in S_+ \mid \text{there exists } e \in E(S_+) \text{ such that } se = s \text{ and } se^\omega \in P\}$$
$$Y = \bigcup_{s \in T} \varphi^{-1}(s)A^\omega$$

The set Y is open by construction, and we claim that $X = Y$. If $u \in X$, there exists by Proposition II.2.3 a linked pair (s, e) of S_+ such that $\varphi(u) = se^\omega$. Therefore $se^\omega \in P$ and $s \in T$. In particular, $u \in \varphi^{-1}(s)A^\omega$, and thus X is a subset of Y. Conversely, if $u \in Y$, there exists $s \in T$ such that $u = u_0u'$ with $\varphi(u_0) = s$ and $u' \in A^\omega$. Since $s \in T$, there exists $e \in E(S_+)$ such that $se = s$ and $se^\omega \in P$. Furthermore, there exists by Proposition II.2.3 a linked pair (s', e') of S_+ such that $\varphi(u') = s'e'^\omega$. Since $e^\pi s'e'^\omega \leq e^\omega$, the condition $se^\omega \in P$ implies $se^\pi s'e'^\omega \in P$. But $\varphi(u) = \varphi(u_0)\varphi(u') = ss'e'^\omega = se^\pi s'e'^\omega$, since $s = se = se^\pi$. Therefore $\varphi(u) \in P$ and $u \in X$, proving the claim. □

Since a set is closed if and only if its complement is open, Theorem 3.1 leads to a characterization of the recognizable closed sets (compare with the statement of Exercise V.12).

Corollary 3.2. *Let X a recognizable subset of A^ω. The following conditions are equivalent:*

(1) *X is closed,*

(2) *$S(X)$ satisfies the identity $x^\omega \leq x^\pi yz^\omega$.*

Example 3.1. Let $A = \{a, b\}$ and let $X = a^\omega$. The syntactic ordered ω-semigroup of X is $S = (S_+, S_\omega)$, with $S_+ = \{a, b\}$, $S_\omega = \{a^\omega, b^\omega\}$, ordered by $a \leq b$ and $a^\omega \leq b^\omega$ and satisfying the relations $a^2 = a$, $ab = ba = b^2 = b$, $ab^\omega = bb^\omega = ba^\omega = b^\omega$, $aa^\omega = a^\omega$. One recognizes the ω-semigroup of Example II.6.1. This ordered ω-semigroup satisfies the identity $x^\omega \leq x^\pi yz^\omega$. Consequently, X is closed. However, the identity $x^\pi yz^\omega \leq x^\omega$ is not satisfied, since $b^\omega = a^\omega bb^\omega \not\leq a^\omega$, showing that X is not an open set.

Note that the complement of X, a^*bA^ω, is open but not closed. Its syntactic ordered ω-semigroup is obtained from the one of X by reversing the order.

We derive from Theorem 3.1 and Corollary 3.2 a characterization of the clopen recognizable subsets.

Corollary 3.3. *Let X be a recognizable subset of A^ω. The following conditions are equivalent:*

(1) *X is a clopen set,*

(2) *$S(X)$ satisfies the identity $x^\pi yz^\omega = x^\omega$.*

We have already given several characterizations of the deterministic subsets but it is interesting to reformulate Theorem II.9.4 in terms of varieties (see also Exercise V.13).

Theorem 3.4. *Let X be a recognizable subset of A^ω. The following conditions are equivalent:*

(1) *X is deterministic (class Π_2),*

(2) *$S(X)$ satisfies the identity $(x^\pi y)^\omega \leq (x^\pi y)^\pi x^\omega$.*

Taking the complement yields a characterization of the recognizable subsets whose complement is deterministic

Theorem 3.5. *Let X be a recognizable subset of A^ω. The following conditions are equivalent:*

(1) *X is of complement deterministic (class Σ_2),*

(2) *$S(X)$ satisfies the identity $(x^\pi y)^\pi x^\omega \leq (x^\pi y)^\omega$.*

Example 3.2. Let $A = \{a, b\}$ and let $X = (A^*a)^\omega = (b^*a)^\omega$. The syntactic ordered ω-semigroup of X was computed in Example II.6.4. It is $S = (S_+, S_\omega)$ with $S_+ = \{a, b\}$, $S_\omega = \{a^\omega, b^\omega\}$, ordered by $a \leq b$ and $a^\omega \leq b^\omega$ and satisfying the relations $a^2 = a$, $ab = ba = b^2 = b$, $ab^\omega = bb^\omega = b^\omega$, $aa^\omega = ba^\omega = a^\omega$. This ordered ω-semigroup satisfies the identity $(x^\pi y)^\omega \leq (x^\pi y)^\pi x^\omega$. Consequently, X is deterministic. However, the identity $(x^\pi y)^\pi x^\omega \leq (x^\pi y)^\omega$ is not satisfied for $x = b$ and $y = a$, showing that the complement of X, A^*a^ω is not deterministic. In particular, X is neither open, nor closed.

Finally, a little computation of identities leads to a characterization of the recognizable subsets that are simultaneously deterministic and of complement deterministic.

Proposition 3.6. *The three identities*

$$(x^\pi y^\pi)^\pi x^\omega = (x^\pi y^\pi)^\pi y^\omega, \tag{3.1}$$

$$(x^\pi y)^\omega = (x^\pi y)^\pi x^\omega, \tag{3.2}$$

$$(yx^\pi)^\omega = (yx^\pi)^\pi x^\omega \tag{3.3}$$

are equivalent. If an ω-semigroup S satisfies these identities and if (s, e) and (t, f) are two linked pairs of S_+ such that $s \mathrel{\mathcal{R}} t$, then $se^\omega = tf^\omega$.

Proof. Let us start with (3.1). Since $x^\omega = x^\pi x^\omega$, $(x^\pi y)^\pi x^\omega = (x^\pi y)^\pi x^\pi x^\omega$. Now, since

$$\left(x^\pi (yx^\pi)^\pi\right)^\pi = \left((x^\pi y)^\pi x^\pi\right)^\pi = \left((x^\pi y)^\pi\right)^\pi x^\pi = (x^\pi y)^\pi x^\pi$$

we have $(x^\pi y)^\pi x^\omega = (x^\pi (yx^\pi)^\pi)^\pi x^\omega$. Therefore, by (3.1),

$$\left(x^\pi (yx^\pi)^\pi\right)^\pi x^\omega = \left(x^\pi (yx^\pi)^\pi\right)^\pi (yx^\pi)^\omega = x^\pi (yx^\pi)^\omega = (x^\pi y)^\omega$$

which proves (3.2).

Let us start now with (3.2). We get immediately $(x^\pi y^\pi)^\pi x^\omega = (x^\pi y^\pi)^\omega$, which also gives by permuting x and y, and then multiplying on the left by x^π, $x^\pi (y^\pi x^\pi)^\pi y^\omega = x^\pi (y^\pi x^\pi)^\omega = (x^\pi y^\pi)^\omega$. Observing that $y^\omega = y^\pi y^\omega$ and $x^\pi (y^\pi x^\pi)^\pi y^\pi = (x^\pi y^\pi)^{\pi+1}$, we obtain $(x^\pi y^\pi)^{\pi+1} y^\omega = (x^\pi y^\pi)^\omega$, whence, by multiplying on the left by $(x^\pi y^\pi)^{\pi-1}$, $(x^\pi y^\pi)^{2\pi} y^\omega = (x^\pi y^\pi)^{\pi-1}(x^\pi y^\pi)^\omega = (x^\pi y^\pi)^\omega$, which gives (3.1), since $(x^\pi y^\pi)^{2\pi} = (x^\pi y^\pi)^\pi$.

We also have from (3.2), since $x^\pi x^\omega = x^\omega$, $(x^\pi y)^\omega = (x^\pi y)^\pi x^\pi x^\omega$, whence, by multiplying on the left by $y(x^\pi y)^{\pi-1}$,

$$y(x^\pi y)^{\pi-1}(x^\pi y)^\omega = y(x^\pi y)^{\pi-1}(x^\pi y)^\pi x^\pi x^\omega$$

Now, $y(x^\pi y)^{\pi-1}(x^\pi y)^\omega = (yx^\pi)^\omega$ and $y(x^\pi y)^{\pi-1}(x^\pi y)^\pi x^\pi x^\omega = (yx^\pi)^{2\pi} x^\omega = (yx^\pi)^\pi x^\omega$, which gives (3.3). A symmetric argument would show that (3.3) implies (3.2).

Let S be an ω-semigroup satisfying (3.1) and (3.2), and let (s, e) and (t, f) be two linked pairs of S such that $s \mathcal{R} t$. Then there exist $u, v \in S^1$ such that $t = su$ and $s = tv$. Applying the identity (3.1) with $x = ufv$ and $y = e$ yields

$$\left((ufv)^\pi e\right)^\pi e^\omega = \left((ufv)^\pi e\right)^\pi (ufv)^\omega$$

whence, by multiplying on the left by s and observing that $sufv = tfv = tv = s$ and $se = s$,

$$se^\omega = s(ufv)^\omega = su(fvu)^\omega = t(fvu)^\omega$$

Now, by (3.2), we have $(fvu)^\omega = (f^\pi vu)^\omega = (f^\pi vu)^\pi f^\omega$. It follows, since $t = tf = tvu$,

$$se^\omega = t(f^\pi vu)^\pi f^\omega = tf^\omega$$

completing the proof. □

The following result is close to Corollary V.5.12.

Theorem 3.7. *Let X be a recognizable subset of A^ω. The following conditions are equivalent:*

(1) *X and its complement are deterministic (class Δ_2),*

(2) *X is a finite boolean combination of recognizable open sets (class $\mathcal{E}_1$),*

(3) *$S(X)$ satisfies the identity* $(x^\pi y)^\omega = (x^\pi y)^\pi x^\omega$,

(4) *$S(X)$ satisfies the identity* $(x^\pi y^\pi)^\pi x^\omega = (x^\pi y^\pi)^\pi y^\omega$.

Proof. The equivalence of (1) and (3) follows from Theorems 3.5 and 3.4 and that of (3) and (4) of Proposition 3.6. Since every open set is deterministic and has a deterministic complement, we also have (2) implies (1). In order to establish that (3) implies (2), it suffices to show that if two words $u, v \in A^\omega$ are such that, for each $s \in S_+(X)$

$$u \in \varphi^{-1}(s)A^\omega \Leftrightarrow v \in \varphi^{-1}(s)A^\omega$$

then $\varphi(u) = \varphi(v)$. Indeed, this property will show that X is a finite boolean combination of open sets of the form $\varphi^{-1}(s)A^\omega$. Let (s, e) and (t, f) be linked pairs of u and v respectively. Then $u \in \varphi^{-1}(s)A^\omega$ and hence $v \in \varphi^{-1}(s)A^\omega$, showing that $t \leq_{\mathcal{R}} s$. Similarly, $s \leq_{\mathcal{R}} t$ and hence $s \mathrel{\mathcal{R}} t$. Finally, Proposition 3.6 shows that $se^\omega = tf^\omega$ and hence $\varphi(u) = \varphi(v)$. □

One can summarize the previous results in a slightly different form. Let us recall the terminology introduced in Chapter III: a subset of A^∞ is said to be Σ_1 (resp. Π_1, Δ_1, Π_2, Σ_2, Δ_2) if it is open (resp. closed, open and closed, deterministic, of deterministic complement, deterministic and of deterministic complement).

Theorem 3.8.

(1) *The recognizable Σ_1-sets form a positive variety, which corresponds to the variety* $[\![x^\pi yz^\omega \leq x^\omega]\!]$,

(2) *The recognizable Π_1-sets form a positive variety, which corresponds to the variety* $[\![x^\omega \leq x^\pi yz^\omega]\!]$,

(3) *The recognizable Δ_1-sets form a variety, which corresponds to the variety* $[\![x^\pi yz^\omega = x^\omega]\!]$,

(4) *The recognizable Π_2-sets form a positive variety, which corresponds to the variety* $[\![(x^\pi y)^\omega \leq (x^\pi y)^\pi x^\omega]\!]$,

(5) *The recognizable Σ_2-sets form a positive variety, which corresponds to the variety* $[\![(x^\pi y)^\pi x^\omega \leq (x^\pi y)^\omega]\!]$,

(6) *The recognizable Δ_2-sets form a variety, which corresponds to the variety* $[\![(x^\pi y)^\omega = (x^\pi y)^\pi x^\omega]\!]$, *which is also equal to the variety* $[\![(x^\pi y^\pi)^\pi x^\omega = (x^\pi y^\pi)^\pi y^\omega]\!]$.

Recall that the restriction of the topology of A^∞ to A^+ is trivial: every subset of A^+ is clopen. This explains why the identities characterizing the ordered ω-semigroup of various topological classes concern only the ω-part of the ordered ω-semigroup: there are no restrictions at all on the $+$-part.

Theorem 3.8 provides an algorithm for deciding whether a given recognizable set of A^∞ is open (resp. closed, deterministic, etc.). The decidability itself is a special case of the corresponding property for each Wadge class of ω-rational sets.

Corollary 3.9. *One can decide whether a given recognizable set of A^ω is Σ_1 (open), Π_1 (closed), Δ_1 (open and closed), Π_2 (deterministic), Σ_2 (of deterministic complement), Δ_2 (deterministic and of deterministic complement).*

Proof. Let X be a recognizable set of A^ω, given by a finite Büchi automaton. We have seen in Chapter II how to compute the syntactic ordered ω-semigroup $S(X)$. There remains to test whether the identities (1) to (6) of Theorem 3.8 are satisfied in $S(X)$. □

4 Weak recognition

We introduce in this section a notion of weak recognition that will help us to obtain several difficult results on ∞-varieties.

4.1 Weak recognition by a morphism of ordered semigroups. Let S be a finite ordered semigroup and $\varphi : A^+ \to S$ be a surjective morphism of ordered semigroups. A set of infinite words is called *φ-simple* if it is of the form $X_{(\downarrow s, \downarrow e)} = \varphi^{-1}(\downarrow s)\varphi^{-1}(\downarrow e)^\omega$, where (s, e) is a linked pair of S. Here is a more convenient characterization:

Lemma 4.1. *Let (s, e) be a linked pair of S. Then $u \in X_{(\downarrow s, \downarrow e)}$ if and only if there exists a linked pair (s', e') of S such that $s' \leq s$, $e' \leq e$ and $u \in X_{(s', e')}$.*

Proof. If $u \in X_{(\downarrow s, \downarrow e)}$, there exists by definition elements x and y of S such that $x \leq s$, $y \leq e$ and $u \in X_{(x, y)}$. Thus there exists a factorization $u = u_0 u_1 \cdots$ such that $\varphi(u_0) = x$ and $\varphi(u_n) = y$ for each $n > 0$. Set $s' = xy^\pi$ and $e' = y^\pi$, where π is the exponent of S. The pair (s', e') is a linked pair which satisfies $s' = xy^\pi \leq se^\pi = se = s$ and $e' = y^\pi \leq e^\pi = e$. Furthermore, setting $v_0 = u_0 u_1 \cdots u_\pi$ and $v_n = u_{n\pi+1} \cdots u_{(n+1)\pi-1}$ yields $u = v_0 v_1 \cdots$ with $\varphi(v_0) = s'$ and $\varphi(v_n) = e'$ for each $n > 0$. Therefore $u \in X_{(s', e')}$. □

The morphism φ *weakly recognizes* a subset X of A^ω if X is a finite union of φ-simple subsets. In this case

$$X = \bigcup_{(s,e) \in P} X_{(\downarrow s, \downarrow e)}$$

where P is *the trace of X by φ*, defined by the formula

$$P = \{(s, e) \mid (s, e) \text{ is a linked pair such that } X_{(\downarrow s, \downarrow e)} \neq \emptyset \text{ and } X_{(\downarrow s, \downarrow e)} \subset X\}$$

By extension, an ordered semigroup S is said to weakly recognize a subset X of A^ω if there exists a morphism of ordered semigroups $\varphi : A^+ \to S$ which weakly recognizes X. The next result justifies this terminology.

Proposition 4.2. *Each subset of A^ω recognized by a finite ordered ω-semigroup $S = (S_+, S_\omega)$ is weakly recognized by S_+.*

Proof. It is a direct consequence of Formula II.7.2. □

The converse of Proposition 4.2 does not hold in general: if φ_+ weakly recognizes a subset of A^ω, φ does not necessarily recognize this subset, even in the case where $\varphi_+ : A^+ \to S_+$ and $\varphi = \bar{\varphi}_+ : A^\infty \to \bar{S}_+$, as shown in Example 4.1 (see also Exercise 7).

Example 4.1. Set $A = \{a, b\}$, and denote by $B(1, 2)$ the semigroup $\{e, f\}$, given by its multiplication table: $ee = fe = e$, $ff = ef = f$. Let $\varphi : A^+ \to B(1, 2)$ be a semigroup morphism defined by $\varphi(a) = e$ and $\varphi(b) = f$. The linked pairs of $B(1, 2)$ are (e, e) and (f, f). The corresponding simple subsets are

$$X_{(e,e)} = (A^*a)^\omega \text{(set of infinite words containing infinitely many } a\text{'s)},$$
$$X_{(f,f)} = (A^*b)^\omega \text{(set of infinite words containing infinitely many } b\text{'s)}.$$

The morphism $\bar{\varphi}$ does not recognize any of these sets. Indeed, the pairs (e, e) and (f, f) are conjugate since $(ab)^\omega \in X_{(e,e)} \cap X_{(f,f)}$. In fact, the only subsets of A^ω recognized by $\bar{\varphi}$ are A^ω and the empty set.

There exists a rather simple connection between the sets of finite words recognized by a morphism and the sets of infinite words weakly recognized by the same morphism.

Proposition 4.3. *Let S be a finite ordered semigroup, I an order ideal of S and $\varphi : A^+ \to S$ a surjective morphism. Let $L = \varphi^{-1}(I)$. Then the set $\vec{L}$ is weakly recognized by φ and the following formula holds*

$$\vec{L} = \bigcup_{(s,e) \in \vec{I}} X_{(\downarrow s, \downarrow e)}$$

with $\vec{I} = \{(s, e) \mid (s, e)$ is a linked pair such that $s \in I\}$.

Proof. Let $(s, e) \in \vec{I}$ and let $u \in X_{(\downarrow s, \downarrow e)}$. Then we have $u = u_0 u_1 \cdots$ with $\varphi(u_0) \le s$ and $\varphi(u_n) \le e$ for each $n > 0$. It follows $\varphi(u_0 u_1 \cdots u_n) \le se = s$ for each $n > 0$ and as $s \in I$, we also have $\varphi(u_0 u_1 \cdots u_n) \in I$ and hence $u_0 u_1 \cdots u_n \in L$. Consequently, u has infinitely many prefixes in L and $u \in \vec{L}$.

Conversely, let $u \in \overrightarrow{L}$. Then u factorizes as $u = u_0 u_1 \cdots$ where, for each $n \geq 0$, $u_0 u_1 \cdots u_n \in L$. By Proposition II.2.3, there exists a linked pair (s, e) and a strictly increasing sequence of integers $(k_n)_{n \geq 0}$ such that $\varphi(u_0 u_1 \cdots u_{k_0 - 1}) = s$ and $\varphi(u_{k_n} u_{k_n+1} \cdots u_{k_{n+1}-1}) = e$ for each $n \geq 0$. Since $u_0 u_1 \cdots u_{k_0-1} \in L$, we have $s = \varphi(u_0 u_1 \cdots u_{k_0-1}) \in \varphi(L) = I$ and hence $(s, e) \in \overrightarrow{I}$. Consequently, $u \in \bigcup_{(s,e) \in \overrightarrow{I}} X_{(\downarrow s, \downarrow e)}$. $\square$

A finite union of subsets of A^ω weakly recognized by some ordered semigroups is weakly recognized by the product of these semigroups. More precisely:

Proposition 4.4. *Let $(X_i)_{i \in I}$ be a finite family of subsets of A^ω weakly recognized by morphisms $\varphi_i : A^+ \to (S_i, \leq_i)$. Let $\varphi : A^+ \to \prod_{i \in I} S_i$ be the morphism defined by $\varphi(u) = (\varphi_i(u))_{i \in I}$ and let S be the ordered subsemigroup of $\prod_{i \in I} S_i$ defined by $S = \varphi(A^+)$. Then $\varphi : A^+ \to S$ weakly recognizes the union of the X_i's.*

Proof. Let, for each $i \in I$, E_i be the trace of X_i by φ_i. Then we have

$$X_i = \bigcup_{(s_i, e_i) \in E_i} \varphi_i^{-1}(\downarrow s_i) \varphi_i^{-1}(\downarrow e_i)^\omega$$

Set

$$E = \Big\{ \big((s_i)_{i \in I}, (e_i)_{i \in I}\big) \mid \text{for each } i \in I,\ (s_i, e_i) \text{ is linked in } S_i \\ \text{and there exists } i \in I \text{ such that } (s_i, e_i) \in E_i \Big\}$$

and let us show that $\bigcup_{i \in I} X_i = X$ with

$$X = \bigcup_{(s,e) \in E} \varphi^{-1}(\downarrow s) \varphi^{-1}(\downarrow e)^\omega$$

If $u \in X_i$ for some $i \in I$, there exists $(s_i, e_i) \in E_i$ such that

$$u \in \varphi_i^{-1}(\downarrow s_i) \varphi_i^{-1}(\downarrow e_i)^\omega$$

Therefore, there is a factorization $u = u_0 u_1 u_2 \cdots$ with $\varphi_i(u_0) \leq s_i$ and $\varphi(u_n) \leq e_i$ for each $n > 0$. Furthermore, by Proposition II.2.3, there exists a linked pair (s', e') of S and a strictly increasing sequence of positive integers $(k_n)_{n \geq 0}$ such that

$$\varphi(u_0 u_1 \cdots u_{k_0 - 1}) = s' \text{ and } \varphi(u_{k_n} u_{k_n+1} \cdots u_{k_{n+1}-1}) = e' \text{ for each } n \geq 0.$$

Then we have necessarily

$$s'_i = \varphi_i(u_0 u_1 \cdots u_{k_0-1}) = \varphi_i(u_0) \varphi_i(u_1) \cdots \varphi_i(u_{k_0-1}) \leq s_i e_i \cdots e_i = s_i$$

and similarly $e'_i \le e_i$. Since $u \in X_{(s',e')}$, $X_{(s',e')}$ is nonempty. Furthermore, since $s' \in {\downarrow s}$ and $e' \in {\downarrow e}$, we have $X_{(s',e')} \subset X_{({\downarrow s},{\downarrow e})} \subset X$. Therefore $(s', e') \in E$ and as $u \in X_{(s',e')}$, $u \in X$.

Conversely, if $u \in X$, there exists $(s, e) \in E$ and a factorization $u = u_0u_1u_2\cdots$ such that $\varphi(u_0) \le s$ and $\varphi(u_n) \le e$ for each $n > 0$. By definition of E, there exists i such that $(s_i, e_i) \in E_i$ and since $\varphi_i(u_0) \le s_i$ and $\varphi_i(u_n) \le e_i$ for each $n > 0$, we have $u \in \varphi_i^{-1}({\downarrow s_i})\varphi_i^{-1}({\downarrow e_i})^\omega$ and hence $u \in X_i$. □

We are now ready to state the main result of this section, which motivates the definition of weak recognition.

Theorem 4.5. *Let* $\mathbf{V}$ *be a variety of finite ordered semigroups,* $\mathcal{V}$ *the corresponding* $+$*-positive variety and* Z *a subset of* A^ω. *The following conditions are equivalent:*

(1) *Z is recognized by a Büchi ordered automaton whose ordered transition semigroup belongs to* $\mathbf{V}$,

(2) *Z is weakly recognized by an ordered semigroup of* $\mathbf{V}$,

(3) *Z is a finite union of subsets of the form XY^ω with $XY^* \in \mathcal{V}(A^+)$ and $Y^+ \in \mathcal{V}(A^+)$.*

Proof. (1) implies (2). Let $\mathcal{A} = (Q, \le, A, E, I, F)$ be an ordered Büchi automaton whose ordered transition semigroup S belongs to $\mathbf{V}$. Denote by $\varphi : A^+ \to S$ the corresponding surjective morphism. We claim that

$$L^\omega(\mathcal{A}) = \bigcup_{(s,e)\in E} \varphi^{-1}({\downarrow s})\varphi^{-1}({\downarrow e})^\omega \tag{4.1}$$

where E is the set of linked pairs (s, e) of S such that there exist $i \in I$ and $f \in F$ with $(i, f) \in s$ and $(f, f) \in e$. If $u \in \varphi^{-1}({\downarrow s})\varphi^{-1}({\downarrow e})^\omega$, where $(s, e) \in E$, there exists a linked pair (s', e') of S with $s' \le s$ and $e' \le e$ such that $u \in \varphi^{-1}({\downarrow s'})\varphi^{-1}({\downarrow e'})^\omega$. Since $(s, e) \in E$, there exists $i \in I$ and $f \in F$ such that $(i, f) \in s$ and $(f, f) \in e$. Now, since $s' \le s$, there exists a state $f_0 \le f$ such that $(i, f_0) \in s'$. Since $e' \le e$, $f_0 \le f$ and $(f, f) \in e$, one can find by induction, using Lemma 4.7, a sequence $(f_n)_{n>0}$ of states such that, for each n, $f_n \le f$ and $(f_{n+1}, f_n) \in e'$. Let f' be a state occurring at least twice in this sequence. Since $s'e' = s'$ and $e'e' = e'$, we have $(i, f') \in s'$ and $(f', f') \in e'$ and as $f' \le f$, $f' \in F$. Therefore $u \in L^\omega(\mathcal{A})$.

Conversely, let $u \in L^\omega(\mathcal{A})$. Then there exists an initial state i, a final state f and a factorization $u = u_0u_1u_2\cdots$ such that $(i, f) \in \varphi(u_0)$ and $(f, f) \in \varphi(u_n)$ for each $n > 0$. By Proposition II.2.3, one can, up to the taking of a superfactorization, assume that there exists a linked pair (s, e) of S such that $\varphi(u_0) = s$ and $\varphi(u_n) = e$ for each $n > 0$. In particular $(s, e) \in E$ and $u \in \varphi^{-1}({\downarrow s})\varphi^{-1}({\downarrow e})^\omega$, which proves 4.1 and shows that $L^\omega(\mathcal{A})$ is weakly recognized by S.

(2) implies (3). Let (s, e) be a linked pair of S. Then $\downarrow s$ is a subsemigroup of S since if $x \leq e$ and $y \leq e$, then $xy \leq e^2 = e$. Similarly, if $s' \leq s$ and $e' \leq e$, $s'e' \leq se = s$. Therefore, setting $X = \varphi^{-1}(\downarrow s)$ and $Y = \varphi^{-1}(\downarrow e)$, we have on the one hand $Y^+ = Y = \varphi^{-1}(\downarrow e)$ and on the other hand, since $se = s$,

$$\varphi^{-1}(\downarrow s) = X \subset XY^* = \varphi^{-1}(\downarrow s)\varphi^{-1}(\downarrow e)^* \subset \varphi^{-1}(\downarrow s)$$

and hence $XY^* = \varphi^{-1}(\downarrow s)$. Finally as $S \in \mathbf{V}$, one has $\varphi^{-1}(\downarrow e) \in \mathcal{V}(A^+)$ and $\varphi^{-1}(\downarrow s) \in \mathcal{V}(A^+)$.

(3) implies (2). By Proposition 4.4, the set of subsets of A^ω weakly recognized by a semigroup of $\mathbf{V}$ is closed under finite union. Thus we may assume that $Z = XY^\omega$ with XY^* recognized by a morphism of ordered semigroups $\varphi_1 : A^+ \to S_1$ and Y^+ recognized by a morphism of ordered semigroup $\varphi_2 : A^+ \to S_2$, with $S_1, S_2 \in \mathbf{V}$. Let $\varphi : A^+ \to S = S_1 \times S_2$ the morphism defined by $\varphi(u) = (\varphi_1(u), \varphi_2(u))$ and let $S = \varphi(A^+)$. Then $S \in \mathbf{V}$ since S is a ordered subsemigroup of a product of ordered semigroups of $\mathbf{V}$. Let I_1 (resp. I_2) be an order ideal of S_1 (resp. S_2) such that $XY^* = \varphi_1^{-1}(I_1)$ (resp. $Y^+ = \varphi_2^{-1}(I_2)$). Let $R = \varphi(XY^*)$, $T = \varphi(Y^+)$ and let E be the set of idempotents of T. We claim that $XY^\omega = U$, with

$$U = \bigcup_{\substack{r \in R \\ e \in E}} \varphi^{-1}(\downarrow re)\varphi^{-1}(\downarrow e)^\omega \tag{4.2}$$

If $u \in Z$, there exists $r \in R$ and $e \in E$ and a factorization $u = u_0u_1u_2\cdots$ such that $\varphi(u_0) \leq re$ and $\varphi(u_n) \leq e$ for each $n > 0$. Setting $r = (r_1, r_2)$ and $e = (e_1, e_2)$, we have $\varphi_1(u_0) \leq r_1e_1$. Now $r_1e_1 \in \varphi_1(XY^*)\varphi_1(Y^+) \subset \varphi_1(XY^*) = I_1$ and hence $\varphi(u_0) \in I_1$. It follows $u_0 \in \varphi_1^{-1}(I_1) = XY^*$. Similarly, for each $n > 0$, $\varphi_2(u_n) \leq e_2$ and $u_n \in Y^+$ by the same type of argument. Therefore $u \in XY^\omega$.

Conversely, if $u \in XY^\omega$, u can be factorized as $u = u_0u_1\cdots$ with $u_0 \in XY^*$ and $u_n \in Y^+$ for each $n > 0$. Therefore $\varphi(u_n) \in T$ for each $n > 0$ and by Proposition II.2.3, one can find a linked pair (t, e) such that $u_1u_2\cdots \in \varphi^{-1}(t)(\varphi^{-1}(e))^\omega$. Let $s = \varphi(u_0)$ and $r = st$. It comes $re = r$, $r = \varphi(u_0)\varphi(u_1) \in \varphi(XY^*) = R$ and $u \in \varphi^{-1}(re)(\varphi^{-1}(e))^\omega$, whence $u \in U$, proving 4.2. It follows that XY^ω is weakly recognized by a semigroup of $\mathbf{V}$.

(2) implies (1). Let X be a subset of A^ω weakly recognized by a morphism $\varphi : A^+ \to S$ with $S \in \mathbf{V}$. Then

$$X = \bigcup_{(s,e) \in P} \varphi^{-1}(\downarrow s)\varphi^{-1}(\downarrow e)^\omega$$

where P is the trace of X on S. We shall now build a Büchi automaton recognizing X the transition semigroup of which is a product of quotients of S.

Set $Q = S \cup \{f\}$, where f is a new element. We equip Q with a structure of ordered monoid with identity f by extending the product of S by the rules $fq = qf = q$ for each $q \in Q$ and by extending the order on S by setting

$$p \leq q \text{ if and only if } p, q \in S \text{ and } p \leq q \text{ or if } p = q = f.$$

Finally, we denote by $\leq_Q$ the dual order to the order on Q:

$$p \leq_Q q \quad \text{if and only if} \quad q \leq p.$$

Consider now, for each idempotent e of S, the ordered Büchi automaton

$$\mathcal{A}_e = (Q, \leq_Q, A, E_e, I_e, \{f\})$$

with $I_e = \{s \in S \mid (s, e) \in P\}$ and

$$E_e = \{(p, a, q) \in S^1 \times A \times S^1 \mid \varphi(a)q \leq p \text{ or } \varphi(a)q \leq pe\}$$

A few verifications are in order to make sure that $\mathcal{A}_e$ is indeed an ordered automaton. First of all, the set of final states $\{f\}$ is an order ideal of Q. Next, if $(p, a, q) \in E_e$ and if $p' \leq_Q p$, we have on the one hand $\varphi(a)q \leq p$ or $\varphi(a)q \leq pe$ and on the other hand $p \leq p'$. Therefore $\varphi(a)q \leq p'$ or $\varphi(a)q \leq p'e$ and $(p', a, q) \in E_e$.

Let $\varphi_e : A^+ \to S(\mathcal{A}_e)$ be the surjective morphism from A^+ onto the transition semigroup of $\mathcal{A}_e$. Showing that $S(\mathcal{A}_e)$ is a quotient of S, amounts to verifying, by Proposition II.4.3, that the congruence of ordered semigroup associated with φ refines than the one associated with φ_e:

Lemma 4.6. *Let $u, v \in A^+$. If $\varphi(u) \leq \varphi(v)$, then $\varphi_e(u) \preceq_{\mathcal{A}_e} \varphi_e(v)$.*

Proof. Suppose that $\varphi(u) \leq \varphi(v)$ and let $(p, q) \in \varphi_e(v)$. Then we have $\varphi_e(v)q \leq p$ or $\varphi_e(v)q \leq pe$. It follows $\varphi(u)q \leq p$ or $\varphi(u)q \leq pe$ and hence $(p, q) \in \varphi_e(u)$, which shows that $\varphi_e(u) \preceq_{\mathcal{A}_e} \varphi_e(v)$. □

Furthermore, a word u is accepted by $\mathcal{A}_e$ if and only if there exists a state $s \in I_e$ and a factorization $u = u_0u_1u_2\cdots$ such that $(s, f) \in \varphi_e(u_0)$ and $(f, f) \in \varphi_e(u_n)$ for each $n > 0$. The latter condition can also be written as $\varphi(u_n)f \leq f$ or $\varphi(u_n)f \leq fe$, or, since $\varphi(u_n)f = \varphi(u_n)$ and $fe = f$, as $\varphi(u_n) \leq f$ or $\varphi(u_n) \leq e$. The first possibility is excluded, and the second is actually equivalent to $\varphi(u_n) \leq e$, i.e. $u_n \in \varphi^{-1}(\downarrow e)$. Similarly, the condition $(s, f) \in \varphi_e(\varphi(u_0))$ can be written $\varphi(u_0) \leq s$ or $\varphi(u_0) \leq se$. But for $s \in I_e$, one has $se = e$ and the condition reduces to $\varphi(u_0) \leq s$, i.e. $u_0 \in \varphi^{-1}(\downarrow s)$. Therefore

$$L^\omega(\mathcal{A}_e) = \bigcup_{s \in I_e} \varphi^{-1}(\downarrow s)\varphi^{-1}(\downarrow e)^\omega$$

It follows that X is recognized by the automaton $\mathcal{A}$ formed by the disjoint union of the $\mathcal{A}_e$. The ordered transition semigroup T of $\mathcal{A}$ is equal to the product of the S_e's. Now, we have $S_e \in \mathbf{V}$ since each S_e is a quotient of S and hence $T \in \mathbf{V}$. Consequently, X is weakly recognized by an ordered semigroup of $\mathbf{V}$. □

4.2 Recognition versus weak recognition. We have seen that if $\varphi : A^+ \to S$ weakly recognizes a subset X of A^ω, the morphism $\bar{\varphi} : A^\infty \to \bar{S}$ does not necessarily recognize this subset. It is however possible, given a morphism $\varphi : A^+ \to S$, to find a morphism of ω-semigroups recognizing X. To this end, one defines a relation $\preceq_\varphi$ on A^+ by setting $u \preceq_\varphi v$ if and only if, $\varphi(u) \leq \varphi(v)$ and, for each $x \in A^*$, and for each linked pair (s, e) of S,

$$(vx)^\omega \in X_{(\downarrow s, \downarrow e)} \text{ implies } (ux)^\omega \in X_{(\downarrow s, \downarrow e)}$$

a property that can be reformulated as follows:

Lemma 4.7. *We have $u \preceq_\varphi v$ if and only if $\varphi(u) \leq \varphi(v)$ and, for each $x \in A^*$ and for each linked pair (s, e) of S such that $(vx)^\omega \in X_{(s,e)}$, there exists a linked pair (s', e') of S such that $s' \leq s$, $e' \leq e$ and $(ux)^\omega \in X_{(s',e')}$.*

Proof. Suppose that $u \preceq_\varphi v$. If $(vx)^\omega \in X_{(s,e)}$, then $(vx)^\omega \in X_{(\downarrow s, \downarrow e)}$, and hence $(ux)^\omega \in X_{(\downarrow s, \downarrow e)}$. By Lemma 4.1, there exists a linked pair (s', e') of S such that $s' \leq s$, $e' \leq e$ and $(ux)^\omega \in X_{(s',e')}$.

Conversely, suppose that u and v satisfy the condition of the second part of the statement. If $(vx)^\omega \in X_{(\downarrow s, \downarrow e)}$, there exists by Lemma 4.1 a linked pair (s', e') such that $s' \leq s$, $e' \leq e$ and $(vx)^\omega \in X_{(s',e')}$. Then there exists a linked pair (s'', e'') such that $s'' \leq s'$, $e'' \leq e'$ and $(ux)^\omega \in X_{(s'',e'')}$. Since $s'' \leq s$ and $e'' \leq e$ by transitivity, we have $(ux)^\omega \in X_{(s,e)}$. Therefore $u \preceq_\varphi v$. □

We now extend the relation $\preceq_\varphi$ to A^ω by setting, for all $u, v \in A^\omega$, $u \preceq_\varphi v$ if and only if, for each linked pair (s, e) of S,

$$v \in X_{(\downarrow s, \downarrow e)} \text{ implies } u \in X_{(\downarrow s, \downarrow e)}$$

Before proving that $\preceq_\varphi$ is a congruence, we need an other algebraic tool.

Given a semigroup S, we define a new semigroup

$$T = \left\{ \begin{pmatrix} s & P \\ 0 & s \end{pmatrix} \;\middle|\; s \in S,\ P \text{ is a subset of } S \times S \right\}$$

with multiplication defined by

$$\begin{pmatrix} s & P \\ 0 & s \end{pmatrix} \begin{pmatrix} t & Q \\ 0 & t \end{pmatrix} = \begin{pmatrix} st & sQ \cup Pt \\ 0 & st \end{pmatrix}$$

where

$$sQ = \{(sq_1, q_2) \mid (q_1, q_2) \in Q\} \quad \text{and} \quad Pt = \{(p_1, p_2 t) \mid (p_1, p_2) \in P\}$$

Let now φ be a morphism from A^+ into S and let $\psi : A^+ \to T$ be the map defined by

$$\psi(u) = \begin{pmatrix} \varphi(u) & \tau(u) \\ 0 & \varphi(u) \end{pmatrix} \quad \text{with} \quad \tau(u) = \{(\varphi(u_1), \varphi(u_2)) \mid u = u_1 u_2\}$$

Lemma 4.8. *The map ψ is a semigroup morphism.*

Proof. We claim that

$$\tau(uv) = \varphi(u)\tau(v) \cup \tau(u)\varphi(v)$$

First, if $v = v_1v_2$, then $uv = (uv_1)v_2$ and hence $(\varphi(u)\varphi(v_1), \varphi(v_2)) \in \tau(uv)$. It follows that $\varphi(u)\tau(v)$ is contained in $\tau(uv)$. A dual proof would show that $\tau(u)\varphi(v)$ is contained in $\tau(uv)$.

To prove the opposite inclusion, consider a factorization of uv as w_1w_2. Then either u is a prefix of w_1 or v is a suffix of w_2. In the first case, $w_1 = uv_1$ for some word v_1 such that $v_1w_2 = v$, and thus

$$(\varphi(w_1), \varphi(w_2)) = (\varphi(uv_1), \varphi(w_2)) = (\varphi(u)\varphi(v_1), \varphi(w_2)) \in \varphi(u)\tau(v)$$

A dual argument works in the latter case, which proves the claim. It follows that

$$\begin{aligned}\psi(u)\psi(v) &= \begin{pmatrix}\varphi(u) & \tau(u)\\ 0 & \varphi(u)\end{pmatrix}\begin{pmatrix}\varphi(v) & \tau(v)\\ 0 & \varphi(v)\end{pmatrix}\\ &= \begin{pmatrix}\varphi(u)\varphi(v) & \varphi(u)\tau(v)\cup\tau(u)\varphi(v)\\ 0 & \varphi(u)\varphi(v)\end{pmatrix}\\ &= \begin{pmatrix}\varphi(uv) & \tau(uv)\\ 0 & \varphi(uv)\end{pmatrix} = \psi(uv)\end{aligned}$$

and thus ψ is a semigroup morphism. □

Let us denote by $\Diamond(S)$ the semigroup $\psi(A^+)$. The linked pairs of $\Diamond(S)$ have a very special property.

Proposition 4.9. *Let m be an idempotent of $\Diamond(S)$ and let U be a set of words of A^+ such that, for each $u \in U$, $\psi(u) = m$. If a linked pair of S is associated with a word of U^ω, then it is associated with each word of U^ω.*

Proof. Let (s, e) be a linked pair of S. Let $(u_n)_{n\in\mathbb{N}}$ and $(v_n)_{n\in\mathbb{N}}$ be two arbitrary sequences of words of U. Put $u = u_0u_1\cdots$ and $v = v_0v_1\cdots$. It suffices to show that if $u \in \varphi^{-1}(s)(\varphi^{-1}(e))^\omega$, then $v \in \varphi^{-1}(s)(\varphi^{-1}(e))^\omega$. Put $m = \begin{pmatrix} f & P\\ 0 & f\end{pmatrix}$. Since m is idempotent, so is f, and $\varphi(u_n) = \varphi(v_n) = f$ for each $n \geq 0$. If (s, e) is a linked pair of u, there exists by Proposition II.2.8, a factorization of u associated with (s, e) and conjugate of $(u_n)_{n\in\mathbb{N}}$. Then there exist a strictly increasing sequence of positive integers $(k_n)_{n\geq 0}$, and factorizations $x_0 = u_0\cdots u_{k_0-1}u'_{k_0}$ and, for each $n > 0$, $x_n = u''_{k_{n-1}}u_{k_{n-1}+1}\cdots u_{k_n-1}u'_{k_n}$ such that $u_{k_n} = u'_{k_n}u''_{k_n}$ for each $n \geq 0$, $\varphi(x_0) = s$ and $\varphi(x_n) = e$ for $n > 0$. Now, as $\psi_{1,2}(v_{k_n}) = \psi_{1,2}(u_{k_n}) = P$, there exists a factorization $v_{k_n} = v'_{k_n}v''_{k_n}$ such that $\varphi(v'_{k_n}) = \varphi(u'_{k_n})$ and $\varphi(v''_{k_n}) = \varphi(u''_{k_n})$. Setting

$y_0 = v_0 \cdots v_{k_0-1} v'_{k_0}$ and, for each $n > 0$, $y_n = v''_{k_{n-1}} v_{k_{n-1}+1} \cdots v_{k_n-1} v'_{k_n}$, we obtain $\varphi(y_0) = \varphi(v_0) \cdots \varphi(v_{k_0-1})\varphi(v'_{k_0}) = \varphi(u_0) \cdots \varphi(u_{k_0-1})\varphi(u'_{k_0}) = \varphi(x_0) = s$ and similarly $\varphi(y_n) = \varphi(x_n) = e$ for each $n > 0$. Therefore $v \in \varphi^{-1}(s)(\varphi^{-1}(e))^\omega$. □

We use the previous result to show that $\preceq_\varphi$ is stable under infinite product.

Proposition 4.10. *Let $(u_n)_{n\in\mathbb{N}}$ and $(u'_n)_{n\in\mathbb{N}}$ be two sequences of words of A^+ such that, for each $n \geq 0$, $u'_n \preceq_\varphi u_n$. Then $u'_0 u'_1 \cdots \preceq_\varphi u_0 u_1 \cdots$.*

Proof. Suppose that, for each $n \geq 0$, $u'_n \preceq_\varphi u_n$ and put $u = u_0 u_1 \cdots$ (resp. $u' = u'_0 u'_1 \cdots$). Let $\psi : A^+ \to \Diamond(S)$ be the morphism induced by φ. By Proposition II.2.3 applied to the sequence $(\psi(u_n), \psi(u'_n))$ of $\Diamond(S) \times \Diamond(S)$, there exists a linked pair (h, m) (resp. (h', m')) of $\Diamond(S) \times \Diamond(S)$ and a superfactorization $(v_0, v_1, \ldots)$ (resp. $(v'_0, v'_1, \ldots)$ of $(u_0, u_1, \ldots)$ (resp. $(u'_0, u'_1, \ldots)$) such that $\psi(v_0) = h$ (resp. $\psi(v'_0) = h'$) and, for each $n > 0$, $\psi(v_n) = m$ (resp. $\psi(v'_n) = m'$). Since $u'_n \preceq_\varphi u_n$ for each $n \geq 0$, and since the restriction of $\preceq$ to A^+ is stable, we have $v'_n \preceq_\varphi v_n$ for each $n \geq 0$.

Let (s, e) be a linked pair of S such that $u \in X_{(s,e)}$. Setting $v = v_1 v_2 \cdots$ (resp. $v' = v'_1 v'_2 \cdots$), we have $u = v_0 v$ and by Proposition II.2.5, there exists an element $t \in S$ such that $s = \varphi(v_0)t$ and $v \in X_{(t,e)}$. Now Proposition 4.9 shows that $v_1^\omega \in X_{(t,e)}$ and as $v'_1 \leq v_1$, there exists a linked pair (t', e') with $t' \leq t$ and $e' \leq e$ such that $v_1'^{\omega} \in X_{(t',e')}$. It follows, again by Proposition 4.9, that $v' \in X_{(t',e')}$. Finally, one has $s' = \varphi(v'_0)t' \leq \varphi(v_0)t = s$ and (s', e') is a linked pair such that $s' \leq s$, $e' \leq e$ and $v \in X_{(s',e')}$. Therefore $v' \preceq_\varphi v$ and $u' \preceq_\varphi u$. □

We are now ready to show that $\preceq_\varphi$ is a congruence.

Proposition 4.11. *The relation $\preceq_\varphi$ is a congruence of ordered ω-semigroup of finite index on $(A^\infty, =)$.*

Proof. To start with, consider the restriction of $\preceq_\varphi$ to A^+. The relation $\preceq_\varphi$ is reflexive and transitive. Let us verify it is stable. Suppose $u \preceq_\varphi v$ and let $w \in A^+$. If $(uwx)^\omega \in X_{(\downarrow s, \downarrow e)}$, we have $(vwx)^\omega \in X_{(\downarrow s, \downarrow e)}$: this follows from the relation $u \preceq_\varphi v$, by taking wx as right context. Since $\varphi(u) \leq \varphi(v)$, we also have $\varphi(uw) \leq \varphi(vw)$ and hence $uw \preceq_\varphi vw$.

If now $(wvx)^\omega \in X_{(\downarrow s, \downarrow e)}$, we have $(wvx)^\omega \in X_{(t,f)}$ with $t \leq s$ and $f \leq e$. Moreover, since $(wvx)^\omega = w(vxw)^\omega$, there exists by Proposition II.2.5 an element $r \in S$ such that $(vxw)^\omega \in X_{(r,f)}$ and $t = \varphi(w)r$. In particular, $(vxw)^\omega \in X_{(\downarrow r, \downarrow f)}$ and hence $(uxw)^\omega \in X_{(\downarrow r, \downarrow f)}$, since $u \preceq_\varphi v$. Therefore $(uxw)^\omega \in X_{(r',f')}$ with $r' \leq r$ and $f' \leq f$. It follows $w(uxw)^\omega = (wux)^\omega \in X_{(\varphi(w)r',f')}$. Now $\varphi(w)r' \leq \varphi(w)r = t \leq s$ and $f' \leq f \leq e$. Therefore $(wux)^\omega \in X_{(\downarrow s, \downarrow e)}$. Since $\varphi(wu) \leq \varphi(wv)$, we have $wu \preceq_\varphi wv$.

The restriction of $\preceq_\varphi$ to $(A^+, =)$ is therefore a congruence of ordered semigroup. The restriction of $\preceq_\varphi$ to $(A^\omega, =)$ is, by construction, reflexive and transitive. It could be proved it is stable under product on the left by using, as above, Proposition II.2.5. Finally, Proposition 4.10 shows that the relation is stable under infinite product. □

Denote by $\hat{S}$ the quotient $(A^\infty, =)/\preceq_\varphi$ and by $\hat{\varphi} : A^\infty \to \hat{S}$ the natural morphism. The ordered ω-semigroup $\hat{S}$ is called the *strong expansion* of S.

Since, for each $u, v \in A^+$, $\hat{\varphi}(u) = \hat{\varphi}(v)$ implies $\varphi(u) \leq \varphi(v)$, there exists by Proposition A.1.2 a surjective morphism $\pi : \hat{S}_+ \to S_+$ such that the following diagram commutes

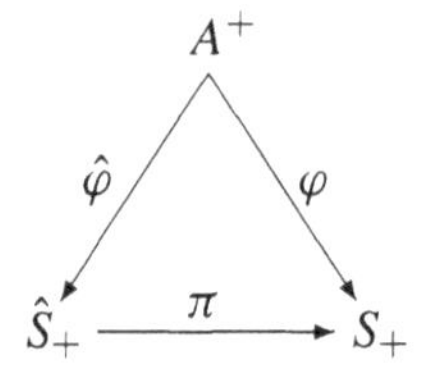

Then the following property holds:

Theorem 4.12. *Let X be a subset of A^ω and let S be a finite ordered semigroup. If $\varphi : A^+ \to S$ weakly recognizes X, then $\hat{\varphi} : A^\infty \to \hat{S}$ recognizes X.*

Proof. Since X is weakly recognized by φ, it is union of φ-simple subsets. As the subsets recognized by a morphism of ω-semigroups are closed under finite union, it suffices to verify that for each linked pair (s, e) of S, $X_{(\downarrow s, \downarrow e)}$ is recognized by $\hat{\varphi}$. To this extend, we show that the syntactic ordered ω-semigroup $X_{(\downarrow s, \downarrow e)}$ is a quotient of $\hat{S}$. This amounts to proving that if two words u and v of A^+ (resp. A^ω) verify $\bar{\varphi}(u) = \bar{\varphi}(v)$, then they are equivalent modulo the syntactic congruence of $X_{(s,e)}$. Suppose first that $u, v \in A^+$. By symmetry, it suffices to verify the two following properties

(1) if $xuyz^\omega \in X_{(\downarrow s, \downarrow e)}$, then $xvyz^\omega \in X_{(\downarrow s, \downarrow e)}$,

(2) if $x(uy)^\omega \in X_{(\downarrow s, \downarrow e)}$, then $x(vy)^\omega \in X_{(\downarrow s, \downarrow e)}$

Suppose that $xuyz^\omega \in X_{(\downarrow s, \downarrow e)}$. By Proposition II.2.5, there exists a linked pair (t, e) of yz^ω such that $s = \varphi(xu)t$. Since $\varphi(u) = \varphi(v)$, we also have $s = \varphi(xv)t$ and hence (s, e) is a linked pair of $xvyz^\omega$, proving (1).

Suppose now that $x(uy)^\omega \in X_{(\downarrow s, \downarrow e)}$. By Proposition II.2.5, there exists a linked pair (t, e) of $(uy)^\omega$ such that $s = \varphi(x)t$. Since $u \equiv_\varphi v$, (t, e) is also associated with $(vy)^\omega$, which proves (2). □

4.3 Properties of the strong expansion. The goal of the next statements is to compare the algebraic properties of S and $\hat{S}$. The proofs rely on the following combinatorial lemma.

Lemma 4.13. *Let u be a finite word and let n be a positive integer such that $\varphi(u)^n$ is idempotent. If (s, e) is a linked pair of u^ω, there exists a factorization $u = u'u''$ and an integer $k > n$ such that $\varphi(u^k u') = s$ and $\varphi(u''u^{2n-1}u') = e$.*

Proof. By hypothesis, u^ω admits a factorization $u^\omega = u_0 u_1 u_2 \cdots$ such that $\varphi(u_0) = s$ and $\varphi(u_i) = e$ for each $i > 0$. Up to a regrouping of the u_i's, we may assume that $|u_i| > |u|$ for each $i > 0$. Each u_i can then be written as $u_i = s_i u^{n_i} p_i$, where p_i (resp. s_i) is a prefix (resp. suffix) of u and n_i is a nonnegative integer. Then we have by construction the relations $p_i s_{i+1} = u$ for each $i \geq 0$. Since u only has a finite number of suffixes, there exist two indices i and j such that $1 \leq i < j$ and $s_i = s_j$. Setting $u' = p_{j-1}$ and $u'' = s_i$, we have $u'u'' = p_{j-1}s_i = p_{j-1}s_j = u$. On the other hand $u_i \cdots u_{j-1} = s_i u^{r-1} p_{j-1} = u''u^{r-1}u'$ for some $r \geq 2$ and thus $(u''u^{r-1}u')^n = u''u^{rn-1}u'$. It follows, since $\varphi(u)^n$ is idempotent,

$$\begin{aligned}\varphi(u''u^{2n-1}u') &= \varphi(u'')\varphi(u)^n\varphi(u)^{n-1}\varphi(u') = \varphi(u'')\varphi(u)^{(r-1)n}\varphi(u)^{n-1}\varphi(u')\\ &= \varphi(u'')\varphi(u)^{rn-1}\varphi(u') = \varphi(u''u^{rn-1}u') = \varphi((u''u^{r-1}u')^n)\\ &= \varphi((u_i \cdots u_{j-1})^n) = e^n = e\end{aligned}$$

Finally, we have $u_0 \cdots u_{j-1}(u''u^{2n-1}u') = u^k u'$ with $k > n$ and hence

$$\begin{aligned}\varphi(u^k u') &= \varphi(u_0 \cdots u_{j-1}(u''u^{2n-1}u'))\\ &= \varphi(u_0) \cdots \varphi(u_{j-1})\varphi(u''u^{2n-1}u') = se = s\end{aligned}$$

which concludes the proof. □

Lemma 4.14. *Let $u \in A^+$ and $v \in A^*$. If $\bar\varphi(u)$ is idempotent, the words $(uv)^\omega$ and $(uvu)^\omega$ have the same linked pairs.*

Proof. From $\bar\varphi(u) = \bar\varphi(u)^2$, follows in particular $u \equiv_\varphi u^2$ and the words $(uv)^\omega$ and $(u^2v)^\omega$ have the same linked pairs. Now $(u^2v)^\omega = u(uvu)^\omega$ and by Proposition II.2.5, the linked pairs associated with $(u^2v)^\omega$ are of the form $(\varphi(u)s, e)$ where (s, e) is a linked pair associated with $(uvu)^\omega$. But if (s, e) is associated with $(uvu)^\omega$, then $s = \varphi(uvup)$ for some prefix p of $(uvu)^\omega$ and hence $\varphi(u)s = \varphi(u)\varphi(u)\varphi(vup) = \varphi(u)\varphi(vup) = \varphi(uvup) = s$. Therefore $(uv)^\omega$ and $(uvu)^\omega$ have the same associated linked pairs. □

Proposition 4.15. *If S is an ordered $\mathcal{J}$-trivial semigroup, then $\hat{S}_+ = S$.*

Proof. Let $u \in A^*$ and let π be the exponent of S. We claim that if $u^\omega \in X_{(\downarrow s, \downarrow e)}$, where (s, e) is a linked pair of S, then $\varphi(u)^\pi \leq s$ and $\varphi(u)^\pi \leq e$. Indeed, there exists by the hypothesis a linked pair (t, f) of u^ω such that $t \leq s$ and $f \leq e$. Since $(\varphi(u)^\pi, \varphi(u)^\pi)$ is another linked pair of u^ω, it is conjugate from (t, f). But as S is $\mathcal{J}$-trivial, two conjugate pairs are necessarily equal, proving the claim.

Consider now two words u and v such that $\varphi(u) \leq \varphi(v)$ and let $x \in A^*$. If $(vx)^\omega \in X_{(\downarrow s, \downarrow e)}$, we have $\varphi(vx) \leq s$ and $\varphi(vx) \leq e$ by the first part of the proposition and as $\varphi(u) \leq \varphi(v)$, we also have $\varphi(ux) \leq s$ and $\varphi(ux) \leq e$. Consequently, $(ux)^\omega \in X_{(\downarrow s, \downarrow e)}$, showing that $u \preceq_\varphi v$, and therefore $\hat{\varphi}(u) \leq \hat{\varphi}(v)$. Thus $\varphi(u) \leq \varphi(v)$ if and only if $\hat{\varphi}(u) \leq \hat{\varphi}(v)$. □

It is worth stating separately the most important corollaries of Proposition 4.15.

Corollary 4.16. *If S is nilpotent, then $\hat{S} = S$.*

Corollary 4.17. *If S is aperiodic and commutative, then $\hat{S} = S$.*

Our last example relies on a manipulation of identities.

Lemma 4.18. *A semigroup which satisfies the identities $x^\pi y \leq x^\pi$ and $yx^\pi \leq x^\pi$ is $\mathcal{J}$-trivial.*

Proof. Indeed, these identities imply $x^\pi x \leq x^\pi$ and $x^\pi = x^\pi x^\pi = xx^\pi x^{\pi-1} \leq xx^\pi$ whence $x^\pi = x^{\pi+1}$. Therefore

$$(xy)^\pi = (xy)^{\pi+1} = x(yx)^\pi y \leq x(yx)^\pi = (xy)^\pi x \leq (xy)^\pi$$

whence $(xy)^\pi x = (xy)^\pi$ and similarly $y(xy)^\pi = (xy)^\pi$. The latter identities indeed define the variety of $\mathcal{J}$-trivial semigroups. □

Corollary 4.19. *If S satisfies the identities $x^\pi y \leq x^\pi$ and $yx^\pi \leq x^\pi$, then $\hat{S} = S$.*

We investigate now the properties of the morphism $\pi : \hat{S} \to S$.

Proposition 4.20. *For each idempotent f of S, the semigroup $\pi^{-1}(f)$ satisfies the identity $x^3 = x^4$.*

Proof. Let $u \in A^+$ be such that $\varphi(u) = \varphi(u)^2 = f$. We need to show that $u^3 \equiv_\varphi u^4$, in other words that, for each $x \in A^*$, the linked pairs of the words $(u^3x)^\omega$ and $(u^4x)^\omega$ are the same. Set $v = u^3x$ and $w = u^4x$ and let n be the exponent of the semigroup S. Since $\varphi(u) = \varphi(u)^2$, we have $\varphi(v) = \varphi(w)$.

If (s, e) is a linked pair of v^ω, there exists by Lemma 4.13 a factorization $v = v'v''$ and an integer $k > n$ such that $\varphi(v^k v') = s$ and $\varphi(v''v^{2n-1}v') = e$. Now, the factorization $u^3x = v'v''$ can take one of the following forms:

(1) u is prefix of v' and $v' = ut$ with $t \in A^*$ and $tv'' = u^2x$,

(2) v' is prefix of u and $u = v't$ with $t \in A^*$ and $v'' = tu^2x$.

We put $w' = u^2t$, $w'' = v''$ in the first case, and $w' = v'$, $w'' = tu^3x$ in the second case. In both cases, we have $w = w'w''$, $\varphi(w') = \varphi(v')$ and $\varphi(w'') = \varphi(v'')$. It follows

$$\varphi(w^k w') = \varphi(v^k v') = s \text{ and } \varphi(w'' w^{2n-1} w') = \varphi(v'' v^{2n-1} v') = e$$

showing that (s, e) is a linked pair of w^ω.

Conversely, let (s, e) be a linked pair of w^ω. By Lemma 4.13, there exists a factorization $w = w'w''$ and an integer $k > n$ such that $\varphi(w^k w') = s$ and $\varphi(w''w^{2n-1}w') = e$. Now, the factorization $u^4x = w'w''$ can take one of the following forms:

(1) u^2 is a prefix of w' and $w' = u^2t$ with $t \in A^*$ and $tw'' = u^2x$,

(2) w' is prefix of u^2 and $u^2 = w't$ with $t \in A^*$ and $w'' = tu^2x$.

We put $v' = ut$, $v'' = w''$ in the first case, and $v' = w'$, $v'' = tux$ in the second case.[1] Then in both cases $v = v'v''$ and by an easy computation $\varphi(v^k v') = s$ and $\varphi(v''v^{2n-1}v') = e$, showing that (s, e) is a linked pair of v^ω. □

Proposition 4.20 shows that $\hat{S}$ is an aperiodic extension of S. Since an aperiodic extension of an aperiodic semigroup is aperiodic, we derive the two following corollaries.

Corollary 4.21. *If S is aperiodic, then so is $\hat{S}$.*

Corollary 4.22. *Let $\mathbf{V}$ be a variety of semigroups closed under aperiodic extension. If S is a semigroup of $\mathbf{V}$, then so is $\hat{S}$.*

Here are two results involving expansions of ordered semigroups.

Proposition 4.23. *If S satisfies the identity $xy \leq x$, then so does $\hat{S}$.*

Proof. It suffices of establish that, for each $x, y \in A^+$, for each $z \in A^*$ and for each linked pair (s, e),

$$(xz)^\omega \in X_{(\downarrow s, \downarrow e)} \text{ implies } (xyz)^\omega \in X_{(\downarrow s, \downarrow e)}$$

Put $u = xz$. If $u^\omega \in X_{(\downarrow s, \downarrow e)}$, there exists a factorization $u = u_1u_2$, integers $n > 0$ and $m \geq 0$ and a linked pair (s', e') such that $\varphi(u^n u_1) = s' \leq s$ and $\varphi(u_2 u^m u_1) = e' \leq e$.

If $x = u_1u_3$ and $u_2 = u_3z$ with $u_3 \neq 1$, we have $\varphi(u_3y) \leq \varphi(u_3)$ since S satisfies the identity $xy \leq x$. Similarly, $\varphi(xy) \leq \varphi(x)$. Therefore

$$(xyz)^\omega = ((xyz)^n u_1)(u_3yz(xyz)^m u_1)^\omega$$

$\varphi((xyz)^n u_1) \leq \varphi((xz)^n u_1) = s'$ and $\varphi(u_3yz(xyz)^m u_1) \leq \varphi(u_3z(xz)^m u_1) = \varphi(u_2u^m u_1) = e'$, showing that $(xyz)^\omega \in X_{(\downarrow s, \downarrow e)}$.

[1] This is the reason why the identity $x^3 = x^4$ is used instead of $x^2 = x^3$: one needs to be able to replace a factor u^2 by u in every case.

Otherwise, we have $u_1 = xu_3$ with $u_3 \in A^*$. In this case, we can write

$$(xyz)^\omega = ((xyz)^n xyu_3)(u_2(xyz)^m xyu_3)^\omega$$

and, since $\varphi(xy) \leq \varphi(x)$, it comes

$$\varphi((xyz)^n xyu_3) \leq \varphi((xz)^n xu_3) = \varphi(u^n u_1) = s'$$

and

$$\varphi(u_2(xyz)^m xyu_3) \leq \varphi(u_2(xz)^m xu_3) = \varphi(u_2 u^m u_1) = e'$$

and therefore $(xyz)^\omega \in X_{(\downarrow s, \downarrow e)}$. □

Proposition 4.24. *If S satisfies the identity $x^\pi yx^\pi \leq x^\pi$, then so does $\hat{S}$.*

Proof. Let π be the exponent of $\hat{S}$. It suffices to establish that, for every $x, y \in A^+$, for each $z \in A^*$ and for each linked pair (s, e),

$$(x^\pi z)^\omega \in X_{(\downarrow s, \downarrow e)} \text{ implies } (x^\pi yx^\pi z)^\omega \in X_{(\downarrow s, \downarrow e)}$$

Put $u = x^{2\pi}z$. If $(x^\pi z)^\omega \in X_{(\downarrow s, \downarrow e)}$, then $u^\omega \in X_{(\downarrow s, \downarrow e)}$, since $\hat{\varphi}(x^{2\pi}) = \hat{\varphi}(x^\pi)$, and there exists a factorization $u = u_1u_2$, integers $n > 0$ and $m \geq 0$ and a linked pair (s', e') such that $\varphi(u^n u_1) = s' \leq s$ and $\varphi(u_2 u^m u_1) = e' \leq e$. Now, the factorization $u = u_1u_2$ can take one of the following forms:

(1) $u_1 = x^\pi u_3$, with $u_3u_2 = x^\pi z$,
(2) $u_2 = u_3x^\pi z$, with $u_1u_2 = x^\pi$.

In the first case, we have

$$(x^\pi yx^{2\pi}z)^\omega = ((x^\pi yx^{2\pi}z)^n x^\pi yx^\pi u_3)(u_2(x^\pi yx^{2\pi}z)^m x^\pi yx^\pi u_3)^\omega$$

and, since $\varphi(x^\pi yx^\pi) \leq \varphi(x^\pi)$, we have

$$\varphi((x^\pi yx^{2\pi}z)^n x^\pi yx^\pi u_3) \leq \varphi(x^{2\pi}z)^n x^\pi u_3) = \varphi(u^n u_1) = s' \leq s \text{ and}$$
$$\varphi(u_2(x^\pi yx^{2\pi}z)^m x^\pi yx^\pi u_3)\varphi(u_2(x^{2\pi}z)^m x^\pi u_3) = \varphi(u_2u^m u_1) = e' \leq e.$$

In the second case, we have

$$(x^{2\pi}yx^\pi z)^\omega = ((x^{2\pi}yx^\pi z)^n u_1)(u_3x^\pi yx^\pi z(x^{2\pi}yx^\pi z)^m u_1)^\omega$$

with

$$\varphi((x^{2\pi}yx^\pi z)^n u_1) \leq \varphi((x^{2\pi}z)^n u_1) = \varphi(u^n u_1) = s' \leq s$$

and

$$\varphi(u_3x^\pi yx^\pi z(x^{2\pi}yx^\pi z)^m u_1) \leq \varphi(u_3x^\pi z(x^{2\pi}z)^m u_1) = \varphi(u_2u^m u_1) = e' \leq e.$$

Therefore $(x^\pi yx^\pi z)^\omega \in X_{(\downarrow s, \downarrow e)}$ in both cases, since $\hat{\varphi}(x^\pi) = \hat{\varphi}(x^{2\pi})$. □

5 Extensions of McNaughton's theorem

McNaughton's Theorem (Theorem II.9.3) shows that every recognizable subset of A^ω is a boolean combination of deterministic subsets. The question raises whether this result can be relativized to varieties. In other words, if $\mathcal{V}$ is a ∞-variety, can every subset of $\mathcal{V}(A^\omega)$ be written as a boolean combination of deterministic subsets of $\mathcal{V}(A^\omega)$? One can also consider positive varieties: if $\mathcal{V}$ is an ∞-positive variety, is it possible to express every subset of $\mathcal{V}(A^\omega)$ as a positive boolean combination of deterministic subsets of $\mathcal{V}(A^\omega)$? The answer to these two questions is generally negative, but it is positive in some particular cases that are worth being studied.

It may also seem natural to use deterministic subsets to pass from a (positive) +-variety to a (positive) ∞-variety. More precisely, if $\mathcal{V}$ is a (positive) +-variety, we consider, for each alphabet A, the set $\overrightarrow{\mathcal{V}}(A^\omega)$ of the (positive) boolean combinations of subsets of the form $\overrightarrow{L}$ with $L \in \mathcal{V}(A^+)$. One may think that the classes $\mathcal{V}(A^+)$ and $\overrightarrow{\mathcal{V}}(A^\omega)$ define a (positive) ∞-variety. In fact, this is not the case and one can even prove that there exist some ∞-varieties for which the class $\overrightarrow{\mathcal{V}}$ is the ω-component of no ∞-variety (Exercise 8). But once again, if this counter-example ruins any hope of general result, much more satisfying statements can be obtained in some particular cases.

We have seen in Chapter II how to associate with each finite semigroup S an ω-semigroup $\bar{S}$. We now present another construction, which works only for the $\mathcal{R}$-trivial semigroups, to help us in our study of the operation $L \to \overrightarrow{L}$.

Let S be a finite $\mathcal{R}$-trivial ordered semigroup. Define on $S^{\mathcal{R}} = (S, SE(S))$ a structure of Wilke algebra by taking for mixed product $S \times SE(S) \to SE(S)$ the product in S and by setting, for each $s \in S$, $s^\omega = s^\pi$. The identities $(x^n)^\omega = x^\omega$ (for each $n > 0$) and $x(yx)^\omega = (xy)^\omega$ hold, since, for each $x, y \in S$,

$$(x^n)^\pi = x^\pi \quad \text{and} \quad x(yx)^\pi = (xy)^\pi x = (xy)^\pi$$

because, in a $\mathcal{R}$-trivial semigroup, $(xy)^\pi \mathrel{\mathcal{R}} (xy)^\pi x$. Finally, the order of S defines an order on $S^{\mathcal{R}}$.

Proposition 5.1. *Let S be a finite $\mathcal{R}$-trivial ordered semigroup. If a subset L of A^+ is recognized by S, then $\overrightarrow{L}$ is recognized by $S^{\mathcal{R}}$.*

Proof. We also denote by φ the morphism from A^∞ into $S^{\mathcal{R}}$ induced by φ. Let I be an order ideal of S such that $L = \varphi^{-1}(I)$. We claim that

$$\overrightarrow{L} = \varphi^{-1}(I \cap SE(S))$$

First of all, we have

$$\overrightarrow{L} = \overrightarrow{\varphi^{-1}(I)} = \overrightarrow{\bigcup_{s\in I} \varphi^{-1}(\downarrow s)} = \bigcup_{s\in I} \overrightarrow{\varphi^{-1}(\downarrow s)}$$

Therefore, if $u \in \overrightarrow{L}$, there exists a $s \in I$ such that $u \in \overrightarrow{\varphi^{-1}(\downarrow s)}$. Consequently, u admits a factorization $u = u_0 u_1 \cdots$ such that, for each $n \geq 0$, $\varphi(u_0 \cdots u_n) \leq s$. Then there exists a linked pair (t, f) and a superfactorization of $u_0 u_1 \cdots$, say $u = v_0 v_1 \cdots$, such that $\varphi(v_0) = t$ and $\varphi(v_n) = f$ for each $n > 0$. Then we have by definition $\varphi(u) = tf^\omega = tf^\pi = tf = t = \varphi(v_0)$ and $\varphi(v_0) \leq s$, since $v_0 = u_0 u_1 \cdots u_k$ for some $k > 0$. Therefore $\varphi(u) \leq s$ and $\varphi(u) \in I$.

Conversely, if $u \in \varphi^{-1}(I \cap SE(S))$, there exists $s \in I$ such that $\varphi(u) = s$. Let (t, f) be a linked pair of u and let $u = u_0 u_1 \cdots$ be a factorization such that $\varphi(u_0) = t$ and $\varphi(u_n) = f$ for each $n > 0$. We have $\varphi(u) = tf^\omega = tf^\pi = tf = t$ and hence $t = s$. It follows that $\varphi(u_0 u_1 \cdots u_n) = s$ for each $n > 0$ and hence $u \in \overrightarrow{L}$, proving the claim. □

Let $\mathbf{V}$ be a variety of semigroups and let $\mathcal{V}$ be the corresponding +-variety.

Theorem 5.2. *Let Z be a recognizable subset of A^ω. If $\mathbf{V}$ is a variety of semigroups contained in the variety $\mathbf{R}$, the following conditions are equivalent:*

(1) *$Z = \overrightarrow{L}$, with $L \in \mathcal{V}(A^+)$,*

(2) *Z is a boolean combination of subsets of the form $\overrightarrow{L}$, with $L \in \mathcal{V}(A^+)$,*

(3) *$S_+(Z) \in \mathbf{V}$ and $S(Z)$ satisfies the identity $(x^\pi y^\pi)^\pi x^\omega = (x^\pi y^\pi)^\pi y^\omega$.*

Proof. It is clear that (1) implies (2).

(2) implies (3). Let $\mathbf{V}'$ be the variety of all ω-semigroups S which satisfy $S_+ \in \mathbf{V}$ and the identity $(x^\pi y^\pi)^\pi x^\omega = (x^\pi y^\pi)^\pi y^\omega$, and let $\mathcal{V}'$ be the corresponding variety. In particular, for each alphabet A, $\mathcal{V}(A^+)$ is a boolean algebra. Therefore it suffices to verify that if $Z = \overrightarrow{L}$, with $L \in \mathcal{V}(A^+)$, then $S(Z) \in \mathbf{V}'$. Let S be the syntactic semigroup of L. Since $L \in \mathcal{V}(A^+)$, $S \in \mathbf{V}$. Furthermore, by Proposition 5.1, $\overrightarrow{L}$ is recognized by $S^{\mathcal{R}}$ and hence $S(Z)$ is a quotient of $S^{\mathcal{R}}$. Now $S^{\mathcal{R}}$ satisfies the identity $(x^\pi y^\pi)^\pi x^\omega = (x^\pi y^\pi)^\pi y^\omega$, because, for each $x, y \in S$, we have, since S is $\mathcal{R}$-trivial,

$$(x^\pi y^\pi)^\pi x^\omega = (x^\pi y^\pi)^\pi x^\pi = (x^\pi y^\pi)^\pi = (x^\pi y^\pi)^\pi y^\pi = (x^\pi y^\pi)^\pi y^\omega$$

Therefore $S^{\mathcal{R}} \in \mathbf{V}'$ and $S(Z) \in \mathbf{V}'$.

(3) implies (1). Let now $\varphi : A^\infty \to S(Z)$ be the syntactic morphism of Z. By Proposition 3.6, if (s, e) and (s, f) are two linked pairs of $S(Z)$, then $se^\omega = sf^\omega$. Let $u \in A^\omega$ and let (s, e) be a linked pair of u. By Proposition II.2.8, all the linked pairs of u are conjugate of (s, e) and subsequently, since $S(Z)$ is $\mathcal{R}$-trivial, are of the form (s, f) with $se^\omega = sf^\omega$. We claim that

$$\varphi^{-1}(se^\omega) = \overrightarrow{\varphi^{-1}(s)} \tag{5.1}$$

If $u \in \varphi^{-1}(se^\omega)$, there exists a linked pair (s, f) and a factorization $u = u_0 u_1 \cdots$ such that $\varphi(u_0) = s$ and $\varphi(u_n) = f$ for each $n > 0$. Therefore $u_0 \cdots u_n \in \varphi^{-1}(s)$ for each $n > 0$ and hence $u \in \overrightarrow{\varphi^{-1}(s)}$. Conversely, if $u \in \overrightarrow{\varphi^{-1}(s)}$, u has infinitely many prefixes in $\varphi^{-1}(s)$ and hence, by Proposition II.2.10, $R(u) = s$. It follows that every linked pair of u is of the form (s, f). Therefore $\varphi(u) = sf^\omega = se^\omega$ and $u \in \varphi^{-1}(se^\omega)$, proving (5.1). Finally, Z is a finite union of subsets of the form $\varphi^{-1}(se^\omega)$ and hence a finite union of subsets of the form $\overrightarrow{\varphi^{-1}(s)}$. Since $\varphi^{-1}(s)$ is recognized by $S_+(Z)$ by construction, we have $\varphi^{-1}(s) \in \mathcal{V}(A^+)$. Finally, the formula

$$\overrightarrow{\bigcup_{1\le i\le n} L_i} = \bigcup_{1\le i\le n} \overrightarrow{L_i}$$

shows that Z is indeed of the form $\overrightarrow{L}$ with $L \in \mathcal{V}(A^+)$. □

For the positive varieties, another result is available. Let $\mathbf{V}$ be a variety of ordered semigroups and let $\mathcal{V}$ be the corresponding positive +-variety.

Theorem 5.3. *Let Z be a recognizable subset of A^ω. If $\mathbf{V}$ is contained in the variety $[\![x^\pi y^\pi \le x^\pi,\ (xy)^\pi x = (xy)^\pi]\!]$, the following conditions are equivalent:*

(1) *$Z = \overrightarrow{L}$, with $L \in \mathcal{V}(A^+)$,*

(2) *Z is positive boolean combination of subsets of the form $\overrightarrow{L}$, with $L \in \mathcal{V}(A^+)$,*

(3) *$S_+(Z) \in \mathbf{V}$ and $S(Z)$ satisfies the identity $x^\pi y^\omega \le x^\omega$.*

Proof. It is clear that (1) implies (2).

(2) implies (3). Let $\mathbf{V}'$ be the variety of all ordered ω-semigroups S which satisfy $S_+ \in \mathbf{V}$ and the identity $x^\pi y^\omega \le x^\omega$. It suffices to verify that if $Z = \overrightarrow{L}$, with $L \in \mathcal{V}(A^+)$, then $S(Z) \in \mathbf{V}'$. Let S be the syntactic semigroup of L. Since $L \in \mathcal{V}(A^+)$, we have $S \in \mathbf{V}$. Furthermore, by hypothesis, $\mathbf{V}$ is contained in the variety $[\![x^\pi y^\pi \le x^\pi,\ (xy)^\pi x = (xy)^\pi]\!]$, which is itself contained in $\mathbf{R}$. Applying Proposition 5.1, we conclude that $\overrightarrow{L}$ is recognized by $S^\mathcal{R}$ and hence $S(Z)$ is a quotient of $S^\mathcal{R}$. Now $S^\mathcal{R}$ satisfies the identity $x^\pi y^\omega \le x^\omega$, since, for each $x, y \in S$,

$$x^\pi y^\omega = x^\pi y^\pi \le x^\pi = x^\omega$$

Therefore $S^\mathcal{R} \in \mathbf{V}'$ and $S(Z) \in \mathbf{V}'$.

(3) implies (1). We claim that the identity

$$x^\pi y^\omega \le x^\omega \tag{5.2}$$

implies

$$(x^\pi y)^\omega = (x^\pi y)^\pi x^\omega \tag{5.3}$$

Indeed we have by 5.2, $(x^\pi y)^\pi x^\omega \le (x^\pi y)^\omega$ and $x^\pi (x^\pi y)^\omega \le x^\omega$. The latter identity gives $(x^\pi y)^\omega = (x^\pi y)^\pi x^\pi (x^\pi y)^\omega \le (x^\pi y)^\pi x^\omega$, proving the claim. Let $\varphi : A^\infty \to S(Z)$ the syntactic morphism of Z. By Proposition 3.6, if (s, e) and (s, f) are two linked pairs of $S(Z)$, then $se^{\omega} = sf^{\omega}$. Moreover, if (s, e) and (s', e') are two linked pairs such that $s' \leq s$, then $s'e'^{\omega} \leq se^{\omega}$. We have indeed by 5.2 $ee'^{\omega} = e^{\pi}e'^{\omega} \leq e^{\omega}$ and hence, since $s = se$, $s'e'^{\omega} \leq se'^{\omega} = see'^{\omega} \leq se^{\omega}$.

Let (s, e) be a linked pair of S. Put

$$T = \{t \in S \mid \text{ there exists a idempotent } f \text{ such that } tf = t \text{ and } tf^{\omega} \leq se^{\omega}\}$$

We first establish the formula

$$\varphi^{-1}(\downarrow se^{\omega}) = \bigcup_{t \in T} \overrightarrow{\varphi^{-1}(\downarrow t)} = \overrightarrow{\bigcup_{t \in T} \varphi^{-1}(\downarrow t)} \tag{5.4}$$

The second equality is obvious. Let us prove the first. If $u \in \varphi^{-1}(\downarrow se^{\omega})$, there exists a linked pair (t, f) of u such that $tf^{\omega} \leq se^{\omega}$. Therefore $t \in T$. Moreover $u \in \overrightarrow{\varphi^{-1}(t)}$ and hence $u \in \overrightarrow{\varphi^{-1}(\downarrow t)}$. Conversely, suppose there exists $t \in T$ such that $u \in \overrightarrow{\varphi^{-1}(\downarrow t)}$. Then there exists $t' \leq t$ such that $u \in \overrightarrow{\varphi^{-1}(t')}$ and, as we have seen in the proof of Theorem 5.2, u admits a linked pair of the form (t', f'). Then we have $t'f'^{\omega} \leq tf^{\omega} \leq se^{\omega}$ and hence $u \in \varphi^{-1}(\downarrow se^{\omega})$, thereby proving the formula 5.4. Since $\varphi^{-1}(\downarrow t)$ is recognized by $S_{+}(Z)$ by construction, we have $\varphi^{-1}(\downarrow t) \in \mathcal{V}(A^{+})$, showing that Z is indeed of the form $\overrightarrow{L}$ with $L \in \mathcal{V}(A^{+})$. □

The identity $x^{\pi}y^{\omega} \leq x^{\omega}$ is a consequence of the identity $x^{\pi}yz^{\omega} \leq x^{\omega}$ that characterizes, by Theorem 3.1, the syntactic ordered ω-semigroups of the open recognizable sets of A^{ω}. There exists however nonopen recognizable subsets whose syntactic ordered ω-semigroup satisfies the identity $x^{\pi}y^{\omega} \leq x^{\omega}$ (cf. Exercise 6).

We shall conclude this section with a formula which, given a morphism recognizing a subset L of A^{+}, produces an expression of the complement of $\overrightarrow{L}$.

Theorem 5.4. *Let S be a finite semigroup, P a subset of S and $\varphi : A^{+} \to S$ a surjective morphism. Let $L = \varphi^{-1}(P)$. Then the formula holds*

$$A^{\omega} \setminus \overrightarrow{L} = \bigcup_{t \in S} \varphi^{-1}(t)(A^{\omega} \setminus \varphi^{-1}(t^{-1}P)A^{\omega}) \tag{5.5}$$

Proof. Denote by Z the right hand side of (5.5). If $u \in \overrightarrow{L}$, we have $\varphi(u) = se^{\omega}$, where (s, e) is a linked pair such that $s \in P$. If $u \in Z$, there exists a $t \in S$ such that $u = vx$ with $\varphi(v) = t$ and $x \notin \varphi^{-1}(t^{-1}P)A^{\omega}$. Since v is a prefix of u, there exists a prefix p of x such that $\varphi(vp) = s$. Therefore $\varphi(vp) = \varphi(v)\varphi(p) = t\varphi(p) = s$. Thus $\varphi(p) = t^{-1}P$ and $x \in \varphi^{-1}(t^{-1}P)A^{\omega}$, a contradiction. Therefore $u \notin Z$.

Suppose now that $u \notin \overrightarrow{L}$. Then the word u has only finitely many prefixes in L. Let $u = u_0u_1\cdots$ be a Ramseyan factorization of u. Up to replacing u_0 by a product

$u_0 u_1 \cdots u_k$, we may assume that all the prefixes of u that are in L are shorter than u_0. Put $\varphi(u_0) = t$. If $u_1 u_2 \cdots \in \varphi^{-1}(t^{-1}P)A^\omega$, the word $u_1 u_2 \cdots$ has a prefix p in $\varphi^{-1}(t^{-1}P)$ and hence $\varphi(u_0 p) = \varphi(u_0)\varphi(p) = t\varphi(p) \in P$, which means that $u_0 p \in L$, a contradiction. Therefore $u_1 u_2 \cdots \in A^\omega \setminus \varphi^{-1}(t^{-1}P)A^\omega$ and $u \in Z$. □

6 Varieties closed under aperiodic extension

With each variety of finite semigroups **V**, we have associated three classes of subsets of A^ω: the subsets Z such that $S_+(Z) \in \mathbf{V}$, the subsets weakly recognized by a semigroup of **V** and the boolean combinations of subsets of the form $\overrightarrow{L}$, where L is recognized by a semigroup of **V**. Up to now, we didn't encounter any nontrivial ∞-variety for which these three classes were the same. We shall see however that these three classes coincide when **V** is closed under aperiodic extension (or, equivalently, if the $+$-variety corresponding to **V** is closed under marked product).

Theorem 6.1. *Let* **V** *be a variety of semigroups closed under aperiodic extension and let Z be a recognizable subset of A^ω. The following conditions are equivalent:*

(1) *Z is boolean combination of subsets of the form $\overrightarrow{L}$, where L is recognized by a semigroup of* **V**,

(2) *Z is a finite union of subsets of the form XY^ω, where XY^* and Y^+ are recognized by a semigroup of* **V**,

(3) *Z is weakly recognized by a semigroup of* **V**,

(4) $S_+(Z) \in \mathbf{V}$.

Proof. The equivalence of (2) and (3) follows from Theorem 4.5 and that of (3) and (4) from Corollary 4.22 and Theorem 4.12. By Proposition 4.3, the subsets of the form $\overrightarrow{L}$, where L is recognized by a semigroup of **V**, satisfy (3) and hence also (4). Since Condition (4) defines a variety of ω-semigroups, the subsets Z satisfying (4) form a boolean algebra. Consequently (1) implies (4). Finally, the implication (4) implies (1) is a direct consequence of Theorem II.9.3. Indeed, Theorem B.4.16 shows that the $+$-variety $\mathcal{V}$ corresponding to **V** is closed under concatenation product. Consequently, if $X \in \mathcal{V}(A^+)$, we have $P = X \setminus XA^+ \in \mathcal{V}(A^+)$ and the subsets $U_{s,e}$ and $V_{s,e}$ occurring in Theorem II.9.3 are also $\mathcal{V}(A^+)$. □

This result can be applied in particular in the case of the variety **A** of aperiodic semigroups. A subset of A^ω is said to be *star-free* if it belongs to the smallest boolean algebra $\mathcal{S}$ of subsets of A^ω such that if X is a star-free subset of A^* and if $Y \in \mathcal{S}$, then $XY \in \mathcal{S}$.

In other words, the star-free sets of A^ω form the smallest boolean algebra closed under product on the left by a star-free subset of A^*.

Theorem 6.2. *Let Z be a recognizable subset of A^ω. The following conditions are equivalent:*

(1) *Z is a star-free subset of A^ω,*

(2) *Z is boolean combination of subsets of the form $\overrightarrow{L}$, where L is a star-free subset of A^+,*

(3) *Z is a finite union of subsets of the form XY^ω, where XY^* and Y^+ are star-free subsets of A^+,*

(4) *Z is weakly recognized by an aperiodic semigroup,*

(5) *$S_+(Z)$ is aperiodic.*

Proof. The equivalence of conditions (2) to (5) follows directly from Theorem 6.1. The subsets satisfying these conditions form a boolean algebra (by (2)) — and are closed under product on the left by a subset star-free of A^* (by (3)). It follows that the star-free sets of A^ω, that form the smallest boolean algebra closed under product on the left by a subset star-free of A^*, satisfy Conditions (2)–(5). Therefore (1) implies (2).

Theorem 5.4 and Theorem B.4.1 show that if L is a star-free subset of A^+, then $\overrightarrow{L}$ is a star-free subset of A^ω. It follows that (2) implies (1). □

7 Concatenation hierarchies for infinite words

As for finite words, one can define concatenation hierarchies inside the star-free sets of infinite words. Let A be a finite alphabet.

(1) The subsets of level 0 of A^ω are $\emptyset$ and A^ω.

(2) For each integer $n \geq 0$, the subsets of level $n+1/2$ of A^ω are the positive boolean combinations of subsets of the form XaY where X is a subset of level $n+1/2$ of A^*, $a \in A$ and Y is a subset of level n of A^ω.

(3) For each integer $n \geq 0$, the subsets of level $n+1$ of A^ω are the boolean combinations of subsets of level n.

For each integer n, we denote by $\mathcal{V}_n(A^\omega)$ (resp. $\mathcal{V}_{n+1/2}(A^\omega)$) the set of subsets of A^ω of level n (resp. $n+1/2$).

Here is the syntactic characterization of the subsets of level $1/2$.

Theorem 7.1. *Let Z be a recognizable subset of A^ω. The following conditions are equivalent:*

(1) *Z is of level $1/2$,*

(2) *Z is a finite union of subsets of the form $A^*a_1A^*a_2\cdots A^*a_kA^\omega$, where $a_1,\ldots,a_k \in A$,*

(3) *Z is of the form $\overrightarrow{L}$, where L is a subset of level $1/2$ of A^+,*

(4) *$S(Z)$ satisfies the identities $xy \leq x$, $yx \leq x$ and $x^\pi y^\omega \leq x^\omega$.*

Proof. The equivalence of (1) and (2) follows immediately from the definitions and the equivalence of (2) and (3) results from the formula

$$\overrightarrow{A^*a_1A^*a_2\cdots A^*a_kA^*} = A^*a_1A^*a_2\cdots A^*a_kA^\omega$$

and

$$\bigcup_{i\in I}\overrightarrow{L_i} = \overrightarrow{\bigcup_{i\in I} L_i}$$

for each finite family $(L_i)_{i\in I}$ of recognizable subsets of A^+.

(2) implies (4). The ordered ω-semigroups satisfying Condition (4) form a variety of ordered ω-semigroups. By Proposition II.6.4, it suffices to verify Condition (4) when $Z = A^*a_1A^*a_2\cdots A^*a_kA^\omega$. In this case, Z is the set of infinite words admitting $a_1\cdots a_k$ as a subword. Consequently, every infinite word admitting a word of Z as a subword is still in Z. In particular, for each $x, y \in A^+$, $u, v \in A^*$ and $z \in A^\omega$, the condition $uxw \in Z$ (resp. $u(xv)^\omega \in Z$) implies $uxyw \in Z$ and $uyxw \in Z$ (resp. $u(xyv)^\omega \in Z$ and $u(yxv)^\omega \in Z$). Therefore $S(Z)$ satisfies the identities $xy \leq x$ and $yx \leq x$. Similarly, if $ux^\omega \in Z$, there exists an integer n such that, for each $m \geq n$, $a_1\cdots a_k$ is a subword of ux^m. Therefore, $ux^my^\omega \in Z$ for all $m \geq n$, showing that $S(Z)$ satisfies the identity $x^\pi y^\omega \leq x^\omega$.

(4) implies (2). Let $\varphi : A^\infty \to S$ be the syntactic morphism of Z and let I be an order ideal of S such that $Z = \varphi^{-1}(I)$. Associate with each linked pair (s, e) of S_+ such that $se^\omega \in I$, the set $F(s, e)$ of minimal words of $\varphi^{-1}(s)\varphi^{-1}(e)$ with respect to the subword order. By Theorem B.4.6, this set is finite. We claim that

$$Z = \bigcup_{\{(s,e)\mid se^\omega\in I\}}\ \bigcup_{a_1\cdots a_k\in F(s,e)} A^*a_1A^*a_2\cdots A^*a_kA^\omega \tag{7.1}$$

First of all, if $u \in Z$, there exists by Theorem II.2.2 a linked pair (s, e) of S_+ and a factorization $u = u_0u_1\cdots$ such that $\varphi(u_0) = s$ and $\varphi(u_n) = e$ for each $n > 0$. Therefore $u_0u_1 \in \varphi^{-1}(s)\varphi^{-1}(e)$ and hence there exists a subword $a_1\cdots a_k$ of u_0u_1 in $F(s, e)$. Therefore $u_0u_1 \in A^*a_1A^*a_2\cdots A^*a_kA^*$ and $u \in A^*a_1A^*a_2\cdots A^*a_kA^\omega$, thereby proving the inclusion $Z \subset Y$, where Y is the right hand side of 7.1.

Conversely, if $u \in Y$, we have $u = x_0a_1x_1\cdots x_{k-1}a_kx$ with $a_1\cdots a_k \in F(s, e)$, $se^\omega \in I$, $x_0, \ldots, x_{k-1} \in A^+$ and $x \in A^\omega$. Setting $\varphi(x) = tf^\omega$, we have

$$\begin{aligned}\varphi(u) &= \varphi(x_0)\varphi(a_1)\varphi(x_1)\cdots\varphi(x_{k-1})\varphi(a_k)tf^\omega\\ &\leq \varphi(a_1)\varphi(a_2)\cdots\varphi(a_k)f^\omega = sef^\omega = se^\pi f^\omega \leq se^\omega\end{aligned}$$

Therefore $\varphi(u) \in I$ and $u \in Z$, which proves the formula 7.1. Consequently, Z is of level 1/2. □

Theorem 7.1 should be compared with the following theorem:

Theorem 7.2. *Let Z be a recognizable subset of A^ω. The following conditions are equivalent:*

(1) *Z is positive boolean combination of sets of the form $A^*a_1A^* \cdots A^*a_kA^\omega$, with $k > 0$, $a_1, \ldots, a_k \in A$, or $(A^*a)^\omega$, with $a \in A$,*
(2) *Z is a finite union of subsets of the form XY^ω, where XY^* and Y^+ are of level $1/2$,*
(3) *Z is weakly recognized by a semigroup satisfying the identities $xy \le x$ and $yx \le x$,*
(4) *Z is weakly recognized by a semigroup satisfying the identities $x^\pi y \le x^\pi$ and $yx^\pi \le x^\pi$,*
(5) *$S(Z)$ satisfies the identities $xy \le x$ and $yx \le x$,*
(6) *$S(Z)$ satisfies the identities $x^\pi y \le x^\pi$ and $yx^\pi \le x^\pi$.*

Proof. The equivalence of (2) and (3) follows from Proposition B.4.8 and Theorem 4.5. The equivalence of (3) and (5) (resp. of (4) and (6)) follows from Corollary 4.19 and Theorem 4.12. It is clear also that (5) implies (6).

(6) implies (5). Suppose that $S(Z)$ satisfies the identities $x^\pi y \le x^\pi$ and $yx^\pi \le x^\pi$. By Lemma 4.18, $S_+(Z)$ is $\mathcal{J}$-trivial and by Proposition B.1.16, it satisfies the identities $(xy^\pi)^\pi = (xy)^\pi = (x^\pi y)^\pi$. Proving that $S(Z)$ satisfies the identities $xy \le x$ and $yx \le x$, amounts to verifying that, for each word $x, y \in A^+$ and $u \in A^*$, the condition $(xu)^\omega \in Z$ implies $(xyu)^\omega \in Z$ and $(yxu)^\omega \in Z$. Let π be the exponent of $S_+(Z)$. If $(xu)^\omega \in Z$, we have $((xu)^\pi)^\omega \in Z$ and by the identity $yx^\pi \le x^\pi$, we have $(y(xu)^\pi)^\omega \in Z$, and thus $((y(xu)^\pi)^\pi)^\omega \in Z$. We deduce from the identity $(xy^\pi)^\pi = (xy)^\pi$, that $((yxu)^\pi)^\omega = (yxu)^\omega \in Z$. One would show in a similar way that $(xyu)^\omega \in Z$ by using the identity $x^\pi y \le x^\pi$.

We now have established the equivalence of conditions (2) to (6). It follows that the subsets of A^ω satisfying (2) are closed under finite intersection. To prove that (1) implies (5), it suffices to show that the subsets of the form $A^*a_1A^* \cdots A^*a_kA^\omega$, with $k > 0$, $a_1, \ldots, a_k \in A$, or $(A^*a)^\omega$, with $a \in A$, satisfy (5), which follows immediately from Theorems 7.1 and Example II.8.5.

There remains to show that (2) implies (1). Let Z be a subset of the form XY^ω, where XY^* and Y^+ are subsets of A^+ level $1/2$. The proof requires an elementary combinatorial result.

Lemma 7.3. *Let $L_1, \ldots, L_k$ be ideals of A^+. Then*

$$\Big(\bigcup_{1 \le i \le k} L_i\Big)^\omega = \bigcup_{1 \le i \le k} L_i^\omega$$

Proof. The inclusion $\bigcup_{1 \le i \le k} L_i^\omega \subset (\bigcup_{1 \le i \le k} L_i)^\omega$ is clear. To establish the opposite inclusion, consider a word $u \in (\bigcup_{1 \le i \le k} L_i)^\omega$. Then $u = u_0u_1 \cdots$ with, for each $n \ge 0$,

$u_n \in \bigcup_{1 \le i \le k} L_i$. Let i be an index such that the set of n's satisfying $u_n \in L_i$ is infinite. Since $L_i = A^* L_i A^*$, we have $u \in L_i^\omega$ and hence $u \in \bigcup_{1 \le i \le k} L_i^\omega$. □

By Lemma 7.3, Y^ω is a finite union of subsets of the form

$$(A^* b_1 A^* \cdots A^* b_k A^*)^\omega$$

with $k > 0$ and $b_1, \ldots, b_k \in A$. It follows that Z can be written as a finite union of subsets of the form

$$\begin{aligned} A^* a_1 A^* \cdots A^* a_p (A^* b_1 A^* \cdots A^* b_k A^*)^\omega \\ = A^* a_1 A^* \cdots A^* a_p A^* b_1 A^* \cdots A^* b_k A^\omega \cap \bigcap_{1 \le i \le k} (A^* b_i)^\omega \end{aligned}$$

showing that Z satisfies (1). □

At first sight, Theorem 7.2 seems to contradict Theorem 2.8, since the varieties of ordered ω-semigroups $\mathbf{V}_1 = [\![xy \le x,\ yx \le x]\!]$ and $\mathbf{V}_2 = [\![x^\pi y \le x^\pi,\ yx^\pi \le x^\pi]\!]$, which are distinct, correspond to the same sets of infinite words. In fact, we have indeed $\mathcal{V}_1(A^\omega) = \mathcal{V}_2(A^\omega)$ for each alphabet A, but Propositions B.4.8 and B.4.10 show that $\mathcal{V}_1(A^+) \ne \mathcal{V}_2(A^+)$.

The syntactic characterization of level 1 is also known.

Theorem 7.4. *Let Z be a recognizable subset of A^ω. The following conditions are equivalent:*

(1) *Z is of level 1,*

(2) *Z is boolean combination of subsets of the form $A^* a_1 A^* a_2 \cdots A^* a_k A^\omega$, where $a_1, \ldots, a_k \in A$,*

(3) *Z is boolean combination of subsets of the form $\overrightarrow{L}$, where L is a subset of level 1/2 of A^+,*

(4) *$Z = \overrightarrow{L}$, where L is a subset of level 1 of A^+,*

(5) *Z is boolean combination of subsets of the form $\overrightarrow{L}$, where L is a subset of level 1 of A^+,*

(6) *$S_+(Z) \in \mathbf{J}$ and $S(Z)$ satisfies the identity $(x^\pi y^\pi)^\pi x^\omega = (x^\pi y^\pi)^\pi y^\omega$.*

Recall that, by Proposition 3.6, the identity $(x^\pi y^\pi)^\pi x^\omega = (x^\pi y^\pi)^\pi y^\omega$ characterizes the subsets Δ_2 of A^ω.

Proof. The equivalence of (1), (2) and (3) follows from Theorem 7.2 and that of (4), (5) and (6) from Theorem 5.2. Since (3) implies (5), it only remains to verify that (4) implies (2).

Let $Z = \overrightarrow{L}$, where L is a subset of level 1 of A^+. Then there exists an integer $n \geq 0$ and a set E of words of length n such that $L = U \setminus V$ with

$$L = \bigcap_{a_1 \cdots a_n \in E} A^*a_1A^* \cdots A^*a_nA^* \setminus \bigcup_{a_1 \cdots a_n \in A^n \setminus E} A^*a_1A^* \cdots A^*a_nA^*$$

We claim that $\overrightarrow{L} = U \setminus V$, where:

$$U = \bigcap_{a_1 \cdots a_n \in E} A^*a_1A^* \cdots A^*a_nA^\omega$$

$$V = \bigcup_{a_1 \cdots a_n \in A^n \setminus E} A^*a_1A^* \cdots A^*a_nA^\omega$$

Let $Y = U \setminus V$. If $u \in Y$, there exists an integer N such that all the prefixes of u of length $\geq N$ admit all the words of E as subwords. Moreover, no word of $A^N \setminus E$ is a subword of such a prefix. Therefore all the prefixes of u of length $\geq N$ are in L and $u \in \overrightarrow{L}$.

Conversely, if $u \in \overrightarrow{L}$, u has infinitely many prefixes in L and hence in particular, every word of E is a subword of u, showing that $u \in U$. If $u \in V$, there exists a word $a_1 \cdots a_n \in A^n \setminus E$ which is a subword of u. Then there exists an integer N such that $a_1 \cdots a_n$ is a subword of all the prefixes of u of length $\geq N$. Thus there would be only finitely many prefixes of u in L, a contradiction. Therefore $u \in Y$. □

Theorem 7.5. *Let Z be a recognizable subset of A^ω. The following conditions are equivalent:*

1. *Z is boolean combination of subsets of the form $A^*a_1A^* \cdots A^*a_kA^\omega$, with $a_1, \ldots, a_k \in A$, or $(A^*a)^\omega$, with $a \in A$,*
2. *Z is boolean combination of subsets of the form XY^ω, where XY^* and Y^+ are of level $1/2$,*
3. *Z is a finite union of subsets of the form XY^ω, where XY^* and Y^+ are of level 1,*
4. *Z is boolean combination of subsets of the form XY^ω, where XY^* and Y^+ are of level 1,*
5. *Z is weakly recognized by a semigroup of* **J**,
6. $S_+(Z) \in \mathbf{J}$.

Proof. The equivalence of (1) and (2) follows from Theorem 7.2. The equivalence of (3) and (5) follows from Theorems 4.5 and B.4.12 and (6) implies (5) follows from Proposition 4.2.

(5) implies (6). Let S be a $\mathcal{J}$-trivial semigroup weakly recognizing Z. By Proposition 4.15, we have $\hat{S}_+ = S$ and hence $S(Z)$, which is a quotient of $\hat{S}$, satisfies $S_+(Z) \in \mathbf{J}$.

It follows now from the equivalence of (3) and (6) that the subsets Z satisfying (3) form a boolean algebra. Consequently, (3) and (4) are equivalent.

(2) implies (4) is trivial.

(6) implies (1). Suppose that $S_+(Z) \in \mathbf{J}$. By Theorem B.4.14, $(S_+(Z), =)$ is quotient of a finite ordered semigroup of the form $T = S \times \check{S}$, where S satisfies the identities $xy \leq x$ and $yx \leq x$. By Corollary II.5.3, S(Z) is quotient of $\bar{T}$, which is itself isomorphic to $\bar{S} \times (\bar{S})\check{}$. Proposition II.6.5 then shows that Z is positive boolean combination of subsets recognized by $\bar{S}$ or by $(\bar{S})\check{}$. But, by Proposition II.6.3, the complement of the subsets recognized by $(\bar{S})\check{}$ is recognized by $\bar{S}$. Finally, Theorem 7.2 shows that the subsets recognized by $\bar{S}$ are positive boolean combination of subsets of the form $A^*a_1A^* \cdots A^*a_kA^\omega$, with $k > 0, a_1, \ldots, a_k \in A$, or $(A^*a)^\omega$, with $a \in A$. Therefore Z satisfies (1). □

The preceding results lead to consider the variety of ordered ω-semigroups defined by the identity $xy \leq x$.

Theorem 7.6. *Let Z be a recognizable subset of A^ω. The following conditions are equivalent:*

(1) *Z is a positive boolean combination of subsets of the form $(A^*a)^\omega$ or*

$$a_0A^*a_1A^* \cdots A^*a_kA^\omega,$$

with $a, a_0, \ldots, a_n \in A$

(2) *Z is a union of subsets of the form XY^ω, where $S(XY^*), S(Y^+) \in [\![xy \leq x]\!]$,*

(3) *Z is weakly recognized by an ordered semigroup of $[\![xy \leq x]\!]$,*

(4) *$S(Z)$ satisfies the identity $xy \leq x$.*

Proof. Let $\mathbf{V}$ be the variety of semigroups defined by the identity $xy \leq x$ and let $\mathcal{V}$ be the corresponding positive variety.

(2) is equivalent to (3) by Theorem 4.5.

(3) implies (4). Let S be an ordered semigroup of $[\![xy \leq x]\!]$ weakly recognizing Z. By Proposition 4.23, $\hat{S}$ also satisfies the identity $[\![xy \leq x]\!]$. Furthermore, $\hat{S}$ recognizes Z by Theorem 4.12 and by Corollary II.8.10, $S(Z)$ is a quotient of $\hat{S}$. Therefore $S(Z)$ also satisfies the identity $[\![xy \leq x]\!]$.

(4) implies (3). Indeed, $S(Z)$ recognizes Z and Proposition 4.2 shows that $S_+(Z)$ weakly recognizes Z.

(1) implies (2). The equivalence of (2) and (4) shows that the subsets of the form XY^ω, where $XY^*, Y^+ \in \mathcal{V}(A^+)$, define a positive variety and are in particular closed under finite union and finite intersection. Showing that (1) implies (2) amounts to verifying that the subsets of the form $a_0A^*a_1A^* \cdots A^*a_kA^\omega$ or $(A^*a)^\omega$, with $a \in A$, are of the form XY^ω, with $XY^*, Y^+ \in \mathcal{V}(A^+)$. For the subsets of the form $(A^*a)^\omega$, it suffices to take $X = A^*aA^*$ and $Y = A$. For the subsets of the form

$Z = a_0A^*a_1A^* \cdots A^*a_kA^\omega$, set $X = a_0A^*a_1A^* \cdots A^*a_kA^*$ and $Y = A$, which gives $XY^* = a_0A^*a_1A^* \cdots A^*a_kA^* \in \mathcal{V}(A^+)$, $Y^+ = A^+ \in \mathcal{V}(A^+)$ and $XY^\omega = Z$.

(2) implies (1). By Proposition B.4.9, $\mathcal{V}(A^+)$ consists of the finite unions of subsets of the form $a_0A^*a_1A^* \cdots A^*a_kA^*$. The next lemma describes the subsets of the form Y^ω with $Y^+ \in \mathcal{V}(A^+)$.

Lemma 7.7. *Let $L_1, \ldots, L_k$ be right ideals of A^+. Then*

$$\Big(\bigcup_{1\le i\le k} L_i\Big)^\omega = \bigcup_{1\le i,j\le k} L_iL_j^\omega$$

Proof. The inclusion $\bigcup_{1\le i,j\le k} L_iL_j^\omega \subset (\bigcup_{1\le i\le k} L_i)^\omega$ is obvious. For the opposite inclusion, consider a word $u \in (\bigcup_{1\le i\le k} L_i)^\omega$. Then $u = u_0u_1 \cdots$ with, for each $n \ge 0$, $u_n \in \bigcup_{1\le i\le k} L_i$. Choose i such that $u_0 \in L_i$ and j such that the set of all n satisfying $u_n \in L_j$ is infinite. Since $L_j = L_jA^*$, $u \in L_iL_j^\omega$, proving the lemma. □

It follows from Lemma 7.7 that the subsets of the form XY^ω, where XY^* and Y^+ are in $\mathcal{V}(A^+)$, are finite unions of subsets of the form

$$a_0A^*a_1A^* \cdots A^*a_pA^*(b_0A^* \cdots b_qA^*)^\omega$$

The formula

$$a_0A^*a_1A^* \cdots A^*a_pA^*(b_0A^* \cdots b_qA^*)^\omega = a_0A^*a_1A^* \cdots A^*a_pA^\omega \cap \bigcap_{1\le i\le q} (A^*b_i)^\omega$$

gives the conclusion. □

Again, the subsets of the form $\overrightarrow{L}$, where $S(L) \in [\![xy \le x]\!]$, define a ∞-positive variety.

Theorem 7.8. *Let Z be a recognizable subset of A^ω. The following conditions are equivalent:*

(1) *Z is a finite union of sets of the form*

$$a_0A^*a_1A^* \cdots A^*a_kA^\omega$$

with $a_0, \ldots, a_n \in A$,

(2) *Z is of the form $\overrightarrow{L}$, where $S(L) \in [\![xy \le x]\!]$,*

(3) *$S(Z)$ satisfies the identities $xy \le x$ and $x^\pi y^\omega \le x^\omega$.*

Proof. To establish the equivalence of (2) and (3), it suffices to apply Theorem 5.3 to the variety of ordered semigroups $\mathbf{V} = [\![xy \le x]\!]$, which is indeed contained in the variety $[\![x^\pi y^\pi \le x^\pi,\ (xy)^\pi x = (xy)^\pi]\!]$.

The equivalence of (1) and (2) is obvious, since, by Proposition B.4.9, $\mathcal{V}(A^+)$ consists of the finite unions of sets of the form $a_0A^*a_1A^* \cdots A^*a_kA^*$. □

No syntactic characterization of the other levels is known. We shall see the logical counterpart on this hierarchy in Chapter VIII.

8 Exercises

8.1 Varieties of finite or infinite words.

Exercise 1. Show that if an ordered semigroup satisfies the identity $(xy)^\pi \leq (xy)^\pi x$, then S is $\mathcal{R}$-trivial, and hence satisfies the identity $(xy)^\pi = (xy)^\pi x$.

Exercise 2. Show that, for each alphabet A, the monoid U_1 recognizes the subset $\{1\}$. Show that the commutative monoid $\{1, s, 0\}$ defined by $s^2 = 0$ is the syntactic monoid of the subsets of the form $\{a\}$, with $a \in A$. Verify that this monoid is aperiodic.

Exercise 3. Let **V** be a variety of monoids and let $\mathcal{V}$ be the corresponding $*$-variety. Show that the following conditions are equivalent:

(1) **V** is locally finite,

(2) For each alphabet A, $\mathcal{V}(A^*)$ is finite,

(3) For each alphabet A, the monoid $\hat{F}_{\mathbf{V}}(A)$ is finite.

Exercise 4. For $0 \leq k \leq n$, denote by $L(a, k, n)$ the set of words of A^* in which the number of occurrences of a's is congruent to k modulo n:

$$L(a, k, n) = \{u \in A^* \mid |u|_a \equiv k \mod n\}$$

Let $\mathcal{G}com$ be the $*$-variety corresponding to the variety **Gcom** of commutative groups. Show that for each alphabet A, $\mathcal{G}com(A^*)$ is the boolean algebra generated by the subsets of the form $L(a, k, n)$ where $a \in A$ and $0 \leq k < n$.

Exercise 5. Let $\mathcal{C}om$ be the $*$-variety corresponding to the variety **Com** of commutative monoids. Show that for each alphabet A, $\mathcal{C}om(A^*)$ is the boolean algebra generated by the subsets of the form

$$L(a, k) = \{u \in A^* \mid |u|_a = k\} \quad \text{or} \quad L(a, k, n) = \{u \in A^* \mid |u|_a \equiv k \bmod n\}$$

where $a \in A$ and $0 \leq k < n$.

8.2 Varieties and topology.

Exercise 6. Let $A = \{a, b, c\}$ and let $X = a^*cb^*\{a, c\}A^\omega \cup a^*bA^\omega \cup a^\omega$. Show that the syntactic ordered ω-semigroup of X satisfies the identity $x^\pi y^\omega \leq x^\omega$ but neither the identity $x^{\pi}zy^{\omega} \leq x^{\omega}$ nor $x^{\omega} \leq x^{\pi}zy^{\omega}$. Deduce that X is deterministic but is neither open nor closed.

8.3 Weak recognition.

Exercise 7. Let S be a finite semigroup and let $\varphi : A^+ \to (S, =)$ be a semigroup morphism weakly recognizing a subset X of A^{ω}. Show that $\bar{\varphi} : A^{\infty} \to \bar{S}$ recognizes X if and only if, for every linked pair (s, e) of S, $X_{(\downarrow s, \downarrow e)} \cap X \neq \emptyset$ implies $X_{(\downarrow s, \downarrow e)} \subset X$.

8.4 Boolean combinations of deterministic sets.

Exercise 8. For a $+$-variety $\mathcal{V}$, we denote, for each alphabet A, $\overrightarrow{\mathcal{V}}(A^{\omega})$ the boolean algebra generated by the subsets of the form $\overrightarrow{L}$, with $L \in \mathcal{V}(A^+)$. The aim of this problem is to show that the class $\overrightarrow{\mathcal{V}}$ is not necessarily the ω-component of an ∞-variety.

(1) Let $A = \{a, b\}$ and let $X = (ab)^+$. Show that the syntactic semigroup S of X is the semigroup of functions given by the table below

	a	b	ab	ba	a^2
1	2	–	1	–	–
2	–	1	–	2	–

Deduce that S is the semigroup with zero presented by the relations $aba = a$, $bab = b$ and $a^2 = b^2 = 0$.

(2) Prove that the variety of semigroups $\mathbf{V}$ generated by S is defined by the identities $xyx = xyxyx$, $x^2y^2 = y^2x^2$ and $x^2 = x^3$.

(3) Show that $\mathbf{V}$ satisfies the identities $xyxzx = xzxyx$, $x^2y^2 = x^2y^2x = x^2y^3$ and $x^2yx = x^2y^2$.

(4) Deduce that the *relatively free semigroup* $F_{\mathbf{V}}(A)$ of the variety $\mathbf{V}$ on the alphabet A, that is, the quotient of A^+ by the relations $xyx = xyxyx$, $x^2y^2 = y^2x^2$ and $x^2 = x^3$, for each $x, y \in A^+$, is the 17 element semigroup whose right representation is given in Figure 8.1. In this graph, the arcs are of the form $s \xrightarrow{a} sa$ with $s \in F_{\mathbf{V}}(A)$ and $a \in A$, but the edges arriving in 0 are omitted. Therefore $aaba = aabb = abbaa = abbab = abaa = ababb = babb = babaa = baaba = baabb = bbaa = bbab = 0$. For example, $babab = bab$ since there is an edge labelled by b from $baba$ to bab.

(5) Deduce from (4) that every element of $\mathcal{V}(A^+)$ can be written as a union of subsets of one of the following types:

(a) $aa^+, bb^+, abb^+, baa^+, ab(ab)^+, (ab)^+a, ba(ba)^+, (ba)^+b$,

(b) a finite subset,

(c) $A^*(a^2ba + a^2b^2 + aba^2 + bab^2 + b^2a^2 + b^2ab + b^3)A^*$.

(6) Show that if $L \in \mathcal{V}(A^+)$, then $(aab)^{\omega} \in \overrightarrow{L}$ if and only if $aab^{\omega} \in \overrightarrow{L}$.

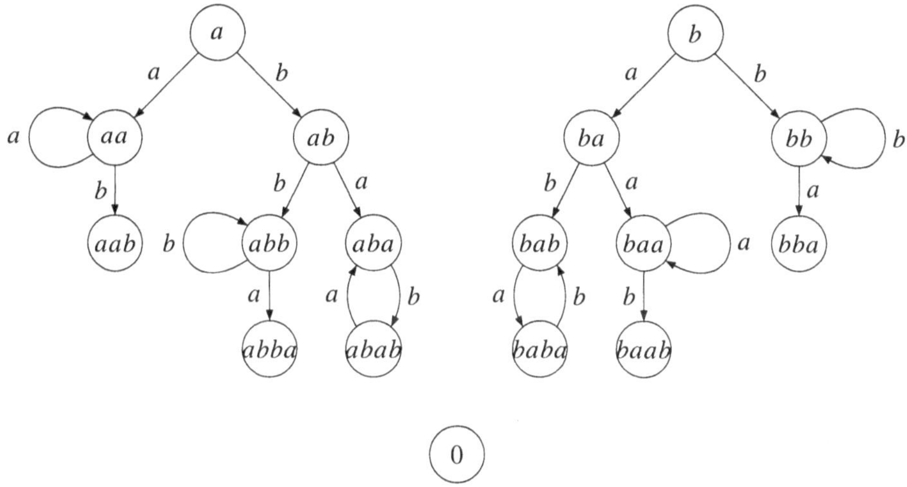

Figure 8.1. The relatively free semigroup $F_{\mathbf{V}}(A)$.

(7) Let $B = \{a, b, c\}$ and let $\varphi : A^+ \to B^+$ be the morphism of semigroups defined by $\varphi(a) = ac$ and $\varphi(b) = b$. Let $L = \{b, ac\}^*a$. Show that the syntactic semigroup of L is BA_2. Deduce that $L \in \mathcal{V}(B^+)$.

(8) Let $X = \varphi^{-1}(\overrightarrow{L})$. Show that X is the set of infinite words on the alphabet A containing an infinite number of a's. Deduce that $X = \overrightarrow{A^*a}$.

(9) Show, by using (6), that X is not a boolean combination of subsets of the form $\overrightarrow{K}$, with $K \in \mathcal{V}(A^+)$.

8.5 Varieties closed under aperiodic extension.

Exercise 9. A subset X of A^* is called *pure* if $u^n \in X$ for some $n > 0$ implies $u \in X$. Show that if X is star-free and pure, then X^* is star-free.

Exercise 10. Show that, in Theorem 6.2, condition (3) can be replaced by

(3′) Z is a finite union of subsets of the form XY^ω, where X and Y^+ are star-free subsets of A^+.

9 Notes

The notion of variety was introduced by Birkhoff [32] in the more general setting of universal algebra. In this chapter, we called them varieties in Birkhoff's sense. Their characterization in terms of identities is also due to Birkhoff. The "ordered" version is due to Bloom [35]. The study of varieties of finite algebras seems to be much more recent. Their characterization by identities is due to Reiterman [263] and, in the ordered case, to Pin and Weil [248]. Reiterman's theorem and its consequences are treated in detail in the book of Almeida [7].

The variety theorem, settled by Eilenberg [98], gave a common framework to several earlier results. For instance, Kleene's theorem [160] can be considered as the first example of correspondence between varieties. But the most important example remains Schützenberger's characterization of star-free sets by aperiodic monoids [274]. Another proof of this result, using wreath product decomposition of semigroups, was found independently by Brzozowski-Cohen [42] and Meyer [193]. The characterization of piecewise testable subsets is due to Simon [285]. The other examples given in this chapter are presented in the treatise of Eilenberg [98]. One may also refer to the books by Lallement [167] or Pin [241] or to Chapter 10 of the Handbook of Language Theory [246].

The original version of the variety theorem dealt only with semigroup varieties. Its extension to ordered semigroups is due to Pin [244]. The examples of positive varieties come from the same article.

The semigroup $\Diamond(S)$ is a variant of the Schützenberger product [274]. A proof of Higman's theorem can be found in chapter 6 of Lothaire [183].

The introduction of concatenation hierarchies goes back to Cohen and Brzozowski [74] and there exists an abundant literature on this topic. See the articles [288, 242, 250] for more details. The characterization of the varieties closed under product is due to Straubing [309].

The main result on varieties of infinite words is Theorem 6.2, which summarizes results of Ladner [166], Thomas [324, 325] and Perrin [233]. Theorem 6.1 is due to Perrin [234]. The systematic study of varieties of infinite words was initiated by Pécuchet [227, 225], who introduced in particular the three classes associated with a variety of finite semigroups, and proved Theorems 4.5 and 7.5. But the final definition of a variety and the suitable extension of the variety theorem are due to Wilke [355, 356]. Wilke [356, 357] also proved Theorem 5.3. The extension of these results to ordered semigroups and positive varieties is due to Pin [247]. The case of varieties of finite words is from Pin [244].

VII

LOCAL PROPERTIES

1 Introduction

In this chapter, we continue on the study of varieties of finite or infinite words. The varieties introduced in this chapter share the property of being defined by "local" properties on words. A word property is said to be *local* if it can be tested by a *scanner*, that is a machine equipped with a finite memory and a sliding window of fixed size n to scan the input word. The window can also be moved before the first and beyond the last

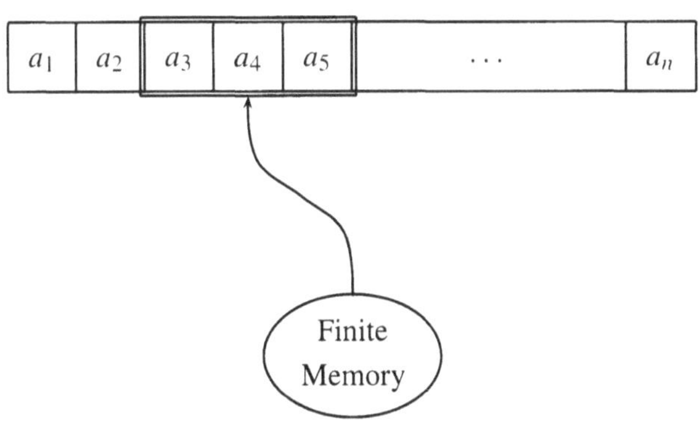

Figure 1.1. A scanner.

letter of the word, in order to read the prefixes and suffixes of length $< n$. For instance, if $n = 3$, and if $u = abbaaabab$, the various positions of the window are represented in the following diagram:

$$\boxed{a}bbaaabab \quad \boxed{ab}baaabab \quad \boxed{abb}aaabab \quad a\boxed{bba}aabab \quad \cdots$$

$$\cdots \quad abbaaa\boxed{bab} \quad abbaaab\boxed{ab} \quad abbaaaba\boxed{b}$$

At the end of the scan, the scanner memorizes the prefixes and suffixes of length $< n$ and the set of factors of length n of the input word. The memory of the scanner contains a table of possible lists of prefixes (resp. suffixes, factors). A word is accepted by the scanner if the list of prefixes (resp. suffixes, factors) obtained after the scan matches one of the lists of the table. Another possibility is to take into account the number of occurrences of the factors of the word.

The chapter is organised as follows. In Section 2, we come back to the strong extension defined in Section VI.4.2. We prove two new properties of this expansion which are crucial for proving the results of Section 3, devoted to local properties. We shall first treat properties defined by letters, which correspond to the case where the width of the window is equal to 1, then we shall treat the general case, in which the width of the window is unbounded.

2 Weak recognition

In this section, we return to the properties of the expansion $\hat{S}$. The variety of locally idempotent and commutative semigroups is not stable by aperiodic extension, but it is nevertheless stable by the operation $S \to \hat{S}$.

Proposition 2.1. *If S is locally idempotent and commutative, then so is $\hat{S}$.*

Proof. Let u be a word such that $\hat{\varphi}(u) = \hat{\varphi}(u)^2$. We need to show that, for each word $v, w \in A^+$, $\hat{\varphi}(uvuvu) = \hat{\varphi}(uvu)$ and $\hat{\varphi}(uvuwu) = \hat{\varphi}(uwuvu)$. We already know that $\varphi(u) = \varphi(u)^2$ since $\hat{\varphi}(u) = \hat{\varphi}(u)^2$ and hence $\varphi(uvuvu) = \varphi(uvu)$ and $\varphi(uvuwu) = \varphi(uwuvu)$ since S is locally idempotent and commutative. Let $x \in A^*$. We shall prove successively the following properties:

(1) each linked pair of $(uvux)^\omega$ is also a linked pair of $(uvuvux)^\omega$,

(2) each linked pair of $(uvuvux)^\omega$ is also a linked pair of $(uvux)^\omega$,

(3) each linked pair of $(uvuwux)^\omega$ is also a linked pair of $(uwuvux)^\omega$.

Since $\hat{\varphi}(u) = \hat{\varphi}(u)^2$, two words of the form xuy and xu^2y have the same linked pairs. We shall implicitly use this result several times without any further warning in the sequel. Let n be an integer such that $\varphi(uvux)^n$ and $\varphi(uvuwux)^n$ are idempotent.

(1) Put $f = u^2vux$. If (s, e) is a linked pair of f^ω, there exists by Lemma VI.4.13 a factorization $f = f'f''$ and an integer $k > n$ such that $\varphi(f^k f') = s$ and

$\varphi(f''f^{2n-1}f') = e$. Let us consider the factorization $u^2vux = f'f''$. If u is a prefix of f', then $f' = ut$ with $t \in A^*$ and $tf'' = uvux$. Setting $g' = uvut$ and $g'' = f''$, we have $g = g'g'' = uvutf'' = uvu^2vux$. On the other hand, since S is locally idempotent and commutative,

$$\varphi(fg') = \varphi(uvuxuvut) = \varphi(uvuxut) = \varphi(ff')$$

Therefore

$$\varphi(g^kg') = \varphi(f^kg') = \varphi(f^kf') = s \quad \text{and}$$
$$\varphi(g''g^{2n-1}g') = \varphi(f''f^{2n-1}g') = \varphi(f''f^{2n-1}f') = e$$

If f' is a prefix of u, we have $u = f't$ with $t \in A^*$ and $f'' = tuvux$. Setting $g' = f'$ and $g'' = tuvuvux$, we get $g = g'g'' = u^2vuvux$, $\varphi(g') = \varphi(f')$ and $\varphi(g'') = \varphi(f'')$ whence $\varphi(f) = \varphi(g)$, $\varphi(g^kg') = \varphi(f^kf') = s$ and $\varphi(g''g^{2n-1}g') = \varphi(f''f^{2n-1}f') = e$. We conclude in both cases that (s, e) is a linked pair of $(uvuvux)^\omega$.
(2) It suffices to mimic the proof of (1) by arguing on the word $f = uvu^2vux$. We consider next the factorization $f = f'f''$. If uvu is a prefix of f', then $f' = uvut$ and one may take $g' = ut$ and $g'' = f''$ to get rid of the factor vu. If f' is a prefix of uvu, then $f't = uvu$ and $f'' = tuvux$. Then we choose $g' = f'$ and $g'' = tux$ to get rid of the factor vu. In both cases it is easy to see that (s, e) is a linked pair of $(uvux)^\omega$.
(3) Again, we mimic the proof of (1), arguing this time on the word $f = u^2vu^2wu^2x$. There are several cases according to the factorization $f'f''$ of f.

(a) $f' \le u$. Then there exists $t \in A^*$ such that $u = f't$ and $f'' = tuvu^2wu^2x$. Then one may take $g' = f'$ and $g'' = tuwuvux$.

(b) $u < f' \le u^2vu$. Then there exist $r, t \in A^*$ such that $ur = f'$, $rt = uvu$ and $f'' = tuwu^2x$. In this case, we take $g' = uwur$ and $g'' = tux$.

(c) $u^2vu < f' \le u^2vu^2wu$. Then there exists $r, t \in A^*$ such that $u^2vur = f'$, $rt = uwu$, $f'' = tux$. In this case, we take $g' = ur$ and $g'' = tuvux$.

(d) $u^2vu^2wu < f'$. Then there exists $t \in A^*$ such that $u^2vu^2wut = f'$ and $tf'' = ux$. In this case, we take $g' = uwuvut$ and $g'' = f''$.

We claim that, in all cases, (s, e) is a linked pair of g^ω, where $g = g'g''$. The cases (a) and (d) are easy and the case (b) is essentially dual to the case (c). Consider this last case. First, we have $g = g'g'' = urtuvux = u^2wu^2vux$, whence $\varphi(g) = \varphi(f)$. On the other hand, since S is locally idempotent and commutative, we have $\varphi(gg') = \varphi(fg') = \varphi(u^2vu^2wu^2xur) = \varphi(uvuwuxur) = \varphi(uvuwuxuvur) = \varphi(u^2vu^2wu^2xu^2vur) = \varphi(ff')$ and similarly $\varphi(g''g) = \varphi(f''f)$. It follows that $\varphi(g^kg') = \varphi(f^kg') = \varphi(f^kf') = s$ and $\varphi(g''g^{2n-1}g') = \varphi(f''f^{2n-1}f') = e$, proving the claim.

Properties (1) and (2) show that $uvu \simeq_\varphi uvuvu$ and Property (3) shows that $uvuwu \simeq_\varphi uwuvu$. Therefore $\hat{S}$ is locally idempotent and commutative. □

Lemma 2.2. *Let S be an aperiodic semigroup satisfying Thérien's condition and let n be the exponent of S. Let e and f be two idempotents of S and let $p \in eSf$, $q \in fSe$ and $r \in eSf$. Then $(pqrq)^n p = (pqrq)^n r$, $p(qrqp)^n = r(qrqp)^n$ and $(pqrq)^n pq = (pqrq)^n rq = (pqrq)^n$.*

Proof. Thérien's condition can be written as $pqr = rqp$. It follows by induction $(pqrq)^n = (pq)^n(rq)^n = (rq)^n(pq)^n$. Now, since S is aperiodic, we have $(rq)^n = (rq)^{n+1}$. Consequently $(pqrq)^n p = (rq)^n(pq)^n p = (rq)^{n+1}(pq)^n p = (rq)^n r(qp)^{n+1} = (rq)^n r(qp)^n = (pqrq)^n r$. The second formula can be proved in an analogous way. Finally, we have $(pqrq)^n rq = (pqrq)^n pq = (pq)^{n+1}(rq)^n = (pq)^n(rq)^n = (pqrq)^n$. □

Proposition 2.3. *If S is aperiodic and satisfies Thérien's condition, then so does $\hat{S}$.*

Proof. Corollary VI.4.21 already shows that if S is an aperiodic semigroup, then so is $\hat{S}$. Let u and v be two words such that $\hat{\varphi}(u) = \hat{\varphi}(u)^2$ and $\hat{\varphi}(v) = \hat{\varphi}(v)^2$. We need to show that for each $p, q, r \in A^+$, $\hat{\varphi}(upvqurv) = \hat{\varphi}(urvqupv)$. We have $\varphi(u) = \varphi(u)^2$ and $\varphi(v) = \varphi(v)^2$ since $\varphi = \pi \circ \hat{\varphi}$ and since $\hat{\varphi}(u) = \hat{\varphi}(u)^2$ and $\hat{\varphi}(v) = \hat{\varphi}(v)^2$. It follows, since S satisfies Thérien's condition, that $\varphi(upvqurv) = \varphi(urvqupv)$. It remains to show that $upvqurv \simeq_\varphi urvqupv$. By symmetry, it suffices to verify that, for each $x \in A^*$, every linked pair of $(upvqurvx)^\omega$ is also a linked pair of $(urvqupvx)^\omega$. Since $\bar{\varphi}(u) = \bar{\varphi}(u)^2$, it suffices by Lemma VI.4.14, to show that the linked pairs of the words $(upvqurvxu)^\omega$ and $(urvqupvxu)^\omega$ are the same. Since $\bar{\varphi}(u) = \bar{\varphi}(u)^2$ and $\bar{\varphi}(v) = \bar{\varphi}(v)^2$, two words of the form xuy and xu^2y (resp. xvy and xv^2y) have the same linked pairs. We shall use implicitly this result in the sequel. Let n be the exponent of S and let $f = u^2pv^2qu^2rv^2xu^2$. If (s, e) is a linked pair of f^ω, there exists by Lemma VI.4.13 a factorization $f = f'f''$ and an integer $k > n$ such that $\varphi(f^k f') = s$ and $\varphi(f'' f^{2n-1} f') = e$. Several cases arise according to the position of the scansion of f induced by the factorization $f'f''$. The various possibilities are represented in Figure 2.1.

The cases (1′), (2′) and (3′) being symmetric of (1), (2) and (3) respectively, we are left with three cases only.

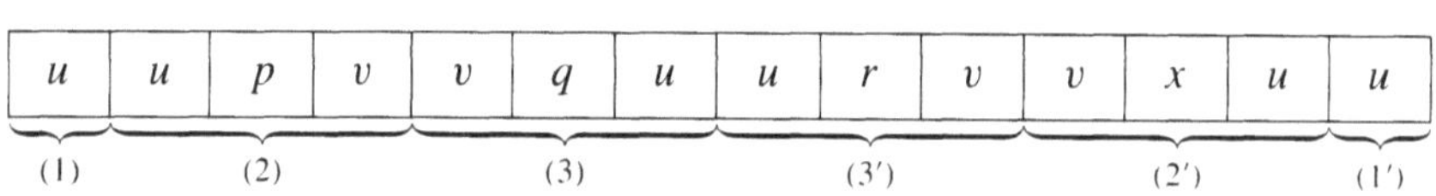

Figure 2.1. The various possibilities.

(1) $f' \le u$. Then there exists $y \in A^*$ such that

$$u = f'y \text{ and } f'' = yupv^2qu^2rv^2xu^2$$

Setting $g' = f'$, $g'' = yurvqupvxu$ and $g = g'g''$, we have $g = u^2rvqupvxu$ and thus $\varphi(g) = \varphi(f)$ since $\varphi(u) = \varphi(u^2)$ and since S satisfies Thérien's condition. Therefore, $\varphi(g^k g') = \varphi(f^k f') = s$ and

$$\varphi(g'') = \varphi(y)\varphi(urvqupvxu) = \varphi(y)\varphi(upvqurvxu) = \varphi(f'')$$

Thus $\varphi(g''g^{2n-1}g') = \varphi(f''f^{2n-1}f') = e$ and (s, e) is a linked pair of g^ω.

(2) $u < f' \le u^2pv$. Then there exists $y, z \in A^*$ such that $uy = f'$, $yz = upv$ and $f'' = zvqu^2rv^2xu^2$. Setting $g' = urvquy$, $g'' = zvxu$ and $g = g'g''$, we have $g = urvqu^2pv^2xu$ and thus $\varphi(g) = \varphi(f)$. On the other hand, we have by Lemma 2.2, $\varphi(furvqu) = \varphi(f)$ and hence $\varphi(g^k g') = \varphi(f^k g') = \varphi(f^n g') = \varphi(f^n y) = \varphi(f^n f')$. Similarly,

$$\begin{aligned}\varphi(f''f^{2n-1}) &= \varphi(z)\varphi((vqurv)(vxupvqurv)^{2n-1}(vxu)) \\ &= \varphi(z)\varphi((vxupvqurv)^{2n-1}(vxu)) \\ &= \varphi(z)\varphi(vxu)\varphi(upvqurvxu)^{2n-1} \\ &= \varphi(g''g^{2n-1})\end{aligned}$$

whence $\varphi(g''g^{2n-1}g') = \varphi(f''f^{2n-1}f') = e$. Thus (s, e) is a linked pair of g^ω.

(3) $u^2pv < f' \le u^2pv^2qu$. Then there exists $y, z \in A^*$ such that $u^2pvy = f'$, $yz = vqu$ and $f'' = zurv^2xu^2$. Setting $g' = urvy$, $g'' = zupvxu$ and $g = g'g''$, we have $g = urv^2qu^2pvxu$ and hence $\varphi(g) = \varphi(f)$. On the other hand, we have by Lemma 2.2,

$$\begin{aligned}\varphi(g^k g') = \varphi(f^k g') = \varphi(f^n g') &= \varphi((upvqurvxu)^n urvy) \\ &= \varphi((upvqurvxu)^n upvy) = \varphi(f^n f') = s\end{aligned}$$

and similarly,

$$\begin{aligned}\varphi(g''g^n) = \varphi(g''f^n) &= \varphi(zupvxu(upvqurvxu)^n) \\ &= \varphi(zurvxu(upvqurvxu)^n) = \varphi(f''f^n)\end{aligned}$$

Thus $\varphi(g''g^{2n-1}g') = \varphi(f''f^{2n-1}f') = e$ and (s, e) is a linked pair of g^ω.

It follows from this discussion that each linked pair of $(upvqurvx)^\omega$ is also a linked pair of $(urvqupvx)^\omega$. □

3 Local properties of infinite words

3.1 Properties defined by letters. In this section, we characterize sets of infinite words u satisfying conditions of the type

(a) A given letter occurs in u,

(b) A given letter does not occur in u,

(c) A given letter occurs infinitely often in u.

Three varieties will occur in the forthcoming statements: the variety $\mathbf{J_1}$ of idempotent and commutative semigroups, the variety of ordered $\mathbf{J}_1^+$, defined by the identities $x^2 = x$, $xy = yx$ and $xy \leq x$, and its dual, the variety $\mathbf{J}_1^-$, defined by the identities $x^2 = x$, $xy = yx$ and $x \leq xy$.

Theorem 3.1. *Let Z be a recognizable subset of A^ω. The following conditions are equivalent:*

(1) *Z is positive boolean combination of subsets of the form A^*aA^ω or $(A^*a)^\omega$, with $a \in A$,*

(2) *Z is union of subsets of the form XY^ω, where $XY^*, Y^+ \in \mathcal{J}_1^+(A^+)$,*

(3) *Z is weakly recognized by an ordered semigroup of $\mathbf{J}_1^+$,*

(4) *$S(Z)$ belongs to $\mathbf{J}_1^+$.*

Proof. (2) and (3) are equivalent by Theorem VI.4.5.

(3) implies (4). Let S be an ordered semigroup of $\mathbf{J}_1^+$ weakly recognizing Z. By Proposition VI.4.15, we have $\hat{S}_+ = S$ and hence $\hat{S}$ satisfies the identities of $\mathbf{J}_1^+$. On the other hand, $\hat{S}$ recognizes Z by Theorem VI.4.12 and by Corollary II.8.10, $S(Z)$ is a quotient of $\hat{S}$. Hence $S(Z)$ satisfies also the identities of $\mathbf{J}_1^+$.

(4) implies (3). Indeed, $S(Z)$ recognizes Z and Proposition VI.4.2 shows that $S_+(Z)$ weakly recognizes Z.

(1) implies (2). The equivalence of (2) and (4) shows that the subsets of the form XY^ω, where $XY^*, Y^+ \in \mathcal{J}_1^+(A^+)$, define a positive variety and are in particular closed under finite union and finite intersection. Showing that (1) implies (2) amounts to verifying that the subsets of the form A^*aA^ω or $(A^*a)^\omega$, with $a \in A$, are indeed of the form XY^ω, with $XY^*, Y^+ \in \mathcal{J}_1^+(A^+)$. Now, setting $X = A^*aA^*$ and $Y = A$, we get $XY^* = A^*aA^* \in \mathcal{J}_1^+(A^+)$, $Y^+ = A^+ \in \mathcal{J}_1^+(A^+)$ and $XY^\omega = A^*aA^\omega$. Similarly, setting $X = A^+$ and $Y = A^*aA^*$, we obtain $XY^* = A^+ \in \mathcal{J}_1^+(A^+)$, $Y^+ = A^*aA^* \in \mathcal{J}_1^+(A^+)$ and $XY^\omega = A^+(A^*aA^*)^\omega = (A^*a)^\omega$.

(2) implies (1). By Proposition B.3.6, $\mathcal{J}_1^+(A^+)$ consists of the finite unions of subsets of the form $F(B) = \bigcap_{a \in B} A^*aA^*$, where $B \subset A$. If $B_1, \ldots, B_k$ are subsets of A. Lemma VI.7.3 now shows that

$$\Big(\bigcup_{1 \leq i \leq k} F(B_i)\Big)^\omega = \bigcup_{1 \leq i \leq k} F(B_i)^\omega \tag{3.1}$$

Let Z be a finite union of subsets of the form XY^ω with $XY^*, Y^+ \in \mathcal{J}_1^+(A^+)$. Formula 3.1 shows that Z can be written as a finite union of subsets of the form $F(B)F(C)^\omega$, with $B, C \subset A$. Finally, the formula

$$F(B)F(C)^\omega = \Big(\bigcap_{a \in B} A^*aA^\omega\Big) \cap \Big(\bigcap_{a \in C} (A^*a)^\omega\Big)$$

permits us to conclude. □

A dual version of Theorem 3.1 also holds.

Theorem 3.2. *Let Z be a recognizable subset of A^ω. The following conditions are equivalent:*

(1) *Z is a positive boolean combination of subsets of the form B^ω or A^+B^ω,*
(2) *Z is union of subsets of the form XY^ω, where $XY^*, Y^+ \in \mathcal{J}_1^-(A^+)$,*
(3) *Z is weakly recognized by an ordered semigroup of $\mathbf{J}_1^-$,*
(4) *$S(Z)$ belongs to $\mathbf{J}_1^-$.*

Proof. The equivalence of (2), (3) and (4) is proved as in the proof of Theorem 3.1. Furthermore the formulas

$$B^\omega = A^\omega \setminus \bigcup_{a \in A \setminus B} A^*aA^\omega \quad \text{and} \quad A^+B^\omega = A^\omega \setminus \bigcup_{a \in A \setminus B} (A^*a)^\omega$$

show that the positive boolean combinations of subsets of the form B^ω or A^+B^ω form a positive boolean algebra $\mathcal{V}(A^\omega)$, dual of that constituted by the positive boolean combinations of subsets of the form A^*aA^ω or $(A^*a)^\omega$, with $a \in A$. This latter positive boolean algebra is associated, according to Theorem 3.1, with the variety of ordered ω-semigroups defined by the identities $xy = yx$, $x^2 = x$ and $xy \leq x$. It follows that $\mathcal{V}(A^\omega)$ is associated with the variety of ordered ω-semigroups defined by the identities $xy = yx$, $x^2 = x$ and $x \leq xy$, thereby establishing the equivalence of (1) and (4). □

Finally, for the variety $\mathbf{J}_1$, the following theorem holds, whose proof is similar to that of Theorem 3.1.

Theorem 3.3. *Let Z be a recognizable subset of A^ω. The following conditions are equivalent:*

(1) *Z is boolean combination of subsets of the form A^*aA^ω or $(A^*a)^\omega$, with $a \in A$,*
(2) *Z is union of subsets of the form XY^ω, where $XY^*, Y^+ \in \mathcal{J}_1(A^+)$,*
(3) *Z is weakly recognized by a semigroup of $\mathbf{J}_1$,*
(4) *$S(Z)$ belongs to $\mathbf{J}_1$.*

Theorems 3.1 and 3.3 should be compared with the next two statements.

Theorem 3.4. *Let Z be a recognizable subset of A^ω. The following conditions are equivalent:*

(1) *Z is positive boolean combination of subsets of the form A^*aA^ω, where $a \in A$,*
(2) *$Z = \overrightarrow{L}$, with $L \in \mathcal{J}_1^+(A^+)$,*
(3) *Z is positive boolean combination of subsets of the form $\overrightarrow{L}$, where $L \in \mathcal{J}_1^+(A^+)$,*
(4) *$S(Z)$ satisfies the identities $x^2 = x$, $xy = yx$, $xy \leq x$ and $xyx^\omega = yxy^\omega$.*

Proof. The equivalence of (2), (3) and (4) follows from Theorem VI.5.3, applied to $\mathbf{V} = \mathbf{J}_1^+$. Indeed, in the presence of the identities $x^2 = x$ and $xy = yx$, the identity $(x^\pi y^\pi)^\pi x^\omega = (x^\pi y^\pi)^\pi y^\omega$ reduces to $xyx^\omega = yxy^\omega$. Finally, (1) implies (3) follows from the formula

$$A^*aA^\omega = \overrightarrow{A^*aA^*}$$

and (2) implies (1) from the formula

$$\overrightarrow{\bigcap_{a \in B} A^*aA^*} = \bigcap_{a \in B} A^*aA^\omega$$

□

Theorem 3.5. *Let Z be a recognizable subset of A^ω. The following conditions are equivalent:*

(1) *Z is boolean combination of subsets of the form A^*aA^ω, where $a \in A$,*
(2) *Z is boolean combination of subsets of the form $\overrightarrow{L}$, where $L \in \mathcal{J}_1(A^+)$,*
(3) *$S(Z)$ satisfies the identities $x^2 = x$, $xy = yx$ and $xyx^\omega = yxy^\omega$.*

Proof. The proof mimics the proof of Theorem 3.4, starting this time from the variety of semigroups $\mathbf{J}_1$. □

3.2 Prefixes and suffixes of infinite words. As we did for finite words, we call *local* a property of infinite words defined by its factors of finite length. A subset Z of A^ω is said to be *prefix-testable* if Z is a boolean combination of subsets of the form uA^ω, with $u \in A^+$, which amounts to saying that $Z = XA^\omega$ with X finite.

Let us adapt to infinite words the notation introduced in Section B.5.2. For an infinite word u, denote by $F_n(u)$ the set of all words of length n which are factors of u and by $F_n^\infty(u)$ the set of all words of length n which occur infinitely often in u.

Theorem 3.6. *Let Z be a recognizable subset of A^ω. The following conditions are equivalent:*

(1) *Z is prefix-testable,*
(2) *$Z = \overrightarrow{L}$, with L prefix-testable,*

(3) *Z is boolean combination of subsets of the form* $\overrightarrow{L}$, *where L is prefix-testable,*

(4) $S(Z)$ *satisfies the identity* $x^\pi y = x^\pi$.

Proof. By Proposition B.5.4, the prefix-testable subsets are of the form $L = XA^* \cup Y$, with X and Y finite. Then we have $\overrightarrow{L} = XA^\omega$, establishing the equivalence of (1) and (2). Observing that the subsets satisfying (1) (or (2)) are closed under boolean operations gives the equivalence of (2) and (3).

For the equivalence of (2) and (4), we apply Theorem VI.5.3 to the variety of ordered semigroups $\mathbf{V} = [\![x^\pi y = x^\pi]\!]$, which is obviously contained in the variety $[\![x^\pi y^\pi \leq x^\pi, (xy)^\pi x = (xy)^\pi]\!]$. This theorem shows that (2) is equivalent to

(4′) $S(Z)$ satisfies the identities $x^\pi y = x^\pi$ and $x^\pi y^\omega \leq x^\omega$.

Now the second identity follows from the identity $x^\pi y^\omega = x^\omega$, which is itself a consequence of $x^\pi y = x^\pi$, since, if $x^\pi y = x^\pi$, then $x^\pi y^\pi = x^\pi$ and

$$x^\omega = (x^\pi)^\omega = (x^\pi y^\pi)^\omega = x^\pi (y^\pi x^\pi)^\omega = x^\pi (y^\pi)^\omega = x^\pi y^\omega$$

Consequently, conditions (4) and (4') are equivalent. □

A subset Z of A^ω is said to be *suffix-testable* if Z is a boolean combination of subsets of the form $(A^*u)^\omega$, with $u \in A^+$.

Theorem 3.7. *Let Z be a recognizable subset of A^ω. The following conditions are equivalent:*

(1) *Z is suffix-testable,*

(2) *Z is a boolean combination of subsets of the form* $\overrightarrow{L}$, *where L is suffix-testable,*

(3) *Z is a boolean combination of subsets weakly recognized by semigroup of* $\mathbf{L'I}$,

(4) $S(Z)$ *is locally idempotent and commutative and satisfies the identity* $xy^\omega = y^\omega$.

Proof. The equivalence of (1) and (2) follows from the formula $(A^*u)^\omega = \overrightarrow{A^*u}$.

(1) implies (3). By Theorem VI.4.5, a subset Z of A^ω is weakly recognized by a semigroup of $\mathbf{L'I}$ if it can be written as XY^ω, where XY^* and Y^+ are recognized by semigroups of $\mathbf{L'I}$. This condition is indeed satisfied for $Z = (A^*u)^\omega$, by taking $X = A^*$ and $Y = A^*u$.

(3) implies (1). If Z is weakly recognized by a semigroup S of $\mathbf{L'I}$, Z can be written as a finite union of subsets of the form $X_s X_e^\omega$, where (s, e) is a linked pair of S. Then $se = s$ by definition of a linked pair, but also $se = e$ since $S \in \mathbf{L'I}$. Thus $s = e$ and there exists a finite subset F of A^+ such that $\varphi^{-1}(e) = A^*F$. Therefore

$$X_s X_e^\omega = (\varphi^{-1}(e))^\omega = (A^*F)^\omega = \bigcap_{u \in F} (A^*u)^\omega$$

(1) implies (4). Let $Z = (A^*u)^\omega$. By (3), Z is weakly recognized by a semigroup S of $\mathbf{L'I}$. Since $S \in \mathbf{L'I}$, S is a fortiori idempotent and commutative, and so is $\hat{S}$, by

Proposition 2.1. Now Corollary II.8.10 shows that $S(Z)$ is a quotient of $\hat{S}$ and hence $S_+(Z)$ is also locally idempotent and commutative. Finally $S(Z)$ satisfies the identity $xy^\omega = y^\omega$. Indeed, it is easy to see that for each $x, y \in A^+$ and $z \in A^*$, $zxy^\omega \in (A^*u)^\omega$ if and only if $zy^\omega \in (A^*u)^\omega$.

(4) implies (2). By Proposition B.5.8, each locally idempotent and commutative semigroup which is a quotient of A^+ is quotient of a semigroup of the form $S_k = A^+/{\sim_k}$. Consequently, there exists an integer $k > 0$ such that Z is recognized by $\bar{\psi}$, where $\psi : A^+ \to S_k$ is the quotient morphism. Let $\equiv_{2k}$ be the equivalence on A^ω defined by $u \equiv_{2k} v$ if and only if $F_{2k}^\infty(u) = F_{2k}^\infty(v)$. The class of a word u for this equivalence is

$$\Big(\bigcap_{x \in F_{2k}^\infty(u)} (A^*x)^\omega \setminus \bigcup_{x \notin F_{2k}^\infty(u)} (A^*x)^\omega\Big)$$

and thus it is a suffix-testable subset. We claim that if $u \equiv_{2k} v$, then $\varphi(u) = \varphi(v)$, which will show that Z is union of $\equiv_{2k}$-classes.

Suppose that $u \equiv_{2k} v$. Let us fix a factor $h \in F_{2k}^\infty(u)$ and set $h = aspb$, where a and b are (possibly equal) letters and s and p are words of length $k-1$. As the factor h occurs infinitely often in u and v, there exist factorizations $u = u_0u_1\cdots$ and $v = v_0v_1\cdots$ such that, for each $n > 0$, $|u_n| \geq 2k$, $|v_n| \geq 2k$ and

$$p_{k-1}(u_n) = p = p_{k-1}(v_n),\ s_{k-1}(u_n) = s = s_{k-1}(v_n) \text{ and}$$
$$F_{2k}(u_n) = F_{2k}(v_n) = F_{2k}^\infty(u)$$

As each factor of length k of u_n (resp. v_n) is factor of a factor of length $2k$, we have, for each $n > 0$, $u_n \sim_k u_{n+1} \sim_k v_n \sim_k v_{n+1}$ and thus $\psi(u_n) = \psi(v_n) = \psi(u_{n+1}) = \psi(v_{n+1})$. Therefore, there exists $y \in S_+(Z)$ such that $\varphi(u_n) = \varphi(v_n) = y$ for each $n > 0$. It follows, by the identity $xy^\omega = y^\omega$, that $\varphi(u) = \varphi(u_0)y^\omega = y^\omega = \varphi(v)$, proving the claim. □

One can also consider only positive boolean combinations. A subset Z of A^ω is said to be *positively suffix-testable* if Z is positive boolean combination of subsets of the form $(A^*u)^\omega$, with $u \in A^+$. The following theorem, whose proof is left to the reader, is the "positive counterpart" of Theorem 3.7.

Theorem 3.8. *Let Z be a recognizable subset of A^ω. The following conditions are equivalent:*

(1) *Z is positively suffix-testable,*
(2) *Z is intersection of subsets of the form $\overrightarrow{L}$, where L is suffix-testable,*
(3) *Z is positive boolean combination of subsets weakly recognized by a semigroup of $\mathbf{L}^r\mathbf{I}$,*
(4) *$S(Z)$ is locally idempotent and commutative and satisfies the identities $x^\pi y x^\pi \leq x^\pi$ and $xy^\omega = y^\omega$.*

A subset Z of A^ω is said to be *prefix-suffix testable* if Z is a boolean combination of subsets of the form uA^ω or $(A^*u)^\omega$, with $u \in A^+$.

Theorem 3.9. *Let Z be a recognizable subset of A^ω. The following conditions are equivalent:*

(1) *Z is prefix-suffix testable,*

(2) *Z is boolean combination of subsets of the form $\overrightarrow{L}$, where L is a prefix-suffix testable subset of A^+,*

(3) *Z is boolean combination of subsets weakly recognized by a semigroup of* $\mathbf{L}^\ell\mathbf{I}$,

(4) *Z is boolean combination of subsets weakly recognized by a semigroup of* $\mathbf{LI}$,

(5) *$S(Z)$ is locally idempotent and commutative and satisfies the identities $x^\pi yz^\omega = x^\pi z^\omega$.*

Proof. (1) implies (2). This follows immediately from the formula

$$uA^\omega = \overrightarrow{uA^*} \qquad \text{and} \qquad (A^*u)^\omega = \overrightarrow{A^*u}$$

(2) implies (1). If L is prefix-suffix testable, then $L = XA^*Y \cup Z$ with X, Y and Z finite and thus

$$\overrightarrow{L} = \overrightarrow{XA^*Y} = XA^*(YA^*)^\omega = \Big(\bigcup_{u\in X} uA^*\Big) \cap \Big(\bigcup_{u\in Y} (A^*u)^\omega\Big)$$

(1) implies (3). By Theorem VI.4.5, a subset Z of A^ω is weakly recognized by a semigroup of $\mathbf{L}^\ell\mathbf{I}$ if it can be written as XY^ω, where XY^* and Y^+ are recognized by semigroups of $\mathbf{L}^\ell\mathbf{I}$. This condition is indeed verified for $Z = uA^\omega$ and for $Z = (A^*u)^\omega$. It suffices to take $X = uA^*$ and $Y = A$ in the first case, $X = A^*$ and $Y = uA^*$ in the second case.

(3) implies (4) since $\mathbf{L}^\ell\mathbf{I}$ is contained in $\mathbf{LI}$.

(4) implies (2). If Z is weakly recognized by a semigroup S of $\mathbf{LI}$, Z can be written as a finite union of subsets of the form $X_sX_e^\omega$, where (s, e) is a linked pair of S. Since $se = s$ by definition of a linked pair, s and e are two $\mathcal{L}$-equivalent idempotents of the minimal ideal of S. Therefore there exist finite subsets F, F' and G of A^+ such that $\varphi^{-1}(s) = FA^*G$ and $\varphi^{-1}(e) = F'A^*G$. Consequently,

$$X_sX_e^\omega = FA^*G(F'A^*G)^\omega = F(A^*GF')^\omega = \overrightarrow{FA^*GF'}$$

(1) implies (5). It suffices to verify that $S(Z)$ satisfies (5) for $Z = uA^\omega$ and for $Z = (A^*u)^\omega$. For $Z = (A^*u)^\omega$, this follows from Theorem 3.7 and for $Z = uA^\omega$, from Theorem 3.6, since the identities $x^\pi y = x^\pi$ and $x^\pi y^\omega = x^\omega$ give $x^\pi yz^\omega = x^\pi z^\omega = x^\omega$.

(5) implies (1). By Proposition B.5.8, every locally idempotent and commutative semigroup which is a quotient of A^+ is quotient of a semigroup of the form $S_k =$

$A^+/\sim_k$. Therefore, there exists an integer $k > 0$ such that Z is recognized by $\bar{\psi}$, where $\psi : A^+ \to S_k$ is the quotient morphism. Set $K = \max\{2k-2, |S_+(Z)|\}$ and let $\equiv_K$ be the equivalence on A^ω defined by $u \equiv_K v$ if and only if $p_K(u) = p_K(v)$ and $F_K^\infty(u) = F_K^\infty(v)$. The class of a word u for this equivalence is

$$p_K(u)A^\omega \cap \Big(\bigcap_{x \in F_K^\infty(u)} (A^*x)^\omega \setminus \bigcup_{x \notin F_K^\infty(u)} (A^*x)^\omega \Big)$$

and thus it is a prefix-suffix testable subset. We claim that if $u \equiv_K v$, then $\varphi(u) = \varphi(v)$, which will prove that Z is union of $\equiv_K$-classes.

Suppose that $u \equiv_K v$. Then in particular $u = wu'$ and $v = wv'$, where w is the common prefix of length K of u and v. Since $|w| \geq |S_+(Z)|$, there exists by Proposition A.1.12 a factorization $w = w_1w_2$ and an idempotent $e \in S_+(Z)$ such that $\varphi(<_1) = \varphi(w_1)e$. Let us fix a factor $h \in F_K^\infty(u)$ and set $h = sph'$ with $|s| = |p| = k-1$. Since the factor h occurs infinitely often in u and v, there exist factorizations $u = u_0u_1\cdots$ and $v = v_0v_1\cdots$ such that, for each $n > 0$,

$$\begin{aligned} p_{k-1}(u_n) &= p = p_{k-1}(v_n),\ s_{k-1}(u_n) = s = s_{k-1}(v_n) \text{ and} \\ F_K(u_n) &= F_K(v_n) = F_K^\infty(u) \end{aligned}$$

Then we have, for each $n > 0$, $u_n \sim_k u_{n+1} \sim_k v_n \sim_k v_{n+1}$ and hence $\psi(u_n) = \psi(v_n) = \psi(u_{n+1}) = \psi(v_{n+1})$. Therefore there exist $z \in S_+(Z)$ such that $\varphi(u_n) = \varphi(v_n) = z$ for each $n > 0$. It follows that $\varphi(u) = \varphi(w)\varphi(u_0)z^\omega = \varphi(w)e\varphi(u_0)z^\omega$. Now the identity $x^\pi yz^\omega = x^\pi z^\omega$ shows that $e\varphi(u_0)z^\omega = ez^\omega$. It follows that $\varphi(u) = \varphi(w)ez^\omega$. A similar computation holds for $\varphi(v)$ and thus $\varphi(u) = \varphi(v)$, proving the claim. □

If we restrict ourselves to positive boolean combinations, we obtain the following statement. Once again, we state without proof the positive version of Theorem 3.9. A subset Z of A^ω is said to be *positively prefix-suffix testable* if Z is a positive boolean combination of subsets of the form uA^ω or $(A^*u)^\omega$, with $u \in A^+$.

Theorem 3.10. *Let Z be a recognizable subset of A^ω. The following conditions are equivalent:*

(1) *Z is positively prefix-suffix testable,*
(2) *Z is a finite intersection of subsets of the form $\overrightarrow{L}$, where L is prefix-suffix testable,*
(3) *Z is a positive boolean combination of subsets weakly recognized by a semigroup of* $\mathbf{L}^\ell\mathbf{I}$,
(4) *Z is a positive boolean combination of subsets weakly recognized by a semigroup of* $\mathbf{LI}$,
(5) *$S_+(Z)$ is locally idempotent and commutative and $S(Z)$ satisfies the identities $x^\pi yx^\pi \leq x^\pi$ and $x^\pi yz^\omega = x^\pi z^\omega$.*

3.3 Factors of infinite words. A subset X of A^ω is said to be *locally testable* if it is a boolean combination of subsets of the form xA^ω, A^*xA^ω or $(A^*x)^\omega$, where $x \in A^+$. We shall give several characterizations of these subsets, in a form that reminds the characterizations of recognizable subsets of Chapter I.

Theorem 3.11. *Let Z be a subset of A^ω. The following conditions are equivalent:*

(1) *Z is locally testable,*
(2) *Z is a boolean combination of subsets of the form $\overrightarrow{L}$, where L is a locally testable subset of A^+,*
(3) *Z is a finite union of subsets of the form XY^ω such that XY^* and Y^+ are locally testable subsets of A^+,*
(4) *Z is weakly recognized by a locally idempotent and commutative semigroup,*
(5) *$S_+(Z)$ is locally idempotent and commutative.*

Proof. (1) implies (2). This follows immediately from the formulas

$$uA^\omega = \overrightarrow{uA^*} \qquad A^*uA^\omega = \overrightarrow{A^*uA^\omega} \qquad \text{and} \qquad (A^*u)^\omega = \overrightarrow{A^*u}$$

(3) is equivalent to (4) by Theorem VI.4.5.
(4) implies (5). If Z is weakly recognized by S, Z is recognized by $\hat{S}$ by Theorem VI.4.12. Now, by Proposition 2.1, if S is locally idempotent and commutative, then so is $\hat{S}$. Finally, by Corollary II.8.10, $S(Z)$ is a quotient of $\hat{S}$ and hence $S_+(Z)$ is also locally idempotent and commutative.
(5) implies (4). Set $S(Z) = (S_+, S_\omega)$. Then $S(Z)$ is a quotient of $\bar{S}_+$ by Corollary II.5.4. Thus, by Proposition II.8.9, $\bar{S}_+$ recognizes Z and by Theorem VI.4.12, S_+ recognizes weakly Z. Now, if $S(Z)$ is locally idempotent and commutative, then so is S_+ by definition.
(2) implies (4). It follows from the equivalence of (3), (4) and (5) that the subsets of A^ω satisfying condition (4) form a boolean algebra. Now by Proposition VI.4.3, the subsets of the form $\overrightarrow{L}$, where L is a locally testable subset of A^+, are weakly recognized by a locally idempotent and commutative semigroup.
(5) implies (1). By Proposition B.5.8, $S_+(Z)$ is a quotient of a semigroup of the form $S_k = A^+/{\sim_k}$. Consequently, there exists an integer $k > 0$ such that Z is recognized by $\bar{\varphi}$, where $\varphi : A^+ \to S_k$ is the quotient morphism. Let (s, e) be a linked pair of S_k and let x and y be words of A^+ such that $\varphi(x) = s$ and $\varphi(y) = e$. As $se = s$ and $e^2 = e$, we may suppose, up to replacing x by xy^k (or y by y^k), that $|x|, |y| \geq k$. The next lemma uses the notation introduced at the beginning of Section 3.2.

Lemma 3.12. *The following formula holds:*

$$\begin{aligned}\bar{\varphi}^{-1}([s,e]) \;=\; \big\{u \in A^\omega \mid\; & p_{k-1}(u) = p_{k-1}(x),\; F_k(u) = F_k(x),\\ & F_k^\infty(u) = F_k(y),\; \textit{and}\; s_{k-1}(y)p_{k-1}(y) = F_{2k-2}^\infty(u)\big\}\end{aligned}$$

Proof. Denote by Z the right hand side of the equality to be proved. If $\bar{\varphi}(u) = [s, e]$, u can be factorized as $u = u_0 u_1 u_2 \cdots$, with $\varphi(u_0) = s$ and $\varphi(u_1) = \varphi(u_2) = \cdots = e$. It follows, since $s = se$,

$$u_0 \sim_k u_0 u_1 \sim_k u_0 u_1 u_2 \sim_k \cdots \sim_k x \text{ and } u_1 \sim_k u_1 u_2 \sim_k \cdots \sim_k y$$

Consequently, $p_{k-1}(u) = p_{k-1}(u_0) = p_{k-1}(x)$ and $F_k(x) = F_k(u_0) \subset F_k(u)$. Furthermore, every factor of length k of u is factor of $u_0 \cdots u_n$ for some n. Now, since $u_0 \cdots u_n \sim_k u_0$, we have $F_k(u_0 \cdots u_n) = F_k(u_0) = F_k(x)$. Consequently, $F_k(u) = F_k(x)$. On the other hand, $F_k(u_1) = F_k(u_2) = \cdots = F_k(y)$ and hence each factor of length k of y belongs to $F_k^\infty(u)$. Conversely, each factor of $F_k^\infty(u)$ is a factor of $u_1 \cdots u_n$ for some n. As $u_1 \cdots u_n \sim_k y$, it is also a factor of y and hence $F_k^\infty(u) = F_k(y)$. Finally, $s_{k-1}(y) = s_{k-1}(u_1) = s_{k-1}(u_2) = \cdots$ and $p_{k-1}(y) = p_{k-1}(u_1) = p_{k-1}(u_2) = \cdots$. Consequently, $s_{k-1}(y)p_{k-1}(y) \in F_{2k-2}^\infty(u)$, which proves the inclusion $\bar{\varphi}^{-1}([s, e]) \subset Z$.

Conversely, let $u \in Z$. As $s_{k-1}(y)p_{k-1}(y)$ occurs infinitely often in u, one can build a sequence $(v_n)_{n \geq 0}$ such that

$$u = p_{k-1}(u)v_0 s_{k-1}(y)p_{k-1}(y)v_1 s_{k-1}(y)p_{k-1}v_2 \cdots$$

Furthermore, one can choose $v_0, v_1, v_2, \ldots$ sufficiently long so that

$$F_k(p_{k-1}(u)v_0) = F_k(u) \text{ and } F_k(v_1) = F_k(v_2) = \ldots = F_k^\infty(u)$$

We claim that $p_{k-1}(u)v_0 s_{k-1}(y) \sim_k x$. Indeed, $p_{k-1}(u) = p_{k-1}(x)$ by assumption and $s_{k-1}(y) = s_{k-1}(xy) = s_{k-1}(x)$ since $xy \sim_k x$. On the other hand, $F_k(p_{k-1}(u)v_0) = F_k(u) = F_k(x)$ and hence

$$F_k(x) = F_k(p_{k-1}(u)v_0) \subset F_k(p_{k-1}(u)v_0 s_{k-1}(xy)) \subset F_k(u) = F_k(x)$$

which proves the claim.

One would prove in a similar way that

$$p_{k-1}(y)v_1 s_{k-1}(y) \sim_k p_{k-1}(y)v_2 s_{k-1}(y) \sim_k \cdots \sim_k y$$

and hence $u \in \bar{\varphi}^{-1}([s, e])$. Thus $Z = \bar{\varphi}^{-1}([s, e])$. □

Lemma 3.12 shows that $\bar{\varphi}^{-1}([s, e])$ is locally testable. More precisely,

$$\bar{\varphi}^{-1}([s, e]) = X_1 \cap X_2 \cap X_3 \cap X_4$$

with

$$\begin{aligned}
X_1 &= p_{k-1}(x)A^\omega \\
X_2 &= \bigcap_{v\in F_k(x)} A^*vA^\omega \setminus \bigcup_{v\notin F_k(x)} A^*vA^\omega \\
X_3 &= \bigcap_{v\in F_k(y)} (A^*v)^\omega \setminus \bigcup_{v\notin F_k(y)} (A^*v)^\omega \\
X_4 &= (A^*s_{k-1}(y)p_{k-1}(y))^\omega
\end{aligned}$$

As Z is recognized by $\bar{\varphi}$, Z is union of subsets of the form $\bar{\varphi}^{-1}([s, e])$ and hence is locally testable. □

A subset of A^ω is said to be positively locally testable if it is positive boolean combination of subsets of the form uA^ω, A^*uA^ω or $(A^*u)^\omega$, with $u \in A^+$. We give below a characterization of this restricted class of locally testable sets, for which the proof is omitted.

Theorem 3.13. *Let Z be a subset of A^ω. The following conditions are equivalent:*

(1) *Z is positively locally testable,*
(2) *Z is positive boolean combination of subsets of the form $\overrightarrow{L}$, where L is a positively locally testable subset of A^+,*
(3) *Z is a finite union of subsets of the form XY^ω such that XY^* and Y^+ are positively locally testable subsets of A^+,*
(4) *Z is weakly recognized by a locally idempotent and commutative ordered semigroup satisfying the identity $x^\pi yx^\pi \leq x^\pi$,*
(5) *$S(Z)$ is locally idempotent and commutative and satisfies the identity $x^\pi yx^\pi \leq x^\pi$.*

We shall now establish a similar result for the threshold locally testable subsets. Recall that, if x is a finite word and u is a finite or infinite word, $\begin{bmatrix} u \\ x \end{bmatrix}$ denotes the number of occurrences of x in u. Set, for each word $x \in A^+$ and for each integer $t \geq 0$,

$$F(x, t) = \left\{u \in A^+ \mid \begin{bmatrix} u \\ x \end{bmatrix} \geq t\right\} \quad \text{and} \quad F_\omega(x, t) = \left\{u \in A^\omega \mid \begin{bmatrix} u \\ x \end{bmatrix} \geq t\right\}$$

and also

$$F_\omega(x, \infty) = \left\{u \in A^\omega \mid \begin{bmatrix} u \\ x \end{bmatrix} = \infty\right\}$$

Note that, in general, $F(x,t)$ is not equal to $(A^*x)^t A^\omega$, because overlapping factors may occur. However, $F_\omega(x,\infty) = (A^*x)^\omega$, since, if there are infinitely many factors equal to x, one can select an infinite sequence of nonoverlapping factors.

A subset of A^ω is said to be *threshold locally testable* (TLT) if it is equal to a boolean combination of subsets of the form xA^ω, $F_\omega(x,t)$ or $F_\omega(x,\infty)$ with $x \in A^+$ and $t \geq 0$.

Theorem 3.14. *Let Z be a subset of A^ω. The following conditions are equivalent:*

(1) *Z is threshold locally testable,*

(2) *Z is a boolean combination of subsets of the form $\overrightarrow{L}$, where L is a threshold locally testable subset of A^+,*

(3) *Z is a finite union of subsets of the form XY^ω such that XY^* and Y^+ are threshold locally testable subsets of A^+,*

(4) *Z is weakly recognized by an aperiodic semigroup satisfying Thérien's condition,*

(5) *$S_+(Z)$ is aperiodic and satisfies Thérien's condition.*

Proof. The proof is inspired by the proof of Theorem 3.11, which just requires a few adjustments.

(1) implies (2) follows immediately from the formulas

$$xA^\omega = \overrightarrow{xA^*}, \quad F_\omega(x,t) = \overrightarrow{F(x,t)}, \quad F_\omega(x,\infty) = \overrightarrow{A^*x}$$

The equivalence of (3) and (4) follows from Theorem VI.4.5.

(4) implies (5). If Z is weakly recognized by S, Z is recognized by $\hat{S}$ by Theorem VI.4.12. Now, by Proposition 2.3, if S is aperiodic and satisfies Thérien's condition, then so is $\hat{S}$. Finally, by Corollary II.8.10, $S(Z)$ is a quotient of $\hat{S}$ and hence $S_+(Z)$ is aperiodic and satisfies Thérien's condition.

(5) implies (4). Set $S(Z) = (S_+, S_\omega)$. Then $S(Z)$ is a quotient of $\bar{S}_+$ by Corollary II.5.4. Thus, by Proposition II.8.9, $\bar{S}_+$ recognizes Z and by Theorem VI.4.12, S_+ recognizes weakly Z. Now, if $S(Z)$ is aperiodic and satisfies Thérien's condition, then so is S_+ by definition.

(2) implies (4). It follows from the equivalence of (3), (4) and (5) that the subsets of A^ω satisfying condition (4) form a boolean algebra. Now by Proposition VI.4.3 and B.5.10, the subsets of the form $\overrightarrow{L}$, where L is a locally threshold testable subset of A^+, are weakly recognized by an aperiodic semigroup satisfying Thérien's condition.

(5) implies (1). By Proposition B.5.11, $S_+(Z)$ is quotient of a semigroup of the form $S_{k,t} = A^+/{\sim_{k,t}}$. Consequently, there exist two positive integers k and t such that Z is recognized by $\bar{\varphi}$, where $\varphi : A^+ \to S_{k,t}$ is the quotient morphism. Let (s,e) be a linked pair of $S_{k,t}$ and let x and y be words of A^+ such that $\varphi(x) = s$ and $\varphi(y) = e$. As $se = s$ and $e^2 = e$, we may suppose, up to replacing x by xy^k (or y by y^k), that $|x|, |y| \geq k$.

For the counterpart of Lemma 3.12, we introduce an auxiliary notation. Given an infinite word u, and two positive integers k and r, we denote by $F_k^r(u)$ the set of words of length k which occur at least r times as a factor of u. Note that, since $y \sim_{k,t} y^2$, $F_k(y) = F_k^2(y) = \ldots = F_k^t(y)$. We also denote by $F_n^\infty(u)$ the set of factors of length n occurring infinitely often in u.

Lemma 3.15. *The following formula holds:*

$$\bar{\varphi}^{-1}([s,e]) = \{u \in A^\omega \mid p_{k-1}(u) = p_{k-1}(x),\ s_{k-1}(y)p_{k-1}(y) = F_{2k-2}^\infty(u),$$
$$F_k(u) = F_k(x),\ F_k^2(u) = F_k^2(x),\ \ldots,\ F_k^t(u) = F_k^t(x) \text{ and } F_k^\infty(u) = F_k(y)\}$$

Proof. Denote by Z the right hand side of the equality to be proved. If $\bar{\varphi}(u) = [s,e]$, u can be factorized as $u = u_0u_1u_2\cdots$, with $\varphi(u_0) = s$ and $\varphi(u_1) = \varphi(u_2) = \cdots = e$. It follows, since $s = se$,

$$u_0 \sim_{k,t} u_0u_1 \sim_{k,t} u_0u_1u_2 \sim_{k,t} \cdots \sim_{k,t} x \text{ and } u_1 \sim_{k,t} u_1u_2 \sim_{k,t} \cdots \sim_{k,t} y$$

Consequently, $p_{k-1}(u) = p_{k-1}(u_0) = p_{k-1}(x)$. Furthermore, by choosing n sufficiently large, we may assume that every factor of length k of u occurs as a factor of $u_0\cdots u_n$ for some n, with the same multiplicity counted threshold t. Now, since $u_0\cdots u_n \sim_{k,t} x$, we have, for every word z of length k,

$$\begin{bmatrix} u \\ z \end{bmatrix} \equiv_{k,t} \begin{bmatrix} u_0\cdots u_n \\ z \end{bmatrix} \equiv_{k,t} \begin{bmatrix} x \\ z \end{bmatrix}$$

Consequently, $F_k(u) = F_k(x)$, $F_k^2(u) = F_k^2(x)$, $\ldots$, $F_k^t(u) = F_k^t(x)$. On the other hand, $F_k(u_1) = F_k(u_2) = \cdots = F_k(y)$ and hence each factor of length k of y belongs to $F_k^\infty(u)$. Conversely, each factor of $F_k^\infty(u)$ is a factor of $u_1\cdots u_m$ for some m. As $u_1\cdots u_m \sim_{k,t} y$, it is also a factor of y and hence $F_k^\infty(u) = F_k(y)$. Finally, $s_{k-1}(y) = s_{k-1}(u_1) = s_{k-1}(u_2) = \cdots$ and $p_{k-1}(y) = p_{k-1}(u_1) = p_{k-1}(u_2) = \cdots$. Consequently, $s_{k-1}(y)p_{k-1}(y) \in F_{2k-2}^\infty(u)$, which proves the inclusion $\bar{\varphi}^{-1}([s,e]) \subset Z$.

Conversely, let $u \in Z$. As $s_{k-1}(y)p_{k-1}(y)$ occurs infinitely often in u, one can build a sequence $(v_n)_{n\geq 0}$ such that

$$u = p_{k-1}(u)v_0s_{k-1}(y)p_{k-1}(y)v_1s_{k-1}(y)p_{k-1}v_2\cdots$$

Furthermore, one can choose v_0 sufficiently long so that

(1) all factors of length k of $p_{k-1}(y)v_1s_{k-1}(y)p_{k-1}v_2\cdots$ are in $F_k^\infty(u)$,

(2) for every word z of length k, $\left[\begin{smallmatrix} p_{k-1}(u)v_0 \\ z \end{smallmatrix}\right] \equiv_{k,t} \left[\begin{smallmatrix} u \\ z \end{smallmatrix}\right]$.

One can also choose $v_1, v_2, \ldots$ sufficiently long so that

(3) $F_k(v_1) = F_k(v_2) = \ldots = F_k^2(v_1) = F_k^2(v_2) = \ldots = F_k^t(v_1) = F_k^t(v_2) = \ldots = F_k^\infty(u)$.

We claim that $p_{k-1}(u)v_0s_{k-1}(y) \sim_{k,t} x$. Indeed, $p_{k-1}(u) = p_{k-1}(x)$ by assumption and $s_{k-1}(y) = s_{k-1}(xy) = s_{k-1}(x)$ since $xy \sim_{k,t} x$. Now, it follows from (2) that $F_k(p_{k-1}(u)v_0) = F_k(u) = F_k(x)$, $F_k^2(p_{k-1}(u)v_0) = F_k^2(u) = F_k^2(x)$, ..., $F_k^t(p_{k-1}(u)v_0) = F_k^t(u) = F_k^t(x)$, which proves the claim.

Furthermore, for every $n > 0$, and for $1 \leq i \leq t$, we have by (1) and (3)

$$F_k^\infty(u) = F_k^i(v_n) \subset F_k^i(p_{k-1}(y)v_ns_{k-1}(y) \subset F_k^\infty(u) = F_k^i(y)$$

and hence $F_k^i(p_{k-1}(y)v_is_{k-1}(y)) = F_k^\infty(u) = F_k^i(y)$. It follows that

$$p_{k-1}(y)v_1s_{k-1}(y) \sim_{k,t} p_{k-1}(y)v_2s_{k-1}(y) \sim_{k,t} \cdots \sim_{k,t} y$$

and hence $u \in \bar{\varphi}^{-1}([s,e])$. Thus $Z = \bar{\varphi}^{-1}([s,e])$. Lemma 3.15 shows that $\bar{\varphi}^{-1}([s,e])$ is threshold locally testable. More precisely,

$$\bar{\varphi}^{-1}([s,e]) = X_1 \cap X_2 \cap X_3 \cap Z_1 \cap \cdots \cap Z_t$$

with

$$X_1 = p_{k-1}(x)A^\omega$$

$$X_2 = \bigcap_{v \in F_k(y)} (A^*v)^\omega \setminus \bigcup_{v \notin F_k(y)} (A^*v)^\omega$$

$$X_3 = (A^*s_{k-1}(y)p_{k-1}(y))^\omega$$

and, for $1 \leq r \leq t$,

$$Z_r = \bigcap_{v \in F_k(x)} \left\{u \in A^\omega \mid \begin{bmatrix} u \\ v \end{bmatrix} \geq r\right\} \setminus \bigcup_{v \notin F_k(x)} \left\{u \in A^\omega \mid \begin{bmatrix} u \\ v \end{bmatrix} \geq r\right\}$$

As Z is recognized by $\bar{\varphi}$, Z is union of subsets of the form $\bar{\varphi}^{-1}([s,e])$ and hence is threshold locally testable. □

4 Exercises

4.1 Local properties of finite words.

Exercise 1. Let $A = \{a, b, c\}$. Denote by $\mathcal{F}$ the set of subsets of A^ω which are finite union of subsets of the form $XA^*(YA^*)^\omega$, with X and Y finite. Set $X = aA^*(bA^*)^\omega$ and $Y = aA^*(cA^*)^\omega$. Show that $X \in \mathcal{F}$ and $Y \in \mathcal{F}$, but that $X \cap Y$ is weakly recognized by no semigroup of $\mathbf{L}^\ell\mathbf{I}$. Derive that $\mathcal{F}$ is not closed under intersection.

4.2 Factors of infinite words.

Exercise 2. A subset of A^ω is said to be *prefix-factors testable* if it is equal to a boolean combination of subsets of the form xA^ω or A^*xA^ω with $x \in A^+$. Let Z be a subset of A^ω. Prove that the following conditions are equivalent:

(1) Z is prefix-factors testable,

(2) Z is boolean combination of subsets of the form $\overrightarrow{L}$, where L is a prefix-factors testable subset of A^+,

(3) Z is finite union of subsets of the form XY^ω such that XY^* and Y^+ are prefix-factors testable subsets of A^+,

(4) $S(Z)$ is locally idempotent and commutative and satisfies the identity

$$(x^\pi y^\pi)^\pi x^\omega = (x^\pi y^\pi)^\pi y^\omega.$$

Exercise 3. A subset of A^ω is said to be *positively prefix-factors testable* if it is equal to a positive boolean combination of subsets of the form xA^ω or $A * xA^\omega$ with $x \in A^+$. Let Z be a subset of A^ω. Prove that the following conditions are equivalent:

(1) Z is positively prefix-factors testable,

(2) Z is positive boolean combination of subsets of the form $\overrightarrow{L}$, where L is a positively prefix-factors testable subset of A^+,

(3) Z is finite union of subsets of the form XY^ω such that XY^* and Y^+ are positively prefix-factors testable subsets of A^+,

(4) $S(Z)$ is locally idempotent and commutative and satisfies the identities

$$(x^\pi y^\pi)^\pi x^\omega = (x^\pi y^\pi)^\pi y^\omega \quad \text{and} \quad x^\pi y x^\pi \leq x^\pi$$

Exercise 4. A subset of A^ω is said to be *positively threshold locally testable* if it is equal to a positive boolean combination of subsets of the form xA^ω, $F_\omega(x, t)$ or $F_\omega(x, \infty)$ with $x \in A^+$ and $t \geq 0$.

Let Z be a subset of A^ω. Show that the following conditions are equivalent:

(1) Z is positively locally threshold testable,

(2) Z is positive boolean combination of subsets of the form $\overrightarrow{L}$, where L is a positively locally threshold testable subset of A^+,

(3) Z is a finite union of subsets of the form XY^ω such that XY^* and Y^+ are positively locally threshold testable subsets of A^+,

(4) Z is weakly recognized by an aperiodic ω-semigroup satisfying Thérien's condition and the identity $x^\pi y x^\pi \leq x^\pi$,

(5) $S_+(Z)$ is aperiodic and satisfies Thérien's condition and $S(Z)$ satisfies the identity $x^\pi y x^\pi \leq x^\pi$.

Exercise 5. Let Z be a subset of A^ω. Show that the following conditions are equivalent:

(1) Z is positively threshold prefix-factors testable,

(2) Z is positive boolean combination of subsets of the form $\overrightarrow{L}$, where L is a positively threshold prefix-factors testable subset of A^+,

(3) $S_+(Z)$ is aperiodic and satisfies Thérien's condition and $S(Z)$ satisfies the identities

$$(x^\pi y)^\omega = (x^\pi y)^\pi x^\omega \quad \text{and} \quad x^\pi y x^\pi \leq x^\pi$$

Exercise 6. A subset of A^ω is said to be *threshold prefix-factors testable* (TPFT) if it is equal to a boolean combination of subsets of the form xA^ω or $F_\omega(x, t)$ with $x \in A^+$ and $t \geq 0$. Let Z be a subset of A^ω. Show that the following conditions are equivalent:

(1) Z is threshold prefix-factors testable,

(2) Z is a boolean combination of subsets of the form $\overrightarrow{L}$, where L is a threshold prefix-factors testable subset of A^+,

(3) $S_+(Z)$ is aperiodic and satisfies Thérien's condition and $S(Z)$ satisfies the identity

$$(x^\pi y)^\omega = (x^\pi y)^\pi x^\omega$$

5 Notes

The characterization of locally testable subsets was obtained independently by Brzozowski-Simon [44] and McNaughton [190] and that of threshold locally testable subsets follows from the work of Straubing [311] and Thérien-Weiss [321].

Theorems 3.6 and 3.11 and Exercise 1 are due to Pécuchet [227, 225]. Theorems 3.14 and Exercise 6 are due to Wilke [355, 357]. The extension of these results to ordered semigroups and positive varieties is presented here for the first time in the framework of infinite words.

VIII

AN EXCURSION INTO LOGIC

1 Introduction

This chapter is devoted to the presentation of the links between finite automata and logic. Büchi was the first to set up a logical language equivalent to finite automata. This logical formalism is rather unusual since it makes use of second-order variables. The reader will find in the notes at the end of this chapter the motivations that drove Büchi in this direction and a description of the genesis of this aspect of the theory of automata. Here we shall simply emphasize two aspects of this question.

It is remarkable that all the logical theories occurring in this framework are decidable theories. Such behaviour is typical of the theory of finite automata in which most of the usual problems are not only decidable but even of low complexity. Exhibiting a decidable logical theory equivalent to finite automata is, in a sense, a confirmation of this general principle.

Furthermore, the logical formalism is the same for finite words and for infinite words. The only switch occurs in the interpretation of formulas, in contrast to the formalism of automata or rational expressions, in which the finite and the infinite case have to be distinguished. In a sense, the material of this chapter constitutes a justification of the unity of the theories of automata on finite or infinite words.

We first present in Section 2 the elements of formal logic necessary to understand the sequel. This concerns essentially very elementary notions which can be skipped by a reader with some background in logic. In particular, we introduce monadic second-order logic, which is the logical language equivalent to automata, as is shown later. We detail the interpretation of formulas on words. This amounts mainly to considering a word as a structure by associating with each index i the letter in position i. The relations that are used are the order relation between the indices and the relation $\mathbf{a}(i)$ expressing that the letter in position i is an a.

In Section 3, we prove the equivalence between finite automata and monadic second-order logic (Büchi's theorem). We prove this result for finite words and for infinite words. One of the consequences of this result is the decidability of the monadic second-order logic on natural integers (also called monadic second order logic of one successor). As it happens, the proof of this latter result was one motivation of the work of Büchi. The extension of Büchi's decidability result to two successors is Rabin's theorem, which will be presented in Chapter X (Theorem X.4.3).

The conversion from automata to logic can be made by writing a formula stating that a word is the label of a successful path in an automaton. This proof brings us a supplementary clue: every formula of monadic second-order logic is equivalent to a formula of low complexity in terms of quantifier alternation. In particular, the hierarchy based on the number of quantifier alternations collapses. We shall see later on that this is a major difference between first and second order.

We also investigate the expressive power of the logical language obtained by replacing the order relation on integers by the successor function. We shall see that, for infinite words, the monadic second-order logics of these two languages are equivalent. This should not be a surprise to the reader familiar with logic, since the order relation is definable in monadic second-order logic of the successor function.

Furthermore, we show that monadic second-order logic is, on the integers, equivalent to *weak* monadic second-order logic. The weak logic utilizes the same formulas as the "strong" one but the interpretation is more restrictive: set variables are always interpreted as finite sets of integers. The equivalence between weak and strong logic is a consequence of McNaughton's theorem. Indeed, one can deduce from this theorem that each rational set $X \subset A^\omega$ is a boolean combination of subsets of the form $\overrightarrow{Y}$ with $Y \subset A^*$. And such a set is definable by the formula

$$\forall x\ \exists y\ (y > x \wedge \varphi(y))$$

where $\varphi(y)$ is the formula expressing the fact that the first y letters of a word are in Y. Since Y is a set of finite words, we can restrict ourselves to weak logic to interpret $\varphi(y)$. This interdependence of results expressible in logic and in terms of automata reflects the deep and real connection between the two approaches. We shall see other examples in the sequel when we shall deal with first-order logic.

In Section 4, we present the corresponding theory for first-order logic. We start with the first-order theory of the linear order since, in contrast to the monadic case, the logic of the linear order and that of the successor are no longer equivalent.

The first-order logic of the linear order is shown to be equivalent to aperiodic automata or, as we have seen in Chapter VI, to star-free sets. This is true for finite words and for infinite words.

In Section 4.2, we describe the hierarchy of the first-order logic of the linear order, corresponding to the alternate use of existential and universal quantifiers. We show that this hierarchy corresponds, on words, to the concatenation hierarchy described in Chapter VI. This leads to a doubly satisfying situation: first-order formulas not only correspond globally to a natural class of recognizable sets (the star-free sets), but this correspondence holds level by level.

In Section 5, we study a more restricted language, the first-order logic of the successor. We give an effective characterization of the sets of infinite words definable in this logic: they are the threshold locally testable sets (defined in Chapter VII) which are in the class Δ_2 (defined in Chapter III). We shall introduce at this point a technique of proof, called a *Fraïssé-Ehrenfeucht* game, and we shall use it to prove the characterization theorem.

In the last section, we introduce the formalism of *temporal logic* and show that this logic is equivalent to first-order logic. Finally, we describe the expressive power of the temporal logic without the until operator.

In Chapters IX and X, we shall see how the logical framework presented in this chapter extends to the more general cases of bi-infinite words and infinite trees.

2 The formalism of logic

In this section, we review the basic definitions of logic that will be used in this book. We shall define successively first-order logic, second-order, monadic second-order and weak monadic second-order logic.

2.1 Syntax. We shall first define the syntax of logical formulas. Let us start by *first-order logic*.

The basic ingredients are the *logical symbols* which encompass the logical connectives: $\wedge$ (and), $\vee$ (or), $\neg$ (not), $\rightarrow$ (implies), the equality symbol $=$, the *quantifiers* $\exists$ (there exists) and $\forall$ (for all), an infinite set of variables (most often denoted by x, y, z, or $x_0, x_1, x_2, \ldots$) and parenthesis (to ensure legibility of the formulas).

In addition to these logical symbols, we make use of a set $\mathcal{L}$ of nonlogical symbols. These auxiliary symbols can be of three types: relation symbols (for instance $<$), function symbols (for instance min), or constant symbols (for instance 0, 1). Expressions are built from the symbols of $\mathcal{L}$ by obeying the usual rules of the syntax, then first-order formulas are built by using the logical symbols, and are denoted by $F_1(\mathcal{L})$. We now give a detailed description of the syntactic rules to obtain the logical formulas in three steps, passing successively, following the standard terminology, from terms to atomic formulas and subsequently to formulas.

We shall illustrate the formation rules by taking as an example

$$\mathcal{L} = \{ < , \min, 0\}$$

in which $<$ is a binary relation symbol, min is a two-variable function symbol and 0 is a constant.

We first define the set of $\mathcal{L}$-*terms*. It is the least set of expressions containing the variables, the constant symbols of $\mathcal{L}$ (if there are some) which is closed under the following formation rule: if $t_1, t_2, \ldots, t_n$ are terms and if f is a function symbol with n arguments, then the expression $f(t_1, t_2, \ldots, t_n)$ is a term. In particular, if $\mathcal{L}$ does not contain any function symbol, the only terms are the variables and the constant symbols. In our example, the following expressions are terms:

$$x_0 \qquad \min(0,0) \qquad \min(x_1, \min(0, x_2)) \qquad \min(\min(x_0, x_1), \min(x_1, x_2))$$

The *atomic formulas* are formulas either of the form

$$(t_1 = t_2)$$

where t_1 and t_2 are terms, or of the form

$$R(t_1, \ldots, t_n)$$

where $t_1, \ldots, t_n$ are terms and where R is a n-ary relation symbol of $\mathcal{L}$. On our example, the following expressions are atomic formulas:

$$(\min(x_1, \min(0, x_2)) = x_1) \qquad (\min(0,0) < 0)$$
$$(\min(\min(x_0, x_1), \min(x_1, x_2)) < \min(x_3, x_1))$$

Notice that, in order to improve legibility, we didn't take literally the definition of atomic formulas: indeed, since $<$ is a symbol of binary relation, one should write $<(t_1, t_2)$ instead of $t_1 < t_2$.

Finally, the *first-order formulas* on $\mathcal{L}$ form the least set of expressions containing the empty formula and the atomic formulas and closed under the following formation rules:

(i) If $(\varphi_i)_{i \in I}$ is a finite family of first-order formulas, so are the expressions

$$\Big(\bigwedge_{i \in I} \varphi_i\Big) \quad \text{and} \quad \Big(\bigvee_{i \in I} \varphi_i\Big)$$

(ii) If φ and ψ are first-order formulas, so are the expressions

$$\neg\varphi \quad \text{and} \quad (\varphi \rightarrow \psi)$$

(iii) If φ is a first-order formula and if x is a variable, then the expressions

$$(\exists x \varphi) \quad \text{and} \quad (\forall x \varphi)$$

are first-order formulas.

To make notations easier, we set

$$\textbf{true} = \bigwedge_{i \in \emptyset} \varphi_i \quad \text{and} \quad \textbf{false} = \bigvee_{i \in \emptyset} \varphi_i$$

The following expressions are first-order formulas of our example language:

$$(\exists x\ (\forall y\ ((y < \min(z, 0)) \wedge (x < 0)))) \qquad\qquad (\forall x\ (y = x))$$

Again, it is convenient to simplify notation by suppressing some of the parenthesis and we shall write the previous formulas under the form

$$\exists x\ \forall y\ (y < \min(x, 0)) \wedge (z < 0) \qquad\qquad \forall x\ y = x$$

In a first-order formula, some variables occur after a quantifier (existential or universal): the occurrences of these variables are said to be *bounded* and the other occurrences are said to be *free*. For example, in the formula

$$\exists x\ (\underline{\underline{y}} < h(\underline{x}, 0)) \wedge \forall y\ (\underline{\underline{z}} < \underline{y})$$

the simply underlined occurrences of x and y are bounded and the occurrences of z and y doubly underlined are free. A variable is free if at least one of its occurrences is free. The set $FV(\varphi)$ of free variables of a formula φ can be defined inductively as follows:

(1) If φ is an atomic formula, $FV(\varphi)$ is the set of variables occurring in φ,
(2) $FV(\neg\varphi) = FV(\varphi)$,
(3) $FV(\bigwedge_{i\in I} \varphi_i) = FV(\bigvee_{i\in I} \varphi_i) = \bigcup_{i\in I} FV(\varphi_i)$,
(4) $FV(\varphi \to \psi) = FV(\varphi) \cup FV(\psi)$,
(5) $FV(\exists x\varphi) = FV(\forall x\varphi) = FV(\varphi) \setminus \{x\}$.

A *statement* is a formula in which all occurrences of variables are bounded. For example, the formula

$$\exists x\ \forall y\ (y < \min(x, 0))$$

is a statement.

We shall denote by $\varphi(x_1, x_2, \ldots, x_n)$ a formula φ in which the set of free variables is contained in $\{x_1, \ldots, x_n\}$ (but is not necessarily equal to $\{x_1, \ldots, x_n\}$).

The variables used in first-order logic, or *first order variables*, are interpreted, as we shall see, as the elements of a set. In *second-order logic*, one makes use of another type of variables, called *second-order variables*, which represent relations. These variables are denoted traditionally by capital letters: X_0, X_1, etc. One builds in this way the set of second-order formulas on $\mathcal{L}$, denoted by $F_2(\mathcal{L})$. The set of terms is the same as for first-order. The *atomic formulas* are either of the form

$$(t_1 = t_2)$$

where t_1 and t_2 are terms, or of the form

$$R(t_1, \dots, t_n) \qquad \text{or} \qquad X(t_1, \dots, t_n)$$

where $t_1, \dots, t_n$ are terms, R is a n-ary relation symbol of $\mathcal{L}$ and X is a variable representing an n-ary relation.

Finally, *second-order formulas* on $\mathcal{L}$ form the least set of expressions containing the atomic formulas and closed under the following formation rules:

(i) If φ and ψ are second-order formulas, then so are

$$\neg\varphi, \quad (\varphi \wedge \psi), \quad (\varphi \vee \psi), \quad (\varphi \to \psi)$$

(ii) If φ is a second-order formula, if x is a variable and if X is a variable of relation, then the expressions

$$(\exists x \varphi) \qquad (\forall x \varphi) \qquad (\exists X \varphi) \qquad (\forall X \varphi)$$

are second-order formulas.

We call *monadic second-order logic* the fragment of second order logic in which the only relation variables are unary relation variables, in other words, variables representing subsets of the domain. By convenience, they are called *set variables*. We denote by $MF_2(\mathcal{L})$ the set of monadic second-order logic on $\mathcal{L}$. We shall also use the notation $x \in X$ instead of $X(x)$.

2.2 Semantics. We have adopted so far a syntactic point of view to define formulas with no reference to semantics. But of course, formulas would be uninteresting if they were meaningless. Giving a precise meaning to formulas requires to specify the domain on which we want to interpret each of the symbols of the language $\mathcal{L}$. Formally, a *structure* $\mathcal{S}$ on $\mathcal{L}$ is given by a nonempty set D, called *domain* and by a map defined on $\mathcal{L}$, called an *assignment* which associates

(1) with each n-ary relation symbol of $\mathcal{L}$, a n-ary relation defined on D,

(2) with each n-ary function symbol f of $\mathcal{L}$, a n-ary function defined on D,

(3) with each constant symbol c of $\mathcal{L}$, an element of D.

To improve notations, we shall use the same notation for the relation (resp. function, constant) symbols and for the relations (resp. functions, constants) represented by these symbols. The context will allow us to determine easily what the notation stands for. For example, we shall always employ the symbol $<$ to designate the usual order relation on a set of integers, independently of the domain ($\mathbb{N}$, $\mathbb{Z}$, or a subset of $\mathbb{N}$).

We still have to interpret variables. Let us start by first-order variables. Given a fixed structure $\mathcal{S}$ on $\mathcal{L}$, with domain D, a *valuation* on $\mathcal{S}$ is a map ν from the set of variables into the set D. It is then easy to extend ν into a function of the set of terms of $\mathcal{L}$ into D, by induction on the formation rules of terms:

(1) if c is a constant symbol, we put $\nu(c) = c$,

(2) if f is a n-ary function symbol and if $t_1, \dots, t_n$ are terms,

$$\nu\big(f(t_1, \dots, t_n)\big) = f(\nu(t_1) \dots \nu(t_n))$$

If ν is a valuation and a an element of D, we denote by $\nu\binom{a}{x}$ the valuation ν' defined by

$$\nu'(y) = \begin{cases} \nu(y) & \text{if } y \neq x \\ a & \text{if } y = x \end{cases}$$

The notion of interpretation can be now easily formalized. Define, for each first-order formula φ and for each valuation ν, the expressions "the valuation ν satisfies φ in $\mathcal{S}$", or "$\mathcal{S}$ satisfies $\varphi[\nu]$", denoted by $\mathcal{S} \models \varphi[\nu]$, as follows:

(1) $\mathcal{S} \models (t_1 = t_2)[\nu]$	if and only if $\nu(t_1) = \nu(t_2)$
(2) $\mathcal{S} \models R(t_1, \dots, t_n)[\nu]$	if and only if $\big(\nu(t_1), \dots, \nu(t_n)\big) \in R$
(3) $\mathcal{S} \models \neg\varphi[\nu]$	if and only if not $\mathcal{S} \models \varphi[\nu]$
(4) $\mathcal{S} \models (\bigwedge_{i \in I} \varphi)[\nu]$	if and only if for each $i \in I,\ \mathcal{S} \models \varphi_i[\nu]$
(5) $\mathcal{S} \models (\bigvee_{i \in I} \varphi)[\nu]$	if and only if there exists $i \in I,\ \mathcal{S} \models \varphi_i[\nu]$
(6) $\mathcal{S} \models (\varphi \to \psi)[\nu]$	if and only if $\mathcal{S} \not\models \varphi[\nu]$ or $\mathcal{S} \models \psi[\nu]$
(7) $\mathcal{S} \models (\exists x \varphi)[\nu]$	if and only if $\mathcal{S} \models \varphi[\nu\binom{a}{x}]$ for some $a \in D$
(8) $\mathcal{S} \models (\forall x \varphi)[\nu]$	if and only if $\mathcal{S} \models \varphi[\nu\binom{a}{x}]$ for each $a \in D$

Note that, actually, the truth of the expression "the valuation ν satisfies φ in $\mathcal{S}$" only depends on the values taken by the free variables of φ. In particular, if φ is a statement, the choice of the valuation is irrelevant. Therefore, if φ is a statement, one says that φ is satisfied by $\mathcal{S}$ (or that $\mathcal{S}$ satisfies φ), and denote by $\mathcal{S} \models \varphi$, if, for each valuation ν, $\mathcal{S} \models \varphi[\nu]$.

Next we move to the interpretation of second-order formulas. Given a structure $\mathcal{S}$ on $\mathcal{L}$, with domain D, a *second-order valuation* on $\mathcal{S}$ is a map ν which associates with each first order variable an element of D and to each n-ary relation variable a subset of D^n (i.e. a n-ary relation on D).

If ν is a valuation and R a subset of D^n, $\nu\binom{R}{X}$ denotes the valuation ν' defined by

$$\nu'(x) = \nu(x) \text{ if } x \text{ is a first-order variable}$$

$$\nu'(Y) = \begin{cases} \nu(Y) & \text{if } Y \neq X \\ R & \text{if } Y = X \end{cases}$$

The notion of interpretation, already defined for first order, is supplemented by the following rules:

(9) $\mathcal{S} \models (X(t_1, \ldots, t_n))[\nu]$ if and only if $\big(\nu(t_1), \ldots, \nu(t_n)\big) \in \nu(X)$

(10) $\mathcal{S} \models (\exists X\varphi)[\nu]$ if and only if there exists $R \subseteq D^n$, $\mathcal{S} \models \varphi[\nu\binom{R}{X}]$

(11) $\mathcal{S} \models (\forall X\varphi)[\nu]$ if and only if for each $R \subseteq D^n$, $\mathcal{S} \models \varphi[\nu\binom{R}{X}]$

Weak monadic second-order logic is composed with the same formulas than monadic second-order logic, but the interpretation is even more restricted: are only considered valuations which associate with set variables *finite* subsets of the domain D.

Two formulas φ and ψ are said to be *logically equivalent* if, for each structure $\mathcal{S}$ on $\mathcal{L}$, we have $\mathcal{S} \models \varphi$ if and only if $\mathcal{S} \models \psi$.

It is easy to see that the following formulas are logically equivalent:

(1)	$\varphi \wedge \psi$	and	$\neg(\neg\varphi \vee \neg\psi)$
(2)	$\varphi \to \psi$	and	$\neg\varphi \vee \psi$
(3)	$\forall x\varphi$	and	$\neg(\exists x\ \neg\varphi)$
(4)	$\varphi \vee \psi$	and	$\psi \vee \varphi$
(5)	$\varphi \wedge \psi$	and	$\psi \wedge \varphi$
(6)	$\varphi \wedge false$	and	$false$
(7)	$\varphi \vee false$	and	φ

Consequently, up to logical equivalence, we may assume that the formulas are built without the symbols $\wedge$, $\to$ and $\forall$.

Logical equivalence also permits to give a more structured form to formulas. A formula is said to be under *disjunctive normal form* if it can be written as disjunction of conjunctions of atomic formulas or of negations of atomic formulas, in other words under the form

$$\bigvee_{i \in I} \bigwedge_{j \in J_i} (\varphi_{ij} \vee \neg\psi_{ij})$$

where I and the J_i are finite sets, and where the φ_{ij} and the ψ_{ij} are atomic formulas. The next result is standard and easily proved.

Proposition 2.1. *Every quantifier-free formula is logically equivalent to a quantifier free formula in disjunctive normal form.*

One can set up a hierarchy inside first-order formulas as follows. Let $\Sigma_0 = \Pi_0$ be the set of quantifier-free formulas. Next, for each $n \geq 0$, denote by Σ_{n+1} the least set Δ of formulas such that

(1) Δ contains the boolean combinations of formulas of Σ_n,

(2) Δ is closed under disjunctions and finite conjunctions,
(3) if $\varphi \in \Delta$ and if x is a variable, $\exists x\varphi \in \Delta$.

Similarly, denote by Π_{n+1} the least set Γ of formulas such that

(1) Γ contains the boolean combinations of formulas of Π_n,
(2) Γ is closed under disjunctions and finite conjunctions,
(3) if $\varphi \in \Gamma$ and if x is a variable, $\forall x\varphi \in \Gamma$.

In particular, Σ_1 is the set of *existential* formulas — that is of the form

$$\exists x_1\ \exists x_2\ \ldots\ \exists x_k\ \varphi$$

where φ is quantifier-free.

Finally, we denote by $\mathcal{B}\Sigma_n$ the set of boolean combinations of formulas of Σ_n.

A first-order formula is said to be in *prenex normal form* if it is of the form

$$\psi = Q_1x_1\ Q_2x_2\ \ldots\ Q_nx_n\ \varphi$$

where the Q_i are existential or universal quantifiers ($\exists$ or $\forall$) and φ is quantifier-free. The sequence $Q_1x_1\ Q_2x_2\ \ldots\ Q_nx_n$, which can be considered as a word on the alphabet

$$\{\ \exists x_1, \exists x_2, \ldots, \forall x_1, \forall x_2, \ldots\ \},$$

is called the quantification prefix of ψ. The interest of these formulas in prenex normal form comes from the following result.

Proposition 2.2. *Every first-order formula is logically equivalent to a formula in prenex normal form.*

Proof. It suffices to verify that if the variable x does not occur in the formula ψ then

$$\exists x(\varphi \wedge \psi) \equiv (\exists x\varphi) \wedge \psi$$
$$\exists x(\varphi \vee \psi) \equiv (\exists x\varphi) \vee \psi$$

and the same formulas hold for the quantifier $\forall$. Hence it is possible, by renaming the variables, to throw back the quantifiers to the outside. □

Proposition 2.2 can be improved to take into account the level of the formula in the Σ_n hierarchy.

Proposition 2.3. *For each integer $n \geq 0$,*

(1) *Every formula of Σ_n is logically equivalent to a formula in prenex normal form in which the quantification prefix is a sequence of n (possibly empty) alternating blocks of existential and universal quantifiers, starting with a block of existential quantifiers.*

(2) *Every formula of Π_n is logically equivalent to a formula in prenex normal form in which the quantification prefix is a sequence of n (possibly empty) alternating blocks of existential and universal quantifiers, starting with a block of universal quantifiers.*

For example, the formula

$$\underbrace{\exists x_1 \, \exists x_2 \, \exists x_3}_{\text{block 1}} \, \underbrace{\forall x_4 \, \forall x_5}_{\text{block 2}} \, \underbrace{\exists x_6 \, \exists x_7}_{\text{block 3}} \, \varphi(x_1, \ldots, x_6)$$

belongs to Σ_3 (and also to all Σ_n's such that $n \geq 3$). Similarly the formula

$$\underbrace{}_{\text{block 1}} \, \underbrace{\forall x_4 \, \forall x_5}_{\text{block 2}} \, \underbrace{\exists x_6 \, \exists x_7}_{\text{block 3}} \, \varphi(x_1, \ldots, x_7)$$

belongs to Σ_3 and to Π_2, but not to Σ_2, since the counting of blocks of a Σ_n-formula should always begin by a possibly empty block of existential quantifiers.

One can also introduce normal forms and a hierarchy for second-order monadic formulas. Thus, one can show that every monadic second-order formula is logically equivalent to a formula of the form

$$\psi = Q_1 X_1 \, Q_2 X_2 \, \ldots \, Q_n X_n \, \varphi$$

where the Q_i are existential or universal quantifiers and φ is a first-order formula.

2.3 Logic on words. The logical language that we shall use now was introduced by Büchi under the name of "sequential calculus". This permits us to formalize certain properties of words. To interpret formulas on words, we shall consider a word $u = u(0)u(1)\ldots$ as a map associating a letter with each index. The set of indices will itself be considered as a subset of the set

$$\mathcal{N} = \mathbb{N} \cup \{\infty\}$$

the set of natural numbers with a new maximal element, denoted by ∞.

Let A be an alphabet. For an infinite word $u \in A^\omega$, we set $|u| = \infty$, so that the length of a word is always an element of the set $\mathcal{N}$. For each word $u \in A^\infty$, one defines the *domain* of u, denoted by $\mathrm{Dom}(u)$ as

$$\mathrm{Dom}(u) = \{i \in \mathcal{N} \mid 0 \leq i \leq |u|\}$$

So, if u is a finite word, $\mathrm{Dom}(u) = \{0, \ldots, |u|\}$ and if u is an infinite word, $\mathrm{Dom}(u) = \mathcal{N}$. Define for each letter $a \in A$ a unary relation $\mathbf{a}$ on the domain of u by

$$\mathbf{a} = \{i < |u| \mid u(i) = a\}.$$

Finally, let us associate with each word u the structure

$$\mathcal{M}_u = \big(\mathrm{Dom}(u), (\mathbf{a})_{a \in A}\big),$$

Beware that, with these definitions, the domain of a finite or infinite word comprises a supplementary position at the end. One can imagine that this position is occupied by an end symbol (the $ prized by lexical analyzers...).

For example, if $u = abbaab$, then $\mathrm{Dom}(u) = \{0, 1, \ldots, 6\}$, $\mathbf{a} = \{0, 3, 4\}$ and $\mathbf{b} = \{1, 2, 5\}$. If $u = (aba)^\omega$, then

$$\mathbf{a} = \{n \in \mathbb{N} \mid n \equiv 0 \mod 3 \text{ or } n \equiv 2 \mod 3\} \text{ and}$$
$$\mathbf{b} = \{n \in \mathbb{N} \mid n \equiv 1 \mod 3\}$$

From now on, we shall interpret logical formulas on words, that is, on a structure of the form $\mathcal{M}_u$ as explained above. Let φ be a statement. A finite or infinite word $u \in A^\infty$ *satisfies* φ if the structure $\mathcal{M}_u$ satisfies φ. This is denoted by $u \models \varphi$. We also say that u is a model of φ. Define the *spectrum* of φ as the set

$$S(\varphi) = \{u \in A^\infty \mid u \text{ satisfies } \varphi\}$$

We also set $S^*(\varphi) = S(\varphi) \cap A^*$, $S^+(\varphi) = S(\varphi) \cap A^+$ and $S^\omega(\varphi) = S(\varphi) \cap A^\omega$.

From now on, all the variables will be interpreted as natural integers or ∞. Therefore, we shall use logical equivalence restricted to interpretations of domain $\mathcal{N}$.

The various logical languages that we shall consider all contain, for each $a \in A$, a unary relation symbol denoted $\mathbf{a}$ when no confusion arises. We shall also use two other nonlogical symbols, $<$ and S, that will be interpreted as the usual order and as the successor relation on $\mathrm{Dom}(u)$: $S(x, y)$ if and only if $y = x + 1$. In the sequel, we shall mainly consider two logical languages: the language

$$\mathcal{L}_< = \{<\} \cup \{\mathbf{a} \mid a \in A\}$$

will be called the *language of the linear order* and the language

$$\mathcal{L}_S = \{S\} \cup \{\mathbf{a} \mid a \in A\}$$

will be called the *language of the successor.* The atomic formulas of the language of the linear order are of the form

$$\mathbf{a}(x), \qquad x = y, \qquad x < y$$

and those of the language of the successor are of the form

$$\mathbf{a}(x), \qquad x = y, \qquad S(x, y).$$

We shall denote respectively by $F_1(<)$ and $MF_2(<)$ the set of first-order and monadic second-order formulas of signature $\{<, (\mathbf{a})_{a \in A}\}$. Similarly, we denote by $F_1(S)$ and $MF_2(S)$ the same sets of formulas of signature $\{S, (\mathbf{a})_{a \in A}\}$. Inside $F_1(<)$ (resp. $F_1(S)$), the $\Sigma_n(<)$ (resp. $\Sigma_n(S)$) hierarchy is based on the number of quantifier alternations.

We shall now start the comparison between the various logical languages we have introduced. First of all, the distinction between the signatures S and $<$ is only relevant for first order in view of the following proposition. A relation $R(x_1, \dots, x_n)$ on the integers is said to be *defined* by a formula $\varphi(x_1, \dots, x_n)$ if, for each $i_1, \dots, i_n \geq 0$, $R(i_1, \dots, i_n)$ if and only if $\varphi(i_1, \dots, i_n)$ is true.

Proposition 2.4. *The successor relation can be defined in $F_1(<)$, and the order relation on integers can be defined in $MF_2(S)$.*

Proof. The successor relation can be defined by the formula

$$(i < j) \wedge \forall k \Big((i < k) \rightarrow ((j = k) \vee (j < k))\Big)$$

which states that $j = i + 1$ if i is smaller than j and there exist no element between i and j. Conversely, $i < j$ can be expressed in $MF_2(S)$ as follows:

$$\exists X \left[\forall x \forall y \Big(\big((x \in X) \wedge S(x, y)\big) \rightarrow (y \in X)\Big)\right] \wedge (j \in X) \wedge (i \notin X)$$

which intuitively means that there exists an interval of the form $[k, +\infty[$ containing j but not i. □

On the contrary, we shall see later on (Corollary 5.10) that the relation $<$ cannot be defined in $F_1(S)$.

Given two sets of formulas F and G, we use the notation $F \leq G$ if for each formula of F, there exists an equivalent formula in G (we remind the reader that the variables are always interpreted as integers). Denote by $F \equiv G$ the associated equivalence and by $F < G$ the relation $F \leq G$ but $F \not\equiv G$. The next result summarizes the expressive power of the various logical languages.

Proposition 2.5. *The following relations hold:*

$$F_1(S) < F_1(<) < MF_2(S) \equiv MF_2(<)$$

Proof. Proposition 2.4 shows the double inequality $F_1(S) \leq F_1(<) \leq MF_2(S)$. The fact that these inequalities are strict follow respectively from Corollary 5.10 and Corollary 4.4, which will be proved later on.

Finally, we have $MF_2(S) \leq MF_2(<)$ since S can be expressed in $F_1(<)$, which in turn is contained in $MF_2(<)$. But we also have $MF_2(<) \leq MF_2(S)$ since, by Proposition 2.4, the relation $<$ can be defined in $MF_2(S)$. □

It will be convenient to enrich our logical languages by adding auxiliary formulas which can be considered as abbreviations. We shall see an example a few lines below with the predicate Finite(X) which expresses the fact that a set (of integers) is finite.

If x is a variable, it is convenient to write $x + 1$ (resp. $x - 1$) to replace a variable y submitted to the condition $S(x, y)$ (resp. $S(y, x)$). However, the reader should be aware of the fact that $x + y$ is not definable in $MF_2(<)$ (Exercise 3).

We can also use the symbols $\leq$ and $\neq$ with their usual interpretations: $x \leq y$ stands for $(x < y) \vee (x = y)$ and $x \neq y$ for $\neg(x = y)$.

Another example, to be used below, are the symbols min, max, which designate the first and the last element of the domain. These symbols can be defined in $F_1(S)$ with two alternations of quantifiers:

$$\begin{aligned} \min : &\quad \exists \min \ \forall x \ \neg S(x, \min) \\ \max : &\quad \exists \max \ \forall x \ \neg S(\max, x) \end{aligned}$$

These symbols permit to express that u is the empty word ($\max = \min$) or that u is finite ($(\max = \min) \vee \exists\, x S(x, \max)$). We shall denote by FINITE the latter formula.

We shall sometimes need a parametrized, or relative, notion of satisfaction for a formula. Let indeed $\varphi(x, y)$ be a formula with only two free variables x and y. Let $u \in A^\infty$ be a word and $i, j \in \mathrm{Dom}(u)$. A word $u \in A^\infty$ is said to satisfy the formula φ between i and j if $u_i \dots u_{j-1} \models \varphi$ (or $u_i u_{i+1} \dots \models \varphi$ if $j = \infty$).

Proposition 2.6. *For each statement φ, there exists a formula $\varphi(x, y)$ with the same signature, the same order (and, in the case of a first-order formula, the same level in the hierarchy Σ_n), having x and y as unique free variables and which satisfies the following property: for each finite or infinite word u and for each $s, t \in \mathrm{Dom}(u)$, $u \models \varphi(s, t)$ if and only if u satisfies φ between s and t.*

Proof. The formulas $\varphi(x, y)$ are built by induction on the formation rules as follows: if φ is an atomic formula, we set $\varphi(x, y) = \varphi$. Otherwise, we set

$$\begin{aligned} (\neg\varphi)(x, y) &= \neg\varphi(x, y) \\ (\varphi \vee \psi)(x, y) &= \varphi(x, y) \vee \psi(x, y) \\ (\exists z\varphi)(x, y) &= \exists z \ ((x \leq z) \wedge (z < y) \wedge \varphi(x, y)) \\ (\exists X\varphi)(x, y) &= \exists X \ ((\forall\, z(X(z) \rightarrow (x \leq z) \wedge (z < y)) \wedge \varphi(x, y)) \end{aligned}$$

It is easy to verify that the formulas $\varphi(x, y)$ built in this way have the required properties. □

We conclude this section by examining the place of the weak theories of the linear order and the successor. First of all, one can verify that the formulas used in the proof of Proposition 2.4 actually define the relation $<$ in $WMF_2(S)$ (Exercise 1). It follows that

$$WMF_2(S) \equiv WMF_2(<)$$

and that the weak monadic theories of the successor and of the linear order are the same.

We shall see on the other hand that

$$WMF_2(<) \equiv MF_2(<)$$

so that the weak and strong theories of the linear order coincide. We can also note immediately the next result, which is true in a much more general setting (actually as soon as one can write a formula expressing that a set is finite).

Proposition 2.7. *The formula* $WMF_2(<) \leq MF_2(<)$ *holds.*

Proof. One can write in $MF_2(<)$ a formula Finite(X) expressing that a set X is finite. For instance

$$\text{Finite}(X) = \exists x\ (\forall y\ X(y) \rightarrow y < x)$$

Every formula of $WMF_2(<)$ is equivalent to a formula of the form

$$\psi = Q_1 X_1 \dots Q_n X_n \varphi$$

where φ is a first-order formula. The formula ψ is then equivalent to the formula

$$Q_1 X_1 \dots Q_n X_n \varphi \wedge \bigwedge_{1 \leq i \leq n} \text{Finite}(X_i)$$

of $MF_2(<)$. □

The formulas in which the predicates **a** do not occur pertain uniquely to integers. We shall see in the next section that the words have nevertheless the same role to play, as soon as formulas with free variables have to be interpreted.

3 Monadic second-order logic on words

This section is devoted to the proof of a result of Büchi stating that the subsets of A^∞ definable in monadic second-order logic are exactly the rational (resp. ω-rational) sets.

Theorem 3.1. *Let* $X \subset A^\infty$ *be a set of finite or infinite words. The following conditions are equivalent:*

(1) *X is definable by a formula of $MF_2(<)$,*

(2) *X is recognizable.*

The proof of this result can be decomposed into two parts: passing from words to formulas, and from formulas to words.

To pass from words to formulas, we simulate the behaviour of an automaton by a formula.

Proposition 3.2. *For each Büchi automaton $\mathcal{A} = (Q, A, E, I, F)$, there exist formulas $\varphi_+, \varphi_\omega$ of $MF_2(<)$ such that $S^+(\varphi_+) = L^+(\mathcal{A})$, $S^\omega(\varphi_\omega) = L^\omega(\mathcal{A})$ and $L^\infty(\mathcal{A}) = S(\varphi_+ \vee \varphi_\omega)$.*

Proof. Suppose that $Q = \{1, \dots, n\}$. We first write a formula ψ expressing the existence of a path of label u. To this purpose, we associate with each state q a set variable X_q which encodes the set of positions in which a given path visits the state q. The formula states that the X_q's are pairwise disjoint and that if a path is in state q in position x, in state q' in position $x+1$ and if the x-th letter is an a, then $(q, a, q') \in E$. This gives

$$\psi = \Big(\bigwedge_{q \neq q'} \neg \exists x (X_q(x) \wedge X_{q'}(x)) \Big) \wedge$$
$$\Big(\forall x \forall y\ S(x, y) \rightarrow \bigvee_{(q,a,q') \in E} \big(X_q(x) \wedge \mathbf{a}(x) \wedge X_{q'}(y) \big) \Big)$$

There remains to state that the path is successful. For a finite path, it suffices to know that 0 belongs to one of the X_q's such that $q \in I$ and that max belongs to one of the X_q's such that $q \in F$. Therefore, we set

$$\psi_+ = \psi \wedge \text{FINITE} \wedge \Big(\bigvee_{q \in I} X_q(0) \Big) \wedge \Big(\bigvee_{q \in F} X_q(\max) \Big)$$

An infinite path is successful if 0 belongs to one of the X_q's such that $q \in I$ and if, for each position x, there exists a position $y > x$ in which the path visits a final state. In summary, we set:

$$\psi_\omega = \psi \wedge \Big(\bigvee_{q \in I} X_q(0) \Big) \wedge \Big(\bigvee_{q \in F} \forall x\ \exists y\ (x < y\ \wedge\ X_q(y)) \Big)$$

The formulas

$$\varphi_+ = \exists X_1 \exists X_2 \dots \exists X_n (\psi_+)$$
$$\varphi_\omega = \exists X_1 \exists X_2 \dots \exists X_n (\psi_\omega)$$

now entirely encode the automaton. □

To pass from statements to sets of words, a natural idea is to argue by induction on the formation rules of formulas. The problem is that the set $S(\varphi)$ is only defined when φ is a statement. The traditional solution in this case consists of adding constants to interpret free variables to the structure in which the formulas are interpreted. For the sake of homogeneity, we proceed in a slightly different way, so that these structures remain words.

u_0	a	b	a	a	b	a	b	$\cdots$
u_1	0	1	0	0	0	0	0	$\cdots$
u_2	0	0	0	0	1	0	0	$\cdots$
u_3	1	0	0	0	0	0	0	$\cdots$
u_4	0	1	1	0	0	1	1	$\cdots$
u_5	1	1	0	1	0	1	0	$\cdots$

Figure 3.1. A word of $B^{\omega}_{3,2}$.

The idea is to use an extended alphabet of the form

$$B_{p,q} = A \times \{0,1\}^p \times \{0,1\}^q$$

such that p (resp. q) is greater than or equal to the number of first-order (resp. second-order) variables of φ. An infinite word on the alphabet $B_{p,q}$ can be identified with the sequence

$$(u_0, u_1, \ldots, u_p, u_{p+1}, \ldots, u_{p+q})$$

where $u_0 \in A^\omega$ and $u_1, \ldots, u_p, u_{p+1}, \ldots, u_{p+q} \in \{0,1\}^\omega$. We are actually interested in the set $K_{p,q}$ of words of $B^{\infty}_{p,q}$ in which each of the components $u_1, \ldots, u_p$ contain exactly one occurrence of 1. If the context permits, we shall note B instead of $B_{p,q}$ and K instead of $K_{p,q}$. We also set

$$K_* = K \cap B^*, \qquad K_\omega = K \cap B^\omega$$

For example, if $A = \{a, b\}$, a word of $B^{\omega}_{3,2}$ is represented in Figure 3.1. The elements of K are called *marked words* on A. This terminology expresses the fact that the elements of K are (finite or infinite) words in which labels marking certain positions have been added. Each of the p first rows only marks one position and the last q ones an arbitrary number of positions.

Recall that the *star-free* subsets of B^∞ are obtained from the sets B^∞ and $\emptyset$ by using the boolean operations and the *marked product*

$$(X, b, Y) \to XbY$$

which associates with $X \subset A^*$, $b \in B$ and $Y \subset B^\infty$, the subset XbY of B^∞. Star-free sets are in particular rational subsets of B^∞. Then we have the following property.

Proposition 3.3. *For each $p, q \geq 0$, the set $K_{p,q}$ is a star-free, and hence rational, subset of $B^{\infty}_{p,q}$.*

Proof. Set, for $1 \leq i \leq p$,

$$C_i = \{\, (b_0, b_1, \ldots, b_{p+q}) \in B \mid b_i = 1 \,\}$$

Then K is the set of words of B^∞ containing exactly one letter of each C_i, for $1 \leq i \leq p$. Now the formula

$$K = \bigcap_{1 \leq i \leq p} B^* C_i B^\infty \setminus \bigcup_{1 \leq i \leq p} B^* C_i B^* C_i B^\infty$$

shows that K is a star-free subset of B^∞. □

The interpretation of formulas on the words of $B^\infty_{p,q}$ follows the main lines of the interpretation described in Section 2.3, but the interpretation of **a** is slightly modified by setting

$$\mathbf{a} = \{i < |u| \mid u_0(i) = a\}.$$

Let $\varphi(x_1, \ldots, x_r, X_1, \ldots, X_s)$ be a formula in which the first-order (resp. second-order) free variables are $x_1, \ldots, x_r$ (resp. $X_1, \ldots, X_s$), with $r \leq p$ and $s \leq q$. Let $u = (u_0, u_1, \ldots, u_{p+q})$ be a word of $K_{p,q}$ and, for $1 \leq i \leq p$, denote by n_i the position of the unique 1 of the word u_i. In the example above, one would have $n_1 = 1$, $n_2 = 4$ and $n_3 = 0$. A word u is said to satisfy φ if u_0 satisfies $\varphi[\nu]$, where ν is the valuation defined by

$$\nu(x_j) = n_j \qquad (1 \leq j \leq r)$$
$$\nu(X_j) = \{\, i \in \mathbb{N} \mid u_{p+j,i} = 1 \,\} \qquad (1 \leq j \leq s)$$

In other words, each X_j is interpreted as the set of positions of 1's in u_{p+j}, and each x_j as the unique position of 1 in u_j. Note that for $p = q = 0$, we recover the customary interpretation of statements.

Set

$$S_{p,q}(\varphi) = \{\, u \in K_{p,q} \mid u \text{ satisfies } \varphi(x_1, \ldots, x_p, X_1, \ldots, X_q) \,\}$$

Again, we shall sometimes simply use the notation $S(\varphi)$ instead of $S_{p,q}(\varphi)$. We also note that $S^*(\varphi) = S(\varphi) \cap A^*$ and $S^\omega(\varphi) = S(\varphi) \cap A^\omega$. Conjunctions and disjunctions are easily converted into boolean operations.

Proposition 3.4. *For each finite family of formulas $(\varphi_i)_{i \in I}$, the following equalities hold:*

(1) $S(\bigvee_{i \in I} \varphi_i) = \bigcup_{i \in I} S(\varphi_i)$,
(2) $S(\bigwedge_{i \in I} \varphi_i) = \bigcap_{i \in I} S(\varphi_i)$,
(3) $S(\neg\varphi) = K_{p,q} \setminus S(\varphi)$.

Proof. This is an immediate consequence of the definitions. □

To conclude the proof of Theorem 3.1, it remains to prove by induction on the formation rules of formulas that the sets $S(\varphi)$ are rational. Let us start with the atomic formulas. As for the set K, we prove a slightly more precise result that will be used later on in this chapter.

Proposition 3.5. *For each variable x, y, for each set variable X and for each letter $a \in A$, the sets of the form $S(\mathbf{a}(x))$, $S(x = y)$, $S(x < y)$ and $S(X(x))$ are star-free, and hence rational, subsets of B^∞.*

Proof. Set, for $i, j \in \{1, \dots, p+q\}$

$$\begin{aligned} C_{j,a} &= \{b \in B_{p,q} \mid b_j = 1 \text{ and } b_0 = a\} \\ C_{i,j} &= \{b \in B_{p,q} \mid b_i = b_j = 1\} \\ C_i &= \{b \in B_{p,q} \mid b_i = 1\} \end{aligned}$$

Then we have, by setting $B = B_{p,q}$

$$\begin{aligned} S(\mathbf{a}(x_i)) &= K \cap B^* C_{i,a} B^\infty \\ S(x_i = x_j) &= K \cap B^* C_{i,j} B^\infty \\ S(x_i < x_j) &= K \cap B^* C_i B^* C_j B^\infty \\ S(X_i(x_j)) &= K \cap B^* C_{i+p,j} B^\infty \end{aligned}$$

which establishes the proposition. □

Proposition 3.4 allows one to treat logical connectives. There remains to treat the case of the formulas of the form $\exists x\varphi$ and $\exists X\varphi$. Denote by π_i the function that erases the i-th component, defined by

$$\pi_i(b_0, b_1, \dots, b_{p+q}) = (b_0, b_1, \dots, b_{i-1}, b_{i+1}, \dots, b_{p+q})$$

Thus π_i should be considered as a function from $B_{p,q}$ into $B_{p-1,q}$ if $i \le p$ and into $B_{p,q-1}$ if $p < i \le p+q$.

Proposition 3.6. *For each formula φ, the following formulas hold*

(1) $S_{p-1,q}(\exists x_p\varphi) = \pi_p(S_{p,q}(\varphi))$,
(2) $S_{p,q-1}(\exists X_q\varphi) = \pi_{p+q}(S_{p,q}(\varphi))$.

Proof. This follows from the definition of existential quantifiers. □

We are now ready to show that if $S(\varphi)$ is rational, then so are $S(\exists x\varphi)$ and $S(\exists X\varphi)$. We may assume that $x = x_p$ and $X = X_q$. Then, by Proposition 3.6, we have $S(\exists x_p\varphi) = \pi_p(S(\varphi))$ and $S(\exists X_q\varphi) = \pi_{p+q}(S(\varphi))$. Since morphisms preserve rationality, the result follows.

This concludes the proof of Büchi's theorem. The reader will find in Exercise 4 the sketch of another proof to pass from words to formulas based on a translation of rational expressions.

We now show, as was announced in Section 2.3 that strong and weak logic coincide on words.

Theorem 3.7. *The theories $MF_2(<)$ and $WMF_2(<)$ are equivalent on words.*

Proof. Proposition 2.7 shows that $WMF_2(<) \leq MF_2(<)$. To establish the opposite direction, it suffices, by Büchi's theorem, to show that each rational subset of A^∞ can be defined by a formula of $WMF_2(<)$. There is no problem for a set of finite words, since set variables are necessarily interpreted as finite sets. Therefore it suffices to consider the case of a recognizable set X of infinite words.

By McNaughton's theorem (Theorem I.9.1), X is a boolean combination of sets of the form $\overrightarrow{Y}$. We claim that such a set is definable in weak monadic second-order logic. Indeed, let $\varphi(x, y)$ be a formula of $WMF_2(<)$ defining Y (which consists of finite words). Then we have $u \models \varphi(0, n)$ if and only if $u_0 u_1 \dots u_{n-1} \in Y$ by Proposition 2.6 and the formula

$$\psi = \forall x \big(\exists y (x < y) \wedge \varphi(0, y)\big)$$

guarantees that u has infinitely many prefixes in Y. Thus $S(\psi) = \overrightarrow{Y}$. □

Another proof of this result is given in Exercise 5. It consists in coding directly a Muller automaton by a weak monadic second-order formula. We conclude this section by making two observations of logical nature.

First of all, Büchi's theorem allows one to show that the monadic second-order theory of nonnegative integers is decidable. This contrasts with the undecidability of more expressive logical theories of integers (see the Notes Section).

Theorem 3.8. *There is an algorithm to decide whether a given statement of $MF_2(<)$ is true on $\mathbb{N}$.*

Proof. Let $\varphi \in MF_2(<)$. One can effectively calculate the set $S(\varphi)$ and decide whether $S(\varphi) = A^\omega$ or $S(\varphi) = \emptyset$. □

Next, the proof of Proposition 3.2 shows that the hierarchy on formulas of $MF_2(<)$ based on the number of alternations of second-order quantifiers collapses to the first level. We have indeed the following result.

Proposition 3.9. *Every formula of $MF_2(<)$ is equivalent to a formula of the form*

$$\exists X_1 \dots \exists X_k \varphi(X_1, \dots, X_k)$$

where φ is a first-order formula.

In fact, one can even show that every formula of $MF_2(<)$ is equivalent to a formula of the form $\exists X \varphi(X)$ where φ is a first-order formula (Exercise 2).

4 First-order logic of the linear order

We now study the language $F_1(<)$ of the first-order logic of the linear order. We shall, in a first step, characterize the sets of words definable in this logic, which happen to be the star-free sets. Next, we shall see in Section 4.2 how this result can be refined to establish a correspondence between the levels the Σ_n-hierarchy of first-order logic and the concatenation hierarchy of star-free sets.

4.1 First order and star-free sets. We shall prove the following key result.

Theorem 4.1. *A set of finite or infinite words is star-free if and only if it is definable by a formula of* $F_1(<)$.

The first part of the proof consists in converting the sets of words into formulas. To start with, the equalities

$$S(\mathbf{true}) = A^\infty \qquad S(\mathbf{false}) = \emptyset$$

show that the basic star-free sets are definable by a formula of $F_1(<)$. The boolean operations are easily converted into connectives as in Proposition 3.4.

We now treat the marked product. We shall use Proposition 2.6 to replace each formula $\varphi \in F_1(<)$ by a formula $\varphi(x, y) \in F_1(<)$ such that for each word $u \in A^*$ and each $s, t \in \mathrm{Dom}(u)$, we have $u \models \varphi(s, t)$ if and only if $u_s u_{s+1} \dots u_{t-1} \models \varphi$. The next result, whose proof is immediate, shows that if $X_1 \subset A^*$ and $X_2 \subset A^\infty$ are definable in $F_1(<)$, then so is the marked product $X_1 a X_2$ for each $a \in A$.

Proposition 4.2. *Let* $X_1 \subset A^*$ *and* $X_2 \subset A^\infty$. *If* $X_1 = S(\varphi_1)$ *and* $X_2 = S(\varphi_2)$, *then* $X_1 a X_2 = S(\varphi)$ *with* $\varphi = \exists y\ \varphi_1(0, y) \wedge \mathbf{a}(y) \wedge \varphi_2(y + 1, \max)$.

For the opposite implication, from formulas to words, we show by induction on the formation rules of formulas that $S(\varphi)$ is star-free for every formula $\varphi \in F_1(<)$. In order to treat formulas with free variables, we shall work with the extended alphabets $B_{p,q}$ introduced in Section 3. But since there are no set variables any more, we may assume that $q = 0$, which permits to eliminate the q indices in the notation. Therefore, we simply set $B_p = A \times \{0, 1\}^p$ and we denote by K_p (or simply K) the set of the words u of B_p^∞ in which the components $u_1, \dots, u_p$ contain exactly one occurrence of 1. Note that $K_0 = A^*$, but that K_p does not contain the empty word for $p > 0$. Denote by $K' = K \cap B^\omega$ the set of infinite words of K.

We also set, for each formula $\varphi = \varphi(x_1, \dots, x_p)$,

$$S_p(\varphi) = \{u \in K_p \mid u \models \varphi\}$$

By Proposition 3.5, the sets $S(\mathbf{a}(x))$, $S(x = y)$ and $S(x < y)$ are star-free. By Proposition 3.3, the set K is also star-free. The boolean operations are translated in the usual way by connectives. There remains to treat the case of quantifiers. We shall use the following result.

Proposition 4.3. *Let B be an alphabet and let $B = C \cup D$ be a partition of B. Let $X \subset B^\infty$ be a star-free subset of B^∞ such that every word of X has exactly one letter in C. Then X can be written under the form*

$$X = \bigcup_{1 \le i \le n} Y_i b_i Z_i$$

with $n \ge 0$, $b_i \in C$, and where the $Y_i \subset D^*$ and $Z_i \subset D^\infty$ are star-free sets.

Proof. Let $\mu : B^\infty \to S$ be the syntactic morphism of X. Since X is star-free, Theorem VI.6.2 shows that S is aperiodic. Then

$$X = \bigcup \mu^{-1}(s) b \mu^{-1}(u)$$

where the union runs over all the triples (s, b, u) such that $s \in S_+$, $u \in S$, $b \in C$ and $s\mu(b)u \in \mu(X)$. Since S is an aperiodic ω-semigroup, the sets $\mu^{-1}(s)$ and $\mu^{-1}(u)$ are star-free by Theorem VI.6.2. □

We now treat the case of the existential quantifiers. Suppose that $X = S(\varphi)$ is star-free. Denote by $\pi : B_p \to B_{p-1}$ the projection which erases the component corresponding to x_p (we may assume $x = x_p$). Let us apply Proposition 4.3 with $X = S(\varphi)$, $B = B_p$, $C = \{b \in B_p \mid b_p = 1\}$ and $D = \{b \in B_p \mid b_p = 0\}$. The fact that each word of X contains exactly one occurrence in C follows from the inclusion $X \subset K$. Therefore

$$\pi(X) = \bigcup_{1 \le i \le n} \pi(Y_i)\pi(b_i)\pi(Z_i)$$

Since the restriction of π to D^∞ is an isomorphism, the sets $\pi(Y_i)$ and $\pi(Z_i)$ are all star-free. Therefore $S(\exists x \varphi) = \pi(X)$ is a star-free subset. This concludes the proof of Theorem 4.1. □

The proof of Theorem 4.1 given above makes use of the syntactic characterization of star-free sets. One can also show directly that $S(\varphi)$ is aperiodic for $\varphi \in F_1(<)$ (Exercise 8).

We can now prove the result announced before.

Corollary 4.4. *One has $F_1(<) < MF_2(<)$.*

Proof. We have seen in Proposition 2.5 that $F_1(<) \le MF_2(<)$. Let φ be a formula of $MF_2(<)$ such that $S(\varphi)$ is not a star-free set. Then, by Theorem 4.1, there are no formula of $F_1(<)$ equivalent to φ. □

Example 4.1. Let φ be the formula

$$\varphi = \exists X\ \big(X(\min) \wedge \forall x\ (X(x) \leftrightarrow \neg X(x+1)) \wedge X(\max)\big)$$

where $\leftrightarrow$ is the connective designating logical equivalence.

A finite word u satisfies φ if and only if $|u|$ is even. Thus $X = S(\varphi)$ is not star-free and hence φ cannot be equivalent to any formula of $F_1(<)$ or, equivalently, X is not definable in $F_1(<)$.

4.2 Logical hierarchy. We shall see in this section how to refine the characterization of the first-order definable sets of the logic of the linear order (Theorem 4.1). We shall actually establish a bijection between the concatenation hierarchy of star-free sets defined in Chapter VI and the logical hierarchy $\Sigma_n(<)$. Since the concatenation hierarchy is infinite, it will show that the hierarchy defined on formulas by quantifier alternation is also infinite.

Let us recall the notation introduced in Section 2: we denote by $\Sigma_n(<)$ the set of formulas of the language $F_1(<)$ with at most n alternating blocks of quantifiers and by $\mathcal{B}\Sigma_n(<)$ the set of boolean combinations of formulas of $\Sigma_n(<)$.

Theorem 4.5. *For each integer $n \ge 0$,*

(1) *A set of finite or infinite words is definable by a formula of $\mathcal{B}\Sigma_n(<)$ if and only if it is a star-free set of level n.*

(2) *A set of finite or infinite words is definable by a formula of $\Sigma_{n+1}(<)$ if and only if it is a star-free set of level $n + 1/2$.*

Proof. The first part of the proof consists in converting sets of words to formulas. To start with, the equalities

$$S(\textbf{true}) = A^\infty \qquad S(\textbf{false}) = \emptyset$$

show that each star-free subset of level 0 is definable by a formula of $\mathcal{B}\Sigma_0$.

By induction, suppose that the star-free sets of level $n+1/2$ are definable by a formula of $\Sigma_{n+1}(<)$. Proposition 3.4 shows now that the star-free sets of level $n+1$ are definable by a formula of $\mathcal{B}\Sigma_{n+1}(<)$.

Suppose now that the star-free sets of level n are definable by a formula of $\mathcal{B}\Sigma_n(<)$. If X is a star-free subset of A^∞ of level $n + 1/2$, X can be written as a finite union of subsets of the form

$$X_0 a_1 X_1 a_2 \cdots a_k X_k$$

where $k \geq 0$, $X_0, X_1, \ldots, X_{k-1}$ are star-free sets of level n of A^*, X_k is a star-free subset of level $n + 1/2$ of A^∞ and $a_1, \ldots, a_k$ are letters.

We now use again the "relativised" version of the first-order formula given by Proposition 2.6. If $X_0 = S(\varphi_0)$, $X_1 = S(\varphi_1), \ldots, X_k = S(\varphi_k)$, we have $X_0 a_1 X_1 \cdots a_k X_k = S(\varphi)$, where φ is the formula

$$\exists x_1 \cdots \exists x_k \bigwedge_{1 \leq i \leq k-1} (x_i < x_{i+1}) \bigwedge_{1 \leq i \leq k} \mathbf{a}_i(x_i) \wedge \varphi_1(0, x_1 - 1)$$
$$\bigwedge_{1 \leq i \leq k-1} \varphi(x_i + 1, x_{i+1} - 1) \wedge \varphi(x_k, \max)$$

showing that $X_0 a_1 X_1 \cdots a_k X_k$ is definable by a formula of Σ_{n+1}. It follows, by Proposition 3.4, that if the star-free subsets of level n are definable by a formula of $\mathcal{B}\Sigma_n(<)$, the star-free subsets of level $n + 1/2$ are definable by a formula of $\Sigma_{n+1}(<)$.

We now arrive at the second part of the proof, the conversion from formulas to star-free sets. We recall the scheme of the proof of Theorem 4.1, which consists in arguing by induction on the formation rules of formulas. For this purpose, we build, for each alphabet B, a hierarchy of star-free subsets of B^∞ which differs from the usual hierarchy, in particular at level 0.

The induction that follows forces very precise requirements on the subsets of B^∞ of level 0. In particular, they have to contain the subsets of the form $S(\mathbf{a}(x_i))$, $S(x_i = x_j)$ and $S(x_i < x_j)$ and to be closed under quotient. This leads us to the following definition, which will be subsequently justified by the lemmas 4.7 and 4.8. For each $\varepsilon = (\varepsilon_1, \ldots, \varepsilon_p) \in \{0, 1\}^p$, we set

$$K(\varepsilon) = \{u \in B_p^* \mid \text{for } 1 \leq i \leq p,\ u_i \text{ contains exactly } \varepsilon_i \text{ occurrences of } 1\}$$
$$K'(\varepsilon) = \{u \in B_p^\omega \mid \text{for } 1 \leq i \leq p,\ u_i \text{ contains exactly } \varepsilon_i \text{ occurrences of } 1\}$$

Note that $K_p = K(1, \ldots, 1)$ and $K'_p = K'(1, \ldots, 1)$. For $a \in A$, $i \in \{1, \ldots, p\}$ and $\varepsilon \in \{0, 1\}^p$, we set

$$K(a, i, \varepsilon) = \{u \in K(\varepsilon) \mid \text{if } u_{i,n_i} = 1, \text{ then } u_{0,n_i} = a\}$$
$$K'(a, i, \varepsilon) = \{u \in K'(\varepsilon) \mid \text{if } u_{i,n_i} = 1, \text{ then } u_{0,n_i} = a\}$$

This definition is interesting only if $\varepsilon_i = 1$. In this case, $K(a, i, \varepsilon)$ is the set of the words of $K(\varepsilon)$ such that, if n_i is the unique position of 1 in u_i, the letter of u_0 in position n_i is an a. Finally, if $i, j \in \{1, \ldots, p\}$, we set

$$K_=(i, j, \varepsilon) = \{u \in K(\varepsilon) \mid u_{i,n} = 1 \text{ if and only if } u_{j,n} = 1\}$$
$$K'_=(i, j, \varepsilon) = \{u \in K'(\varepsilon) \mid u_{i,n} = 1 \text{ if and only if } u_{j,n} = 1\}$$
$$K_<(i, j, \varepsilon) = \{u \in K(\varepsilon) \mid \text{if } u_{i,n_i} = 1 \text{ and } u_{j,n_j} = 1, \text{ then } n_i < n_j\}$$
$$K'_<(i, j, \varepsilon) = \{u \in K'(\varepsilon) \mid \text{if } u_{i,n_i} = 1 \text{ and } u_{j,n_j} = 1, \text{ then } n_i < n_j\}$$

Again, these definitions are only interesting if $\varepsilon_i = \varepsilon_j = 1$. In this case, $K_=(i, j, \varepsilon)$ (resp. $K_<(i, j, \varepsilon)$) is the set of the words of $K(\varepsilon)$ such that the unique position of 1 in u_i is equal to (resp. precedes) the unique position of 1 in u_j.

The subsets of B_p^* (resp. B_p^ω) of level 0 form the least set of subsets closed under finite intersection and finite union and containing the subsets of the form $K(\varepsilon)$, $K(a, i, \varepsilon)$, $K_=(i, j, \varepsilon)$ and $K_<(i, j, \varepsilon)$ (resp. $K'(\varepsilon)$, $K'(a, i, \varepsilon)$, $K'_=(i, j, \varepsilon)$ and $K'_<(i, j, \varepsilon)$) for each $\varepsilon \in \{0, 1\}^p$, $a \in A$ and $i, j \in \{1, \dots, p\}$.

Next, level $n + 1/2$ is the least set of subsets closed under finite intersection and finite union and containing the subsets of the form

$$L_0 b_1 L_1 b_2 \cdots b_k L_k$$

where $k \geq 0$, $b_1, \dots, b_k \in B_p$ and $L_0, L_1, \dots, L_k$ are subsets of level n of B_p^* (resp. $L_0, L_1, \dots, L_{k-1}$ are subsets of level n of B_p^* and L_k is a subset of level n of B_p^ω). The level $n + 1$ is the closure of level $n + 1/2$ under finite union, finite intersection and the operations $X \to Y \setminus X$ where Y is a subset of level 0.

For $p = 0$, we recover the usual hierarchy.

Lemma 4.6. *The hierarchy of subsets of B_0^* (resp. B_0^ω) coincides with the concatenation hierarchy.*

Proof. We have $B_0 = A$ and the only subsets of level 0 are $\emptyset$ and A^* (resp. A^ω). It follows that the two hierarchies coincide. □

The subsets of level n (resp. $n + 1/2$) are closed under quotient.

Lemma 4.7. *For each $n \geq 0$, the following properties hold:*

(1) *Let X be a subset of B_p^* of level n (resp. $n + 1/2$) and let $u, v \in B_p^*$. Then the subsets $u^{-1}X$ and Xv^{-1} are of level n (resp. $n + 1/2$).*

(2) *Let X a subset of B_p^ω of level n (resp. $n + 1/2$), let $u \in B_p^*$ and $v \in B_p^\omega$. Then the subsets $u^{-1}X$ and Xv^{-1} are of level n (resp. $n + 1/2$).*

Proof. We treat the case of the sets of the form Xu^{-1}, where X is a subset of B_p^ω and $u \in B_p^\omega$. The other cases are analogous.

We already know that the residuals commute with boolean operations: if X_1 and X_2 are subsets of B_p^ω, then

$$(X_1 \cup X_2)u^{-1} = X_1u^{-1} \cup X_2u^{-1}$$
$$(X_1 \cap X_2)u^{-1} = X_1u^{-1} \cap X_2u^{-1}$$
$$(X_1 \setminus X_2)u^{-1} = X_1u^{-1} \setminus X_2u^{-1}$$

Moreover, if $b_1, \dots, b_k$ are letters of B_p, $L_0, L_1, \dots, L_{k-1}$ are subsets of B_p^* and L_k a subset of B_p^ω, we have the formula

$$(L_0 b_1 L_1 b_2 \cdots b_k L_k)u^{-1} = \bigcup_{0 \le i \le k} \bigcup_{x_i \in E_i} L_0 b_1 L_1 \ \cdots \ L_{i-1} b_i (L_i x_i^{-1}) \qquad (4.1)$$

where $E_i = \{x_i \mid u \in x_i b_{i+1} L_{i+1} \ \cdots \ b_k L_k\}$. Indeed, we have

$$x \in (L_0 b_1 L_1 b_2 \cdots b_k L_k)u^{-1}$$

if and only if $xu \in L_0 b_1 L_1 b_2 \cdots b_k L_k$, which amounts to saying there exists i such that $0 \le i \le k$ and two words y_i and x_i such that $y_i x_i \in L_i$, $x \in L_0 b_1 L_1 b_2 \cdots L_{i-1} b_i y_i$ and $u \in x_i b_{i+1} L_{i+1} \ \cdots \ b_k L_k$. Formula (4.1) follows.

We now conclude the proof by induction. The subsets of level 0 are closed under quotient. Let indeed K' be one of the subsets $K'(\varepsilon)$, $K'(a, i, \varepsilon)$, $K'_=(i, j, \varepsilon)$ or $K'_<(i, j, \varepsilon)$. If u is not a suffix of any word of K', then of course $K'u^{-1} = \emptyset$. Otherwise, we have in particular, denoting by ε'_i the number of occurrences of 1 in u_i, $\varepsilon'_i \le \varepsilon_i$ and

$$\begin{aligned}
K'(\varepsilon)u^{-1} &= K'(\varepsilon - \varepsilon') \\
K'(a, i, \varepsilon)u^{-1} &= K'(a, i, \varepsilon - \varepsilon') \\
K'_=(i, j, \varepsilon)u^{-1} &= K'_=(i, j, \varepsilon - \varepsilon') \\
K'_<(i, j, \varepsilon)u^{-1} &= K'_<(i, j, \varepsilon - \varepsilon')
\end{aligned}$$

If the subsets of level n are closed under quotient, the formula 4.1 shows that every residual of a marked product of such subsets is of level $n + 1/2$. It follows that the subsets of level $n + 1/2$ are closed under quotient since the residuals commute with boolean operations. Next, the same argument shows that the subsets of level $n + 1$ are closed under quotient, completing the induction. □

We come back to the conversion of formulas into star-free sets. The induction starts by the formulas of $\mathcal{B}\Sigma_0$.

Lemma 4.8. *If $\varphi \in \mathcal{B}\Sigma_0$, $S_p^*(\varphi)$ (resp. $S_p^\omega(\varphi)$) is a subset of level 0 of B_p^* (resp. B_p^ω).*

Proof. We may assume that φ is in disjunctive normal form. Then we get rid of the negations of atomic formulas by replacing

$$\begin{aligned}
\neg(x = y) &\quad \text{by} \quad (x < y) \vee (y < x) \\
\neg(x < y) &\quad \text{by} \quad (x = y) \vee (y < x) \\
\neg\mathbf{a}(x) &\quad \text{by} \quad \bigvee_{b \ne a} \mathbf{b}(x)
\end{aligned}$$

Thus φ is now a disjunction of conjunctions of atomic formulas, and by Proposition 3.4, we are left with the case of atomic formulas. Finally, the formulas

$$\begin{array}{ll} S_p^*(\mathbf{a}(x_i)) = K(a, i, 1, \ldots, 1) & S_p^\omega(\mathbf{a}(x_i)) = K'(a, i, 1, \ldots, 1) \\ S_p^*(x_i = x_j) = K_=(i, j, 1, \ldots, 1) & S_p^\omega(x_i = x_j) = K'_=(i, j, 1, \ldots, 1) \\ S_p^*(x_i < x_j) = K_<(i, j, 1, \ldots, 1) & S_p^\omega(x_i < x_j) = K'_<(i, j, 1, \ldots, 1) \end{array}$$

conclude the proof since these subsets are of level 0 by definition. □

The key step of the induction consists to convert existential quantifiers into concatenation products.

Lemma 4.9. *Let φ be a formula of Σ_n and $x_{i_1}, \ldots, x_{i_k}$ be free variables of φ. If $S_p^*(\varphi)$ (resp. $S_p^\omega(\varphi)$) is a subset of level n of B_p^* (resp. B_p^ω), then $S_{p-k}^*(\exists x_{i_1} \cdots \exists x_{i_k} \varphi)$ (resp. $S_{p-k}^\omega(\exists x_{i_1} \cdots \exists x_{i_k} \varphi)$) is a subset of level $n + 1/2$ of B_p^* (resp. B_p^ω).*

Proof. We shall treat the case of infinite words, but the other case is similar. Up to a permutation of the free variables, we may assume that $x_{i_1} = x_1, \ldots, x_{i_k} = x_k$. We first establish the formula

$$S_{p-k}^\omega(\exists x_{i_1} \cdots \exists x_{i_k} \varphi) = \pi(S_p^\omega(\varphi)) \tag{4.2}$$

where $\pi : B_p \to B_{p-k}$ is the function erasing the components of index 1 to k, defined by:

$$\pi_i(b_0, b_1, \ldots, b_p) = (b_0, b_{k+1}, \ldots, b_p)$$

Indeed, if $u = (u_0, u_1, \ldots, u_p) \in S_p^\omega(\varphi)$, we have in particular $u \in K_p$. For $1 \le i \le p$, denote by n_i the position of the unique 1 of the word u_i. Then we have $u_0 \models \varphi[\nu]$, where ν is the valuation defined by $\nu(x_i) = n_i$ for $1 \le i \le p$. Now $\nu = \nu'[\binom{n_1}{x_1} \cdots \binom{n_k}{x_k}]$, where ν' is the valuation defined by $\nu'(x_i) = n_i$ for $k+1 \le i \le p$. It follows $u_0 \models (\exists x_1 \cdots \exists x_k \varphi)[\nu']$ and as $\pi(u)_0 = u_0$, it follows $\pi(u) \in S_{p-k}^\omega(\exists x_1 \cdots \exists x_k \varphi)$.

Conversely, if $v \in S_{p-k}^\omega(\exists x_1 \cdots \exists x_k \varphi)$, there exist positions $n_1, \ldots, n_k$ such that

$$v_0 \models \varphi[\nu'\tbinom{n_1}{x_1} \cdots \tbinom{n_k}{x_k}].$$

Let u be the unique word of K_p such that $\pi(u) = v$ and such that, for $1 \le i \le k$, the position of the unique 1 of u_i is n_i. Then $u_0 = v_0$, and as $\nu = \nu'[\binom{n_1}{x_1} \cdots \binom{n_k}{x_k}]$, we have $u_0 \models \varphi[\nu]$, whence $u \in S_p^\omega(\varphi)$, proving Formula (4.2).

We shall now express $S_p^\omega(\varphi)$ as a finite union of marked products of subsets of level n. Set $L = S_p^\omega(\varphi)$, $D = \{ b \in B_p \mid \text{for } 1 \le i \le k, b_i = 0 \}$ and, for $I \subseteq \{1, \ldots, p\}$,

$$C_I = \{ b \in B_p \mid \text{for each } i \in I \ b_i = 1 \}$$

If $u \in L$, there exists, for $1 \leq i \leq k$, a unique position n_i such that $u_{i,n_i} = 1$. Then there exists an ordered partition $P = (I_1, \ldots, I_r)$ of $\{1, \ldots, k\}$ such that u is uniquely factorized as $u = u_0 b_1 u_1 b_2 \cdots b_r u_k$, where $b_j \in C_{I_j}$, $u_0, \ldots, u_{r-1} \in D^*$ and $u_r \in D^\omega$.

		a_1			a_2				a_3		$\cdots$
0	0	1	0	0	0	0	0	0	0	0	$\cdots$
0	0	0	0	0	0	0	0	0	1	0	$\cdots$
0	0	0	0	0	1	0	0	0	0	0	$\cdots$
0	0	1	0	0	0	0	0	0	0	0	$\cdots$
					$\vdots$						$\cdots$

Next we obtain the formula below, in which $Q(x) = \{u \in B_p^* \mid u^{-1}L = x^{-1}L\}$ and $\mathcal{P}$ denotes the set of ordered partitions of $\{1, \ldots, k\}$:

$$L = \bigcup_{P \in \mathcal{P}} \bigcup_{(x_0, b_1, x_1, \ldots, b_r, x_r) \in E_P} [Q(x_0)]\, b_1\, [(x_0 b_1)^{-1} Q(x_0 b_1 x_1)]\, b_2 \cdots$$
$$\cdots b_r\, [(x_0 b_1 x_1 \cdots b_r)^{-1} Q(x_0 b_1 x_1 \cdots b_r x_r)] \qquad (4.3)$$

in which if $P = (I_1, \ldots, I_r)$ is an ordered partition, E_P is the set of

$$(x_0, b_1, x_1, \ldots, b_r, x_r)$$

such that $x_0 b_1 x_1 \cdots b_r x_r \in L$, with $x_0, x_1, \ldots, x_{p-1} \in B_p^*$, $x_p \in B_p^\omega$ and, for $1 \leq j \leq r$, $b_j \in C_{I_j}$.

Let L' be the right hand side of the Formula (4.3). First of all, if $u \in L$, there exists an ordered partition $(I_1, \ldots, I_r)$ of $\{1, \ldots, k\}$ and a factorization $u = u_0 b_1 u_1 \cdots b_r u_r$, with, for $1 \leq j \leq r$, $b_j \in C_{I_j}$. Now, by construction, $x \in Q(x)$ for each x, and hence

$$u_0 \in Q(u_0),\ u_1 \in (u_0 b_1)^{-1} Q(u_0 b_1 u_1),\ \ldots,$$
$$u_r \in (u_0 b_1 \cdots u_{r-1} b_r)^{-1} Q(u_0 b_1 \cdots u_{r-1} b_r u_r)$$

thereby establishing the inclusion $L \subseteq L'$. Conversely, if there exists an ordered partition P of $\{1, \ldots, k\}$ and if

$$u \in [Q(x_0)]\, b_1\, [(x_0 b_1)^{-1} Q(x_0 b_1 x_1)]\, b_2 \cdots$$
$$b_k\, [(x_0 b_1 x_1 \cdots b_r)^{-1} Q(x_0 b_1 x_1 \cdots b_r x_r)]$$

for some $(x_0, b_1, x_1, \ldots, b_k, x_r) \in E_P$, we have $u = u_0 b_1 u_1 \cdots b_r u_r$, with $b_j \in C_{I_j}$ and

$$u_0 \in Q(x_0),\ u_1 \in (x_0 b_1)^{-1} Q(x_0 b_1 x_1), \ldots,$$
$$u_r \in (x_0 b_1 x_1 \cdots b_r)^{-1} Q(x_0 b_1 x_1 \cdots b_r x_r)$$

Therefore, for $0 \le i \le r$, $x_0 b_1 x_1 \cdots b_i u_i \in Q(x_0 b_1 x_1 \cdots b_i x_i)$, whence

$$(x_0 b_1 x_1 \cdots b_i u_i)^{-1} L = (x_0 b_1 x_1 \cdots b_i x_i)^{-1} L$$

Next we show that $x_0 b_1 x_1 \cdots b_i x_i b_{i+1} u_{i+1} \cdots b_r u_r \in L$ by induction on $r - i$. For $i = r$, the result follows from the fact that $(x_0, b_1, x_1, \ldots, b_r, x_r) \in E_P$. Next, if

$$x_0 b_1 x_1 \cdots b_i x_i b_{i+1} u_{i+1} \cdots b_r u_r \in L$$

we have $b_{i+1} u_{i+1} \cdots b_r u_r \in (x_0 b_1 x_1 \cdots b_i x_i)^{-1} L$ and hence

$$b_{i+1} u_{i+1} \cdots b_r u_r \in (x_0 b_1 x_1 \cdots b_i u_i)^{-1} L$$

whence $x_0 b_1 x_1 \cdots b_{i-1} x_{i-1} b_i u_i \cdots b_r u_r \in L$, allowing the induction. Consequently

$$u = u_0 b_1 u_1 \cdots b_r u_r \in L$$

and the inclusion $L' \subseteq L$ is proved.

A second formula gives an expression of $Q(x)$ in terms of the residuals of L.

$$Q(x) = \Big(\bigcap_{y \in x^{-1}L} Ly^{-1} \Big) \setminus \Big(\bigcup_{y \notin x^{-1}L} Ly^{-1} \Big) \tag{4.4}$$

Let indeed $Q'(x)$ be the right hand side of (4.4). If $u \in Q(x)$, then $u^{-1}L = x^{-1}L$. Therefore the following sequence of equivalences hold

$$y \in x^{-1}L \Longleftrightarrow y \in u^{-1}L \Longleftrightarrow uy \in L \Longleftrightarrow u \in Ly^{-1}$$

showing that $Q(x) \subseteq Q'(x)$. Let now $u \in Q'(x)$. If $y \in x^{-1}L$, then $u \in Ly^{-1}$ and hence $uy \in L$ and $y \in u^{-1}L$. Similarly, if $y \notin x^{-1}L$, then $y \notin u^{-1}L$. Therefore $u^{-1}L = x^{-1}L$ and $u \in Q(x)$, proving (4.4).

Since L is ω-rational, Corollary II.7.7 shows that L has only finitely many residuals on the right and on the left. It follows that only finitely many unions and intersections occur in Formulas (4.3) and (4.4). Besides, Lemma 4.7 shows that all the residuals of L are of level n. Consequently, the subsets of the form $Q(x)$ and their residuals are of level n. Formula (4.3) now gives an expression of $S_p^\omega(\varphi)$ as a finite union of products of the form $X_0 b_1 X_1 \cdots b_r X_r$ where $b_j \in C_{I_j}$ and the $X_0, \ldots, X_{r-1}$ (resp. X_r) are level n subsets of D^* (resp. D^ω). By Formula (4.2), $S_{p-r}^\omega(\exists x_{i_1} \cdots \exists x_{i_r} \varphi)$ is thus a finite union of subsets of the form $\pi(X_0 b_1 X_1 b_2 \cdots b_r X_r)$. Therefore, proving that $S_{p-r}^\omega(\exists x_{i_1} \cdots \exists x_{i_k} \varphi)$ is of level $n + 1/2$, amounts to proving that if X is a subset of B_p^* (resp. B_p^ω) of level n contained in D^* (resp. D^ω), then $\pi(X)$ is a subset of B_{p-k}^* (resp. B_{p-k}^ω) of level n.

Let $\delta : B_{p-k} \to B_p$ be the function defined by

$$\delta(b_0, \ldots, b_{p-k}) = (b_0, 0, \ldots, 0, b_1, \ldots, b_{p-k})$$

Then we have, for each subset X of D^* (resp. $D^{(\omega)}$),

$$\pi(X) = \delta^{-1}(X)$$

Thus, it suffices to verify by induction that the subsets of level n (resp. $n + 1/2$) are stable under δ^{-1}. For level 0, this follows from the formulas

$$\begin{aligned}
\delta^{-1}(K(\varepsilon_1, \ldots, \varepsilon_{p-k})) &= K(0, \ldots, 0, \varepsilon_{k+1}, \ldots, \varepsilon_p) \\
\delta^{-1}(K(a, i, \varepsilon_1, \ldots, \varepsilon_{p-k})) &= K(a, k+i, 0, \ldots, 0, \varepsilon_{k+1}, \ldots, \varepsilon_p) \\
\delta^{-1}(K_=(i, j, \varepsilon_1, \ldots, \varepsilon_{p-k})) &= K_=(k+i, k+j, 0, \ldots, 0, \varepsilon_{k+1}, \ldots, \varepsilon_p) \\
\delta^{-1}(K_=(i, j, \varepsilon_1, \ldots, \varepsilon_{p-k})) &= K_<(k+i, k+j, 0, \ldots, 0, \varepsilon_{k+1}, \ldots, \varepsilon_p)
\end{aligned}$$

and from the fact that δ^{-1} commutes with the boolean operations. This latter property also permits to pass from level $n + 1/2$ to level $n + 1$ in the induction step. Passing from level n to level $n + 1/2$, requires the formula

$$\delta^{-1}(X_0 b_1 X_1 \cdots b_k X_k) = \bigcup_{\substack{c_1 \in \delta^{-1}(b_1) \\ \vdots \\ c_k \in \delta^{-1}(b_k)}} \delta^{-1}(X_0) c_1 \delta^{-1}(X_1) \cdots c_k \delta^{-1}(X_k)$$

which follows from the fact that δ is length preserving. □

We can now conclude the second part of the proof of Theorem 4.5. Since the only variable-free formulas of $\mathcal{B}\Sigma_0$ are **true** and **false**, the only subsets of A^* (resp. A^{ω}) definable by a formula of $\mathcal{B}\Sigma_0(<)$ are $\emptyset$ and A^* (resp. A^{ω}), which are indeed of level 0.

By Lemma 4.8, for each formula φ of $\mathcal{B}\Sigma_0(<)$, $S_p^*(\varphi)$ (resp. $S_p^{\omega}(\varphi)$) is a subset of B_p^* (resp. B_p^{ω}) of level 0. Suppose by induction that, for each formula φ of $\mathcal{B}\Sigma_n(<)$, $S_p^*(\varphi)$ (resp. $S_p^{\omega}(\varphi)$) is a subset of B_p^* (resp. B_p^{ω}) of level n. Consider the set Δ of formulas φ such that, for each $p \geq 0$, $S_p^*(\varphi)$ (resp. $S_p^{\omega}(\varphi)$) is a subset of B_p^* (resp. B_p^{ω}) of level $n + 1/2$. The set Δ contains $\mathcal{B}\Sigma_n(<)$ by the induction hypothesis and, by Proposition 3.4, it is closed under finite disjunction and finite conjunction. Finally, Lemma 4.9 shows that if $\varphi \in \Delta$ and if $x_1, \ldots, x_k$ are free variables of φ, then $\exists x_1 \cdots \exists x_k \varphi \in \Delta$. Therefore Δ contains $\Sigma_{n+1}(<)$ and each formula of $\Sigma_{n+1}(<)$ defines a subset B_p^* (resp. B_p^{ω}) of level $n + 1/2$.

To conclude, suppose by induction that for each formula φ of $\Sigma_{n+1}(<)$, $S_p^*(\varphi)$ (resp. $S_p^{\omega}(\varphi)$) is a subset of B_p^* (resp. B_p^{ω}) of level $n + 1/2$. The proof of Proposition 3.3 shows that K_p is of level 1 and thus Proposition 3.4 shows that for each formula φ of $\mathcal{B}\Sigma_{n+1}(<)$, $S_p^*(\varphi)$ (resp. $S_p^{\omega}(\varphi)$) is a subset of B_p^* (resp. B_p^{ω}) of level $n + 1$. □

5 First-order logic of the successor

We now broach the study of a more restricted logical language, the first-order logic of the successor. We shall characterize the sets definable in this language. This will show in particular that this logic is strictly less powerful than the first-order logic of the linear order. The characterization utilizes the technique of Fraïssé-Ehrenfeucht games, that we first describe.

5.1 Fraïssé-Ehrenfeucht Games. Here we consider the first-order formulas obtained by using a set $\mathcal{V}$ of variables and a finite set $\mathcal{S}$ of relational symbols (including the symbols $\mathbf{a}$ for $a \in A$) but *no function symbol*. Denote by $F_1(\mathcal{S})$ the set of these formulas.

In this section, there will be no difference between the cases of finite and infinite words. Therefore, we shall consider indifferently finite or infinite words of A^∞, and they will be simply called words.

It is convenient to introduce the following notion for interpreting formulas on words. Let A and $\mathcal{V}$ be two finite sets. A $\mathcal{V}$-structure on A is a word of $(A \times 2^{\mathcal{V}})^\infty$

$$(a_0, U_0)(a_1, U_1)\cdots$$

with $u = a_0a_1\cdots \in A^\infty$ and $U_i \subset \mathcal{V}$ such that, for each j, k with $0 \leq j < k \leq |u|$,

$$U_j \cap U_k = \emptyset.$$

The $\mathcal{V}$-structures are of course the same objects as the words of the set K_p built on the alphabet $B_p = A \times \{0, 1\}^p$ of Section 4. Indeed, set $\mathcal{V} = \{x_1, \ldots, x_p\}$. Then the $\mathcal{V}$-structure $(a_0, U_0)(a_1, U_1)\cdots$ corresponds to the word $(u_0, u_1, \ldots, u_p)$ where $u_0 = a_0\cdots$ and $U_i = \{x_k \mid u_{k,i} = 1\}$.

Define the *quantifier height* $c(\varphi)$ of $\varphi \in F_1(\mathcal{S})$ by setting $c(\varphi) = 0$ if φ is atomic, and then

$$c(\neg\varphi) = c(\varphi)$$
$$c(\varphi \vee \psi) = c(\varphi \wedge \psi) = \max(c(\varphi), c(\psi))$$
$$c(\exists x\ \varphi) = c(\forall x\ \varphi) = c(\varphi) + 1$$

This quantifier height should not be mixed up with the level of the formula in the Σ_n hierarchy of quantifier alternations.

The next proposition depends of course on our hypothesis that the alphabet A is finite and that the set of symbols only contains a finite number of relation symbols (and no function symbol). It is also assumed that the set of variables is a fixed finite set, which prevents one to convert all formulas in prenex normal form.

Proposition 5.1. *Given a finite set of variables, there are, up to logical equivalence, only a finite number of formulas of quantifier height* $\leq r$.

Proof. The atomic formulas are of the form $u = v$ or $R(u, v, \ldots)$ with $R \in \mathcal{S}$ and $u, v, \ldots \in \mathcal{V}$. Therefore there are only finitely many atomic formulas, and hence only finitely many boolean combinations of them, which proves the result for $r = 0$. Next, the formulas of quantifier height r are boolean combinations of formulas φ, $\exists x\varphi$ or $\forall x\varphi$ with $c(\varphi) \leq r - 1$. Arguing by induction, we may assume there are finitely many such formulas, up to logical equivalence. The proposition follows, since there are now only finitely many nonequivalent boolean combinations of formulas taken from a finite set. □

Given two (finite or infinite) words w_1 and w_2 on A, we write $w_1 \equiv_r w_2$ if w_1 and w_2 satisfy the same formulas of quantifier height at most r.

It follows from Proposition 5.1 that the equivalence $\equiv_r$ has finite index.

To ensure that $w_1 \equiv_r w_2$, it suffices that w_2 satisfies all the formulas of quantifier height $\leq r$ satisfied by w_1, and conversely. Indeed, if φ is a formula of quantifier height $\leq r$ not satisfied by w_1, then $\neg\varphi$ is a formula of the same quantifier height satisfied by w_1 and hence by w_2. Thus w_2 does not satisfy φ and it follows that $w_1 \equiv_r w_2$.

A *Fraïssé-Ehrenfeucht game* is a game with two players playing on two fixed words w_1 and w_2 of A^∞. The players are called Player I and Player II. Player I tries to show that the two words are different while Player II, on the contrary, tries to show they are the same for the rules of the game. Each player owns r tokens numbered $z_1, z_2, \ldots, z_r$. At round i, Player I puts her token z_i on one of the letters of one of the two words. Player II puts his token z_i on one of the letters of the other word. Once put down, a token cannot be moved anymore. The game ends after r rounds when the two players have depleted all their tokens.

At the end of the game, we are left with two $\{z_1, z_2, \ldots, z_r\}$-structures. Player II wins if the two structures satisfy the same atomic formulas. Otherwise, Player I wins.

Denote by $G(w_1, w_2, r)$ the r round game defined by the words w_1 and w_2. It is obvious that the games $G(w_1, w_2, r)$ and $G(w_2, w_1, r)$ are identical.

Moreover, such a game is *determined*, that is, one of the two players has a winning strategy. This is easy to verify by induction on r. It is obvious if $r = 0$. Suppose next that the property holds for $r - 1$. Then, either Player I can play a move that will give her a winning strategy after any answer of Player II, or, for each first move played by Player I, there exists a move of Player II giving him a winning strategy. Player I has a winning strategy in the first case, and Player II in the second case.

Example 5.1. Consider the game defined by $w_1 = ab$ and $w_2 = baa$, with equality as the unique predicate. If $r = 1$, Player II can win by playing on a letter identical to that played by Player I. If $r \geq 2$, Player I can win by showing that w_2 contains two occurrences of letter a. Indeed, she first plays on the first a of baa. Then Player II is forced to answer on the unique a of ab. Then Player I puts her token on the second a of baa. If Player II answers on a, the formula $z_1 = z_2$ is satisfied on the structure associated with ab, but not on the other structure. If he plays on b, $\mathbf{a}(z_2)$ is not true on

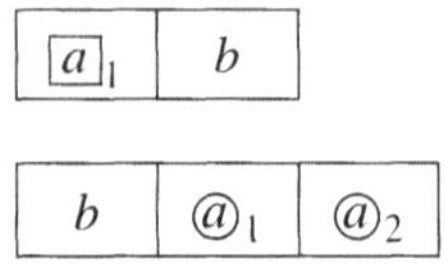

Figure 5.1. A winning game for Player I.

ab. Therefore, Player I wins. In Figure 5.1, the moves of Player I are represented by circled letters, and that of Player II by letters in a square.

A Fraïssé-Ehrenfeucht game $G(w_1, w_2, r)$ starts with two words w_1 and w_2. At the end, we are left with two structures w'_1 and w'_2. Assuming that w_1 and w_2 are also structures would not change the definitions, but will be quite useful to argue by induction (we just did it implicitly to prove that Fraïssé-Ehrenfeucht games are determined).

The next result is a very useful characterization of the equivalence $\equiv_r$.

Theorem 5.2 (Fraïssé, Ehrenfeucht). *Let w_1, w_2 be two words on A and let $r \geq 0$. Then $w_1 \equiv_r w_2$ if and only if Player II has a winning strategy in the game $G(w_1, w_2, r)$.*

Proof. Let us first assume that $w_1 \equiv_r w_2$. We show by induction on r that Player II has a winning strategy. If $r = 0$, Player II wins by definition since w_1 and w_2 satisfy the same atomic formulas. Let now $r > 0$ and suppose the result holds for $r - 1$ but not for r. Since the game is determined, Player I has necessarily a winning strategy. Suppose that Player I plays the first move of this strategy (say, on w_1). This defines a structure w'_1 such that, for each move played by Player II, Player I has a winning strategy in the resulting game $(w'_1, w'_2, r - 1)$. By induction hypothesis, we thus have $w'_1 \not\equiv_r w'_2$. Let ψ be the conjunction of all the formulas of quantifier height $< r$ satisfied by w'_1 (up to logical equivalence, to ensure finiteness). Then $w'_2 \not\models \psi$. Since this hold for any structure w'_2 obtained from w_2 by playing the token z_1, we have

$$w_2 \not\models \exists z_1 \psi \quad \text{and} \quad w_1 \models \exists z_1 \psi$$

a contradiction with the hypothesis, since $c(\exists z_1 \psi) = r$.

In the other direction, we also argue by induction on r. If $r = 0$ and if Player II has a winning strategy, then w_1 and w_2 satisfy the same atomic formulas and hence $w_1 \equiv_0 w_2$. Next suppose that $r > 0$ and that the implication is proved for $r - 1$.

Suppose that Player II has a winning strategy for the game $G(w_1, w_2, r)$. Then he also has one for the game $G(w_1, w_2, r - 1)$. Therefore $w_1 \equiv_{r-1} w_2$ by the induction hypothesis. To show that $w_1 \equiv_r w_2$, it suffices to consider a formula of the form $\exists x \psi$ with $c(\psi) = r - 1$. Suppose that

$$w_1 \models \exists x \psi \quad \text{and} \quad w_2 \not\models \exists x \psi$$

Up to a renaming of x, we may assume that $x = z_1$. Then Player I can play her token z_1 on w_1, in such a way that the resulting structure w_1' satisfies ψ. But since Player II is supposed to have a winning strategy, he can answer on w_2 to obtain a winning strategy on $G(w_1', w_2', r-1)$. By induction hypothesis, this implies that $w_1' \equiv_{r-1} w_2'$. In particular, $w_2' \models \psi$, whence $w_2 \models \exists x\, \psi$, a contradiction. □

Fraïssé-Ehrenfeucht games can be used to give a direct proof (i.e. without using star-free sets) that $F_1(<) < MF_2(<)$ (see Exercise 9).

5.2 Characterization of $F_1(S)$. We shall give in this section a syntactic characterization of the sets definable in $F_1(S)$, i.e. definable by a first-order formula using the successor function, but not the relation $<$.

This characterization relies on the definition of a threshold locally testable (TLT) subset of A^+ given in Section B.5.2. For the convenience of the reader, we briefly recall this definition: a subset of A^+ is TLT if and only if it is a finite union of $\sim_{k,t}$-classes of A^+ for some positive integers k and t. This equivalence is defined on A^+ by setting $u \sim_{k,t} v$ if and only if

(a) for each $n < k$, $p_n(u) = p_n(v)$,

(b) for each $n < k$, $s_n(u) = s_n(v)$,

(c) $u \equiv_{k,t} v$.

Equivalently, a subset of A^+ is TLT if and only if it is a boolean combination of subsets of the form uA^*, A^*u or $F(x, t)$ where $x \in A^+$, and $t > 0$.

The notion of a TLT subset of A^ω was introduced in Section VII.3.3, but a slight variation of this definition is more suitable for our purpose.

A subset of A^ω is said to be *threshold prefix-factors testable* (TPFT) if it is equal to a boolean combination of subsets of the form xA^ω or $F_\omega(x, t)$ with $x \in A^+$ and $t \geq 0$. It is easy to give a description of these sets in terms of equivalence relations. The idea is to drop the second condition of the definition of $\sim_{k,t}$, which does not make sense for infinite words. Thus, we define an equivalence $\sim_{k,t}$ on A^ω by setting $u \sim_{k,t} v$ if and only if

(a) for each $n < k$, $p_n(u) = p_n(v)$,

(b) $u \equiv_{k,t} v$.

Observe that $\simeq_{k+1,t}$ refines $\simeq_{k,t}$.

Proposition 5.3. *A subset of A^ω is threshold prefix-factors testable if and only if it is a finite union of $\sim_{k,t}$-classes of A^ω for some positive integers k and t.*

Proof. Let x be a finite word of length k. Then xA^ω is saturated modulo $\sim_{k+1,1}$. Indeed, if $u \sim_{k+1,1} v$ and $u \in xA^\omega$, then $p_k(u) = p_k(v) = x$. In the same way, $F_\omega(x, t)$ is saturated modulo $\sim_{k,t}$. Indeed, if $u \sim_{k,t} v$, then $\left[{u \atop x}\right] \equiv_t \left[{v \atop x}\right]$. Thus, if $u \in F_\omega(x, t)$, $\left[{u \atop x}\right] \geq t$ and thus $\left[{v \atop x}\right] \geq t$, that is, $v \in F_\omega(x, t)$. □

The characterization of the class of threshold prefix-factors testable sets within the largest class of threshold locally testable sets is a much more difficult result, whose proof is omitted.

Theorem 5.4. *A subset of A^ω is threshold prefix-factors testable if and only if it is threshold locally testable and Δ_2.*

We are now ready to state the announced characterization, which makes a subtle distinction between the finite and the infinite words.

Theorem 5.5. *A subset of A^+ is definable in $F_1(S)$ if and only if it is threshold locally testable. A subset of A^ω is definable in $F_1(S)$ if and only if it is threshold prefix-factors testable.*

Proof. The conversion of a set of words into a $F_1(S)$-formula relies on two straightforward lemmas. The first one takes care of subsets of the form uA^*, A^*u and uA^ω.

Lemma 5.6. *Let $u = a_1 \cdots a_k$. Then $uA^* = L^+(\varphi \wedge x_1 = 0)$, $A^*u = L^+(\varphi \wedge x_k = \max)$, and $uA^\omega = L^\omega(\varphi \wedge x_1 = 0)$ where*

$$\varphi = \exists x_1 \cdots \exists x_k \left(\bigwedge_{i=1}^{k-1} ((x_{i+1} = x_i + 1) \wedge \bigwedge_{i=1}^{k} \mathbf{a}_i(x_i) \right)$$

The second one covers the subsets of the form $F_\omega(x, t)$.

Lemma 5.7. *Let $x = a_1 \cdots a_k$ be and let t be a positive integer. Then $F_\omega(x, t) = L^\omega(\varphi)$ where*

$$\varphi = \exists x_{1,1} \cdots \exists x_{s,k} \Big(\bigwedge_{\substack{1 \le i \le s \\ 1 \le j \le k-1}} (x_{i,j+1} = x_{i,j} + 1))$$

$$\bigwedge_{1 \le i < j \le s} (x_{i,1} \neq x_{j,1}) \qquad \bigwedge_{\substack{1 \le i \le s \\ 1 \le j \le k}} \mathbf{a}_j(x_{i,j}) \Big)$$

It view of Lemmas 5.6 and 5.7, it follows from the definitions that a threshold locally testable subset of A^+ and a threshold prefix-factors testable subset of A^ω are $F_1(S)$-definable.

Passing from formulas to sets of words is a bit more difficult, and requires an auxiliary result on Fraïssé-Ehrenfeucht games.

Lemma 5.8. *Let w_1 and w_2 be two words, both simultaneously finite or infinite. Let r be a positive integer and let $R = 3^r$. If $w_1 \sim_{R,3R} w_2$, then $w_1 \equiv_r w_2$.*

Proof. We prove the lemma when w_1 and w_2 are infinite. For w_1 and w_2 finite, the proof should be slightly modified to take into account possible moves in the last R positions of each word.

We shall prove that if $w_1 \sim_{R,3R} w_2$, then Player II has a winning strategy in the game (w_1, w_2, r). The key idea is that each position defines a region around it. Intuitively, the strategy of Player II consists to reply to each move of Player I, defining some region S, by a move played in a region isomorphic to S, but on the other word. The size of these regions will be large enough to ensure that, at the end, the possible relations between all the played positions are the same on the two words.

We now give the details of the strategy. Suppose that, after i moves, the tokens $z_1, \ldots, z_i$ have been played on each word. For each position m carrying a token on w_j $(j = 1, 2)$, consider the factor of length 2.3^{r-i} centered in m

$$w_j[m - 3^{r-i} + 1, m + 3^{r-i}]$$

If the bounds happens to be negative or larger than the length of w_j, we simply reduce the interval accordingly. We now form the union of the supports of all these factors. We obtain in this way a union of disjoint connected subsets which are the supports of the factors

$$\{u_{i,j,1}, \ldots, u_{i,j,k_{i,j}}\}$$

In this way, we have for $i = r$, $|u_{r,j,h}| \geq 2$ and for $1 \leq i \leq r$, $|u_{i-1,j,h}| \geq 2.3^{r-i}$.

We shall verify by induction on i that Player II can always apply the following strategy: at each round p, he plays in such a way that the factors $u_{i,1,h}$ and $u_{i,2,h}$ associated with the two tokens z_p are equal, and that the two tokens occupy the same relative position within the two factors.

This strategy is winning. Indeed, for $i = r$, if w_j satisfies $z_p = z_q + 1$, the tokens z_p and z_q are placed in the same factor $u_{r,j,h}$ of w_i (which is of length at least equal to 2) and hence the other word satisfies the same formula.

For $i = 0$, there is nothing to prove. Suppose the assertion has been proved for some $i \geq 0$. Then Player I plays the token z_{i+1} in one of the words, say w_1. If the factor

$$v = w_1[m - 3^{r-i-1} + 1, m + 3^{r-i-1}]$$

is entirely contained in one of the factors $u_{i,1,h}$, then Player II can play at the corresponding position of the factor $u_{i,2,h}$ and the condition will hold for the set of disjoint factors at the end of the next round. If, on the contrary, the above factor v is not entirely contained in one of the factors $u_{i,1,h}$, it is necessarily disjoint of all the factors $u_{i+1,1,h}$ associated with the tokens $z_1, \ldots, z_i$ (see Figure 5.2). To conclude the proof, it suffices now to show that w_2 contains a factor equal to v and disjoint of the factors $u_{i+1,1,h}$

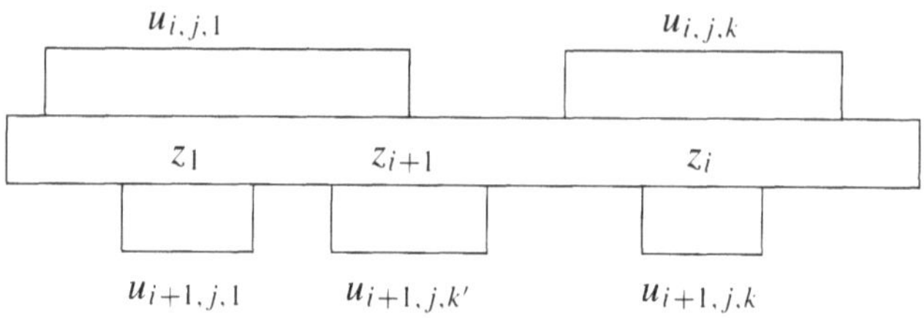

Figure 5.2. The strategy of Player II for the $i+1$-th move.

associated with the first i tokens. Player II can then place her token z_{i+1} in the middle of this factor. First observe that the sum of the lengths of the factors $u_{i,1,h}$ is at most $2i.3^{r-i} < 3^{r+1}$, and therefore the factor v occurs at least 3^{r+1} times in the factors $u_{i,1,h}$. Since $w_1 \sim_{R,3R} w_2$ and since the factors $u_{i,1,h}$ occur in the same way in the two words, there is at least one occurrence of v in w_2 out of the factors $u_{i,2,h}$. □

We can now complete the proof of Theorem 5.5. Suppose indeed that a subset of A^+ (resp. A^ω) is definable in $F_1(S)$ by a formula of quantifier height r. Then it is a finite union of $\equiv_r$-classes and Lemma 5.8 shows that it is a finite union of $\sim_{R,3R}$-classes. Therefore it is threshold locally testable (resp. threshold prefix-factors testable). □

Corollary 5.9. *It is decidable to know whether or not a recognizable subset of A^+ (resp. A^ω) is definable in $F_1(S)$.*

Proof. By Theorem 5.5, a subset of A^+ is definable in $F_1(S)$ if and only if it is TLT. Now, by Theorem B.5.10, one can decide whether a recognizable subset of A^+ is TLT.

By Theorem 5.5, a subset of A^ω is definable in $F_1(S)$ if and only if it is TPFT. Now, by Theorem 5.4, a subset of A^ω is TPFT if and only if it is threshold locally testable and Δ_2. Finally, by Theorem VII.3.14 (resp. VI.3.7), one can decide whether a recognizable subset of A^ω is TPFT (resp. Δ_2). □

Example 5.2. The set $X = (ab)^+$ is defined by the formula

$$\varphi = \mathbf{a}(0) \wedge \mathbf{b}(\max -1) \wedge \neg\exists x\ (\mathbf{a}(x) \wedge \mathbf{a}(x+1)) \wedge \neg\exists x\ (\mathbf{b}(x) \wedge \mathbf{b}(x+1))$$

and the set of words having two occurrences of a is defined by the formula

$$\exists x \exists y\ (\mathbf{a}(x) \wedge \mathbf{a}(y) \wedge \neg(x = y))$$

We also deduce from Theorems 5.5 that the logic of the successor is less expressive than the logic of the linear order.

Corollary 5.10. *The relation $<$ cannot be defined in $F_1(S)$.*

Proof. We have seen above that $X = a^*ba^*ca^*$ is not threshold locally testable. Therefore, by Theorem 5.5, it is not definable in $F_1(S)$. However, it is clearly definable in $F_1(<)$, whence the result. □

An analysis of the proof of Theorem 5.5 also shows that the hierarchy of $F_1(S)$ built on the number of quantifier alternations collapses.

Proposition 5.11. *Every formula of $F_1(S)$ is equivalent to a boolean combination of formulas of Σ_2.*

Proof. This is clear by Lemmas 5.6 and 5.7, since the constants 0 and max are definable by formulas of Σ_2. □

One can also characterize (Exercise 11) the sets definable by boolean combinations of Σ_1-formulas in the language of the successor (without the constants 0 and max). They are the *threshold strongly locally testable* sets, which are the unions of classes of the equivalence $\equiv_{r,k}$ defined as follows: $w_1 \equiv_{r,k} w_2$ if, for each $v \in A^+$ of length $\leq k$, $\binom{w_1}{v} \equiv_r \binom{w_2}{v}$.

6 Temporal logic

Another way to express properties on words is supplied by *temporal logic*. This logic makes use of a formalism slightly different from that of the customary logic. It avoids the use of variables to the price of extraneous operators. The ingredients are:

(1) a proposition p_a for each letter $a \in A$,

(2) the connectives $\vee$, $\wedge$ and $\neg$,

(3) the temporal operators **X** ("next"), **F** ("eventually") and **U** ("until")

and the formation rules of formulas are quite simple:

(1) for each $a \in A$, p_a is a formula,

(2) if φ and ψ are formulas, so are $\varphi \vee \psi$, $\varphi \wedge \psi$, $\neg\varphi$, $\mathbf{X}\varphi$, $\mathbf{F}\varphi$, $\varphi\,\mathbf{U}\,\psi$.

Denote by TL the set of formulas obtained in this way. They are called *temporal formulas*. Some authors also use the notation $\circ$ instead of **X** and $\diamond$ instead of **F**.

Semantics is defined by induction on the formation rules of temporal formulas. Given a word $w \in A^\infty$, and $n \in \mathrm{Dom}(w)$, we say that w satisfies φ at the instant n, denoted by $(w, n) \models \varphi$, in the following way:

(1) $(w, n) \models p_a$ if the n-th letter of w is an a.

(2) $(w, n) \models \varphi \vee \psi$ (resp. $\varphi \wedge \psi$, $\neg\varphi$) if $(w, n) \models \varphi$ or $(w, n) \models \psi$ (resp. if $(w, n) \models \varphi$ and $(w, n) \models \psi$, if (w, n) does not satisfy φ).

(3) $(w, n) \models \mathbf{X}\varphi$ if $(w, n+1)$ satisfies φ.

(4) $(w, n) \models \mathbf{F}\varphi$ if there exists m such that $n \leq m \leq |w|$ and $(w, m) \models \varphi$.

(5) $(w, n) \models \varphi \mathbf{U} \psi$ if there exists m such that $n \le m \le |w|$, $(w, m) \models \psi$ and, for each k such that $n \le k < m$, $(w, k) \models \varphi$.

Whether w satisfies φ or not at the instant n only depends on the word $w = w_n w_{n+1} \cdots$. If φ is a temporal formula, w is said to satisfy φ if $(w, 0) \models \varphi$.

Example 6.1. Let $w = abbababcba$. Then $(w, 4) \models p_a$ since the fourth letter of w is an a, $(w, 4) \models \mathbf{X} p_b$ since the fifth letter of w is a b and $(w, 4) \models \mathbf{F}(p_c \wedge \mathbf{X} p_b)$ since cb is a factor of $babcba$.

With each formula of temporal logic φ, is associated the set of words satisfying φ:

$$S(\varphi) = \{u \in A^\infty \mid u \text{ satisfies } \varphi\}$$

We also set $S^+(\varphi) = S(\varphi) \cap A^+$ and $S^\omega(\varphi) = S(\varphi) \cap A^\omega$.

Example 6.2. The set $X = (a^*b)^\omega$ of the words having infinitely many b's is defined by the formula $\neg\mathbf{F}\neg\mathbf{F} p_b$

We have just defined "future" temporal formulas but one can define in the same way "past" temporal formulas by reversing time: it suffices to replace "next" by "previous" (symbol **Y**), "eventually" by "sometimes" (symbol **G**) and "until" by "since" (symbol **S**). The corresponding semantics are modified as follows.

(3′) $(w, n) \models \mathbf{Y}\varphi$ if $n > 1$ and $(w, n-1)$ satisfies φ.

(4′) $(w, n) \models \mathbf{G}\varphi$ if there exists $m \le n$ such that $(w, m) \models \varphi$.

(5′) $(w, n) \models \varphi \mathbf{S} \psi$ if there exists $m \le n$ such that $(w, m) \models \psi$ and for every k such that $m < k \le n$, $(w, k) \models \varphi$.

Figure 6.1 illustrates the symmetry between the operators "until" and "since". If φ is a past temporal formula, we say that w satisfies φ if $(w, |w|) \models \varphi$. The language defined by a formula φ is the set $L(\varphi)$ of all words of A^+ that satisfy φ.

We shall see (Theorem 6.7) that temporal logic and first-order logic of the linear order have the same expressive power. More precisely, we shall see that a set of finite (resp.

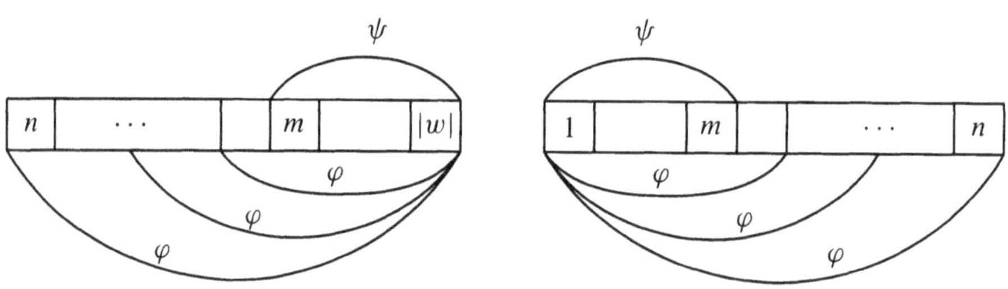

Figure 6.1. A diagram for $(w, n) \models \varphi \mathbf{U} \psi$ and for $(w, n) \models \varphi \mathbf{S} \psi$.

infinite) words is expressible by a temporal logic formula if and only if it is star-free. This result requires a new characterization of star-free sets.

6.1 Another characterization of star-free sets. A prefix-free set X of A^+ is said to have *bounded synchronization delay* if there is an integer $d \geq 0$ such that for all $u, w \in A^*$ and $v \in X^d$, one has

$$uvw \in X^* \text{ implies } uv, w \in X^*.$$

The smallest integer such that the above property is true is called the synchronization delay of X.

Example 6.3. Any subset B of A is prefix-free and has synchronization delay 0. For $A = \{a, b\}$, the prefix-free set a^*b has synchronization delay 1. On the same alphabet, the prefix-free set $X = \{ab, ba\}$ does not have a bounded synchronization delay. Indeed, for any integer $d \geq 0$, $a(ba)^d b \in X^*$ although $b \notin X^*$, contradicting the definition for $u = a$, $v = (ba)^d$ and $w = b$.

A word w is called a *constant* for a subset S of A^* if for all $u_1, u_2, u_3, u_4 \in A^*$, the following diamond property holds.

$$u_1wu_2, u_3wu_4 \in S \text{ implies } u_1wu_4, u_3wu_1 \in S$$

The set of constants is clearly an ideal of A^*. The following property will be used below.

Lemma 6.1. *If X is prefix-free with synchronizing delay d, any word having a factor in X^d is a constant for X^*.*

Proof. Let $u, v \in A^*$, $x \in X^d$ and let $w = uxv$. Suppose that u_1wu_2 and u_3wu_4 are in X^*. Since X has delay d, $u_1uxvu_2 \in X^*$ implies $u_1ux, vu_2 \in X^*$ and $u_3uvwu_4 \in X^*$ implies $u_3ux, vu_4 \in X^*$. Thus $u_1wu_4, u_3wu_2 \in X^*$, proving that w is a constant. □

Let $\mathcal{F}$ be a class of sets. A mapping $\alpha : B \to \mathcal{P}(A^*)$ is called an *$\mathcal{F}$-substitution* if the following conditions are satisfied:

(1) There exists a partition of A into two subsets A_0, A_1 such that, for every $b \in B$, $\alpha(b) \subset A_0^*A_1$.

(2) For all $b, c \in B$ with $b \neq c$, $\alpha(b) \cap \alpha(c) = \emptyset$.

(3) For all $b \in B$, $\alpha(b) \in \mathcal{F}$.

A mapping α satisfying only (1) and (2) is called a *marked substitution*. If $\alpha : B \to A^*$ is a marked substitution, we also denote by α the mapping from B^* into $\mathcal{P}(A^*)$ defined by $\alpha(1) = \{1\}$ and by

$$\alpha(x) = \alpha(b_1)\alpha(b_2)\cdots\alpha(b_n) \text{ for } x = b_1b_2\cdots b_n$$

Note that if $\alpha : B \to A^*$ is a marked substitution, the set $Z = \alpha(B)$ is prefix-free with synchronization delay 1. Indeed, if $uvw \in Z^*$ for $v \in A_0^* A_1$, then $uv, w \in Z^*$. More generally, the following property holds.

Proposition 6.2. *Let $\alpha : B^* \to A^*$ be a marked substitution. If $X \subset B^*$ is prefix-free with bounded synchronization delay, then $\alpha(X)$ is also prefix-free with bounded synchronization delay.*

Proof. We first verify that $Y = \alpha(X)$ is prefix-free. Let $u = a_1 \cdots a_n$ and $v = b_1 \cdots b_m$ be two elements of A^* with $a_i, b_j \in A$. Suppose that an element x of $\alpha(u)$ is a prefix of an element y of $\alpha(v)$. Then there exist $x_1 \in \alpha(a_1), \ldots, x_n \in \alpha(a_n)$, $y_1 \in \alpha(a_1)$, $\ldots, y_m \in \alpha(a_m)$ and $z \in A^*$ such that $x = x_1 \cdots x_n$, $y = y_1 \cdots y_m$ and $xz = y$. Since $\alpha(B)$ is prefix-free, it follows that $x_1 = y_1, \ldots, x_n = y_n$ and by (2), this forces u to be a prefix of v. Thus the image of a prefix-free set is prefix-free.

Let d be the synchronization delay of X. Suppose that for $u, v \in A^*$, $y \in Y$ and $z \in Y^d$, we have $uyzv \in Y^*$. We use the fact that $\alpha(B)$ is prefix-free with synchronization delay 1. Since $y, z \in A^* A_1$, we have $uy, z, v \in \alpha(B)^*$. Thus there exist $r, s \in A^*$ and $x \in X^d$ such that $uy \in \alpha(r)$, $z \in \alpha(x)$ and $s \in \alpha(v)$. Since X has delay d, we deduce from $rxs \in X^*$ that $rx, s \in X^*$. This implies $uyz, v \in Y^*$, showing that Y has synchronization delay at most $d + 1$. □

The following result gives a description of the class of star-free sets in terms of rational operations with a restricted star operation. Note that complementation is not used.

Theorem 6.3. *The class of star-free sets in A^* is the smallest class of subsets of A^* containing $\emptyset$ and $\{a\}$ for each $a \in A$ and which is closed under union, concatenation and star operation restricted to prefix-free sets with bounded synchronization delay.*

Proof. Let $\mathcal{F}$ be the class of sets described in the statement of the theorem. We first prove that any element of $\mathcal{F}$ is star-free. For a set $X \subset A^*$, we set

$$W(X) = \{w \in A^* \mid A^* w A^* \cap X = \emptyset\}$$

It is the set of words which are not a factor of a word in X. As usual, the set of words which are factors of some word in X is denoted by $F(X)$. The next lemma gives an expression of $W(X^*)$ in a particular case.

Lemma 6.4. *Let X be a star-free, prefix-free set with bounded synchronization delay. Then $W(X^*) = A^* V A^*$ where $V = (A^* \setminus A^* X^{d+1} A^*) \cap (A^* \setminus F(X^{d+2}))$.*

Proof. The set V defined in the above statement can be described as follows: it is the set of words which are not a factor of a word in X^{d+2}, and which do not contain a word of X^{d+1} as a factor.

We first show that $A^*VA^* \subset W(X^*)$. It is enough to show that $V \subset W(X^*)$. Let $v \in V$. We argue by contradiction. Let n be the least integer such that v is a factor of a word in X^n. Then $uvw = x_1x_2\cdots x_n$ for some $u, v \in A^*$ and $x_i \in X$. By the definition of V, we have $n > d+2$ and by the minimality of n, u is a prefix of x_1 and w is a suffix of x_n. Thus $x_2\cdots x_{n-1}$ is a factor of v, a contradiction with the fact that v does not have a factor in X^{d+1}.

Let us show the opposite inclusion. Let V' be the set of words in $W(X^*)$ without a proper factor in $W(X^*)$. It suffices to show that $V' \subset V$. Let $v \in V'$. The only thing to prove is that v does not have a factor in X^{d+1}. If $v \in A$, it is clear. Otherwise, let $v = ahb$ with $a, b \in A$ and $h \in A^*$. Suppose first that h has a factor in X^d. By the definition of V', the words ah and hb are not in $W(X^*)$ and thus there are words u_1, u_2, u_3, u_4 such that

$$u_1ahu_2, u_3hbu_4 \in X^*$$

Since h has a factor in X^d, it is a constant by Lemma 6.1 and hence $u_1ahbu_4 \in X^*$, a contradiction. Let us now suppose that v has a factor in X^{d+1}. Thus $ahb = uxv$ with $u, v \in A^*$ and $x \in X^{d+1}$. Since h does not have a factor in X^d, the only possibility is that $u = v = 1$. But this contradicts the fact that $v \in W(X^*)$. □

The next lemma completes the proof that any element of $\mathcal{F}$ is star-free.

Lemma 6.5. *If X is a star-free, prefix-free set with bounded synchronization delay, then X^* is star-free.*

Proof. Let d be the synchronization delay of X and let

$$U = (X^dA^* \cap A^*X^d) \setminus W(X^*)$$

We claim that $U = X^dX^*$. It is clear that X^dX^* is a subset of U. To prove the opposite inclusion, consider an element u of U. There exists $x, y \in X^d$ and $r, s \in A^*$ such that $u = xr = sy$. Since $u \notin W(X^*)$, there exist some $p, q \in A^*$ such that $puq \in X^*$. Since X has synchronization delay d, it follows from $pxrq \in X^*$ that $rq \in X^*$. Since $x \in X^*$, this implies $xrq \in X^*$. Then, from $xrq = syq \in X^*$, we conclude that $u = sy$ is in X^*, proving the claim.

We finally obtain the formula

$$X^* = 1 \cup X \cup \ldots \cup X^{d-1} \cup U$$

and since the set $W(X^*)$ is star-free by Lemma 6.4, the set U is also star-free. Hence X^* itself is star-free. □

To prove conversely that any star-free set is in $\mathcal{F}$, we use a decomposition result similar to Theorem A.4.16, based on the following lemma.

Lemma 6.6. *The class $\mathcal{F}$ is closed under $\mathcal{F}$-substitution.*

Proof. Let $\alpha : B^* \to A^*$ be an $\mathcal{F}$-substitution and let $X \in \mathcal{F}$. If $X = \emptyset$, then $\alpha(X) = \emptyset$. If $X = \{a\}$ for some $a \in A$, then $\alpha(X) \in \mathcal{F}$ since α is an $\mathcal{F}$-substitution. Finally, let Y and Z be sets of $\mathcal{F}$ such that $\alpha(Y), \alpha(Z) \in \mathcal{F}$. If $X = Y \cup Z$, then $\alpha(X) \in \mathcal{F}$ since $\alpha(X \cup Y) = \alpha(X) \cup \alpha(Y)$. Next, if $X = YZ$, we have again $X \in \mathcal{F}$ since $\alpha(XY) = \alpha(X)\alpha(Y)$. Finally, assume that $X = Y^*$ where $Y \in \mathcal{F}$ is a prefix-free set with bounded synchronization delay such that $\alpha(Y) \in \mathcal{F}$. By Proposition 6.2, $\alpha(Y)$ is prefix-free with bounded synchronization delay. Thus $\alpha(X)^*$ is in $\mathcal{F}$. □

End of the proof of Theorem 6.3. Let $X \subset A^*$ and let $\varphi : A^* \to M$ be the morphism onto the syntactic monoid of X. We prove by induction on Card(M) that if M is aperiodic, then $X \in \mathcal{F}$. According to Proposition A.4.14, three cases may now occur.

First assume that $M = \{1, x, x^2, \ldots, x^n\}$ is a cyclic monoid, defined by the relation $x^{n+1} = x^n$. Let $\varphi : A^* \to M$ be the syntactic morphism of X and let $B = \varphi^{-1}(1)$. For $0 \leq k < n$, the set $\varphi^{-1}(x^k)$ is a finite union of sets of the form $B^*a_1B^*a_2B^* \cdots B^*a_rB^*$, where $r \leq k$ and $a_1, \ldots, a_r \in A \setminus B$. Furthermore, the set $\varphi^{-1}(x^n)$ can be written as a finite union of sets of the form $B^*a_1B^*a_2B^* \cdots B^*a_rA^*$, where $r \leq n$ and $a_1, \ldots, a_r \in A \setminus B$. All these sets belongs to $\mathcal{F}$, since B is a prefix-free set with bounded synchronization delay. Thus X itself is in $\mathcal{F}$.

If M is isomorphic to $\tilde{U}_n$ for some $n > 0$, then by Proposition B.3.7, X is a disjoint union of sets of the form $a_1\{a_1\}^*a_2\{a_1, a_2\}^* \cdots a_n\{a_1, a_2, \ldots, a_n\}^*$ where the a_i's are distinct letters of A. Thus X is in $\mathcal{F}$, since each subset of A is a prefix-free set with bounded synchronization delay.

Otherwise, there is a partition $A = A_0 \cup A_1$ such that $\varphi(A_0^*A_1)^*$ and $\varphi(A_0^*)$ are proper submonoids of M. Without loss of generality, we may assume that $X = \varphi^{-1}(m)$ for some $m \in M$. We claim that

$$X = \bigcup_{rs=m} X_rY_s$$

where

$$X_r = \varphi^{-1}(r) \cap A_0^* \quad \text{and} \quad Y_s = \varphi^{-1}(s) \cap (A_0^*A_1)^*$$

By the induction hypothesis, we have $X_r \in \mathcal{F}$. We now show that $Y_s \in \mathcal{F}$.

Consider an alphabet S in bijection with the set $\varphi(A_0^*A_1)$. Let $\beta : S \to \varphi(A_0^*A_1)$ be a bijection. Define a mapping $\alpha : S \to \mathcal{P}(A^*)$ by

$$\alpha(x) = \varphi^{-1}(\beta(x)) \cap A_0^*A_1$$

for each $x \in S$. In other terms, $\alpha(x) = \{u \in A_0^*A_1 \mid \varphi(u) = \beta(x)\}$. One can verify that for every $x \in S^*$, $\alpha(x) = \varphi^{-1}(\beta(x)) \cap (A_0^*A_1)^*$. The mapping α is an $\mathcal{F}$-substitution. Indeed, conditions (1) and (2) of the definition clearly hold. For each $s \in S$, the set $\alpha(s)$ is in $\mathcal{F}$ by induction hypothesis since it is recognized by $\beta(S^*) = \varphi(A_0^*A_1)^*$ which is strictly included in M. Thus condition (3) is also satisfied.

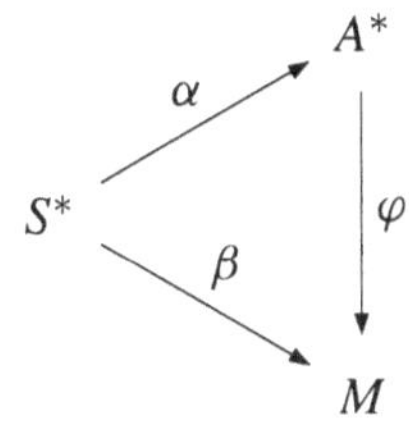

For every $r \in \varphi(A_0^* A_1)^*$, we have $Y_r = \alpha(\beta^{-1}(r))$. The set $\beta^{-1}(r)$ is in $\mathcal{F}$ by induction hypothesis since it is recognized by $\beta(S^*)$. Thus, by Lemma 6.6, we have $Y_r \in \mathcal{F}$ and this concludes the proof. □

6.2 Star-free sets and temporal logic. Temporal logic is actually equivalent to first-order logic of the linear order.

Theorem 6.7. *Temporal logic is equivalent to $F_1(<)$ on finite words.*

We first show that temporal logic is weaker that $F_1(<)$, by induction on the formation rules of temporal formulas. This is clear for the predicates p_a and for the use of connectives. Suppose next that a formula $\varphi \in TL$ is equivalent to a formula $\psi \in F_1(<)$. Then we make use of the relativised formula $\psi(x, y)$ of Proposition 2.6 to obtain

$$\begin{aligned} \mathbf{X}\varphi &\equiv \psi(x+1, \max) \\ \mathbf{F}\varphi &\equiv \exists z \psi(z, \max) \\ \varphi \, \mathbf{U} \, \varphi' &\equiv \exists x \big[\psi'(x, \max) \wedge \forall y \big(y < x \rightarrow \psi(y, \max)\big)\big] \end{aligned}$$

It follows by Theorem 4.1 that a set of finite or infinite words definable by a temporal logic formula is star-free. The converse statement follows from the following result.

Theorem 6.8. *Every subset of A^+ recognized by an aperiodic semigroup is definable by a temporal logic formula.*

Proof. Since the reverse of an aperiodic semigroup is also aperiodic, it suffices to prove the dual version of the theorem, obtained by using *past* temporal logic.

Let $\mathbf{C}$ be the class of all semigroups S such that *every* language recognized by S is definable in past temporal logic (PTL). By Schützenberger's theorem (Theorem B.4.1), it suffices to show that each aperiodic semigroup belongs to $\mathbf{C}$. The class $\mathbf{C}$ is certainly closed under division, because if S divides T, every language recognized by S is also recognized by T. Next, the trivial semigroup $\{1\}$ belongs to $\mathbf{C}$, since the languages of A^+ recognized by $\{1\}$ are A^+ and the empty set. Now, by Theorem A.4.16, it remains to show that if $T \in \mathbf{C}$, then $U_2 \circ T \in \mathbf{C}$.

Consider a subset K of A^+ recognized by $\eta : A^+ \rightarrow U_2 \circ T$. Denote by π the natural projection from $U_2 \circ T$ onto T and let $\gamma = \pi \circ \eta$. Set $B = T^1 \times A$. By Theorem A.5.4,

K is a finite union of languages of the form $U \cap \sigma^{-1}(V)$ where $\sigma : A^+ \to B^+$ is the sequential function associated with γ, $U \subset A^+$ is recognized by T and $V \subset B^+$ is recognized by U_2. First, the formulas $L(\neg\varphi) = A^+ \setminus L(\varphi)$ and $L(\varphi \vee \psi) = L(\varphi) \cup L(\psi)$ show that PTL-definable languages are closed under boolean operations. Thus it suffices to show that every language of the form $U \cap \sigma^{-1}(V)$ is PTL-definable. Since $T \in \mathbf{C}$, U is PTL-definable by definition. Furthermore, by the dual version of Proposition B.3.7, V is a boolean combination of languages of the form B^*bC^*, where $b \in B$ and $C \subset B$. Since σ^{-1} commutes with boolean operations, it remains to show that languages of the form $\sigma^{-1}(B^*bC^*)$ are PTL-definable. We claim that

$$\sigma^{-1}(B^*bC^*) = \sigma^{-1}(B^*C)\,\mathbf{S}\,\sigma^{-1}(B^*b) \tag{6.1}$$

Indeed, let $u = a_1 \cdots a_n$ be a word of A^+ and let $\sigma(a_1 \cdots a_n) = b_1 \cdots b_n$. Then $\sigma(u) \in B^*bC^*$ if and only if there exists an i such that $b_i = b$ and, for every $j > i$, $b_j \in C$. This is equivalent to saying that $\sigma(a_1 \cdots a_i) \in B^*b$ and for every $j > i$, $\sigma(a_1 \cdots a_j) \in B^*C$, and this proves (6.1). Now

$$\sigma^{-1}(B^*C) = \bigcup_{b \in C} \sigma^{-1}(B^*b)$$

and therefore it suffices to show that the languages of the form $\sigma^{-1}(B^*b)$ are PTL-definable. Set $b = (t, a)$. Then we have

$$\sigma(a_1 \cdots a_n) = (1, a_1)(\gamma(a_1), a_2) \cdots (\gamma(a_1 \cdots a_{n-1}), a_n).$$

It follows that $\sigma(a_1 \cdots a_n) \in B^*b$ if and only if $\gamma(a_1 \cdots a_{n-1}) = t$ and $a_n = a$. Therefore $\sigma^{-1}(B^*b) = La$, where $L = \{u \in A^+ \mid \gamma(u) = t\}$. Thus L is recognized by T and since $T \in \mathbf{C}$, is PTL-definable. Now, since $L(\varphi)a = L(\mathbf{Y}\varphi \wedge p_a)$, $La = \sigma^{-1}(B^*b)$ is PTL-definable and this concludes the proof. □

We now extend Theorem 6.8 to infinite words.

Theorem 6.9. *The class of subsets of A^ω definable in temporal logic is the class of star-free sets.*

Proof. We first prove that any set definable in temporal logic is star-free. We have first for any $a \in A$, $S(p_a) = aA^\omega$, which is star-free. Next, consider a formula built from two formulas φ, ψ such that $X = S(\varphi)$ and $Y = S(\psi)$ are star-free. Since the star-free sets over A^* form a boolean algebra, the sets $S(\varphi \vee \psi) = X \cup Y$ and $S(\neg\varphi) = A^\omega \setminus X$ are star-free. We also have

$$S(\mathbf{X}\varphi) = AX, \quad S(\mathbf{F}\varphi) = A^*X$$

which are also star-free. The until operator is treated as in the case of finite words. For $X, Y \subset A^\omega$, we set

$$X \mathbf{U} Y = \{w \in A^* \mid x = uv \text{ where } v \in Y \text{ and } u'v \in X \text{ for each suffix } u' \text{ of } u\}$$

Actually, we have $S(\varphi \mathbf{U} \psi) = S(\varphi) \mathbf{U} S(\psi)$.

For a set Z of words, we set $\rho(Z) = A^* \setminus A^*(A^* \setminus Z)$. It is the set of words with all their suffixes in Z. If Z is star-free, so is $\rho(Z)$. The following lemma proves that $X \mathbf{U} Y$ is star-free if X and Y are star-free.

Lemma 6.10. *There is a partition $Y = Y_1 \cup \ldots \cup Y_n$ of Y in a finite number of subsets such that*

$$X \mathbf{U} Y = \bigcup_{i=1}^{n} \rho(Z_i) Y_i$$

with $Z_i = \{u \in A^* \mid uY_i \subset X\}$.

Proof. Let $\alpha : A^\infty \to S$ be a morphism onto a finite ω-semigroup $S = (S_+, S_\omega)$ recognizing X. We have the partition $Y = \bigcup_{s \in S} Y_s$ with $Y_s = (\alpha^{-1}(s) \cap Y)$. Let for each $s \in S$, $Z_s = \{u \in A^* \mid uY_i \subset X\}$. Since α recognizes X, we have for any word $u \in A^*$ and $s \in S$ the three conditions $uY_s \cap X \neq \emptyset$, $uY_s \subset X$ and $u \in Z_s$ are equivalent. This proves that $X \mathbf{U} Y = \bigcup_{i=1}^{n} \rho(Z_i) Y_i$. □

To prove conversely that any star-free set is definable, we prove the following lemma.

Lemma 6.11. *Let X be a prefix-free set with bounded synchronization delay. Let $u \in A^*$, $v \in X^{2d}$, and $w \in Y$ be such that*

(1) *$uvw \in X^d A^\omega$ and*

(2) *$u'vw \in X^{d+1} A^\omega \cup (A^\omega \setminus X^d A^\omega)$ for each suffix u' of u.*

Then $uvw \in X^* Y$.

Proof. By induction on the length of u. The result obviously holds for $|u| = 0$. Suppose that the result holds for $|u| \leq n - 1$ and let us suppose that $|u| = n$. Since condition (ii) holds with $u' = u$, we have $uvw = x_1 x_2 \cdots x_{d+1} r$ with $x_1, \ldots, x_{d+1} \in X$ and $r \in A^*$. We distinguish three cases.

(1) The word x_1 is a prefix of u. Let $u = x_1 q$. We can then apply the induction hypothesis to (q, v, w). We conclude that $qvw \in X^* Y$ and thus $uvw \in X^* Y$.

(2) The word uv is a prefix of $x_1 \cdots x_{d+1}$. Let $x_1 \cdots x_{d+1} = uvr$. Since X has delay d, we obtain that $uv \in X^*$. Thus $uvw \in X^* Y$ again.

(3) Otherwise, we have $|u| < |x_1|$ and $|x_1 \cdots x_d| < |uv|$. Then we have $x_1 = up$ and $v = px_2 \cdots x_{d+1} q$ for some $p, q \in A^*$. Since X has delay d, we have $q \in X^*$. Thus $uvw = x_1 \cdots x_d q w$ is in $X^* Y$ in this case also.

This completes the proof of the lemma. □

Lemma 6.12. *Let X be a prefix-free set with bounded synchronization delay $d \geq 1$ and let $Y \subset A^\omega$. Then*

$$X^*Y = Y \cup XY \cup \ldots \cup X^{2d-1}Y \cup U$$

with

$$U = X^d A^\omega \cap \Big(\big(X^{d+1}A^\omega \cup (A^\omega \setminus X^d A^\omega)\big) \mathbf{U}\, X^{2d}Y\Big).$$

Proof. Let $Z = Y \cup XY \cup \ldots \cup X^{2d-1}Y \cup U$. To prove that $Z \subset X^*Y$, we only need to prove that $U \subset X^*Y$. Let $z \in U$. Then $z \in X^d A^\omega$ and $z = uvw$ with $u \in A^*$, $v \in X^{2d}$ and $w \in Y$ with $u'vw \in X^{d+1}A^\omega \cup (A^\omega \setminus X^d A^\omega)$ for each suffix u' of u. By Lemma 6.11, we have $z \in X^*Y$.

We now consider the opposite inclusion $X^*Y \subset Z$. Let $z \in X^n Y$ for some $n \geq 0$. If $n < 2d$, then $z \in Z$. Let then $z = uvw$ with $u \in X^*$, $v \in X^{2d}$ and $w \in Y$.

Clearly $z \in X^d A^\omega$. There remains to prove that for each suffix u' of u, $u'vw \in X^{d+1}A^\omega \cup (A^\omega \setminus X^d A^\omega)$. This is equivalent to saying that $u'vw \in X^d A^\omega$ implies $u'vw \in X^{d+1}A^\omega$. Suppose that $u'vw \in X^d A^\omega$. Then $u'vw = xq$ with $x = x_1 x_2 \cdots x_d$, $x_i \in X$ and $q \in A^\omega$. There are two cases.

(1) x is a proper prefix of $u'v$. Let $u = pu'$. Since $z = pu'vw = pxq$, there is a word $s \in A^+$ such that $pxs = pu'v = uv \in X^*$. Since $x \in X^d$, we have $s \in X^*$. Thus $u'v = xs$ is in $X^d A^*$ and $u'vw$ is in $X^{d+1}A^\omega$.

(2) $u'v$ is a prefix of x. Then $x = u'vs$ for some $s \in A^*$. Let $v = v_1 v_2$ with $v_1, v_2 \in X^d$. Since $x \in X^*$ and $v_1 \in X^d$, we have $u'v_1 \in X^+$. Hence

$$z = uv_1v_2w \in X^{d+1}A^\omega.$$

This concludes the proof. □

In the case $d = 0$, X is a subset of the alphabet. One may either use the formula above (a prefix-free set of delay 0 is also of delay 1) or write a simple ad hoc formula, which can be written $X^*Y = Y \cup (XA^\omega \,\mathbf{U}\, XY)$.

End of the proof of Theorem 6.9. By Theorem VI.6.2, the class of star-free sets of infinite words is the least class $\mathcal{C}$ of subsets of A^ω containing $\emptyset$ and A^ω, closed under boolean operations and such that for any star-free set $X \subset A^*$ and any $Y \in \mathcal{C}$, one has $XY \in \mathcal{C}$. By Theorem 6.3, the class $\mathcal{A}$ of star-free subsets of A^* is the smallest class of subsets of A^* containing $\emptyset$, $\{1\}$ and a for each $a \in A$ and which is closed under union, concatenation and star operation restricted to prefix-free sets with bounded synchronization delay.

Let $\mathcal{L}$ be the class of subsets of A^ω definable in temporal logic. We have to prove that $\mathcal{C} \subset \mathcal{L}$. Clearly, the only thing to prove is that for any star-free set $X \in \mathcal{A}$ and each $Y \in \mathcal{L}$ we have $XY \in \mathcal{L}$.

Let $\mathcal{S}$ be the class of all subsets of A^* such that for any $Y \in \mathcal{L}$, we have $XY \in \mathcal{L}$. We prove that $\mathcal{A} \subset \mathcal{S}$. Let first $a \in A$. Then for any $Y \in \mathcal{L}$ such that $Y = S(\varphi)$, we have

$$aY = S(p_a \wedge \mathbf{X}\varphi)$$

Thus $aY \in \mathcal{L}$, hence $\{a\} \in \mathcal{S}$. Let then $X, Y \in \mathcal{S}$ and let $Z \in \mathcal{L}$. Since $XZ, YZ \in \mathcal{L}$, there are formulas φ, ψ such that $XZ = S(\varphi)$, $YZ = S(\psi)$. Then

$$(X \cup Y)Z = S(\varphi \vee \psi)$$

Thus $X \cup Y \in \mathcal{S}$. In the same way, since $YZ \in \mathcal{L}$, we have $(XY)Z = X(YZ) \in \mathcal{L}$. Thus $XY \in \mathcal{S}$. Finally, let $X \in \mathcal{S}$ be a prefix-free set with finite synchronization delay. For any $Y \in \mathcal{L}$, we have by Lemma 6.12,

$$X^*Y = Y \cup XY \cup \ldots \cup X^{2d-1}Y \cup U$$

with

$$U = X^d A^\omega \cap \Big(\big(X^{d+1}A^\omega \cup (A^\omega \setminus X^d A^\omega)\big)\, \mathbf{U}\, X^{2d}Y\Big).$$

Clearly $U \in \mathcal{L}$ and thus $X^*Y \in \mathcal{L}$. This shows that $X^* \in \mathcal{S}$ and completes the proof that $\mathcal{A} \subset \mathcal{S}$, whence $\mathcal{C} \subset \mathcal{L}$. □

7 Restricted temporal logic

If the operator "Until" is omitted, we obtain the *restricted temporal logic* . The sets of finite words definable in this logic are described in the following proposition. The subtle distinction between conditions (2) and (3) will be used in the proof of the main theorem below.

Proposition 7.1. *Let L be a subset of A^+. The following conditions are equivalent:*

(1) *L is RTL-definable,*

(2) *L belongs to the smallest boolean algebra containing the sets aA^* and closed under the operations $L \to A^*L$ and $L \to aL$ for every $a \in A$,*

(3) *L belongs to the smallest boolean algebra containing the sets aA^* and closed under the operations $L \to A^*aL$ and $L \to aL$ for every $a \in A$.*

Proof. Let $\mathcal{C}$ (respectively $\mathcal{C}'$) be the smallest boolean algebra closed under the operations $L \to A^*L$ (respectively $L \to A^*aL$) and $L \to aL$ for every letter $a \in A$. In particular the sets $\emptyset$ and A^+ belong to $\mathcal{C}$ and $\mathcal{C}'$ by definition. We first prove that $\mathcal{C} = \mathcal{C}'$. The inclusion $\mathcal{C}' \subset \mathcal{C}$ follows directly from the formula $A^*aL = A^*(aL)$. The opposite inclusion follows from the formula

$$A^*L = L \cup \bigcup_{a \in A} A^*aL.$$

Thus (2) and (3) are equivalent.

(1) implies (2). We show by induction on the formation rules that $L(\varphi) \in \mathcal{C}$ for every RTL-formula φ. First, if $\varphi = p_a$, then

$$L(p_a) = aA^* \in \mathcal{C}.$$

If φ and ψ are formulas such that $L(\varphi)$ and $L(\psi)$ belong to $\mathcal{C}$, then

$$\begin{aligned} L(\varphi \vee \psi) &= L(\varphi) \cup L(\psi) \in \mathcal{C}, \\ L(\varphi \wedge \psi) &= L(\varphi) \cap L(\psi) \in \mathcal{C}, \\ L(\neg\varphi) &= A^+ \setminus L(\varphi) \in \mathcal{C}, \\ L(\mathbf{X}\varphi) &= AL(\varphi) = \bigcup_{a \in A} aL(\varphi) \in \mathcal{C}, \\ L(\mathbf{F}\varphi) &= A^*L(\varphi) \in \mathcal{C}. \end{aligned}$$

(2) implies (1). Let $\mathcal{F}$ be the set of RTL-definable sets. Then $\mathcal{F}$ contains $aA^* = L(p_a)$, for every $a \in A$. The formulas $L(\varphi) \cup L(\psi) = L(\varphi \vee \psi)$ and $A^+ \setminus L(\varphi) = L(\neg\varphi)$ show that $\mathcal{F}$ is a boolean algebra and the formula $A^*L(\varphi) = L(\mathbf{F}\varphi)$ shows that $\mathcal{F}$ is closed under the operation $L \to A^*L$. Finally, the formula $aL(\varphi) = L(p_a \wedge \mathbf{X}\varphi)$ shows that $\mathcal{F}$ is closed under the operation $L \to aL$, for every letter $a \in A$. Therefore $\mathcal{F}$ contains $\mathcal{C}$. □

We can now state our main result.

Theorem 7.2. *Let L be a subset of A^+. The following conditions are equivalent:*

(1) *L is RTL-definable,*

(2) *the syntactic semigroup of L is locally $\mathcal{L}$-trivial.*

Proof. As for Theorem 6.8, we prove the dual version of the theorem, which states that L is definable in past restricted temporal logic if and only if its syntactic semigroup is $\mathcal{R}$-trivial. Consider the smallest boolean algebra $\mathcal{B}$ containing the sets A^*a and closed under the operations $L \to LaA^*$ and $L \to La$ for every $a \in A$. By Proposition 7.1 and duality, it suffices now to prove that a set of finite words belongs to $\mathcal{B}$ if and only if its syntactic semigroup belongs to $\mathbf{LR}$.
First, $S(A^*a) \in \mathbf{LR}$. Now, by Propositions A.5.2 and A.5.3, $S(LaA^*)$ divides $U_1 \circ S(L)$, and $S(La)$ divides $\bar{2} \circ S(L)$. It follows by Proposition A.4.18, that if $S(L) \in \mathbf{LR}$, then $S(LaA^*) \in \mathbf{LR}$ and $S(La) \in \mathbf{LR}$. Therefore, if $L \in \mathcal{B}$, then $S(L) \in \mathbf{LR}$.

In the other direction, the proof mimics the proof of Theorem 6.8. Let $\mathbf{C}$ be the class of all transformation semigroups X such that *every* set recognized by X belongs to $\mathcal{B}$. The class $\mathbf{C}$ contains the trivial semigroup and is closed under division. Therefore, to show that $\mathbf{C}$ contains $\mathbf{LR}$, it suffices, by Proposition A.4.18, to verify that if $Y \in \mathbf{C}$, then $U_1 \circ Y \in \mathbf{C}$ and $\bar{2} \circ Y \in \mathbf{C}$.

By Theorem A.5.4, every subset of A^+ recognized by $U_1 \circ Y$ (respectively $\bar{\mathbf{2}} \circ Y$) is a finite union of sets of the form $U \cap \sigma^{-1}(V)$ where $\sigma : A^+ \to B^+ = (Q \times A)^+$ is a certain sequential function, $U \subset A^+$ is recognized by Y and $V \subset B^+$ is recognized by U_1 (respectively $\bar{\mathbf{2}}$). Since $Y \in \mathbf{C}$, U belongs to $\mathcal{B}$ by definition. Furthermore, by Proposition A.3.4 (resp. Proposition A.3.2), V is a boolean combination of sets of the form B^*bB^*, (respectively B^*b) where $b \in B$. Since σ^{-1} commutes with boolean operations, it remains to show that the sets of the form $\sigma^{-1}(B^*bB^*)$ (respectively $\sigma^{-1}(B^*b)$) belong to $\mathcal{B}$. We take again the notations used in the definition of σ (cf. Theorem A.5.4). Set $b = (q, a)$ (recall that $B = Q \times A$). Then we have

$$\sigma(a_1 \cdots a_n) = (q_0, a_1)(q_0 \cdot \pi(\eta(a_1)), a_2) \cdots (q_0 \cdot \pi(\eta(a_1 \cdots a_{n-1})), a_n).$$

First assume $q \neq q_0$. Then $\sigma(a_1 \cdots a_n) \in B^*bB^*$ if and only if there exists an index i such that $q_0 \cdot \pi(\eta(a_1 \cdots a_{i-1})) = q$ and $a_i = a$. Therefore $\sigma^{-1}(B^*bB^*) = LaA^*$, where

$$L = \{u \in A^+ \mid q_0 \cdot \pi(\eta(u)) = q\}.$$

If $q = q_0$, then $\sigma^{-1}(B^*bB^*) = LaA^* \cup aA^*$. But L is recognized by Y and since $Y \in \mathbf{C}$, L belongs to $\mathcal{B}$. Furthermore, aA^* also belongs to $\mathcal{B}$, since

$$\{a\} = A^*a \setminus \big((A^*a)aA^* \cup (A^*a)bA^* \cup (A^*b)aA^* \cup (A^*b)bA^*\big),$$

and

$$aA^* = \{a\} \cup \{a\}aA^* \cup \{a\}bA^*$$

It follows that $\sigma^{-1}(B^*bB^*)$ belongs to $\mathcal{B}$. Similarly, $\sigma(a_1 \cdots a_n) \in B^*b$ if and only if $q_0 \cdot \pi(\eta((a_1 \cdots a_{n-1})) = q$ and $a_n = a$. Therefore $\sigma^{-1}(B^*b) = La$ or $La \cup \{a\}$ (if $q = q_0$) and $\sigma^{-1}(B^*b)$ also belongs to $\mathcal{B}$. □

Corollary 7.3. *Given a rational set L, one can effectively decide whether it is RTL-definable.*

Proof. The set L can be given either by a rational expression or by a finite automaton. In both cases, there are well-known algorithms to compute its minimal automaton $\mathcal{A}(L)$, and then its syntactic semigroup $S(L)$, which is also the transformation semigroup of $\mathcal{A}(L)$. Then it suffices, by Theorem 7.2 to verify that $S(L)$ is locally $\mathcal{L}$-trivial. □

Say that two PTL-formulas φ and ψ are *equivalent* if $L(\varphi) = L(\psi)$, that is, if they agree when interpreted on finite words.

Corollary 7.4. *Given a PTL-formula, one can effectively decide whether it is equivalent to some RTL-formula.*

We conclude this section with three examples.

Example 7.1. Let $A = \{a, b\}$ and let $L = (ab)^+$.
The syntactic semigroup of L is the semigroup S with zero presented by the relations

$$a^2 = b^2 = 0, \quad aba = a, \quad bab = b.$$

Thus $S = \{a, b, ab, ba, 0\}$. There are three idempotents ab, ba, and 0. The corresponding local semigroups are

$$abSab = \{ab, 0\}, \quad baSba = \{ba, 0\} \quad \text{and} \quad 0S0 = \{0\},$$

all of which are $\mathcal{L}$-trivial. Therefore L is expressible in restricted temporal logic. Indeed, we have $L = L(\varphi)$, where

$$\varphi = p_a \wedge \mathbf{F}(p_b \wedge \neg\mathbf{X}p_a \wedge \neg\mathbf{X}p_b) \wedge \neg\mathbf{F}(p_a \wedge \mathbf{X}p_a) \wedge \neg\mathbf{F}(p_b \wedge \mathbf{X}p_b).$$

Example 7.2. Let $A = \{a, b, c\}$ and let $L = A^*a\{a, c\}^*$.
The syntactic semigroup of L is U_2, which is locally $\mathcal{L}$-trivial. Therefore L is expressible in restricted temporal logic. Indeed, we have $L = L(\varphi)$, where

$$\varphi = \mathbf{F}(p_a \wedge \mathbf{X}\neg\mathbf{F}p_b).$$

Example 7.3. Let $A = \{a, b, c\}$ and let $L = a^*b\{a, b, c\}^*$.
The syntactic semigroup of L is the monoid S presented by the relations

$$a = 1, \quad bb = bc = b, \quad cb = cc = c.$$

This is the reverse of U_2, and it is aperiodic, but not locally $\mathcal{L}$-trivial. Therefore, any formula ψ such that $L = L(\psi)$ makes use of the Until operator. In fact, $L = L(\varphi)$, where $\varphi = p_a \,\mathbf{U}\, p_b$.

This does not give, however, a polynomial algorithm to check whether a given PTL-formula is equivalent to a RTL formula. We presently do not know any reasonable bound on the complexity of this problem.

8 Exercises

8.1 The formalism of logic.

Exercise 1. Show that the relation $<$ can be expressed in $WMF_2(S)$.

8.2 Monadic second-order logic on words.

Exercise 2. Show that the formulas with only one second-order existential quantifier suffice to define the rational subsets.

Exercise 3. Let $A = \{a, b\}$ and let φ be the formula

$$\exists x\big((x + x + 1 = \max) \wedge (\forall y\ y \leq x \rightarrow \mathbf{a}(y)) \wedge (\forall z\ x < z < \max \rightarrow \mathbf{b}(z))\big)$$

Show that $S(\varphi) = \{a^n b^n \mid n \geq 0\}$. Deduce that addition is not definable in $MF_2(<)$.

Exercise 4. Given the monadic second-order formulas defining two rational sets X and Y, give the formulas defining $X \cup Y$, XY, X^* and X^ω. *Hint*: use Proposition 2.6

Exercise 5. Find a formula φ of $WMF_2(<)$ such that $S(\varphi)$ is the set recognized by a given Muller automaton.

Exercise 6. The sentence

$$\forall X \exists x \forall y (X(x) \rightarrow x \leq y)$$

states that every nonempty subset of $\mathbb{N}$ has a lower bound. Use the interpretation of formulas by automata to give a formal proof of this property.

Exercise 7. *Presburger arithmetic* is the set of first-order formulas which are true in the structure $(\mathbb{N}, +)$ formed by the nonnegative integers with the addition.

Formulas of $WMF_2(S)$ can be interpreted as first-order formulas on $\mathbb{N}$ by interpreting a second order variable X as the set of positions of the 1's in the binary representation of an integer x.

(1) Find a formula $\varphi(X, Y, Z) \in WMF_2(S)$ to express the fact that the integers x, y, z associated with the variables X, Y, Z satisfy the relation $x + y = z$.
(2) Deduce that every formula of Presburger arithmetic can be translated into a formula of $MF_2(<)$.
(3) Deduce that Presburger arithmetic is decidable.

8.3 First-order logic of the linear order.

Exercise 8. Show by induction on the formation rules of first-order formulas that, for each formula $\varphi \in F_1(<)$, the set $S(\varphi)$ is aperiodic.

8.4 First-order logic of the successor.

Exercise 9. Show by using Fraïssé-Ehrenfeucht games that the set $X = (a^2)^*$ is not definable in $F_1(<)$. *Hint*: show that Player II has a winning strategy in the game (a^k, a^{k+1}, r) for $k \geq 2^r - 1$.

Exercise 10. Consider the theory $F_1(=)$, in which the unique relation symbol, besides the relations $\mathbf{a}$ for $a \in A$, is the equality. Show that the sets definable in $F_1(=)$ are the threshold 1-locally testable sets.

Exercise 11. Show that a subset of A^+ is definable by a boolean combination of existential formulas of $F_1(S)$ if and only if it is strongly threshold locally testable.

Exercise 12. Show that the transition semigroup of a deterministic automaton $\mathcal{A}$ is locally $\mathcal{L}$-trivial if and only if there exist no configurations of $\mathcal{A}$ of the form

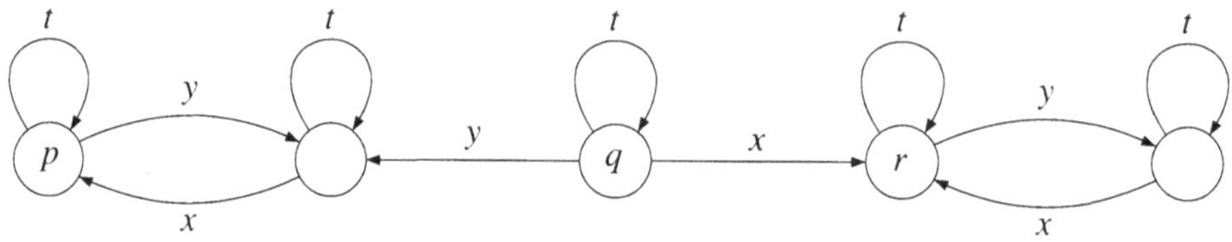

in which x, y and t are words, with $p \neq r$.

Use this result to give a polynomial time algorithm for testing whether the set accepted by an n-state deterministic automaton is RTL-definable.

8.5 Temporal logic.

Exercise 13. Let φ and ψ be past temporal formulas and let $L = L(\varphi)$ and $K = L(\psi)$. Let $\eta : A^+ \to S$ be a semigroup morphism recognizing L, and let $P = \eta(L)$. Prove the following formula

$$\begin{aligned} L(\varphi \,\mathbf{S}\, \psi) &= \{uv \in A^+ \mid u \in K, v \in A^* \text{ and for each} \\ &\qquad\qquad \text{left factor } v' \neq 1 \text{ of } v, uv' \in L\} \\ &= \bigcup_{s \in S} (\eta^{-1}(s) \cap K)\big(A^* \setminus (A^+ \setminus \eta^{-1}(s^{-1}P))A^*\big). \end{aligned}$$

Show that if L and K are star-free, then $L(\varphi \,\mathbf{S}\, \psi)$ is star-free.

9 Notes

Our introduction to logic is inspired by various classical treatises including Ebbinghaus et al. [95] and Barwise [17]. The topic treated in this chapter is seldom covered in books on automata, with the notable exceptions of the handbook of Trakhtenbrot and Barzdin [337] and the more recent book of Straubing [313]. Let us also mention the survey of Thomas [332] which contains some supplements to our presentation. The multi-author survey edited by Grädel, Thomas and Wilke [127] contains a presentation of many of the topics treated here and extensions to others. In particular three chapters are devoted to monadic second-order logic [354, 262, 266].

The standard reference about Büchi's theorem (Theorem 3.1) is the 1962 paper of Büchi [46]. In fact, Büchi had previously published, in answer to a question raised by Tarski, a version concerning the weak logic [45]. The first-order theory of integers with $+$ and $\times$ is well known to be undecidable and the question of finding decidable subfragments has been raised at the same epoch. Other contributions to this field include the decidability of the first-order theory of integers, known as Presburger arithmetic (Exercise 7 shows how to deduce this result from Büchi's theorem). The more important new idea is the one which led to the proof of the closure under complement, a necessary step to handle negation in logic. The method employed by Büchi amounts to using the definition of recognizable sets by ω-semigroups, as seen in Chapter II.

The formalism of second-order logic can be used beyond the monadic case. This leads to a logical description of the standard complexity classes. For instance, Stockmeyer [306] and Immerman [145], pursuing the seminal work of Fagin [108], established that famous classes like P, NP, PSPACE and others can be described by various fragments of the second-order logic.

The decidability of the monadic second-order logic of the successor has been generalized in many ways. First, as we shall see in Chapter X, the result still holds if $MF_2(S)$ is replaced by the corresponding logic with two successors, a deep result due to Rabin [253]. Formulas are then interpreted on infinite trees. In contrast with the case of words, the equivalence between weak and strong monadic logic does not hold for the logic with two successors (Thomas [327]). The software MONA designed by Klarlund and Moller [158] translates formulas in the logics with one or two successors into finite-state automata.

Next, the result can also be generalized to ordinals. Büchi [49], and also Shelah [283], have extended the decidability result of the monadic theory of the ordinal ω to transfinite ordinals.

First-order definable sets and star-free expressions were introduced by R. McNaughton (see [192]). The equivalence between these notions and that of an aperiodic semigroup relies on Schützenberger's theorem [274]. The extension of these notions to infinite words is due to Ladner [166], Thomas [324] and Perrin and Pin [236].

The correspondence between the concatenation hierarchy and the Σ_n hierarchy was established by Thomas [326]. Our presentation follows our article [236].

Our presentation of the first-order logic of the successor follows the main lines of the presentation given in the book of Straubing [313]. Theorem 5.5 is due to Thomas [326]. Lemma 5.8 is due to Hanf [133] and Theorem 5.4 to Wilke [357].

Temporal logic is a special case of what is called "modal logic". It was introduced by Pnueli [186] as a model to specify and verify concurrent systems. The logic we have defined, is, properly speaking, linear temporal logic (versus branching logic, which allows branching in the evolution of time). The equivalence with first-order logic was established by Kamp [151] (see also Gabbay and al. [113]) in a much more general setting. The characterization of the restricted temporal logic is due to Cohen, Perrin and Pin [73] for finite words, and to Thérien and Wilke [322] for infinite words. Our presentation of the equivalence between temporal logic and star-free sets follows the work of Cohen [77]. This approach using a restricted star operator was initiated by

Peikert [229]. It relies on Theorem 6.3, a result of Schützenberger [275] (see also Lallement [167, Chapter 7]). The notion of prefix-free set with bounded synchronization delay was originally introduced by Golomb and Gordon [125] (see also [31]).

Various fragments of temporal logic have been studied in the literature, notably by Etessami, Peled, Thérien and Wilke [106, 105, 107, 230, 322, 323, 360].

IX

BI-INFINITE WORDS

1 Introduction

In this chapter, we consider two-sided infinite words instead of the one-sided infinite words studied in the rest of the book. This notion is particularly relevant to the study of sets invariant under the shift transformation. Indeed, the shift becomes an invertible transformation on bi-infinite words.

We shall see in this chapter that most of the results proved in the one-sided case can be naturally extended to the two-sided case. There are, however, some difficulties, in particular, as we shall see, for the notion of determinism.

In Section 2, we introduce the basic terminology and results about recognizable sets of bi-infinite words. In Sections 3 and 4, we prove the two-sided counterpart of the statements relating to determinism (in particular Theorem 4.10, analogous to Landweber's theorem). However, in this case, we lack a two-sided version of Safra's construction which would be needed to build directly a deterministic Muller automaton. In Section 4, we introduce the notion of ζ-semigroup, which is the counterpart for bi-infinite words of the notion of ω-semigroup.

In Section 5, we prove an unambiguity theorem which relies on Theorem II.10.3.

In Section 6, we study a notion which is specific to the two-sided case: the notion of words inseparable by a finite automaton. We show that two words are inseparable if and only if they are either shift equivalent or if they are both recurrent with the same set of factors (Theorem 6.1).

In Section 7, we reconsider the previous questions from a logical point of view.

2 Bi-infinite words

Let $A^{\mathbb{Z}}$ be the set of all functions from $\mathbb{Z}$ into A, denoted

$$u = \cdots u(-1)u(0)u(1)\cdots$$

with $u(i) \in A$ or also

$$u = \cdots u_{-1}u_0u_1\cdots$$

and called *bi-infinite words*. For $i, j \in \mathbb{Z}$ with $i \leq j$, we set

$$u[i, j] = u_iu_{i+1}\cdots u_j$$

The set $A^{\mathbb{Z}}$ is equipped with the *shift* function $\sigma : A^{\mathbb{Z}} \to A^{\mathbb{Z}}$ defined by

$$\sigma(w)_n = w_{n-1} \qquad (n \in \mathbb{Z})$$

For a subset U of $A^{\mathbb{Z}}$, we set

$$\sigma(U) = \{\sigma(u) \mid u \in U\}$$

and the sets $\sigma^n(U)$ are called the *shifts* of U. A set U is said to be *shift invariant* if $\sigma(U) = U$.

As in the one-sided case, one gets an ultrametric by setting for $u, v \in A^{\mathbb{Z}}$,

$$d(u, v) = 2^{-n} \text{ where } n = \sup\{i \geq 0 \mid u[-i, i] = v[-i, i]\}$$

Of course, we adopt the usual convention that $n = +\infty$ and $d(u, v) = 0$ if $u = v$ and $n = -1$ if $u_0 \neq v_0$. The open sets of the corresponding topology are this time the shifts of the sets of the form

$$c(X) = \{u \in A^{\mathbb{Z}} \mid u[0, +\infty[\, \in XA^{\omega}\}$$

for some subset X of A^*.

One can build a bi-infinite word from two infinite one-sided words in the following way. We consider, in addition to the set $A^{\mathbb{N}} = A^{\omega}$ of one-sided infinite words, the set $A^{-\mathbb{N}}$ of *left infinite words* of the form

$$u = \cdots u_{-2}u_{-1}u_0$$

with $u_i \in A$. For each $u \in A^{\mathbb{N}}$, we denote by $\tilde{u}$ the left infinite word defined by

$$\tilde{u}_{-n} = u_n \qquad (n \in \mathbb{N})$$

The map $u \mapsto \tilde{u}$ is a bijection from $A^{\mathbb{N}}$ onto $A^{-\mathbb{N}}$. If X is a subset of $A^{\mathbb{N}}$, we set

$$\tilde{X} = \{\tilde{u} \mid u \in X\}$$

The same notation is used for the inverse map from $A^{-\mathbb{N}}$ onto $A^{\mathbb{N}}$.

For a subset X of A^+, we denote by $\tilde{X}$ the set of reversals $\tilde{x}$, for $x \in X$. We set

$$X^{\tilde{\omega}} = \{u \in A^{-\mathbb{N}} \mid \tilde{u} \in (\tilde{X})^{\omega}\}$$
$$\overleftarrow{X} = \{u \in A^{-\mathbb{N}} \mid \tilde{u} \in \overrightarrow{Y}, \text{ where } Y = \tilde{X}\}$$

We have in particular $A^{\tilde{\omega}} = A^{-\mathbb{N}}$.

For $x \in A^{-\mathbb{N}}$ and $X \subset A^{-\mathbb{N}}$, we set

$$x^{-1}X = \{y \in A^* \mid xy \in X\}$$

Given $u \in A^{-\mathbb{N}}$ and $v \in A^{\mathbb{N}}$, the bi-infinite word w defined by

$$w_n = \begin{cases} u_n & \text{if } n \leq 0 \\ v_{n-1} & \text{if } n \geq 1 \end{cases}$$

is denoted by $[u, v]$. Given two subsets $U \subset A^{-\mathbb{N}}$ and $V \subset A^{\mathbb{N}}$, we put

$$[U, V] = \{[u, v] \in A^{\mathbb{Z}} \mid u \in U, v \in V\}$$

and we denote by UV the closure of $[U, V]$ under the shift operator, that is,

$$UV = \bigcup_{k \in \mathbb{Z}} \sigma^k([U, V])$$

Now, since for $U \subset A^{-\mathbb{N}}$, $V \subset A^*$ and $W \subset A^{\mathbb{N}}$, the associativity formula

$$(UV)W = U(VW)$$

holds, this set will be simply denoted by UVW.

Example 2.1. The set

$$a^{\tilde{\omega}}(a + b)^* a^{\omega}$$

is formed by the words containing finitely many occurrences of b.

For each subset X of A^+, we put

$$X^{\zeta} = X^{\tilde{\omega}} X^{\omega}$$

Thus X^{ζ} consists of the concatenations $\cdots x_{-1}x_0x_1 \cdots$ of the words of X.

Example 2.2. The following equality holds

$$(ab)^{\zeta} = [(ab)^{\tilde{\omega}}, (ab)^{\omega}] \cup [(ba)^{\tilde{\omega}}, (ba)^{\omega}]$$

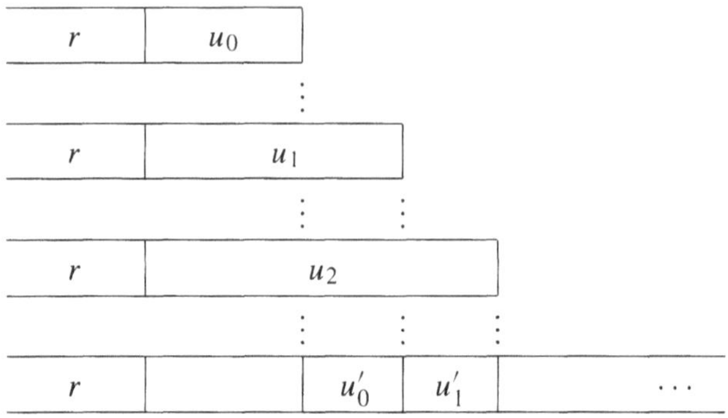

Figure 2.1. An element of $\overrightarrow{U}$.

The set A^{ζ} is the quotient of $A^{\mathbb{Z}}$ by the equivalence $\equiv$ defined by

$$x \equiv y \text{ if and only if there exists } n \in \mathbb{Z} \text{ such that } x = \sigma^n(y)$$

The $\equiv$-classes are called the *orbits* of σ.

Let $U \subset A^{-\mathbb{N}}$ be a set of left infinite words. We denote by $\overrightarrow{U}$ the set of bi-infinite words $x \in A^{\mathbb{Z}}$ for which there exists a sequence $(u_n)_{n\geq 0}$ of elements of U of the form $u_{n+1} = u_n u'_n$ such that (cf. Figure 2.1)

$$\begin{aligned} x &= u_0 u'_0 u'_1 u'_2 \cdots \\ &= u_1 u'_1 u'_2 \cdots \\ &\vdots \\ &= u_n u'_n u'_{n+1} \cdots \end{aligned}$$

Equivalently, $x \in \overrightarrow{U}$ if and only if there exists a word $r \in A^{-\mathbb{N}}$ and a sequence $(u_n)_{n\geq 0}$ of elements of A^* such that $ru_n \in U$ and $x = r(\lim_{n\to\infty} u_n)$ (cf. Figure 2.1). In a more concise form, we have

$$\overrightarrow{U} = \bigcup_{r \in A^{-\mathbb{N}}} r(\overrightarrow{r^{-1}U})$$

Similarly, the set $\overleftarrow{V}$ can be defined for $V \subset A^{\mathbb{N}}$. Finally, if $X \subset A^*$ is a set of finite words, $\overleftrightarrow{X}$ is defined as the set of bi-infinite words $x \in A^{\mathbb{Z}}$ for which there exists a sequence (u_n, v_n) of pairs of words of the form

$$u_{n+1} = u'_n u_n, \quad v_{n+1} = v_n v'_n$$

with $u_n v_n \in X$, such that $x = \cdots u'_2 u'_1 u'_0 u_0 v_0 v'_1 v'_2 \cdots$ (cf. Figure 2.2).

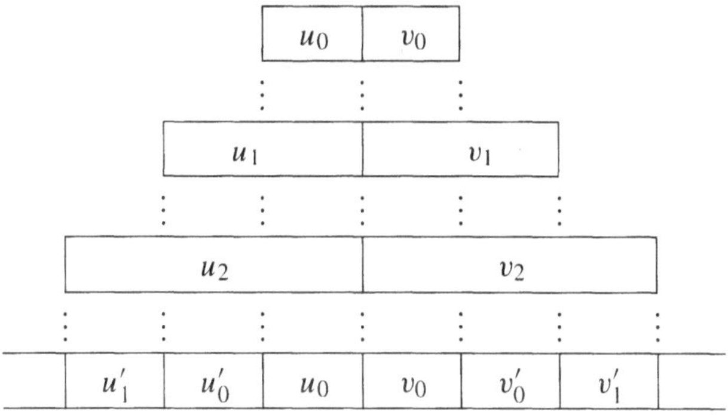

Figure 2.2. An element of $\overleftrightarrow{X}$.

Equivalently, $w \in \overleftrightarrow{X}$ if and only if $w = [u, v]$ with $\tilde{u} = \lim_{n\to\infty} \tilde{r}_n$, $v = \lim_{n\to\infty} s_n$ and $r_n s_n \in X$ for each $n \geq 0$.

Example 2.3. Let $X = a^*b^*$. Then $\overleftrightarrow{X} = a^\zeta + b^\zeta + a^{\tilde{\omega}}b^\omega$.

Proposition 2.1. *If U and V are two recognizable subsets of A^+, the following formula holds:*

$$\overrightarrow{U^{\tilde{\omega}}V} = U^{\tilde{\omega}}(\overrightarrow{U^*V})$$

Proof. The inclusion from right to left is clear. To prove the opposite inclusion, consider a word $w \in \overrightarrow{U^{\tilde{\omega}}V}$. There exists by definition $r \in A^{-\mathbb{N}}$ and a sequence $(s_n)_{n\geq 0}$ of words of A^* such that $w = r(\lim_{n\to\infty}(s_n))$ and $rs_n \in U^{\tilde{\omega}}V$ for each $n \geq 0$. For each $n \geq 0$, two cases arise:

(1) either $r = r'_n r''_n$ and $s_n = s'_n s''_n$ with $r'_n \in U^{\tilde{\omega}}$, $r''_n s'_n \in U$, $s''_n \in U^*V$
(2) or $r = r'_n r''_n$ with $r'_n \in U^{\tilde{\omega}}$, $r''_n s_n \in V$

One of these cases must arise for infinitely many n's. Up to taking a subsequence, we may assume that one of the two cases holds for all n. We now consider these two cases successively.

(1) We have $s'_n \in r''^{-1}_n U$. Since U is recognizable, there are only finitely many distinct sets $r''^{-1}_n U$ for all $n \geq 0$. Therefore, we may assume, up to taking a subsequence, that $r''_0 s'_n \in U$ for each $n \geq 0$. Then we have $w = r'_0 \lim(r''_0 s_n)$, and the result follows since $r''_0 s_n \in U^*V$.
(2) Since V is recognizable, we may assume, up to taking a subsequence, that $r''_0 s_n \in V$ for each $n \geq 0$. Therefore $w = r'_0 \lim(r''_0 s_n)$, and the result follows. □

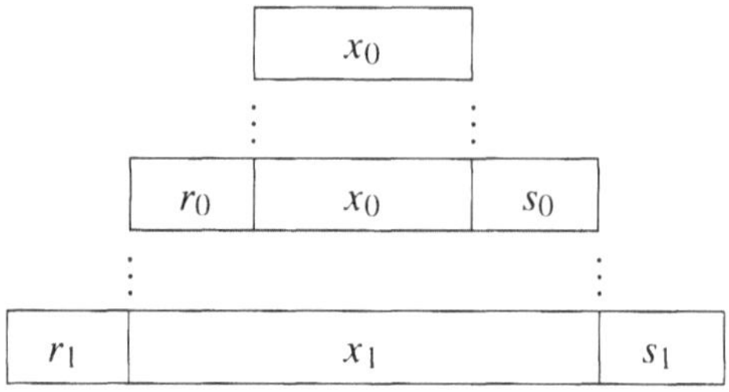

Figure 2.3. The sequence $(x_n)_{n\geq 0}$.

Example 2.4. The formula $\overrightarrow{A^{\tilde{\omega}}b} = A^{\tilde{\omega}}(\overrightarrow{A^*b})$ holds.

Proposition 2.2. *If X is a prefix (resp. suffix) recognizable set, then* $\overleftrightarrow{X} = \emptyset$.

Proof. Let $(x_n)_{n\geq 0}$ be a sequence of words of X such that $x_{n+1} = r_n x_n s_n$ with $r_n, s_n \in A^+$, as shown in Figure 2.3. Let $t_n = r_n r_{n-1} \cdots r_0$. Since X is recognizable, the set $\{t_n^{-1}X \mid n \geq 0\}$ is finite. Therefore, one can find two integers $m < n$ such that $t_n^{-1}X = t_m^{-1}X$. It follows that

$$r_m r_{m-1} \cdots r_0 x_0 s_n \in X,$$

a contradiction, since X is prefix. □

Let $\mathcal{A} = (Q, A, E, I, F)$ be an automaton. A *bi-infinite path* in the automaton is a bi-infinite sequence of consecutive arrows, i.e. of the form $(p_n, a_n, p_{n+1})_{n\in\mathbb{Z}}$. Its *label* is the bi-infinite word

$$u = \cdots a_{-1}a_0a_1\cdots$$

A path is *successful* if $p_n \in I$ holds for infinitely many $n < 0$ and $p_n \in F$ for infinitely many $n > 0$. The set of bi-infinite words recognized by the automaton, denoted by $L^{\zeta}(\mathcal{A})$, is the set of labels of the successful bi-infinite paths.

In a similar way a left infinite path $(p_{n-1}, a_n, p_n)_{n\leq 0}$ is successful if $p_n \in I$ holds for infinitely many n. The set of left infinite words recognized by the automaton, denoted by $L^{\tilde{\omega}}(\mathcal{A})$, is the set of labels of the successful left infinite paths.

A subset W of $A^{\mathbb{Z}}$ (resp. $A^{-\mathbb{N}}$) is said to be *recognizable* if there exists a finite automaton $\mathcal{A}$ such that $W = L^{\zeta}(\mathcal{A})$. We shall denote by $\mathrm{Rec}(A^*)$, $\mathrm{Rec}(A^{\mathbb{N}})$, $\mathrm{Rec}(A^{-\mathbb{N}})$ and $\mathrm{Rec}(A^{\mathbb{Z}})$, the classes of recognizable subsets of A^*, $A^{\mathbb{N}}$, $A^{-\mathbb{N}}$ and $A^{\mathbb{Z}}$, respectively. Clearly, a set X of infinite words belongs to $\mathrm{Rec}(A^{\mathbb{N}})$ if and only if $\tilde{X}$ belongs to $\mathrm{Rec}(A^{-\mathbb{N}})$.

A subset W of $A^{\mathbb{Z}}$ is said to be ζ*-rational* if W is a finite union of sets of the form $X^{\tilde{\omega}}YZ^{\omega}$ with $X, Y, Z \in \mathrm{Rec}(A^*)$.

Note that, by definition, a recognizable subset of $A^{\mathbb{Z}}$ is shift invariant. Indeed, the set of bi-infinite successful paths of an automaton is shift invariant and so is the set of their labels. The following proposition relates $\mathrm{Rec}(A^*)$, $\mathrm{Rec}(A^{\mathbb{N}})$ and $\mathrm{Rec}(A^{\mathbb{Z}})$.

Proposition 2.3. *For each subset W of $A^{\mathbb{Z}}$, the following conditions are equivalent:*

(1) $W \in \mathrm{Rec}(A^{\mathbb{Z}})$,

(2) *W is shift invariant and is a finite union of sets of the form $[\tilde{U}, V]$ for $U, V \in \mathrm{Rec}(A^{\mathbb{N}})$,*

(3) *W is a finite union of sets of the form $\tilde{U}V$ for $U, V \in \mathrm{Rec}(A^{\mathbb{N}})$,*

(4) *W is ζ-rational.*

Proof. (1) implies (2). Let $\mathcal{A} = (Q, I, F)$ be a finite automaton such that $W = L^{\zeta}(\mathcal{A})$. For each $q \in Q$, define the automata $\mathcal{A}_q = (Q, I, q)$ and $\mathcal{B}_q = (Q, q, F)$. Setting

$$U_q = L^{\tilde{\omega}}(\mathcal{A}_q), \quad V_q = L^{\omega}(\mathcal{B}_q)$$

it is easy to verify the formula

$$W = \bigcup_{q \in Q} [U_q, V_q]$$

This shows that W satisfies Condition (2) since, besides, W is shift invariant.
(2) implies (3). If $W = \bigcup_{i=1}^{n} [U_i, V_i]$ is shift invariant, we have

$$W = \bigcup_{k \in \mathbb{Z}} \sigma^k(W) = \bigcup_{i=1}^{n} \Big(\bigcup_{k \in \mathbb{Z}} \sigma^k([U_i, V_i]) \Big) = \bigcup_{i=1}^{n} U_i V_i$$

(3) implies (4). It suffices to show that each UV is of the required form. Since $\tilde{U}$ and V are in $\mathrm{Rec}(A^{\mathbb{N}})$, we have

$$\tilde{U} = \bigcup_{i=1}^{n} X_i Y_i^{\omega}, \quad V = \bigcup_{j=1}^{m} Z_j T_j^{\omega}$$

whence

$$UV = \bigcup_{i=1}^{n} \bigcup_{j=1}^{m} \tilde{Y}_i^{\tilde{\omega}} \tilde{X}_i Z_j T_j^{\omega}$$

and the result follows.
(4) implies (1). Consider $X, Y, Z \in \mathrm{Rec}(A^*)$. Then one can build a finite automaton $\mathcal{A} = (Q, i, f)$ with the following properties:

$$X^* = L(Q, i, i), \quad X^* Y Z^* = L(Q, i, f), \quad Z^* = L(Q, f, f).$$

We have then $X^{\tilde{\omega}} Y Z^{\omega} = L^{\zeta}(\mathcal{A})$. This shows that $X^{\tilde{\omega}} Y Z^{\omega} \in \mathrm{Rec}(A^{\mathbb{Z}})$. Since, on the other hand, $\mathrm{Rec}(A^{\mathbb{Z}})$ is closed under union, this concludes the proof. □

A nontrivial consequence of Proposition 2.3 is that $\text{Rec}(A^{\mathbb{Z}})$ is closed under complement. The proof relies on the analogous property of $\text{Rec}(A^{\mathbb{N}})$.

Corollary 2.4. *The class* $\text{Rec}(A^{\mathbb{Z}})$ *is a boolean algebra.*

Proof. We have already seen that $\text{Rec}(A^{\mathbb{Z}})$ is closed under union. For the complement, we use Condition (2) of Proposition 2.3. Let us consider

$$W = \bigcup_{1 \le i \le n} [\tilde{U}_i, V_i]$$

with $U_i, V_i \in \text{Rec}(A^{\mathbb{N}})$. Since $\text{Rec}(A^{\mathbb{N}})$ is a boolean algebra, we may assume that the family U_i consists of pairwise disjoint sets. Then we have

$$A^{\mathbb{Z}} \setminus W = [\tilde{Z}, A^{\mathbb{N}}] \cup \Big(\bigcup_{1 \le i \le n} [\tilde{U}_i, A^{\mathbb{N}} \setminus V_i] \Big).$$

with $Z = A^{\mathbb{N}} \setminus \bigcup_{1 \le i \le n} U_i$, showing that the complement of W is recognizable. □

Example 2.5. Let $\mathcal{A} = (Q, A, E, I, F)$ be the automaton given in Figure 2.4, with $Q = \{1, 2, 3\}$, $I = \{1\}$ and $F = \{3\}$.

Let $W = L^{\zeta}(\mathcal{A})$. Then $W = a^{\tilde{\omega}}(a+b)^* a^{\omega}$ and also

$$W = [a^{\tilde{\omega}}(a+b)^*, (a+b)^* a^{\omega}]$$

The complement of W can be written as

$$W^c = [(ba^*)^{\tilde{\omega}}, (a+b)^{\omega}] \cup [a^{\tilde{\omega}}(a+b)^*, (a^*b)^{\omega}]$$

which is the union of the sets of words having infinitely many b's respectively on the left and on the right. This can be rewritten as an unambiguous expression

$$W^c = (ba^*)^{\tilde{\omega}}(a^*b)^{\omega} \cup (ba^*)^{\tilde{\omega}} a^{\omega} \cup a^{\tilde{\omega}}(ba^*)^{\omega}$$

by using the sets of words with infinitely many b's on the left (respectively on the right).

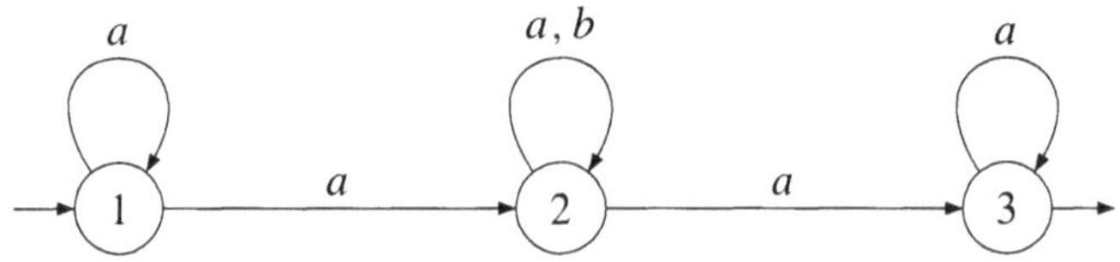

Figure 2.4. The automaton of W.

The set $A^{\mathbb{Z}}$ can be naturally equipped with three topologies.

(i) The two-sided topology in which the open sets are the shifts of the cylinders

$$c(X) = \{u \in A^{\mathbb{Z}} \mid u[0, +\infty[\in XA^{\omega}\}$$

for $X \subset A^*$.

(ii) The right topology in which the open sets are the shifts of the cylinders

$$\gamma(X) = \{u \in A^{\mathbb{Z}} \mid u[0, +\infty[\in X\}$$

for $X \subset A^{\omega}$.

(iii) The left topology, defined in a symmetric way.

Each of these topologies can be defined by a metric. Indeed, if we set

$$d_0(x, y) = 2^{-n}$$

with $n = \max\{i \geq 0 \mid x[-i, i] = y[-i, i]\}$ (with the usual conventions when $x = y$ and when $x_0 \neq y_0$), the corresponding topology is the two-sided topology since the open balls are the sets

$$B(x, 2^{-n}) = \{y \in A^{\mathbb{Z}} \mid y[-n, n] = x[-n, n]\}$$

and thus we have $B(x, 2^{-n}) = C(x[-n, n])$.

Next, the right topology is associated with the metric

$$d_1(x, y) = 2^{-n}$$

with $n = \max\{i \geq 0 \mid x] - \infty, i] = y] - \infty, i]\}$ (and the usual conventions).

Every open set of the two-sided topology is open for the right topology. It is a consequence of the inequality on the metrics

$$d_0(x, y) \leq d_1(x, y)$$

The next example shows that these two topologies are nevertheless distinct.

Example 2.6. Let

$$X = b^{\tilde{\omega}}\{a^n b^n \mid n \geq 0\}a^{\omega} \cup a^{\zeta} \cup b^{\zeta}$$

This set is closed for the right and for the left topologies. However, it is not closed for the two-sided topology. Indeed, the word $a^{\tilde{\omega}}b^{\omega}$ belongs to the closure of X, but does not belong to X.

This example shows that a set can be closed (or open) for each of the right and left topologies without being closed for the two-sided topology. This is not possible for a recognizable set since if $\mathcal{A} = (Q, A, E, I, F)$ is a finite automaton, the following equivalences hold.

$$L^{\zeta}(\mathcal{A}) \text{ is "right" closed} \iff L^{\zeta}(\mathcal{A}) = L^{\zeta}(I, F).$$
$$L^{\zeta}(\mathcal{A}) \text{ is "left" closed} \iff L^{\zeta}(\mathcal{A}) = L^{\zeta}(Q, F).$$
$$L^{\zeta}(\mathcal{A}) \text{ is closed} \iff L^{\zeta}(\mathcal{A}) = L^{\zeta}(Q, Q).$$

We shall see later on that this connection between the left, the right and the two-sided topologies can be extended, for the recognizable sets, to the upper levels of the Borel hierarchy.

3 Determinism

The next result is, in appearance, quite analogous to the corresponding result for the one-sided case (Theorem I.6.2). Nevertheless, its proof is, as we shall see, much more involved. A subset U of $A^{-\mathbb{N}}$ is recognizable if $\tilde{U}$ is a recognizable subset of $A^{\mathbb{N}}$.

Theorem 3.1. *A subset W of $A^{\mathbb{Z}}$ is recognizable by a finite deterministic automaton if and only if there exists a recognizable subset U of $A^{-\mathbb{N}}$ such that $W = \overrightarrow{U}$.*

One can summarize the proposition by the formula

$$\overrightarrow{\text{Rec}(A^{-\mathbb{N}})} = \text{Det}(A^{\mathbb{Z}})$$

with obvious notation. We shall see in Section 4 a more general result given by the formula

$$\overrightarrow{\mathcal{P}(A^{-\mathbb{N}})} \cap \text{Rec}(A^{\mathbb{Z}}) = \text{Det}(A^{\mathbb{Z}})$$

which is the two-sided counterpart of Landweber's theorem (Theorem I.9.9).

Before proceeding to the proof of Theorem 3.1, we give some examples.

Example 3.1. Let $W = (a+b)^{\tilde{\omega}}a(a+b)^{\omega}$, a set recognized by the deterministic automaton drawn in Figure 3.1. In accordance with Theorem 3.1, we have $W = \overrightarrow{U}$ with

$$U = (ab^*)^{\tilde{\omega}} \cup b^{\tilde{\omega}}a(a+b)^*$$

We now establish the following result, which proves the inclusion $\text{Det}(A^{\mathbb{Z}}) \subset \overrightarrow{\text{Rec}(A^{-\mathbb{N}})}$.

Proposition 3.2. *Let $\mathcal{A}$ be a deterministic automaton. Then $\overrightarrow{L^{\tilde{\omega}}(\mathcal{A})} = L^{\zeta}(\mathcal{A})$.*

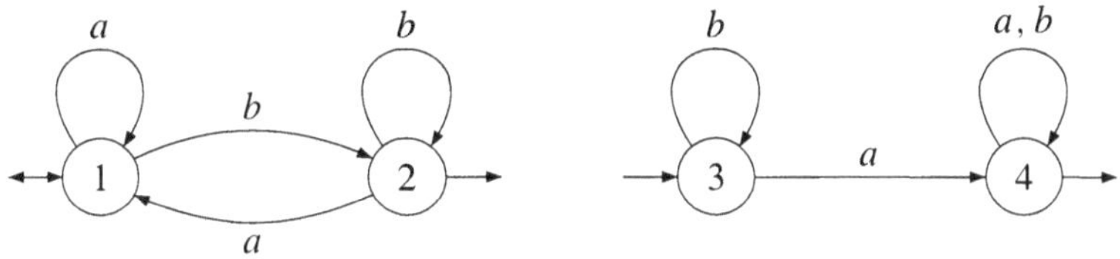

Figure 3.1. A deterministic automaton.

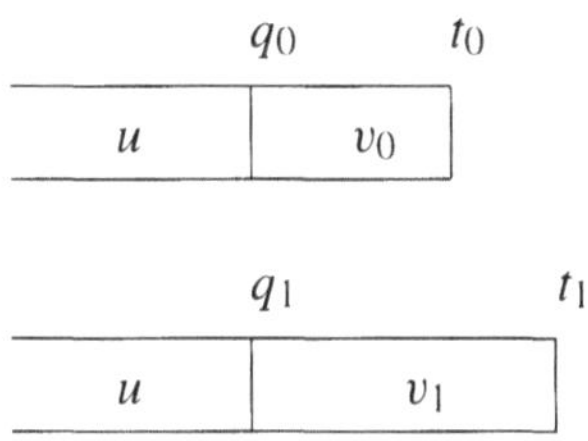

Figure 3.2. The paths c_n.

Proof. If $x \in \overrightarrow{L^{\tilde{\omega}}(\mathcal{A})}$, there exists a word $u \in A^{-\mathbb{N}}$ such that $x = u \lim(v_n)$ and each uv_n is the label of a successful path c_n. Since $\mathcal{A}$ is deterministic, the number of paths of label u is bounded by Card(Q). Thus we may assume that this path is the same for all the c_n's. One obtains in this way a successful path of label x. Thus $x \in L^{\zeta}(\mathcal{A})$. The inclusion $L^{\zeta}(\mathcal{A}) \subset \overrightarrow{L^{\tilde{\omega}}(\mathcal{A})}$ is immediate. □

We shall now prove the missing inclusion $\overrightarrow{\mathrm{Rec}(A^{-\mathbb{N}})} \subset \mathrm{Det}(A^{\mathbb{Z}})$. It relies on the following result, which is interesting in itself.

Theorem 3.3. *Every recognizable subset U of $A^{-\mathbb{N}}$ can be recognized by a deterministic finite automaton, i.e.* $\mathrm{Rec}(A^{-\mathbb{N}}) = \mathrm{Det}(A^{-\mathbb{N}})$.

Proof. Let φ be a morphism from A^{∞} onto an ω-semigroup S recognizing $\tilde{U}$. Since $\mathrm{Det}(A^{-\mathbb{N}})$ is closed under union, we may assume that $\tilde{U} = \varphi^{-1}(s)\varphi^{-1}(e)^{\omega}$ for some linked pair (s, e) of S_+.

Let $X = \varphi^{-1}(s)$, $Y = \varphi^{-1}(e)$ and $L = Y \setminus YA^+$. Let $\mathcal{A} = (Q, i, I)$ be a deterministic automaton recognizing YL and such that, for some subset F of Q, the automaton (Q, i, F) recognizes X. Thus

$$L^*(Q, i, I) = YL, \qquad L^*(Q, i, F) = X.$$

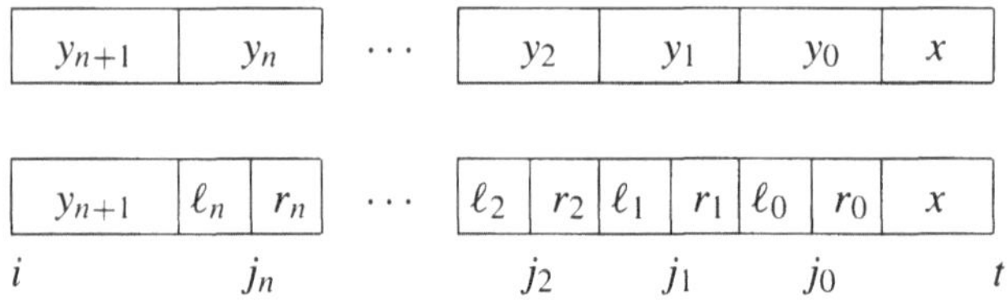

Figure 3.3. The path c_n.

We also assume that each state $q \in Q$ is accessible from i. We claim that

$$L^{\tilde{\omega}}(Q, I, F) = Y^{\tilde{\omega}} X = U$$

First consider $u = \cdots y_2 y_1 y_0 x$ with $y_k \in Y$ and $x \in X$. For each integer $k \geq 0$, the word y_k has a unique prefix ℓ_k in L. Setting $y_k = \ell_k r_k$, there is, for each $n \geq 0$, a path c_n of the form represented in Figure 3.3 with $j_k \in I$ and $t \in F$. Since Q is finite, there exists a converging subsequence of the sequence (c_n) and thus u is accepted by the automaton (Q, I, F).

Conversely, consider $u \in L^{\tilde{\omega}}(Q, I, F)$. We have

$$u = \cdots r_2 r_1 r_0 v$$

with $j \xrightarrow{r_k} j$ for some $j \in I$ and $j \xrightarrow{v} t$ for some $t \in F$. Furthermore, we may assume, up to taking a subsequence, that all the $\varphi(r_k)$ are equal to a fixed idempotent f of S. We shall prove that for each $k \geq 0$, one can write

$$r_{2k+1} r_{2k} = g_k d_k \tag{3.1}$$

with $d_k \in Y$ and $Y g_k \subset Y$, which gives the conclusion. Indeed, we have then

$$u = \cdots g_2 d_2 g_1 d_1 g_0 d_0 v$$

with $d_k g_{k-1} \in Y$. We also have $d_0 v \in X$. Indeed, let w be a word such that $i \xrightarrow{w} j$. Since $j \in I$, we have $w \in YL$ and hence $\varphi(w) = e$. Thus

$$\varphi(d_0 v) = \varphi(d_0)\varphi(v) = e\varphi(s) = \varphi(w)\varphi(v) = \varphi(wv)$$

Since $i \xrightarrow{wv} t$, we have $wv \in X$ and hence $\varphi(wv) = s$ whence $\varphi(d_0 v) = s$, or, equivalently, $d_0 v \in X$.

Finally, let us prove Formula (3.1). Set $r = r_{2k+1}$ and $r' = r_{2k}$ and let w be such that $i \xrightarrow{w} j$. Since $j \in I$, we have $wrr' \in YL$ and hence $wrr' = y\ell$ with $y \in Y, \ell \in L$. Let us verify that y cannot be a prefix of w. Otherwise, one would have

$$w = yw', \qquad w'rr' = \ell.$$

But we have

$$\varphi(w'r) = \varphi(w')f = \varphi(w')\varphi(rr') = \varphi(\ell) = e$$

which gives $w'r \in Y$, a contradiction since ℓ has no prefix in L.

Thus we have $y = ww'$ and $rr' = w'\ell$. Put $g_k = w'$ and $d_k = \ell$. It remains to verify that $Yw' \subset Y$. But since $w = yw'$ and $\varphi(w) = \varphi(y) = e$, the equality $e = e\varphi(w')$ proves the inclusion. This concludes the proof of the theorem. □

Proof of Theorem 3.1. We have already seen that $\mathrm{Det}(A^{\mathbb{Z}}) \subset \overrightarrow{\mathrm{Rec}(A^{-\mathbb{N}})}$ (Proposition 3.2). We now establish now the opposite inclusion. Let $U \subset A^{-\mathbb{N}}$ be a recognizable set. By Theorem 3.3, there exists a deterministic automaton $\mathcal{A}$ which recognizes U. Now $\overrightarrow{U} = L^{\zeta}(\mathcal{A})$ by Proposition 3.2, whence $\overrightarrow{U} \in \mathrm{Det}(A^{\mathbb{Z}})$. □

We shall now introduce a new type of deterministic automata to recognize the subsets of $A^{\mathbb{Z}}$ of the form $\overleftrightarrow{L}$, where L is a recognizable subset of A^+. This class of sets will be denoted by $\overleftrightarrow{\mathrm{Rec}(A^*)}$. We shall see in Section 4 that the following formula holds

$$\overrightarrow{\mathrm{Rec}(A^{-\mathbb{N}})} \cap \overleftarrow{\mathrm{Rec}(A^{\mathbb{N}})} = \overleftrightarrow{\mathrm{Rec}(A^*)}$$

a stronger result than our first characterization.

A *bi-automaton* $\mathcal{A} = (\mathcal{A}_-, \mathcal{A}_+, I)$ consists of a pair of automata

$$\mathcal{A}_- = (Q_-, I_-, F_-), \qquad \mathcal{A}_+ = (Q_+, I_+, F_+)$$

and a subset $I \subset I_- \times I_+$ submitted to a certain condition specified below. The automaton $\mathcal{A}_-$ is used to read the words from right to left and $\mathcal{A}_+$ from left to right. We shall use a backwards notation for the transitions of $\mathcal{A}$, that is $q \xleftarrow{x} p$ instead of $p \xrightarrow{\tilde{x}} q$. The condition on I is the following: the set of pairs of words (x, y) such that

$$f_- \xleftarrow{x} i_-, \qquad i_+ \xrightarrow{y} f_+$$

with $f_- \in F_-$, $(i_-, i_+) \in I$, $f_+ \in F_+$ is a union of classes for the equivalence $\equiv$ defined by $(x, yz) \equiv (xy, z)$. We denote by $L^*(\mathcal{A})$ the set of pairs satisfying these conditions. The set of bi-infinite words recognized by $\mathcal{A}$ is the set

$$L^{\zeta}(\mathcal{A}) = \bigcup_{(i_-, i_+) \in I} [L^{\tilde{\omega}}(Q_-, i_-, F_-), L^{\omega}(Q_+, i_+, F_+)]$$

It follows from the definitions that $L^{\zeta}(\mathcal{A})$ is shift invariant.

Example 3.2. Let $\mathcal{A}$ be the bi-automaton given in Figure 3.4, where $\mathcal{A}_-$ and $\mathcal{A}_+$ are separated by a vertical line and $I = \{(1, 2), (6, 7), (11, 12)\}$. For this bi-automaton,

$$L^*(\mathcal{A}) = [a^*, a^+b^+] \cup [a^+, b^+] \cup [a^+b^+, b^*] \quad \text{and} \quad L^{\zeta}(\mathcal{A}) = a^{\tilde{\omega}}b^{\omega}$$

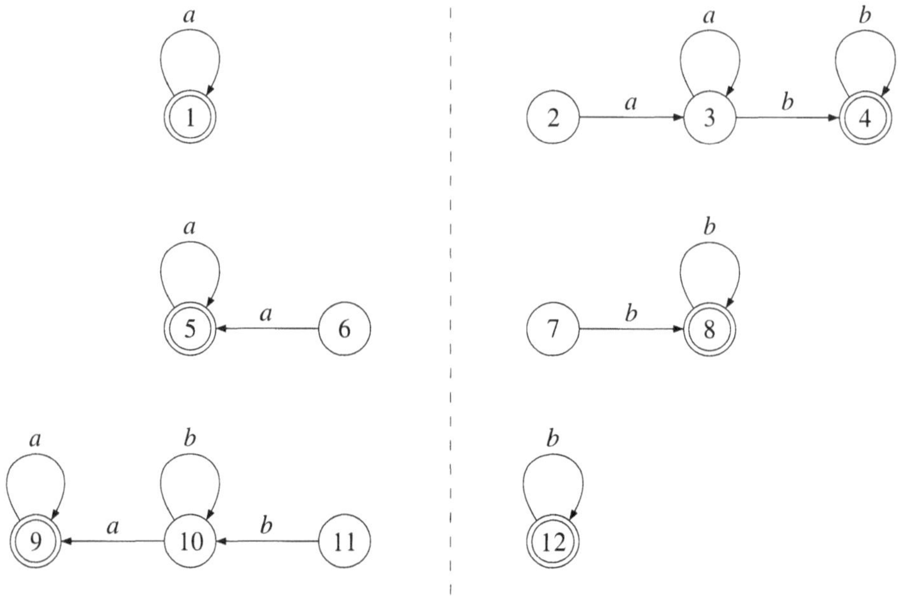

Figure 3.4. A bi-automaton.

A bi-automaton $\mathcal{A} = (\mathcal{A}_-, \mathcal{A}_+)$ is said to be deterministic if $\mathcal{A}_-$ and $\mathcal{A}_+$ are deterministic.

Theorem 3.4. *A subset W of $A^{\mathbb{Z}}$ is recognizable by a finite deterministic bi-automaton if and only if there exists a recognizable subset X of A^* such that $W = \overleftrightarrow{X}$.*

Proof. Let $\mathcal{A} = (\mathcal{A}_-, \mathcal{A}_+, I)$ be a deterministic bi-automaton and let

$$X = \{xy \mid (x, y) \in L^*(\mathcal{A})\}$$

Let us verify that $L^\zeta(\mathcal{A}) = \overleftrightarrow{X}$. First, if $w \in L^\zeta(\mathcal{A})$, w is equal to $[u, v]$ for some $\tilde{u} \in L^\omega(Q_-, i_-, F_-)$, $v \in L^\omega(Q_+, i_+, F_+)$ and $(i_-, i_+) \in I$. There exists a strictly increasing sequence of prefixes $(x_n)_{n \geq 0}$ of $\tilde{u}$ such that $\tilde{x_n} \in L^*(Q_-, i_-, F_-)$. There is also a strictly increasing sequence of prefixes (y_n) of v such that $y_n \in L^*(Q_+, i_+, F_+)$. Thus $(x_n, y_n) \in L^*(\mathcal{A})$, and since $u = \lim_{n\to\infty} x_n y_n$, it follows that $u \in \overleftrightarrow{X}$.

Conversely, if $w \in \overleftrightarrow{X}$, $w = [u, v]$ where $\tilde{u} = \lim_{n\to\infty} \tilde{x_n}$, $v = \lim_{n\to\infty} y_n$ and $x_n y_n \in X$. Since $\mathcal{A}$ is a bi-automaton, we have $(x_n, y_n) \in L^*(\mathcal{A})$ for each $n \geq 0$. Since Q_- and Q_+ are finite, we may assume that the same pair (i_-, i_+) is used to recognize each (x_n, y_n), which gives, for each $n \geq 0$, $\tilde{x}_n \in L^*(Q_-, i_-, F_-)$, $y_n \in L^*(Q_+, i_+, F_+)$. Therefore, since the automata are deterministic, $\tilde{u} \in L^\omega(Q_-, i_-, F_-)$ and $v \in L^\omega(Q_+, i_+, F_+)$. This means, by definition, that $w \in L^\zeta(\mathcal{A})$.

Consider now a recognizable subset X of A^* and let $\mathcal{A} = (Q, i, F)$ be a finite deterministic automaton recognizing X. For each state $q \in Q$, there exists a deterministic automaton $\mathcal{A}_{q-}$ which recognizes the set of words x such that

$$q \xleftarrow{x} i$$

Set $\mathcal{A}_{q-} = (Q_{q-}, i_{q-}, F_{q-})$ and let $\mathcal{A} = (\mathcal{A}_+, \mathcal{A}_-, I)$ be the bi-automaton obtained by taking for $\mathcal{A}_+$ the union of the automata $\mathcal{A}_q$, for $\mathcal{A}_-$ the union of the automata $\mathcal{A}_{q-}$ and for I the set of pairs (i_{q-}, q). It is easy to verify that $L^\zeta(\mathcal{A}) = \overleftrightarrow{X}$. □

4 Morphisms

We shall now introduce a notion showing how to recognize bi-infinite words with morphisms. For this purpose, we introduce the notion of ζ-semigroup which is the counterpart, for the two-sided case, of that of ω-semigroup. We first define the notion dual to that of an ω-semigroup.

An $\tilde{\omega}$-semigroup is a pair $(S_+, S_{\tilde{\omega}})$ where S_+ is a semigroup and $S_{\tilde{\omega}}$ is a set equipped with an infinite product $\tilde{\pi} : S_+^{-\mathbb{N}} \to S_{\tilde{\omega}}$. The associativity rules are the same as for an ω-semigroup, but should be applied backwards. They allow to use the notation $\cdots s_{-2}s_{-1}s_0$ instead of $\tilde{\pi}(\ldots, s_{-1}, s_0)$. There is also a notion of Wilke $\tilde{\omega}$-algebra with an operator $s \mapsto s^{\tilde{\omega}}$ satisfying the reversed version of the identities of a Wilke ω-algebra.

A *ζ-semigroup* is an algebra $S = (S_+, S_\omega, S_{\tilde{\omega}}, S_\zeta)$ where:

(1) (S_+, S_ω) is an ω-semigroup.

(2) $(S_+, S_{\tilde{\omega}})$ is a $\tilde{\omega}$-semigroup

with a surjective map $\rho : S_{\tilde{\omega}} \times S_\omega \to S_\zeta$ such that for each $s \in S_{\tilde{\omega}}$, $t \in S_+$ and $u \in S_\omega$, we have

$$\rho(st, u) = \rho(s, tu).$$

These rules allow one to replace the notation

$$\rho(\cdots s_{-1}s_0, s_1s_2\cdots),$$

for $(s_n)_{n\in\mathbb{Z}} \in S_+^{\mathbb{Z}}$, by $\cdots s_{-1}s_0s_1\cdots$.

Example 4.1. The algebra $F_\zeta(A) = (A^+, A^{\mathbb{N}}, A^{-\mathbb{N}}, A^\zeta)$ with the obvious operations is a ζ-semigroup. It is in fact the free ζ-semigroup on A, in the same way as A^∞ is the free ω-semigroup on A (Exercise 2).

Let S and T be two ζ-semigroups. A *morphism* of ζ-semigroups from S into T is a 4-tuple $\varphi = (\varphi_+, \varphi_{\tilde{\omega}}, \varphi_\omega, \varphi_\zeta)$ such that $(\varphi_+, \varphi_{\tilde{\omega}})$ (resp. φ_ω) is a morphism of $\tilde{\omega}$-semigroups (resp. ω-semigroups) and $\varphi_\zeta : S_\zeta \to T_\zeta$ is a map such that, for each $s \in S_{\tilde{\omega}}, t \in S_\omega$,

$$\varphi_\zeta(st) = \varphi_{\tilde{\omega}}(s)\varphi_\omega(t).$$

Just as in the one-sided case, where finite ω-semigroups can be replaced by Wilke algebra, there is a notion in which all operations are of finite arity. A *Wilke ζ-algebra* is a 4-tuple $S = (S_+, S_{\tilde{\omega}}, S_\omega, S_\zeta)$ such that $S_{\tilde{\omega}}$ (resp. S_ω) is a Wilke $\tilde{\omega}$-algebra (resp. a Wilke ω-algebra) with a surjective mapping $\rho : S_{\tilde{\omega}} \times S_\omega \to S_\zeta$ such that, for each $s \in S_{\tilde{\omega}}, t \in S_+$ and $u \in S_\omega$,

$$\rho(st, u) = \rho(s, tu)$$

Thus, Wilke ζ-algebras are to ζ-semigroups what Wilke algebras are to ω-semigroups. In particular, finite ζ-semigroups can be replaced by finite Wilke ζ-algebras.

A triple (e, s, f) of elements of a semigroup S is said to be *linked* if

$$e = e^2, \quad es = s, \quad sf = s, \quad f^2 = f$$

This implies that the pair (s, f) is right linked. The pair (e, s) is said to be *left linked*. Two linked triples (e, s, f) and (e', s', f') are said to be *conjugate* if there exists $x, y, z, t \in S^1$ such that

$$e = xy, \quad e' = yx, \quad s' = ysz, \quad f = zt, \quad f' = tz$$

Then we also have $s = xs't$. The conjugacy relation is an equivalence on the linked triples.

For every finite semigroup S, one can build a finite universal ζ-semigroup such that $S_+ = S$. Indeed, it suffices to take the ω-semigroup $\bar{S}$ constructed in Proposition II.5.2. Recall that each element of $\bar{S}_\omega$ is a conjugacy class $[s, e]$ of the linked pair (s, e) and that $s^\omega = [se, e]$ where $e = s^\pi$ is the unique idempotent of s^+. We construct in the same way a $\tilde{\omega}$-semigroup by taking for $\bar{S}_{\tilde{\omega}}$ the set of conjugacy classes of the left linked pairs and by setting $s^{\tilde{\omega}} = [e, es]$. Finally, we choose for $\bar{S}_\zeta$ the set of conjugacy classes of linked triples. The map ρ is defined by

$$\rho([e, s], [t, f]) = [e, st, f].$$

The following result is an immediate consequence of Theorem II.5.1.

Theorem 4.1. *Let S be a finite ζ-semigroup. For each $x \in S_\zeta$, there exists a linked triple (e, s, f) such that $x = e^{\tilde{\omega}} s f^\omega$.*

Furthermore, the following result holds:

Proposition 4.2. *Let S be a finite semigroup. Two linked triples (e, s, f) and (e', s', f') are conjugate if and only if, for each ζ-semigroup (S_+, S_ζ) such that $S = S_+$, the equality $e^{\tilde{\omega}} s f^{\omega} = e'^{\tilde{\omega}} s' f'^{\omega}$ holds.*

Proof. The condition is clearly necessary. It is also sufficient since, in the ζ-semigroup $\bar{S}$, the two conditions are equivalent. □

We say that a morphism from $F_\zeta(A)$ onto a finite ζ-semigroup S *recognizes* a subset X of $A^{\mathbb{Z}}$ if there exists a subset U of S_ζ such that $X = \varphi^{-1}(U)$.

Theorem 4.3. *A subset of $A^{\mathbb{Z}}$ is recognizable if and only if it can be recognized by a morphism from $F_\zeta(A)$ onto a finite ζ-semigroup.*

Proof. The condition is sufficient since if φ is a morphism from $F_\zeta(A)$ onto a finite ζ-semigroup S and if $X = \varphi^{-1}(U)$ with $U \subset S_\zeta$, one can write

$$X = \bigcup_{s^{\tilde{\omega}} t u^{\omega} \in U} \varphi^{-1}(s)^{\tilde{\omega}} \varphi^{-1}(t) \varphi^{-1}(u)^{\omega}$$

and the right hand side is a finite union of recognizable subsets.

Conversely, it suffices to treat the case of a set of the form $X = UV$ with $U \subset A^{\tilde{\omega}}$ and $V \subset A^{\omega}$ recognizable. One can choose the same finite semigroup S_+ to recognize U and V, and then form the $\tilde{\omega}$-semigroup $(S_+, \bar{S}_{\tilde{\omega}})$ recognizing U and the ω-semigroup $(S_+, \bar{S}_\omega)$ recognizing V. Then it is immediate that the ζ-semigroup $\bar{S}_\zeta$ recognizes X. □

We now establish some results which extend to the two-sided case the fundamental results of McNaughton and of Landweber. See Theorems II.9.3 and II.9.4. If φ is a morphism from A^+ onto a finite semigroup S and if $s \in S$, we set

$$X_s = \varphi^{-1}(s) \qquad D_s = X_s \setminus X_s A^+ \qquad G_s = X_s \setminus A^+ X_s$$

If (e, s, f) is a linked triple, we set

$$\begin{aligned} K(e, s, f) &= \{(e', f') \in Se \times fS \mid (e', s, f') \text{ is linked}\} \\ L(e, s) &= \{e' \in Se \mid (e', s) \text{ is left linked}\} \\ R(s, f) &= \{f' \in fS \mid (s, f') \text{ is linked}\} \end{aligned}$$

Note that $K = R \times L$.

Proposition 4.4. *The following formula holds, with $R = R(s, f)$:*

$$X_e^{\tilde{\omega}} X_s X_f^{\omega} \subset \overrightarrow{X_e^{\tilde{\omega}} X_s D_f} \subset \bigcup_{f' \in R} X_e^{\tilde{\omega}} X_s X_{f'}^{\omega} \tag{4.1}$$

Proof. By Proposition 2.1, and since $es = s$, we have

$$\overrightarrow{X_e^{\tilde{\omega}} X_s D_f} = X_e^{\tilde{\omega}}(\overrightarrow{X_s D_f})$$

The formula is thus a direct consequence of the formula

$$X_s X_f^{\omega} \subset \overrightarrow{X_s D_f} \subset \bigcup_{f' \in R} X_s X_{f'}^{\omega}$$

which has been proved in the one-sided case (Proposition II.9.1). □

A direct consequence of this statement is a generalization of the results pertaining to the one-sided case (Theorem II.9.3). Let us recall that, by Theorem 3.1, $Det(A^{\mathbb{Z}}) = \overrightarrow{Rec(A^{-\mathbb{N}})}$.

Theorem 4.5. *Every recognizable subset of $A^{\mathbb{Z}}$ is a boolean combination of subsets of the form $\overrightarrow{X}$ where X is a recognizable subset of $A^{-\mathbb{N}}$.*

The proof is very similar to the corresponding proof in the one-sided case (see Theorem II.9.4), but makes use of Formula (4.1). We omit to enter into all the details.

Theorem 4.6. *The equality $\overrightarrow{\mathcal{P}(A^{-\mathbb{N}})} \cap \mathrm{Rec}(A^{\mathbb{Z}}) = \mathrm{Det}(A^{\mathbb{Z}})$ holds.*

Again, the proof is exactly the same as in the one-sided case. It relies on the following lemma.

Lemma 4.7. *Let U be a recognizable subset of $A^{\mathbb{Z}}$. Let φ be a morphism from $F_{\zeta}(A)$ onto a finite ζ-semigroup S recognizing U. Then $U \in \mathrm{Det}(A^{\mathbb{Z}})$ if and only if, for each linked triple (e, s, f) of S_+ such that $X_e^{\tilde{\omega}} X_s X_f^{\omega} \subset U$, we also have $X_e^{\tilde{\omega}} X_s X_{f'}^{\omega} \subset U$ for each $f' \in R(s, f)$.*

We consider now the more symmetric case of two-sided limits.

Proposition 4.8. *The following formula holds, with $K = K(e, s, f)$:*

$$X_e^{\tilde{\omega}} X_s X_f^{\omega} \subset \overleftrightarrow{G_e X_s D_f} \subset \bigcup_{(e', f') \in K} X_{e'}^{\tilde{\omega}} X_s X_{f'}^{\omega} \tag{4.2}$$

Proof. The first inclusion is clear. To establish the second one, let us consider a sequence (u_n, v_n) of pairs of words such that $u_n v_n \in G_e X_s D_f$ and let $w = [u, v]$ with $\tilde{u} = \lim_{n\to\infty} \tilde{u_n}$, $v = \lim_{n\to\infty} v_n$. Since G_e is suffix and recognizable, we have $\overleftrightarrow{G_e} = \emptyset$ by Proposition 2.2. Similarly, since D_f is prefix and recognizable, $\overleftrightarrow{D_f} = \emptyset$.

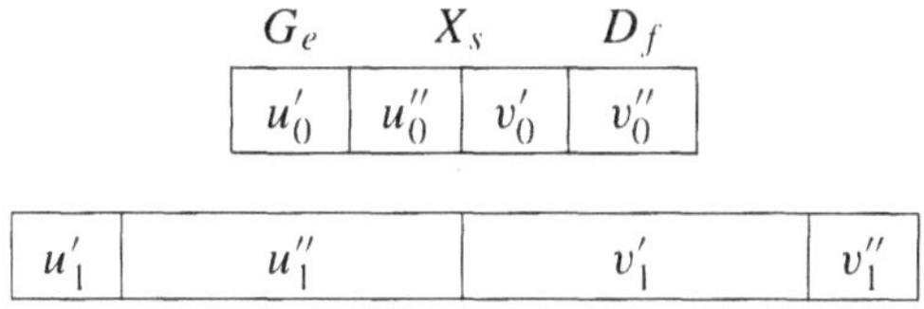

Up to taking a subsequence, we may assume that, for each $n \geq 0$, one has $u_n = u'_n u''_n$, $v_n = v'_n v''_n$ with $u'_n \in G_e$, $u''_n v'_n \in X_s$, $v''_n \in D_f$. We may also assume (again at the price of taking a subsequence), that $\varphi(u''_n) = s'$, $\varphi(v'_n) = s''$ with $s = s's''$. Setting $R = R(e, s)$ and $L = L(s, f)$, we have, by Proposition 4.4

$$\begin{aligned}\overleftrightarrow{G_e X_s D_f} &= \bigcup_{s=s's''} (\overleftarrow{G_e X_{s'}})(\overrightarrow{X_{s''} D_f}) \\ &\subset \bigcup_{s=s's''} \Big(\bigcup_{e' \in L} X_{e'}^{\tilde{\omega}} X_{s'}\Big)\Big(\bigcup_{f' \in R} X_{s''} X_{f'}^{\omega}\Big) \\ &\subset \bigcup_{e' \in L} \bigcup_{f' \in R} X_{e'}^{\tilde{\omega}} X_s X_{f'}^{\omega}\end{aligned}$$

Formula (4.2) follows, since $K = L \times R$. □

The next theorem generalizes McNaughton's theorem to the two-sided case.

Theorem 4.9. *Every recognizable subset of $A^{\mathbb{Z}}$ is a boolean combination of sets of the form $\overleftrightarrow{X}$, where X is a recognizable subset of A^*.*

Proof. Let $U \in \mathrm{Rec}(A^{\mathbb{Z}})$. Let S be a finite ζ-semigroup and let $\varphi : F_\zeta(A) \to S$ be a morphism of ζ-semigroups recognizing U. Let (e, s, f) be a linked triple of S_+ such that $X_e^{\tilde{\omega}} X_s X_f^{\omega} \subset U$. Then we have, with the notation of Proposition 4.8

$$X_e^{\tilde{\omega}} X_s X_f^{\omega} = \overleftrightarrow{G_e X_s D_f} \setminus \Big(\bigcup_{e' \in L \setminus E} X_{e'}^{\tilde{\omega}} X_s X_f^{\omega} \cup \bigcup_{f \in R \setminus F} X_e^{\tilde{\omega}} X_s X_{f'}^{\omega} \Big) \qquad (4.3)$$

where E denotes the $\mathcal{D}$-class of e and F that of f. Formula 4.3 suffices to establish the result by induction on $\mathrm{Card}(L) + \mathrm{Card}(R)$. To verify this formula, it suffices to observe that, on the one hand we have $X_e^{\tilde{\omega}} X_s X_f^{\omega} = X_{e'}^{\tilde{\omega}} X_{s'} X_{f'}^{\omega}$ for each $(e', f') \in E \times F$, since φ recognizes U. On the other hand, we have $X_e^{\tilde{\omega}} X_s X_f^{\omega} \cap X_{e'}^{\tilde{\omega}} X_{s'} X_{f'}^{\omega} = \emptyset$ as soon as $e' \notin E$ or $f' \notin F$. Therefore, Formula (4.3) can be directly deduced from Formula (4.2). □

Example 4.2. If $U = a^{\tilde{\omega}}(a+b)^* a^{\omega}$, then by Formula (4.3),

$$U = \overleftrightarrow{a(a+b)^* a} \setminus \big((ba^*)^{\tilde{\omega}}(a+b)^{\omega} \cup (a+b)^{\tilde{\omega}}(a^* b)^{\omega}\big)$$

The next theorem generalizes Landweber's theorem to the two-sided case.

Theorem 4.10. *The formula* $\overleftrightarrow{\mathcal{P}(A^*)} \cap \mathrm{Rec}(A^{\mathbb{Z}}) = \overleftrightarrow{\mathrm{Rec}(A^*)}$ *holds.*

Proof. Let U be a recognizable subset of $A^{\mathbb{Z}}$ of the form $\overleftrightarrow{X}$, with $X \subset A^*$. Let φ be a morphism of ζ-semigroups from $F_\zeta(A)$ onto a finite ζ-semigroup S recognizing U. The theorem is now a consequence of the following lemma.

Lemma 4.11. *We have* $U \in \overleftrightarrow{\mathcal{P}(A^*)}$ *if and only if, for each linked triple* (e, s, f) *of* S_+ *such that* $X_e^{\tilde{\omega}} X_s X_f^{\omega} \subset U$*, we also have* $X_{e'}^{\tilde{\omega}} X_s X_{f'}^{\omega} \subset U$ *for each* $(e', f') \in K$.

Proof. The condition is sufficient since, according to Formula 4.2, one has

$$\bigcup_{(e',f') \in K} X_{e'}^{\tilde{\omega}} X_s X_{f'}^{\omega} = \bigcup_{(e',f') \in K} \overleftrightarrow{G_{e'} X_s D_{f'}}$$

The argument used in the one-sided case can be readily adapted to show that the condition is also necessary. □

The last result of this section sets up a curious link between the three topologies introduced on $A^{\mathbb{Z}}$, and their restriction to recognizable subsets. Roughly speaking, it means that, for a recognizable subset of $A^{\mathbb{Z}}$, its left and right topological properties are independent.

Theorem 4.12. *The following equality holds:*

$$\overleftarrow{\mathrm{Rec}(A^{-\mathbb{N}})} \cap \overrightarrow{\mathrm{Rec}(A^{\mathbb{N}})} = \overleftrightarrow{\mathrm{Rec}(A^*)}$$

Proof. The inclusion from right to left is obvious. The opposite inclusion relies on the previous results. Let $U \in \overleftarrow{\mathrm{Rec}(A^{-\mathbb{N}})} \cap \overrightarrow{\mathrm{Rec}(A^{\mathbb{N}})}$ and let φ be a morphism of ζ-semigroups from $F_\zeta(A)$ onto a finite ζ-semigroup S recognizing U. Let (e, s, f) be a linked triple of S_+ such that $X_e^{\tilde{\omega}} X_s X_f^{\omega} \subset U$. By Lemma 4.7, the inclusion $X_e^{\tilde{\omega}} X_s X_{f'}^{\omega} \subset U$ holds for each $f' \in R$. Symmetrically, we have $X_{e'}^{\tilde{\omega}} X_s X_f^{\omega} \subset U$ for each $e' \in L$. Since $K = L \times R$, the result follows from Lemma 4.11. □

5 Unambiguous automata on bi-infinite words

Theorem 4.5 shows that, in some sense, McNaughton's theorem holds in the bi-infinite case. However, there is no direct formulation of the theorem in terms of automata, because no counterpart of the notion of Muller automaton is known in the two-sided

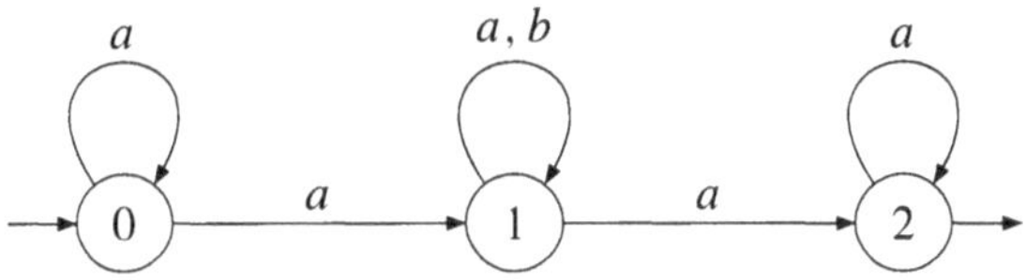

Figure 5.1. A ζ-ambiguous automaton.

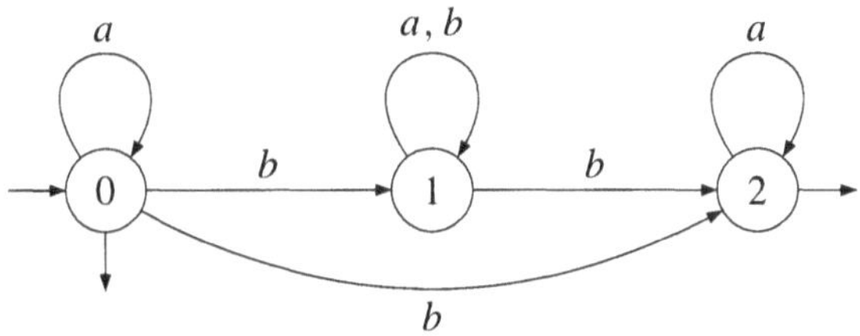

Figure 5.2. A ζ-unambiguous automaton.

case. We introduce in this section a notion of unambiguity which, in some sense, fills this gap.

An automaton $\mathcal{A} = (Q, A, E, I, F)$ is said to be *ζ-unambiguous* if and only if any bi-infinite word is the label of at most one successful path.

Example 5.1. The automaton pictured in Figure 5.1 recognizes the set $a^{\tilde{\omega}}(a+b)^*a^{\omega}$ of bi-infinite words having finitely many b's. This automaton is ζ-ambiguous since the word a^{ζ} is the label of the two successful paths $0^{\tilde{\omega}}12^{\omega}$ and $0^{\tilde{\omega}}112^{\omega}$. The automaton pictured in Figure 5.2 recognizes the same set but is ζ-unambiguous.

The next example shows that it is not always so easy to find a ζ-unambiguous automaton recognizing a given set of bi-infinite words.

Example 5.2. The set $(ba^*)^{\tilde{\omega}}(a+b)^{\omega} + (a+b)^{\tilde{\omega}}(ba^*)^{\omega}$ is the of bi-infinite words having infinitely many b's. It is recognized by the automaton pictured in Figure 5.3.

Recall that a Büchi automaton $\mathcal{A} = (Q, A, E, I, F)$ is ω-unambiguous if every infinite word is the label of at most one infinite successful path. An easy way to obtain a ζ-unambiguous automaton is to combine two ω-unambiguous automata as follows.

Let $\mathcal{B} = (Q_{\mathcal{B}}, A, E_{\mathcal{B}}, I_{\mathcal{B}}, F_{\mathcal{B}})$ and $\mathcal{C} = (Q_{\mathcal{C}}, A, E_{\mathcal{C}}, I_{\mathcal{C}}, F_{\mathcal{C}})$ be two Büchi automata. The *synchronised product* $\tilde{\mathcal{B}} \times \mathcal{C}$ is the automaton $\mathcal{A} = (Q, A, E, I, F)$ where the set of states is $Q = Q_{\mathcal{B}} \times Q_{\mathcal{C}}$, the sets of initial and final states are $F_{\mathcal{B}} \times Q_{\mathcal{C}}$ and $Q_{\mathcal{B}} \times F_{\mathcal{C}}$,

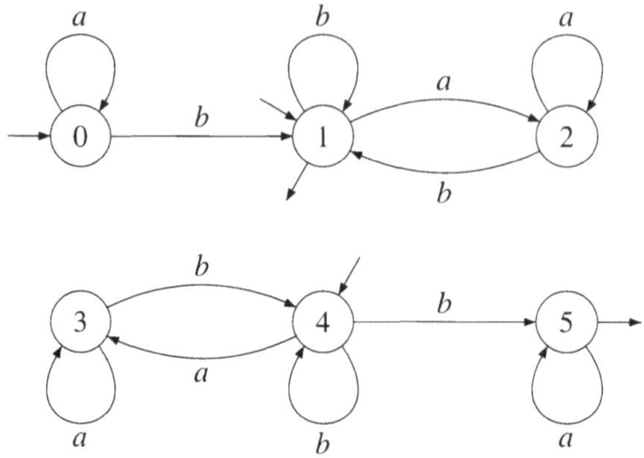

Figure 5.3. Another ζ-unambiguous automaton.

respectively, and the set E of transitions is defined by

$$E = \{((p,q), a, (p',q')) \mid (p', a, p) \in E_{\mathcal{B}} \text{ and } (q, a, q') \in E_{\mathcal{C}}\}.$$

Proposition 5.1. *The synchronised product of two ω-unambiguous automata is a ζ-unambiguous automaton.*

Proof. Let $\mathcal{B}$ and $\mathcal{C}$ be two ω-unambiguous automata and let $\mathcal{A}$ be their synchronised product. Let u be a bi-infinite word and let

$$\cdots \ (p_{k-1}, q_{k-1}) \xrightarrow{u_{k-1}} (p_k, q_k) \xrightarrow{u_k} (p_{k+1}, q_{k+1}) \ \cdots$$

be a successful bi-infinite path of $\mathcal{A}$. For each $n \in \mathbb{Z}$, let $x_n = \cdots u_{n-2}u_{n-1}$ and $y_n = u_n u_{n+1} \cdots$. By definition of $\mathcal{A}$, the path

$$p_n \xrightarrow{u_{n-1}} p_{n-1} \xrightarrow{u_{n-2}} p_{n-2} \ \cdots \tag{5.1}$$

is a final path in $\mathcal{B}$ labelled by x_n and

$$q_n \xrightarrow{u_n} q_{n+1} \xrightarrow{u_{n+1}} q_{n+2} \ \cdots \tag{5.2}$$

is a final path in $\mathcal{C}$ labelled by y_n. Since $\mathcal{B}$ (resp. $\mathcal{C}$) is ω-unambiguous, p_n (resp. q_n) is the first state of the unique final path in $\mathcal{B}$ (resp. $\mathcal{C}$) labelled by x_n (resp. y_n). This shows that the state (p_n, q_n) is fully determined by u. Therefore u labels at most one successful path in $\mathcal{B} \times \mathcal{C}$. □

The proof of the main theorem of this section relies on a slight generalisation of Corollary II.10.7.

Proposition 5.2. *Let $X_1, \ldots, X_n$ be n ω-rational sets. There is a prophetic Büchi automaton $\mathcal{A} = (Q, A, E, F)$ and, for $1 \leq k \leq n$, a subset I_k of Q such that the automaton $\mathcal{A}_k = (Q, A, I_k, E, F)$ recognizes X_k.*

Proof. For $n = 1$, the result follows directly from Corollary II.10.7. Let $n \geq 2$, and suppose that the result holds for $n - 1$. There is a prophetic automaton $\mathcal{B} = (Q_\mathcal{B}, A, E_\mathcal{B}, F_\mathcal{B})$ and $n-1$ subsets $J_1, \ldots, J_{n-1}$ of $Q_\mathcal{B}$ such that, for $1 \leq k \leq n-1$, the automaton $\mathcal{B}_k = (Q_\mathcal{B}, A, I_k, E_\mathcal{B}, F_\mathcal{B})$ recognizes X_k. By Corollary II.10.7, there is also a prophetic automaton $\mathcal{C} = (Q_\mathcal{C}, A, E_\mathcal{C}, I_\mathcal{C}, F_\mathcal{C})$ recognizing the set X_n. We now combine these two automata to obtain, for $1 \leq k \leq n$, an automaton $\mathcal{A}_k = (Q, A, I_k, E, F)$ recognizing X_k. Our construction is reminiscent of the construction used in the proof of Proposition I.6.4. Let us set

$$Q = Q_\mathcal{B} \times Q_\mathcal{C} \times \{0, 1\}$$

$$I_k = \begin{cases} J_k \times Q_\mathcal{C} \times \{0, 1\} & \text{if } 1 \leq k \leq n-1 \\ Q_\mathcal{B} \times I_\mathcal{C} \times \{0, 1\} & \text{if } k = n \end{cases}$$

$$F = Q_\mathcal{B} \times F_\mathcal{C} \times \{1\}$$

$$E = \left\{ \big((p, q, \varepsilon), a, (p', q', \varepsilon')\big) \mid (p, a, p') \in E_\mathcal{B}, (q, a, q') \in E_\mathcal{C} \text{ and} \right.$$

$$\left. \varepsilon = \begin{cases} 1 & \text{if } p' \in F_\mathcal{B} \\ 0 & \text{if } p' \notin F_\mathcal{B} \text{ and } q' \in F_\mathcal{C} \\ \varepsilon' & \text{otherwise} \end{cases} \right\}$$

Consider an infinite path

$$(p_0, q_0, \varepsilon_0) \xrightarrow{u_0} (p_1, q_1, \varepsilon_1) \xrightarrow{u_1} (p_2, q_2, \varepsilon_2) \quad \cdots$$

The ε-bit acts as a semaphore, whose value can only change in two circumstances. First, ε_n is reset to 0 each time q_n is final and p_{n+1} is not final. Next, ε_n is reset to 1 if p_{n+1} is final. In particular, if the path is final, then the path $q_0 \xrightarrow{u_0} q_1 \xrightarrow{u_1} q_2 \cdots$ is final in $\mathcal{C}$, because q_n has to visit $F_\mathcal{C}$ infinitely often, and the path $p_0 \xrightarrow{u_0} p_1 \xrightarrow{u_1} p_2 \cdots$ is final in $\mathcal{B}$, because ε_n has to take the value 1 infinitely often.

By construction, the automaton $\mathcal{A}$ has co-deterministic transitions. Let us verify that it is co-deterministic. Assume that a word u is the label of two final paths of $\mathcal{A}$, say

$$(p_0, q_0, \varepsilon_0) \xrightarrow{u_0} (p_1, q_1, \varepsilon_1) \xrightarrow{u_1} (p_2, q_2, \varepsilon_2) \quad \cdots$$

$$(p'_0, q'_0, \varepsilon'_0) \xrightarrow{u_0} (p'_1, q'_1, \varepsilon'_1) \xrightarrow{u_1} (p'_2, q'_2, \varepsilon'_2) \quad \cdots$$

Then $p_0 \xrightarrow{u_0} p_1 \xrightarrow{u_1} p_2 \cdots$ and $p'_0 \xrightarrow{u_0} p'_1 \xrightarrow{u_1} p'_2 \cdots$ are two final paths of $\mathcal{B}_1$ with the same label, and since $\mathcal{B}_1$ is co-deterministic, $p_n = p'_n$ for all $n \geq 0$. Next, $q_0 \xrightarrow{u_0} q_1 \xrightarrow{u_1} q_2 \cdots$ and $q'_0 \xrightarrow{u_0} q'_1 \xrightarrow{u_1} q'_2 \cdots$ are two final paths of $\mathcal{C}$ and since $\mathcal{C}$ is co-deterministic, $q_n = q'_n$ for all $n \geq 0$. Finally, for the infinitely many values n such that $q_{n+1} \in F_{\mathcal{C}}$, we have

$$\varepsilon_n = \varepsilon'_n = \begin{cases} 1 & \text{if } p_{n+1} \in F_{\mathcal{B}} \\ 0 & \text{if } p_{n+1} \notin F_{\mathcal{B}} \end{cases}$$

Since the transitions of $\mathcal{A}$ are co-deterministic, this implies that $\varepsilon_n = \varepsilon'_n$ for all $n \geq 0$ and thus the two paths coincide.

We claim that $\mathcal{A}$ is co-complete. Let u be an infinite word. Since $\mathcal{B}_1$ is co-complete, there is a final path $p_0 \xrightarrow{u_0} p_1 \xrightarrow{u_1} p_2 \cdots$ of $\mathcal{B}_1$ labelled by u. Similarly, since $\mathcal{C}$ is co-complete, there is a final path $q_0 \xrightarrow{u_0} q_1 \xrightarrow{u_1} q_2 \cdots$ of $\mathcal{C}$ labelled by u. We now just have to adjust the ε_n to obtain a final path of label u. Let us set, in a first round

$$\varepsilon_n = \begin{cases} 1 & \text{if } p_{n+1} \in F_{\mathcal{B}} \\ 0 & \text{if } p_{n+1} \notin F_{\mathcal{B}} \text{ and } q_{n+1} \in F_{\mathcal{C}} \end{cases}$$

Since p_n is in $F_{\mathcal{B}}$ for infinitely many values of n, it suffices now to fill the holes by setting $\varepsilon_n = \varepsilon_m$ where m is the smallest integer larger than n such that ε_m was already defined in the first round. The path

$$(p_0, q_0, \varepsilon_0) \xrightarrow{u_0} (p_1, q_1, \varepsilon_1) \xrightarrow{u_1} (p_2, q_2, \varepsilon_2) \cdots$$

is now well-defined. It is a little bit more tricky to show that this path is final. Let n_0 be an integer. Choose $n_1 > n_0$ such that $q_{n_1+1} \in F_{\mathcal{C}}$. Next, let n_2 be the smallest integer larger than n_1 such that $p_{n_2+1} \in F_{\mathcal{B}}$. Finally, let n_3 be the largest integer smaller that n_2 such that $q_{n_3+1} \in F_{\mathcal{C}}$. By construction $n_0 \leq n_1 \leq n_3 < n_2$. Furthermore, $\varepsilon_{n_2} = 1$ since $p_{n_2+1} \in F_{\mathcal{B}}$ and, for $n_3 < m < n_2$, $\varepsilon_m = 1$ since $p_{m+1} \notin F_{\mathcal{B}}$ and $q_{m+1} \notin F_{\mathcal{C}}$. In particular, the state $(p_{n_3+1}, q_{n_3+1}, \varepsilon_{n_3+1})$ is final and thus the path is final.

Let us show that, for $1 \leq k \leq n$, $\mathcal{A}_k$ recognizes X_k. Since $\mathcal{A}$ is prophetic, there is a unique final path of label u in $\mathcal{A}$, say

$$(p_0, q_0, \varepsilon_0) \xrightarrow{u_0} (p_1, q_1, \varepsilon_1) \xrightarrow{u_1} (p_2, q_2, \varepsilon_2) \cdots$$

Furthermore, $p_0 \xrightarrow{u_0} p_1 \xrightarrow{u_1} p_2 \cdots$ is, for $1 \leq k \leq n-1$, the unique final path of $\mathcal{B}_k$ of label u. It follows that $u \in X_k$ if and only if $p_0 \in J_k$ and thus $\mathcal{A}_k$ recognizes X_k.

Finally, $q_0 \xrightarrow{u_0} q_1 \xrightarrow{u_1} q_2 \cdots$ is the unique final path of $\mathcal{C}$ of label u. Thus $u \in X_n$ if and only if $q_0 \in I_{\mathcal{C}}$ and thus $\mathcal{A}_n$ recognizes X_n. □

Theorem 5.3. *Any rational set of A^ζ can be recognized by a ζ-unambiguous automaton.*

Proof. Let Z be a ζ-rational set of bi-infinite words. By Corollary 2.4, its complement is also ζ-rational, and hence, by Proposition 2.3, is equal to a finite union $\bigcup_{k=1}^{n} \tilde{U}_n V_n$ where the sets U_k and V_k are ω-rational subsets of A^ω. By Proposition 5.2, there is a prophetic automaton $\mathcal{B} = (Q_\mathcal{B}, A, E_\mathcal{B}, F_\mathcal{B})$ and n subsets $I_1, \ldots, I_n$ of $Q_\mathcal{B}$ such that for $1 \leq k \leq n$, the automaton $\mathcal{B}_k = (Q_\mathcal{B}, A, E_\mathcal{B}, I_k, F_\mathcal{B})$ recognizes X_k. Similarly, there is a prophetic automaton $\mathcal{C} = (Q_\mathcal{C}, A, E_\mathcal{C}, F_\mathcal{C})$ and n subsets $J_1, \ldots, J_n$ of $Q_\mathcal{C}$ such that for $1 \leq k \leq n$, the automaton $\mathcal{C}_k = (Q_\mathcal{C}, A, E_\mathcal{C}, J_k, F_\mathcal{C})$ recognizes X_k.

By Proposition 5.1, the automaton $\mathcal{A} = \tilde{\mathcal{B}} \times \mathcal{C}$ is ζ-unambiguous. Note that a bi-infinite word z belongs to Z^c if and only if it can be written as $z = xy$ with $x \in X_k$ and $y \in Y_k$ for some integer $1 \leq k \leq n$. This means that z belongs to Z^c if a state (p, q) in $F_k \times I_k$ occurs in the unique successful path of $\mathcal{A}$ labelled by z. Let P be the subset $\bigcup_{i=1}^{n} F_k \times I_k$. Let $\mathcal{A}'$ be the automaton obtained from $\mathcal{A}$ by suppressing all states of P. More formally the automaton $\mathcal{A}'$ is the automaton (Q', A, E', I', F') where Q', I' and F' are respectively equal to $Q \setminus P$, $I \setminus P$ and $F \setminus Q$ and E' is equal to

$$E' = \{((p, q), a, (p', q')) \in E \mid (p, q) \notin P \text{ and } (p', q') \notin P\}.$$

The automaton $\mathcal{A}'$ is still ζ-unambiguous and recognizes Z. □

The construction used in the proof of Theorem 5.3 is illustrated in the following example.

Example 5.3. The set $X = A^{\tilde{\omega}} b A^\omega$ of bi-infinite words having at least one occurrence of b is recognized by the automaton pictured in Figure 5.4 but this automaton is ζ-ambiguous. The construction described in the proof of Theorem 5.3 can be used to get a ζ-unambiguous automaton recognizing X. The complement of X is the set a^ζ. Automata $\tilde{\mathcal{B}}$ and $\mathcal{C}$ pictured in Figure 5.5 recognize the sets $a^{\tilde{\omega}}$ and a^ω. One gets an ζ-unambiguous automata recognizing X by removing the state 02 of the product automaton $\tilde{\mathcal{B}} \times \mathcal{C}$.

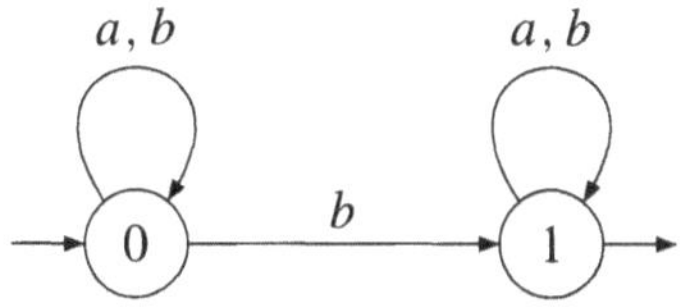

Figure 5.4. A ζ-ambiguous automaton.

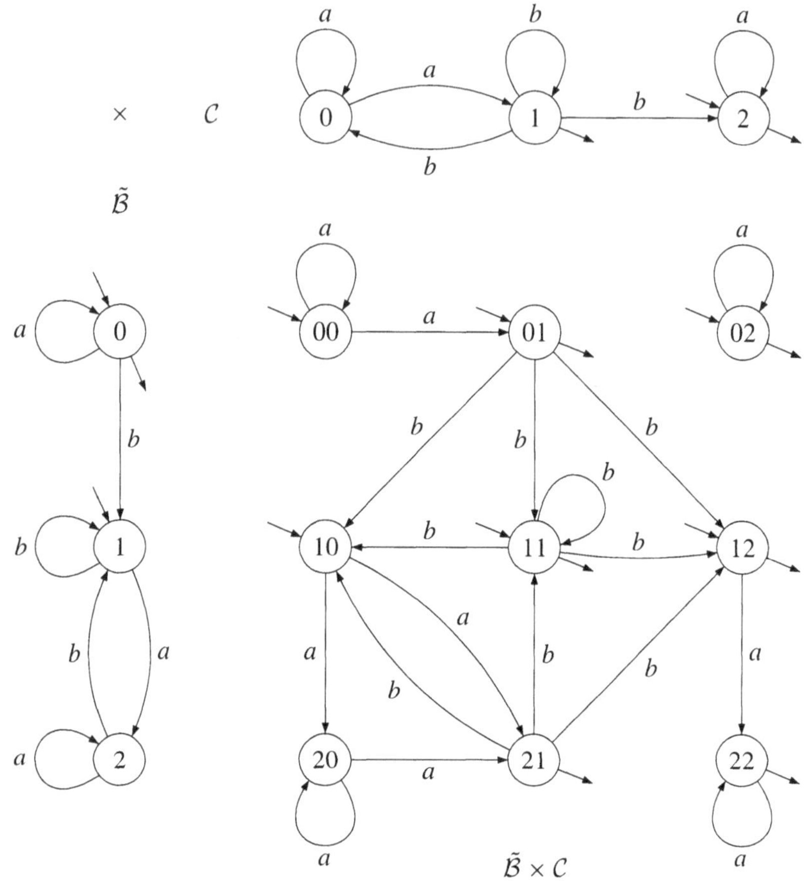

Figure 5.5. The product automaton.

6 Discrimination

Two bi-infinite words $x, y \in A^{\mathbb{Z}}$ are said to be *distinguishable* if there exists a finite Büchi automaton $\mathcal{A}$ accepting x but not y. Otherwise, x and y are said to be *indistinguishable*. Two shifted words are certainly indistinguishable. It is clear that the existence of distinct indistinguishable words relies on our definition of recognition by finite automaton, which does not refer to an origin. This phenomenon is specific to bi-infinite words since finite or one-sided infinite words are always discriminated by a finite automaton.

We shall study this notion in this section, and give a characterization of indistinguishable words.

First observe that, if two words $x, y \in A^{\mathbb{Z}}$ are indistinguishable, then for each ζ-rational subset $X \subset A^{\mathbb{Z}}$, $x \in X$ if and only if $y \in X$.

A bi-infinite word $x \in A^{\mathbb{Z}}$ is *recurrent* if every word w occurring in x occurs infinitely often on the left and on the right. The set $F(x)$ of *factors* of x is the set of finite words occurring in x.

Theorem 6.1. *Two bi-infinite words are indistinguishable if and only if one of the following conditions is satisfied:*

(1) *they are shift equivalent,*

(2) *they are both recurrent and have the same set of factors.*

It is easy to see that the condition is necessary. Indeed, two indistinguishable words have necessarily the same set of factors since $w \in F(x)$ is equivalent to $x \in A^{\tilde{\omega}}wA^{\omega}$ and that $A^{\tilde{\omega}}wA^{\omega}$ is a ζ-recognizable set. Next, if x is not recurrent, there exists a word w which does not occur infinitely often in both directions. Let us suppose there is a last occurrence of w on the right of x. Intuitively, this last occurrence can be used to define an origin, thus rendering x unique. The set $wU \subset A^{\omega}$, which consists of the infinite words starting by w but not containing any other occurrence of w, is ω-recognizable. Thus $A^{\tilde{\omega}}wU$ is ζ-recognizable and if y is indistinguishable from x, it has to be equal to x on the right of w. A similar argument shows that in fact, $x = y$.

Let $S = F(x)$ be the set of factors of a recurrent word x on the alphabet A. It is clear that S has the following property: if $u, v \in S$, there exists a $w \in S$ such that uw, wv and uwv are in S. Conversely, if S is a set with this property, one can construct a recurrent word x such that $F(x) = S$. More generally, we introduce the following definition. A subset T of a semigroup S is *transitive* if, for each $t_1, t_2 \in T$, there exists an element $t \in T$ such that $t_1t \in T$, $tt_2 \in T$ and $t_1tt_2 \in T$.

We prove the following lemma:

Lemma 6.2. *Let S be a finite semigroup and let T be a transitive subset of S. The intersection K of all the ideals of S meeting T is a principal ideal which meets T. Furthermore, $StS = K$ for each $t \in T \cap K$.*

Proof. Since S is finite, it suffices to prove that if I and J are two ideals meeting T, then $I \cap J$ meets T. Let $r \in I \cap T$, $s \in J \cap T$. Since T is transitive, there exists $t \in T$ such that $rts \in T$. But since I and J are ideals, we have $rts \in I \cap J$. Therefore $I \cap J$ meets T, and hence K meets T. Furthermore, K is a principal ideal since, for each $t \in T \cap K$, $StS \subset K$ and thus $StS = K$. □

The intersection of all the ideals of S meeting T is called the *T-minimal ideal* of S.

Lemma 6.3. *Let S be a finite semigroup and let K be a principal ideal of S. Let D be the set of generators of K, i.e.*

$$D = \{s \in S \mid K = SsS\}$$

Let $\ell, r \in D$ be such that $\ell r \in D$. Then the set $G = S\ell \cap rS \cap D$ is a finite group.

Proof. By definition, D is a $\mathcal{D}$-class of S. Next G is an $\mathcal{H}$-class containing an idempotent since $lr \in D$. Therefore, by Proposition A.2.4, it is a group. □

Lemma 6.4. *Let $\varphi : A^+ \to S$ be a morphism from A^+ onto a finite semigroup S. Let T be a transitive subset of S, let K be the T-minimal ideal of S, let D be the set of generators of K and let $x \in A^{\mathbb{Z}}$ be a recurrent word such that $\varphi(F(x)) = T$. Then, for all $u, v \in F(x)$, we have*

$$x \in (\varphi^{-1}(e))^{\zeta}$$

where e is the idempotent of the group $G = \varphi(v)S \cap S\varphi(u) \cap D$.

Proof. Since $\varphi(u), \varphi(v) \in T \cap K$, we have by Lemma 6.2 that $\varphi(u), \varphi(v) \in D$. By Lemma 6.3, the set $G = \varphi(v)S \cap S\varphi(u) \cap D$ is a group. Note for further use that each word $w \in vA^* \cap A^*u \cap F(x)$ satisfies $\varphi(w) \in G$.

Since $uv \in F(x)$, one has $x \in yuvz$ with $z \in A^\omega$. We claim that $vz \in (\varphi^{-1}(e))^\omega$. One would show, symmetrically, that $yu \in (\varphi^{-1}(e))^{\tilde{\omega}}$.

For each $w \in F(x)$, set

$$R(w) = \{\varphi(vsu) \mid w \in A^*uvsu\}$$

by the remarks above, $R(w)$ is contained in the group G. We also observe that $R(w) \subset R(xw)$ for each $x \in A^*$.

Consider a word $w \in F(x)$ such that Card$(R(w))$ is maximal. Since $uv \in F(x)$ and since x is recurrent, we have Card$(R(w)) \geq 1$. In particular, $w \in A^*u$.

Since x is recurrent, one can write $z = twz'$ with $t \in A^+$. Now $R(w) \subset R(uvtw)$ by the previous remark. Since $R(w)$ is maximal, we have thus $R(w) = R(uvtw)$. As $w \in A^*u$, we have (see Figure 6.1) $\varphi(vtw) \in R(uvtw)$ and hence $\varphi(vtw) \in R(w)$. Set $w = ruvsu$ with $\varphi(vtw) = \varphi(vsu)$ and let $x = vtru$ and $y = vsu$. Then we have $\varphi(x), \varphi(y) \in G$ and $\varphi(xy) = \varphi(y)$. This implies $\varphi(x) = e$, whence

$$vz \in \varphi^{-1}(e)vA^\omega.$$

The conclusion follows from the fact that $vz \in (\varphi^{-1}(e))^\omega$ by repeating the argument on the word $vsuz'$ instead of vz. □

We conclude now the proof of Theorem 6.1. Two shift equivalent words are clearly indistinguishable. Let $x, y \in A^{\mathbb{Z}}$ be two recurrent words such that $F(x) = F(y)$. Let $U \subset A^{\mathbb{Z}}$ be a recognizable set. Let $\varphi : F_\zeta(A) \to S$ be a morphism of ζ-semigroups from $F_\zeta(A)$ onto a finite ζ-semigroup S recognizing U. Let $T = \varphi(F(x))$ and let K be the T-minimal ideal of S_+. Let $u \in \varphi^{-1}(K \cap T)$. Since $F(x)$ is transitive, there exists $t \in A^*$ such that $utu \in F(x)$. Let $v = tu$. Then $u, v \in \varphi^{-1}(K \cap T)$ and $uv \in F(x) = F(y)$. Let e be the identity of G. By Lemma 6.4, we have $x, y \in \varphi^{-1}(e)^\zeta$. Therefore $x \in U$ if and only if $y \in U$. Thus x and y are indistinguishable. □

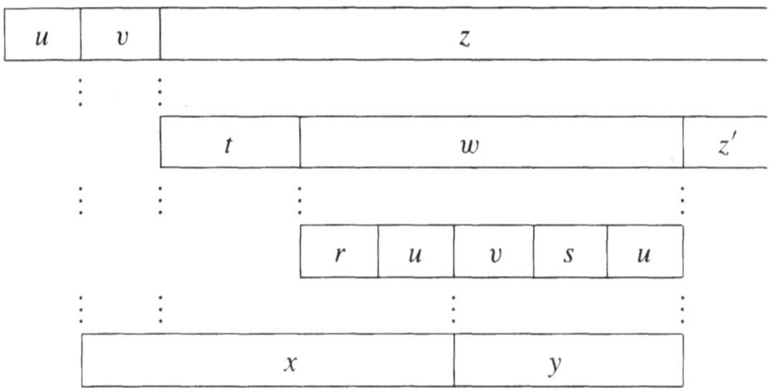

Figure 6.1. The factorization of vz.

7 Logic on $\mathbb{Z}$

The considerations developed in Chapter VIII on the link between logic and automata can be transposed to the case of bi-infinite words with the remarkable feature, due to shift invariance, that the formulas do not use the zero.

We start with the generalization of Büchi's theorem. Consider the monadic second-order logic of the successor $MF_2(S)$, which makes use of the successor relation and the relation symbols **a** for $a \in A$. Variables are now interpreted as elements of $\mathbb{Z}$ (and not $\mathbb{N}$ as in Chapter VIII).

We set

$$S^{\zeta}(\varphi) = \{x \in A^{\mathbb{Z}} \mid x \models \varphi\}$$

and we say that φ defines the set $S^{\zeta}(\varphi)$.

The constant 0 cannot, this time, be defined in $MF_2(S)$. Indeed, all the subsets of $\mathbb{Z}$ definable in $MF_2(S)$ are shift invariant. We still have $MF_2(S) = MF_2(<)$ since the order relation of $\mathbb{Z}$ can be defined in $MF_2(S)$ as that of $\mathbb{N}$.

The following theorem shows that the logic $MF_2(S)$ is decidable.

Theorem 7.1. *A subset of $A^{\mathbb{Z}}$ is definable by a sentence of $MF_2(S)$ if and only if it is recognizable.*

Proof. If X is recognized by an automaton $\mathcal{A} = (Q, I, T, F)$, we build, as in the proof of Proposition VIII.3.2, a formula expressing the existence of a successful path. In the opposite direction, we proceed by induction on the formation rules of formulas to show that the set of words satisfying a formula is recognizable. □

Example 7.1. The set of words on $A = \{a, b\}$ with infinitely many b's on the left and on the right is defined by the formula

$$\forall x\ (\exists y < x\ \mathbf{b}(y) \wedge \exists z > x\ \mathbf{b}(z))$$

The equivalence between weak and strong monadic second-order logic still holds.

Theorem 7.2. *On $\mathbb{Z}$, $WMF_2(S)$ and $MF_2(S)$ are equivalent.*

Proof. By Theorem 4.9, every recognizable subset of $A^{\mathbb{Z}}$ is a boolean combination of subsets of the form $\overleftrightarrow{X}$ for some recognizable set $X \subset A^+$. Now such a set is definable in $WMF_2(S)$. Indeed, let $\varphi(x, y)$ be a formula of $WMF_2(S)$ such that $w \models \varphi(n, m)$ if and only if $w_n w_{n+1} \ldots w_n \in X$. Let ψ be the formula of $WMF_2(S)$

$$\psi = \forall x\ \forall y\ \exists z\ \exists t\ \ (z < x \wedge t < y \wedge \varphi(z, t))$$

Then $\overleftrightarrow{X} = S^{\zeta}(\psi)$ and hence $\overleftrightarrow{X}$ is definable in $WMF_2(S)$. □

The next theorem shows that, as in the case of $\mathbb{N}$, monadic second-order logic is, on $\mathbb{Z}$, strictly more expressive that first-order logic. The notion of star-free subset of $A^{\mathbb{Z}}$ is entirely analogous to the corresponding notion in $A^{\mathbb{N}}$. The star-free subsets are the finite unions of subsets of the form $\tilde{U}V$ for some star-free subsets U and V of $A^{\mathbb{N}}$. They still correspond to the subsets of $A^{\mathbb{Z}}$ recognized by an aperiodic ζ-semigroup (see Exercise 4).

Theorem 7.3. *A subset X of $A^{\mathbb{Z}}$ is definable in $F_1(<)$ if and only it is star-free.*

Proof. To show that the subsets definable in $F_1(<)$ are star-free, it suffices to proceed by induction on the formation rules of formulas as in the proof of Theorem VIII.4.1.

In the opposite direction, it suffices to consider a set of the form $\tilde{U}V$ where U and V are star-free subsets of $A^{\mathbb{N}}$. Now, if $\varphi, \psi \in F_1(<)$ are formulas such that $U = S^{\omega}(\varphi)$, $V = S^{\omega}(\psi)$, one can build two formulas $\varphi(x)$ and $\psi(x)$ such that, for each $w \in A^{\mathbb{Z}}$, $w \models \varphi(n)$ if and only if $w_n w_{n+1} \cdots \models \varphi$ and $w \models \psi(n)$ if and only if $w_{-n-1} w_{-n-2} \cdots \models \psi$. Then $\tilde{U}V$ is defined by the formula $\exists x\ \tilde{\psi}(x) \wedge \varphi(x)$. □

8 Exercises

8.1 Bi-infinite words.

Exercise 1. Show that the class of recognizable subsets of $A^{\mathbb{Z}}$ consists of the finite unions of subsets of the form UV for some recognizable subset $\tilde{U}$, $V \subset A^{\omega}$.

8.2 Morphisms.

Exercise 2. Show that for each map $\varphi : A \to S$ from a set A into a ζ-semigroup S, there exists a unique extension of φ into a morphism of ζ-semigroups from $F_\zeta(A)$ into S.

Exercise 3. Show that one can define a notion of syntactic ζ-semigroup (extending that of ω-semigroup) to the case of bi-infinite words by considering for a word $x \in A^*$ the three sets

$$\{(u, v, w, t) \in A^* \mid (uxv)^{\tilde{\omega}} wt^{\omega} \in X\}$$
$$\{(u, v, w, t) \in A^* \mid u^{\tilde{\omega}}(vxw)t^{\omega} \in X\}$$
$$\{(u, v, w, t) \in A^* \mid u^{\tilde{\omega}}v(wxt)^{\omega} \in X\}$$

Exercise 4. The class $SF(A^{\mathbb{Z}})$ of star-free subsets of $A^{\mathbb{Z}}$ can be defined as follows: it consists of the finite unions of subsets of the form UV where $\tilde{U}$, V are star-free subsets of $A^{\mathbb{N}}$. A finite ζ-semigroup S is *aperiodic* if the semigroup S_+ is aperiodic.

Show that a recognizable subset is star-free if and only if it can be recognized by an aperiodic ζ-semigroup.

8.3 Discrimination.

Exercise 5. Let $A = \{a, b\}$ and let $\pi : A^+ \to A^+$ be the morphism defined by $\pi(a) = ab$ and $\pi(b) = ba$. There exists a unique word $x \in A^\omega$ such that $x \in aA^\omega$ and $\pi(x) = x$. This word, $x = abbabaab\cdots$, is called the infinite *word of Thue-Morse*. The bi-infinite word of Thue-Morse is the word $y = [\tilde{x}, x]$. Show that y is recurrent.

8.4 Logic on $\mathbb{Z}$.

Exercise 6. Show that the three following conditions are equivalent for $x, y \in A^{\mathbb{Z}}$:

(i) x and y have the same set of factors and are both recurrent,

(ii) x and y satisfy the same formulas of $MF_2(S)$,

(iii) x and y satisfy the same formulas of $MF_2(<)$.

Exercise 7. Show that the three following conditions are equivalent for $x, y \in A^{\mathbb{Z}}$.

(i) x and y have the same set of factors,

(ii) x and y satisfy the same formulas of $F_1(S)$,

(iii) x and y satisfy the same existential formulas of $F_1(S)$.

9 Notes

There is a relatively small number of articles dealing with automata on bi-infinite words [18, 21, 19, 20]. The counterpart of McNaughton's theorem for bi-infinite words (Theorem 4.9) was proved in [220] (see also the final version in [221]). The proof presented in this chapter is simpler.

Theorem 6.1 and Exercise 6 are proved in Perrin and Schupp [238]. The proof is inspired by an argument due to M.P. Schützenberger which is used in Blanchard and Perrin [34]. The theorem was obtained independently by A. Semenov [282]. It extends a result of K. Compton [78], who introduced the notion of *rich word* to designate a recurrent bi-infinite word $x \in A^{\mathbb{Z}}$ such that $F(x) = A^*$. Compton proves that two rich words cannot be discriminated by a finite automaton and that the monadic second-order theory of a rich word is decidable.

The results presented in Section 5 are due to Carton (unpublished).

Exercise 7 is due to Beauquier and Pin [22, 23].

Bruyère and Carton [41] have extended the notion of automata to countable linear orders. Their main result is an extension of Kleene's theorem to words indexed by countable scattered orders.

X

INFINITE TREES

1 Introduction

In this chapter, we treat a generalization from words to trees. Trees are actually functions from words to a set of labels. Thus they are an object of higher order than words.

In applications, trees arise in several forms. One of them is the trees linked to syntax analysis, which are the *parse trees* obtained in the process of parsing a string of symbols according to a context-free grammar. Another slightly different context in which trees arise is the definition of *terms* in an algebra. In practice, these terms can describe functional objects or elements of a type system. The infinite trees studied here appear in particular as recursively defined types. Indeed, a recursively defined type in a high-level programming language can be described by an infinite tree. Infinite trees also appear as describing the behaviour of discrete systems. In particular, the semantics of a concurrent program is given by a computation tree.

We shall see here that many of the notions seen in the previous chapters can be generalized to trees; In particular, the connexion between automata and logic, provided one uses the monadic second-order logic of two successors instead of just one successor.

There are however many differences between words and trees. First of all, there is a distinction between Büchi recognizability and Muller recognizability, the latter being more general. Also, the notion of determinism is quite different. Indeed, deterministic top-down tree automata are not equivalent to nondeterministic ones. This corresponds to an intuitive explanation: deterministic top-down automata can only specify properties of paths in a tree. For infinite trees, the definition of acceptance by a finite automaton is based on a top-down approach. This leads to a distinction between Büchi and Muller (or Rabin) automata, which are not equivalent in the case of infinite trees.

Nevertheless, an important property of automata on infinite words is preserved for infinite trees. In both cases, there is an alternative definition in terms of a decidable logical theory. For infinite trees, a fundamental result states that the monadic second-order theory of two successors (denoted $S2S$) is decidable (Rabin's theorem). Since Muller recognizable sets of trees are exactly those which are definable in $S2S$, this is also linked to the fact that the family of recognizable sets of infinite trees is closed under complement.

The chapter is organized as follows. We first define in Section 2 the basic notions concerning trees, including the definition of rational and ω-rational sets of trees. In Section 3, we introduce tree automata with a version of Kleene's theorem for finite trees (Theorem 3.2). Tree automata on infinite trees are introduced with the variants of Büchi and Muller automata. We give a proof of Rabin's basis theorem, according to which any rational set of trees contains a rational tree. In Section 4.2, we give a proof of Rabin's complementation theorem. It relies on the Theorem on forgetful determinacy proved in Chapter IV. In Section 5, we consider the topological aspects of infinite trees. Here again, the situation is more complex than in the case of words. Indeed, the recognizable sets of trees do not fall into the family of Borel or even Suslin sets. The last section (Section 6) deals with the logical aspects of the theory of infinite trees. We define the monadic theory of two successors and prove its decidability (Theorem 6.2). We also state without proof a result, also due to Rabin, characterizing the sets of trees definable in the weak monadic theory of two successors.

2 Finite and Infinite trees

We begin by defining the trees studied in this chapter. These are valued trees and the definitions given here are special to the chapter. In particular, although the objects are essentially the same, they do not refer to the notions on trees introduced in Chapter I.

Let D and A be two nonempty alphabets. The elements of D are called *directions* and those of A *labels*. A *D-tree* on A is a map $t : P \to A$ where P is a nonempty subset of D^* which is prefix-closed. We denote by $T(D, A)$ the set of all D-trees on A.

The set P is called the *domain* of the tree t, denoted $\mathrm{Dom}(t)$. The elements of P are called the *nodes* of the tree. If $x \in P$ is a node, any node of the form xi for $i \in D$ is called a *child* of x. The tree is said to be finite if its domain is a finite set.

The *height* of a finite tree t is the maximal length of the elements of its domain.

Let A and B be two alphabets and let $\varphi : A \to B$ be a map. It induces a map, also denoted by φ, from $T(D, A)$ into $T(D, B)$ defined by $t \to \varphi \circ t$. Such a map is usually called a *projection*.

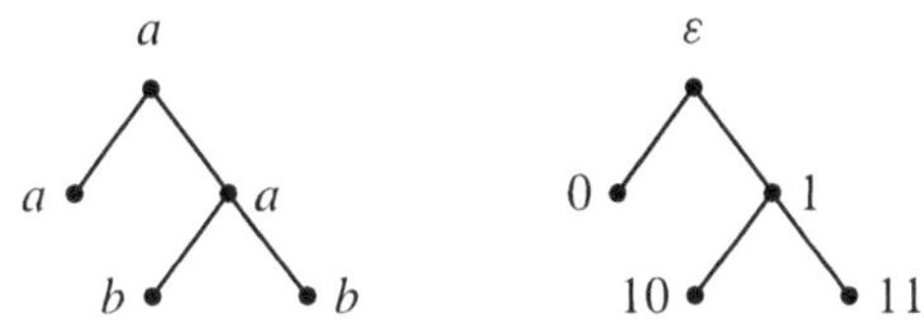

Figure 2.1. A finite tree and its domain.

Example 2.1. We have represented in Figure 2.1 a finite tree on $A = \{a, b\}$ with its domain on the right.

When D has only one symbol, a tree is the same as a prefix-closed subset of A^*. Trees often arise as *syntax trees* of expressions using symbols from a *ranked alphabet* $A = A_0 \cup A_1 \cup \ldots \cup A_k$ where A_i contains i-ary function symbols. We shall use freely this correspondence and denote $f(t_1, \ldots, t_n)$ a tree with a root labeled f with n children which are roots of the trees $t_1, \ldots, t_n$.

Example 2.2. The tree represented in Figure 2.2 is the syntax tree of the arithmetic expression written in infix notation $2+5*4$. It is a finite binary tree on the ranked alphabet formed of $A_0 = \{2, 4, 5\}$ and $A_2 = \{+, *\}$. Its domain is the set $\{\varepsilon, 0, 1, 10, 11\}$.

For a tree t and a word $u \in D^*$, we denote by t_u the tree defined by $t_u(x) = t(ux)$.

Let t be a tree of domain P. The *frontier* of t, denoted $\mathrm{Fr}(t)$ is the set

$$\mathrm{Fr}(t) = \{x \in P \mid xD \cap P = \emptyset\}$$

The elements of $\mathrm{Fr}(t)$ are usually called the *leaves* of the tree. The *outer frontier* of t, denoted by $\mathrm{Fr}^+(t)$ is the set $PD - P$ formed of all $xi \notin P$ such that $x \in P$ and $i \in D$. We set $\mathrm{Dom}^+(t) = \mathrm{Dom}(t) \cup \mathrm{Fr}^+(t)$.

A *path* π through the tree t is a maximal chain for the prefix order in $\mathrm{Dom}(t)$. Thus, a path is the sequence of prefixes of a finite or infinite word in D^∞. If π is a path, we

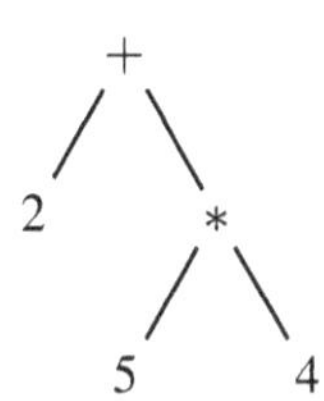

Figure 2.2. A syntax tree.

denote by $t \mid \pi$ the restriction of t to the set π. Clearly, $t \mid \pi$ is a finite or infinite word on the alphabet A.

An *infinite tree* on A is a tree t such that $\mathrm{Dom}(t) = D^*$. Thus, by definition, $\mathrm{Fr}(t)$ is empty for an infinite tree. We note that an infinite tree is actually a 'purely' infinite one and that there are trees which are neither finite nor infinite with these definitions, like $f(a, f(a, f(a, \ldots) \cdots)$ for example.

An infinite tree is a generalization of an infinite word: when $\mathrm{Card}(D) = 1$, we obtain an infinite word on the alphabet A.

To simplify the notation, we restrict ourselves to complete binary trees. This means that there are only two directions, i.e. $D = \{0, 1\}$ and that for every node x of the tree, there are either two children or no children. We denote by T_A the set of finite binary trees on A and by T_A^ω the set of corresponding infinite trees. Let $T_A^\infty = T_A \cup T_A^\omega$ be the set of finite or infinite trees.

Let $S \subset T_A$ be a set of finite trees, let $T \subset T_A^\infty$ be a set of finite or infinite ones and let $c \in A$. The concatenation product $S \cdot_c T$ is the set of all trees obtained by replacing in a tree $s \in S$ each occurrence of c on $\mathrm{Fr}(t)$ by a tree from T (different trees from T may be used for different occurrences of c). This operation is associative, in the sense that, if $S, T \subset T_A$ are sets of finite trees, $U \subset T_A^\infty$ is a set of finite or infinite ones and $c, d \in A$, then $(S \cdot_c T) \cdot_d U = S \cdot_c (T \cdot_d U)$.

For a fixed symbol $c \in A$, a set $T \subset T_A$ and an integer $n \geq 0$, we denote by $T^{n,c}$ the set defined by induction by

$$T^{0,c} = \{c\}, \qquad \text{and} \qquad T^{n,c} = T \cdot_c T^{n-1,c} \cup T^{n-1,c}$$

Accordingly, we define

$$T^{*,c} = \bigcup_{n \geq 0} T^{n,c}$$

A set of trees $T \subset T_A$ is *rational* if there is a finite set C containing A, such that T can be obtained from finite subsets of T_C by a finite number of unions, concatenations $\cdot_c$ and stars ^{*,c} for $c \in C$.

This notion generalizes the notion of rational set of words since a word $a_0 a_1 \cdots a_k$ can be seen as the unary tree $a_0(a_1(\ldots a_k(c)\ldots)$.

Example 2.3. The set T_A itself is rational since

$$T_A = T^{*,a}$$

where a is an arbitrary element of A and T is formed of all trees of height at most one. Actually, it is enough to have in T the trees of height zero (i.e. reduced to one node) and the trees of height one with all the leaves labeled by a.

Example 2.4. The set of finite trees labeled in $A \cup c$ with symbols c on their frontier and nowhere else can be written $T^{*,c}$ where T is the set of trees of this form of height at most one. It is therefore also rational.

We shall need to consider a variant of the definition of the product where a tuple $\mathbf{c} = (c_1, \ldots, c_m)$ is used instead of a single symbol c. For $T, T_1, \ldots, T_m \subset T_A$, we denote by $T \cdot_{\mathbf{c}} (T_1, \ldots, T_m)$ the set of all trees obtained from a tree $t \in T$ by substituting each c_i, for $i = 1, \ldots, m$, by a tree in T_i. For $m = 1$, the definition coincides with the previous one.

We now come to the definition of the infinite product. Let $\mathbf{c} = (c_1, \ldots, c_m)$ and let $T_1, \ldots, T_m \subset T_A$. We define $(T_1, \ldots, T_m)^{\omega,\mathbf{c}}$ as the set of all infinite trees t obtained as follows. We consider infinite sequences $(t_0, t_1, \ldots)$ of trees such that $t_0 \in \{c_1, \ldots, c_m\}$ and for $n \geq 0$, $t_{n+1} \in t_n \cdot_{\mathbf{c}} (T_1, \ldots, T_m)$. It is clear that the tree t_n is, except for the c_i at the frontier, a restriction of t_{n+1}. Thus there is a common extension of all trees t_n (except perhaps at their frontier). We let t be this extension.

A set $T \subset T_A^{\omega}$ of infinite trees is ω-rational if there is a finite set $C = \{c_1, \ldots, c_m\}$ and rational sets $T_0, T_1, \ldots, T_m \subset T_{A \cup C}$ such that

$$T = T_0 \cdot_{\mathbf{c}} (T_1, \ldots, T_m)^{\omega,\mathbf{c}}$$

As usual, the singleton $\{t\}$ is simply denoted by t. Consequently, we use the notation $t_1 \cdot_c t_2$ or $t \cdot_{\mathbf{c}} (t_1, \ldots, t_m)$ instead of $\{t_1\} \cdot_c \{t_2\}$ or $\{t_0\} \cdot_{\mathbf{c}} (\{t_1\}, \ldots, \{t_m\})$. We use the same notation when the t_i's are infinite trees with the convention that, instead of frontier occurrences of the c_i, the first occurrences of the c_i are used for replacement.

Example 2.5. The set T_A^{ω} itself is ω-rational since

$$T_A^{\omega} = T \cdot_c (T)^{\omega,c}$$

where T is the set of finite trees in $T_{A \cup c}$ with all leaves labeled by c (see Example 2.4).

Example 2.6. The set of infinite trees on $\{a, b\}$ with a finite number of b can be written

$$T = T_A \cdot_a (T_a)^{\omega,a}$$

3 Tree automata

In this section, we introduce tree automata. We shall first define automata working on finite trees and subsequently on infinite trees, with several possible forms including Büchi and Muller tree automata.

3.1 Automata on finite trees. A *tree automaton* on the alphabet A is given by a finite set Q of states, a set $E \subset Q \times A \times Q \times Q$ of edges, a set $I \subset Q$ of initial states and a set $F \subset Q$ of final states.

A *run* of the automaton $\mathcal{A}$ on a tree t is map $r : \mathrm{Dom}^+(t) \to Q$ with $r(\varepsilon) \in I$ such that $(r(x), t(x), r(x0), r(x1)) \in E$ for all $x \in \mathrm{Dom}(t)$. The run is *successful* if $r(w) \in F$ for all w on the outer frontier $\mathrm{Fr}^+(t)$ of t. The set of finite trees recognized

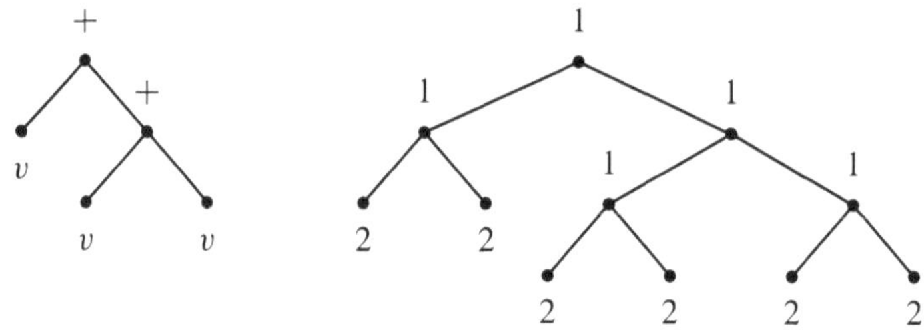

Figure 3.1. A tree t and a run on t.

by the automaton is formed of all trees t such that there is a successful run of $\mathcal{A}$ on t. A set $T \subset T_A$ of finite trees is *recognizable* if there is a tree automaton $\mathcal{A}$ recognizing T.

Example 3.1. The set T of syntax trees of arithmetic expressions formed using an operation symbol $+$ and a variable symbol v is a recognizable set of trees. Let indeed $\mathcal{A}$ be the tree automaton defined by $Q = \{1, 2\}$, $I = \{1\}$, $F = \{2\}$ and

$$E = \{(1, +, 1, 1), (1, v, 2, 2)\}$$

Then $\mathcal{A}$ recognizes the set T. An example of a run of $\mathcal{A}$ is represented in Figure 3.1.

A tree automaton $\mathcal{A} = (Q, E, I, F)$ is said to be *top-down deterministic* if $\text{Card}(I) = 1$ and for each pair $(p, a) \in Q \times A$, there is at most one pair $(q, r) \in Q \times Q$ such that $(p, a, q, r) \in E$. The following example shows that top-down deterministic automata are not equivalent to deterministic ones.

Example 3.2. Let $T = \{f(a, b), f(b, a)\}$ where f is a 2-ary function symbol. The set T being finite, it is certainly recognizable. However, a top-down deterministic tree automaton recognizing T would also accept $f(a, a)$ and $f(b, b)$.

The notion of a *bottom-up deterministic* tree automaton is symmetric, with the requirement that $\text{Card}(F) = 1$ and that for each triple $(a, q, r) \in A \times Q \times Q$, there is at most one state $p \in Q$ such that $(p, a, q, r) \in Q$.

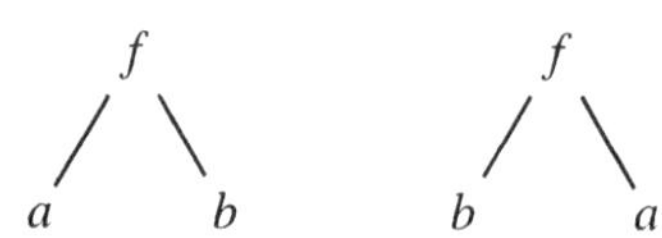

Figure 3.2. The set $T = \{f(a, b), f(b, a)\}$.

A bottom-up tree automaton is *complete* if, for any triple $(a, q, r) \in A \times Q \times Q$, there is at least one (and thus exactly one) state $p \in Q$ such that $(p, a, q, r) \in Q$.

We note that on a tree representing an expression, a bottom-up computation corresponds to a bottom-up evaluation of the expression.

It can be shown by a simple subset construction that any tree automaton is equivalent to a complete, bottom-up, deterministic automaton (see Exercise 1). As a consequence, we have the following statement, which shows that the family of recognizable sets is closed under all boolean operations.

Proposition 3.1. *The family of recognizable sets of trees is closed under complement.*

Proof. We use complete bottom-up deterministic tree automata. Taking the complement just consists in changing the set I with its complement $Q \setminus I$. □

The following is a version of Kleene's theorem for finite trees. The proof is quite similar to that of the corresponding result for words and we leave it as an exercise (Exercise 5).

Theorem 3.2. *A set $T \subset T_A$ is recognizable if and only if it is rational.*

Example 3.3. The set of syntax trees of arithmetic expressions formed using an operation symbol $+$ and a variable symbol v is a recognizable set of trees (see Example 3.1). It is also rational since it can be written $T^{*,v}$ where T is the set of trees of this form of height at most one (see Example 2.4).

3.2 Büchi tree automata. We now come to automata on infinite trees.

A *Büchi tree automaton* is a tree automaton $\mathcal{A} = (Q, A, E, i, F)$ with $E \subset Q \times A \times Q \times Q$ as set of edges, $i \in Q$ as initial state and $F \subset Q$ as set of final states.

A *run* of the automaton $\mathcal{A}$ on a tree t is a map $r : \{0, 1\}^* \to Q$ with $r(\varepsilon) = i$ such that $(r(x), t(x), r(x0), r(x1)) \in E$ for all $x \in \{0, 1\}^*$. The run is *successful* if on each path some final state occurs infinitely often, i.e.

$$\mathrm{Inf}(r \mid \pi) \cap F \neq \emptyset$$

for all paths π in the tree t. An infinite tree t is recognized by the automaton $\mathcal{A}$ if there is a successful run of $\mathcal{A}$ on t. The set recognized by $\mathcal{A}$ is the set of all infinite trees recognized by $\mathcal{A}$. A set T of infinite trees is called Büchi recognizable if it is recognized by some Büchi automaton. The class of Büchi recognizable trees is closed under finite union and projection (see Exercise 2).

Example 3.4. Let T be the set of infinite trees t on $A = \{a, b\}$ such that some path of t carries infinitely many a. A Büchi automaton recognizing T is obtained by selecting in a nondeterministic way a path on which an infinite number of occurrences of a is checked. Formally, the set of states is $Q = \{1, 2, 3\}$ with $I = \{1\}$, $F = \{1, 3\}$ and E consists in the quadruples $(3, x, 3, 3)$ for $x = a, b$ and all quadruples $(p, x, q, 3)$ or $(p, x, 3, q)$ with $q = 1$ if $x = a$ and $q = 2$ if $x = b$. Thus, the states in a run are all equal to 3 except on a path on which state 1 corresponds to an occurrence of a and state 2 to an occurrence of b. It is clear that an infinite tree is accepted if and only if there is a path with an infinite number of a. Thus T is Büchi recognizable. We shall see (Example 3.6) that the complement of T is not Büchi recognizable.

Theorem 3.3. *A set $T \subset T_A^\omega$ is Büchi recognizable if and only if it is ω-rational.*

Proof. It is easy to see that any ω-rational set of trees is Büchi recognizable (see Exercise 5).
Conversely, let $\mathcal{A} = (Q, E, i, F)$ be a Büchi automaton with $F = \{q_1, \ldots, q_m\}$. For each $q \in Q$, let T_q be the set of finite trees t on $A \cup Q$ such that

(1) they have values in F on their frontier but elsewhere in A,

(2) there is a run r starting at q on the tree obtained by deleting the frontier of t and r is such that $r(x) = t(x)$ for all $x \in \mathrm{Fr}(t)$.

Each set T_q is recognizable. For each $p \in Q$, let S_p be the set recognized by the Büchi tree automaton (Q, E, p, F). We can write each S_p as

$$S_p = T_p \cdot_{\mathbf{q}} (T_{q_1}, \ldots, T_{q_m})^{\omega, \mathbf{q}}$$

where $\mathbf{q} = (q_1, \ldots, q_m)$. Indeed, for any tree t which belongs to the right-hand side of the formula, there is a run of $\mathcal{A}$ on t starting at p and passing infinitely often in F. Thus $t \in S_p$.

Conversely, if $t \in S_p$, let r be a successful run of (Q, E, p, F) on t. Any path π in r passes infinitely often in F. Thus, there is a tree $s \in T_p$ such that $t \in s \cdot_{\mathbf{q}} (S_{q_1}, \ldots, S_{q_m})$. Continuing in this way, we obtain the desired decomposition. □

Example 3.5. Let T be the set of trees on $\{a, b\}$ such that there is a path with an infinite number of a. We have seen in Example 3.4 that T is Büchi recognizable. An expression as an ω-rational set is

$$T = U \cdot_{(c,d)} (U, V)^{\omega, (c,d)}$$

where $U, V \subset T_{a,b,c,d}$ is formed of trees having values in $\{c, d\}$ exactly on their frontier. The trees of U have exactly one c and the father of the node with label c has label a. The trees of V have only d on their frontier.

3.3 Muller tree automata. We now come to the definition of Muller tree automata. Note that, in contrast to the definition on infinite words, we allow this time nondeterminism.

A *Muller tree automaton* $\mathcal{A} = (Q, E, i, \mathcal{F})$ is the same as a Büchi tree automaton except that the set of final states is replaced by a family $\mathcal{F}$ of subsets of Q. A run of $\mathcal{A}$ on a tree t is called successful if for every path π of r, the set of states that occur infinitely often is in $\mathcal{F}$, i.e.

$$\mathrm{Inf}(r \mid \pi) \in \mathcal{F}$$

It is clear that Büchi tree automata are a particular case of Muller tree automata. Indeed, one can use the family $\mathcal{F}$ of sets of states containing a final state.

An important difference with the case of infinite words is the fact that Büchi automata are less powerful than Muller automata, as shown by the following example.

Example 3.6. Let T be the set of infinite trees on $A = \{a, b\}$ such the number of a along any path is finite. The set T is the complement of the set of Example 3.4. It is recognizable by a Muller tree automaton with two states 1 and 2 and all the edges (p, x, q, q) with $q = 1$ if $x = a$ and $q = 2$ otherwise. The family $\mathcal{F}$ is formed of the singleton $\{2\}$.

Let us show that T is not Büchi recognizable. Let $\mathcal{A}$ be a Büchi tree automaton accepting all the trees of T. Let n be the number of states of $\mathcal{A}$. We shall show that $\mathcal{A}$ also accepts trees in the complement of T. For $i \geq 0$, consider the set $U_i = \bigcup_{k=0}^{i-1}(1^+0)^k$ (see Figure 3.3) and let t_i be the infinite tree defined by $t_i(x) = a$ for $x \in U_i$ and $t_i(x) = b$ otherwise. We have $t_n \in T$ and thus there is a successful run r of $\mathcal{A}$ on t_n. We show by induction on n that in an n-state automaton accepting all trees of T, there is a path in t_n with three nodes $u < v < w$ such that $r(u) = r(w) \in F$ and $t_n(v) = a$.

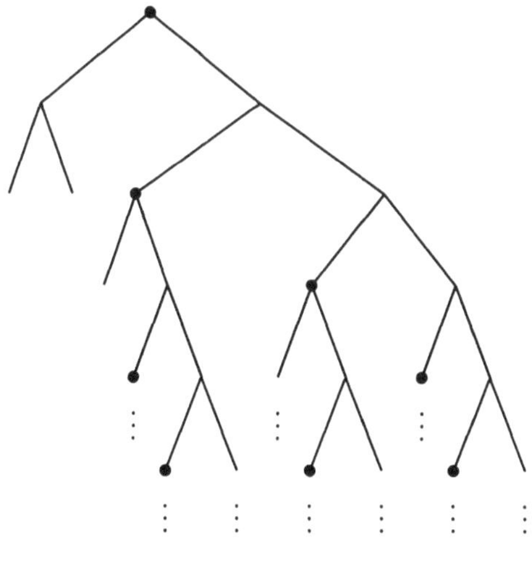

Figure 3.3. The set U_i.

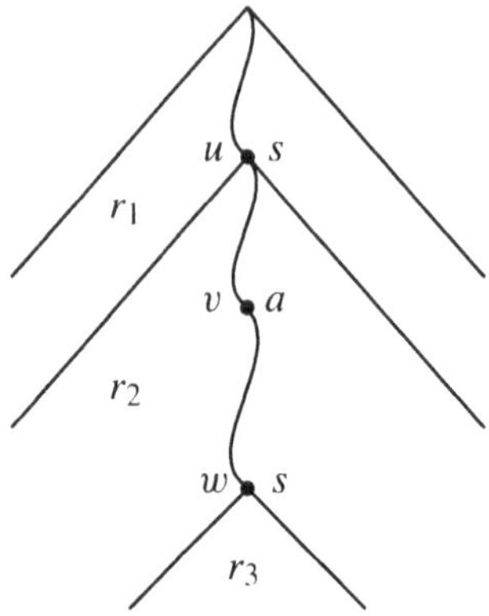

Figure 3.4. A path in t_n.

It is true for $n = 2$ since in this case, $\mathcal{A} = (\{i, f\}, E, i, \{f\})$ and the only possible run has $r(\varepsilon) = i$ and $r(x) = f$ otherwise. Let us assume the property true for $n - 1$. The path 1^* of r passes infinitely often through F and thus, there is an $u \in 1^*$ such that $r(u) \in F$. By the definition of t_n, we have $(t_n)_{u0} = t_{n-1}$. The tree r_{u0} is a run of the automaton $(Q, r(u0), F)$ on t_{n-1} which is successful. If this run uses $r(u)$, the claim is proved. Otherwise, it is a run on an automaton with $n - 1$ states accepting all trees of T. The claim then follows by induction.

Let r_1 be the tree obtained from r by deleting r_u and setting $r_1(u) = c$. Let r_2 be the tree obtained from r_u by deleting r_w and let $r_3 = r_w$. Then $r = r_1 \cdot r_2 \cdot r_3$ and similarly $t_n = s_1 \cdot s_2 \cdot s_3$. The tree $r_1 \cdot r_2^{\omega}$ is a successful run of $\mathcal{A}$ on $s_1 \cdot s_2^{\omega}$. Since $s_1 \cdot s_2^{\omega}$ has an infinite number of symbols a, we have proved the assertion.

The following property is easy to prove.

Proposition 3.4. *The class of Muller recognizable sets of trees is closed under union and projection.*

Proof. It is clear, since we use nondeterministic automata that if T and T' are Muller recognizable, then $T \cup T'$ is Muller recognizable. For projection, consider a function $f : A \to B$ and let $T \subset T_A$ be recognized by a Muller automaton $\mathcal{A} = (Q, E, i, \mathcal{F})$. Let $\mathcal{B} = (Q, E', i, \mathcal{F})$ be the tree automaton which is the same as $\mathcal{A}$ with the set of edges $E' = \{(p, b, q, r) \mid (p, a, q, r) \in E\}$. Then $\mathcal{B}$ recognizes $f(T)$. □

We shall see later (Theorem 4.3) that the class of Muller recognizable sets of trees is closed under complement. As for automata on infinite words, variants of Rabin and Streett automata can be defined. A Rabin tree automaton is the same as a Muller tree automaton except that the family $\mathcal{F}$ is replaced by a set $\mathcal{P}$ of pairs $(L_1, R_1), \ldots, (L_n, R_n)$. A run r is successful if there is an index k with $1 \leq k \leq n$ such that

$$\mathrm{Inf}(r \mid \pi) \cap L_i = \emptyset \quad \text{and} \quad \mathrm{Inf}(r \mid \pi) \cap R_i \neq \emptyset$$

for all paths π of r. Likewise, a run is successful in a Streett automaton if the negation of this condition holds.

It is clear that Rabin and Streett tree automata are particular cases of Muller tree automata. Conversely, a simple construction transforms a Muller tree automaton into a Rabin (or Streett) tree automaton (see Exercise 7).

The notion of a Rabin chain tree automaton is also a simple generalization of the corresponding notion for infinite words (see Section IV.6.2). Indeed, it is an automaton of the form $\mathcal{A} = (Q, q_0, E, \mathcal{S})$ where $\mathcal{S}$ is a strictly increasing sequence

$$E_1 \subset F_1 \subset E_2 \subset F_2 \subset \ldots \subset E_n \subset F_n$$

of subsets of Q. A run r is successful if for each path π, there is an integer k such that

$$\mathrm{Inf}(r \mid \pi) \cap E_k = \emptyset \text{ and } \mathrm{Inf}(r \mid \pi) \cap F_k \neq \emptyset$$

The notion of the *memory extension* of a Muller tree automaton is also a straightforward extension of the corresponding construction for automata on words. The only difference is that we allow nondeterminism. Formally, the memory extension of the Muller tree automaton $\mathcal{A} = (Q, E, i, \mathcal{F})$ is the Rabin chain tree automaton $\mathcal{B} = (S, F, j, \mathcal{S})$. The set of states S is defined in the same way as for ordinary automata. It is the set

$$S = \{(u, v) \mid uv \in \mathrm{Perm}(Q)\}$$

where $\mathrm{Perm}(Q)$ is the set of permutations of elements of Q. The initial state j is any element of S of the form xi. The chain $\mathcal{S}$ is also defined identically. The only difference is the set F of transitions. It is the set of all

$$((u, v), a, (x, yq), (z, tr))$$

such that $uv = xqy = zrt$ and $(p, a, q, r) \in E$. The idea is thus the same: in the memory extension, we keep track along each path of the run of the order of last appearance of each state.

The following result holds true for tree automata with the same proof as for ordinary automata.

Proposition 3.5. *The automata $\mathcal{A}$ and its memory extension $\mathcal{B}$ are equivalent. Consequently, any Muller tree automaton is equivalent to a Rabin chain tree automaton.*

The following result gives an interesting connection between the notion of Muller recognizability for trees and for words. For a set $X \subset (\{0, 1\} \times A)^\omega$, we denote by $T(X)$ the set of all trees $t \in T_A^\omega$ such that for all $\pi \in \{0, 1\}^\omega$, one has $(\pi, t \mid \pi) \in X$.

Proposition 3.6. *Let $X \subset (\{0,1\} \times A)^\omega$ be a recognizable set of words on the alphabet $\{0,1\} \times A$. Then, the set $T(X)$ is Muller recognizable.*

If X is recognizable by a Büchi deterministic automaton, then T is Büchi recognizable.

Proof. Let $\mathcal{A} = (Q, E, i, \mathcal{F})$ be a (deterministic) Muller automaton recognizing X. We transform $\mathcal{A}$ into a tree automaton $\mathcal{B} = (Q, F, i, \mathcal{F})$ where F is the set of all (p, a, q, r) for $p, q, r \in Q$ and $a \in A$ such that $p \cdot (0, a) = q$ and $p \cdot (1, a) = r$ by the definition of $\mathcal{B}$. If $t \in T_A^\omega$ is recognized by $\mathcal{B}$, then $t \in T$. Conversely, if $t \in T$, there is for each path π in t a successful path in $\mathcal{A}$ labeled $(\pi, t \mid \pi)$. Since $\mathcal{A}$ is deterministic, these paths can be merged into a run of $\mathcal{B}$ and thus t is recognized by $\mathcal{B}$. Thus $\mathcal{B}$ recognizes T and this proves the first statement. To prove the second one, we replace $\mathcal{A}$ by a deterministic Büchi automaton and the rest of the argument is the same. □

Actually, the converse of Proposition 3.6 holds. See the notes section.

Example 3.7. Let $A = \{a, b\}$ and let $X \subset (\{0,1\} \times A)^\omega$ be the set of infinite words on $\{0,1\} \times A$ having a finite number of a's. The corresponding set of trees on $A = \{a, b\}$ is the Muller recognizable set of trees such that the number of a along any path is finite of Example 3.6.

Example 3.8. Let T be the set of infinite trees on $\{a, b\}$ such that the number of a on any path is infinite. Since the set X of words with an infinite number of a is recognizable by a deterministic Büchi automaton, the set T is Büchi recognizable.

3.4 Rabin basis theorem. A tree t is said to be *rational* if there is only a finite number of different subtrees t_x for $x \in \{0,1\}^*$.

The following result extends to trees the result that any nonempty recognizable set of infinite words contains an ultimately periodic word (Lemma I.5.1). We shall see farther on in this chapter another proof of this result using games.

Theorem 3.7. *Any nonempty Muller recognizable set of infinite trees contains a rational tree.*

The proof uses the following lemma showing that one may reduce the problem to 'input-free' tree automata, i.e. an automaton with a set of edges $E \subset Q \times Q \times Q$.

Lemma 3.8. *For any Muller tree automaton $\mathcal{A} = (Q, E, i, \mathcal{F})$, one can build an input-free tree automaton $\mathcal{A}' = (Q', E', i', \mathcal{F}')$ with a function $f : Q' \to A$ such that r' is a successful run of $\mathcal{A}'$ if and only if $r = f \circ r'$ is a successful run of $\mathcal{A}$.*

In particular, the set recognized by $\mathcal{A}$ contains a rational tree if and only if the set recognized by $\mathcal{A}'$ does.

Proof. Indeed, it is enough to define $Q' = Q \times A$ and to choose

$$E' = \{((q,a),(q,a'),(q'',a'')) \mid (q,a,q',q'') \in E\}$$

The result then follows with f being the projection $f(q,a) = a$. □

Proof of Theorem 3.7. By the above lemma, it is enough to prove that an input-free tree automaton $\mathcal{A} = (Q, E, i, \mathcal{F})$ with a successful run has a rational successful run. Call a state $q \in Q$ *live* if $q \neq i$ and if there are transitions (q, a, q', q'') with $q' \neq q$ or $q'' \neq q$. Thus a state is live if we are not forced to stay forever in this state. We proceed by induction on the number of live states.

If there are no live states, all runs will be stationary from the children of the root onwards and hence will also be rational.

Let us distinguish three cases for the induction step.

First, if some live state q is not used in the successful run r, we apply the induction hypothesis to the automaton obtained by cancelling the state q.

Second, suppose that there is a node u in r such that the state $q = r(u)$ is live but that some live state q' does not appear beyond u. Let $\mathcal{A}_1$ be the automaton obtained from $\mathcal{A}$ by replacing all transitions from state q by the unique transition (q, q, q). Thus q is no longer live in $\mathcal{A}_1$ and by induction hypothesis, there is a successful rational run r_1 of $\mathcal{A}_1$. Let $\mathcal{A}_2$ be the tree automaton obtained from $\mathcal{A}$ by choosing q as initial state and deleting q'. By the induction hypothesis, there is a rational successful run r_2 of $\mathcal{A}_2$. Then it is clear that $r_1 \cdot_q r_2$ is a successful rational run of $\mathcal{A}$.

There remains to treat the case where all live states appear in r beyond any given node. In this case, the set of states $\mathrm{Inf}(\pi)$ is, for each path π in r, equal to the set of all live states. Let us choose an arbitrary live state q. First, any path in r passes by q. Second, there is a finite run s of $\mathcal{A}$ such that

(1) s starts at q, i.e. $s(\varepsilon) = q$

(2) s ends with q, i.e. $s(x) = q$ for all $x \in \mathrm{Fr}(s)$

(3) any path in s passes through all live states.

Then the tree $r \cdot_q s^{\omega \cdot q}$ is a rational successful run of $\mathcal{A}$. □

Corollary 3.9. *The emptiness problem for Muller tree automata is decidable.*

Proof. The proof consists in following the proof of Theorem 3.7 to build a rational tree recognized by the automaton. □

4 Tree automata and games

In this section, we introduce a game played on trees which simulates the possible runs of an automaton on the tree. We shall use this game to prove Rabin's complementation theorem.

4.1 Automaton and Pathfinder. Let $\mathcal{A} = (Q, q_0, E, \mathcal{F})$ be a Muller tree automaton. For each infinite tree $t \in T^{\omega}_{A}$, we define a game $G_{\mathcal{A},t}$ which is defined as follows. In this game, Player I, called *Automaton*, chooses transitions from the set E and Player II, called *Pathfinder* chooses directions from the set $\{0, 1\}$. Intuitively, the game describes all the possible runs of the automaton, with the first player choosing the transitions of the automaton and the second one the path to follow in the tree. More precisely, Automaton first chooses a transition $(q_0, a_0, q_0', q_0'') \in E$ with $a_0 = t(\varepsilon)$. Then Pathfinder chooses a direction $\delta_0 \in \{0, 1\}$. Automaton chooses in turn a transition $(q_1, a_1, q_1', q_1'') \in E$ with $a_1 = t(\delta_0)$ such that $q_1 = q_0'$ if $\delta_0 = 0$ and $q_1 = q_0''$ otherwise. Continuing in this way, there results a sequence of states $(q_0, q_1, \ldots)$. If this sequence satisfies the Muller condition defined by $\mathcal{F}$, then Automaton wins the play, otherwise Pathfinder wins.

We shall actually, in order to be able to use the results of Chapter IV, present the game $G_{\mathcal{A},t}$ as a game on a graph. We consider as set of vertices the set $V = V_1 \cup V_2$ with $V_1 = \{0, 1\}^* \times Q$ and $V_2 = \{0, 1\}^* \times E$. Thus we identify the positions of Automaton with a node of the tree t labeled by a state of $\mathcal{A}$ and the positions of Pathfinder by nodes of t labeled by a transition. Formally, there is an edge of the graph from (w, q) to (w, e) for all $e = (q, t(w), q', q'')$ and for all such $e \in E$, there are edges of the graph from (w, e) to $(w0, q')$ and to $(w1, q'')$. The game begins at (ε, q_0) and the winning condition is given by the Muller condition of the automaton $\mathcal{A}$ applied to the second component of the vertices of V_1.

Viewed as an abstract game on $Q \cup E$, the game $G_{\mathcal{A},t}$ is a rational game. Thus, by Theorem IV.4.9, the game $G_{\mathcal{A},t}$ is determined.

The following result is the basis of the proof of Theorem 4.3.

Proposition 4.1. *The automaton $\mathcal{A}$ accepts the tree t if and only if Automaton wins the game $G_{\mathcal{A},t}$. It does not accept t if and only if Pathfinder wins the game $G_{\mathcal{A},t}$.*

If $\mathcal{A}$ is moreover a Rabin chain tree automaton, the winning strategies can be chosen to be memoryless.

Proof. Since the game $G_{\mathcal{A},t}$ is determined, we only have to prove one of the two first statements. Let us consider the first one. By definition, $\mathcal{A}$ accepts t if and only if there is a successful run r of $\mathcal{A}$ on t, i.e. such that for each path π, the sequence $r \mid \pi$ satisfies $\mathcal{F}$. Thus, if $\mathcal{A}$ accepts t, Automaton has a winning strategy consisting in choosing for each position (w, q) the transition $(q, t(w), r(w0), r(w1))$. Conversely, a winning strategy for Automaton is a function $f : \{0, 1\}^* \times Q \to E$. We define a run by induction on the length of the word $w \in \{0, 1\}^*$ by $r(\varepsilon) = q_0$ and assuming $r(w)$ already defined,

$$r(w0) = q', \quad r(w1) = q''$$

when $f(w, r(w)) = (q, t(w), q', q'')$. Clearly r is a successful run. This proves the first statement.

Finally, if $\mathcal{A}$ is a Rabin chain tree automaton, the game $G_{\mathcal{A},t}$ is a parity game and the result follows from Theorem IV.4.6. □

A special case of the game $G(\mathcal{A}, t)$ occurs when one considers an input-free Muller tree automaton $\mathcal{A} = (Q, q_0, \mathcal{F})$. In this case, the game $G(\mathcal{A}, t)$ becomes a game $G(\mathcal{A})$ which is a game on the finite graph (V, E) with $V_1 = Q$, $V_2 = F$ and where E is formed of all

$$(q, (q, q', q'')), \quad ((q, q', q''), q'), \quad ((q, q', q''), q'')$$

for $(q, q', q'') \in F$. The winning set consists of the state sequences which satisfy $\mathcal{F}$.

Proposition 4.1 takes the following form.

Proposition 4.2. *Let $\mathcal{A} = (Q, q_0, \mathcal{F})$ be an input-free Muller automaton. Then Automaton wins the game $G(\mathcal{A})$ if and only if $\mathcal{A}$ has at least one successful run.*

If $\mathcal{A}$ happens to be a Rabin chain automaton, then $G(\mathcal{A})$ is a parity game. If Automaton moreover wins the game $G(\mathcal{A})$, it has thus a memoryless winning strategy. Such a strategy is a successful run of $\mathcal{A}$ which is a rational tree. Indeed, the subtree at a node $x \in \{0, 1\}^*$ only depends on the label $r(x)$.

4.2 Rabin's tree theorem. The aim of this section is to prove the following result, using games.

Theorem 4.3 (Rabin tree theorem). *The family of Muller recognizable sets of trees is closed under complement.*

Proof. By Proposition 3.5, any Muller recognizable set of trees can be recognized by a Rabin chain tree automaton. Let $\mathcal{A} = (Q, q_0, E, S)$ be a Rabin chain tree automaton. We have to prove that the set X of trees $t \in T_A^{\omega}$ which are not accepted by $\mathcal{A}$ is recognizable. According to Proposition 4.1, if t is not accepted by $\mathcal{A}$, then Pathfinder has a memoryless winning strategy in the game $G_{\mathcal{A},t}$. Our aim is to obtain the trees $t \in X$ as the projection of a tree s labeled in $A \times I$ where I (the set of *instructions*) is composed of the functions from E into $\{0, 1\}$. Indeed, a memoryless winning strategy of Pathfinder is a function $f : \{0, 1\}^* \times E \to \{0, 1\}$. It can be viewed as a function $s : \{0, 1\}^* \to I$ which defines a tree (s, t) with $(s, t)(w) = (s(w), t(w))$.

Formally, let Z be the set of infinite words on $I \times E \times \{0, 1\}$ formed of all (x, y, π) such that

(1) if the sequence of instructions x is applied to the sequence of transitions y, the result is the path π.

(2) The corresponding sequence of states is not successful in $\mathcal{A}$.

It is clear that Z is recognizable. By Proposition 3.6, the set of trees

$$Y = \{s \in T_{I \times E}^{\omega} \mid \text{for all } \pi \in \{0, 1\}^{\omega}, (s(\pi), \pi) \in Z\}$$

is recognizable. Since X is obtained from Y by projection, this concludes the proof. □

4.3 A second proof of Rabin's basis theorem. We shall now see how one can use games to give a second proof of Rabin's basis theorem asserting that any nonempty Muller recognizable set of trees contains a rational tree (Theorem 3.7).

By Proposition 3.5, any Muller tree automaton is equivalent to a Rabin chain tree automaton. Thus we may restrict ourselves to considering a set T recognized by a Rabin chain tree automaton $\mathcal{A} = (Q, A, q_0, F, \mathcal{S})$. We replace $\mathcal{A}$ by an input-free automaton $\mathcal{A}' = (Q \times A, \{q_0\} \times A, F', \mathcal{S}')$ in such a way that $T \neq \emptyset$ if and only if $\mathcal{A}'$ has some successful run. One may imagine that $\mathcal{A}'$ guesses its input. We consider the game $G(\mathcal{A}')$. By Proposition 4.2, Automaton wins the game if and only if $\mathcal{A}'$ admits at least one successful run. Since $\mathcal{A}'$ is a Rabin chain automaton, $G(\mathcal{A}')$ is a parity game. Thus, the winner of the game has a memoryless winning strategy. Thus, if $T \neq \emptyset$, Automaton has a memoryless winning strategy, which defines a run of $\mathcal{A}'$ which is a rational tree. Thus, if $T \neq \emptyset$, T contains a rational tree, which concludes the proof.

One also deduces easily from this proof the fact that the emptiness problem is decidable for Muller tree automata. Indeed, $G(\mathcal{A}')$ is a parity game on a finite graph and thus one may effectively find which player has a memoryless winning strategy (see Chapter IV).

5 Topology

In this section, we study the topological aspects of the theory of trees. We shall first see how the topology is defined on the set of trees, as a generalization of the topology on the space of infinite words.

5.1 The topological space of infinite trees. The set T_A^ω of infinite trees on A is a topological space which is a generalization of the space A^ω of infinite words. The topology can be defined in several ways, just as in the case of words.

First of all, since a tree is a function from $\{0, 1\}^*$ into A, we can obtain T_A^ω as a product space, considering the discrete topology on A. This means that a basis of open sets is formed by the sets of trees that have a specified value on each of the words $x \in F$ where $F \subset \{0, 1\}^*$ is a finite set. Thus, the open sets have the form $S \cdot_c T_A^\omega$ where S is a set of finite trees.

This topology can also be defined by a distance, as in the case of words. Let us indeed define the distance of two trees $s, t \in T_A^\omega$ as $d(s, t) = 2^{-e(s,t)}$ where

$$e(s, t) = \min\{|x| \mid s(x) \neq t(x)\}$$

with the usual conventions that $e(s, s) = \infty$ and $2^{-\infty} = 0$.

The set $T(D, A)$ of all D-trees on a finite alphabet A is also a topological space which is a generalization of the space A^∞ of finite or infinite words.

The topology is simply defined as follows. By adjoining an extra element to the set of labels, any tree can be viewed as a function from D^* into A. The set of functions from

D^* into A is a topological space for the product topology, considering A as a discrete topological space. The set $T(D, A)$ of trees is a closed subset of this space.

The open sets of the topology are the sets of the form $S \cdot_c T(D, A)$ where S is a set of finite trees.

We denote by Σ_1^0 the class of open sets of trees and by Π_1^0 the class of closed sets (the upper index was not used before and we shall see its use shortly).

5.2 Suslin sets. Since trees are more complicated objects than words, we have to introduce more complicated topological classes. The classes of Borel and Suslin sets were defined in Chapter III for any metric spaces. We denote by Σ_{n+1}^0 the closure of Π_n^0 by countable union and by Π_{n+1}^0 the closure of Σ_n^0 by countable intersection. These are the *finite Borel classes*.

The class of Suslin sets is formed by the continuous images of Borel sets. We have seen in Chapter III (Proposition III.6.3 and Corollary III.6.6) that the class of Suslin subsets of $A^{\mathbb{N}}$ is closed under countable union and intersection. One can show that this property holds for any Polish space. Consequently, any countable union or intersection of Suslin sets is again Suslin. The class of Suslin sets contains strictly the class of Borel sets, as seen in Chapter III.

The general properties of these classes are also the same. In particular, any Suslin set can be obtained as a continuous image of a closed set.

The analogue of Theorem III.7.1 (Suslin's separation theorem) holds also for trees with a very similar proof. Thus a set of trees which is Suslin as well as its complement is a Borel set.

We define the *Suslin classes* Σ_n^1, Π_n^1 and Δ_n^1 of subsets of T_A^ω as follows. The class Σ_1^1 is the class of Suslin sets and the class Π_1^1 is the class of co-Suslin sets formed by the complements of Σ_1^1-sets. Then, for $n \geq 1$, we define Σ_{n+1}^1 (resp. Π_{n+1}^1) as the class formed by the continuous images of Π_n^1-sets (resp. Σ_n^1-sets). We also set

$$\Delta_n^1 = \Sigma_n^1 \cap \Pi_n^1.$$

We shall use the following result.

Proposition 5.1. *Let $X \subset A^\omega$ be a Borel subset. Then the set*

$$T = \{t \in T_A^\omega \mid \textit{for all } \pi \in \{0,1\}^*,\ t|\pi \in X\}$$

is in the class Π_1^1 of co-Suslin sets.

Proof. Let

$$U = \{(t, x) \in T_A^\omega \times A^\omega \mid \text{there exists } \pi \in \{0,1\}^*\ t|\pi = x\}$$

The set U is a Π_2^0-set. Indeed, let U_n be the set of pairs (t, x) satisfying the following property: there is a word $a_1 a_2 \ldots a_n \in \{0,1\}^*$ such that $t(\varepsilon)t(a_1)\ldots t(a_n)$ is a prefix of x. Then $U = \bigcap_{n \geq 1} U_n$ which proves the claim since the sets U_n are clearly open.

The set $V = U \cap \bar{X}$ obtained by intersecting U with the complement of X is therefore a Borel set. Finally, the projection of V on its first component is the complement of T, which is thus co-Suslin. □

Example 5.1. The set T of trees on $\{a, b\}$ having an infinite number of a on some path of Example 3.4 is Büchi recognizable. By Proposition 5.1, it is Suslin. Its complement is the set of trees such that the number of a along any path is finite. We have seen in Example 3.6 that it is Muller recognizable but not Büchi recognizable. It is co-Suslin.

It can be shown that T is Σ_1^1-complete (Exercise 10). Thus the set $\bar{T}$ is not Suslin since, otherwise, T would be Borel by Suslin's theorem.

5.3 Recognizable sets. We now consider the topological complexity of recognizable sets of trees. We begin with Büchi recognizable sets of trees.

Proposition 5.2. *Any Büchi recognizable set of trees is Suslin.*

Proof. Let $\mathcal{A}$ be a Büchi tree automaton recognizing a set T. The set of pairs (r, t) where r is a successful run of $\mathcal{A}$ on the tree t is a Borel set (and even a Π_2-set). Thus T is Suslin. □

The following result reflects the fact that the topological complexity of Muller recognizable sets of trees is even higher than the one of Büchi recognizable sets.

Proposition 5.3. *A Muller recognizable set of trees belongs to the class Δ_2^1.*

Proof. We shall prove that any Muller recognizable set T of trees is in Σ_2^1. The dual statement $T \in \Pi_2^1$ will then follow from the fact that the class of recognizable sets is closed under complement.

Let $\mathcal{A}$ be a Muller tree automaton recognizing a set T. By definition, a tree t is in T if there is a successful run r of $\mathcal{A}$ on t. Since the map $(r, t) \mapsto t$ is continuous, it is enough to prove that the set of successful runs of $\mathcal{A}$ is in Π_1^1. This follows actually directly from Proposition 5.1. □

6 Monadic second order logic of two successors

We denote by $S2S$ the logical language of the *second-order monadic theory of two successors*. It uses individual variables $x, y, \ldots$ as well as set variables $X, Y, \ldots$. The terms are formed from the individual variables and the constant ε by application of two symbols of unary functions succ_0 and succ_1. We write $x0$ and $x1$ instead of $\mathrm{succ}_0(x)$ and $\mathrm{succ}_1(x)$. The atomic formulas are of the form

$$t = t', \quad t < t', \quad t \in X$$

where t, t' are terms and X is a set variable. Formulas are formed as usual from atomic formulas using logical connectives and quantifiers.

We interpret the formulas on trees by considering a tree as a model

$$(\mathrm{Dom}(t), \varepsilon, \mathrm{succ}_0, \mathrm{succ}_1, <, (\mathbf{a})_{a\in A})$$

where succ_0, succ_1 are the two successor functions from $\mathrm{Dom}(t)$ into $\mathrm{Dom}^+(t)$ defined by $\mathrm{succ}_0(x) = x0$ and $\mathrm{succ}_1(x) = x1$. The order $<$ is the proper prefix relation on $\{0, 1\}^*$. The unary relation denoted $\mathbf{a}$ is the set of $x \in \mathrm{Dom}(t)$ such that $t(x) = a$.

If φ is an $S2S$-formula, we say that a tree t satisfies the formula φ if φ is true on t.

The set of trees satisfying a formula φ is denoted by $\mathcal{M}(\varphi)$. We also set

$$\mathcal{M}^*(\varphi) = \mathcal{M}(\varphi) \cap T_A, \quad \mathcal{M}^\omega(\varphi) = \mathcal{M}(\varphi) \cap T_A^\omega$$

A set T of finite (resp. infinite) trees is *definable in* $S2S$ if there is a formula $\varphi \in S2S$ such that $T = \mathcal{M}^*(\varphi)$ (resp. $T = \mathcal{M}^\omega(\varphi)$).

Example 6.1. The set of infinite trees satisfying the formula

$$\forall x \exists y \; (x < y \wedge (\forall z \; (y < z \rightarrow \mathbf{b}(z)))$$

is the set of trees with a finite number of a of Example 3.6.

The monadic second-order logic $MF_2(S)$ defined in Chapter VIII results from $S2S$ by deleting succ_1. Similarly to the case of $MF_2(S)$, the order $<$ (and the constant ε) can be defined in terms of succ_0 and succ_1. We use them to simplify our notation. Similarly, other abbreviations can be introduced. For example, the following notation (in which some abbreviations like $\forall x \in X$ or $X \subset Y$ are already used) will be used.

$$\mathrm{Chain}(X)\colon \forall x \in X \; \forall y \in X \; (x < y \vee x = y \vee y < x)$$
$$\mathrm{Path}(X)\colon \mathrm{Chain}(X) \wedge \neg \, \exists Y (X \subset Y \wedge X \neq Y \wedge \mathrm{Chain}(Y))$$
$$\mathrm{Inf}(X)\colon \exists Y \subset X \; \forall y \in Y \; \exists z \in Y \; (y < z)$$

The following theorem is a generalization of the corresponding result for words (Theorem VIII.3.1). The proof is also quite similar and we give it with less details.

Theorem 6.1. *A set of infinite trees is definable in $S2S$ if and only if it is Muller recognizable. More precisely, for every formula $\varphi(X_1, X_2, \ldots, X_n) \in S2S$, one can construct effectively a Muller tree automaton $\mathcal{A}$ which accepts t if and only if t satisfies φ.*

Proof. To prove that a Muller recognizable set of infinite trees is definable in $S2S$, one has to express by a formula of $S2S$ the existence of a successful run.

Conversely, one first reduces $S2S$ to a pure second-order formalism $S2S_0$ with atomic formulas of the form $\mathrm{succ}_0(X, Y)$, $\mathrm{succ}_1(X, Y)$, $X \subset Y$ only. We then proceed by induction on the formulas of $S2S_0$ by interpreting a formula $\varphi(X_1, \dots X_n)$ on the alphabet $A = \{0, 1\}^n$. Induction over formulas $\varphi \in S2S_0$ show the Muller recognizability of $\mathcal{M}^\omega(\varphi)$. For $\vee$ and $\exists$, the proof is easy since the family of Rabin recognizable sets is closed under union and projection (Proposition 3.4). For negation, we use Theorem 4.3. □

As a consequence of Theorem 6.1, we obtain the following result.

Theorem 6.2. *The theory $S2S$ is decidable.*

Proof. Let φ be an $S2S$-sentence. By Theorem 6.1, φ is true if and only if there is an input-free tree automaton which admits a successful run. The existence of such a run can be decided effectively by Corollary 3.9. □

A result similar to Theorem 6.1 also holds for finite trees (Exercise 8).

As we did for $MF_2(S)$, we introduce the *weak* monadic second-order logic $WS2S$, in which the set variables are interpreted as finite sets.

We quote without proof the following result which shows a close connexion between $WS2S$ and Büchi recognizability.

Theorem 6.3. *A set of infinite trees is Büchi recognizable if and only if it can be defined by a formula*

$$\exists X_1 \dots \exists X_n \ \varphi(X_1, \dots, X_n)$$

where $\varphi(X_1, \dots, X_n)$ is a $WS2S$-formula.

A set T of infinite trees is definable in $WS2S$ if and only if T and its complement T_A^ω are Büchi recognizable.

7 Effective algorithms

We have seen that the emptiness problem for Muller tree automata is decidable (Corollary 3.9). The complexity of the emptiness problem for Muller tree automata is high. The problem can indeed be shown to be NP-complete for Rabin tree automata. The emptiness problem for Büchi tree automata is of lower complexity. Indeed, it can be shown that the complexity of the emptiness problem for Büchi tree automata is polynomial (see Exercise 9).

8 Exercises

Exercise 1. A tree automaton $\mathcal{A} = (Q, E, I, F)$ is called bottom-up deterministic if $\text{Card}(F) = 1$ and for each triple $(a, q, r) \in A \times Q \times Q$, there is at most one state $p \in Q$ such that $(p, a, q, r) \in E$.

Show that any tree automaton is equivalent (on finite trees) to a bottom-up deterministic one.

Exercise 2. Show that the class of Büchi recognizable trees is closed under finite union and projection.

Exercise 3. Show that if $S, T \subset T_A$ are recognizable, then $S \cdot_c T$ and $S^{*,c}$ are recognizable.

Exercise 4. Let $\mathcal{A} = (Q, E, I, F)$ be tree automaton with $Q = \{1, 2, \ldots, n\}$. Let $R_i^{(k)}$ be the set of runs r of $\mathcal{A}$ such that

(1) $r(\varepsilon) = i$,

(2) $r(x) \leq k$ for all $x \in \text{Dom}(r) \setminus (\{\varepsilon\} \cup \text{Fr}(r))$.

Show that, for $1 \leq k \leq n$,

$$R_i^{(k)} = R_i^{(k-1)} \cup R_i^{(k-1)} \cdot_k (R_k^{(k-1)})^{*,k} \cdot_k R_k^{(k-1)}.$$

Conclude that a recognizable set of trees is rational.

Exercise 5. Use Exercises 3 and 4 to prove Theorem 3.3.

Exercise 6. Show that if $T_0, \ldots, T_m$ are recognizable sets of trees of $T_{A \cup C}$, where $C = \{c_1, \ldots, c_m\}$ then for $c = (c_1, \ldots, c_m)$, $T_0 \cdot_c (T_1, \ldots, T_m)^{\omega,c}$ is Büchi recognizable.

Exercise 7. Prove by a construction analogous to the one used for ordinary automata (see Exercise I.17) that for any Muller tree automaton, there exists an equivalent Rabin tree automaton.

Exercise 8. Prove the a set of finite trees is recognizable if and only if it is definable in $S2S$.

Exercise 9. Show that there is a polynomial time algorithm to decide whether a Büchi tree automaton recognizes a nonempty set (hint: use the proof of Theorem 3.3).

Exercise 10. A tree $t \in T(D, A)$ is said to be *well-founded* if there are no infinite paths in t. Otherwise, it is said to be *ill-founded*.

(1) Show that the set of ill-founded unlabeled trees on $D = \mathbb{N}$ is Σ_1^1-complete.

(2) Deduce from (1) that the set of infinite trees on $\{a, b\}$ having an infinite number of a's on some path is Σ^1_1-complete.

9 Notes

Our presentation of automata on infinite trees follows closely the two surveys of Wolfgang Thomas [330, 332]. The multi-author monograph edited by Grädel, Thomas and Wilke [127] presents also most of this material. This includes a general presentation of tree automata [217], alternating tree automata [154] and the link with μ-calculus [368].

The theory of automata on infinite trees was first studied by Rabin [253, 254]. Theorem 3.2 dealing with finite trees appears in Thatcher and Wright [320] and in Doner [91]. Example 3.6 appears in Rabin [254]. Theorem 3.3 is implicit in Rabin [254]. A stronger statement, where also rational expressions for tree languages are introduced, appears in [317]. The converse of Proposition 3.6 has been proved by Kupferman, Safra and Vardi [163]. Theorem 3.7 appeared in Rabin [255].

The idea of using games in connexion with tree automata appears in Büchi [50, 51] and Gurevitch, Harrington [131], who introduced the names Automaton and Pathfinder for the two players.

The topological aspects of Section 5 were considered by Mostowski [206] and Skurczynski [291]. It is proved in [291] that the Borel hierarchy of recognizable sets of infinite trees is infinite.

The results of Section 6 are due to Rabin. Theorem 6.2 appeared originally in [253]. A proof of Theorem 6.3 can be found in [254]. The NP-completeness of the emptiness problem for Rabin tree automata is due to Emerson and Jutla [102, 103].

Muller and Schupp have developed in [215] a theory of *alternating tree automata*. The basic idea is to replace the states of the automaton by propositional formulas on the set of states. The transition function of the automaton is then a function

$$\delta : Q \times A \to \mathcal{L}(\{0, 1\} \times Q)$$

from $Q \times A$ to $\mathcal{L}(\{0, 1\} \times Q)$ where $\mathcal{L}(S)$ is the free distributive lattice on a set S. It is then easy to pass from an automaton to its *dual*, recognizing the complement, by taking the negation of the formulas. One has however to prove that the model is equivalent to ordinary tree automata and is not stronger. This is the difficult point of this approach. A characterization in terms of alternating automata of weakly definable sets of infinite trees appears in [211].

For an approach using μ-calculus, see the book by Arnold and Niwinski [14].

ANNEX A

FINITE SEMIGROUPS

This annex is a brief introduction to the algebraic structures used in this book: semigroups, monoids, semirings, and their ordered counterparts. Basic definitions are given in Section 1. Green relations, which give important structural properties of semigroups, are introduced in Section 2. Section 3 is devoted to transformation semigroups, that is, semigroups acting on a set. Semidirect and wreath products are defined in Section 4 and used in Section 5.

1 Monoids, semigroups and semirings

We first briefly recall some definitions from elementary algebra.

1.1 Definitions. A *relation* on a set E is a subset of $E \times E$. If $\mathcal{R}$ is a relation on E, two elements x and y of E are said to be *related by* $\mathcal{R}$ if $(x, y) \in \mathcal{R}$, which is also denoted by $x \mathrel{\mathcal{R}} y$. The relation $\mathcal{R}$ is *reflexive* if, for each $x \in E$, $x \mathrel{\mathcal{R}} x$, *symmetric* if, for each $x, y \in E$, $x \mathrel{\mathcal{R}} y$ implies $y \mathrel{\mathcal{R}} x$, *antisymmetric* if, for each $x, y \in E$, $x \mathrel{\mathcal{R}} y$ and $y \mathrel{\mathcal{R}} x$ imply $x = y$ and *transitive* if, for each $x, y, z \in E$, $x \mathrel{\mathcal{R}} y$ and $y \mathrel{\mathcal{R}} z$ implies $x \mathrel{\mathcal{R}} z$. It is a *preorder* if it is reflexive and transitive, an *order* (or *partial order*) if it is reflexive, transitive and antisymmetric and an *equivalence relation* (or an *equivalence*) if it is reflexive, transitive and symmetric. If $\mathcal{R}$ is a preorder, the relation $\sim$ defined by $x \sim y$ if and only if $x \mathrel{\mathcal{R}} y$ and $y \mathrel{\mathcal{R}} x$ is an equivalence relation, called the *equivalence relation associated* with $\mathcal{R}$.

Relations are ordered by inclusion. More precisely, if $\mathcal{R}_1$ and $\mathcal{R}_2$ are two relations on a set S, $\mathcal{R}_1$ *refines* $\mathcal{R}_2$ (or $\mathcal{R}_2$ is *coarser* than $\mathcal{R}_1$) if and only if, for each $s, t \in S$, $s \mathrel{\mathcal{R}_1} t$ implies $s \mathrel{\mathcal{R}_2} t$. Equality is thus the thinnest equivalence relation and the *universal* relation, in which all elements are related, is the coarsest.

Let S be a set. A mapping $S \times S \to S$ is called an *operation*. The image of (x, y) under this mapping is called the composition of x and y. In many cases, the composition is denoted multiplicatively (xy) or additively $(x + y)$ and is called the product or the sum, respectively. An operation is associative if, for every x, y, z in S, $(xy)z = x(yz)$. An element 1 such that $1x = x1 = x$ for each $x \in S$ is called an *identity*. It is an easy exercise to verify that there can be at most one identity in S, and thus it is referred as *the* identity. When the composition is written additively, the identity, if it exists, is denoted by 0.

A *monoid* is a set equipped with an associative operation and an identity.

We already met several examples of monoids, in particular the free monoid A^* on an alphabet A, equipped with the concatenation product. Another example is the set $\mathcal{R}(Q)$ of all binary relations on a set Q, equipped with the composition of relations: if $\mathcal{R}$ and $\mathcal{S}$ are two relations on Q, $\mathcal{RS}$ denotes the relation defined by

$$(q, q'') \in \mathcal{RS} \text{ if and only if there exists } q' \in Q \text{ such that } (q, q') \in \mathcal{R} \text{ and } (q', q'') \in \mathcal{S}.$$

A *semigroup* is a set equipped with an associative operation which does not necessarily admit an identity. The free semigroup A^+, which is the set of nonempty words on an alphabet A, is an example.

A relation $\mathcal{R}$ on a semigroup S is said to be *stable on the right* (resp. *stable on the left*) if, for each $x, y, z \in S$, $x \mathrel{\mathcal{R}} y$ implies $xz \mathrel{\mathcal{R}} yz$ (resp. $zx \mathrel{\mathcal{R}} zy$). A relation which is stable on the right and on the left is said to be *stable*.

An *ordered semigroup* is a semigroup equipped with a stable partial order $\leq$: for each $s, s', t, t' \in S$, $s \leq t$ and $s' \leq t'$ implies $ss' \leq tt'$. An *order ideal* of an ordered semigroup S is a subset I of S such that $s \in I$ and $t \leq s$ implies $t \in I$. The *order ideal generated* by an element x is the set $\downarrow x$ of all the elements y of S such that $y \leq x$.

If S is an ordered semigroup, the *dual* of S is the ordered semigroup $\check{S}$ obtained by replacing the order $\leq$ on S by *the dual order* $\leq_r$, defined by $s \leq_r t$ if and only if $t \leq s$.

Note that any semigroup, equipped with the equality relation is an ordered semigroup. Roughly speaking, this allows the theory of semigroups to be embedded in that of ordered semigroups. In order to avoid cumbersome repetitions, we shall take advantage of this observation and use, from now onwards, the word "semigroup" for "ordered semigroup". The theory of unordered semigroups can be recovered by simply assuming that the order is always the equality relation.

A semigroup (or a monoid) S is said to be *commutative* if its operation is commutative, that is, if $xy = yx$ for each $x, y \in S$.

If S is a semigroup, S^1 denotes the monoid equal to S if S is a monoid, and to $S \cup \{1\}$ if S is not a monoid. In the latter case, the operation of S is completed by the rules

$$1s = s1 = s$$

for each $s \in S^1$.

A *semiring* is a set S equipped with two operations, denoted resp. additively and multiplicatively, and with a special element, denoted 0 and called the zero, satisfying the following conditions:

(1) $(S, +, 0)$ is a commutative monoid,

(2) $(S, .)$ is a semigroup,

(3) addition is distributive over multiplication, that is, for each $s, t_1, t_2 \in S$, $s(t_1 + t_2) = st_1 + st_2$ and $(t_1 + t_2)s = t_1 s + t_2 s$,

(4) for each $s \in S$, $0s = s0 = 0$.

The only difference with the definition of a ring is thus the lack of inverse for the addition.

A semiring is said to be *commutative* (resp. *unitary*) if the multiplication is commutative (resp. has an identity). If S is a semigroup, the set $\mathcal{P}(S)$ of subsets of S is a semiring for the operations

$$X + Y = X \cup Y \quad \text{and} \quad XY = \{xy \mid x \in X \text{ and } y \in Y\},$$

the empty set being the zero. Similarly, the set of rational subsets of A^*, equipped with the same operations (union and product), is a unitary semiring. Another standard example is the boolean semiring $\mathbb{B}$. Recall that $\mathbb{B} = \{0, 1\}$, with the operations $0 + 0 = 0$, $1 + 0 = 0 + 1 = 1 + 1 = 1$; $0 \times 0 = 0 \times 1 = 1 \times 0 = 0$, $1 \times 1 = 1$. Note that the only equality which is not required by the definition of an unitary semiring is $1 + 1 = 1$.

On a general level, a morphism between two algebraic structures is a map preserving the operations. Therefore a *semigroup morphism* is a map φ from a semigroup S into a semigroup T such that, for each $s_1, s_2 \in S$,

(1) $\varphi(s_1 s_2) = \varphi(s_1)\varphi(s_2)$.

Similarly, a *monoid morphism* is a map φ from a monoid S into a monoid T satisfying (1) and such that

(2) $\varphi(1) = 1$.

A *morphism of ordered semigroups* must also be order compatible: for each $s_1, s_2 \in S$,

(3) if $s_1 \leq s_2$ then $\varphi(s_1) \leq \varphi(s_2)$.

Finally, a map φ from a semiring S in a semiring T is a *semiring morphism* if, for each $s_1, s_2 \in S$,

(1) $\varphi(s_1 + s_2) = \varphi(s_1) + \varphi(s_2)$.

(2) $\varphi(s_1 s_2) = \varphi(s_1)\varphi(s_2)$.

(3) $\varphi(0) = 0$.

A *subsemigroup* of a semigroup S is a subset T of S such that $s_1 \in T$ and $s_2 \in T$ imply $s_1 s_2 \in T$. A *submonoid* of a monoid is a subsemigroup containing the identity. An ordered subsemigroup (resp. monoid) of an ordered semigroup (resp. monoid) S is a subsemigroup (resp. submonoid), equipped with the restriction of the order on S.

Given a subset R of a semigroup S, the subsemigroup of S *generated* by R is the smallest subsemigroup of S containing R. It consists of all products $r_1 \cdots r_n$ of elements of R. If S is a monoid, the submonoid generated by R is defined in a similar way, but it always contains the identity of S. A semigroup (resp. monoid) is called *cyclic* if it is generated by a singleton.

Let S and T be two semigroups (resp. monoids, ordered semigroups, ordered monoids, semirings). Then T is a *quotient* of S if there exists a surjective morphism from S onto T. Finally, a semigroup (resp. monoid) S *divides* a semigroup (resp. monoid) T if S is quotient of a subsemigroup of T. It is easy to see that division is a transitive relation.

A semigroup (resp. monoid, semiring, ordered semigroup) morphism $\varphi : S \to T$ is an *isomorphism* if there exists a morphism $\psi : T \to S$ such that $\varphi \circ \psi = Id_T$ and $\psi \circ \varphi = Id_S$. A semigroup morphism (resp. monoid, semiring) is an isomorphism if and only if it is a bijective morphism. This is no longer true for morphisms of ordered semigroup as shown in the next example:

Example 1.1. Let $U_1 = \{0, 1\}$ be the semigroup with two elements defined by $1.1 = 1$ and $0.0 = 0.1 = 1.0 = 0$ and let $\leq$ be the order relation defined by $0 \leq 1$. The identity on U_1 defines a bijective morphism of ordered semigroups from $(U_1, =)$ into $(U_1, \leq)$ which is not an isomorphism.

In fact, a morphism of ordered semigroups $\varphi : S \to T$ is an isomorphism if and only if it is bijective and if, for each $x, y \in S$, $x \leq y$ is equivalent to $\varphi(x) \leq \varphi(y)$.

Example 1.1 leads us to make precise the relations between semigroups and ordered semigroups. First of all, each semigroup S can be ordered by the equality relation, which is of course stable. Moreover, every ordered semigroup $(S, \leq)$ with underlying semigroup S is a quotient of $(S, =)$, since the identity on S defines a surjective morphism from $(S, =)$ onto $(S, \leq)$. In the sequel, we shall simply denote by S the ordered semigroup $(S, =)$. Note that in this ordered semigroup, every subset is an order ideal.

Given a family $(S_i)_{i\in I}$ of semigroups, the *product* $\Pi_{i\in I}$ is the semigroup defined on the Cartesian product of the sets S_i by the operation

$$(s_i)_{i\in I}(s'_i)_{i\in I} = (s_i s'_i)_{i\in I}$$

The product of a family of ordered semigroups is defined analogously.

1.2 Congruences. A *semigroup congruence* is a stable equivalence relation. Thus, an equivalence relation $\sim$ on a semigroup S is a congruence if, for each $m, n \in S$ and

$u, v \in S^1$, we have

$$m \sim n \text{ implies } umv \sim unv.$$

The set $S/{\sim}$ of equivalence classes of the elements of S is naturally equipped with a structure of semigroup, and the function which maps every element onto its equivalence class is a semigroup morphism from S onto $S/{\sim}$.

Let S be an ordered semigroup. A *congruence of ordered semigroup* on S is a stable preorder coarser than $\leq$. If $\preceq$ is a congruence of ordered semigroup, the equivalence $\sim$ associated with $\preceq$ is a semigroup congruence and the preorder $\preceq$ induces a stable order relation on the quotient semigroup $S/{\sim}$, that will also be denoted by $\leq$. Finally, the function which maps each element onto its equivalence class is a morphism of ordered semigroups from S onto $(S/{\sim}, \leq)$.

For each semigroup morphism $\varphi : S \to T$, the equivalence $\sim_\varphi$ defined on S by

$$x \sim_\varphi y \text{ if and only if } \varphi(x) = \varphi(y)$$

is a congruence. Similarly, for each morphism of ordered semigroups $\varphi : S \to T$, the relation $\preceq_\varphi$ defined on S by

$$x \preceq_\varphi y \text{ if and only if } \varphi(x) \leq \varphi(y)$$

is a congruence of ordered semigroup. In both cases, this congruence, called the *nuclear congruence* of φ, has the following standard property.

Proposition 1.1. *Let $\varphi : S \to T$ be a morphism of semigroups and let $\pi : S \to S/{\sim_\varphi}$ be the quotient morphism. Then there exists a unique semigroup morphism $\bar{\varphi} : S/{\sim_\varphi} \to T$ such that $\varphi = \bar{\varphi} \circ \pi$. Moreover, $\bar{\varphi}$ is an isomorphism from $S/{\sim_\varphi}$ onto $\varphi(S)$.*

When two congruences are comparable, the quotient structures associated with them can also be compared. We give separately the statements corresponding to semigroups and to ordered semigroups.

Proposition 1.2. *Let $\sim_1$ and $\sim_2$ be two congruences on a semigroup S and π_1 (resp. π_2) the canonical morphism from S onto $S/{\sim_1}$ (resp. $S/{\sim_2}$). If $\sim_2$ is coarser than $\sim_1$, there exists a unique surjective morphism $\pi : S/{\sim_1} \to S/{\sim_2}$ such that $\pi \circ \pi_1 = \pi_2$.*

Proposition 1.3. *Let $\preceq_1$ and $\preceq_2$ two congruences on an ordered semigroup S and π_1 (resp. π_2) the canonical morphism from S onto $S/{\preceq_1}$ (resp. $S/{\preceq_2}$). If $\preceq_1$ is thinner than $\preceq_2$, there exists a unique surjective morphism of ordered semigroups $\pi : S/{\preceq_1} \to S/{\preceq_2}$ such that $\pi \circ \pi_1 = \pi_2$.*

The free semigroup and monoid on A are respectively A^+ and A^*. Indeed, they satisfy the universal property characterizing free structures.

Proposition 1.4. *If φ is a function from A into a semigroup (resp. monoid) S, there exists a unique semigroup (resp. monoid) morphism $\bar{\varphi} : A^+ \to S$ (resp. $A^* \to S$) such that, for each $a \in A$, $\bar{\varphi}(a) = \varphi(a)$. Moreover, $\bar{\varphi}$ is surjective if and only if the set $\varphi(A)$ generates S.*

Proof. Define a mapping $\bar{\varphi} : A^+ \to S$ by setting, for each word $a_0 a_1 \cdots a_n$,

$$\bar{\varphi}(a_0 a_1 \cdots a_n) = \varphi(a_0)\varphi(a_1) \cdots \varphi(a_n)$$

One can easily verify that $\bar{\varphi}$ is the required morphism. On the other hand, any morphism $\bar{\varphi}$ such that $\bar{\varphi}(a) = \varphi(a)$ for each $a \in A$ must satisfy these two equalities, which shows it is unique.

By construction, the set $\varphi(A)$ generates $\bar{\varphi}(A)$. Consequently, the morphism $\bar{\varphi}$ is surjective if and only if the set $\varphi(A)$ generates S. □

Similarly, $(A^+, =)$ and $(A^*, =)$ are respectively the free ordered semigroup and monoid on A. Indeed, it suffices to imitate the proof of Proposition 1.4 to obtain the following statement.

Proposition 1.5. *If φ is a function from A into an ordered semigroup (resp. monoid) $(S, \leq)$, there exists a unique morphism of ordered semigroups (resp. monoids) $\bar{\varphi} : (A^+, =) \to (S, \leq)$ (resp. $(A^*, =) \to S$) such that, for each $a \in A$, $\bar{\varphi}(a) = \varphi(a)$. Moreover, $\bar{\varphi}$ is surjective if and only if the set $\varphi(A)$ generates S.*

These results have several frequently used corollaries. To avoid multiple statements, one for nonordered structures, and another one for ordered structures, we have written the word "ordered" in brackets.

Corollary 1.6. *Let S be an [ordered] semigroup and let A be a subset of S generating S. The identity map from A into S induces a morphism of [ordered] semigroups from A^+ onto S.*

This morphism is called the *natural* morphism from A^+ onto S.

Corollary 1.7. *Let $\eta : A^+ \to S$ and $\beta : T \to S$ be surjective morphisms of [ordered] semigroups. Then there exists a surjective morphism of [ordered] semigroups $\varphi : A^+ \to T$ such that $\eta = \beta \circ \varphi$.*

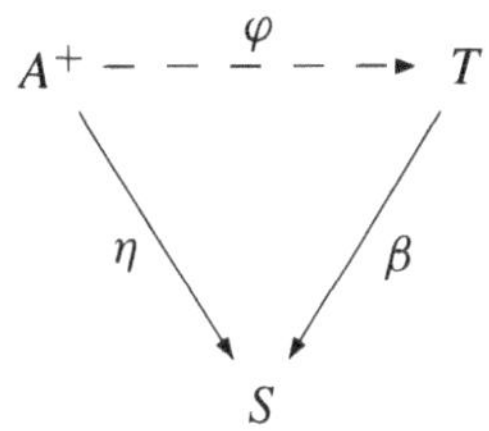

Proof. Let us associate with each letter $a \in A$ an element $\varphi(a)$ of $\beta^{-1}(\eta(a))$. We thus define a function $\varphi : A \to T$, which, by Proposition 1.5, can be extended to a morphism $\varphi : A^+ \to T$ such that $\eta = \beta \circ \varphi$. $\square$

Proposition 1.8. *Let S be an [ordered] semigroup, $(\sim_i)_{i \in I}$ be a family of congruences on S and $\sim$ be the intersection of these congruences. Then the [ordered] semigroup $S/\sim$ is a subsemigroup of the product $\Pi_{i \in I} S/\sim_i$.*

Proof. We give the proof in the unordered case, which can be readily extended to the ordered case.

Denote by $\pi_i : S \to S_i$ the natural projections and by $\pi : S \to \Pi_{i \in I} S/\sim_i$ the morphism defined by $\pi(s) = (\pi_i(s))_{i \in I}$ for every $s \in S$. The nuclear congruence of π is none other than $\sim$ and thus by Proposition 1.1, $S/\sim$ is isomorphic to $\pi(S)$. $\square$

1.3 Structure of finite semigroups. An *idempotent* in a semigroup S is an element e such that $e = e^2$. The set of idempotents of a semigroup is denoted by $E(S)$.

If S is a cyclic semigroup, generated by a single element x, the set S consists of the successive powers of x. If S is infinite, it is isomorphic to the additive semigroup of strictly positive integers. If S is finite, there exist integers i, $p > 0$ such that

$$x^{i+p} = x^i.$$

The minimal i and p with this property are called respectively the *index* and the *period* of x. The semigroup S then has $i + p - 1$ elements and its multiplicative structure is represented in Figure 1.1.

The next result is well known.

Proposition 1.9. *In a finite semigroup, each element has an idempotent power.*

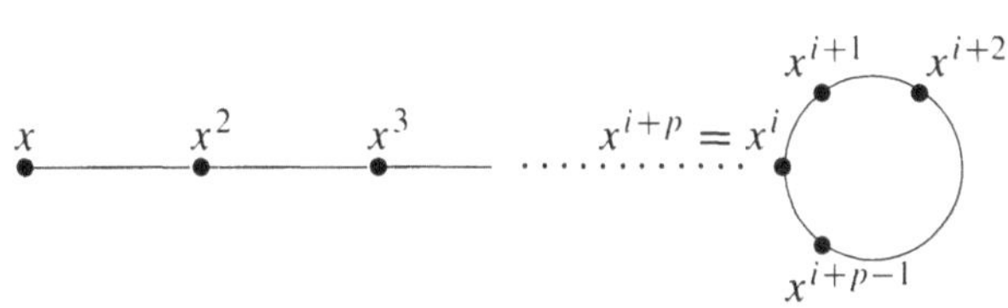

Figure 1.1. The semigroup generated by x.

Proof. Let i and p be the index and the period of an element x. Observe that, for $k \geq i$, $x^k = x^{k+p}$. In particular, if k is a multiple of p, we have

$$(x^k)^2 = x^{2k} = x^{k+qp} = x^k$$

and thus x^k is idempotent. In fact, it is easy to see that the subsemigroup $\{x^i, \dots, x^{i+p-1}\}$ is isomorphic to the additive group $\mathbb{Z}/p\mathbb{Z}$. □

Proposition 1.9 has two important consequences.

Corollary 1.10. *Every nonempty finite semigroup contains at least one idempotent.*

Proposition 1.11. *For each finite semigroup S, there exists an integer π such that, for each $s \in S$, s^π is idempotent.*

Proof. By Proposition 1.9, every element s of S has an idempotent power s^{n_s}. Let n be the least common multiple of the n_s, for $s \in S$. Then s^n is idempotent for each $s \in S$. □

The least integer π satisfying the latter property stated in Proposition 1.11 is called the *exponent* of S.

Here is another elementary property connected with idempotents.

Proposition 1.12. *Let S be a finite semigroup and let $n = |S|$. For every sequence $s_1, \dots, s_n$ of n elements of S, there exists an index $i \in \{1, \dots, n\}$ and an idempotent $e \in S$ such that $s_1 \cdots s_i e = s_1 \cdots s_i$.*

Proof. Consider the sequence $s_1, s_1 s_2, \dots, s_1 \cdots s_n$. If these elements are all distinct, the sequence exhausts the elements of S and one of them, say $s_1 \cdots s_i$, is idempotent. The result is thus clear in this case. Otherwise, two elements of the sequence are equal, say $s_1 \cdots s_i$ and $s_1 \cdots s_j$ with $i < j$. Then we have

$$s_1 \cdots s_i = s_1 \cdots s_i (s_{i+1} \cdots s_j) = s_1 \cdots s_i (s_{i+1} \cdots s_j)^\pi$$

where π is the exponent of S. The proposition follows, since $(s_{i+1} \cdots s_j)^\pi$ is idempotent. □

Let S be a semigroup. For each idempotent e of S, the set

$$eSe = \{ese \mid s \in S\}.$$

is a subsemigroup of S, called the *local* semigroup associated with e. It is a monoid with identity e, and actually, eSe is the set of elements s of S such that $es = se = s$.

An element s of a semigroup S is a *zero* (resp. *zero on the right, zero on the left*) if, for each $t \in S$, $ts = s = st$ (resp. $ts = s$, $st = s$).

Let S be a semigroup. An *right ideal* of S is a subset R of S such that

$$RS^1 = R.$$

Thus R is a right ideal if, for each $r \in R$ and $s \in S^1$, $rs \in R$. Symmetrically, a *left ideal* is a subset L of S such that

$$S^1 L = L.$$

An *ideal* is a subset I of S such that

$$S^1 I S^1 = I.$$

Therefore a subset of S is an ideal if and only if it is simultaneously a right and a left ideal.

2 Green relations

Define a preorder $\leq_{\mathcal{R}}$ on S by setting $s \leq_{\mathcal{R}} s'$ if the right ideal generated by s is contained in the right ideal generated by s', that is if

$$sS^1 \subset s'S^1,$$

or, equivalently, if there exist $p \in S^1$ such that $s = s'p$. Two elements s and s' of S are $\mathcal{R}$-equivalent if they generate the same right ideal, that is if

$$sS^1 = s'S^1,$$

or, equivalently, if there exist $p, q \in S^1$ such that $s = s'p$ and $s' = sq$. The equivalence classes of this relation are called the *$\mathcal{R}$-classes* of S. Denote by $R(s)$ the $\mathcal{R}$-class of s.

Symmetrically, define a preorder $\leq_{\mathcal{L}}$ on S by setting $s \leq_{\mathcal{L}} s'$ if and only if the left ideal generated by s is contained in the left ideal generated by s'. The elements s and s' are $\mathcal{L}$-equivalent if they generate the same left ideal, that is if

$$S^1 s = S^1 s'.$$

The equivalence classes of this relation are called the *$\mathcal{L}$-classes* of S, and $L(s)$ denotes the $\mathcal{L}$-class of s. Finally, the preorder $\leq_{\mathcal{J}}$ is defined by $s \leq_{\mathcal{J}} s'$ if and only if the ideal generated by s is contained in the ideal generated by s'. The elements s and s' are $\mathcal{J}$-equivalent if they generate the same ideal, that is if

$$S^1 s S^1 = S^1 s' S^1.$$

The equivalence classes of this relation are called the *$\mathcal{J}$-classes* of S, and $J(s)$ denotes the $\mathcal{J}$-class of s.

Proposition 2.1. *In each semigroup S, the relations $\leq_{\mathcal{R}}$ and $\mathcal{R}$ are stable on the left and the relations $\leq_{\mathcal{L}}$ and $\mathcal{L}$ are stable on the right.*

Proof. Indeed, if $s \leq_{\mathcal{R}} s'$, there exists $p \in S^1$ such that $s = s'p$. Therefore the equality $ts = ts'p$ holds for each $t \in S$, and thus $ts \leq_{\mathcal{R}} ts'$. The other cases are analogous. □

Proposition 2.2. *Let e be an idempotent of a semigroup S, and let x be an element of S such that $x \leq_{\mathcal{R}} e$ (resp. $x \leq_{\mathcal{L}} e$). Then $ex = x$ (resp. $xe = x$).*

Proof. If $x \leq_{\mathcal{R}} e$, there exists $p \in S^1$ such that $ep = x$. It follows $ex = eep = ep = x$. The proof for the relation $\leq_{\mathcal{L}}$ is symmetric. □

Proposition 2.3. *In each semigroup S, the relations $\mathcal{R}$ and $\mathcal{L}$ commute.*

Proof. Suppose that $s \mathrel{\mathcal{R}} s'$ and $s' \mathrel{\mathcal{L}} s''$. Then there exist m, n, p, q such that $s = s'm$, $s' = sn$, $s' = ps''$ and $s'' = qs'$. Consider the element $t = qs'm = s''m = qs$. Then we have on one hand $s \mathrel{\mathcal{L}} t$ since $t = qs$ and $s = s'm = ps''m = pt$, and on the other hand $t \mathrel{\mathcal{R}} s''$ since $t = s''m$ and $s'' = qs' = qsn = tm$. For instance $s \mathrel{\mathcal{LR}} s''$, showing that $\mathcal{RL} \subset \mathcal{LR}$. The proof of the inclusion $\mathcal{LR} \subset \mathcal{RL}$ is symmetric. □

Denote by $\mathcal{D}$ the equivalence $\mathcal{RL}$ (or $\mathcal{LR}$) and by $\mathcal{H}$ the equivalence $\mathcal{R} \cap \mathcal{L}$. The equivalence classes of these relations are called the *$\mathcal{D}$-classes* and *$\mathcal{H}$-classes* of S, respectively. The next proposition describes the structure of a $\mathcal{D}$-class.

Proposition 2.4. *Let D be a $\mathcal{D}$-class of a semigroup S, and let m, m' and m'' be elements of S.*

(1) *If $m \mathrel{\mathcal{R}} m'$, and if $m = m'p$ and $m' = mq$, the maps $x \to xp$ and $x \to xq$ define inverse bijections between $L(m)$ and $L(m')$, and these bijections preserve the $\mathcal{H}$-classes.*
(2) *One has $mm'' \in R(m) \cap L(m'')$ if and only if $R(m'') \cap L(m)$ contains an idempotent.*
(3) *An $\mathcal{H}$-class is a group if and only if it contains the product of two of its elements.*
(4) *If D contains an idempotent, it contains at least one idempotent in each $\mathcal{R}$-class and in each $\mathcal{L}$-class.*

Proof. (1) Let $n \in L(m)$ (cf. Figure 2.1). Then there exist $s \in S^1$ such that $n = sm$, whence $nqp = smqp = sm'p = sm = n$. Symmetrically, if $n' \in L(m')$, there exists $r \in S^1$ such that $n' = rm'$ and thus $n'pq = rm'pq = rmq = rm' = n'$. Moreover, Proposition 2.1 shows that the maps $x \to xp$ and $x \to xq$ preserve the $\mathcal{H}$-classes. This proves (1).

(2) If $R(m'') \cap L(m)$ contains an idempotent e, put

$$e = m''p, \quad m'' = eq, \quad e = rm, \quad m = se.$$

Then

$$mm''p = me = (se)e = se = m$$

whence $mm'' \mathcal{R} m$. The relation $mm'' \mathcal{L} m''$ is proved in a similar way.

Conversely, if $mm'' \in R(m) \cap L(m'')$, the multiplication on the right by m'' is, by (1), a bijection from $L(m)$ onto $L(m'')$. Thus there exists an element $e \in L(m)$ such that $em'' = m''$. Furthermore $e \in R(m'')$ since left multiplication preserves the $\mathcal{R}$-classes. Therefore there exists $v \in S^1$ such that $e = m''v$. For instance

$$ee = em''v = m''v$$

and thus e is idempotent.

(3) An $\mathcal{H}$-class which is a group obviously contains the product of two of its elements. Conversely, let H be a $\mathcal{H}$-class containing the product of two of its elements m and m'. Property (2) shows, as $mm' \in H$, that H contains an idempotent e. The same property shows that H is a semigroup: indeed, if $x, x' \in H$, we have $xx' \in R(x) \cap L(x') = R(x) \cap L(x) = H$. Moreover, if $x \in H$, we have $x \mathcal{R} e$ and hence $ex = x$ by Proposition 2.2. Finally, for each $x \in H$, the property (1) shows that the map $n \to nx$ is a permutation on H. In particular, there exists x' such that $x'x = e$, and thus H is a group with identity e.

(4) Suppose that D contains an idempotent e and let $m \in D$. Then $e \mathcal{R} r$ and $r \mathcal{L} m$ for some $r \in D$. Therefore there exists $s \in S^1$ such that $rs = e$. It follows that ser is idempotent, since

$$serser = se(rs)er = seer = ser.$$

Furthermore, the equality $r = er$ holds by Proposition 2.2, and thus $r(ser) = eer = er = r$. Consequently $r \mathcal{L} ser$, $L(m) = L(r)$ and this $\mathcal{L}$-class contains an idempotent. □

An $\mathcal{R}$-class (resp. $\mathcal{L}$-class, $\mathcal{D}$-class) which contains an idempotent is called *regular*. By Proposition 2.4, a regular $\mathcal{D}$-class contains at least one idempotent in each $\mathcal{R}$-class and in each $\mathcal{L}$-class. It follows that the $\mathcal{R}$-classes and $\mathcal{L}$-classes contained in a regular $\mathcal{D}$-class are regular.

The $\mathcal{D}$-classes are traditionally represented by the classical "egg-box picture" of Figure 2.1. Each row represents an $\mathcal{R}$-class, each column an $\mathcal{L}$-class and each cell an $\mathcal{H}$-class. The possible presence of an idempotent within an $\mathcal{H}$-class is indicated by a star: these $\mathcal{H}$-classes are thus groups, and all the groups occurring in a given $\mathcal{D}$-class are isomorphic. We now state a few results about finite semigroups.

Proposition 2.5. *Let S be a finite semigroup.*

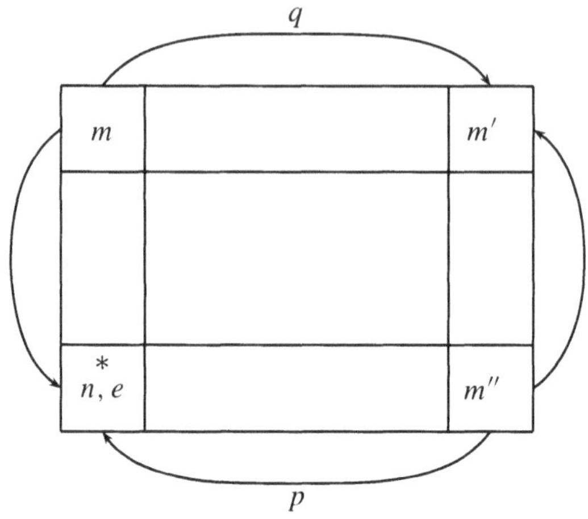

Figure 2.1. A $\mathcal{D}$-class.

(1) (Simplification lemma). *Let* $m \in S$ *and* $r, s \in S^1$. *If* $m = rms$, *then* $rm \in L(m)$ *and* $ms \in R(m)$.

(2) *Let* $m, m' \in S$. *Then* $m \mathcal{D} m'$ *if and only if* $m \mathcal{J} m'$.

Proof. (1) By Proposition 1.9, there exist integers i and j such that $e = r^i$ and $f = s^j$ are idempotent. Then we have

$$m = rms = r^{ij}ms^{ij} = emf = e(emf) = em = r^{i-1}(rm).$$

Thus $m \mathcal{L} rm$. Likewise $m \mathcal{R} ms$.

(2) Suppose that m and m'' generate the same ideal, and let u, v, r, s be such that $m = um''v$ and $m'' = rms$. Then $m = urmsv$ and thus, by the simplification lemma, $urm \mathcal{L} m$ and $msv \mathcal{R} m$. It follows $m \leq_{\mathcal{L}} rm \leq_{\mathcal{L}} urm \leq_{\mathcal{L}} m$ and thus $m \mathcal{L} rm$. Similarly, $ms \mathcal{R} m$, and, by Proposition 2.1, $rms \mathcal{R} rm$. Finally, $m'' = rms \mathcal{R} rm$ and $rm \mathcal{L} m$ and thus $m \mathcal{D} m''$.

Conversely, if $m \mathcal{D} m''$, there exists an element $m' \in S$ such that $m \mathcal{R} m'$ and $m' \mathcal{L} m''$. Therefore $mS^1 = m'S^1$ and $S^1m' = S^1m''$, whence $S^1mS^1 = S^1m''S^1$. □

Corollary 2.6. *Let* S *be a finite semigroup and let* $s, t \in S$.

(1) *If* $s \mathcal{J} t$ *and* $s \leq_{\mathcal{R}} t$, *then* $s \mathcal{R} t$.

(2) *If* $s \mathcal{J} t$ *and* $s \leq_{\mathcal{L}} t$, *then* $s \mathcal{L} t$.

Proof. By symmetry, it suffices to prove the first assertion. If $s \mathcal{J} t$, there exist u and v such that $t = usv$. Furthermore, as $s \leq_{\mathcal{R}} t$, there exists $p \in S^1$ such that

$s = tp$. Therefore $t = utpv$ and, by the simplification lemma, $tS^1 = tpvS^1$. It follows $tS^1 = tpvS^1 \subset tpS^1 \subset tS^1$ and thus $t \mathcal{R} tp$. Therefore $t \mathcal{R} s$. □

A nonempty ideal I of a semigroup S is said to be *minimal* if the only nonempty ideal contained in I is I itself. If such an ideal exists — this is not always the case if S is infinite — it is unique and contained in all the nonempty ideals of S. Indeed, if J is another nonempty ideal, the ideal $I \cap J$ is nonempty, as it contains the nonempty set IJ, it is contained in I and is therefore equal to I. The existence of a minimal ideal is guaranteed in two important particular cases: if S has a zero or if S is finite.

Proposition 2.7. *A finite nonempty semigroup has a unique minimal ideal.*

Proof. Let S be a finite nonempty semigroup. Since S is itself a nonempty ideal, there exists at least a minimal ideal. Let I_1 and I_2 be two minimal ideals of S and let $s_1 \in I_1$ and $s_2 \in I_2$. Then $s_1 s_2 \in I_1 \cap I_2$, showing that $I_1 \cap I_2$ is a nonempty ideal contained in I_1 and in I_2. Since I_1 and I_2 are minimal, it follows $I_1 = I_1 \cap I_2 = I_2$. □

In a semigroup, the preorder $\leq_{\mathcal{J}}$ induces an order on the set of $\mathcal{R}$-classes ($\mathcal{L}$-classes, $\mathcal{J}$-classes). If the semigroup is finite, the minimal $\mathcal{R}$-classes (resp. $\mathcal{L}$-classes, $\mathcal{J}$-classes) are contained in the minimal ideal and are always regular. The maximal classes are described in the next proposition.

Proposition 2.8. *In a finite semigroup, the maximal $\mathcal{R}$-classes (respectively $\mathcal{L}$-classes, $\mathcal{J}$-classes) for the order $\leq_{\mathcal{J}}$ are either regular or reduced to a single element.*

Proof. We give the proof for $\mathcal{R}$-classes, but the other cases would be similar. Let R be a maximal $\mathcal{R}$-class. Suppose that R is not regular and let a and b be two elements of R. Since $a \mathcal{R} b$, there exist $s, t \in S^1$ such that $b = as$ and $a = bt$. It follows $a = a(st)$ and hence $a = a(st)^n$ for all $n \geq 0$. By Proposition 1.9, we may choose n such that $e = (st)^n$ is idempotent. Now $a = ae$ and $es(t(st)^{n-1}) = e$, whence $a \leq_{\mathcal{J}} e \mathcal{J} es$. But since R is not regular, a is not $\mathcal{J}$-equivalent to an idempotent and thus $a <_{\mathcal{J}} e$. Since R is maximal in S, the only possibility is that $e = es = 1$. Thus $a = b$ and R is a singleton. □

A finite semigroup S is said to be *$\mathcal{J}$-trivial* (resp. *$\mathcal{R}$-trivial*, *$\mathcal{L}$-trivial*, *$\mathcal{H}$-trivial*) if and only if the Green relation $\mathcal{J}$ (resp. $\mathcal{R}$, $\mathcal{L}$, $\mathcal{H}$) is the equality relation.

It is *aperiodic* if there is an integer $n > 0$ such that, for each $x \in S$, $x^n = x^{n+1}$. Since we assume finiteness, quantifiers can be inverted in the definition: S is aperiodic if there is, for each $x \in S$, there is an integer $n > 0$ such that $x^n = x^{n+1}$. It turns out that aperiodic semigroups are exactly the $\mathcal{H}$-trivial ones. Other characterizations are given in the next proposition.

Proposition 2.9. *Let S be a finite semigroup. The following conditions are equivalent*

(1) *S is aperiodic,*

(2) *each element generates a subsemigroup of period 1,*

(3) *each $\mathcal{H}$-class of S is trivial,*

(4) *every group in S is trivial.*

Proof. (1) implies (2). If S is aperiodic, then for every $x \in S$, there exists a positive integer n such that $x^n = x^{n+1}$. Thus the subsemigroup generated by x is of period 1.

(2) implies (3). Suppose that $x \mathcal{H} y$. Then there exists $a, b, c, d, \in S^1$ such that $ax = y$, $by = x$, $xc = y$ and $yd = x$, whence $x = axd$ and $x = a^n x d^n$ for every n. Since the subsemigroup generated by a has period 1, one can choose n such that $a^n = a^{n+1}$. It follows that $y = ax = a(a^n x d^n) = a^{n+1} x d^n = a^n x d^n = x$ and thus S is $\mathcal{H}$-trivial.

(3) implies (4). This is trivial, since every maximal group in S is an $\mathcal{H}$-class.

(4) implies (1). Let s be an element of S, i its index and p its period. Then the subsemigroup $\{x^i, \dots, x^{i+p-1}\}$ is a group and thus, by (4), is trivial. It follows that $x^i = x^{i+1}$ and hence S is aperiodic. □

3 Transformation semigroups

Let P be a set and S be a semigroup. A *right action* from S on P is a map $P \times S \to P$, denoted $(p, s) \mapsto p \cdot s$, such that, for each $s, t \in S$ and $p, q \in P$,

$$(p \cdot s) \cdot t = p \cdot (st)$$

This condition implies that one may use the notation $p \cdot st$ in the place of $(p \cdot s) \cdot t$ or $p \cdot (st)$ without any ambiguity. We will follow this convention in the sequel.

The action is *faithful* if, given $s, t \in S$, the condition $p \cdot s = p \cdot t$ for all $p \in P$ implies $s = t$. A *transformation semigroup* (P, S) is a semigroup S equipped with a faithful action of S on P.

Example 3.1. Each semigroup S defines a transformation semigroup (S^1, S), given by the faithful action $q \cdot s = qs$.

Example 3.2. For each nonnegative integer n, let $U_n = \{1, a_1, \dots, a_n\}$ be the monoid defined by the multiplication $a_i a_j = a_j$ for each $i, j \in \{0, \dots, n\}$. In particular U_1 is the monoid considered in Example 1.1 with a slightly different notation.

The monoid $\tilde{U}_n$ has the same underlying set, but the multiplication is defined in the opposite way: $a_i a_j = a_i$ for each $i, j \in \{0, \dots, n\}$.

Example 3.3. With each finite set R, one can associate a transformation semigroup (R, R) defined by the action $r \cdot s = s$ for every $r, s \in R$. In particular, if $R = \{1, \dots, n\}$, this transformation semigroup is usually denoted by $\bar{\mathbf{n}}$.

Let $(P_i, S_i)_{1 \leq i \leq n}$ be a family of transformation semigroups. The direct product of this family is the transformation semigroup

$$\Big(\prod_{1 \leq i \leq n} P_i, \prod_{1 \leq i \leq n} S_i\Big).$$

The action is defined componentwise:

$$(p_1, \dots, p_n) \cdot (s_1, \dots, s_n) = (p_1 \cdot s_1, \dots, p_n \cdot s_n).$$

A transformation semigroup (P, S) *divides* a transformation semigroup (Q, T) if there exists a surjective partial function $\pi : Q \to P$ and, for every $s \in S$, an element $\hat{s} \in T$, called a *cover* of s, such that, for each $q \in Q$, $\pi(q) \cdot s = \pi(q \cdot \hat{s})$. The chosen terminology is justified by the following result.

Proposition 3.1. *If (P, S) divides (Q, T), then S divides T. If S divides T, then (S^1, S) divides (T^1, T).*

Proof. If (P, S) divides (Q, T), there exists a surjective partial function $\pi : Q \to P$ such that every element $s \in S$ has at least one cover. Furthermore, if $\hat{s}_1$ is a cover of s_1 and $\hat{s}_2$ is a cover of s_2, then $\hat{s}_1 \hat{s}_2$ is a cover of $s_1 s_2$, since, for each $q \in \mathrm{Dom}(\pi)$,

$$\pi(q) \cdot s_1 s_2 = \pi(q \cdot \hat{s}_1) \cdot s_2 = \pi((q \cdot \hat{s}_1) \cdot \hat{s}_2) = \pi(q \cdot \hat{s}_1 \hat{s}_2).$$

It follows that the set R of all covers of an element of S form a subsemigroup of S form a subsemigroup of T. Furthermore, if two elements s_1 and s_2 have the same cover $\hat{s}$, then, for each $q \in \mathrm{Dom}(\pi)$,

$$\pi(q) \cdot s_1 = \pi(q \cdot \hat{s}) = \pi(q) \cdot s_2$$

Since π is surjective and the action of S is faithful, it follows $s_1 = s_2$. Therefore, there is a well-defined map $\pi : \hat{s} \to s$ from R onto S and this map is a morphism.

Suppose now that S divides T. Then there exists a subsemigroup R of T and a surjective morphism π from R onto S, which can be extended to a surjective partial function from T^1 onto S^1, by setting $\pi(1) = 1$ if R is not a monoid. For each $s \in S$, choose an element $\hat{s} \in \pi^{-1}(s)$. Then, for every $q \in R^1$, $\pi(q \cdot \hat{s}) = \pi(q)s$ and thus (S^1, S) divides (T^1, T). □

Transformation semigroups are somewhere between automata and semigroups and, not surprisingly, there is a natural notion of recognizability attached to them. Let (P, S) be a transformation semigroup and let R be a semigroup. A subset L of R is recognized by (P, S) if there exists a semigroup morphism $\varphi : R \to S$, a state $p_0 \in P$, and a subset F of P, such that

$$L = \{r \in R \mid p_0 \cdot \varphi(r) \in F\}$$

Let us give a few examples to illustrate this notion.

Proposition 3.2. *A subset of A^+ is recognized by $\bar{\mathbf{2}}$ if and only if it is of the form A^*B, with $B \subset A$.*

Proof. Let B be a subset of A and let $\varphi : A^+ \to \{1, 2\}$ be the semigroup morphism defined by

$$\varphi(u) = \begin{cases} 1 & \text{if the last letter of } u \text{ is in } B \\ 2 & \text{otherwise} \end{cases}$$

Then $A^*B = \{u \in A^+ \mid 1 \cdot \varphi(u) = 1\}$ and thus $\bar{\mathbf{2}}$ recognizes A^*B.

Conversely, suppose that a subset L of A^+ is recognized by $\bar{\mathbf{2}}$. Then there exists a semigroup morphism $\varphi : A^+ \to \{1, 2\}$, a state $p_0 \in \{1, 2\}$ and a subset F of $\{1, 2\}$ such that $L = \{u \in A^+ \mid p_0 \cdot \varphi(u) \in F\}$. Let B be the set of all letters of A such that $\varphi(a) \in F$. Then $p_0 \cdot \varphi(u)$ is in F if and only if the last letter of u is in B and thus $L = A^*B$. □

If the transformation semigroup is attached to a semigroup, we recover the usual notion of recognition.

Lemma 3.3. *A set is recognized by a semigroup S if and only if it is recognized by the transformation semigroup (S^1, S).*

Proof. If L is a subset of R recognized by S, there exists a semigroup morphism $\varphi : R \to S$ and a subset P of S such that $L = \varphi^{-1}(P)$. It follows that

$$L = \{r \in R \mid 1 \cdot \varphi(r) \in P\}$$

and thus L is recognized by (S^1, S). Conversely, if L is recognized by (S^1, S), there exists a semigroup morphism $\varphi : R \to S$, a state $p_0 \in S^1$, and a subset F of S^1, such that $L = \{r \in R \mid p_0 \cdot \varphi(r) \in F\}$. Setting $P = \{s \in S \mid p_0 \cdot s \in F\}$, we have $L = \varphi^{-1}(P)$ and thus L is recognized by S. □

Proposition 3.4. *A subset of A^* is recognized by U_1 if and only if it is of the form A^*BA^* or B^*, for some $B \subset A$.*

Proof. Let B be a subset of A and let $\varphi : A^+ \to U_1$ be the monoid morphism defined by

$$\varphi(a) = \begin{cases} 1 & \text{if } a \in B \\ 0 & \text{otherwise} \end{cases}$$

Then, setting $C = A \setminus B$, we have

$$\begin{aligned} B^* &= \{u \in A^* \mid \varphi(u) = 1\} \\ A^*CA^* &= \{u \in A^* \mid \varphi(u) = 0\} \end{aligned}$$

and the sets of the form A^*BA^* or B^* are recognized by U_1. Conversely, suppose that a subset L of A^* is recognized by U_1. Then there exists a monoid morphism $\varphi : A^* \to U_1$, and a subset P of U_1 such that $L = \varphi^{-1}(P)$. Let B be the set of all letters of A such that $\varphi(a) = 0$ and let $C = A \setminus B$. Then

$$L = \begin{cases} A^* & \text{if } P = U_1 \\ C^* & \text{if } P = \{1\} \\ A^*BA^* & \text{if } P = \{0\} \\ \emptyset & \text{if } P = \emptyset \end{cases}$$

In all cases, L has the required form. □

4 Semidirect product and wreath product

4.1 Definitions. Let S and T be semigroups. We write the product in S additively to provide a more transparent notation, but it is not meant to suggest that S is commutative. A *left action* of T on S is a map $(t, s) \mapsto t \cdot s$ from $T^1 \times S$ into S such that, for all $s, s_1, s_2 \in S$ and $t, t_1, t_2 \in T$,

(1) $(t_1t_2) \cdot s = t_1(t_2 \cdot s)$

(2) $t \cdot (s_1 + s_2) = t \cdot s_1 + t \cdot s_2$

(3) $1 \cdot s = s$

If S is a monoid with identity 0, the action is *unitary* if it satisfies, for all $t \in T$,

(4) $t \cdot 0 = 0$

The *semidirect product* of S and T (with respect to the given action) is the semigroup $S * T$ defined on $S \times T$ by the multiplication

$$(s, t)(s', t') = (s + t \cdot s', tt')$$

Let $X = (P, S)$ and $Y = (Q, T)$ be two transformation semigroups. To make the notation more readable, we shall denote the semigroup S and its action on P additively and the semigroup T and its action on Q multiplicatively. The *wreath product* of X and Y, denoted $X \circ Y$, is the transformation semigroup $(P \times Q, W)$ where W consists of all pairs (f, t), with f is a function from Q into S and $t \in T$. Since we are thinking of f as acting on the right on Q, we will use the more suitable notation $q \cdot f$ in place of $f(q)$. The action of W on $P \times Q$ is given by

$$(p, q) \cdot (f, t) = (p + q \cdot f, q \cdot t) \tag{4.1}$$

We claim that this action is faithful. Indeed, if $(p, q) \cdot (f, t) = (p, q) \cdot (f', t')$ for all $(p, q) \in P \times Q$, then $q \cdot t = q \cdot t'$ for all $q \in Q$ and thus $t = t'$ since T acts faithfully on Q. On the other hand, $p + q \cdot f = p + q \cdot f'$ for all $p \in P$ and thus $q \cdot f = q \cdot f'$ since S acts faithfully on P. Thus $f = f'$, proving the claim. In particular W can be considered as a subset of the semigroup of all transformations on $P \times Q$. We leave it to the reader to verify that W is closed under composition and that the product on W is defined by

$$(f, t)(f', t') = (g, tt')$$

where g is defined, for each $q \in Q$ by

$$q \cdot g = q \cdot f + (q \cdot t) \cdot f'$$

Let us now verify that Formula (4.1) really defines an action of W on $P \times Q$. If $(p, q), (p', q') \in P \times Q$ and $(f, t), (f', t') \in W$, we have

$$\big((p, q) \cdot (f, t)\big) \cdot (f', t') = (p + q \cdot f, q \cdot t) \cdot (f', t') = (p + q \cdot f + (q \cdot t) \cdot f', q \cdot tt') \\ = (p, q)\big((f, t)(f', t')\big)$$

Given two semigroups S and T, consider the wreath product $(S^1, S) \circ (T^1, T) = (S^1 \times T^1, W)$. The semigroup W is called the *wreath product* of S and T and is denoted $S \circ T$. The connections with the semidirect product and the direct product are given in the next propositions.

Proposition 4.1. *Let S and T be semigroups. Then every semidirect product of S and T is a subsemigroup of $S \circ T$. Furthermore, $S \circ T$ is a semidirect product of S^{T^1} and T.*

Proof. Let $S * T$ be a semidirect product of S and T. Let $\varphi\colon S * T \to S \circ T$ be the function defined by $\varphi(s, t) = (f, t)$ where $f : T^1 \to S$ is given by $t \cdot f = t \cdot s$ for every $t \in T^1$. It is easy to verify that φ is a semigroup morphism.

For the second part of the statement, define an action $(t, f) \mapsto t \cdot f$ of S^{T^1} on T by setting $t' \cdot (t \cdot f) = (t't) \cdot f$. Then the semidirect product defined by this action is isomorphic to $S \circ T$. □

Proposition 4.2. *Let X and Y be transformation semigroups. Then $X \times Y$ divides $X \circ Y$.*

Proof. Let $X = (P, S)$ and $Y = (Q, T)$. Since the transformation semigroups $X \times Y$ and $X \circ Y$ have the same set of states, $P \times Q$, it suffices to show that $S \times T$ can be embedded into $S^Q \times T$. With each pair (s, t), associate the pair (f, t), where f is the constant map onto s. Then, for every pair $(p, q) \in P \times Q$, $(p, q) \cdot (s, t) = (p + s, q \cdot t) = (p + q \cdot f, q \cdot t) = (p, q) \cdot (f, t)$, which concludes the proof. □

A routine computation shows that the wreath product on transformation semigroups is associative. The wreath product also preserves division.

Proposition 4.3. *If (P_1, S_1) divides (Q_1, T_1) and (P_2, S_2) divides (Q_2, T_2), then $(P_1, S_1) \circ (P_2, S_2)$ divides $(Q_1, T_1) \circ (Q_2, T_2)$.*

Proof. Let $\pi_1 : Q_1 \to P_1$ and $\pi_2 : Q_2 \to P_2$ be the surjective mappings defining the divisions. Let $\pi = \pi_1 \times \pi_2 : Q_1 \times Q_2 \to P_1 \times P_2$. For $(f, s_2) \in (P_1, S_1) \circ (P_2, S_2)$, define $\widehat{(f, s_2)} = (g, \hat{s}_2)$ by choosing a cover $\hat{s}_2$ of s_2 and, for each $q_2 \in Q_2$, a cover $g(q_2)$ of $f(\pi_2(q_2))$. Now, for each $(q_1, q_2) \in Q_1 \times Q_2$,

$$\begin{aligned}\pi(q_1, q_2) \cdot (f, s_2) &= (\pi_1(q_1), \pi_2(q_2)) \cdot (f, s_2) = (\pi_1(q_1) \cdot f(\pi_2(q_2)), \pi_2(q_2) \cdot s_2) \\ &= (\pi_1(q_1 \cdot g(q_2)), \pi_2(q_2 \cdot \hat{s}_2)) = \pi(q_1 \cdot g(q_2), q_2 \cdot \hat{s}_2) \\ &= \pi((q_1, q_2) \cdot (g, \hat{s}_2))\end{aligned}$$

and this computation concludes the proof. □

In view of Proposition 4.3, we have the following corollary.

Corollary 4.4. *If S_1 divides T_1 and S_2 divides T_2, then $S_1 \circ S_2$ divides $T_1 \circ T_2$.*

If $X = (P, S)$ is a transformation semigroup, then X^1 denotes the transformation semigroup obtained by adjoining to S the identity map 1_P on P. If p is a state, we denote by c_p the constant map defined, for all $q \in P$, by $c_p(q) = p$. The transformation semigroup obtained by adjoining to S all the constant maps c_p is denoted by $\bar{X}$.

Proposition 4.5. *Let X and Y be transformation semigroups. Then $(X \circ Y)^1$ divides $X^1 \circ Y^1$ and $\overline{X \circ Y}$ divides $\bar{X} \circ \bar{Y}$.*

Proof. Let $X = (P, S)$ and $Y = (Q, T)$. First note that the four transformation semigroups $\overline{X \circ Y}$, $\bar{X} \circ \bar{Y}$, $(X \circ Y)^1$ and $X^1 \circ Y^1$ have the same set of states, $P \times Q$. Next, $1_{P \times Q}$ has the same action as $(f, 1_Q) \in (S^1)^Q \times T$, where $f(q) = 1_P$ for all $q \in Q$. Thus $(X \circ Y)^1$ embeds into $X^1 \circ Y^1$.

Finally, if $(p, q) \in P \times Q$, the constant map $c_{(p,q)}$ has exactly the same action as the pair $(g, c_q) \in \bar{S}^Q \times \bar{T}$ where $g(x) = c_p$ for all $x \in Q$. Thus $\overline{X \circ Y}$ embeds into $\bar{X} \circ \bar{Y}$. □

4.2 Basic decomposition results. In this section, we give some useful decomposition results. Let us start with commutative monoids.

Proposition 4.6. *Every commutative monoid divides the product of its cyclic submonoids.*

Proof. Let M be a commutative monoid and let N be the product of its cyclic submonoids. Let $\varphi : N \to M$ be the morphism which transform each element on N into the product of its coordinates. Then φ is clearly surjective and thus M is a quotient of N. □

We now study decompositions involving $\tilde{U}_n$ and U_n.

Proposition 4.7. *For every $n > 0$, U_n divides U_2^n and $\tilde{U}_n$ divides $\tilde{U}_2^n$.*

Proof. Arguing by induction on n, it suffices to verify that U_n divides $U_{n-1} \times U_2$. But a simple computation shows that U_n is isomorphic to the submonoid N of $U_{n-1} \times U_2$ defined as follows:

$$N = \{(1, 1)\} \cup \{(a_i, a_1) \mid 0 \leq i \leq n-1\} \cup \{(a_1, a_2)\}$$

A dual proof works for $\tilde{U}_n$. □

A more precise result follows from Proposition B.3.7: a monoid is idempotent and $\mathcal{R}$-trivial if and only if it divides $\tilde{U}_2^n$ for some $n > 0$. Dually, a monoid is idempotent and $\mathcal{L}$-trivial if and only if it divides U_2^n for some $n > 0$.

Proposition 4.8. *For every $n > 0$, $\tilde{U}_n$ divides $U_n \circ U_2$.*

Proof. Let $\pi : U_n \times U_2 \to \tilde{U}_n$ be the surjective partial map defined by $\pi(1, a_1) = 1$ and, for $1 \leq i \leq n$, $\pi(a_i, a_2) = a_i$.

For $1 \leq j \leq n$, we set $\hat{a}_j = (f_j, a_2)$ where $f_j : U_2 \to U_n$ is defined by $1 \cdot f_j = a_2 \cdot f_j = 1$ and $a_1 \cdot f_j = a_j$. We also set $\hat{1} = (f, 1)$ where $f : U_2 \to U_n$ is defined by $1 \cdot f = a_1 \cdot f = a_2 \cdot f = 1$. Now a simple verification shows that π is indeed a cover:

$$\begin{aligned}
\pi(a_i, a_2) \cdot 1 &= \pi(a_i, a_2) = \pi(a_i + a_2 \cdot f, a_2 \cdot 1) = \pi((a_i, a_2) \cdot \hat{1}) \\
\pi(1, a_1) \cdot 1 &= \pi(1, a_1) = \pi(1 + a_1 \cdot f, a_1 \cdot 1) = \pi((1, a_1)(f, 1)) = \pi((1, a_1) \cdot \hat{1}) \\
\pi(a_i, a_2) \cdot a_j &= a_i = \pi(a_i, a_2) = \pi(a_i + a_2 \cdot f_j, a_2 \cdot a_2) = \pi((a_i, a_2) \cdot \hat{a}_j) \\
\pi(1, a_1) \cdot a_j &= a_j = \pi(a_j, a_2) = \pi(1 + a_1 \cdot f_j, a_1 \cdot a_2) = \pi((1, a_1) \cdot \hat{a}_j)
\end{aligned}$$

Thus $\tilde{U}_n$ divides $U_n \circ U_2$. □

It follows now immediately from Propositions 4.8 and 4.7:

Corollary 4.9. *For every $n > 0$, $\tilde{U}_n$ divides $\underbrace{U_2 \circ \cdots \circ U_2}_{n+1 \text{ times}}$.*

For each $n > 0$, let $\mathbf{D_n}$ be the class of finite semigroups S such that, for all $s_0, s_1, \ldots, s_n$ in S, $s_0 s_1 \cdots s_n = s_1 \cdots s_n$. In such a semigroup, a product of more than n elements is determined by the last n elements. More details on the structure of these semigroups is given in Exercise V.1. We shall now give a decomposition result for the semigroups in $\mathbf{D_n}$. As a first step, we decompose $\bar{\mathbf{n}}$ as a direct product of copies of $\bar{\mathbf{2}}$.

Lemma 4.10. *If $2^k > n$, then $\bar{\mathbf{n}}$ divides $\bar{\mathbf{2}}^k$.*

Proof. The result is trivial, since if T is any subset of size n of $\bar{\mathbf{2}}^k$, (T, T) is a sub-transformation semigroup of $\bar{\mathbf{2}}^k$ isomorphic to $\bar{\mathbf{n}}$. □

We now decompose the semigroups of $\mathbf{D_n}$ as an iterated wreath product of transformation semigroups of the form (T, T).

Proposition 4.11. *Let S be a semigroup of $\mathbf{D_n}$ and let $T = S \cup \{t\}$, where t is a new element. Then (S^1, S) divides $\underbrace{(T, T) \circ \cdots \circ (T, T)}_{n \text{ times}}$.*

Proof. Let $\varphi : T^n \to S^1$ be the partial function defined on sequences of the form $(t, \ldots, t, x_i, \ldots, x_1)$, where $x_1, \ldots, x_i \in S$, by

$$\varphi(t, \ldots, t, x_i, \ldots, x_1) = \begin{cases} x_i \cdots x_1 & \text{if } i > 0 \\ 1 & \text{if } i = 0 \end{cases}$$

Clearly φ is surjective. If $s \in S$, we set $\hat{s} = (f_{n-1}, \ldots, f_1, s)$, where, for $1 \leq i \leq n-1$, $f_i : T^i \to T$ is defined by $(t_i, \ldots, t_1) \cdot f_i = t_i$. Thus

$$(t_n, \ldots, t_1)\hat{s} = (t_{n-1}, \ldots, t_1, s)$$

It follows that if $p = (t, \ldots, t, x_i, \ldots, x_1)$ is in the domain of φ, then $p \cdot \hat{s}$ is also in the domain of φ and $\varphi(p \cdot \hat{s}) = \varphi(p) \cdot s$. This proves the proposition. □

Proposition 4.11 and Lemma 4.10 now give immediately.

Corollary 4.12. *Every semigroup of $\mathbf{D_n}$ divides a wreath product of copies of $\bar{\mathbf{2}}$.*

The $\mathcal{R}$-trivial monoids admit also a simple decomposition.

Theorem 4.13. *A monoid is $\mathcal{R}$-trivial if and only if it divides a wreath product of the form $U_1 \circ \cdots \circ U_1$.*

Proof. We first show that every monoid of the form $U_1 \circ \cdots \circ U_1$ is $\mathcal{R}$-trivial. Since U_1 itself is $\mathcal{R}$-trivial, and since, by Proposition 4.1, a wreath product is a special case of semidirect product, it suffices to show that the semidirect product $S * T$ of two $\mathcal{R}$-trivial monoids S and T is again $\mathcal{R}$-trivial. Indeed, consider two $\mathcal{R}$ equivalent elements (s, t) and (s', t') of $S * T$. Then, $(s, t)(x, y) = (s', t')$ and $(s', t')(x, y) = (s, t)$ for some elements (x, y) and (x', y') of $S * T$. Therefore, on one hand $s + tx = s'$ and $s' + t'x = s$ and on the other hand, $ty = t'$ and $t'y' = t$. It follows that $s \mathrel{\mathcal{R}} s'$ and $t \mathrel{\mathcal{R}} t'$. Therefore $s = s'$ and $t = t'$, and $S * T$ is $\mathcal{R}$-trivial.

Let $M = \{s_1, \ldots, s_n\}$ be an $\mathcal{R}$-trivial monoid of size n. We may assume that $s_i \leq_{\mathcal{R}} s_j$ implies $j \leq i$. Let us identify the elements of U_1^n with words of length n on the alphabet $\{0, 1\}$. Let $\varphi : U_1 \times \cdots \times U_1 \to M$ be the onto partial function defined by

$$\varphi(1^{n-j}0^j) = s_j \quad (0 \leq j \leq n)$$

Thus $\varphi(u)$ is not defined if $u \notin 1^*0^*$. For each $s \in M$, let

$$\hat{s} = (f_{n-1}, \ldots, f_2, a_1)$$

with

$$a_1 = \begin{cases} 1 & \text{if } s = 1 \\ 0 & \text{if } s \neq 1 \end{cases}$$

where $f_{i+1} : \underbrace{U_1 \circ \cdots \circ U_1}_{i \text{ times}} \to U_1$ is defined by

$$f_{i+1}(1^{i-j}0^j) = \begin{cases} 1 & \text{if } s_j s = s_k \text{ and } k \leq i \\ 0 & \text{if } s_j s = s_k \text{ and } k > i \end{cases}$$

If $u \notin 1^*0^*$, the value of $f_{i+1}(u)$ can be chosen arbitrarily.

Let $p = 1^{n-j}0^j$ and $s \in M$. Let k be such that $s_k = s_j s$. Since $s_k \leq_{\mathcal{R}} s_j$, $k \geq j$. Then

$$\begin{aligned} p\hat{s} &= (f_{n-1}, \ldots, f_2, a_1)(1^{n-j}0^j) \\ &= 1^{n-k}0^k \end{aligned}$$

whence $\varphi(p\hat{s}) = s_k = s_j s = \varphi(p)s$. Therefore, M divides $U_1 \circ \cdots \circ U_1$. □

As a preparation to the next theorem, we prove another decomposition result, which is important in its own right.

Proposition 4.14. *Let M be a finite aperiodic monoid and let $\pi : A^* \to M$ be a surjective morphism. Then one of the following cases occur:*

(1) *M is a cyclic monoid,*

(2) *M is isomorphic to $\tilde{U}_n$ for some $n > 0$,*

(3) *there is a proper partition $A = B \cup C$ such that $\pi((B^*C)^*)$ and $\pi(B^*)$ are proper submonoids of M.*

Proof. Let $S = M \setminus \{1\}$. Since M is aperiodic, S is a subsemigroup of M. Let L be an $\mathcal{L}$-class of S, maximal for the order $\leq_{\mathcal{J}}$. First assume that S is the semigroup generated by L. If $|L| = 1$, S is cyclic and so is M. Otherwise, $|L| > 1$ and by Proposition 2.8,

L is regular and thus consists of $\mathcal{L}$-equivalent idempotents. Now, Proposition 2.2 shows that if e and f are $\mathcal{L}$-equivalent idempotents, then $ef = e$ and $fe = f$. It follows that L is a semigroup and hence $S = L$. In particular, we are in the second case.

Now assume that the semigroup generated by L is strictly contained in S and put

$$B = \{a \in A \mid \pi(a) \in L\} \quad \text{and} \quad C = A \setminus B$$

As $\pi(A)$ generates S and $\pi(B)$ does not, B is a strict subset of A, so that C is nonempty. We claim that $\pi(B^+)$ and $\pi(B^*C)^+$ are proper subsemigroups of S. For the first part of the claim, we observe that $\pi(B)$ is contained in L and thus $\pi(B^+)$ is contained in the subsemigroup of S generated by L, which is a proper subsemigroup of S. For the second part, we observe that every element of $\pi(B^*C)^+$ is $<_{\mathcal{J}}$-below L and thus $\pi(B^*C)^+$ is contained in $S \setminus L$. It follows now from the claim that we are in the third case of the proposition. □

Proposition 4.15. *Let M be a monoid. Suppose that $M = L \cup N$ where L is a left ideal and N is a submonoid of M. Then M divides $L^1 \circ \bar{N}$.*

Proof. Let $\varphi : L^1 \times N \to M$ be the map defined by $\varphi(l, n) = ln$. Since $M = L \cup N$ and $L \cup N \subseteq L^1N$, $M = L^1N$, and φ is onto.

Let $m \in M$. If $m \in L$, we set $\hat{m} = (g, c_1)$, where $g : N \to L^1$ is defined by $g(n) = nm$ for all $n \in N$. Otherwise, if $m \notin L$, we set $\hat{m} = (f, m)$, where $f(n) = 1$ for all $n \in N$.

Let $(l, n) \in L^1 \times N$. Then

$$\varphi(l, n) \cdot m = (ln) \cdot m = lnm$$

Now, if $m \in L$,

$$(l, n) \cdot \hat{m} = (l, n)(g, c_1) = (l \cdot g(n), 1) = (lnm, 1)$$

and since L is a left ideal, $lnm \in L$. On the other hand, if $m \in N$,

$$(l, n) \cdot \hat{m} = (l, n)(f, m) = (l \cdot f(n), nm) = (l, nm)$$

and since N is a monoid, $nm \in N$. In both cases, $\varphi((l, n) \cdot \hat{m}) = lnm$. It follows that φ is a covering and thus M divides $L^1 \circ \bar{N}$. □

Theorem 4.16. *A monoid is aperiodic if and only if it divides a wreath product of the form $U_2 \circ \cdots \circ U_2$.*

Proof. Let M be an aperiodic monoid. Consider the three cases given by Proposition 4.14. If M is cyclic, then it is $\mathcal{R}$-trivial, and the result follows from Theorem 4.13.

If M is isomorphic to $\tilde{U}_n$ for some $n > 0$, the result follows from Corollary 4.9. Finally, suppose there is a proper partition $A = B \cup C$ such that $L = \pi((B^*C)^*)$ and $M = \pi(B^*)$ are proper submonoids of M. Then L is a left ideal of M, and since $A^* = (B^*C)^* \cup B^*$, $M = L \cup N$. Thus by Proposition 4.15, M divides $L^1 \circ \bar{N}$. Arguing by induction on $|M|$, we may assume that L and N divide wreath products of copies of U_2. It follows, by Proposition 4.5, that L^1 and $\bar{N}$ also divide wreath products of copies of U_2, since $U_2 = U_2^1 = \bar{U}_2$. Finally, M itself divides a wreath product of copies of U_2. □

Proposition 4.17. *Let $X = (P, S)$ be a transformation semigroup such that $P \cdot S = P$. Then $\bar{\mathbf{2}} \circ X$ divides $X \circ (R, R)$, where R is the set $\{1, 2\}^P \times S$.*

Proof. Define $\varphi : P \times R \to \{1, 2\} \times P$ by setting $\varphi(p, f, s) = (p \cdot f, p \cdot s)$ for each $p \in P$, $f \in \{1, 2\}^P$ and $s \in S$. Given a transformation $v = (g, t)$ of $\bar{\mathbf{2}} \circ X$, with $g \in \{1, 2\}^P$ and $t \in S$, define the transformation $\hat{v}$ of $X \circ (R, R)$ by setting

$$(p, f, s) \cdot \hat{v} = (p \cdot s, g, t)$$

then we have

$$\begin{aligned}\varphi(p, f, s) \cdot v = (p \cdot f, p \cdot s)(g, t) &= (p \cdot f + (p \cdot s) \cdot g, p \cdot st)\\ ((p \cdot s) \cdot g, p \cdot st) &= \varphi(p \cdot s, g, t) = \varphi((p, f, s) \cdot \hat{v})\end{aligned}$$

Thus $\bar{\mathbf{2}} \circ X$ divides $X \circ (R, R)$. □

Given a property $\mathcal{P}$, we say that a semigroup S is *locally* $\mathcal{P}$ if the local semigroup of each idempotent satisfies $\mathcal{P}$. For instance, a semigroup S is *locally trivial* if, for each $s \in S$ and $e \in E(S)$, $ese = e$.

We shall admit without proof our last decomposition result (see the Notes section).

Proposition 4.18. *A semigroup is locally $\mathcal{R}$-trivial if and only if it divides a wreath product of the form $U_1 \circ \cdots \circ U_1 \circ \bar{\mathbf{2}} \circ \cdots \circ \bar{\mathbf{2}}$.*

5 The wreath product principle

Wreath products are especially useful in the study of operations on sets of finite words. We first introduce a standard tool of the theory of automata, the sequential transducers.

5.1 Sequential functions. A *sequential transducer* $\mathcal{T} = (Q, A, R, q_0, \cdot, *)$ is given by a finite set of *states* Q, a finite *input alphabet* A, an *output semigroup* R (possibly infinite), an initial state q_0, a transition function $Q \times A \to Q$, denoted $(q, a) \mapsto q \cdot a$, an output function $Q \times A \to R$, denoted $(q, a) \mapsto q * a$.

The transition and the output functions are extended to functions $Q \times A^* \to Q$ (resp. $Q \times A^* \to R$) by setting, for each $u \in A^*$ and each $a \in A$:

$$q \cdot 1 = q \qquad q * 1 = 1$$
$$q \cdot (ua) = (q \cdot u) \cdot a \qquad q * (ua) = (q * u)((q \cdot u) * a)$$

To make this type of formulas simpler, it is convenient to fix some priority rules on the operators. Our choice is to give highest priority to concatenation, then to dot and then to star. For instance, we write $q \cdot ua$ for $q \cdot (ua)$, $q * ua$ for $q * (ua)$ and $q \cdot u * a$ for $(q \cdot u) * a$.

The function $\sigma : A^* \to R$ defined by $\sigma(u) = (q_0 * u)$ is called the sequential function *realized* by the $\mathcal{T}$. A *sequential function* is a function that can be realized by a sequential transducer.

The transformation semigroup of a sequential transducer $\mathcal{T}$ is the transformation semigroup of the underlying automaton $(Q, A, \cdot, q_0, \emptyset)$.

The key result of this section is a standard result on sequential functions. There are actually two versions of this result, one for A^+ and the other one for A^*. We just state the first version, but we shall use the two versions in the next proposition.

Theorem 5.1. *Let $\sigma : A^+ \to R$ be a sequential function realized by a sequential transducer $\mathcal{T}$, and let (Q, T) be the transformation semigroup of $\mathcal{T}$. If L is a subset of R recognized by a transformation semigroup (P, S), then $\sigma^{-1}(L)$ is recognized by $(P, S) \circ (Q, T)$.*

Proof. Let $\mathcal{T} = (Q, A, R, q_0, \cdot, *)$. Since L is recognized by (P, S), there exist a surjective morphism $\varphi \colon R \to S$, a state $p_0 \in P$ and a subset F of P such that

$$L = \{u \in R \mid p_0 \cdot \varphi(u) \in F\}$$

Let $(P, S) \circ (Q, T) = (P \times Q, W)$ and define a morphism $\psi \colon A^+ \to W$ by setting

$$(p, q) \cdot \psi(u) = (p \cdot \varphi(q * u), q \cdot u)$$

By hypothesis, the map $q \mapsto \varphi(q * u)$ is a function from Q into S and thus ψ is well defined. Setting $I = F \times Q$, we have

$$\begin{aligned}
\sigma^{-1}(L) &= \{u \in A^+ \mid \sigma(u) \in L\} \\
&= \{u \in A^+ \mid p_0 \cdot \varphi(\sigma(u)) \in F\} \\
&= \{u \in A^+ \mid p_0 \cdot \varphi(q_0 * u) \in F\} \\
&= \{u \in A^+ \mid (p_0, q_0) \cdot \psi(u) \in I\}
\end{aligned}$$

Therefore, $\sigma^{-1}(L)$ is recognized by $(P \times Q, W)$. $\square$

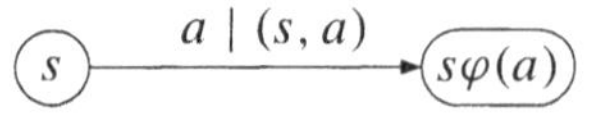

Figure 5.1. The sequential transducer $\mathcal{T}_\varphi$.

5.2 Sets recognized by wreath products. The aim of this section is to characterize the sets recognized by the wreath product of two transformation semigroups. This study motivates the introduction of an auxiliary tool, the sequential transducer of a morphism.

Let T be a semigroup and let $\varphi\colon A^+ \to T$ be a surjective morphism. Set $B_T = T^1 \times A$. Let $\mathcal{T}_\varphi = (T^1, A, B_T^+, \cdot, *, 1)$ be the sequential transducer defined, for each $s \in T^1$ and $a \in A$, by $s \cdot a = s\varphi(a)$ and $s * a = (s, a)$. Note that the transformation semigroup of $\mathcal{T}_\varphi$ is (T^1, T). The sequential function $\sigma_\varphi\colon A^+ \to B_T^+$ defined by $\mathcal{T}_\varphi$ is called the *sequential function associated with* φ. One easily verifies that

$$\sigma_\varphi(a_1 a_2 \cdots a_n) = (1, a_1)(\varphi(a_1), a_2) \cdots (\varphi(a_1 \cdots a_{n-1}), a_n)$$

As a first illustration, we study the operators $L \to LaA^*$ and $L \to La$, where a is a letter.

Proposition 5.2. *Let L be a recognizable subset of A^* and let $a \in A$. Then LaA^* is recognized by $U_1 \circ M(L)$.*

Proof. Let $\varphi : A^+ \to M(L)$ be the syntactic morphism of L. Let $B = B_T$ and let $\sigma_\varphi : A^* \to B^*$ be the sequential function associated with φ. Note that $M(L)$ is the transformation semigroup of the sequential transducer realizing σ_φ. Let $P = \varphi(L)$ and $C = P \times \{a\}$. Then C is a subset of B and we have

$$\begin{aligned}
\sigma_\varphi^{-1}(B^*CB^*) &= \{u \in A^+ \mid \sigma_\varphi(u) \in B^*CB^*\} \\
&= \{a_1 \cdots a_n \in A^+ \mid \exists i\ (\varphi(a_1 \cdots a_{i-1}), a_i) \in C\} \\
&= \{a_1 \cdots a_n \in A^+ \mid \exists i\ a_1 \cdots a_{i-1} \in \varphi^{-1}(P) \text{ and } a_i = a\} \\
&= \varphi^{-1}(P)aA^* = LaA^*.
\end{aligned}$$

It was shown in Proposition 3.4 that B^*CB^* is recognized by U_1. Therefore, by Theorem 5.1, LaA^* is recognized by $U_1 \circ M(L)$.

Proposition 5.3. *Let L be a recognizable subset of A^+ and let $a \in A$. Then La is recognized by $\bar{\mathbf{2}} \circ S(L)$.*

Proof. Let $\varphi : A^+ \to S(L)$ be the syntactic morphism of L. Let $B = B_T$ and let $\sigma_\varphi : A^+ \to B^+$ be the sequential function associated with φ. Note that $S(L)$ is the transformation semigroup of the sequential transducer realizing σ_φ. Let $P = \varphi(L)$ and $C = P \times \{a\}$. Then C is a subset of B and we have

$$\begin{aligned}
\sigma_\varphi^{-1}(B^*C) &= \{u \in A^+ \mid \sigma_\varphi(u) \in B^*C\} \\
&= \{a_1 \cdots a_n \in A^+ \mid (\varphi(a_1 \cdots a_{n-1}), a_n) \in C\} \\
&= \{a_1 \cdots a_n \in A^+ \mid a_1 \cdots a_{n-1} \in \varphi^{-1}(P) \text{ and } a_n = a\} \\
&= \varphi^{-1}(P)a = La.
\end{aligned}$$

It was shown in Proposition 3.2 that B^*C is recognized by $\bar{\mathbf{2}}$. Therefore, by Theorem 5.1, La is recognized by $\bar{\mathbf{2}} \circ S(L)$. □

Let $X = (P, S)$ and $Y = (Q, T)$ be two transformation semigroups, let $Z = X \circ Y = (P \times Q, W)$, and let L be a subset of A^+ recognized by Z. Then there exist a state (p_0, q_0), a subset F of $P \times Q$ and a morphism of semigroups $\eta\colon A^+ \to W$ such that $L = \{u \in A^+ \mid (p_0, q_0)\cdot u \in F\}$. Denote by π the natural projection from W onto T, defined by $\pi(f, t) = t$ and let $\varphi = \pi \circ \eta\colon A^+ \to T$.

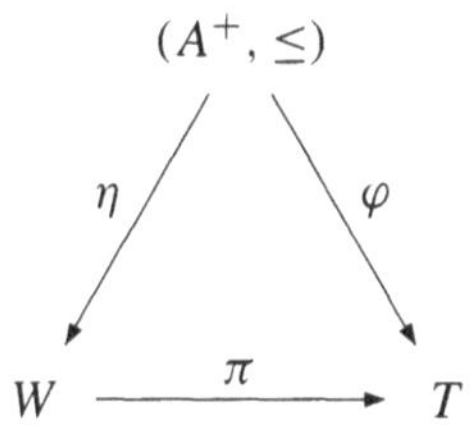

Let $B_Q = Q \times A$ and let $\sigma\colon A^+ \to B_Q^+$ be the function by

$$\sigma(a_1a_2 \cdots a_n) = (q_0, a_1)(q_0\cdot \varphi(a_1), a_2) \cdots ((q_0\cdot \varphi(a_1 \cdots a_{n-1})), a_n)$$

We also define, for each $q \in Q$, a function $\lambda_q\colon B_T^+ \to B_Q^+$ by

$$\lambda_q(t, a) = (q \cdot t, a)$$

Then λ_{q_0} serves as a bridge between σ and σ_φ, since $\sigma = \lambda_{q_0} \circ \sigma_\varphi$, and we have the following diagram:

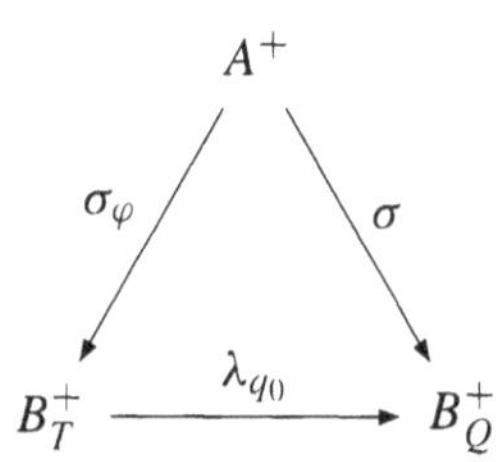

Note also that σ is a sequential function, realized by the transducer $\mathcal{T}_\sigma = (Q, A, B_Q^+, q_0, \cdot, *)$ where $q \cdot a = q \cdot \varphi(a)$ and $q * a = (q, a)$.

Theorem 5.4 (The wreath product principle). *The set L is a finite union of sets of the form $U \cap \sigma_\varphi^{-1}(V)$, where $U \subseteq A^+$ is recognized by φ and $V \subseteq B_T^+$ is recognized by X.*

Proof. Observing that

$$L = \bigcup_{(p,q) \in F} \{u \in A^+ \mid (p_0, q_0) \cdot u = (p, q)\}$$

we may suppose that $F = \{(p, q)$ for some $(p, q) \in P \times Q\}$. For each letter a, set $\eta(a) = (f_a, t_a)$. Note that $\varphi(a) = t_a$. Define a function $\alpha \colon B_Q \to S$ by setting $\alpha(q, a) = q \cdot f_a$ and extend it into a morphism of semigroups $\alpha \colon B_Q^+ \to S$. Setting $\gamma = \lambda_{q_0} \circ \alpha$, we obtain the following commutative diagram:

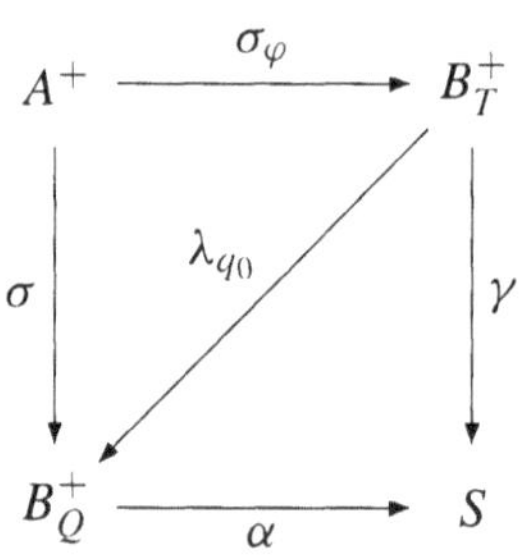

Let $u = a_1 a_2 \cdots a_n$ be a word. Then

$$\begin{aligned}
(p_0, q_0) \cdot u &= (p_0, q_0) \cdot (f_{a_1}, t_{a_1})(f_{a_2}, t_{a_2}) \cdots (f_{a_n}, t_{a_n}) \\
&= \big(p_0 + q_0 \cdot f_{a_1} + \cdots + (q_0 \cdot t_{a_1} \cdots t_{a_{n-1}}) f_{a_n}, q_0 \cdot t_{a_1} \cdots t_{a_n}\big) \\
&= \big(p_0 + \alpha(q_0, a_1) + \cdots + \alpha(q_0 \cdot \varphi(a_1 \cdots a_{n-1}), a_n), q_0 \cdot \varphi(u)\big) \\
&= \big(p_0 + \alpha(\sigma(u)), q_0 \cdot \varphi(u)\big) \\
&= \big(p_0 + \gamma(\sigma_\varphi(u)), q_0 \cdot \varphi(u)\big)
\end{aligned}$$

It follows that $(p_0, q_0) \cdot u \in F$ if and only if the following two conditions are satisfied:

(1) $p_0 + \gamma(\sigma_\varphi(u)) = p$,

(2) $q_0 \cdot \varphi(u) = q$.

Setting $U = \{u \in A^+ \mid q_0 \cdot \varphi(u) = q\}$ and $V = \{v \in B^+ \mid p_0 + \gamma(v) = p\}$, condition (1) can be reformulated as $u \in \sigma_\varphi^{-1}(V)$,and condition (2) as $u \in U$. Thus

$$L = U \cap \sigma_\varphi^{-1}(V)$$

Now, U is recognized by φ and V is recognized by X, which concludes the proof. □

6 Notes

The notions presented in this Annex can be found in most textbooks on semigroups and automata such as Lallement [167], Eilenberg [98] or Pin [241]. Theorem 4.16 is a particular case of the Krohn-Rhodes decomposition theorem. A proof of Proposition 4.18 can be found in [98]. The wreath product principle (Section 5) is due to Straubing [309].

ANNEX B

VARIETIES OF FINITE SEMIGROUPS

This second Annex presents results on the theory of varieties of semigroups with applications to the classification of recognizable languages, mainly used in Chapters VI, VII and VIII.

1 Varieties of algebras

In this section, we establish some general properties of algebras. Here the word *algebra* is taken in a broad sense to cover the algebraic structures of interest to us, namely semigroups, monoids, ω-semigroups and their ordered counterparts. Although algebras could be formally defined, we have preferred to cover in some details the case of ordered semigroups, which is sufficiently generic for our purpose. The other cases can then be readily adapted from this case. As usual, when no order is specified on a semigroup, the equality order is understood.

1.1 Birkhoff varieties. A *Birkhoff variety* of ordered semigroups is a class of ordered semigroups $\mathbf{V}$ such that:

(1) if $S \in \mathbf{V}$ and if T is an ordered subsemigroup of S, then $S \in \mathbf{V}$,

(2) if $S \in \mathbf{V}$ and if T is a quotient of S, then $S \in \mathbf{V}$,

(3) if $(S_i)_{i \in I}$ is a family of ordered semigroups of $\mathbf{V}$, the product $\prod_{i \in I} S_i$ is also in $\mathbf{V}$.

The ordered semigroups of a Birkhoff variety can be finite or infinite.

Example 1.1.

(1) The class of all ordered semigroups forms a Birkhoff variety.

(2) The smallest Birkhoff variety is the trivial variety, consisting only of the empty semigroup and of the semigroup 1.
(3) The class of all commutative semigroups forms a Birkhoff variety.
(4) The class of all groups does not form a Birkhoff variety of monoids. Indeed $\mathbb{Z}$ is a group, but $\mathbb{N}$ is a submonoid of $\mathbb{Z}$ which is not a group.

Let $\mathbf{V}$ be a Birkhoff variety of ordered semigroups and let A be an alphabet. Consider the collection $\mathcal{F}(A, \mathbf{V})$ of all morphisms $\varphi : A^+ \to S$, where S is an ordered semigroup of $\mathbf{V}$. For each such morphism φ, denote by $\sim_\varphi$ the nuclear congruence of φ, and let $\sim$ be the intersection of all these congruences when φ ranges over $\mathcal{F}(A, \mathbf{V})$. Finally, set

$$\mathbf{F}_\mathbf{V}(A) = A^+/{\sim}$$

and let $\pi : A^+ \to \mathbf{F}_\mathbf{V}(A)$ be the natural morphism. Recall that $\mathbf{F}_\mathbf{V}(A)$ is ordered by $u \leq v$ if and only if, for each $\varphi \in \mathcal{F}(A, \mathbf{V})$, $\varphi(u) \leq \varphi(v)$. As a subsemigroup of a product of ordered semigroups of $\mathbf{V}$, this ordered semigroup belongs to $\mathbf{V}$. It is called the $\mathbf{V}$-*free semigroup* on A. This terminology is justified by the following universal property, which relativizes to $\mathbf{V}$ the universal property of A^+ given by Proposition A.1.4.

Proposition 1.1. *If φ is a function from A into an ordered semigroup S of $\mathbf{V}$, there exists a unique semigroup morphism $\bar{\varphi} : F_\mathbf{V}(A) \to S$ such that, for each $a \in A$, $\varphi(a) = \bar{\varphi}(\pi(a))$. Moreover, $\bar{\varphi}$ is surjective if and only if the set $\varphi(A)$ generates S.*

Proof. First, φ can be extended (in a unique way) into a morphism from A^+ into S. Now, since the nuclear congruence of φ is coarser than $\sim$, there exists by Proposition A.1.2 a unique morphism $\bar{\varphi} : F_\mathbf{V}(A) \to S$ such that $\varphi = \bar{\varphi} \circ \pi$.

Clearly, if $\bar{\varphi}$ is surjective, $\varphi : A^+ \to S$ is onto and thus A generates S. Conversely, if $\varphi(A)$ generates S, φ is onto and $\bar{\varphi}$ is also onto. □

Corollary 1.2. *An A-generated semigroup belongs to $\mathbf{V}$ if and only if it is a quotient of $F_\mathbf{V}(A)$.*

Proof. Let S be an A-generated semigroup of $\mathbf{V}$. By definition, there exists a surjective morphism $\varphi : A^+ \to S$. In particular, $\varphi \in \mathcal{F}(A)$ and thus $\pi(u) = \pi(v)$ implies $\varphi(u) = \varphi(v)$. By Proposition A.1.2, S is a quotient of $F_\mathbf{V}(A)$.

Conversely, assume that S is a quotient of $F_\mathbf{V}(A)$. Since $F_\mathbf{V}(A)$ belongs to $\mathbf{V}$, S is also in $\mathbf{V}$. □

Example 1.2. Let $\mathbf{V}$ be the Birkhoff variety of commutative monoids. It is easy to see that the $\mathbf{V}$-free monoid (called, in this case, the free commutative monoid), is the additive monoid $\mathbb{N}^A$, where the addition is defined componentwise: $(n_a)_{a\in A} + (n'_a)_{a\in A} = (n_a + n'_a)_{a\in A}$.

A convenient way to define varieties is to use identities. To avoid any ambiguity, we give the definitions for semigroups and for ordered semigroups. Let A be an alphabet and let $u, v \in A^+$. An ordered semigroup (resp. semigroup) S *satisfies the identity* $u \leq v$ (resp. $u = v$) if and only if, for each morphism of semigroups $\varphi : A^+ \to S$, $\varphi(u) \leq \varphi(v)$ (resp. $\varphi(u) = \varphi(v)$). For example, an ordered semigroup satisfies the identity $xyx \leq x$ if, for each $x, y \in S$, $xyx \leq x$. A semigroup is commutative if and only if it satisfies the identity $xy = yx$; it is idempotent if and only if it satisfies $x = x^2$.

A Birkhoff variety **V** satisfies a given identity if every ordered semigroup of **V** satisfies this identity. We also say in this case that the given identity is *an identity of* **V**. Identities of **V** are closely related to **V**-free semigroups.

Proposition 1.3. *Let $\pi_{\mathbf{V}}$ be the natural morphism from A^+ onto the* **V***-free semigroup on A. Given two words u and v of A^+, $u \leq v$ is an identity of* **V** *if and only if $\pi_{\mathbf{V}}(u) \leq \pi_{\mathbf{V}}(v)$.*

Proof. If $u \leq v$ is an identity of **V**, then $\pi_{\mathbf{V}}(u) \leq \pi_{\mathbf{V}}(v)$ since $F_{\mathbf{V}}(A) \in \mathbf{V}$. Conversely if $\pi_{\mathbf{V}}(u) \leq \pi_{\mathbf{V}}(v)$, then by definition $\varphi(u) \leq \varphi(v)$ for every $\varphi \in \mathcal{F}(A, \mathbf{V})$, and thus $u \leq v$ is an identity of **V**. □

Corollary 1.4. *Let* **V** *and* **W** *be two varieties satisfying the same identities on the alphabet A. Then the* **V***-free semigroup and the* **W***-free semigroup on A are isomorphic.*

Proof. Indeed, by Proposition 1.3, the nuclear congruences of $\pi_{\mathbf{V}}$ and $\pi_{\mathbf{W}}$ are the same. Therefore, the semigroups $\pi_{\mathbf{V}}(A^+)$ and $\pi_{\mathbf{W}}(A^+)$ are equal. □

Let E be a set of identities. The class $[\![E]\!]$ of ordered semigroups is defined to be the class of ordered semigroups satisfying the identities of E. The next result explains the importance of identities.

Theorem 1.5. *A class of ordered semigroups is a Birkhoff variety if and only if it can be defined by a set of identities.*

Proof. We first prove that every class defined by a set of identities is a Birkhoff variety. Since Birkhoff varieties are closed under intersection, it suffices to establish the result when the class is defined by a single identity, say $u \leq v$. Let S be an ordered semigroup satisfying this identity. Then clearly, every subsemigroup of S satisfies the same identity. Furthermore, if $\pi : S \to T$ is an onto morphism, T also satisfies $u \leq v$. Indeed, if $\varphi : A^+ \to T$ is a morphism, there exists by Corollary 4.8 a morphism $\psi : A^+ \to S$ such that $\varphi = \pi \circ \psi$. Now, $\psi(u) \leq \psi(v)$ since S satisfies the identity $u \leq v$ and thus $\pi(\psi(u)) \leq \pi(\psi(v))$. Therefore, T satisfies the identity $u \leq v$. Finally, if $(S_i)_{i \in I}$ is a family of semigroups satisfying the identity $u \leq v$, their product $S = \prod_{i \in I} S_i$ also satisfies this identity. Indeed, let π_i denotes the projection from S onto S_i

and let φ be a morphism from A^+ into S. Since $\pi_i \circ \varphi$ is a morphism from A^+ into S_i, $\pi_i \circ \varphi(u) \leq \pi_i \circ \varphi(v)$. As this holds for each i, we have $\varphi(u) \leq \varphi(v)$.

Let now **V** be a Birkhoff variety. Let E be the class of all identities satisfied by **V** and let $\mathbf{W} = [\![E]\!]$. Clearly $\mathbf{V} \subseteq \mathbf{W}$. Let $S \in \mathbf{W}$ and let Σ be a generating set of S. Then there exists an onto morphism $\varphi : \Sigma^+ \to S$. Let $\pi : \hat{\Sigma}^+ \to F_{\mathbf{V}}(\Sigma)$ be the natural morphism. Let $u, v \in \Sigma^+$. By Proposition 1.3, if $\pi(u) \leq \pi(v)$ then $u \leq v$ is an identity of **V** and thus, is satisfied by S. In particular, $\varphi(u) \leq \varphi(v)$. It follows by Proposition A.1.3 that φ factors through π. Therefore S is a quotient of $F_{\mathbf{V}}(\Sigma)$ and by Corollary 1.2, S is in **V**. Thus $\mathbf{V} = [\![E]\!]$. □

1.2 Varieties of finite semigroups. The semigroups occurring in automata theory are mostly finite, which motivates the following definition.

A *variety of finite ordered semigroups* is a class of finite ordered semigroups **V** such that:

(1) if $S \in \mathbf{V}$ and if T is a subsemigroup of S, then $T \in \mathbf{V}$,

(2) if $S \in \mathbf{V}$ and if T is a quotient of S, then $T \in \mathbf{V}$,

(3) if $(S_i)_{i \in I}$ is a finite family of semigroups of **V**, then $\prod_{i \in I} S_i$ is also in **V**.

Condition (3) can be replaced by the conjunction of conditions (4) and (5):

(4) the trivial semigroup 1 belongs to **V**,

(5) if S_1 and S_2 are semigroups of **V**, then $S_1 \times S_2$ is also in **V**.

Indeed, condition (4) is obtained by taking $I = \emptyset$ in (3).

Varieties of semigroups are defined in the same way. Since a semigroup can be considered as an ordered semigroup, ordered by the equality relation, one can define the variety of ordered semigroups generated by a class $\mathcal{C}$ of semigroups as the variety of ordered semigroups generated by the class $\{(S, =) \mid S \in \mathcal{C}\}$. The context will make clear if we are dealing with varieties of semigroups or of ordered semigroups.

Example 1.3.

(1) The class **S** of all finite semigroups forms a variety of finite semigroups.

(2) The smallest variety of finite semigroups is the trivial variety, consisting of the empty semigroup and of the semigroup with one element 1. This variety is denoted by **1**.

(3) The class of all commutative finite semigroups forms a variety of finite semigroups, denoted by **Com**.

Let us give an important example of variety of finite monoids, which should be compared with Example 1.1 (4).

Proposition 1.6. *The class of all finite groups is a variety of finite monoids, denoted by* **G**.

Proof. It suffices to verify that every submonoid M of a group G is a group. Since M is nonempty, it contains an idempotent, necessarily equal to the identity 1 of G. Furthermore, for a given $s \in M$, the map $\rho_s : M \to M$ defined by $\rho_s(x) = xs$ is injective, and hence bijective since M is finite. In particular, there exists an element $\bar{s}$ of G such that $\bar{s}s = 1$. It follows $s\bar{s}s = s$ whence $s\bar{s} = 1$, showing that s is invertible. Therefore M is a group. □

In the sequel, we shall use the term "variety" as a shorthand for variety of finite semigroups (resp. monoids, etc.).

The *supremum* of two varieties $\mathbf{V}_1$ and $\mathbf{V}_2$ is the smallest variety containing $\mathbf{V}_1$ and $\mathbf{V}_2$. It is denoted by $\mathbf{V}_1 \vee \mathbf{V}_2$.

Let $\mathcal{C}$ be a class of ordered semigroups. The class of all ordered semigroups that divide a finite product of ordered semigroups of $\mathcal{C}$ is a variety, denoted $\langle \mathcal{C} \rangle$ and called the *variety generated* by $\mathcal{C}$. It is the smallest variety containing $\mathcal{C}$.

It is tempting to use identities to obtain a counterpart of Theorem 1.5 for varieties of finite algebras, but an incursion into topology is necessary in order to achieve this goal. This approach is very similar to the introduction of p-adic numbers in number theory and is a good illustration of the following quotation of Marshall Stone [307]: 'A cardinal principle of modern mathematical research may be stated as a maxim: "One must always topologize"'.

1.3 Profinite algebras. The details are provided for semigroups, but the results can be readily adapted to monoids, ω-semigroups and their ordered counterparts.

A *metric semigroup* is a semigroup S equipped with a metric d, such that (S, d) is a complete metric space and the multiplication of S is uniformly continuous. Morphisms between two metric semigroups are required to be uniformly continuous.

If X is a subset of S, the *metric subsemigroup* of S generated by X is by definition the smallest *closed* subsemigroup of S containing X.

Finite semigroups can be considered as metric semigroups, equipped with the discrete metric. More precisely, if S is a finite semigroup, the discrete metric d is defined by

$$d(s, t) = \begin{cases} 0 & \text{if } s = t \\ 1 & \text{otherwise} \end{cases}$$

In this section, we shall systematically consider finite semigroups as metric semigroups without any further warning.

Another important example of metric semigroup is the free pro-$\mathbf{V}$ semigroup on A, that we now define. A semigroup S *separates* two words u and v of the free semigroup A^+ if there exists a morphism φ from A^+ onto S such that $\varphi(u) \neq \varphi(v)$. Let $\mathbf{V}$ be a variety of finite semigroups. We set

$$r_{\mathbf{V}}(u, v) = \min\{\mathrm{Card}(S) \mid S \text{ is a semigroup of } \mathbf{V} \text{ that separates } u \text{ and } v\}$$

and $d_{\mathbf{V}}(u, v) = 2^{-r_{\mathbf{V}}(u,v)}$, with the usual conventions $\min \emptyset = +\infty$ and $2^{-\infty} = 0$. We first establish some general properties of $d_{\mathbf{V}}$.

Proposition 1.7. *The following properties hold for every $u, v, w \in A^+$*

(1) $d_{\mathbf{V}}(u, v) = d_{\mathbf{V}}(v, u)$,

(2) $d_{\mathbf{V}}(uw, vw) \leq d_{\mathbf{V}}(u, v)$ *and* $d_{\mathbf{V}}(wu, wv) \leq d_{\mathbf{V}}(u, v)$,

(3) $d_{\mathbf{V}}(u, w) \leq \max\{d_{\mathbf{V}}(u, v), d_{\mathbf{V}}(v, w)\}$.

Proof. The first assertion is trivial. A semigroup of $\mathbf{V}$ separating uw and vw certainly separates u and v. Therefore $d_{\mathbf{V}}(uw, vw) \leq d_{\mathbf{V}}(u, v)$, and dually, $d_{\mathbf{V}}(wu, wv) \leq d_{\mathbf{V}}(u, v)$.

Let S be a semigroup of $\mathbf{V}$ separating u and w. Then S separates either u and v, or v and w. It follows that $\min(r_{\mathbf{V}}(u, v), r_{\mathbf{V}}(v, w)) \leq r_{\mathbf{V}}(u, w)$ and hence $d_{\mathbf{V}}(u, w) \leq \max\{d_{\mathbf{V}}(u, v), d_{\mathbf{V}}(v, w)\}$. □

If $\mathbf{V}$ is the variety of all finite semigroups, we simplify the notation $d_{\mathbf{V}}$ to d.

Proposition 1.8. *The function d is an ultrametric on A^+.*

Proof. Suppose that $d(u, v) = 0$. In particular, the syntactic semigroup of $\{u\}$ does not separate u from v, showing that $u = v$. Thus by Proposition 1.7, d is an ultrametric. □

For the metric d, the closer are two words, the larger is the semigroup needed to separate them.

In the general case, $d_{\mathbf{V}}$ is not always a metric, because one may have $d_{\mathbf{V}}(u, v) = 0$ even if $u \neq v$. For instance, if $\mathbf{V}$ is the variety of commutative semigroups, $d_{\mathbf{V}}(ab, ba) = 0$, since there is no way to separate ab and ba by a commutative semigroup. To work around this difficulty, we first observe that, by Proposition 1.7, the relation $\sim_{\mathbf{V}}$ defined by

$$u \sim_{\mathbf{V}} v \text{ if and only if } d_{\mathbf{V}}(u, v) = 0$$

is a congruence on A^+. Then Proposition 1.8 can be generalized as follows.

Proposition 1.9.

(1) *The function $d_{\mathbf{V}}$ is an ultrametric on $A^+/{\sim_{\mathbf{V}}}$.*

(2) *The product on $A^+/{\sim_{\mathbf{V}}}$ is uniformly continuous for this metric.*

Proof. (1) follows directly from Proposition 1.7, since $d_{\mathbf{V}}(u, v) = 0$ implies $u \sim_{\mathbf{V}} v$ by definition. We use the same proposition to obtain the relation

$$d_{\mathbf{V}}(uv, u'v') \leq \max\{d_{\mathbf{V}}(uv, uv'), d_{\mathbf{V}}(uv', u'v')\} \leq \max\{d_{\mathbf{V}}(v, v'), d_{\mathbf{V}}(u, u')\}$$

which proves (2). □

The completion of A^+ for d, denoted by $\widehat{A^+}$, is called the *free profinite semigroup* on A. The completion of the metric space $(A^+/{\sim_{\mathbf{V}}}, d_{\mathbf{V}})$, denoted by $\hat{F}_{\mathbf{V}}(A)$, is called the *free pro-V semigroup* on A. It satisfies the following properties:

Proposition 1.10. *Let* $\mathbf{V}$ *be a variety of semigroups and* A *a finite alphabet.*

(1) *The semigroup* $\hat{F}_{\mathbf{V}}(A)$ *is compact.*

(2) *There is a surjective morphism* $\pi_{\mathbf{V}}$ *from* $\widehat{A^+}$ *onto* $\hat{F}_{\mathbf{V}}(A)$.

(3) *Every morphism from* $\widehat{A^+}$ *into a semigroup of* $\mathbf{V}$ *factorizes through* $\pi_{\mathbf{V}}$.

(4) *An* A*-generated finite semigroup belongs to* $\mathbf{V}$ *if and only if it is a quotient of* $\hat{F}_{\mathbf{V}}(A)$.

Proof. (1) Since $\hat{F}_{\mathbf{V}}(A)$ is complete, it suffices to verify that, for every $n > 0$, A^+ is covered by a finite number of open balls of radius $< 2^{-n}$. Consider the congruence $\sim_n$ defined on A^+ by

$$u \sim_n v \text{ if and only if } \varphi(u) = \varphi(v) \text{ for every morphism } \varphi \text{ from } A^+ \text{ onto a semigroup of size} \leq n \text{ of } \mathbf{V}.$$

Since A is finite, there are only finitely many morphisms from A^+ onto a semigroup of size $\leq n$, and thus $\sim_n$ is a congruence of finite index. Furthermore, $d_{\mathbf{V}}(u, v) < 2^{-n}$ if and only if u and v cannot be separated by a semigroup of $\mathbf{V}$ of size $\leq n$, i.e. are $\sim_n$-equivalent. It follows that the $\sim_n$-classes are open balls of radius $< 2^{-n}$ and cover A^+.

(2) Let $\pi_{\mathbf{V}}$ be the natural morphism from A^+ onto $A^+/{\sim_{\mathbf{V}}}$. Since $d_{\mathbf{V}}(u, v) \leq d(u, v)$, $\pi_{\mathbf{V}}$ is uniformly continuous, and can be extended into a uniformly continuous morphism from $\widehat{A^+}$ onto $\hat{F}_{\mathbf{V}}(A)$.

(3) Let φ be a morphism from $\widehat{A^+}$ into a semigroup S of $\mathbf{V}$. Up to replacing S by $\varphi(S)$, we may assume that φ is onto. Since A^+ is dense in $\widehat{A^+}$, and S is discrete, the restriction of φ to A^+ is also surjective. Furthermore, since $u \sim_{\mathbf{V}} v$ implies $\varphi(u) = \varphi(v)$, Proposition A.1.2 shows that there is a surjective morphism π from $A^+/{\sim_{\mathbf{V}}}$ onto S such that $\varphi = \pi \circ \pi_{\mathbf{V}}$. We claim that this morphism is uniformly continuous. Indeed if $d_{\mathbf{V}}(u, v) < 2^{-|S|}$, then u and v cannot be separated by S, and hence $\varphi(u) = \varphi(v)$. Since A^+ is dense in $\hat{F}_{\mathbf{V}}(A)$, π can be extended by continuity into a surjective morphism from $\hat{F}_{\mathbf{V}}(A)$ onto S. Thus S is a quotient of $\hat{F}_{\mathbf{V}}(A)$.

(4) If S is an A-generated semigroup of $\mathbf{V}$, there exists a surjective morphism φ from A^+ onto S. Following the argument used in (3), if $d_{\mathbf{V}}(u, v) < 2^{-|S|}$, then $\varphi(u) = \varphi(v)$, and thus φ is uniformly continuous with respect to $d_{\mathbf{V}}$. Therefore φ can be extended into a uniformly continuous morphism from $\hat{F}_{\mathbf{V}}(A)$ onto S.

Conversely, assume that S is a finite quotient of $\hat{F}_{\mathbf{V}}(A)$ and let $\pi : \hat{F}_{\mathbf{V}}(A) \to S$ be a surjective morphism. The set

$$D = \{(u, v) \in \hat{F}_{\mathbf{V}}(A) \times \hat{F}_{\mathbf{V}}(A) \mid \pi(u) = \pi(v)\}$$

is the inverse image under π of the diagonal of $S \times S$, and since S is discrete and π is continuous, it is a clopen subset of $\hat{F}_{\mathbf{V}}(A) \times \hat{F}_{\mathbf{V}}(A)$. Let $\mathcal{F}$ be the class of all morphisms from $\hat{F}_{\mathbf{V}}(A)$ onto a semigroup of $\mathbf{V}$. For each $\varphi \in \mathcal{F}$, let

$$C_\varphi = \{(u, v) \in \hat{F}_{\mathbf{V}}(A) \times \hat{F}_{\mathbf{V}}(A) \mid \varphi(u) \neq \varphi(v)\}$$

Each C_φ is open by continuity of φ. Furthermore, if (u, v) does not belong to any C_φ, then u and v cannot be separated by any semigroup of $\mathbf{V}$ and hence $d_{\mathbf{V}}(u, v) = 0$, which gives $u = v$ and $\pi(u) = \pi(v)$. It follows that the family $D \cup (C_\varphi)_{\varphi \in \mathcal{F}}$ is a covering of $\hat{F}_{\mathbf{V}}(A) \times \hat{F}_{\mathbf{V}}(A)$ by open sets, and since $\hat{F}_{\mathbf{V}}(A)$ is compact, it admits a finite subcovering, say $D \cup (C_\varphi)_{\varphi \in F}$. Therefore, if $\varphi(u) = \varphi(v)$ for each $\varphi \in F$, then $\pi(u) = \pi(v)$. Consequently S is a quotient of a subsemigroup of the finite semigroup $\prod_{\varphi \in F} \varphi(\hat{F}_{\mathbf{V}}(A))$ and thus belongs to $\mathbf{V}$. □

We now extend the notion of identity as follows. Let A be a finite alphabet and let $u, v \in \widehat{A^+}$. A finite ordered semigroup (resp. semigroup) S *satisfies the identity* $u \leq v$ (resp. $u = v$) if and only if, for each morphism $\varphi : \widehat{A^+} \to S$, $\varphi(u) \leq \varphi(v)$ (resp. $\varphi(u) = \varphi(v)$).

A variety $\mathbf{V}$ satisfies a given identity if every semigroup of $\mathbf{V}$ satisfies this identity. We also say in this case that the given identity is an identity of $\mathbf{V}$. Identities of $\mathbf{V}$ are closely related to free pro-$\mathbf{V}$ semigroups.

Proposition 1.11. *Let A be a finite alphabet. Given two elements u and v of $\widehat{A^+}$, $u = v$ is an identity of $\mathbf{V}$ if and only if $\pi_{\mathbf{V}}(u) = \pi_{\mathbf{V}}(v)$.*

Proof. If $u = v$ is an identity of $\mathbf{V}$, then u and v cannot be separated by any semigroup of $\mathbf{V}$. Thus $d_{\mathbf{V}}(u, v) = 0$, $u \sim_{\mathbf{V}} v$ and $\pi_{\mathbf{V}}(u) = \pi_{\mathbf{V}}(v)$. Conversely if $\pi_{\mathbf{V}}(u) = \pi_{\mathbf{V}}(v)$, then by Proposition 1.10, $\varphi(u) = \varphi(v)$ for every morphism φ from $\widehat{A^+}$ into a semigroup of $\mathbf{V}$, and thus $u = v$ is an identity of $\mathbf{V}$. □

Corollary 1.12. *Let $\mathbf{V}$ and $\mathbf{W}$ be two varieties of finite semigroups satisfying the same identities on the alphabet A. Then $\hat{F}_{\mathbf{V}}(A)$ and $\hat{F}_{\mathbf{W}}(A)$ are isomorphic.*

In particular, an identity of a semigroup of $\mathbf{V}$ can be given as a pair (u, v) of elements of $\hat{F}_{\mathbf{V}}(A)$. We are now ready to state the generalization of Theorem 1.5. Given a set E of identities, we denote by $[\![E]\!]$ the class of finite semigroups satisfying all the identities of E.

Theorem 1.13. *A class of finite semigroups is a variety if and only if it can be defined by a set of identities.*

Proof. The fact that every class of finite semigroups defined by a set of identities is a variety can be proved exactly as in Theorem 1.5.

Let now $\mathbf{V}$ be a variety. Let E be the class of all identities which are satisfied by every semigroup of $\mathbf{V}$ and let $\mathbf{W} = [\![E]\!]$. Clearly $\mathbf{V} \subseteq \mathbf{W}$. Let $S \in \mathbf{W}$. Since S is finite, there exists a finite alphabet A and an onto morphism $\varphi : A^+ \to S$ which can be extended to a uniformly continuous morphism from $\hat{A}^+$ onto S. Let $\pi : \hat{A}^+ \to F_{\mathbf{V}}(A)$ be the natural morphism and let $u, v \in \hat{A}^+$. By Proposition 1.11, if $\pi(u) = \pi(v)$, then $u = v$ is an identity of $\mathbf{V}$ and thus, is satisfied by S. In particular, $\pi(u) = \pi(v)$ implies $\varphi(u) = \varphi(v)$ and by Proposition A.1.2, φ factors through π. Therefore S is a quotient of $\hat{F}_{\mathbf{V}}(A)$ and by Proposition 1.10, S is in $\mathbf{V}$. Thus $\mathbf{V} = \mathbf{W}$. □

For instance, the semigroup $U_1 = \{0, 1\}$, equipped with the usual multiplication of integers, generates the variety of finite idempotent and commutative semigroups, defined by the identities $xy = yx$ and $x = x^2$.

Theorem 1.13 is thus formally similar to Theorem 1.5. The difference lies in the definition of the identities. In Theorem 1.5, an identity is a pair (u, v) of words of A^+ while in Theorem 1.13, an identity is a pair (u, v) of elements of $\widehat{A^+}$.

Each element of $\widehat{A^+}$ is the limit of a Cauchy sequence of (A^+, d). The most important example of a converging sequence is given by the next lemma.

Lemma 1.14. *For each $x \in \widehat{A^+}$, the sequence $(x^{n!})_{n\geq 0}$ is a Cauchy sequence. It converges to an idempotent element of $\widehat{A^+}$.*

Proof. For the first part of the statement, it suffices to show that for $p, q \geq n$, $x^{p!}$ and $x^{q!}$ cannot be separated by a semigroup of size $\leq n$. Let indeed $\varphi : \widehat{A^+} \to S$ be a morphism such that $\text{Card}(S) \leq n$, and put $s = \varphi(x)$. By Proposition A.1.9, s has an idempotent power $e = s^r$, with $r \leq n$. By the choice of p and q, the integer r divides simultaneously $p!$ and $q!$. Consequently, $s^{p!} = s^{q!} = e$, which shows that S cannot separate $x^{p!}$ and $x^{q!}$.

For n large enough, we also have $\varphi(x^{n!})\varphi(x^{n!}) = ee = e = \varphi(x^{n!})$. It follows that the limit of the sequence $(x^{n!})_{n\geq 0}$ is idempotent. □

The limit of the sequence $(x^{n!})_{n\geq 0}$ is denoted x^π. In practice, it suffices to remember that its image under any morphism $\varphi : \widehat{A^+} \to S$ onto a finite semigroup is the unique idempotent of the subsemigroup of S generated by $\varphi(x)$. In particular, one can write $\varphi(x^\pi) = \varphi(x)^\pi$ if the integer π on the right hand side is interpreted as the exponent of S.

Example 1.4. The variety $[\![x^\pi y x^\pi \leq x^\pi]\!]$ is the class of ordered semigroups S such that, for each $s \in S$ and for each idempotent $e \in S$, $ese \leq e$.

Example 1.5. A semigroup belongs to the variety $\mathbf{G_S}$ if and only if it satisfies the identity $x^\pi y = yx^\pi = y$.

A semigroup is *aperiodic* if and only if it satisfies the identity $x^\pi = x^\pi x$, which can also be written, by abuse of notation, $x^\pi = x^{\pi+1}$. Other characterizations of aperiodic semigroups were given in Proposition A.2.9.

A semigroup S is *nilpotent* if and only if, for each idempotent e of S and for each $s \in S$, $es = e = se$. Thus the nilpotent semigroups form a variety, denoted by **Nil**, and defined by the identities $x^\pi y = x^\pi = yx^\pi$. In particular, a nilpotent semigroup is aperiodic. The next proposition summarizes some characteristic properties of the nilpotent semigroups.

Proposition 1.15. *Let S be a nonempty semigroup. The following conditions are equivalent:*

(1) *S is nilpotent,*

(2) *S has a zero which is the unique idempotent of S,*

(3) *there exists an integer $n > 0$ such that S satisfies the identity $x_1 \cdots x_n = y_1 \cdots y_n$.*

Proof. (1) implies (2). Since S is nonempty, Proposition A.1.9 shows that S contains an idempotent e. If S is nilpotent, we have $es = e = se$ for each $s \in S$, showing that e is a zero. Moreover, for each idempotent f of S, we also have $fs = f = sf$, whence in particular $fe = f$. Since e is a zero, it follows that $f = e$.

(2) implies (3). Let $n = \text{Card}(S)$ and let $s_1, \ldots, s_n$ be a finite sequence of elements of S. By Proposition A.1.12, there exists an idempotent $e \in S$ such that $s_1 \cdots s_i e s_{i+1} \cdots s_n = s_1 \cdots s_n$. Since the unique idempotent of S is a zero, we have $s_1 \cdots s_n = 0$. Therefore S satisfies the identity $x_1 \cdots x_n = y_1 \cdots y_n$.

(3) implies (1). Let $s \in S$ and $e \in E(S)$. Setting $x_1 = s$ and $x_2 = \ldots = x_n = y_1 = \ldots = y_n = e$, we find $s = e$ if $n = 1$ and $se = e$ if $n \geq 2$ and hence $se = e$ in all cases. One would show in a similar way that $es = e$ and hence S is nilpotent. □

Denote by $\mathbf{Nil}^+$ (resp. $\mathbf{Nil}^-$) the variety of ordered semigroups defined by the identities $x^\pi y = x^\pi = yx^\pi$ and $x^\pi \leq y$ (resp. $x^\pi y = x^\pi = yx^\pi$ and $y \leq x^\pi$). One can see that $\mathbf{Nil}^+$ is the class of nilpotent ordered semigroups S such that $0 \leq s$ for each $s \in S$. □

A semigroup S is said to be $\mathcal{J}$-trivial (resp. $\mathcal{R}$-trivial, $\mathcal{L}$-trivial, $\mathcal{H}$-trivial) if and only if the Green relation $\mathcal{J}$ (resp. $\mathcal{R}$, $\mathcal{L}$, $\mathcal{H}$) is the equality relation. Proposition A.2.9 shows that a semigroup is $\mathcal{H}$-trivial if and only if it belongs to the variety **A** of aperiodic semigroups. Similarly, the $\mathcal{J}$-trivial (resp. $\mathcal{R}$-trivial, $\mathcal{L}$-trivial) semigroups form a variety, denoted by **J** (resp. **R**, **L**). The identities defining these varieties are given in the next proposition.

Proposition 1.16. *The following equalities hold*

$$\mathbf{R} = [\![(xy)^\pi x = (xy)^\pi]\!]$$

$$\mathbf{L} = [\![y(xy)^\pi = (xy)^\pi]\!]$$
$$\mathbf{J} = [\![y(xy)^\pi = (xy)^\pi = (xy)^\pi x]\!] = [\![x^\pi x = x^\pi, (xy)^\pi = (yx)^\pi]\!]$$
$$\mathbf{A} = [\![x^\pi = x^{\pi+1}]\!]$$

Moreover, the identities $(x^\pi y^\pi)^\pi = (x^\pi y)^\pi = (xy^\pi)^\pi = (xy)^\pi$ *are satisfied by* **J**.

Proof. (1) Let M be a monoid and let $x, y \in M$. If π is interpreted as the exponent of M, we observe that $(xy)^\pi x \mathrel{\mathcal{R}} (xy)^\pi$ since $((xy)^\pi x)(y(xy)^{\pi-1} = (xy)^{2\pi} = (xy)^\pi$. Thus if M is $\mathcal{R}$-trivial, the identity $(xy)^\pi x = (xy)^\pi$ holds in M.

Conversely, assume that M satisfies the identity $(xy)^\pi x = (xy)^\pi$ and let u and v be two $\mathcal{R}$-equivalent elements of M. Then, there exist $x, y \in M$ such that $ux = v$ and $vy = u$. It follows that $u = uxy = u(xy)^\pi$ and thus $v = ux = u(xy)^\pi x$. Now, since $(xy)^\pi x = (xy)^\pi$, $u = v$ and M is $\mathcal{R}$ trivial.

(2) The proof is dual for the variety **L**.

(3) Since $\mathbf{J} = \mathbf{R} \cap \mathbf{L}$, it follows from (1) and (2) that **J** is defined by the identities $y(xy)^\pi = (xy)^\pi = (xy)^\pi x$. Taking $y = 1$, we obtain $x^\pi = x^\pi x$ and also $(xy)^\pi = y(xy)^\pi = (yx)^\pi y = (yx)^\pi$. Conversely, suppose that a monoid satisfies the identities $x^\pi x = x^\pi$ and $(xy)^\pi = (yx)^\pi$. Then we have $(xy)^\pi = (yx)^\pi = (yx)^{\pi+1} = y(xy)^\pi x$, whence $(xy)^\pi = y^\pi (xy)^\pi x^\pi = y^{\pi+1}(xy)^\pi x^\pi = y(xy)^\pi$ and likewise $(xy)^\pi = (xy)^\pi x$. □

Note that the following inclusions hold: $\mathbf{J} \subset \mathbf{R} \subset \mathbf{A}$ and $\mathbf{J} \subset \mathbf{L} \subset \mathbf{A}$.

2 The variety theorem

If **V** is a variety of (resp. ordered) semigroups, denote by $\mathcal{V}(A^+)$ the set of recognizable subsets of A^+ whose syntactic semigroup belongs to **V**. It is also the set of subsets of A^+ recognized by a (resp. ordered) semigroup of **V**.

A *class* of recognizable subsets is a correspondence that associates with each finite alphabet A, a set $\mathcal{C}(A^+)$ of recognizable subsets of A^+. We use here the terms class and correspondence instead of set and function to avoid any paradox of set theory, since it is known, for instance, that the finite sets do not form a set... However, we shall use the term "bijection" instead of "one-to-one and onto correspondence". The correspondence $\mathbf{V} \to \mathcal{V}$ allows one to associate with each variety of semigroup a class of recognizable subsets.

Let E be a set. A class of subsets of E forms a *positive boolean algebra* if it is closed under finite union and finite intersection. It contains in particular the empty subset (obtained as the union of the empty family) and the full subset E (obtained as the intersection of the empty family). A *boolean algebra* is a positive boolean algebra which is also closed under complementation. The smallest positive boolean algebra containing a set $\mathcal{E}$ of subsets of E is called the positive boolean algebra *generated* by $\mathcal{E}$. Its elements are obtained from the elements of $\mathcal{E}$ by using the operations of finite union and finite intersection: they are the *positive boolean combination* of elements of $\mathcal{E}$. The notion of generated boolean algebra can be defined in the same way by adding the complement to the basic operations.

Recall that if X is a subset of A^+ and if $u \in A^*$, the *left* (resp. *right*) *quotient* of X by u is the set

$$u^{-1}X = \{v \in A^+ \mid uv \in X\} \qquad (\text{resp. } Xu^{-1} = \{v \in A^+ \mid vu \in X\})$$

A *$+$-positive variety* is a class of recognizable subsets such that

(1) for each alphabet A, $\mathcal{V}(A^+)$ is a positive boolean algebra,

(2) for each morphism of semigroups $\varphi : A^+ \to B^+$, $X \in \mathcal{V}(B^+)$ implies $\varphi^{-1}(X) \in \mathcal{V}(A^+)$,

(3) if $X \in \mathcal{V}(A^+)$ and $u \in A^+$, $u^{-1}X \in \mathcal{V}(A^+)$ and $Xu^{-1} \in \mathcal{V}(A^+)$.

A *$+$-variety* is a positive $+$-variety closed under complementation. This amounts replacing (1) by (1′) in the previous definition

(1′) for each alphabet A, $\mathcal{V}(A^+)$ is a boolean algebra.

We are now ready to state the variety theorem.

Theorem 2.1. *The correspondence* $\mathbf{V} \to \mathcal{V}$ *defines a bijection between the varieties of ordered semigroups and the $+$-positive varieties on one hand and between the varieties of semigroups and the $+$-varieties on the other hand.*

We shall omit the proof of Theorem 2.1, because it follows essentially the same pattern as the proof of Theorem VI.2.8, which is given in full details. Actually, Theorem VI.2.8 is a natural extension of Theorem 2.1 to infinite words.

Boldface letters are usually used to denote varieties of finite semigroups and the corresponding $+$-varieties are denoted with cursive letters.

One can obtain a parallel theory by considering the recognizable subsets of the free monoid A^*. It suffices to define successively the notions of subset recognized by an (ordered) monoid, of (ordered) syntactic monoid, of variety of (ordered) monoids. The main difference with the corresponding notions for semigroups concerns the identities. Consider for instance the identity $x^\pi yx^\pi = x^\pi$. This identity is actually equivalent, in the monoid case, to the identity $y = 1$. To see this, it suffices to substitute x by 1, to obtain $y = 1$. The only monoid satisfying the identity $x^\pi yx^\pi = x^\pi$ is therefore the trivial monoid.

Continuing the parallelism, one next defines positive $*$-varieties as follows: a *$*$-positive variety* is a class of recognizable subsets such that

(1) for each alphabet A, $\mathcal{V}(A^*)$ is a positive boolean algebra,

(2) for each morphism of monoids $\varphi : A^* \to B^*$, $X \in \mathcal{V}(B^*)$ implies $\varphi^{-1}(X) \in \mathcal{V}(A^*)$,

(3) if $X \in \mathcal{V}(A^*)$ and $u \in A^*$, $u^{-1}X \in \mathcal{V}(A^*)$ and $Xu^{-1} \in \mathcal{V}(A^*)$.

Similarly, a *$*$-variety* is a $*$-variety closed under complementation. Then we have a monoid version of Theorem 2.1.

Theorem 2.2. *The correspondence $\mathbf{V} \to \mathcal{V}$ defines a bijection between the varieties of ordered monoids and the positive $*$-varieties on the one hand and between the varieties of monoids and the $*$-varieties on the other hand.*

The variety theorem allows one to associate a $+$-variety (resp. $*$-variety) with each variety of semigroups (resp. monoids). The most interesting case are of course those for which a combinatorial description of the corresponding $+$-variety (resp. $*$-variety) is known. We shall present several such examples in the next section.

3 Some examples of varieties

The description of the $+$-variety associated with the trivial variety of semigroups $\mathbf{1}$ is immediate.

Proposition 3.1. *For each alphabet A, $\mathcal{I}(A^+)$ consists of the empty and the full subsets.*

We pursue with the varieties **Nil**, $\mathbf{Nil}^+$ and $\mathbf{Nil}^-$. Recall that a subset F of a set E is *cofinite* if the complement of F in E is finite.

Proposition 3.2. *For each alphabet A,*

(1) *$\mathcal{N}il^+(A^+)$ consists of the empty subset and of the cofinite subsets of A^+,*
(2) *$\mathcal{N}il^-(A^+)$ consists of A^+ and of the finite subsets of A^+,*
(3) *$\mathcal{N}il(A^+)$ is the set of finite or cofinite subsets of A^+.*

Proof. (1). Denote by $\varphi : A^+ \to S$ the syntactic morphism of L. If L is empty, S is trivial, and hence in $\mathbf{Nil}^+$. If L is a cofinite subset of A^+, there exists an integer n such that L contains all the words of length $\geq n$. If u is such a word, we have $xuy \in L$ for each $x, y \in A^*$, thereby showing that all the words of A^+ of length $\geq n$ are syntactically equivalent and thus have the same image e under φ. By Proposition 1.15, S is thus nilpotent. There remains to prove that $e \leq s$ for every $s \in S$. Let $v \in \varphi^{-1}(s)$. Then the formal implication

$$(xvy \in L \Rightarrow xuy \in L)$$

shows that $u \leq_L v$, whence $e \leq s$ in S. Therefore $S \in \mathbf{Nil}^+$.

Conversely, let $(S, \leq) \in \mathbf{Nil}^+$, I be an order ideal of S and let $\varphi : A^+ \to S$ be a morphism of semigroups. If I is empty, $\varphi^{-1}(I)$ is empty also. Otherwise, I contains necessarily 0, since 0 is minimal for $\leq$. Let u be a word of length greater than or equal to $\mathrm{Card}(S)$. By Proposition 1.15, $\varphi(u) = 0$ and hence $\varphi(u) \in I$. Therefore $\varphi^{-1}(I)$ is cofinite.

(2) Follows from (1) by taking the complement.

(3) What precedes shows that the syntactic semigroup of a finite or cofinite subset is a nilpotent semigroup. To prove the converse, consider a nilpotent nonempty semigroup S. Let P be a subset of S and let $\varphi : A^+ \to S$ be a morphism of semigroups. Then 0 belongs either to P, or to $S \setminus P$ and the argument above shows that $\varphi^{-1}(P)$ is either finite or cofinite. □

A variety of ordered monoids is called *locally finite* if it is generated by a single ordered monoid. Equivalent characterizations are given in Exercise VI.3. The positive variety corresponding to a locally finite variety is easy to describe.

Proposition 3.3. *Let* $\mathbf{V}$ *be a variety of ordered monoids generated by a single ordered monoid* M *and let* $\mathcal{V}$ *be the corresponding positive variety. Then, for every alphabet* A, $\mathcal{V}(A^*)$ *is the positive boolean algebra generated by the sets of the form* $\varphi^{-1}(\downarrow m)$, *where* $\varphi : A^* \to M$ *is an arbitrary morphism and* $m \in M$.

Proof. It is clear that $\varphi^{-1}(\downarrow m) \in \mathcal{V}(A^*)$ and thus $\mathcal{V}(A^*)$ also contains the positive boolean algebra generated by these sets. Conversely, let $L \in \mathcal{V}(A^*)$. Then there exists an integer $n \geq 0$, a morphism $\varphi : A^* \to M^n$ and an order ideal I of M such that $L = \varphi^{-1}(I)$. Since $\varphi^{-1}(I) = \bigcup_{m \in P} \varphi^{-1}(\downarrow m)$, it is sufficient to establish the result when $L = \varphi^{-1}(\downarrow m)$ where $m \in M^n$. Denote by π_i the i-th projection from M^n onto M. Setting $m = (m_1, \ldots, m_n)$, we have $m = \bigcap_{1 \leq i \leq n} \pi_i^{-1}(m_i)$, whence

$$\varphi^{-1}(\downarrow m) = \bigcap_{1 \leq i \leq n} (\pi_i \circ \varphi)^{-1}(\downarrow m_i)$$

Since $m_i \in M$ and $\pi_i \circ \varphi$ is a morphism from A^* into M, the result follows. □

There is of course a similar result for varieties of monoids, the proof of which is similar.

Proposition 3.4. *Let* $\mathbf{V}$ *be a variety of monoids generated by a single monoid* M *and let* $\mathcal{V}$ *be the corresponding variety of languages. Then, for every alphabet* A, $\mathcal{V}(A^*)$ *is the boolean algebra generated by the sets of the form* $\varphi^{-1}(\downarrow m)$, *where* $\varphi : A^* \to M$ *is an arbitrary morphism and* $m \in M$.

It follows from Propositions 3.3 and 3.4 that if $\mathbf{V}$ is locally finite, the set $\mathcal{V}(A^*)$ is finite for each alphabet A. This explains the terminology.

Let $\mathbf{J}_1$ be the variety $\mathbf{J}_1$ of idempotent and commutative monoids, defined by the identities $xy = yx$ and $x^2 = x$. By Proposition A.4.6, $\mathbf{J}_1$ is generated by its cyclic monoids. But there is only one nontrivial cyclic monoid in $\mathbf{J}_1$, and this is the monoid $U_1 = \{0, 1\}$ of Example A.1.1. The corresponding variety is described as follows.

Proposition 3.5. *For each alphabet A, $\mathcal{J}_1(A^*)$ is the boolean algebra generated by the subsets of the form A^*aA^* where a is a letter. Equivalently, $\mathcal{J}_1(A^*)$ is the boolean algebra generated by the subsets of the form B^* where B is a subset of A.*

Proof. The equality of the two boolean algebras considered in the statement results from the formulas

$$B^* = A^* \setminus \bigcup_{a \in A \setminus B} A^*aA^* \quad \text{and} \quad A^*aA^* = A^* \setminus (A \setminus \{a\})^*$$

Since $\mathbf{J_1}$ is generated by U_1, one can use Proposition 3.4 to describe $\mathcal{J}_1$. Let $\varphi : A^* \to U_1$ be a morphism, and let $B = \{a \in A \mid \varphi(a) = 1\}$. Then $\varphi^{-1}(1) = B^*$ and $\varphi^{-1}(0) = A^* \setminus B^*$, which establishes the proposition. □

This result can be divided into two parts. Denote by $\mathbf{J}_1^+$ (resp. $\mathbf{J}_1^-$) the positive variety of idempotent and commutative semigroups satisfying the identity $xy \leq x$ (resp $x \leq xy$). If B is a subset of A, denote by $F(B)$ the set of words of A^+ containing at least one occurrence of each letter of B. Thus

$$F(B) = \bigcap_{a \in B} A^*aA^*$$

The next proposition is thus a variant of Proposition 3.5 and its proof is left as an exercise to the reader.

Proposition 3.6. *For each alphabet A, $\mathcal{J}_1^+(A^*)$ is the set of finite unions of subsets of the form $F(B)$ where $B \subset A$. Similarly, $\mathcal{J}_1^-(A^*)$ is the set of finite unions of subsets of the form B^+ where $B \subset A$.*

Another interesting example of locally finite variety is the variety $\mathbf{R_1}$ of idempotent and $\mathcal{R}$-trivial monoids. The corresponding $*$-variety admits a combinatorial descriptions which is related to the notion of first appearance record. The dual notion was considered in Section IV.4.4.

Proposition 3.7. *Let L be a recognizable subset of A^* and let M be its syntactic monoid. The following conditions are equivalent:*

(1) *M divides $\tilde{U}_2^n$ for some $n > 0$,*

(2) *M belongs to $\mathbf{R_1}$,*

(3) *M satisfies the identity $xyx = xy$,*

(4) *L is a disjoint union of sets of the form*

$$a_1\{a_1\}^*a_2\{a_1, a_2\}^*a_3\{a_1, a_2, a_3\}^* \cdots a_n\{a_1, a_2, \ldots, a_n\}^*$$

where the a_i's are distinct letters of A,

(5) *L is a boolean combination of sets of the form B^*aA^*, where $a \in A$ and $B \subset A$.*

Proof. (1) implies (2) since $\tilde{U}_2 \in \mathbf{R}_1$.

(2) implies (3). Let $x, y \in M$. Since M is idempotent, $xy = xyxy$ and thus $xy \mathcal{R} xyx$. But M is $\mathcal{R}$-trivial and therefore $xy = xyx$.

(3) implies (4). Let $\rho : A^* \to A^*$ be the function which associates with any word u the sequence of all distinct letters of u in the order in which they first appear when u is read from left to right. For example, if $u = caabacb$, then $\rho(u) = cab$. In fact ρ is sequential function, realized by the sequential transducer $\mathcal{T} = (\mathcal{P}(A), A, \emptyset, ., *)$, where the transition and the output functions are defined by

$$B \cdot a = B \cup \{a\}$$

$$B * a = \begin{cases} 1 & \text{if } a \in B \\ 0 & \text{otherwise} \end{cases}$$

Define an equivalence $\sim$ on A^* by setting $u \sim v$ if $\rho(u) = \rho(v)$. It is easy to see that the equivalence classes of $\sim$ are the disjoint sets

$$L_{(a_1,\ldots,a_n)} = a_1\{a_1\}^* a_2\{a_1, a_2\}^* a_3\{a_1, a_2, a_3\}^* \cdots a_n\{a_1, a_2, \ldots, a_n\}^*$$

where $(a_1, \ldots, a_n)$ is a sequence of distinct letters of A. We claim that $\sim$ is a congruence. If $u \sim v$, then u and v belong to some set $L_{(a_1,\ldots,a_n)}$. Let a be a letter. If $a = a_i$ for some i, then $ua, va \in L_{(a_1,\ldots,a_n)}$, and $au, av \in L_{(a,a_1,\ldots,a_{i-1},a_{i+1},\ldots,a_n}$. Thus $ua \sim va$ and $au \sim av$. If $a \notin \{a_1, \ldots, a_n\}$, then $ua, va \in L_{(a_1,\ldots,a_n,a)}$ and $au, av \in L_{(a,a_1,\ldots,a_n)}$ and thus again $ua \sim va$ and $au \sim av$, which proves the claim.

Let $\eta : A^* \to M$ be the syntactic morphism of L. If $u \in L_{(a_1,\ldots,a_n)}$, then $u = a_1u_1a_2u_2 \cdots a_nu_n$ where $u_i \in \{a_1, \ldots, a_i\}^*$ for $1 \leq i \leq n$ and thus by (3), $\eta(u) = \eta(a_1 \cdots a_n)$. It follows that $u \sim v$ implies $\eta(u) = \eta(v)$ and therefore L is a disjoint union of equivalence classes of $\sim$, that is of sets of the form $L_{(a_1,\ldots,a_n)}$.

(4) implies (5). First observe that

$$L_{(a_1,\ldots,a_n)} = A_n^* \cap \bigcap_{1 \leq i \leq n} A_{i-1}^* a_i A_i^* \text{ where } A_i = \{a_1, \ldots, a_i\} \text{ and } A_0 = \emptyset$$

Condition (5) is now a consequence of the following equalities:

$$A_i^* = A^* \setminus \bigcup_{a \notin A_i} A^*aA^* \qquad A_{i-1}^* a_i A_i^* = A_{i-1}^* a_i A^* \cap A_i^*$$

(5) implies (1). By the variety theorem (Theorem 2.1), it is sufficient to show that, for $B \subset A$ and $a \in A$, B^*aA^* is recognized by $\tilde{U}_2$. Let $\tilde{U}_2 = \{1, a_1, a_2\}$ and let $\varphi : A^* \to \tilde{U}_2$ be the morphism defined by

$$\varphi(a) = a_1$$

$$\varphi(b) = \begin{cases} 1 & \text{if } b \in B \setminus \{a\} \\ a_2 & \text{for } b \in A \setminus (B \cup \{a\}) \end{cases}$$

Then $\varphi^{-1}(a_1) = B^*aA^*$, which concludes the proof. □

We shall conclude this section with a description of the $*$-variety corresponding to the variety **Acom** of aperiodic and commutative monoids.

If a is a letter of an alphabet A, let us denote by $L(a, k)$ the set of words of A^* which contain exactly k occurrences of a

$$L(a, k) = \{u \in A^+ \mid |u|_a = k\}$$

Then the following result holds.

Proposition 3.8. *For each alphabet A, $\mathcal{A}com(A^*)$ is the boolean algebra generated by the sets of the form $L(a, k)$ where $a \in A$ and $k \geq 0$.*

Proof. First, every set of the form $L(a, k)$ is recognized by an aperiodic commutative monoid. Indeed, let $N = \{1, x, x^2, \ldots, x^k, x^{k+1}$ be the cyclic monoid defined by the relation $x^{k+2} = x^{k+1}$, and let $\varphi : A^* \to N$ be the morphism defined by $\varphi(a) = x$ and $\varphi(b) = 1$ if $b \neq a$. Then clearly $L(a, k) = \varphi^{-1}(x^k)$.

By Proposition A.4.6, **Acom** is generated by its cyclic monoids, and Proposition 3.4 can be used to describe $\mathcal{A}com$. Let $M = \{1, x, x^2, \ldots, x^n\}$ be a cyclic monoid, defined by the relation $x^{n+1} = x^n$, and let $\varphi : A^* \to M$ be a morphism. Then for each $a \in A$ there exists an integer n_a such that $\varphi(a) = x^{n_a}$. Let k be an integer such that $0 \leq k < n$. Then

$$\begin{aligned}\varphi^{-1}(x^k) &= \Big\{u \in A^* \mid \sum_{a \in A} n_a |u|_a = k\Big\} \\ &= \bigcup \bigcap_{a \in A} L(a, k_a)\end{aligned}$$

where the union is taken over the set of families $(k_a)_{a \in A}$ such that $\sum_{a \in A} n_a k_a = k$. Finally, for $k = n$, we have

$$\varphi^{-1}(x^n) = A^* \setminus \bigcup_{0 \leq k < n} \varphi^{-1}(x^k)$$

which concludes the proof. □

4 Star-free sets

The description of the $*$-variety associated with the variety **A** of aperiodic semigroups, obtained by Schützenberger in 1965, is, after to Kleene's theorem, the most important result of the theory of finite automata.

Let A be a finite alphabet. The set of *star-free* subsets of A^* is the smallest set $\mathcal{R}$ of subsets of A^* such that

(a) $\mathcal{R}$ contains the empty set, the set $\{1\}$ and, for each $a \in A$, the singleton $\{a\}$.

(b) $\mathcal{R}$ is closed under finite union, finite product and complement.

Thus the definition of the star-free subsets follows the same definition scheme as the one of rational subsets, with the difference that the star operation is replaced by the complement. Since the rational subsets are closed under complement, every star-free subset is rational, but we shall see later on that the converse is not true. It follows also immediately from the definition that every finite set is star-free.

We shall follow the notation of Chapter I. Union will be denoted additively, the empty set will be denoted by 0, the singleton $\{u\}$ will be simply denoted by u, the product will be denoted by simple juxtaposition and L^c will denote the complement of a subset L of A^*. The star-free sets are thus described by expressions using the letters of the alphabet A, the constants 0 and 1 and the three operators union, product and complement. It is not always easy to find such an expression, as is shown in the examples below.

Example 4.1.

(1) A^* is a set star-free, since $A^* = 0^c$

(2) If B is a subset of A, A^*BA^* is star-free by (1). It follows that B^* is star-free, since

$$B^* = A^* \setminus \bigcup_{a \in A \setminus B} A^*aA^* = \Big(\sum_{a \in A \setminus B} 0^c a 0^c\Big)^c$$

(3) If $A = \{a, b\}$, the set $(ab)^*$ is star-free. Indeed

$$(ab)^* = (b0^c + 0^c a + 0^c aa0^c + 0^c bb0^c)^c$$

Schützenberger's theorem states that a subset is star-free if and only if its syntactic monoid is aperiodic. Here is an equivalent formulation in terms of varieties

Theorem 4.1. *For each alphabet A, $\mathcal{A}(A^*)$ is the set of star-free sets of A^*.*

Proof. The easiest part of the proof relies on a syntactic property of the concatenation product. Let L_0 and L_1 be two recognizable subsets of A^* and let $L = L_0L_1$. Let $\eta_0 : A^* \to M_0$, $\eta_1 : A^* \to M_1$ and $\eta : A^* \to M$ be the syntactic morphisms of L_0, L_1 and L.

Lemma 4.2. *If M_0 and M_1 are aperiodic, so is M.*

Proof. Let π_0, π_1 and π be the respective exponents of M_0, M_1 and M and let n be a multiple of π such that $n \geq \pi_0 + \pi_1 + 1$. It suffices to show that for each $x, y, u \in A^*$, $xu^ny \in L$ if and only if $xu^{n+1}y \in L$. One can of course suppose that $u \neq 1$. If $xu^ny \in L$, there exists a factorization $xu^ny = x_0x_1$ with $x_0 \in L_0$ and $x_1 \in L_1$. Two cases are possible. Either $x_0 = xu^{\pi_0}r$ with $rx_1 = u^{n-\pi_0}y$, or $x_1 = su^{\pi_1}y$ with $x_0s = xu^{n-\pi_1}$. Let us consider the first case, since the second case is symmetric. Since M_0 is aperiodic and since $xu^{\pi_0}r \in L_0$, we have $xu^{\pi_0+1}r \in L_0$ and hence $xu^{n+1}y \in L$.

Conversely, if $xu^{n+1}y \in L$, we obtain by the same process a factorization $xu^{n+1}y = x_0x_1$ with $x_0 \in L_0$ and $x_1 \in L_1$. Then we have $x_0 = xu^{\pi_0+1}r$ with $rx_1 = u^{n-\pi_0+1}y$, or $x_1 = su^{\pi_1+1}y$ with $x_0s = xu^{n-\pi_1+1}$. The same argument as above would show that $xu^ny \in L$. □

Let us fix an alphabet A. An elementary computation (cf. Exercise VI.2) shows that the syntactic monoid of the subsets $\{1\}$ and a, for $a \in A$, is aperiodic. Therefore, the set $\mathcal{A}(A^*)$ contains the subsets of this type. It is closed under boolean operations, by definition of a $*$-variety, but also under product, by Lemma 4.2. Consequently, $\mathcal{A}(A^*)$ contains the star-free sets.

To prove the converse, we need two elementary properties of the aperiodic monoids. The first property is a simple reformulation of the simplification lemma (Proposition A.2.5 (1)) in the case of aperiodic monoids.

Lemma 4.3. *Let M an aperiodic monoid and let $p, q, r \in M$. If $pqr = q$, then $pq = q = qr$.*

Proof. Let π the exponent of M. Since $pqr = q$, we also have $p^\pi qr^\pi = q$. Since M is aperiodic, we have $p^\pi = p^{\pi+1}$ and hence $pq = pp^\pi qr^\pi = p^\pi qr^\pi = q$ and, in the same way, $qr = q$. □

The second property leads to a decomposition of each subset of an aperiodic monoid as a boolean combination of right ideals, left ideals, or ideals.

Lemma 4.4. *Let M be an aperiodic monoid and let $p, q, r \in M$. Then $\{q\} = (qM \cap Mq) \setminus J_q$, with $J_q = \{s \in M \mid q \notin MsM\}$.*

It is clear that $q \in (qM \cap Mq) \setminus J_q$. Conversely, if $s \in (qM \cap Mq) \setminus J_q$, there exist $p, q \in M$ such that $s = pq = qr$. Moreover, as $s \notin J_q$, $q \in MsM$ and there exist $u, v \in M$ such that $q = usv$. Since $s = pq$, it follows that $q = upqv$, and, by the simplification rule, $q = (up)q = u(pq) = us$. Since $s = qr$, we also have $q = uqr$ and by the simplification rule, $q = qr = s$. □

We now need proving that if $\varphi : A^* \to M$ is a morphism from A^* into an aperiodic monoid M, the set $\varphi^{-1}(P)$ is star-free for every subset P of M. The formula

$$\varphi^{-1}(P) = \bigcup_{m \in P} \varphi^{-1}(m)$$

allows one to assume that $P = \{m\}$. We shall show that $\varphi^{-1}(m)$ is star-free by induction on the integer $r(m) = \text{Card}(M \setminus MmM)$. Put $B = \{a \in A \mid \varphi(a) = 1\}$.

If $r(m) = 0$, then $M = MmM$ and there exist $u, v \in M$ such that $umv = 1$. The simplification rule applied to $(um)1(v) = 1$ and to $(u)1(mv) = 1$ gives $u = v = 1$

and hence also $m = 1$. Let us show that $\varphi^{-1}(1) = B^*$. If $u \in B^*$, we have of course $\varphi(u) = 1$. Conversely, if $\varphi(u) = 1$, the simplification rule shows that $\varphi(a) = 1$ for each letter a occurring in u, and hence $u \in B^*$. Now, as was shown in example 4.1, (2)), B^* is a star-free set.

Assume now that $r(m) > 0$ and that the property has been established for each element s such that $r(s) < r(m)$. We need to establish the formula

$$\varphi^{-1}(m) = (UA^* \cap A^*V) \setminus (A^*CA^* \cup A^*WA^*) \tag{4.1}$$

where

$$U = \bigcup_{(n,a)\in E} \varphi^{-1}(n)aA^* \qquad V = \bigcup_{(n,a)\in F} A^*a\varphi^{-1}(n)$$

$$C = \{a \in A \mid m \notin M\varphi(a)M\} \quad W = \bigcup_{(a,n,b)\in G} A^*a\varphi^{-1}(n)bA^*$$

with

$$E = \{(n,a) \in M \times A \mid n\varphi(a)\ \mathcal{R}\ m \text{ but } n \notin mM\}$$
$$F = \{(a,n) \in A \times M \mid \varphi(a)n\ \mathcal{L}\ m \text{ but } n \notin Mm\}$$
$$G = \{(a,n,b) \in (A \times M \times A \mid m \in M\varphi(a)nM \cap Mn\varphi(b)M) \setminus M\varphi(a)n\varphi(b)M\}$$

Denote by L the right hand side of (4.1). We first prove the inclusion $\varphi^{-1}(m) \subset L$. Let $u \in \varphi^{-1}(m)$ and let p be the shortest prefix of u such that $\varphi(p)\ \mathcal{R}\ m$. The word p cannot be empty, since otherwise $m\ \mathcal{R}\ 1$, whence $m = 1$ by the simplification rule. Put $p = ra$ with $r \in A^*$ and $a \in A$ and let $n = \varphi(r)$. By construction, $(n,a) \in E$ and $u \in \varphi^{-1}(n)aA^*$. Therefore $u \in UA^*$ and a symmetric argument would show that $u \in A^*V$. If $u \in A^*CA^*$, there exists a letter a of C such that $m = \varphi(u) \in M\varphi(a)M$, a contradiction with the definition of C. Similarly, if $u \in A^*WA^*$, there exist $(a,n,b) \in G$ such that $m \in M\varphi(a)n\varphi(b)M$, a contradiction this time with the definition of G. Therefore $u \in L$.

Conversely, let $u \in L$ and let $s = \varphi(u)$. Since $u \in UA^* \cap A^*V$, we have $s \in mM \cap Mm$. By Lemma 4.4, in order to prove that $s = m$, and hence that $u \in \varphi^{-1}(m)$, it suffices to prove that $s \notin J_m$, i.e. $m \in MsM$. Supposing the contrary, consider a factor f of u of minimal length such that $m \notin M\varphi(f)M$. The word f is necessarily nonempty. If f is a letter, this letter is in C and $u \in A^*CA^*$, which is impossible. We may thus set $f = agb$, with $a, b \in A$. Set $n = \varphi(g)$. Since f is of minimal length, we have $m \in M\varphi(a)nM$ and $m \in Mn\varphi(b)M$. Consequently $n \in G$ and $f \in W$, which is impossible again.

Formula (4.1) is thus established and it suffices now to show that UA^*, A^*V and A^*WA^* are star-free, since we have already seen in Example 4.1 that A^*CA^* is star-free. Let $(n,a) \in E$. Since $n\varphi(a)M = mM$, we have $MmM \subset MnM$ and hence

$r(n) \le r(m)$. Moreover, as $m \le_{\mathcal{R}} n$, Corollary A.2.6 shows that if $MmM = MnM$, we have $n \mathrel{\mathcal{R}} m$, which is not possible since $n \notin mM$. Therefore $r(n) < r(m)$ and we can conclude by applying the induction hypothesis.

A symmetric argument would work for A^*V. There remains to treat the case A^*WA^*. Let $(a, n, b) \in G$. We have $r(n) \le r(m)$ since $m \in MnM$. Suppose that $MmM = MnM$. Then in particular, $n \in MmM$ and as $m \in M\varphi(a)nM$ and $m \in Mn\varphi(b)M$, it follows $n \in M\varphi(a)nM$ and $n \in Mn\varphi(b)M$, whence $n \mathrel{\mathcal{L}} \varphi(a)n$ and $n \mathrel{\mathcal{R}} n\varphi(b)$. By Proposition A.2.4, $n\varphi(b) \mathrel{\mathcal{L}} \varphi(a)n\varphi(b)$, and hence $m \mathrel{\mathcal{J}} \varphi(a)n\varphi(b)$, a contradiction with the definition of G. Consequently $r(n) < r(m)$ we can conclude by applying the induction hypothesis. □

Example 4.2. Let $A = \{a, b\}$ and let $L = (ab)^*$. The minimal automaton of L is represented in Figure 4.1. The syntactic monoid M of L is the monoid consisting of the six matrices

$$I = \begin{pmatrix} 1 & 0 \\ 0 & 1 \end{pmatrix} \qquad a = \begin{pmatrix} 0 & 1 \\ 0 & 0 \end{pmatrix} \qquad b = \begin{pmatrix} 0 & 0 \\ 1 & 0 \end{pmatrix}$$
$$aa = \begin{pmatrix} 0 & 0 \\ 0 & 0 \end{pmatrix} \qquad ab = \begin{pmatrix} 1 & 0 \\ 0 & 0 \end{pmatrix} \qquad ba = \begin{pmatrix} 0 & 0 \\ 0 & 1 \end{pmatrix}$$

and it is defined by the relations $a^2 = b^2 = 0$, $aba = a$ and $bab = b$. Its $\mathcal{D}$-class structure is given in Figure 4.2: This monoid is aperiodic, since $x^2 = x^3$ for each $x \in M$, and hence L is star-free.

Example 4.3. Let $A = \{a, b\}$ and let $L' = (aa)^*$. The minimal automaton of L' is represented in Figure 4.3. The syntactic monoid M' of L' is the monoid consisting of the three matrices

$$I = \begin{pmatrix} 1 & 0 \\ 0 & 1 \end{pmatrix} \quad a = \begin{pmatrix} 0 & 1 \\ 1 & 0 \end{pmatrix} \quad b = \begin{pmatrix} 0 & 0 \\ 0 & 0 \end{pmatrix}$$

and it is defined by the relations $a^2 = 1$ and $b = 0$. Its $\mathcal{D}$-class structure is given in Figure 4.4. This monoid is not aperiodic, since, for each $n > 0$, $a^n \ne a^{n+1}$ and hence L' is not star-free.

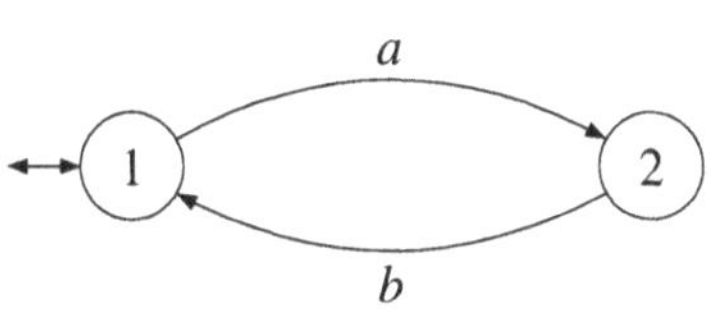

Figure 4.1. The minimal automaton of $(ab)^*$.

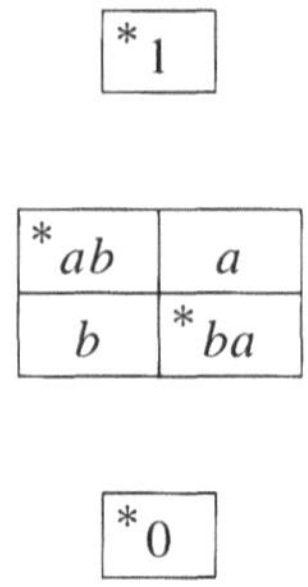

Figure 4.2. The $\mathcal{D}$-class structure of M.

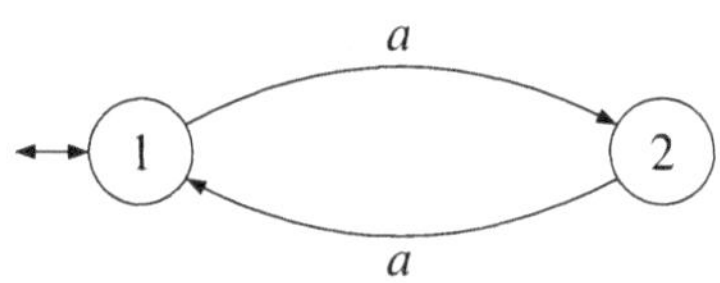

Figure 4.3. The minimal automaton of $(aa)^*$.

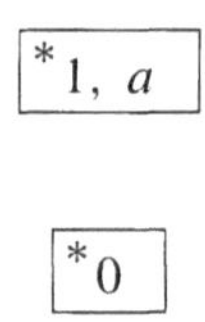

Figure 4.4. The $\mathcal{D}$-class structure of M'.

4.1 Concatenation hierarchies. The definition of star-free sets relies on two types of operations: boolean operations and concatenation product. By alternating the use of the concatenation product and of the boolean closure one can obtain hierarchies of recognizable sets. In fact, to have a better connection with logical hierarchies (see Chapter VIII), it is more advisable to use the marked product instead of the concatenation product.

Here is the precise definition of the hierarchy. Let A be an finite alphabet.

(1) The sets of level 0 are the empty set and A^*.

(2) For each integer $n \geq 0$, the sets of level $n + 1/2$ are the finite unions of subsets of the form

$$L_0 a_1 L_1 \cdots a_k L_k$$

where $L_0, \ldots, L_k$ are sets of level n and $a_1, \ldots, a_k$ are letters of A.

(3) For each integer $n \geq 0$, the sets of level $n+1$ are the finite boolean combinations of the sets of level $n+1/2$.

In particular, every set of level m is also of level n for each $n \leq m$. For each integer n, denote by $\mathcal{V}_n(A^*)$ (resp. $\mathcal{V}_{n+1/2}(A^*)$) the set of subsets of A^* of level n (resp. $n+1/2$).

Before studying in more details the first levels, let us state some general properties of the hierarchy.

Theorem 4.5.

(1) *For each $n \geq 0$, $\mathcal{V}_n$ is a $*$-variety and $\mathcal{V}_{n+1/2}$ is a positive $*$-variety.*

(2) *The union of the $\mathcal{V}_n$ is the $*$-variety of star-free sets.*

(3) *The hierarchy is strict: if A contains at least two letters, there exists for each $n \geq 0$, subsets of level $n+1/2$ that are not of level n and subsets of level $n+1$ that are not of level $n+1/2$.*

We shall denote by $\mathbf{V}_n$ the variety of monoids corresponding to $\mathcal{V}_n$ and by $\mathbf{V}_{n+1/2}$ the variety of ordered monoids corresponding to $\mathcal{V}_{n+1/2}$.

Let us fix an alphabet A. The level 0 consists of the languages $\emptyset$ and A^*. It follows that a subset of A^* is of level 0 if and only if its syntactic monoid is trivial. In other words, the subsets of level 0 define the $*$-trivial variety.

It follows from the definition that the subsets of level $1/2$ are the finite unions of subsets of the form

$$A^*a_1A^*a_2A^* \cdots A^*a_kA^*$$

where $k \geq 0$ and $a_1, \ldots, a_k \in A$. This leads to a few auxiliary definitions. A word $u = a_1a_2 \cdots a_n$ (where the a_i are letters) is said to be a *subword* of a word v if there exist words $v_0, v_1, \ldots, v_n$ of A^* such that $v = v_0a_1v_1a_2 \cdots a_nv_n$. For example, aab is a subword of $baacbc$. A subset L of A^* is a *shuffle ideal* if every word which is a subword of a word of L is also in L. We shall admit the next result, which is a reformulation of a standard result of Higman.

Theorem 4.6. *A subset of A^* is a shuffle ideal if and only if it is a finite union of subsets of the form $A^*a_1A^*a_2A^* \cdots A^*a_kA^*$ with $k \geq 0$ and $a_1, \ldots, a_k \in A$.*

Therefore the subsets of level $1/2$ are exactly the shuffle ideals. These subsets define a $*$-positive variety, as shown in the next result.

Proposition 4.7. *A subset of A^* is of level $1/2$ if and only if its syntactic ordered monoid satisfies the identity $x \leq 1$.*

Proof. Let $L = A^*a_1A^*a_2A^* \cdots A^*a_kA^*$, $\eta : A^* \to (M, \leq)$ the syntactic morphism of L and $I = \eta(L)$. If $xy \in L$, then $xsy \in L$ for each $s \in A^*$. It follows that, for each $x, y, s \in M$, $xy \in I$ implies $xsy \in I$. Therefore $s \leq 1$, showing that M satisfies the identity $x \leq 1$. It follows that the $*$-positive variety associated with the variety defined by this identity contains the subsets of the form $A^*a_1A^*a_2A^* \cdots A^*a_kA^*$ and hence also the finite unions of subsets of this type.

Conversely, let $(M, \leq)$ be a finite ordered monoid satisfying the identity $x \leq 1$ and let $\varphi : A \to M$ a morphism. Let I be an order ideal of S and let $L = \varphi^{-1}(I)$. If $xy \in L$, then $\varphi(xy) = \varphi(x)\varphi(y) \in I$. Let s be a word of A^*. Since $\varphi(s) \leq 1$, we have $\varphi(xsy) = \varphi(x)\varphi(s)\varphi(y) \leq \varphi(x)\varphi(y)$ and hence $\varphi(xsy) \in I$. It follows that $xsy \in L$ and hence L is a shuffle ideal. □

The subsets of level $1/2$ contained in A^+ are also finite unions of subsets of the form

$$A^*a_1A^*a_2A^* \cdots A^*a_kA^*$$

but with $k > 0$. They form a $+$-positive variety, and the corresponding variety is defined by the identities $xy \leq x$ and $yx \leq x$.

Proposition 4.8. *A subset of A^+ is a finite union of subsets of the form*

$$A^*a_1A^* \cdots A^*a_kA^*$$

where $k > 0$ and $a_1, \ldots, a_k \in A$ if and only if its syntactic ordered semigroup satisfies the identities $xy \leq x$ and $yx \leq x$.

Proof. The proof is analogous to that of Proposition 4.7. □

Note the following variations on the same theme, that will be used for infinite words.

Proposition 4.9. *Let $\mathbf{V} = [\![xy \leq x]\!]$. Then, for each alphabet A, $\mathcal{V}(A^+)$ consists of the finite unions of subsets of the form $a_0A^*a_1A^* \cdots A^*a_kA^*$.*

Proposition 4.10. *Let $\mathbf{V} = [\![x^\pi y \leq x^\pi,\ yx^\pi \leq x^\pi]\!]$. Then, for each alphabet A, $\mathcal{V}(A^+)$ is the set of finite unions of subsets of the form F, with F finite, or $A^*a_1A^* \cdots A^*a_kA^*$, where $a_1, \ldots, a_k \in A$.*

Proposition 4.11. *Let $\mathbf{V} = [\![x^\pi y \leq x^\pi]\!]$. Then, for every alphabet A, $\mathcal{V}(A^+)$ is the set of positive boolean combination of subsets of the form u, uA^*, with $u \in A^+$, or $A^*a_1A^* \cdots A^*a_kA^*$, with $a_1, \ldots, a_k \in A$.*

The boolean combinations of subsets of the form $A^*a_1A^*a_2A^* \cdots A^*a_kA^*$, where $k \geq 0$ and $a_1, \ldots, a_k \in A$ form the level 1. These subsets define a $*$-variety and the corresponding variety of monoids is **J**, thanks to a deep result of I. Simon.

Theorem 4.12. *A subset of A^* is of level 1 if and only if its syntactic monoid is finite and $\mathcal{J}$-trivial.*

Proof. The easiest part of the proof consists to verify that the syntactic monoid of a subset of the form $A^*a_1A^*a_2A^* \cdots A^*a_kA^*$ is $\mathcal{J}$-trivial. By Proposition 4.7, its syntactic ordered monoid satisfies the identity $x \leq 1$. Therefore, it suffices to establish the following lemma.

Lemma 4.13. *An ordered monoid satisfying the identity $x \leq 1$ is $\mathcal{J}$-trivial.*

Proof. Indeed, this identity implies $x^\pi x \leq x^\pi$ and $x^\pi = x^\pi x^\pi = xx^\pi x^{\pi-1} \leq xx^\pi$ whence $x^\pi = x^{\pi+1}$. Therefore

$$(xy)^\pi = (xy)^{\pi+1} = x(yx)^\pi y \leq x(yx)^\pi = (xy)^\pi x \leq (xy)^\pi$$

whence $(xy)^\pi x = (xy)^\pi$ and similarly $y(xy)^\pi = (xy)^\pi$. But these two identities define the variety **J**. □

It follows that the $*$-positive variety associated with **J** contains the subsets of the form

$$A^*a_1A^*a_2A^* \cdots A^*a_kA^*$$

and hence the boolean combinations of the subsets of this type.

The hardest part of the proof relies on a result of semigroup theory whose proof is omitted.

Theorem 4.14. *Every finite $\mathcal{J}$-trivial monoid is quotient of a finite ordered monoid satisfying the identity $x \leq 1$.*

We now conclude the proof of Theorem 4.12.

Let L be a recognizable subset of A^* whose syntactic monoid is $\mathcal{J}$-trivial and let $\eta : A^* \to M$ be the syntactic morphism of M. By Theorem 4.14, there is a surjective morphism $\pi : N \to M$, where N is an ordered monoid satisfying the identity $x \leq 1$. By Proposition A.1.7, there exists a morphism $\varphi : A^* \to N$ such that $\pi \circ \varphi = \eta$, and thus φ recognizes L. Setting $P = \varphi(L)$, we observe that

$$L = \varphi^{-1}(P) = \bigcup_{m \in P} \varphi^{-1}(m) \tag{4.2}$$

and that

$$\varphi^{-1}(m) = \varphi^{-1}(\downarrow m) \setminus \bigcup_{x<m} \varphi^{-1}(\downarrow x) \tag{4.3}$$

We claim that, for every $s \in N$, $\varphi^{-1}(\downarrow s) = R_s$, where

$$R_s = \bigcup_{\substack{\varphi(a_1 \cdots a_k) \leq s \\ k \leq |N|}} A^*a_1A^*a_2A^* \cdots A^*a_kA^* \tag{4.4}$$

First assume that $u \in A^*a_1A^*a_2A^* \cdots A^*a_kA^*$, with $k \leq |N|$. Then $\varphi(u) \in N\varphi(a_1)N \cdots N\varphi(a_k)N$, and since the identity $x \leq 1$ holds in N, $\varphi(u) \leq \varphi(a_1) \cdots \varphi(a_k) = \varphi(a_1 \cdots a_k) \leq s$. Thus $u \in \varphi^{-1}(\downarrow s)$. Conversely, let $u \in \varphi^{-1}(\downarrow s)$. Setting $u = a_1 \cdots a_r$, we have

$$1 \leq \varphi(a_1) \leq \varphi(a_1a_2) \leq \ldots \leq \varphi(a_1 \cdots a_r)$$

It follows that u can be factorized as $u = u_0a_1u_1 \cdots a_ku_k$, with $k \leq |N|$ and

$$1 = \varphi(u_0) < \varphi(u_0a_1) = \varphi(u_0a_1u_1) < \ldots < \varphi(u_0a_1 \cdots a_k) = \varphi(u_0a_1 \cdots a_ku_k)$$

It follows that $u \in A^*a_1A^*a_2A^* \cdots A^*a_kA^*$. Furthermore

$$\varphi(a_1 \cdots a_k) \leq \varphi(u_0a_1u_1 \cdots a_ku_k) = \varphi(u) \leq s$$

Thus $u \in R_s$. Formulas 4.2, 4.3 and 4.4 together show that L is a boolean combination of sets of the form $A^*a_1A^*a_2A^* \cdots A^*a_kA^*$. □

Example 4.4. Let $A = \{a, b, c\}$ and let $L = A^*abA^*$. The minimal automaton of L is represented in Figure 4.5. The transitions and the relations defining the syntactic monoid M of L are given in the following tables

1	a	b	c	ab	ba
1	2	1	1	3	2
2	2	3	1	3	3
3	3	3	3	3	3

$a^2 = a$	$b^2 = b$	$cb = c$
$ab = 0$	$bc = b$	$c^2 = c$
$ac = c$	$ca = a$	

The $\mathcal{D}$-class structure of M is represented in Figure 4.6.

One can see in particular that M is not $\mathcal{J}$-trivial. Consequently, the level of L is greater than or equal to 1 (it is actually 3/2).

Example 4.5. Consider now the set $L' = A^*abA^*$ on the alphabet $A = \{a, b\}$. The minimal automaton of L' can be obtained from the one of L by deleting the transitions of label c.

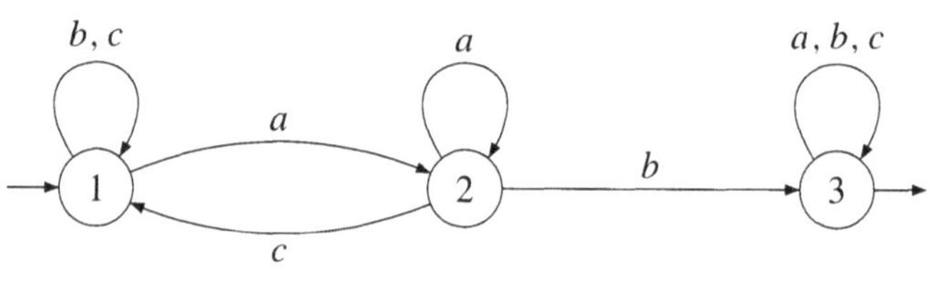

Figure 4.5. The minimal automaton of L.

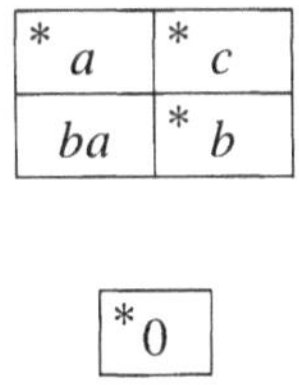

Figure 4.6. The $\mathcal{D}$-classes of M.

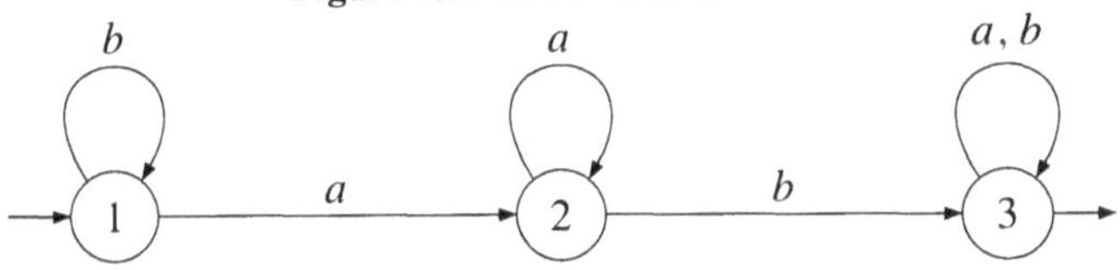

Figure 4.7. The minimal automaton of L'.

The transitions and the relations defining the syntactic monoid M' of L' are given in the following table:

1	a	b	ab	ba
1	2	1	3	2
2	2	3	3	3
3	3	3	3	3

$$a^2 = a$$
$$ab = 0$$
$$b^2 = b$$

The $\mathcal{D}$-class structure of M is represented in Figure 4.8. Therefore M' is $\mathcal{J}$-trivial and L' is of level 1. In fact, we have $L' = A^*aA^*bA^*$.

We show that the sets of level 3/2 are the finite unions of subsets of the form

$$A_0^*a_1A_1^*a_2 \cdots a_kA_k^*$$

where $k \geq 0$, $a_1, \ldots, a_k \in A$ and $A_0, \ldots, A_k$ are subsets of A. Identities defining the variety $\mathbf{V}_{3/2}$ are also known. If x is a word, denote by $c(x)$ the set of letters occurring at least once in x.

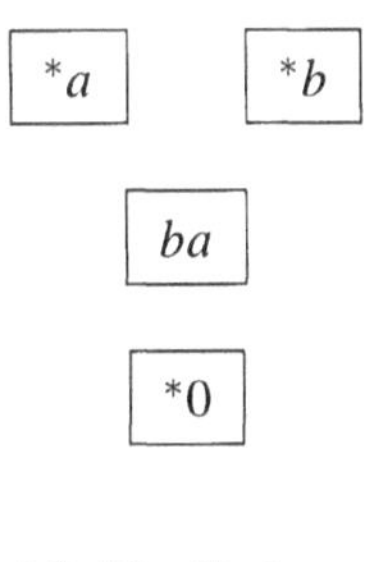

Figure 4.8. The $\mathcal{D}$-classes of M'.

Theorem 4.15. *The variety* $\mathbf{V}_{3/2}$ *is defined by the identities* $x^\pi y x^\pi \leq x^\pi$*, where* x *and* y *are words such that* $c(x) = c(y)$.

This theorem provides an algorithm to decide membership in $\mathbf{V}_{3/2}$. However, the decidability problem of the varieties $\mathbf{V}_n$ is still open for $n \geq 2$.

4.2 Varieties closed under marked product. A $*$-variety is said to be closed under marked product if, for each alphabet A, and for each letter $a \in A$, the marked product XaY of two subsets X, Y of $\mathcal{V}(A^*)$ belongs to $\mathcal{V}(A^*)$.

There is a nice algebraic characterization of these varieties. A semigroup morphism $\pi : S \to T$ is said to be *aperiodic* if, for each idempotent e of T, the semigroup $\pi^{-1}(e)$ is aperiodic. A semigroup S is said to be an *aperiodic extension* of T if there exists a surjective aperiodic morphism from S onto T. Finally a variety of monoids (or semigroups) $\mathbf{V}$ is said to be *closed under aperiodic extension* if and only if every aperiodic extension of a monoid (semigroup) of $\mathbf{V}$ is still in $\mathbf{V}$.

Theorem 4.16. *A $*$-variety is closed under marked product if and only if the corresponding variety of monoids is closed under aperiodic extension.*

Example 4.6. Let $\mathbf{H}$ be a group variety (for instance the variety of commutative groups, nilpotent groups, solvable groups, etc.). Denote by $\bar{\mathbf{H}}$ the variety of monoids all of whose subgroups belong to $\mathbf{H}$. By the results of Annex A, the $\mathcal{H}$-classes containing an idempotent are groups of the variety $\mathbf{H}$. One can show that $\bar{\mathbf{H}}$ is closed under aperiodic extension. It follows that the corresponding $*$-variety is closed under marked product.

5 Local properties of finite words

The simplest local property consists to enforce (or to forbid) the presence of a given letter in a word. The varieties described by these properties were described in Propositions 3.5 and 3.6.

One can also count the number of occurrences of a given letter a. But since we want to count with a finite automaton, we may combine only two types of computation:

(1) Count up to a certain level, like children: 1, 2, 3, 4, 5, many...

(2) Count modulo some integer n.

More formally, let us define for $n > 0$ and $t \geq 0$, a congruence $\equiv_{n,t}$ on $\mathbb{N}$ by $x \equiv_{n,t} y$ if and only if either $x = y \leq t$, or $x > t$, $y > t$ and $x \equiv y \mod n$. It is easy to verify that each congruence of finite index on the additive semigroup $\mathbb{N}$ is of this form. In summary, we shall be able to test with a finite automaton whether two integers are congruent modulo $\equiv_{n,t}$.

As a first step, we shall count only up to a given threshold t, and ignore computation modulo n. We then simply denote by $\equiv_t$ the congruence $\equiv_{t,1}$. If a is a letter of an alphabet A, we denote by $L(a, k)$ the set of words of A^+ which contain exactly k occurrences of a

$$L(a, k) = \{u \in A^+ \mid |u|_a = k\}$$

Denote by **Acom** the variety of commutative and aperiodic semigroups. Then the following result holds.

Proposition 5.1. *For each alphabet A, $\mathcal{A}com(A^+)$ is the boolean algebra generated by the subsets of the form $L(a, k)$ where $a \in A$ and $k \geq 0$.*

Consider now counting modulo n. For $0 \leq k \leq n$, we denote by $L(a, k, n)$ the set of words of A^+ in which the number of occurrences of a's is congruent to k modulo n:

$$L(a, k, n) = \{u \in A^+ \mid |u|_a \equiv k \mod n\}$$

Not surprinsingly, we arrive this time on the variety **Gcom** of commutative groups.

Proposition 5.2. *For each alphabet A, $\mathcal{G}com(A^+)$ is the boolean algebra generated by the subsets of the form $L(a, k, n)$ where $a \in A$ and $0 \leq k < n$.*

Combining the two previous results yields a characterization of the variety **Com** of commutative semigroups.

Proposition 5.3. *For each alphabet A, $\mathcal{C}om(A^+)$ is the boolean algebra generated by the subsets of the form*

$$L(a, k) = \{u \in A^+ \mid |u|_a = k\} \quad \textit{or} \quad L(a, k, n) = \{u \in A^+ \mid |u|_a \equiv k \bmod n\}$$

where $a \in A$ and $0 \leq k < n$.

5.1 Prefixes and suffixes of finite words. A subset L of A^+ is said to be *prefix testable* (resp. *suffix testable*) if L is boolean combination of subsets of the form uA^* (resp. A^*u), where $u \in A^+$. (The term *reverse definite* (resp. *definite*) is also used in the literature.) The syntactic characterization of these sets requires a new definition.

Let S be a semigroup and let e be an idempotent of S. We remind the reader that the local semigroup associated with e, defined in Annex A, is the set eSe of elements of the form ese, with $s \in S$. One defines similarly the *left local semigroup* eS and the *right local semigroup* Se associated with e:

$$eS = \{es \mid s \in S\} \qquad Se = \{se \mid s \in S\}$$

If **V** is a variety of semigroups, the class of semigroups S such that, for each idempotent e of S, the local semigroup (resp. left, right) associated with S belongs to **V** form a variety, called the *local variety* (resp. *left local, right local*), associated with **V** and denoted by **LV** (resp. $\mathbf{L}^{\ell}\mathbf{V}$, $\mathbf{L}^{r}\mathbf{V}$).

Example 5.1. Let us start with the trivial variety **1**.

(1) A semigroup S is locally trivial if and only if, for each idempotent e of S and for each $s \in S$, $ese = e$. The variety **LI** of locally trivial semigroups is defined by the identity $x^\pi yx^\pi = x^\pi$.
(2) A semigroup S is left locally trivial (resp. right) if and only if, for each idempotent e of S and for each $s \in S$, $es = e$ (resp. $se = e$). The variety $\mathbf{L}^\ell\mathbf{I}$ (resp. $\mathbf{L}^r\mathbf{I}$) of left (resp. right) locally trivial semigroups is defined by the identity $x^\pi y = x^\pi$ (resp. $yx^\pi = x^\pi$).

Example 5.2. A semigroup S is locally idempotent and commutative if and only if, for each idempotent e of S and for each $s, t \in S$, $esese = ese$ and $esete = etese$. The variety of locally idempotent and commutative semigroups is defined by the identities $x^\pi yx^\pi = x^\pi yx^\pi yx^\pi$ and $x^\pi yx^\pi zx^\pi = x^\pi zx^\pi yx^\pi$.

The prefix testable subsets are characterized as follows:

Proposition 5.4. *Let L be a recognizable subset of A^+. The following conditions are equivalent:*

(1) *L is prefix testable,*
(2) *$L = XA^* \cup Y$ where X and Y are finite subsets of A^+,*
(3) *$S(L)$ is left locally trivial (i.e. satisfies the identity $x^\pi y = x^\pi$).*

There is of course a dual statement for suffixes instead of prefixes.

Proposition 5.5. *Let L be a recognizable subset of A^+. The following conditions are equivalent:*

(1) *L is suffix testable,*
(2) *$L = A^*X \cup Y$ where X and Y are finite subsets of A^+,*
(3) *$S(L)$ is right locally trivial (i.e. satisfies the identity $yx^\pi = x^\pi$).*

The prefix (resp. suffix) testable subsets hence define a +-variety, and the corresponding variety of semigroups is $\mathbf{L}^\ell\mathbf{I}$ (resp. $\mathbf{L}^r\mathbf{I}$).

If one takes simultaneously prefixes and suffixes into account, one obtains the variety associated with the locally trivial semigroups. A subset L of A^+ is said to be *prefix-suffix testable* if L is boolean combination of subsets of the form uA^* or A^*u, where $u \in A^+$.

Proposition 5.6. *Let L be a recognizable subset of A^+. The following conditions are equivalent:*

(1) *L is prefix-suffix testable,*

(2) *$L = XA^*Y \cup Z$, where X, Y and Z are finite subsets of A^+,*
(3) *$S(L)$ is locally trivial.*

Therefore, the prefix-suffix testable subsets define a +-variety, and the corresponding variety of semigroups is **LI**.

Note that the second condition of the previous propositions does not mention the complement, although, for each alphabet A, the sets $\mathcal{L}^r\mathcal{I}(A^+)$, $\mathcal{L}^\ell\mathcal{I}(A^+)$ and $\mathcal{LI}(A^+)$ are boolean algebra.

5.2 Factors of finite words. The next step is to consider the factors of a word without multiplicity. A subset L of A^+ is said to be *locally testable* if it is boolean combination of subsets of the form uA^*, A^*u or A^*uA^* where $u \in A^+$.

This definition can be reformulated by mean of a congruence. Denote by $p_n(u)$ (resp. $s_n(u)$) the prefix (resp. suffix) of length n of a word u and denote by $F_n(u)$ the set of its factors of length n. Let $\sim_k$ be the congruence on A^+ defined by

$$u \sim_k v \text{ if and only if } F_k(u) = F_k(v) \text{ and, for } n < k,$$
$$p_n(u) = p_n(v) \text{ and } s_n(u) = s_n(v)$$

that is, if u and v have the same prefixes and suffixes of length $< k$ and the same factors of length k. The locally testable subsets admit the following characterization. The proof of this result is omitted.

Theorem 5.7. *Let L be a recognizable subset of A^+. The following conditions are equivalent:*
(1) *L is locally testable,*
(2) *L is union of $\sim_k$-classes for some $k > 0$,*
(3) *$S(L)$ is locally idempotent and commutative.*

Example 5.3. The set $X = (ab)^+$ is locally testable since $X = (aA^* \cap A^*b) \setminus A^*\{aa, bb\}A^*$.

In other words, the locally testable subsets define a +-variety and the corresponding variety of semigroups is the variety of locally idempotent and commutative semigroups.

One of the crucial results for the proof of Theorem 5.7 is a representation result for locally idempotent and commutative semigroups.

Proposition 5.8. *Let S be a finite quotient of A^+. Then S is locally idempotent and commutative if and only if there exists an integer $k > 0$ such that S is a quotient of $A^+/{\sim_k}$.*

Theorem 5.7 can be refined by considering positive varieties.

Theorem 5.9. *Let L be a recognizable subset of A^+. The following conditions are equivalent:*

(1) *L is a positive boolean combination of subsets of the form uA^*, A^*u or A^*uA^* where $u \in A^+$,*

(2) *$S(L)$ is locally idempotent and commutative and satisfies the identity*

$$x^\pi y x^\pi \leq x^\pi$$

Condition (2) means that each ordered local submonoid associated with some idempotent e is an idempotent and commutative monoid in which the identity e is the top element. The recognizable subsets of A^+ verifying the conditions of Theorem 5.7 are called *positively locally testable.*

Counting the number of occurrences of each factor up to a certain threshold t is another possibility. If u and x are words, we denote by $\left[\begin{smallmatrix} u \\ x \end{smallmatrix}\right]$ the number of occurrences of x as a factor of u. For each $k, t > 0$, let $\equiv_{k,t}$ be the equivalence on A^+ defined by setting $u \equiv_{k,t} v$ if and only if, for each word x of length $\leq k$,

$$\begin{bmatrix} u \\ x \end{bmatrix} \equiv_t \begin{bmatrix} v \\ x \end{bmatrix}$$

For instance, $u \equiv_{k,1} v$ if and only if u and v have the same sets of factors of length k, and $u \equiv_{k,5} v$ if and only if u and v have the same factors of length k, counting multiplicities up to threshold 5.

Example 5.4. We have $abababab \equiv_{2,3} abababa$ since $abababab$ contains 4 ($\equiv$ 3 threshold 3) occurrences of ab and 3 ($\equiv$ 3 threshold 3) occurrences of ba, and no occurrences of aa (resp. bb).

Finally, we define a congruence $\sim_{k,t}$ on A^+ by setting $u \sim_{k,t} v$ if and only if

(a) for each $n < k$, $p_n(u) = p_n(v)$,

(b) for each $n < k$, $s_n(u) = s_n(v)$,

(c) $u \equiv_{k,t} v$.

In other words, two words are equivalent modulo $\sim_{k,t}$ if they have the same prefixes and suffixes of length $< k$ and the same factors of length k, counting multiplicity up to threshold t.

A subset L of A^+ is said to be *threshold locally testable* (TLT) if there exist integers $k > 0$ and $t > 0$ such that L is union of classes modulo $\sim_{k,t}$.

Example 5.5. The set $X = A^*aA^*aA^*$ is threshold locally testable since it is the set of words having at least 2 occurrences of a.

Instead, $Y = a^*ba^*ca^*$ is not threshold locally testable. Indeed, we have $a^{kr}ba^{kr}ca^{kr} \sim_{k,r} a^{kr}ca^{kr}ba^{kr}$, although the first word belongs to Y, but not the second one.

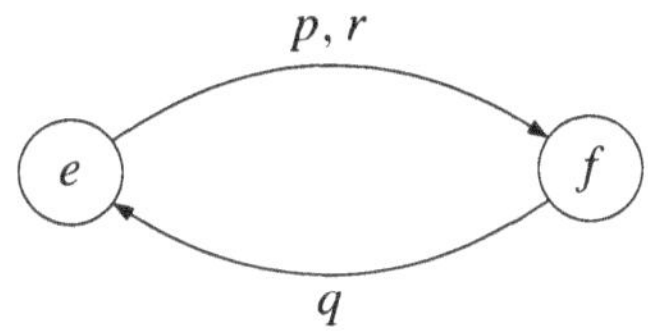

Figure 5.1. The condition $pqr = rqp$.

Set, for $x \in A^+$, and $t \geq 0$,

$$F(x,t) = \left\{ u \in A^+ \mid \begin{bmatrix} u \\ x \end{bmatrix} \geq t \right\}$$

In other words, $F(x,t)$ is the set of words u containing at least t occurrences of the factor x. For instance, $F(ab, 1) = A^*xA^*$, and $F(aba, 2) = A * abaA^*abaA^* \cup A^*ababaA^*$.

Another definition is on order to give the syntactic characterization of the threshold locally testable subsets. The *graph* $G(S)$ of a semigroup S has $E(S)$ as set of vertices and as edges the triples (e, esf, f) such that $e, f \in E(S)$ and $s \in S$ (see Example 5.6 below). Let us say that S satisfies Thérien's condition if, when p and r are the labels of edges from e to f and when q is the label of an edge from f to e, then $pqr = rqp$.

Theorem 5.10. *Let L be a recognizable subset of A^+. The following conditions are equivalent:*

(1) *L is threshold locally testable,*

(2) *L is a boolean combination of subsets of the form uA^*, A^*u or $F(x,t)$ where $x \in A^+$, and $t > 0$,*

(3) *$S(L)$ is aperiodic and satisfies Thérien's condition.*

We shall admit this result, which relies on a proposition of independent interest.

Proposition 5.11. *Let S be a finite quotient of A^+. Then S is aperiodic and satisfies Thérien's condition if and only if there exist two integers $k, t > 0$ such that S is a quotient of $A^+/\sim_{k,t}$.*

Example 5.6. Let $A = \{a, b\}$ and $L = a^*ba^*$. The minimal automaton of L is represented in Figure 5.2. The transitions and relations defining the syntactic semigroup S of L are given in the following tables:

	1	2
a	1	2
b	2	-
bb	-	-

$a = 1$

$b^2 = 0$

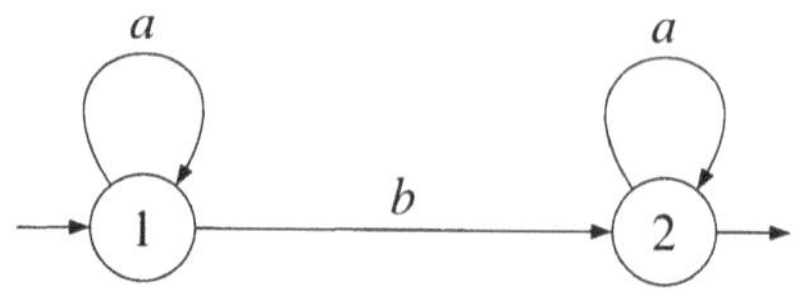

Figure 5.2. The minimal automaton of a^*ba^*.

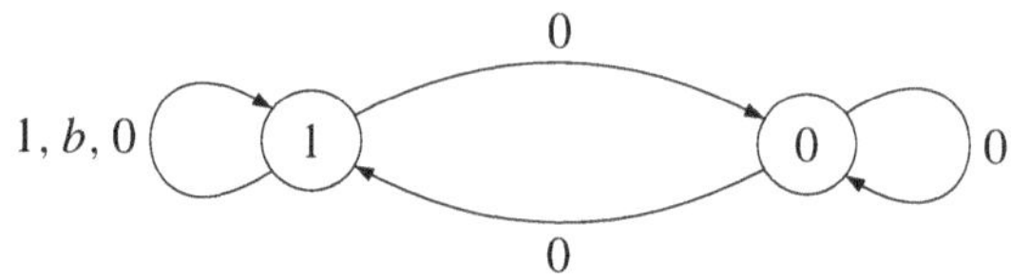

Figure 5.3. The graph of S.

Thus $S = \{a, b, 0\}$, where $a = 1$ is the identity and $E(S) = \{1, 0\}$. The local semigroups are $0S0 = \{0\}$ and $1S1 = S$. The latter local semigroup is not idempotent, since $b^2 \neq b$. Consequently, L is not locally testable. On the other hand, the graph of S satisfies the condition $pqr = rqp$. Therefore L is threshold locally testable.

6 Notes

Most of the results presented in this Annex can be found in Eilenberg [98], Lallement [167], Pin [241], or Almeida [7]. Theorem 1.13 is due to Reiterman [263]. The variety theorem is due to Eilenberg. Theorem 4.6 is due to Higman (see [183] for a proof). Theorem 4.12 is due to Simon (see [98, 241]). A direct proof of Theorem 4.14 can be found in [314]. Theorem 4.16 is due to Straubing [308]. Theorem 5.7 is due to Brzozowski-Simon [44] and to McNaughton independently [190]. The proof is reproduced in [98]. Theorem 5.10 is mainly due to Thérien and Weiss [321].

REFERENCES

[1] M. ABADI, L. LAMPORT AND P. WOLPER, Realizable and unrealizable specifications of reactive systems, in *Automata, languages and programming (Stresa, 1989)*, pp. 1–17, Springer, Berlin, 1989.

[2] R. L. ADLER, D. COPPERSMITH AND M. HASSNER, Algorithms for sliding block codes. An application of symbolic dynamics to information theory, *IEEE Trans. Inform. Theory* **29**,1 (1983), 5–22.

[3] R. L. ADLER AND B. MARCUS, Topological entropy and equivalence of dynamical systems, *Mem. Amer. Math. Soc.* **20**,219 (1979), iv+84.

[4] A. V. AHO, J. E. HOPCROFT AND J. D. ULLMAN, *The design and analysis of computer algorithms*, Addison-Wesley Publishing Co., Reading, Mass.-London-Amsterdam, 1975. Second printing, Addison-Wesley Series in Computer Science and Information Processing.

[5] M. AJTAI, Σ_1^1-formulae on finite structures, *Ann. Pure Appl. Logic* **24**,1 (1983), 1–48.

[6] L. ALBERUCCI, Strictness of the Modal μ-Calculus Hierarchy, in *Automata, Logics, and Infinite Games: A Guide to Current Research [outcome of a Dagstuhl seminar, February 2001]*, E. Grädel, W. Thomas and T. Wilke (éd.), pp. 185–201, *Lecture Notes in Computer Science* vol. 2500, Springer, 2002.

[7] J. ALMEIDA, *Finite semigroups and universal algebra*, World Scientific Publishing Co. Inc., River Edge, NJ, 1994. Translated from the 1992 Portuguese original and revised by the author.

[8] A. ARNOLD, Rational ω-languages are nonambiguous, *Theoret. Comput. Sci.* **26**,1-2 (1983), 221–223.

[9] A. ARNOLD, Topological characterizations of infinite behaviours of transition systems, in *Automata, Languages and Programming*, J. Diaz (éd.), pp. 28–38,

Lecture Notes in Comput. Sci. vol. 154, Springer Verlag, Berlin, Heidelberg, New York, 1983.

[10] A. ARNOLD, Deterministic and nonambiguous rational ω-languages, in *Automata on infinite words (Le Mont-Dore, 1984)*, pp. 18–27, Springer, Berlin, 1985.

[11] A. ARNOLD, A syntactic congruence for rational ω-languages, *Theoret. Comput. Sci.* **39**,2-3 (1985), 333–335.

[12] A. ARNOLD AND D. NIWIŃSKI, Fixed point characterization of Büchi automata on infinite trees, *J. Inf. Process. Cybern. EIK* **26** (1990), 453–461.

[13] A. ARNOLD AND D. NIWIŃSKI, Fixed point characterization of weak monadic logic definable sets of trees, in *Tree automata and languages*, M. Nivat and A. Podelski (éd.), pp. 159–188, Elsevier, Amsterdam, The Netherlands, 1992.

[14] A. ARNOLD AND D. NIWIŃSKI, *Rudiments of μ-Calculus*, Elsevier, Amsterdam, The Netherlands, 2001.

[15] R. J. AUMANN AND S. HART (éd.), *Handbook of game theory with economic applications. Vol. II*, North-Holland Publishing Co., Amsterdam, 1994.

[16] R. BARUA, The Hausdorff-Kuratowski hierarchy of ω-regular languages and a hierarchy of Muller automata, *Theoret. Comput. Sci.* **96**,2 (1992), 345–360.

[17] J. BARWISE (éd.), *Handbook of mathematical logic, Studies in Logic and the Foundations of Mathematics*, North-Holland Publishing Co., Amsterdam, 1977. Edited with the cooperation of H. J. Keisler, K. Kunen, Y. N. Moschovakis and A. S. Troelstra.

[18] D. BEAUQUIER, Bi-limites de langages reconnaissables, *Theoret. Comput. Sci.* **33** (1984), 335–342.

[19] D. BEAUQUIER, Ensembles reconnaissables de mots bi-infinis. Limite et déterminisme, in *Automata on infinite words (Le Mont-Dore, 1984)*, pp. 28–46, Springer, Berlin, 1985.

[20] D. BEAUQUIER AND M. NIVAT, About rational sets of factors of a bi-infinite word, in *Automata, languages and programming (Nafplion, 1985)*, pp. 33–42, Springer, Berlin, 1985.

[21] D. BEAUQUIER AND D. PERRIN, Codeterministic automata on infinite words, *Inform. Process. Lett.* **20**,2 (1985), 95–98.

[22] D. BEAUQUIER AND J.-E. PIN, Factors of words, in *Automata, languages and programming (Stresa, 1989)*, pp. 63–79, Springer, Berlin, 1989.

[23] D. BEAUQUIER AND J.-E. PIN, Languages and scanners, *Theoret. Comput. Sci.* **84** (1991), 3–21.

[24] D. BEAUQUIER AND A. RABINOVICH, Monadic logic of order over naturals has no finite base, Rap. Tech., University Paris XII, 2001.

[25] N. BEDON, Finite automata and ordinals, *Theoret. Comput. Sci.* **156**,1-2 (1996), 119–144.

[26] N. BEDON, Automata, semigroups and recognizability of words on ordinals, *Internat. J. Algebra Comput.* **8**,1 (1998), 1–21.

[27] N. BEDON AND O. CARTON, An Eilenberg theorem for words on countable ordinals, in *LATIN'98: theoretical informatics (Campinas, 1998)*, pp. 53–64, Springer, Berlin, 1998.

[28] N. BEDON AND O. CARTON, An Eilenberg theorem for words on countable ordinals, in *LATIN'98: Theoretical Informatics (Campinas, 1998)*, C. L. Lucchesi and A. V. Moura (éd.), pp. 53–64, *Lecture Notes in Comput. Sci.* vol. 1380, Springer, Berlin, 1998.

[29] E. R. BERLEKAMP, J. H. CONWAY AND R. K. GUY, *Winning ways for your mathematical plays. Vol. 1*, Academic Press Inc. [Harcourt Brace Jovanovich Publishers], London, 1982. Games in general.

[30] E. R. BERLEKAMP, J. H. CONWAY AND R. K. GUY, *Winning ways for your mathematical plays. Vol. 2*, Academic Press Inc. [Harcourt Brace Jovanovich Publishers], London, 1982. Games in particular.

[31] J. BERSTEL AND D. PERRIN, *Theory of codes*, Academic Press Inc., Orlando, Fla., 1985.

[32] G. BIRKHOFF, On the structure of abstract algebras, *Proc. Cambridge Phil. Soc.* **31** (1935), 433–454.

[33] F. BLANCHARD AND G. HANSEL, Languages and subshifts, in *Automata on infinite words (Le Mont-Dore, 1984)*, pp. 138–146, Springer, Berlin, 1985.

[34] F. BLANCHARD AND D. PERRIN, Relèvement d'une mesure ergodique par un codage, *Z. Wahrsch. Verw. Gebiete* **54**,3 (1980), 303–311.

[35] S. L. BLOOM, Varieties of ordered algebras, *J. Comput. System Sci.* **13**,2 (1976), 200–212.

[36] L. BOASSON AND M. NIVAT, Adherences of languages, *J. Comput. System Sci.* **20**,3 (1980), 285–309.

[37] N. BOURBAKI, *Eléments de mathématiques, Topologie Générale*, CCLS, Paris, 1974.

[38] M. BOYLE, B. KITCHENS AND B. MARCUS, A note on minimal covers for sofic systems, *Proc. Amer. Math. Soc.* **95**,3 (1985), 403–411.

[39] M. BOYLE, B. MARCUS AND P. TROW, Resolving maps and the dimension group for shifts of finite type, *Mem. Amer. Math. Soc.* **70**,377 (1987), vi+146.

[40] T. C. BROWN, An interesting combinatorial method in the theory of locally finite semigroups, *Pacific J. Math.* **36** (1971), 285–289.

[41] V. BRUYÈRE AND O. CARTON, Automata on linear orderings, in *MFCS'2001*, J. Sgall, A. Pultr and P. Kolman (éd.), pp. 236–247, *Lecture Notes in Comput. Sci.* vol. 2136, Springer Verlag, Berlin, Heidelberg, New York, 2001.

[42] J. A. BRZOZOWSKI AND R. S. COHEN, On star-free events, in *Proc. Hawaii Int. Conf. on System Science*, pp. 1–4, 1968.

[43] J. A. BRZOZOWSKI AND R. KNAST, The dot-depth hierarchy of star-free languages is infinite, *J. Comput. System Sci.* **16**,1 (1978), 37–55.

[44] J. A. BRZOZOWSKI AND I. SIMON, Characterizations of locally testable events, *Discrete Math.* **4** (1973), 243–271.

[45] J. R. BÜCHI, Weak second-order arithmetic and finite automata, *Z. Math. Logik und Grundl. Math.* **6** (1960), 66–92.

[46] J. R. BÜCHI, On a decision method in restricted second order arithmetic, in *Logic, Methodology and Philosophy of Science (Proc. 1960 Internat. Congr .)*, pp. 1–11, Stanford Univ. Press, Stanford, Calif., 1962.

[47] J. R. BÜCHI, Decision methods in the theory of ordinals, *Bull. Amer. Math. Soc.* **71** (1965), 767–770.

[48] J. R. BÜCHI, Transfinite automata recursions and weak second order theory of ordinals, in *Logic, Methodology and Philos. Sci. (Proc. 1964 Internat. Congr.)*, pp. 3–23, North-Holland, Amsterdam, 1965.

[49] J. R. BÜCHI, The monadic second-order theory of ω_1, in *The Monadic Second-Order Theory of All Countable Ordinals*, J. R. Büchi and D. Siefkes (éd.), pp. 1–127, *Lecture Notes in Math.* vol. 328, Springer Verlag, Berlin, Heidelberg, New York, 1973.

[50] J. R. BÜCHI, Using determinancy of games to eliminate quantifiers, in *Fundamentals of computation theory (Proc. Internat. Conf., Poznań-Kórnik, 1977)*, pp. 367–378, *Lecture Notes in Comput. Sci.* vol. 56, Springer, Berlin, 1977.

[51] J. R. BÜCHI, State-strategies for games in $F_{\sigma\delta} \cap G_{\delta\sigma}$, *J. Symbolic Logic* **48**,4 (1984), 1171–1198.

[52] J. R. BÜCHI, C. ELGOT AND J. B. WRIGHT, The non existence of certain algorithms of finite automata theory, *Notices Amer. Math. Soc.* **5** (1958), 98.

[53] J. R. BÜCHI AND L. H. LANDWEBER, Definability in the monadic second-order theory of successor, *J. Symbolic Logic* **34** (1969), 166–170.

[54] J. R. BÜCHI AND L. H. LANDWEBER, Solving sequential conditions by finite-state strategies, *Trans. Amer. Math. Soc.* **138** (1969), 295–311.

[55] J. R. BÜCHI AND D. SIEFKES, Axiomatization of the monadic second order theory of ω_1, in *The monadic second order theory of all countable ordinals (Decidable theories, II)*, pp. 129–217, *Lecture Notes in Math.* vol. 328, Springer, Berlin, 1973.

[56] O. CARTON, *Mots infinis, ω-semigroupes et topologie*, PhD thesis, University Paris VII, 1993.

[57] O. CARTON, Chain automata, in *Technology and Applications, IFIP*, I. S. B. Pherson (éd.), pp. 451–458, *Information Processing'94, Vol. I*, North Holland, Amsterdam, 1994.

[58] O. CARTON, Chain automata, *Theoret. Comput. Sci.* **161**,1-2 (1996), 191–203.

[59] O. CARTON, Wreath product and infinite words, *J. Pure Appl. Algebra* **153**,2 (2000), 129–150.

[60] O. CARTON AND R. MACEIRAS, Computing the Rabin index of a parity automaton, *Theor. Inform. Appl.* **33**,6 (1999), 495–505.

[61] O. CARTON AND M. MICHEL, Unambiguous Büchi automata, in *LATIN'2000*, G. Gonnet, D. Panario and A. Viola (éd.), Berlin, 2000, pp. 407–416, *Lecture Notes in Comput. Sci.* vol. 1776, Springer.

[62] O. CARTON AND M. MICHEL, Unambiguous Büchi automata, *Theoret. Comput. Sci.* **297** (2003), 37–81.

[63] O. CARTON AND D. PERRIN, Chains and superchains in ω-semigroups, in *Semigroups, automata and languages (Porto, 1994)*, pp. 17–28, World Sci. Publishing, River Edge, NJ, 1996.

[64] O. CARTON AND D. PERRIN, Chains and superchains for ω-rational sets, automata and semigroups, *Internat. J. Algebra Comput.* **7**,6 (1997), 673–695.

[65] O. CARTON AND D. PERRIN, The Wadge-Wagner hierarchy of ω-rational sets, in *Automata, Languages and Programming*, P. Degano, R. Gorrieri and A. Marchetti-Spaccamela (éd.), pp. 17–35, *Lecture Notes in Comput. Sci.* vol. 1256, Springer, 1997.

[66] O. CARTON AND D. PERRIN, The Wagner hierarchy, *Internat. J. Algebra Comput.* **9**,5 (1999), 597–620.

[67] S. CHO AND D. U. T. HUỲNH, Finite-automaton aperiodicity is PSPACE-complete, *Theoret. Comput. Sci.* **88**,1 (1991), 99–116.

[68] Y. CHOUEKA, Theories of automata on ω-tapes: a simplified approach, *J. Comput. System Sci.* **8** (1974), 117–141.

[69] Y. CHOUEKA, Finite automata, definable sets, and regular expressions over ω^n-tapes, *J. Comput. System Sci.* **17**,1 (1978), 81–97.

[70] A. CHURCH, Logic, arithmetic, and automata, in *Proc. Internat. Congr. Mathematicians (Stockholm, 1962)*, pp. 23–35, Inst. Mittag-Leffler, Djursholm, 1963.

[71] E. M. CLARKE AND S. BERND-HOLGER, Model checking, in *Handbook of automated reasoning*, A. Robinson and A. Voronkov (éd.), Elsevier, 1999.

[72] E. M. CLARKE, I. A. DRAGHICESCU AND R. P. KURSHAN, A unified approach for showing language inclusion and equivalence between various types of ω-automata, *Inform. Process. Lett.* **46**,6 (1993), 301–308.

[73] J. COHEN, D. PERRIN AND J.-E. PIN, On the expressive power of temporal logic, *J. Comput. System Sci.* **46**,3 (1993), 271–294.

[74] R. S. COHEN AND J. A. BRZOZOWSKI, Dot-depth of star-free events, *J. Comput. System Sci.* **5** (1971), 1–16.

[75] R. S. COHEN AND A. Y. GOLD, Theory of ω-languages. I. Characterizations of ω-context-free languages, *J. Comput. System Sci.* **15**,2 (1977), 169–184.

[76] R. S. COHEN AND A. Y. GOLD, Theory of ω-languages. II. A study of various models of ω-type generation and recognition, *J. Comput. System. Sci.* **15**,2 (1977), 185–208.

[77] J. COHEN-CHESNOT, On the expressive power of temporal logic for infinite words, *Theoret. Comput. Sci.* **83**,2, Algorithms Automat. Complexity Games (1991), 301–312.

[78] K. J. COMPTON, On rich words, in *Combinatorics on words (Waterloo, Ont., 1982)*, pp. 39–61, Academic Press, Toronto, Ont., 1983.

[79] J. H. CONWAY, *Regular Algebra and Finite Machines*, Chapman and Hall, London, 1971.

[80] J. H. CONWAY, *On numbers and games*, Academic Press [Harcourt Brace Jovanovich Publishers], London, 1976. London Mathematical Society Monographs, No. 6.

[81] B. COURCELLE, Frontiers of infinite trees, *RAIRO Inform. Théor.* **12**,4 (1978), 319–337.

[82] B. COURCELLE, Fundamental properties of infinite trees, *Theoret. Comput. Sci.* **25**,2 (1983), 95–169.

[83] E. M. COVEN AND G. A. HEDLUND, Sequences with minimal block growth, *Math. Systems Theory* **7** (1973), 138–153.

[84] P. DARONDEAU AND L. KOTT, On the observational semantics of fair parallelism, in *Automata, Languages and programming*, J. Diaz (éd.), pp. 147–159, *Lecture Notes in Comput. Sci.* vol. 154, Springer Verlag, Berlin, Heidelberg, New York, 1983.

[85] P. DARONDEAU AND L. KOTT, Towards a formal proof system for ω-rational expressions, *Inform. Process. Lett.* **19**,4 (1984), 173–177.

[86] P. DARONDEAU AND L. KOTT, A formal proof system for infinitary rational expressions, in *Automata on infinite words (Le Mont-Dore, 1984)*, pp. 68–80, Springer, Berlin, 1985.

[87] M. DAUCHET AND E. TIMMERMAN, Decidability of yield's equality for infinite regular trees, in *Automata on Infinite Words*, M. Nivat and D. Perrin (éd.), pp. 118–136, *Lecture Notes in Comput. Sci.* vol. 192, Springer Verlag, Berlin, Heidelberg, New York, 1985.

[88] M. DAVIS, Infinite games of perfect information, in *Advances in game theory*, pp. 85–101, Princeton Univ. Press, Princeton, N.J., 1964.

[89] M. DENKER, C. GRILLENBERGER AND K. SIGMUND, *Ergodic theory on compact spaces*, Springer-Verlag, Berlin, 1976. Lecture Notes in Mathematics, Vol. 527.

[90] J. DEVOLDER, M. LATTEUX, I. LITOVSKY AND L. STAIGER, Codes and infinite words, *Acta Cybernet.* **11**,4 (1994), 241–256.

[91] J. DONER, Tree acceptors and some of their applications, *J. Comput. System Sci.* **4** (1970), 406–451.

[92] J. DONER, Erratum: "Tree acceptors and some of their applications", *J. Comput. System Sci.* **5** (1971), 453.

[93] J. DUPARC AND M. RISS, The missing link for ω-rational sets, automata, and semigroups, *Int. J. Alg. Comput.*, 2003. to appear.

[94] H.-D. EBBINGHAUS AND J. FLUM, *Finite model theory*, Springer-Verlag, New York, 1995.

[95] H.-D. EBBINGHAUS, J. FLUM AND W. THOMAS, *Mathematical logic*, Springer-Verlag, New York, éd. second, 1994. Translated from the German by Margit Meßmer.

[96] A. EHRENFEUCHT, An application of games to the completeness problem for formalized theories, *Fund. Math.* **49** (1960/1961), 129–141.

[97] S. EILENBERG, *Automata, languages, and machines. Vol. A*, Academic Press [A subsidiary of Harcourt Brace Jovanovich, Publishers], New York, 1974. Pure and Applied Mathematics, Vol. 58.

[98] S. EILENBERG, *Automata, languages, and machines. Vol. B*, Academic Press [Harcourt Brace Jovanovich Publishers], New York, 1976. With two chapters ("Depth decomposition theorem" and "Complexity of semigroups and morphisms") by Bret Tilson, Pure and Applied Mathematics, Vol. 59.

[99] C. C. ELGOT, Decision problems of finite automata design and related arithmetics, *Trans. Amer. Math. Soc.* **98** (1961), 21–51.

[100] C. C. ELGOT AND M. O. RABIN, Decidability and undecidability of second (first) order theory of (generalized) successor, *J. of Symbolic Logic* **31** (1966), 169–181.

[101] E. A. EMERSON, Temporal and modal logic, in *Handbook of theoretical computer science, Vol. B*, pp. 995–1072, Elsevier, Amsterdam, 1990.

[102] E. A. EMERSON AND C. JUTLA, Complexity of tree automata and modal logics of programs, in *Proc. 29th Annual IEEE Symp. on Foundations of Computer Science*, pp. 328–337, 1988.

[103] E. A. EMERSON AND C. JUTLA, The complexity of tree automata and logics of programs, *SIAM J. Comput.* **29** (1999), 132–158.

[104] E. A. EMERSON AND C.-L. LEI, Modalities for model checking: branching time logic strikes back, *Sci. Comput. Programming* **8**,3 (1987), 275–306.

[105] K. ETESSAMI, M. Y. VARDI AND TH. WILKE, First-order logic with two variables and unary temporal logic, in *Proceedings 12th Annual IEEE Symposium on Logic in Computer Science*, Warsaw, Poland, 1997, pp. 228–235, IEEE.

[106] K. ETESSAMI AND TH. WILKE, An until hierarchy for temporal logic, in *Proceedings 11th Annual IEEE Symposium on Logic in Computer Science*, New Brunswick, N.J., 1996, pp. 108–117, IEEE.

[107] K. ETESSAMI AND TH. WILKE, An Until Hierarchy and Other Applications of an Ehrenfeucht-Fraïssé Game for Temporal Logic, *Inform. and Computation* **160**,1/2 (Juil. 2000), 88–108.

[108] R. FAGIN, Generalized first-order spectra and polynomial-time recognizable sets, in *Complexity of computation (Proc. SIAM-AMS Sympos. Appl. Math., New York, 1973)*, pp. 43–73. SIAM–AMS Proc., Vol. VII, Amer. Math. Soc., Providence, R.I., 1974.

[109] B. FARWER, ω-Automata, in *Automata, Logics, and Infinite Games: A Guide to Current Research [outcome of a Dagstuhl seminar, February 2001]*, E. Grädel, W. Thomas and T. Wilke (éd.), pp. 3–20, *Lecture Notes in Computer Science* vol. 2500, Springer, 2002.

[110] W. FELLER, *An introduction to probability theory and its applications. Vol. I*, John Wiley & Sons Inc., New York, éd. third, 1968.

[111] R. FISCHER, Sofic systems and graphs, *Monatsh. Math.* **80**,3 (1975), 179–186.

[112] H. FURSTENBERG, *Recurrence in ergodic theory and combinatorial number theory*, Princeton University Press, Princeton, N.J., 1981. M. B. Porter Lectures.

[113] D. GABBAY, A. PNUELI, S. SHELAH AND J. STAVI, On the temporal analysis of fairness, in *Proc. 7th ACM Symp. Princ. Prog. Lang.*, pp. 163–173, Assoc. Comput. Mach., 1980.

[114] D. M. GABBAY, I. HODKINSON AND M. REYNOLDS, *Temporal logic. Vol. 1*, The Clarendon Press Oxford University Press, New York, 1994. Mathematical foundations and computational aspects, Oxford Science Publications.

[115] D. M. GABBAY, I. M. HODKINSON AND M. A. REYNOLDS, Temporal expressive completeness in the presence of gaps, in *Logic Colloquium '90 (Helsinki, 1990)*, pp. 89–121, Springer, Berlin, 1993.

[116] D. M. GABBAY, M. A. REYNOLDS AND M. FINGER, *Temporal logic. Vol. 2*, The Clarendon Press Oxford University Press, New York, 2000. Mathematical foundations and computational aspects, Oxford Science Publications.

[117] D. GALE AND F. STEWART, Infinite games with perfect information, in *Contributions to the Theory of Games*, pp. 245–266, *Ann. Math. Studies*, Princeton Univ. Press, Princeton, N.J., 1953.

[118] M. R. GAREY AND D. S. JOHNSON, *Computers and intractability*, W. H. Freeman and Co., San Francisco, Calif., 1979. A guide to the theory of NP-completeness, A Series of Books in the Mathematical Sciences.

[119] F. GÉCSEG AND M. STEINBY, *Tree automata*, Akadémiai Kiadó (Publishing House of the Hungarian Academy of Sciences), Budapest, 1984.

[120] F. GIRE, Langages rationnels dont la limite est fermée, *C. R. Acad. Sci. Paris Sér. I Math.* **294**,21 (1982), 701–704.

[121] F. GIRE, Une extension aux mots infinis de la notion de transduction, in *Theoretical Computer Science, Proceedings of the 6th GI Conference*, A. Cremers and H. Kriegel (éd.), pp. 123–139, *Lecture Notes in Comput. Sci.* vol. 145, Springer Verlag, Berlin, Heidelberg, New York, 1982.

[122] F. GIRE, Two decidability problems for infinite words, *Inform. Process. Lett.* **22**,3 (1986), 135–140.

[123] F. GIRE AND M. NIVAT, Relations rationnelles infinitaires, *Calcolo* **21**,2 (1984), 91–125.

[124] F. GIRE AND M. NIVAT, Langages algébriques de mots biinfinis, *Theoret. Comput. Sci.* **86**,2 (1991), 277–323.

[125] S. W. GOLOMB AND B. GORDON, Codes with bounded synchronization delay, *Information and Control* **8** (1965), 355–372.

[126] W. H. GOTTSCHALK AND G. A. HEDLUND, *Topological dynamics*, American Mathematical Society, Providence, R. I., 1955. American Mathematical Society Colloquium Publications, Vol. 36.

[127] E. GRÄDEL, W. THOMAS AND T. WILKE, Automata, Logics, and Infinite Games: A Guide to Current Research [outcome of a Dagstuhl seminar, Febru-

ary 2001], in *Automata, Logics, and Infinite Games, Lecture Notes in Computer Science* vol. 2500, Springer, 2002.

[128] R. L. GRAHAM, B. L. ROTHSCHILD AND J. H. SPENCER, *Ramsey theory*, John Wiley & Sons Inc., New York, éd. second, 1990. A Wiley-Interscience Publication.

[129] Y. GUREVICH, Monadic second-order theories, in *Model-theoretic logics*, pp. 479–506, Springer, New York, 1985.

[130] Y. GUREVICH, Games people play, in *The Collected Works of J. Richard Büchi*, S. McLane and D. Siefkes (éd.), pp. 518–524, Springer, 1990.

[131] Y. GUREVICH AND L. HARRINGTON, Trees, automata and games, in *Proc. ACM Symp. on Theory of Computing*, pp. 60–65, Assoc. Comput. Mach., 1982.

[132] Y. GUREVICH, M. MAGIDOR AND S. SHELAH, The monadic theory of ω_2, *J. Symbolic Logic* **48**,2 (1983), 387–398.

[133] W. HANF, Model-theoretic methods in the study of elementary logic, in *Theory of Models (Proc. 1963 Internat. Sympos. Berkeley)*, pp. 132–145, North-Holland, Amsterdam, 1965.

[134] J. HARTMANIS AND R. E. STEARNS, Sets of numbers defined by finite automata, *Amer. Math. Monthly* **74** (1967), 539–542.

[135] F. HAUSDORFF, *Set theory*, Chelsea Publishing Co., New York, 1962. Second edition. Translated from the German by John R. Aumann et al.

[136] T. HEAD, The adherences of languages as topological spaces, in *Automata on infinite words (Le Mont-Dore, 1984)*, pp. 147–163, Springer, Berlin, 1985.

[137] T. HEAD, The topological structure of adherences of regular languages, *RAIRO Inform. Théor. Appl.* **20**,1 (1986), 31–41.

[138] T. HEAD AND B. LANDO, Fixed and Stationary ω-words and ω-languages, in *The Book of L*, A. S. G. Rozenberg (éd.), pp. 147–156, Springer Verlag, Berlin, Heidelberg, New York, 1986.

[139] S. HEILBRUNNER, An algorithm for the solution of fixed-point equations for infinite words, *RAIRO Inform. Théor.* **14**,2 (1980), 131–141.

[140] B. R. HODGSON, Décidabilité par automate fini, *Ann. Sci. Math. Québec* **7**,1 (1983), 39–57.

[141] I. HODKINSON, Expressive completeness of Until and Since over Dedekind complete linear time, in *Modal logic and process algebra (Amsterdam, 1994)*, pp. 171–185, CSLI Publ., Stanford, CA, 1995.

[142] H. J. HOOGEBOOM AND G. ROZENBERG, Infinitary languages: basic theory and applications to concurrent systems, in *Current trends in concurrency (Noordwijkerhout, 1985)*, pp. 266–342, Springer, Berlin, 1986.

[143] J. E. HOPCROFT AND J. D. ULLMAN, *Introduction to automata theory, languages, and computation*, Addison-Wesley Publishing Co., Reading, Mass., 1979. Addison-Wesley Series in Computer Science.

[144] J. M. HOWIE, *An introduction to semigroup theory*, Academic Press [Harcourt Brace Jovanovich Publishers], London, 1976. L.M.S. Monographs, No. 7.

[145] N. IMMERMAN, Languages that capture complexity classes, *SIAM J. Comput.* **16**,4 (1987), 760–778.

[146] A. ISLI, Converting a Büchi alternating automaton to a usual nondeterministic one, *Sādhanā* **21**,2 (1996), 213–228.

[147] H. IZUMI, Y. INAGAKI AND N. HONDA, A complete axiom system for algebra of closed-regular expression, in *Automata, languages and programming (Antwerp, 1984)*, pp. 260–269, Springer, Berlin, 1984.

[148] N. D. JONES, Space-bounded reducibility among combinatorial problems, *J. Comput. System Sci.* **11**,1 (1975), 68–85.

[149] J. JUSTIN AND G. PIRILLO, On a natural extension of Jacob's ranks, *J. Combin. Theory Ser. A* **43**,2 (1986), 205–218.

[150] M. KAMINSKI, A classification of ω-regular languages, *Theoret. Comput. Sci.* **36**,2-3 (1985), 217–229.

[151] J. KAMP, *Tense Logic and the Theory of Linear Order*, PhD thesis, Univ. of California, Los Angeles, 1968.

[152] M. KARPIŃSKI, Almost deterministic ω-automata with existential output condition, *Proc. Amer. Math. Soc.* **53**,2 (1975), 449–452.

[153] A. S. KECHRIS, *Classical descriptive set theory*, Springer-Verlag, New York, 1995.

[154] D. KIRSTEN, Alternating Tree Automata and Parity Games, in *Automata, Logics, and Infinite Games: A Guide to Current Research [outcome of a Dagstuhl seminar, February 2001]*, E. Grädel, W. Thomas and T. Wilke (éd.), pp. 153–167, *Lecture Notes in Computer Science* vol. 2500, Springer, 2002.

[155] F. KLAEDTKE, Complementation of B§chi Automata Using Alternation, in *Automata, Logics, and Infinite Games: A Guide to Current Research [outcome of a Dagstuhl seminar, February 2001]*, E. Grädel, W. Thomas and T. Wilke (éd.), pp. 61–78, *Lecture Notes in Computer Science* vol. 2500, Springer, 2002.

[156] N. KLARLUND, Progress measures for complementation of ω-automata with applications to temporal logic, in *32nd Annual Symposium on Foundations of Computer Science (San Juan, PR, 1991)*, pp. 358–367, IEEE Comput. Soc. Press, Los Alamitos, CA, 1991.

[157] N. KLARLUND, Progress measures, immediate determinacy, and a subset construction for tree automata, *Ann. Pure Appl. Logic* **69**,2-3 (1994), 243–268. Invited papers presented at the 1992 IEEE Symposium on Logic in Computer Science (Santa Cruz, CA).

[158] N. KLARLUND AND A. MØLLER, *MONA Version 1.4 User Manual*, BRICS Notes Series NS-01-1, Department of Computer Science, University of Aarhus, January 2001.

[159] H. KLAUCK, Algorithms for Parity Games, in *Automata, Logics, and Infinite Games: A Guide to Current Research [outcome of a Dagstuhl seminar, February 2001]*, E. Grädel, W. Thomas and T. Wilke (éd.), pp. 107–129, *Lecture Notes in Computer Science* vol. 2500, Springer, 2002.

[160] S. C. KLEENE, Representation of events in nerve nets and finite automata, in *Automata studies*, pp. 3–41, Princeton University Press, Princeton, N. J., 1956. Annals of mathematics studies, no. 34.

[161] K. KOBAYASHI, M. TAKAHASHI AND H. YAMASAKI, Characterization of ω-regular languages by first-order formulas, *Theoret. Comput. Sci.* **28**,3 (1984), 315–327.

[162] K. KOBAYASHI, M. TAKAHASHI AND H. YAMASAKI, Logical formulas and four subclasses of ω-regular languages, in *Automata on infinite words (Le Mont-Dore, 1984)*, pp. 81–88, Springer, Berlin, 1985.

[163] O. KUPFERMAN, S. SAFRA AND M. Y. VARDI, Relating word and tree automata, in *Proceedings, 11th Annual IEEE Symposium on Logic in Computer Science*, pp. 322–332, *Lecture Notes in Comput. Sci.*, IEEE Computer Society Press, 1996.

[164] R. P. KURSHAN, Complementing deterministic Büchi automata in polynomial time, *J. Comput. System Sci.* **35**,1 (1987), 59–71.

[165] R. KÜSTERS, Memoryless Determinacy of Parity Games, in *Automata, Logics, and Infinite Games: A Guide to Current Research [outcome of a Dagstuhl seminar, February 2001]*, E. Grädel, W. Thomas and T. Wilke (éd.), pp. 95–106, *Lecture Notes in Computer Science* vol. 2500, Springer, 2002.

[166] R. E. LADNER, Application of model theoretic games to discrete linear orders and finite automata, *Information and Control* **33**,4 (1977), 281–303.

[167] G. LALLEMENT, *Semigroups and combinatorial applications*, John Wiley & Sons, New York-Chichester-Brisbane, 1979. Pure and Applied Mathematics, A Wiley-Interscience Publication.

[168] L. H. LANDWEBER, Finite state games - A solvability algorithm for restricted second-order arithmetic, *Notices Amer. Math. Soc.* **14** (1967), 129–130.

[169] L. H. LANDWEBER, Decision problems for ω-automata, *Math. Systems Theory* **3** (1969), 376–384.

[170] M. LATTEUX AND E. TIMMERMAN, Finitely generated ω-languages, *Inform. Process. Lett.* **23**,4 (1986), 171–175.

[171] M. LATTEUX AND E. TIMMERMAN, Two characterizations of rational adherences, *Theoret. Comput. Sci.* **46**,1 (1986), 101–106.

[172] M. LATTEUX AND E. TIMMERMAN, Bifaithful starry transductions, *Inform. Process. Lett.* **28**,1 (1988), 1–4.

[173] B. LE SAEC, Saturating right congruences, *RAIRO Inform. Théor. Appl.* **24**,6 (1990), 545–559.

[174] B. LE SAEC, A modular proof of McNaughton's theorem, in *Logic and recognizable sets*, W. Thomas (éd.), pp. 50–55, Kiel Universität, Kiel, 1991.

[175] B. LE SAËC, J.-E. PIN AND P. WEIL, A purely algebraic proof of McNaughton's theorem on infinite words, in *Foundations of software technology and theoretical computer science (New Delhi, 1991)*, pp. 141–151, Springer, Berlin, 1991.

[176] B. LE SAËC, J.-E. PIN AND P. WEIL, Semigroups with idempotent stabilizers and applications to automata theory, *Internat. J. Algebra Comput.* **1**,3 (1991), 291–314.

[177] O. LICHTENSTEIN, A. PNUELI AND L. ZUCK, The glory of the past, in *Logics of programs (Brooklyn, N.Y., 1985)*, pp. 196–218, Springer, Berlin, 1985.

[178] P. A. LINDSAY, Alternation and ω-type Turing acceptors, *Theoret. Comput. Sci.* **43**,1 (1986), 107–115.

[179] P. A. LINDSAY, On alternating ω-automata, *J. Comput. System Sci.* **36**,1 (1988), 16–24.

[180] M. LINNA, On ω-words and ω-computations, *Ann. Univ. Turku. Ser A I* **168** (1975), 53.

[181] M. LINNA, On ω-sets associated with context-free languages, *Information and Control* **31**,3 (1976), 272–293.

[182] I. LITOVSKY AND E. TIMMERMAN, On generators of rational ω-power languages, *Theoret. Comput. Sci.* **53**,2-3 (1987), 187–200.

[183] M. LOTHAIRE, *Combinatorics on words*, Cambridge University Press, Cambridge, 1997. With a foreword by Roger Lyndon and a preface by Dominique Perrin, Corrected reprint of the 1983 original, with a new preface by Perrin.

[184] A. LOUVEAU, Some results in the Wadge hierarchy of Borel sets, in *Cabal seminar 79–81*, pp. 28–55, Springer, Berlin, 1983.

[185] O. MALER AND L. STAIGER, On syntactic congruences for ω-languages, *Theoret. Comput. Sci.* **183**,1 (1997), 93–112.

[186] Z. MANNA AND A. PNUELI, The modal logic of programs, in *Automata, languages and programming (Sixth Colloq., Graz, 1979)*, pp. 385–409, Springer, Berlin, 1979.

[187] D. A. MARTIN, Borel determinacy, *Ann. of Math. (2)* **102**,2 (1975), 363–371.

[188] R. MAZALA, Infinite Games, in *Automata, Logics, and Infinite Games: A Guide to Current Research [outcome of a Dagstuhl seminar, February 2001]*, E. Grädel, W. Thomas and T. Wilke (éd.), pp. 23–42, *Lecture Notes in Computer Science* vol. 2500, Springer, 2002.

[189] R. MCNAUGHTON, Testing and generating infinite sequences by a finite automaton, *Information and Control* **9** (1966), 521–530.

[190] R. MCNAUGHTON, Algebraic decision procedures for local testability, *Math. Systems Theory* **8**,1 (1974), 60–76.

[191] R. MCNAUGHTON, Infinite games played on finite graphs, *Ann. Pure Appl. Logic* **65**,2 (1993), 149–184.

[192] R. MCNAUGHTON AND S. PAPERT, *Counter-free automata*, The M.I.T. Press, Cambridge, Mass.-London, 1971. With an appendix by William Henneman, M.I.T. Research Monograph, No. 65.

[193] A. R. MEYER, A note on star-free events, *J. Assoc. Comput. Mach.* **16** (1969), 220–225.

[194] A. R. MEYER, Weak monadic second order theory of succesor is not elementary-recursive, in *Logic Colloquium (Boston, Mass., 1972–1973)*, pp. 132–154. Lecture Notes in Math., Vol. 453, Springer, Berlin, 1975.

[195] A. R. MEYER AND L. J. STOCKMEYER, The equivalence problem for regular expressions with squaring requires exponential time, in *Proc. 13th IEEE Symp. on Switching and Automata Theory*, pp. 125–129, IEEE Computer Society, 1972.

[196] I. MEZNÍK, On some structural properties of a subclass of ∞-regular languages, *Discrete Appl. Math.* **18**,3 (1987), 315–319.

[197] C. MICHAUX AND F. POINT, Les ensembles k-reconnaissables sont définissables dans $\langle \mathbf{N}, +, V_k \rangle$, *C. R. Acad. Sci. Paris Sér. I Math.* **303**,19 (1986), 939–942.

[198] M. MICHEL, Complementation is more difficult with automata on infinite words. CNET, Paris, 1988.

[199] S. MIYANO AND T. HAYASHI, Alternating finite automata on ω-words, *Theoret. Comput. Sci.* **32**,3 (1984), 321–330.

[200] T. MORIYA AND H. YAMASAKI, Accepting conditions for automata on ω-languages, *Theoret. Comput. Sci.* **61**,2-3 (1988), 137–147.

[201] M. MORSE AND G. A. HEDLUND, Symbolic Dynamics, *Amer. J. Math.* **60** (1938), 815–866.

[202] M. MORSE AND G. A. HEDLUND, Symbolic dynamics II. Sturmian trajectories, *Amer. J. Math.* **62** (1940), 1–42.

[203] M. MORSE AND G. A. HEDLUND, Unending chess, symbolic dynamics and a problem in semigroups, *Duke Math. J.* **11** (1944), 1–7.

[204] Y. N. MOSCHOVAKIS, *Descriptive set theory*, North-Holland Publishing Co., Amsterdam, 1980.

[205] A. W. MOSTOWSKI, Determinancy of sinking automata on infinite trees and inequalities between various Rabin's pair indices, *Inform. Process. Lett.* **15**,4 (1982), 159–163.

[206] A. W. MOSTOWSKI, Regular expressions for infinite trees and a standard form of automata, in *Computation Theory*, A. Skowron (éd.), pp. 157–168, *Lecture Notes in Comput. Sci.* vol. 208, Springer-Verlag, 1984.

[207] A. W. MOSTOWSKI, Hierarchies of weak monadic formulas for two successors arithmetic, *J. Inform. Process. Cybernet.* **23**,10-11 (1987), 509–515.

[208] A. MUCHNIK, Games on infinite trees and automata with dead-end markers - a new proof of the decidability of the monadic theory of two successors, *Semiotics and Information* **24** (1984), 17–40. (in Russian).

[209] A. MUCHNIK, Alternating automata on infinite trees, *Theoret. Comput. Sci.* **54** (1987), 267–276.

[210] D. E. MULLER, Infinite sequences and finite machines, in *Switching Theory and Logical Design, Proc. Fourth Annual Symp. IEEE*, pp. 3–16, IEEE, 1963.

[211] D. E. MULLER, A. SAOUDI AND P. E. SCHUPP, Alternating automata, the weak monadic theory of the tree, and its complexity, in *Automata, languages and programming (Rennes, 1986)*, pp. 275–283, Springer, Berlin, 1986.

[212] D. E. MULLER, A. SAOUDI AND P. E. SCHUPP, Alternating automata, the weak monadic theory of trees and its complexity, *Theoret. Comput. Sci.* **97**,2 (1992), 233–244.

[213] D. E. MULLER AND P. E. SCHUPP, Alternating automata on infinite objects. Determinacy and Rabin's theorem, in *Automata on infinite words (Le Mont-Dore, 1984)*, pp. 100–107, Springer, Berlin, 1985.

[214] D. E. MULLER AND P. E. SCHUPP, The theory of ends, pushdown automata, and second-order logic, *Theoret. Comput. Sci.* **37**,1 (1985), 51–75.

[215] D. E. MULLER AND P. E. SCHUPP, Alternating automata on infinite trees, *Theoret. Comput. Sci.* **54**,2-3 (1987), 267–276.

[216] D. E. MULLER AND P. E. SCHUPP, Simulating alternating tree automata by nondeterministic automata: new results and new proofs of the theorems of Rabin, McNaughton and Safra, *Theoret. Comput. Sci.* **141**,1-2 (1995), 69–107.

[217] F. NIESSNER, Nondeterministic Tree Automata, in *Automata, Logics, and Infinite Games: A Guide to Current Research [outcome of a Dagstuhl seminar, February 2001]*, E. Grädel, W. Thomas and T. Wilke (éd.), pp. 135–152, *Lecture Notes in Computer Science* vol. 2500, Springer, 2002.

[218] M. NIVAT, Sur les ensembles de mots infinis engendrés par une grammaire algébrique, *RAIRO Inform. Théor.* **12**,3 (1978), 259–278, v.

[219] M. NIVAT, Infinite words, infinite trees, infinite computations, in *Foundations of computer science, III (Third Adv. Course, Amsterdam, 1978), Part 2*, pp. 1–52, Math. Centrum, Amsterdam, 1979.

[220] M. NIVAT AND D. PERRIN, Ensembles reconnaissables de mots biinfinis, *14th ACM Symp. of theory of Computing* **XXXVIII** (1982), 47–59.

[221] M. NIVAT AND D. PERRIN, Ensembles reconnaissables de mots biinfinis, *Canad. J. Math.* **38**,3 (1986), 513–537.

[222] D. PARK, Concurrency and automata on infinite sequences, in *Proc. of the 5th GI Conference, Karlsruhe*, P. Deussen (éd.), pp. 167–183, *Lecture Notes in Comput. Sci* vol. 104, Springer Verlag, Berlin, Heidelberg, New York, 1981.

[223] J.-P. PÉCUCHET, Automates boustrophédon et mots infinis, *Theoret. Comput. Sci.* **35**,1 (1985), 115–122.

[224] J.-P. PÉCUCHET, Automates boustrophedon sur des mots infinis, in *Automata on infinite words (Le Mont-Dore, 1984)*, pp. 47–54, Springer, Berlin, 1985.

[225] J.-P. PÉCUCHET, Étude syntaxique des parties reconnaissables de mots infinis, in *Automata, languages and programming (Rennes, 1986)*, pp. 294–303, Springer, Berlin, 1986.

[226] J.-P. PÉCUCHET, On the complementation of Büchi automata, *Theoret. Comput. Sci.* **47**,1 (1986), 95–98.

[227] J.-P. PÉCUCHET, Variétés de semigroupes et mots infinis, in *STACS 86*, B. Monien and G. Vidal-Naquet (éd.), pp. 180–191, *Lecture Notes in Comput. Sci.* vol. 210, Springer Verlag, Berlin, Heidelberg, New York, 1986.

[228] J.-P. PÉCUCHET, Étude syntaxique des parties reconnaissables de mots infinis, *Theoret. Comput. Sci.* **58**,1-3 (1988), 231–248. Thirteenth International Colloquium on Automata, Languages and Programming (Rennes, 1986).

[229] R. PEIKERT, ω-regular languages and propositional temporal logic, Rap. Tech., ETH Zürich, 1985.

[230] D. PELED AND TH. WILKE, Stutter-invariant temporal properties are expressible without the next-time operator, *Inform. Process. Lett.* **63**,5 (1997), 243–246.

[231] D. PERRIN, Variétés de semigroupes et mots infinis, *C. R. Acad. Sci. Paris Sér. I Math.* **295**,10 (1982), 595–598.

[232] D. PERRIN, Variétés de semigroupes et mots infinis, in *Automata, languages and programming (Barcelona, 1983)*, pp. 610–616, Springer, Berlin, 1983.

[233] D. PERRIN, Recent results on automata and infinite words, in *Mathematical foundations of computer science, 1984 (Prague, 1984)*, pp. 134–148, Springer, Berlin, 1984.

[234] D. PERRIN, An introduction to finite automata on infinite words, in *Automata on infinite words (Le Mont-Dore, 1984)*, pp. 2–17, Springer, Berlin, 1985.

[235] D. PERRIN, Finite automata, in *Handbook of theoretical computer science, Vol. B*, pp. 1–57, Elsevier, Amsterdam, 1990.

[236] D. PERRIN AND J.-E. PIN, First-order logic and star-free sets, *J. Comput. System Sci.* **32**,3 (1986), 393–406.

[237] D. PERRIN AND J.-E. PIN, Semigroups and automata on infinite words, in *Semigroups, formal languages and groups (York, 1993)*, pp. 49–72, Kluwer Acad. Publ., Dordrecht, 1995.

[238] D. PERRIN AND P. E. SCHUPP, Automata on the integers, recurrence, distinguishability and the equivalence and decidability of monadic theories, in *Proc. 1st IEEE Symp. on Logic in Computer Science*, pp. 301–304, IEEE, 1986.

[239] J.-E. PIN, Hiérarchies de concaténation, *RAIRO Inform. Théor.* **18**,1 (1984), 23–46.

[240] J.-E. PIN, Star-free ω-languages and first order logic, in *Automata on infinite words (Le Mont-Dore, 1984)*, pp. 56–67, Springer, Berlin, 1985.

[241] J.-E. PIN, *Varieties of formal languages*, Plenum Publishing Corp., New York, 1986. With a preface by M.-P. Schützenberger, Translated from the French by A. Howie.

[242] J.-E. PIN, Finite semigroups and recognizable languages: an introduction, in *Semigroups, formal languages and groups (York, 1993)*, pp. 1–32, Kluwer Acad. Publ., Dordrecht, 1995.

[243] J.-E. PIN, A negative answer to a question of Wilke on varieties of ω-languages, *Inform. Process. Lett.* **56**,4 (1995), 197–200.

[244] J.-E. PIN, A variety theorem without complementation, *Russian Mathematics (Iz. VUZ)* **39** (1995), 80–90.

[245] J.-E. PIN, Logic, semigroups and automata on words, *Ann. Math. Artificial Intelligence* **16**,1-4 (1996), 343–384.

[246] J.-E. PIN, Syntactic semigroups, in *Handbook of formal languages, Vol. 1*, pp. 679–746, Springer, Berlin, 1997.

[247] J.-E. PIN, Positive varieties and infinite words, in *Latin'98*, C. Lucchesi and A. Moura (éd.), pp. 76–87, *Lecture Notes in Comput. Sci.* vol. 1380, Springer Verlag, Berlin, Heidelberg, New York, 1998.

[248] J.-E. PIN AND P. WEIL, Polynomial closure and unambiguous product, in *22th ICALP*, Berlin, 1995, pp. 348–359, *Lecture Notes in Comput. Sci.* n944, Springer.

[249] J.-E. PIN AND P. WEIL, A Reiterman theorem for pseudovarieties of finite first-order structures, *Algebra Universalis* **35**,4 (1996), 577–595.

[250] J.-E. PIN AND P. WEIL, Polynomial closure and unambiguous product, *Theory Comput. Syst.* **30**,4 (1997), 383–422.

[251] W. V. QUINE, Concatenation as a basis for arithmetic, *J. Symbolic Logic* **11** (1946), 105–114.

[252] M. O. RABIN, Decidability of second-order theories and automata on infinite trees., *Bull. Amer. Math. Soc.* **74** (1968), 1025–1029.

[253] M. O. RABIN, Decidability of second-order theories and automata on infinite trees., *Trans. Amer. Math. Soc.* **141** (1969), 1–35.

[254] M. O. RABIN, Weakly definable relations and special automata, in *Mathematical Logic and Foundations of Set Theory*, Y. Bar-Hillel (éd.), pp. 1–23, North Holland, 1970.

[255] M. O. RABIN, *Automata on infinite objects and Church's problem*, American Mathematical Society, Providence, R.I., 1972. Conference Board of the Mathematical Sciences Regional Conference Series in Mathematics, No. 13.

[256] M. O. RABIN, Decidable theories, in *Handbook of Mathematical Logic*, pp. 595–629, North Holland, 1977.

[257] M. O. RABIN AND D. SCOTT, Finite automata and their decision problems, *IBM J. Res. Develop.* **3** (1959), 114–125.

[258] C. W. RACKOFF, *The emptyness and complementation problem for automata on infinite trees*, PhD thesis, MIT, 1972.

[259] F. D. RAMSEY, On a problem of formal logic, *Proc. of the London Math. Soc.* **30** (1929), 338–384.

[260] G. RAUZY, Mots infinis en arithmétique, in *Automata on Infinite Words*, M. Nivat and D. Perrin (éd.), pp. 165–171, *Lecture Notes in Comput. Sci.* vol. 192, Springer Verlag, Berlin, Heidelberg, New York, 1985.

[261] R. R. REDZIEJOWSKI, Infinite-word languages and continuous mappings, *Theoret. Comput. Sci.* **43**,1 (1986), 59–79.

[262] K. REINHARDT, The Complexity of Translating Logic to Finite Automata, in *Automata, Logics, and Infinite Games: A Guide to Current Research [outcome*

of a Dagstuhl seminar, February 2001], E. Grädel, W. Thomas and T. Wilke (éd.), pp. 231–238, *Lecture Notes in Computer Science* vol. 2500, Springer, 2002.

[263] J. REITERMAN, The Birkhoff theorem for finite algebras, *Algebra Universalis* **14**,1 (1982), 1–10.

[264] C. REUTENAUER, Sur les variétés de langages et de monoïdes, in *Theoretical computer science (Fourth GI Conf., Aachen, 1979)*, pp. 260–265, Springer, Berlin, 1979.

[265] M. ROGGENBACH, Determinization of B§chi-Automata, in *Automata, Logics, and Infinite Games: A Guide to Current Research [outcome of a Dagstuhl seminar, February 2001]*, E. Grädel, W. Thomas and T. Wilke (éd.), pp. 43–60, *Lecture Notes in Computer Science* vol. 2500, Springer, 2002.

[266] P. ROHDE, Expressive Power of Monadic Second-Order Logic and Modal ţ-Calculus, in *Automata, Logics, and Infinite Games: A Guide to Current Research [outcome of a Dagstuhl seminar, February 2001]*, E. Grädel, W. Thomas and T. Wilke (éd.), pp. 239–257, *Lecture Notes in Computer Science* vol. 2500, Springer, 2002.

[267] J. G. ROSENSTEIN, *Linear orderings*, Academic Press Inc. [Harcourt Brace Jovanovich Publishers], New York, 1982.

[268] G. ROZENBERG AND A. SALOMAA, *The mathematical theory of L systems*, Academic Press Inc. [Harcourt Brace Jovanovich Publishers], New York, 1980.

[269] S. SAFRA, On the complexity of the ω-automata, in *Proc. 29th Ann. IEEE Symp. on Foundations of Computer Science*, pp. 319–327, IEEE, 1988.

[270] S. SAFRA, Exponential determinization for ω-automata with strong fairness condition, in *Proc. 24th ACM Symp. on the Theory of Computing*, pp. 275–282, ACM, 1992.

[271] A. SAOUDI, D. E. MULLER AND P. E. SCHUPP, Finite state processes, **Z**-temporal logic and the monadic theory of the integers, *Internat. J. Found. Comput. Sci.* **3**,3 (1992), 233–244.

[272] W. J. SAVITCH, Relationships between nondeterministic and deterministic tape complexities, *J. Comput. System. Sci.* **4** (1970), 177–192.

[273] M. SCHÜTZENBERGER, À propos des relations rationelles fonctionnelles, in *Automata, languages and programming (Proc. Sympos., Rocquencourt, 1972)*, pp. 103–114, North Holland, Amsterdam, 1973.

[274] M.-P. SCHÜTZENBERGER, On finite monoids having only trivial subgroups, *Information and Control* **8** (1965), 190–194.

[275] M. P. SCHÜTZENBERGER, Sur certaines opérations de fermeture dans les langages rationnels, in *Symposia Mathematica, Vol. XV (Convegno di Informatica Teorica, INDAM, Roma, 1973)*, pp. 245–253, Academic Press, London, 1975.

[276] S. SCHWOON, Determinization and Complementation of Streett Automata, in *Automata, Logics, and Infinite Games: A Guide to Current Research [outcome of a Dagstuhl seminar, February 2001]*, E. Grädel, W. Thomas and T. Wilke (éd.), pp. 79–91, *Lecture Notes in Computer Science* vol. 2500, Springer, 2002.

[277] V. SELIVANOV, Fine hierarchy of regular ω-languages, *Theoret. Comput. Sci.* **191**,1-2 (1998), 37–59.

[278] V. L. SELIVANOV, Fine hierarchies and Boolean terms, *J. Symbolic Logic* **60**,1 (1995), 289–317.

[279] V. L. SELIVANOV, Fine hierarchies of ω-rational languages, in *TAP-SOFT'95:Theory and Practice of Software Development*, pp. 277–287, *Lecture Notes in Comput. Sci.*, Springer Verlag, Berlin, 1995.

[280] A. SEMENOV, On certain extensions of the arithmetic of addition of natural numbers, *Math. USSR Izvestiya* **15** (1980), 401–418.

[281] A. L. SEMENOV, Logical theories of one-place functions on the natural number series, *Izv. Akad. Nauk SSSR Ser. Mat.* **47**,3 (1983), 623–658.

[282] A. L. SEMENOV, Decidability of monadic theories, in *Mathematical foundations of computer science, 1984 (Prague, 1984)*, pp. 162–175, Springer, Berlin, 1984.

[283] S. SHELAH, The monadic theory of order, *Ann. of Math. (2)* **102**,3 (1975), 379–419.

[284] D. SIEFKES, *Büchi's monadic second order successor arithmetic*, Springer-Verlag, Berlin, 1970.

[285] I. SIMON, Piecewise testable events, in *Proc. 2nd GI Conf.*, H. Brackage (éd.), pp. 214–222, *Lecture Notes in Comp. Sci.* vol. 33, Springer Verlag, Berlin, Heidelberg, New York, 1975.

[286] I. SIMON, Word Ramsey theorems, in *Graph theory and combinatorics (Cambridge, 1983)*, pp. 283–291, Academic Press, London, 1984.

[287] I. SIMON, Factorization forests of finite height, *Theoret. Comput. Sci.* **72**,1 (1990), 65–94.

[288] I. SIMON, The product of rational languages, in *Automata, languages and programming (Lund, 1993)*, pp. 430–444, Springer, Berlin, 1993.

[289] P. SIMONNET, *Automates et Théorie descriptive*, PhD thesis, Université Paris VII, 1992.

[290] A. P. SISTLA, M. Y. VARDI AND P. WOLPER, The complementation problem for Büchi automata with applications to temporal logic, *Theoret. Comput. Sci.* **49**,2-3 (1987), 217–237. Twelfth international colloquium on automata, languages and programming (Nafplion, 1985).

[291] J. SKURCZYŃSKI, The Borel hierarchy is infinite in the class of regular sets of trees, in *Fundamentals of computation theory (Szeged, 1989)*, pp. 416–423, Springer, New York, 1989.

[292] J. SKURCZYŃSKI, The Borel hierarchy is infinite in the class of regular sets of trees, *Theoret. Comput. Sci.* **112**,2 (1993), 413–418.

[293] L. STAIGER, A note on connected ω-languages, *Elektron. Informationsverarb. Kybernet.* **16**,5-6 (1980), 245–251.

[294] L. STAIGER, Finite-state ω-languages, *J. Comput. System Sci.* **27**,3 (1983), 434–448.

[295] L. STAIGER, Projection lemmas for ω-languages, *Theoret. Comput. Sci.* **32**,3 (1984), 331–337.

[296] L. STAIGER, The entropy of finite-state ω-languages, *Problems Control Inform. Theory/Problemy Upravlen. Teor. Inform.* **14**,5 (1985), 383–392.

[297] L. STAIGER, Hierarchies of recursive ω-languages, *Elektron. Informationsverarb. Kybernet.* **22**,5-6 (1986), 219–241.

[298] L. STAIGER, On infinitary finite length codes, *RAIRO Inform. Théor. Appl.* **20**,4 (1986), 483–494.

[299] L. STAIGER, Research in the theory of ω-languages, *J. Inform. Process. Cybernet.* **23**,8-9 (1987), 415–439. Mathematical aspects of informatics (Mägdesprung, 1986).

[300] L. STAIGER, Sequential mappings of ω-languages, *RAIRO Inform. Théor. Appl.* **21**,2 (1987), 147–173.

[301] L. STAIGER, ω-languages, in *Handbook of formal languages, Vol. 3*, pp. 339–387, Springer, Berlin, 1997.

[302] L. STAIGER, Rich ω-words and monadic second-order arithmetic, in *Computer science logic (Aarhus, 1997)*, pp. 478–490, Springer, Berlin, 1998.

[303] L. STAIGER AND K. WAGNER, Automatentheoretische und automatenfreie Charakterisierungen topologischer Klassen regulärer Folgenmengen, *Elektron. Informationsverarbeit. Kybernetik* **10** (1974), 379–392.

[304] J. STERN, Characterizations of some classes of regular events, *Theoret. Comput. Sci.* **35**,1 (1985), 17–42.

[305] J. STERN, Complexity of some problems from the theory of automata, *Inform. and Control* **66**,3 (1985), 163–176.

[306] L. J. STOCKMEYER, The polynomial-time hierarchy, *Theoret. Comput. Sci.* **3**,1 (1976), 1–22 (1977).

[307] M. STONE, The representation of boolean algebras, *Bulletin of the AMS* **44** (19??), 807–816. reviewed in Zentralblatt für Mathematik **20**, 342.

[308] H. STRAUBING, Aperiodic homomorphisms and the concatenation product of recognizable sets, *J. Pure Appl. Algebra* **15**,3 (1979), 319–327.

[309] H. STRAUBING, Families of recognizable sets corresponding to certain varieties of finite monoids, *J. Pure Appl. Algebra* **15**,3 (1979), 305–318.

[310] H. STRAUBING, A generalization of the Schützenberger product of finite monoids, *Theoret. Comput. Sci.* **13**,2 (1981), 137–150.

[311] H. STRAUBING, Finite semigroup varieties of the form $V * D$, *J. Pure Appl. Algebra* **36**,1 (1985), 53–94.

[312] H. STRAUBING, Semigroups and languages of dot-depth two, *Theoret. Comput. Sci.* **58**,1-3 (1988), 361–378. Thirteenth International Colloquium on Automata, Languages and Programming (Rennes, 1986).

[313] H. STRAUBING, *Finite automata, formal logic, and circuit complexity*, Birkhäuser Boston Inc., Boston, MA, 1994.

[314] H. STRAUBING AND D. THÉRIEN, Partially ordered finite monoids and a theorem of I. Simon, *J. Algebra* **119**,2 (1988), 393–399.

[315] H. STRAUBING AND P. WEIL, On a conjecture concerning dot-depth two languages, *Theoret. Comput. Sci.* **104**,2 (1992), 161–183.

[316] R. S. STREETT, Propositional dynamic logic of looping and converse is elementarily decidable, *Inform. and Control* **54**,1-2 (1982), 121–141.

[317] M. TAKAHASHI, The greatest fixed-points and rational omega-tree languages, *Theoret. Comput. Sci.* **44**,3 (1986), 259–274.

[318] M. TAKAHASHI, Brzozowski hierarchy of ω-languages, *Theoret. Comput. Sci.* **49**,1 (1987), 1–12.

[319] M. TAKAHASHI AND H. YAMASAKI, A note on ω-regular languages, *Theoret. Comput. Sci.* **23**,2 (1983), 217–225.

[320] J. W. THATCHER AND J. B. WRIGHT, Generalized finite automata theory with an application to a decision problem of second-order logic, *Math. Systems Theory* **2** (1968), 57–81.

[321] D. THÉRIEN AND A. WEISS, Graph congruences and wreath products, *J. Pure Appl. Algebra* **36**,2 (1985), 205–215.

[322] D. THÉRIEN AND T. WILKE, Temporal logic and semidirect products: an effective characterization of the until hierarchy, in *Proceedings of the 37th Annual Symposium on Foundations of Computer Science*, pp. 256–263, IEEE Computer Science, 1996.

[323] D. THÉRIEN AND T. WILKE, Over words, two variables are as powerful as one quantifier alternation: $FO^2 = \Sigma_2 \cap \Pi_2$, in *Proceedings of the 30th Annual ACM Symposium on Theory of Computing*, pp. 41–47, 1998.

[324] W. THOMAS, Star-free regular sets of ω-sequences, *Inform. and Control* **42**,2 (1979), 148–156.

[325] W. THOMAS, A combinatorial approach to the theory of ω-automata, *Inform. and Control* **48**,3 (1981), 261–283.

[326] W. THOMAS, Classifying regular events in symbolic logic, *J. Comput. System Sci.* **25**,3 (1982), 360–376.

[327] W. THOMAS, A hierarchy of sets of infinite trees, in *Theoretical Computer Science, Proceedings of the 6th GI Conference*, A. Cremers and H. Kriegel (éd.), pp. 335–342, *Lecture Notes in Comput. Sci.* vol. 145, Springer Verlag, Berlin, Heidelberg, New York, 1982.

[328] W. THOMAS, On frontiers of regular trees, *RAIRO Inform. Théor. Appl.* **20**,4 (1986), 371–381.

[329] W. THOMAS, On chain logic, path logic, and first-order logic over infinite trees, in *Proc. 2nd IEEE Symp. on Logic in Comput. Sci. Ithaca, N.Y.*, pp. 245–256, IEEE, 1987.

[330] W. THOMAS, Automata on infinite objects, in *Handbook of Theoretical Computer Science*, J. van Leeuwen (éd.), vol. vol. B, Formal models and semantics, pp. 135–191, Elsevier, 1990.

[331] W. THOMAS, On the synthesis of strategies in infinite games, in *STACS 95 (Munich, 1995)*, pp. 1–13, Springer, Berlin, 1995.

[332] W. THOMAS, Languages, automata, and logic, in *Handbook of formal languages, Vol. 3*, pp. 389–455, Springer, Berlin, 1997.

[333] A. THUE, Über unendliche Zeichenreihen, *Norske Vid. Selsk. Skr. I Math-Nat. Kl.* **7** (1906), 1–22.

[334] A. THUE, Über die gegenseitige Loge gleicher Teile gewisser Zeichenreihen, *Norske Vid. Selsk. Skr. I Math-Nat. Kl. Chris.* **1** (1912), 1–67.

[335] E. TIMMERMANN, The three subfamilies of rational ω-languages closed under ω-transduction, *Theoret. Comput. Sc.* **76** (1988), 243–250.

[336] B. A. TRAKHTENBROT, Finite automata and monadic second order logic (Russian), *Siberian Math. J* **3** (1962), 103–131. (English translation in *Amer. Math. Soc. Transl.* **59**, 1966, 23–55).

[337] B. A. TRAKHTENBROT AND Y. M. BARZDIN′, *Finite automata*, North-Holland Publishing Co., Amsterdam, 1973. Behavior and synthesis, Translated from the Russian by D. Louvish, English translation edited by E. Shamir and L. H. Landweber, Fundamental Studies in Computer Science, Vol. 1.

[338] D. L. VÂN, B. LE SAËC AND I. LITOVSKY, Characterizations of rational ω-languages by means of right congruences, *Theoret. Comput. Sci.* **143**,1 (1995), 1–21.

[339] J. VAN LEEUWEN (éd.), *Handbook of theoretical computer science. Vol. A*, Elsevier Science Publishers B.V., Amsterdam, 1990. Algorithms and complexity.

[340] J. VAN LEEUWEN (éd.), *Handbook of theoretical computer science. Vol. B*, Elsevier Science Publishers B.V., Amsterdam, 1990. Formal models and semantics.

[341] M. Y. VARDI AND P. WOLPER, Automata-theoretic techniques for modal logics of programs, *J. Comput. Syst. Sci.* **32** (1986), 183–221.

[342] M. Y. VARDI AND P. WOLPER, Reasoning about infinite computations, *Inform. and Comput.* **115**,1 (1994), 1–37.

[343] J. VON NEUMANN AND O. MORGENSTERN, *Theory of games and Economic Behavior*, Princeton University Press, Princeton, N. J., 1953.

[344] W. WADGE, *Reducibility and determinateness in the Baire space*, PhD thesis, University of California, Berkeley, 1983.

[345] K. WAGNER, Akzeptierbarkeitsgrade regulären Folgenmengen, *Elektron. Informationsverarb. Kybernet.* **11** (1975), 626–630.

[346] K. WAGNER, A hierarchy of regular sequence sets, in *Mathematical Foundations of Computer Science*, J. Bečvář (éd.), pp. 445–449, *Lecture Notes in Comput. Sci.* vol. 32, Springer, Berlin, 1975.

[347] K. WAGNER, Eine Axiomatisierung der Theorie der regulären Folgenmengen, *Elektron. Informationsverarbeit. Kybernetik* **12**,7 (1976), 337–354.

[348] K. WAGNER, Eine topologische Charakterisierung einiger Klassen regulärer Folgenmengen, *Elektron. Informationsverarbeit. Kybernetik* **13**,9 (1977), 473–487.

[349] K. WAGNER, On ω-regular sets, *Inform. and Control* **43**,2 (1979), 123–177.

[350] K. WAGNER AND L. STAIGER, Recursive ω-languages, in *Fundamentals of Computation Theory*, M. Karpiński (éd.), pp. 532–537, *Lecture Notes in Computer Sci.* vol. 56, Springer Verlag, Berlin, Heidelberg, New York, 1977.

[351] K. WAGNER AND L. STAIGER, Finite automata acceptance of infinite sequences, in *Mathematical Foundations of Computer Science*, A. Blikle (éd.), pp. 69–72, *Lecture Notes in Computer Sci.* vol. 28, Springer Verlag, Berlin, Heidelberg, New York, 1979.

[352] P. WEIL, Closure of varieties of languages under products with counter, *J. Comput. System Sci.* **45** (1992), 316–339.

[353] B. WEISS, Subshifts of finite type and sofic systems, *Monats. Math.* **77** (1973), 462–474.

[354] M. WEYER, Decidability of S1S and S2S, in *Automata, Logics, and Infinite Games: A Guide to Current Research [outcome of a Dagstuhl seminar, February 2001]*, E. Grädel, W. Thomas and T. Wilke (éd.), pp. 207–230, *Lecture Notes in Computer Science* vol. 2500, Springer, 2002.

[355] T. WILKE, An Eilenberg theorem for ∞-languages, in *Automata, Languages and Programming*, pp. 588–599, *Lecture Notes in Computer Sci.* vol. 510, Springer Verlag, Berlin, Heidelberg, New York, 1991.

[356] T. WILKE, An algebraic theory for regular languages of finite and infinite words, *Int. J. Alg. Comput.* **3** (1993), 447–489.

[357] T. WILKE, Locally threshold testable languages of infinite words, in *STACS 93*, P. Enjalbert, A. Finkel and K. Wagner (éd.), pp. 607–616, *Lecture Notes in Comp. Sci.* vol. 665, Springer Verlag, Berlin, Heidelberg, New York, 1993.

[358] T. WILKE AND H. YOO, Computing the Wadge degree, the Lifschitz degree, and the Rabin index of a regular language of infinite words in polynomial time, in *TAPSOFT 95*, P. Mosses, M. Nielsen and M. Schwartzbach (éd.), pp. 288–302, *Lecture Notes in Comp. Sci.* vol. 915, Springer Verlag, Berlin, Heidelberg, New York, 1995.

[359] T. WILKE AND H. YOO, Computing the Rabin index of a regular language of infinite words, *Information and Computation* **130** (1996), 61–70.

[360] TH. WILKE, Classifying discrete Temporal Properties, in *STACS'99*, C. Meinel (éd.), Trier, Germany, 1999, pp. 32–46, *Lecture Notes in Comput. Sci.* vol. 1563, Springer.

[361] J. WOJCIECHOWSKI, Classes of transfinite sequences accepted by finite automata, *Fundamenta Informaticae* **7** (1984), 191–223.

[362] J. WOJCIECHOWSKI, Finite automata on transfinite sequences and regular expressions, *Fundamenta Informaticae* **8** (1985), 379–396.

[363] J. WOJCIECHOWSKI, The ordinals less than ω^ω are definable by finite automata, in *Algebra, combinatorics and logic in computer science, Vol. I, II (Győr, 1983)*, pp. 871–887, North-Holland, Amsterdam, 1986.

[364] P. WOLFE, The strict determinacy of certain infinite games, *Pacific J. Math.* **5** (1955), 841–847.

[365] A. YAKHNIS AND V. YAKHNIS, Extension of Gurevitch-Harrington's restricted determinacy theorem: a criterion for the winning player and an explicit class of winning strategies, *Ann. Pure Appl. Logic* **48** (1990), 277–279.

[366] H. YAMASAKI, Language-theoretical representations of ω-languages, *Theoret. Comput. Sci.* **66**,3 (1989), 247–254.

[367] H. YAMASAKI, M. TAKAHASHI AND K. KOBAYASHI, Characterization of ω-regular languages by monadic second-order formulas, *Theoret. Comput. Sci.* **46**,1 (1986), 91–99.

[368] J. ZAPPE, Modal ţ-Calculus and Alternating Tree Automata, in *Automata, Logics, and Infinite Games: A Guide to Current Research [outcome of a Dagstuhl seminar, February 2001]*, E. Grädel, W. Thomas and T. Wilke (éd.), pp. 171–184, *Lecture Notes in Computer Science* vol. 2500, Springer, 2002.

[369] S. ZEITMAN, Unforgettable forgetful determinacy, *J. Logic Comput.* **4**,3 (1994), 273–283.

[370] W. ZIELONKA, Infinite games on finitely coloured graphs with applications to automata on infinite trees, *Theoret. Comput. Sci.* **200**,1-2 (1998), 135–183.

LIST OF TABLES

LIST OF FIGURES

INDEX

FSC
www.fsc.org
MIX
Papier | Fördert gute Waldnutzung
FSC® C083411

Kolibri 360 GmbH
Ferdinand-Jühlke-Straße 7
99095 Erfurt, Deutschland
produktsicherheit@kolibri360.de